SECOND EDITION

LINEAR ALGEBRA

SECOND EDITION

LINEAR ALGEBRA

John B. Fraleigh
Raymond A. Beauregard
University of Rhode Island

Historical Notes by Victor J. Katz
University of the District of Columbia

Addison-Wesley Publishing Company
Reading, Massachusetts Menlo Park, California New York
Don Mills, Ontario Wokingham, England Amsterdam Bonn
Sydney Singapore Tokyo Madrid San Juan Milan Paris

Acquisitions Editor: David Pallai
Production Administrator: Catherine Felgar
Editorial and Production Services: The Book Company
Text Design: The Book Company
Art Consultant: Loretta Bailey
Manufacturing Supervisor: Roy Logan
Cover Designer: Marshall Henrichs

Library of Congress Cataloging-in-Publication Data

Fraleigh, John B.
  Linear algebra/ by John B. Fraleigh and Raymond A. Beauregard.--
2nd Ed.
    p. cm.
 Includes index.
 ISBN 0-201-11949-8
 1. Algebras, Linear. I. Beauregard, Raymond A. II. Title.
QA184.F75 1989b
512′.5--dc20   89-36-76
               CIP

Reprinted with corrections, August 1991.

4 5 6 7 8 9 10 - DO - 9594939291

PREFACE

Our text is designed to serve as the basis for a first undergraduate course in linear algebra. Because linear algebra provides the tools to deal with many problems in fields ranging from forestry to nuclear physics, it is desirable to make the subject accessible to students from a variety of disciplines. For the mathematics major, a course in linear algebra often serves as a bridge from the typical intuitive treatment of calculus to more rigorous courses such as abstract algebra and analysis. Recognizing this, we have attempted to achieve an appropriate blend of intuition and rigor in our presentation.

FEATURES RETAINED FROM THE FIRST EDITION

- **Applications** first appear early in the text; Chapter 1 concludes with sections on population distribution and on 2-dimensional linear programming. Other applications are presented throughout the text.

- We deal chiefly with the vector spaces $\mathbb{R}^n$. However, many definitions and theorems concerning vectors in $\mathbb{R}^n$ are identical to those for general vector spaces. **Vector spaces** are introduced in Chapter 2, and the basic concepts of generating (or spanning) sets, independent sets, and bases are developed in the vector space context.

- **Eigenvalues and eigenvectors** are introduced, with applications, in Chapter 4. They recur in chapters 5 through 8, so students have the opportunity to continue to work with them.

- Optional **computer exercises** appear throughout the text. We devote a full chapter (Chapter 9) to investigating problems encountered by a **computer** in solving large linear systems. This chapter is of special interest to computer science majors and engineers, and it can be covered at any time after Chapter 1 has been completed.

- Each section concludes with a **summary**, which is very useful for both students and instructors. Students find it a handy reference for formulas and theorems. Instructors can tell by scanning the summary what material is covered and what notation is used in a section.

NEW FEATURES OF THE SECOND EDITION

- The **order of topics** has been altered to follow more closely a standard course in linear algebra. For example, Section 1.1 starts right in with matrix algebra. (The first edition had two initial sections motivating the study of linear algebra.) The discusssion about solving large linear systems with a computer has been moved from the Chapter 2 position to Chapter 9.

- **Exercise sets** have been greatly expanded. In particular, we have added many more nonroutine exercises that require more thought, in order to develop a more thorough understanding of the subject. Most exercise sets also include a 10-part **True-False problem**, designed to emphasize concepts and hypotheses. Answers to many of the odd-numbered exercises that request proofs have been deleted from the answer section at the end of the text, to encourage students to do them independently.

- **Examples from calculus** are now scattered in appropriate places throughout the text. They are all identified with the label *calculus*.

- A new chapter (Chapter 8) is devoted to linear algebra using **complex scalars**; complex scalars are necessary for a thorough treatment of eigenvalues and eigenvectors. The chapter concludes with a section explaining the Jordan canonical form.

- We have adopted a gradual presentation of **linear transformations**. We introduce the concept in the context of the equation $\mathbf{y} = A\mathbf{x}$ in Section 1.5, devote Section 2.9 to linear transformations for general vector spaces, and include the *linear transformation viewpoint* throughout the rest of the text as we introduce new concepts.

- Chapter 5 on **orthogonality**, covering projections, the Gram–Schmidt process, and orthogonal matrices, has been totally reorganized. Now, a more intuitive development of projections is given, and the introduction of the projection matrix is delayed until Section 5.4.

- A **dependence chart** appears immediately after this preface to aid in the construction of a syllabus for the course. The central core consists of just 24 sections from chapters 1 through 6. A great deal of **flexibility** in topic order is possible with our text. For example, chapters 9 and 10 can be done as soon as Chapter 1 has been covered. An instructor who wishes to cover complex scalars might make Section 8.1 on complex numbers the first lesson in the course and follow this by including a few appropriate exercises from Section 8.2 as row reduction of matrices, solution of linear systems, and matrix inversion are discussed. Some of Section 8.3 on diagonalization could be studied in conjunction with the real case in Section 4.2, and the rest of Section 8.3 could be covered after Section 5.3. This would be more efficient than leaving Chapter 8 until the very end of the course, and it would give students more time to work with complex scalars. Section 2.8 on lines and other translates and Section 2.10 on inner product spaces can be omitted without causing trouble later. All of this is indicated by the dependence chart.

SUPPLEMENTS

- An **Instructor's Solutions Manual**, prepared by the authors, is available from the publisher. It contains complete solutions, including proofs, for all of the exercises.

- A **Student's Solutions Manual** is available. It contains the complete solutions from the instructor's solutions manual, including proofs, for every third problem in the text.

- A software package, **LINTEK**, has been designed for use with the text and is available free to the instructor who may make copies for students. Many exercise sets include a few problems that are to be done using LINTEK. This software package includes many interactive programs designed to increase students' understanding of concepts and to provide drill in algorithms. With the exception of the program MATCOMP, which executes matrix algebra, programs have been written to illustrate and instruct as much as possible, rather than just to spit out answers.

Acknowledgments

Reviewers of text manuscripts perform a vital function by keeping authors in touch with reality. We wish to express our appreciation to all the reviewers of the manuscript for this edition, including: Paul Blanchard, Boston University; Henry Cohen, University of Pittsburgh; Sam Councilman, California State University, Long Beach; Murray Eisenberg, University of Massachusetts, Amherst; Mohammed Kazemi, University of North Carolina, Charlotte; Robert Maynard, Tidewater Community College; Robert McFadden, Northern Illinois University; David Meredith, San Francisco State University; Daniel Sweet, University of Maryland; Marvin Zeman, Southern Illinois University.

In addition, we wish to acknowledge the contributions of reviewers of the first edition: Ross A. Beaumont, University of Washington; Lawrence O. Cannon, Utah State University; Daniel Drucker, Wayne State University; Bruce Edwards, University of Florida; Christopher Ennis, University of Minnesota; W. A. McWorter, Jr., Ohio State University; John Morrill, DePauw University; James M. Sobota, University of Wisconsin–LaCrosse.

We are very grateful to Victor Katz for providing the excellent historical notes. His notes are not just biographical information about the contributors to the subject; he actually offers insight into its development.

Finally, we wish to thank Tom Taylor, David Pallai, the mathematics staff at Addison-Wesley, and George Calmenson of The Book Company, which handled production, for their help in preparation of this edition.

DEPENDENCE CHART

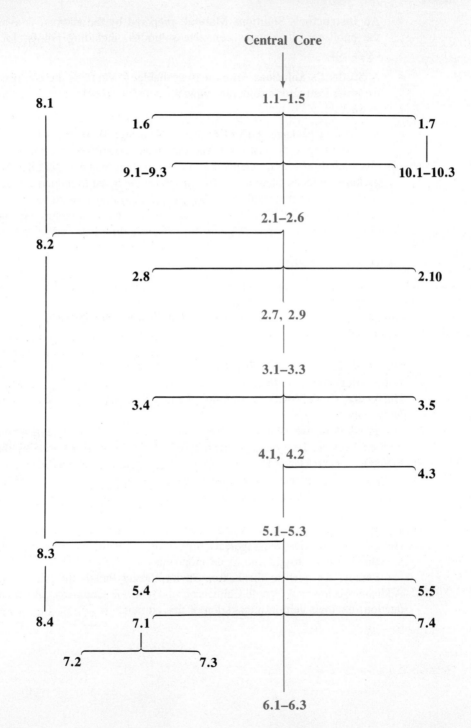

Central Core

8.1

1.1–1.5

1.6   1.7

9.1–9.3   10.1–10.3

2.1–2.6

8.2

2.8   2.10

2.7, 2.9

3.1–3.3

3.4   3.5

4.1, 4.2

4.3

5.1–5.3

8.3

5.4   5.5

8.4   7.1   7.4

7.2   7.3

6.1–6.3

CONTENTS

SECOND EDITION

LINEAR
ALGEBRA

INTRODUCTION TO LINEAR SYSTEMS

We have all solved two simultaneous linear equations in two unknowns—for example,

$$2x + y = 4$$
$$x - 2y = -3.$$

We shall call any such collection of simultaneous linear equations a *linear system*. This chapter is concerned with methods for solving a general linear system of m equations in n unknowns, and with the structure of the solution set of such a system. Section 1.1 introduces matrix algebra, which provides a convenient shorthand for working with linear systems. In Sections 1.2 and 1.3, we discuss the Gauss and Gauss–Jordan methods for solving linear systems and identify properties of their solution sets. Section 1.4 continues the development of matrix algebra by taking up the subject of matrix inversion. More illumination is thrown on linear systems in Section 1.5, where we discuss their relationship to linear transformations. Section 1.6 presents an application of Sections 1.1 through 1.5 to population distribution and Markov chains.

After all this work on linear equalities, we indicate the present-day importance of linear inequalities in Section 1.7, in a discussion of two-dimensional linear programming. Chapter 10, which continues the treatment of linear programming, can be studied at any time after Section 1.7 has been covered.

Before plunging into the mechanics of matrix algebra, we want to indicate briefly why we have started our text with this chapter on linear systems. Finding all solutions of a linear system is fundamental to the study of linear algebra. Indeed, the great practical importance of linear algebra is due to the fact that *linear systems can be solved by algebraic methods*. For example, a *linear* equation in one unknown, such as $3x = 8$, is easy to solve. The solution of every linear equation of the form $ax = b$ is simple to determine. But the nonlinear equations $x^5 + 3x = 1$, $x^x = 100$, and $x - \sin x = 1$ are all tough to solve algebraically.

One often-used technique for dealing with a nonlinear problem consists of *linearizing* the problem—that is, approximating the problem with a linear one that can more easily be solved. Linearization techniques often involve calculus. If you have studied calculus, you may be familiar with Newton's method for approximating a solution to an equation of the form $f(x) = 0$; an example would be $x - 1 - \sin x = 0$. An approximate solution is found by solving sequentially several linear equations of the form $ax = b$, which are obtained by approximating the graph of f with lines. Finding an approximate numerical solution of a partial differential equation may involve solving a linear system consisting of thousands of equations in thousands of unknowns. Computer programs have been developed to implement solving of such systems.

1.1 MATRICES AND THEIR ALGEBRA

The Notion of a Matrix

Matrices provide us with a shorthand method for keeping just the essential data in a linear system. For example, in the linear system

$$2x - 3y = 5$$
$$5x + y = -1,$$

all essential data are contained in the number arrays

$$\begin{bmatrix} 2 & -3 \\ 5 & 1 \end{bmatrix} \quad \text{and} \quad \begin{bmatrix} 5 \\ -1 \end{bmatrix}.$$

The system can be recovered from these arrays, which saves us the trouble of writing variables and equal signs.

A **matrix** is an ordered rectangular array of numbers, usually enclosed in parentheses or square brackets. For example,

$$A = \begin{bmatrix} 2 & -1 & 3 \\ 4 & 2 & 1 \end{bmatrix} \quad \text{and} \quad B = \begin{bmatrix} -1 & 0 & 3 \\ 2 & 1 & 4 \\ 4 & 5 & -6 \\ -3 & -1 & -1 \end{bmatrix}$$

are matrices. We will generally use upper-case letters to denote matrices.

THE TERM *MATRIX* is first mentioned in mathematical literature in an 1850 paper of James Joseph Sylvester (1814–1897). The standard nontechnical meaning of that term is "a place in which something is bred, produced, or developed." For Sylvester, then, a matrix, which was an "oblong arrangement of terms," was an entity out of which one could form various square pieces to produce determinants. These latter quantities, formed from square matrices, were quite well known by this time.

James Sylvester (his original name was James Joseph) was born into a Jewish family in London, and was to become one of the supreme algebraists of the nineteenth century. Despite having studied for several years at Cambridge University, he was not permitted to take his degree there because he "professed the faith in which the founder of Christianity was educated." Therefore, he received his degrees from Trinity College, Dublin. In 1841 he accepted a professorship at the University of Virginia; he remained there only a short time, however, his horror of slavery preventing him from fitting into the academic community. In 1871 he returned to the United States to accept the chair of mathematics at the newly opened Johns Hopkins University. In between these sojourns, he spent about 10 years as an attorney, during which time he met Arthur Cayley (see the note on p. 43), and 15 years as Professor of Mathematics at the Royal Military Academy, Woolwich. Sylvester was an avid poet, prefacing many of his mathematical papers with examples of his work. His most renowned example was the "Rosalind" poem, a 400-line epic, each line of which rhymed with "Rosalind."

The size of a matrix is specified by the number of (horizontal) rows and the number of (vertical) columns that it contains. The matrix A above contains two rows and three columns and is called a 2×3 (read "2 by 3") matrix. Similarly, B is a 4×3 matrix. In writing the notation $m \times n$ to describe the shape of a matrix, we always write the number of rows first. An $m \times m$ matrix has the same number of rows as columns and is said to be a **square matrix**.

Of special interest are matrices having only one row or only one column; these are called **vectors**. The entries in a vector are called its **components**. For example, a $1 \times n$ matrix is a **row vector** with n components, and an $m \times 1$ matrix is a **column vector** with m components. This text uses lower-case boldface letters to denote vectors. For example, we might have

$$\mathbf{v} = \begin{bmatrix} 2 \\ 3 \\ 4 \end{bmatrix}, \qquad \mathbf{w} = [4 \quad 2], \qquad \text{and} \qquad \mathbf{x} = \begin{bmatrix} x_1 \\ x_2 \\ x_3 \end{bmatrix}.$$

Here, $\mathbf{v}$ and $\mathbf{x}$ are column vectors with three components each, while $\mathbf{w}$ is a row vector with two components. In written work, it is customary to place an arrow over a letter to denote a vector, as in $\vec{v}$, $\vec{w}$, and $\vec{x}$. A vector whose components are all 0 is called a **zero vector** and is denoted by **0**.

Notice that we number column vectors from top to bottom, as in the preceding vector $\mathbf{x}$. For a row vector, we number from left to right, as in

$$\mathbf{a} = [a_1 \quad a_2 \quad a_3 \quad a_4].$$

Double subscripts are commonly used to indicate the location of an entry in a matrix that is not a row or column vector. The first subscript gives the number of the row in which the entry appears (counting from the top), and the second subscript gives the number of the column (counting from the left). Thus an $m \times n$ matrix A may be written as

$$A = [a_{ij}] = \begin{bmatrix} a_{11} & a_{12} & a_{13} & \cdots & a_{1n} \\ a_{21} & a_{22} & a_{23} & \cdots & a_{2n} \\ a_{31} & a_{32} & a_{33} & \cdots & a_{3n} \\ & & \vdots & & \\ a_{m1} & a_{m2} & a_{m3} & \cdots & a_{mn} \end{bmatrix}.$$

If we want to express the matrix B on page 2 as $[b_{ij}]$, we would have $b_{11} = -1$, $b_{21} = 2$, $b_{32} = 5$, and so on.

Matrix Multiplication

Matrices are used in this chapter to simplify work with systems of linear equations, saving a lot of writing. Consider, for example, the linear system

$$\begin{aligned} 2x_1 - 3x_2 &= 7 \\ 3x_1 - 4x_2 &= 2. \end{aligned} \tag{1}$$

The important data for this linear system are contained in the *coefficient matrix*

$$A = \begin{bmatrix} 2 & -3 \\ 3 & -4 \end{bmatrix}$$

and in the column vector

$$\mathbf{b} = \begin{bmatrix} 7 \\ 2 \end{bmatrix},$$

which consists of the constants that appear after the equal signs. We let

$$\mathbf{x} = \begin{bmatrix} x_1 \\ x_2 \end{bmatrix}$$

be the column vector of unknowns in system (1). It is customary to abbreviate system (1) as

$$A\mathbf{x} = \mathbf{b},$$

or to express it, written out, as

$$\begin{bmatrix} 2 & -3 \\ 3 & -4 \end{bmatrix}\begin{bmatrix} x_1 \\ x_2 \end{bmatrix} = \begin{bmatrix} 7 \\ 2 \end{bmatrix}.$$

This *matrix equation* suggests the notion of *multiplication* of a matrix A times a column vector $\mathbf{x}$. To prepare for this, we now define the *dot product* of any two vectors with the same number of components.

DEFINITION 1.1 Dot Product

Let **a** and **b** be any two vectors with the same number of components. Each may be either a row vector or a column vector. Let the ordered components of **a** be $a_1, a_2, \ldots, a_n$, and let those of **b** be $b_1, b_2, \ldots, b_n$. The **dot product** of **a** and **b** is

$$\mathbf{a} \cdot \mathbf{b} = a_1 b_1 + a_2 b_2 + \cdots + a_n b_n.$$

This, written in summation notation, is $\mathbf{a} \cdot \mathbf{b} = \sum_{i=1}^{n} a_i b_i$.

It is essential to keep the notions of *vector* and *number* distinct. The term **scalar** is commonly used in linear algebra in place of *number*. The dot product is also called the *scalar product*, since $\mathbf{a} \cdot \mathbf{b}$ is a scalar quantity rather than a vector quantity. We will not use the term *scalar product*, in order to avoid confusion with *scalar multiplication* of a matrix, which we define in a moment.

EXAMPLE 1 Let

$$\mathbf{a} = [1 \quad -3 \quad 4] \quad \text{and} \quad \mathbf{b} = \begin{bmatrix} 2 \\ 0 \\ 8 \end{bmatrix}.$$

Find $\mathbf{a} \cdot \mathbf{b}$.

Solution We have

$$\mathbf{a} \cdot \mathbf{b} = (1)(2) + (-3)(0) + (4)(8) = 2 + 0 + 32 = 34. \quad ■$$

EXAMPLE 2 Let $\mathbf{a} = [1 \quad 5 \quad 6]$ and $\mathbf{b} = [2 \quad 4]$. Find $\mathbf{a} \cdot \mathbf{b}$.

Solution The dot product is not defined, since $\mathbf{a}$ and $\mathbf{b}$ have different numbers of components.

■

Let us return to the matrix expression

$$\begin{bmatrix} 2 & -3 \\ 3 & -4 \end{bmatrix} \begin{bmatrix} x_1 \\ x_2 \end{bmatrix} = \begin{bmatrix} 7 \\ 2 \end{bmatrix} \tag{2}$$

for the linear system

$$2x_1 - 3x_2 = 7 \\ 3x_1 - 4x_2 = 2. \tag{3}$$

We see that the number 7 in this system appears as the dot product of the vectors

$$[2 \quad -3] \quad \text{and} \quad \begin{bmatrix} x_1 \\ x_2 \end{bmatrix}$$

in the first equation of the system, whereas 2 appears as the dot product of

$$[3 \quad -4] \quad \text{and} \quad \begin{bmatrix} x_1 \\ x_2 \end{bmatrix}$$

in the second equation.

The product AB of two matrices A and B is formed by taking all possible dot products of the *row vectors* of A times the *column vectors* of B. In order for such dot products to be defined, the number of components in each row of A must equal the number of components in each column of B. If AB is defined and A is $m \times n$ (so the rows have n components), then B must be $n \times s$ (so the columns of B also have n components).

DEFINITION 1.2 Matrix Multiplication

Let $A = [a_{ik}]$ be an $m \times n$ matrix, and let $B = [b_{kj}]$ be an $n \times s$ matrix. The **matrix product** AB is the $m \times s$ matrix $C = [c_{ij}]$, where c_{ij} is the dot product of the ith row vector of A and the jth column vector of B.

We illustrate the choice of row i from A and column j from B to find the element c_{ij} in AB, according to Definition 1.2, by the equation

$$AB = [c_{ij}] = \begin{bmatrix} a_{11} & \cdots & a_{1n} \\ \vdots & & \vdots \\ a_{i1} & \cdots & a_{in} \\ \vdots & & \vdots \\ a_{m1} & \cdots & a_{mn} \end{bmatrix} \begin{bmatrix} b_{11} & \cdots & b_{1j} & \cdots & b_{1s} \\ \vdots & & \vdots & & \vdots \\ b_{n1} & \cdots & b_{nj} & \cdots & b_{ns} \end{bmatrix},$$

where

$$c_{ij} = (i\text{th row vector of } A) \cdot (j\text{th column vector of } B).$$

In summation notation, we have

$$c_{ij} = a_{i1}b_{1j} + a_{i2}b_{2j} + \cdots + a_{in}b_{nj}$$

$$= \sum_{k=1}^{n} a_{ik}b_{kj}. \tag{4}$$

Notice again that AB is defined only when the second size-number (the number of columns) of A is the same as the first size-number (the number of rows) of B. The product matrix has the shape

(First size-number of A) × (Second size-number of B).

EXAMPLE 3 Let A be a 2×3 matrix, and let B be a 3×5 matrix. Find the size of AB and BA, if they are defined.

Solution Since the second size-number, 3, of A equals the first size-number, 3, of B, we see that AB is defined; it is a 2×5 matrix. However, BA is not defined, because the second size-number, 5, of B is not the same as the first size-number, 2, of A. ■

MATRIX MULTIPLICATION originated in the composition of linear substitutions. Though there are hints of such ideas in the work of Euler and Lagrange, they were first fully discussed by Karl Friedrich Gauss (1777–1855) in his 1801 work *Disquisitiones Arithmeticae* in connection with quadratic forms (functions in two variables of the form $Ax^2 + 2Bxy + Cy^2$). In particular, a linear substitution of the form

$$x = ax' + by' \qquad y = cx' + dy' \tag{i}$$

transforms one such form F in x and y into another form F' in x',y'. If a second substitution

$$x' = ex'' + fy'' \qquad y' = gx'' + hy'' \tag{ii}$$

transforms F' into a form F'' in x'',y'', then the composition of the substitutions, found by replacing x',y' in (i) by their values in (ii), gives a substitution transforming F into F'':

$$x = (ae + bg)x'' + (af + bh)y'' \qquad y = (ce + dg)x'' + (cf + dh)y''. \tag{iii}$$

The coefficient matrix of substitution (iii) is the product of the coefficient matrices of substitutions (i) and (ii). Gauss performed an analogous composition of substitutions for forms in three variables, which gives the multiplication of 3×3 matrices.

Gauss is often called the "prince of mathematicians," since over a long scientific career he contributed greatly to such varied fields as number theory, algebra, geometry, complex analysis, astronomy, geodesy, and mechanics.

EXAMPLE 4   Compute the product

$$\begin{bmatrix} -2 & 3 & 2 \\ 4 & 6 & -2 \end{bmatrix} \begin{bmatrix} 4 & -1 & 2 & 5 \\ 3 & 0 & 1 & 1 \\ -2 & 3 & 5 & -3 \end{bmatrix}.$$

Solution   The product is defined, since the left-hand matrix is 2×3 and the right-hand matrix is 3×4; the product will have size 2×4. The entry in the first row and first column position of the product is obtained by taking the dot product of the first row vector of the left-hand matrix and the first column vector of the right-hand matrix, as follows:

$$(-2)(4) + (3)(3) + (2)(-2) = -8 + 9 - 4 = -3.$$

The entry in the second row and third column of the product is the dot product of the second row vector of the left-hand matrix and the third column vector of the right-hand one:

$$(4)(2) + (6)(1) + (-2)(5) = 8 + 6 - 10 = 4,$$

and so on, through the remaining row and column positions of the product. Eight such computations show that

$$\begin{bmatrix} -2 & 3 & 2 \\ 4 & 6 & -2 \end{bmatrix} \begin{bmatrix} 4 & -1 & 2 & 5 \\ 3 & 0 & 1 & 1 \\ -2 & 3 & 5 & -3 \end{bmatrix} = \begin{bmatrix} -3 & 8 & 9 & -13 \\ 38 & -10 & 4 & 32 \end{bmatrix}. \qquad ▪$$

Example 3 shows that sometimes AB is defined when BA is not. Even if both AB and BA are defined, however, it need not be true that $AB = BA$:

> Matrix multiplication is not commutative.

EXAMPLE 5   Let

$$A = \begin{bmatrix} 0 & 2 \\ 3 & 5 \end{bmatrix} \quad \text{and} \quad B = \begin{bmatrix} 0 & 1 \\ 2 & 5 \end{bmatrix}.$$

Compute AB and BA.

Solution   We compute that

$$AB = \begin{bmatrix} 4 & 10 \\ 10 & 28 \end{bmatrix}, \quad \text{while} \quad BA = \begin{bmatrix} 3 & 5 \\ 15 & 29 \end{bmatrix}. \qquad ▪$$

Of course, for a square matrix A, we denote AA by A^2, AAA by A^3, and so on. It can be shown that matrix multiplication is associative; that is,

$$A(BC) = (AB)C$$

whenever the product is defined. This is not difficult to prove from the definition, although keeping track of subscripts can be a bit challenging. We leave the proof as Exercise 33, whose solution is given in the back of this text.

The $n \times n$ Identity Matrix

Let I be the $n \times n$ matrix $[a_{ij}]$ such that $a_{ii} = 1$ for $i = 1, \ldots, n$ and $a_{ij} = 0$ for $i \neq j$. That is,

$$I = \begin{bmatrix} 1 & 0 & 0 & \cdots & 0 \\ 0 & 1 & 0 & \cdots & 0 \\ 0 & 0 & 1 & \cdots & 0 \\ \vdots & \vdots & \vdots & & \vdots \\ 0 & 0 & 0 & \cdots & 1 \end{bmatrix} = \begin{bmatrix} 1 & & & & \\ & 1 & & \mathbf{0} & \\ & & 1 & & \\ & \mathbf{0} & & \cdot & \\ & & & & 1 \end{bmatrix},$$

where the large zeros above and below the diagonal in the second matrix indicate that each entry of the matrix in those positions is 0. If A is any $m \times n$ matrix and B is any $n \times s$ matrix, we can show that

$$AI = A \qquad \text{and} \qquad IB = B.$$

You will understand why this is so if you think about why it is that

$$\begin{bmatrix} 2 & 3 \\ -1 & 7 \end{bmatrix} \begin{bmatrix} 1 & 0 \\ 0 & 1 \end{bmatrix} = \begin{bmatrix} 2 & 3 \\ -1 & 7 \end{bmatrix} = \begin{bmatrix} 1 & 0 \\ 0 & 1 \end{bmatrix} \begin{bmatrix} 2 & 3 \\ -1 & 7 \end{bmatrix}.$$

Because of the relations $AI = A$ and $IB = B$, the matrix I is called the $n \times n$ **identity matrix**. It behaves for multiplication of $n \times n$ matrices exactly as the scalar 1 behaves for multiplication of scalars. We have one such square identity matrix for each integer $1, 2, 3, \ldots$. To keep notation simple, we denote them all by I, rather than by I_1, $I_2, I_3, \ldots$. The size of I will be clear from the context.

Other Matrix Operations

Although multiplication is the most important matrix operation for our work, we will have occasion to add matrices and to multiply a matrix by a scalar in later chapters. These operations are more natural than matrix multiplication, and consequently they rarely cause students trouble.

DEFINITION 1.3 Matrix Addition

Let $A = [a_{ij}]$ and $B = [b_{ij}]$ be two matrices of the same size $m \times n$. The **sum $A + B$** of these two matrices is the $m \times n$ matrix $C = [c_{ij}]$, where

$$c_{ij} = a_{ij} + b_{ij}.$$

That is, the sum of two matrices of the same size is the matrix of that size obtained by adding corresponding entries.

EXAMPLE 6   Find

$$\begin{bmatrix} 1 & 2 & -4 \\ 0 & 3 & -1 \end{bmatrix} + \begin{bmatrix} -1 & 0 & 2 \\ 1 & -5 & 3 \end{bmatrix}.$$

Solution   The sum is the matrix

$$\begin{bmatrix} 0 & 2 & -2 \\ 1 & -2 & 2 \end{bmatrix}.$$ ■

EXAMPLE 7   Find

$$\begin{bmatrix} 1 & -3 \\ 2 & 4 \end{bmatrix} + \begin{bmatrix} -5 & 4 & 6 \\ 3 & 7 & -1 \end{bmatrix}.$$

Solution   The sum is undefined, since the matrices are not the same size.   ■

Let A be an $m \times n$ matrix, and let O be the $m \times n$ matrix all of whose entries are zero. Then,

$$A + O = O + A = A.$$

The matrix O is called the $m \times n$ **zero matrix**; the size of such a zero matrix is made clear by the context.

DEFINITION 1.4   Scalar Multiplication

Let $A = [a_{ij}]$, and let r be a scalar. The **product** rA of the scalar r and the matrix A is the matrix $B = [b_{ij}]$ having the same size as A, where

$$b_{ij} = ra_{ij}.$$

EXAMPLE 8   Find

$$2\begin{bmatrix} -2 & 1 \\ 3 & -5 \end{bmatrix}.$$

Solution   Multiplying each entry of the matrix by 2, we obtain the matrix

$$\begin{bmatrix} -4 & 2 \\ 6 & -10 \end{bmatrix}.$$ ■

For matrices A and B of the same size, we define the **difference** $A - B$ to be

$$A - B = A + (-1)B.$$

The entries in $A - B$ are obtained by subtracting the entries of B from entries in the corresponding positions in A.

EXAMPLE 9   If

$$A = \begin{bmatrix} 3 & -1 & 4 \\ 0 & 2 & -5 \end{bmatrix} \quad \text{and} \quad B = \begin{bmatrix} -1 & 0 & 5 \\ 4 & -2 & 1 \end{bmatrix},$$

find $2A - 3B$.

Solution   We find that

$$2A - 3B = \begin{bmatrix} 9 & -2 & -7 \\ -12 & 10 & -13 \end{bmatrix}.$$ ■

Taking the *transpose* of a matrix interchanges the roles of row and column. This important operation is used frequently. We mention one use after presenting the formal definition and two examples.

DEFINITION 1.5 Transpose of a Matrix, Symmetric Matrix

The matrix B is the **transpose** of the matrix A, written $B = A^T$, if each entry b_{ij} in B is the same as the entry a_{ji} in A, and conversely. If A is a matrix and if $A = A^T$, then the matrix A is **symmetric**.

EXAMPLE 10   Find A^T if

$$A = \begin{bmatrix} 1 & 4 & 5 \\ -3 & 2 & 7 \end{bmatrix}.$$

Solution   We have

$$A^T = \begin{bmatrix} 1 & -3 \\ 4 & 2 \\ 5 & 7 \end{bmatrix}.$$

Notice that the rows of A become the columns of A^T. ■

Symmetric matrices arise in some applications, as we shall see in Chapter 7.

EXAMPLE 11   Give an example of a 3×3 symmetric matrix.

Solution   An example of a 3×3 symmetric matrix is

$$\begin{bmatrix} 5 & 2 & 1 \\ 2 & -4 & 7 \\ 1 & 7 & 3 \end{bmatrix}.$$

Notice the symmetry in the *main diagonal*, from the upper left-hand corner to the lower right-hand corner. ■

We usually work with column vectors rather than with row vectors, for reasons that are explained in Section 1.5. If $\mathbf{a}$ and $\mathbf{b}$ are two column vectors with n components, the dot product $\mathbf{a} \cdot \mathbf{b}$ can be written in terms of the transpose operation and matrix multiplication—namely,

$$\mathbf{a} \cdot \mathbf{b} = \mathbf{a}^T \mathbf{b} = [a_1 \quad a_2 \quad \cdots \quad a_n] \begin{bmatrix} b_1 \\ b_2 \\ \vdots \\ b_n \end{bmatrix}. \tag{5}$$

Strictly speaking, $\mathbf{a}^T \mathbf{b}$ is a 1×1 matrix, but its sole entry is $\mathbf{a} \cdot \mathbf{b}$. Identifying a 1×1 matrix with its sole entry should cause no difficulty. The use of Eq. (5) makes some formulas later in the text much easier to handle.

Properties of Matrix Operations

For handy reference, we have boxed the algebraic properties of the dot product, of matrix arithmetic, and of the transpose operation. These properties are valid for all vectors, scalars, and matrices for which the indicated quantities are defined. Except for the associative law of matrix multiplication, all of them are easy to check, as we illustrate after stating them. The exercises ask you to prove most of them.

Properties of the Dot Product

$\mathbf{u} \cdot \mathbf{v} = \mathbf{v} \cdot \mathbf{u}$	Commutative property
$\mathbf{u} \cdot (\mathbf{v} + \mathbf{w}) = (\mathbf{u} \cdot \mathbf{v}) + (\mathbf{u} \cdot \mathbf{w})$	Distributive property
$(r\mathbf{u}) \cdot \mathbf{v} = \mathbf{u} \cdot (r\mathbf{v}) = r(\mathbf{u} \cdot \mathbf{v})$	Homogeneous property

---

Properties of Matrix Arithmetic

$A + B = B + A$	Commutativity of addition
$(A + B) + C = A + (B + C)$	Associativity of addition
$A + O = O + A = A$	Identity for addition
$r(A + B) = rA + rB$	A left distributive law
$(r + s)A = rA + sA$	A right distributive law
$(rs)A = r(sA)$	Associativity of scalar multiplication
$(rA)B = A(rB) = r(AB)$	Scalars pull through
$A(BC) = (AB)C$	Associativity of matrix multiplication
$IA = A$ and $BI = B$	Identity for matrix multiplication
$A(B + C) = AB + AC$	A left distributive law
$(A + B)C = AC + BC$	A right distributive law

Properties of the Transpose Operation

$(A^T)^T = A$	Transpose of the transpose
$(A + B)^T = A^T + B^T$	Transpose of a sum
$(AB)^T = B^T A^T$	Transpose of a product

$(XA)^T = A^T X^T$

Proofs of most of these properties involve routine computations with components and entries, as illustrated in the next example.

EXAMPLE 12 Prove that $r(A + B) = rA + rB$ for any two $m \times n$ matrices A and B and any scalar r.

Solution Let $A = [a_{ij}]$ and $B = [b_{ij}]$. The entry in the ith row and jth column position of $r(A + B)$ is

$$r(a_{ij} + b_{ij}) = ra_{ij} + rb_{ij},$$

which is the sum of the entry in the ith row and jth column position of rA and the entry in the corresponding position of rB. But by definition this is the entry in the ith row and jth column position of the matrix $rA + rB$. ■

SUMMARY

1. An $m \times n$ matrix is an ordered rectangular array of numbers containing m rows and n columns.

2. An $m \times 1$ matrix is a column vector with m components, and a $1 \times n$ matrix is a row vector with n components.

3. The dot product of vector **a** having components $a_1, a_2, \ldots, a_n$ and vector **b** having components $b_1, b_2, \ldots, b_n$ is the scalar $\mathbf{a} \cdot \mathbf{b} = a_1b_1 + a_2b_2 + \cdots + a_nb_n$.

4. The product AB of an $m \times n$ matrix A and an $n \times s$ matrix B is the $m \times s$ matrix C, whose entry c_{ij} in the ith row and jth column is the dot product of the ith row vector of A and the jth column vector of B. In general, $AB \neq BA$.

5. If $A = [a_{ij}]$ and $B = [b_{ij}]$ are matrices of the same size, then $A + B$ is the matrix of that size with entry $a_{ij} + b_{ij}$ in the ith row and jth column.

6. For any matrix A and scalar r, the matrix rA is found by multiplying each entry in A by r.

7. The transpose of an $m \times n$ matrix A is the $n \times m$ matrix A^T, which has as its kth row vector the kth column vector of A.

8. Properties of the dot product and matrix operations are given in boxed displays on page 11.

EXERCISES

In Exercises 1–16, let

$$A = \begin{bmatrix} -2 & 1 & 3 \\ 4 & 0 & -1 \end{bmatrix}, \quad B = \begin{bmatrix} 4 & 1 & -2 \\ 5 & -1 & 3 \end{bmatrix}, \quad C = \begin{bmatrix} 2 & -1 \\ 0 & 6 \\ -3 & 2 \end{bmatrix}, \quad and \quad D = \begin{bmatrix} -4 & 2 \\ 3 & 5 \\ -1 & -3 \end{bmatrix}.$$

Compute the indicated quantity, if it is defined.

1. $3A$
2. $0B$
3. $A + B$
4. $B + C$
5. $C - D$
6. $4A - 2B$
7. AB
8. CD
9. $(2A)(5C)$
10. $(5D)(4B)$
11. A^2
12. $(AC)^2$
13. $(2A - B)D$
14. ADB
15. $(A^T)A$
16. $(CD)^T$
17. Let

$$A = \begin{bmatrix} 2 & 0 & 0 \\ 0 & -1 & 0 \\ 0 & 0 & 1 \end{bmatrix}.$$

a) Find A^2. b) Find A^7.

18. Let

$$A = \begin{bmatrix} 0 & 0 & -1 \\ 0 & 2 & 0 \\ 2 & 0 & 0 \end{bmatrix}.$$

a) Find A^2. b) Find A^7.

19. Consider the row and column vectors

$$\mathbf{x} = \begin{bmatrix} -2 & 3 & -1 \end{bmatrix} \quad \text{and} \quad \mathbf{y} = \begin{bmatrix} 4 \\ -1 \\ 3 \end{bmatrix}.$$

Compute the matrix products $\mathbf{xy}$ and $\mathbf{yx}$.

20. Fill in the missing entries in the 4×4 matrix

$$\begin{bmatrix} 1 & -1 & 2 & 5 \\ -1 & 4 & -7 & 8 \\ 2 & -7 & -1 & 6 \\ 5 & 8 & 6 & 3 \end{bmatrix}$$

so that the matrix is symmetric.

21. Mark each of the following True or False. The statements involve matrices A, B, and C that are assumed to have appropriate size.

___ a) If $A = B$, then $AC = BC$.
___ b) If $AC = BC$, then $A = B$.
___ c) If $AB = O$, then $A = O$ or $B = O$.
___ d) If $A + C = B + C$, then $A = B$.
___ e) If $A^2 = I$, then $A = \pm I$.
___ f) If $B = A^2$ and if A is $n \times n$ and symmetric, then $b_{ii} \geq 0$ for $i = 1, 2, \ldots, n$.
___ g) If $AB = C$ and if two of the matrices are square, then so is the third.
___ h) If $AB = C$ and if C is a column vector, then so is B.
___ i) If $A^2 = I$, then $A^n = I$ for all integers $n \geq 2$.

___ j) If $A^2 = I$, then $A^n = I$ for all even integers $n \geq 2$.

22. a) Show that, if A is a matrix and $\mathbf{x}$ is a row vector, then $\mathbf{x}A$ (if defined) is again a row vector.

b) Show that, if A is a matrix and $\mathbf{y}$ is a column vector, then $A\mathbf{y}$ (if defined) is again a column vector.

In Exercises 23–34, show that the given relation holds for all vectors, matrices, and scalars for which the expressions are defined.

23. $\mathbf{u} \cdot \mathbf{v} = \mathbf{v} \cdot \mathbf{u}$
24. $\mathbf{u} \cdot (\mathbf{v} + \mathbf{w}) = (\mathbf{u} \cdot \mathbf{v}) + (\mathbf{u} \cdot \mathbf{w})$
25. $A + B = B + A$
26. $(A + B) + C = A + (B + C)$
27. $(r + s)A = rA + sA$
28. $(rs)A = r(sA)$
29. $A(B + C) = AB + AC$
30. $(A^T)^T = A$
31. $(A + B)^T = A^T + B^T$
32. $(AB)^T = B^T A^T$
33. $(AB)C = A(BC)$
34. $(rA)B = A(rB) = r(AB)$
35. If B is an $m \times n$ matrix and if $B = A^T$, find the size of
   a) A,
   b) AA^T,
   c) $A^T A$.
36. Let A and B be $n \times n$ matrices, and let $\mathbf{v}$ and $\mathbf{w}$ be column vectors with n components each. Express $(A\mathbf{v}) \cdot (B\mathbf{w})$, using just multiplication of certain matrices. [HINT: See Eq. (5).]
37. The Hilbert matrix H_n is the $n \times n$ matrix $[h_{ij}]$, where $h_{ij} = 1/(i + j - 1)$. Show that the matrix H_n is symmetric.
38. Show that, if A is a square matraix, then $A + A^T$ is symmetric.
39. Show that, if A is a matrix, then AA^T is symmetric.
40. a) Show that, if A is a square matrix, then $(A^2)^T = (A^T)^2$ and $(A^3)^T = (A^T)^3$. [HINT: Don't try to show that the matrices have equal entries; instead use Exercise 32.]

b) State the generalization of part (a), and give a proof using mathematical induction (see Appendix A).

41. a) Let A be an $m \times n$ matrix, and let $\mathbf{e}_j$ be the $n \times 1$ column vector whose jth component is 1 and whose other components are 0. Show that $A\mathbf{e}_j$ is the jth column vector of A.

b) Let A and B be matrices of the same size.

   i) Show that, if $A\mathbf{x} = \mathbf{0}$ (the zero vector) for all $\mathbf{x}$, then $A = O$, the zero matrix. [HINT: Use part (a).]

   ii) Show that, if $A\mathbf{x} = B\mathbf{x}$ for all $\mathbf{x}$, then $A = B$. [HINT: Consider $A - B$.]

42. Let A and B be square matrices. Is

$$(A + B)^2 = A^2 + 2AB + B^2?$$

If so, prove it; if not, give a counterexample and state under what conditions the equation is true.

43. Let A and B be square matrices. Is

$$(A + B)(A - B) = A^2 - B^2?$$

If so, prove it; if not, give a counterexample and state under what conditions the equation is true.

■ *The software LINTEK includes a program, MATCOMP, that performs the matrix operations described in this section. Let*

$$A = \begin{bmatrix} 4 & 6 & 0 & 1 & -9 \\ 2 & 11 & 5 & 2 & -5 \\ -1 & 2 & -4 & 5 & 7 \\ 0 & 12 & -8 & 4 & 3 \\ 10 & 4 & 6 & 2 & -5 \end{bmatrix} \quad and$$

$$B = \begin{bmatrix} -8 & 15 & 4 & -11 \\ 3 & 5 & 6 & -2 \\ 0 & -1 & 12 & 5 \\ 1 & 13 & -15 & 7 \\ 6 & -8 & 0 & -5 \end{bmatrix}.$$

Use MATCOMP, or similar software, to enter and store these matrices, and then compute the matrices in Exercises 44–51, if they are defined.

44. $A^4 + A$ 45. A^2B 46. $A^3(A^T)^2$

47. BA^2 48. $B^T(2A)$ 49. $AB(AB)^T$

50. $(2A)^3 - A^5$ 51. $(A^T)^5$

1.2   SOLVING SYSTEMS OF LINEAR EQUATIONS

Solving a system of linear equations is a fundamental problem of linear algebra. Many of the computational exercises in this text involve solving such linear systems. This section presents an algorithm for solving any linear system.

The solution set of any system of equations is the intersection of the solution sets of the individual equations. That is, any solution of a system must be a solution of each equation in the system; and conversely, any solution of every equation in the system is considered to be a solution of the system. Bearing this in mind, we start with an intuitive discussion of the geometry of linear systems; a more detailed study of this geometry appears in Section 2.8.

The Geometry of Linear Systems

Frequently, students are under the impression that a linear system containing the same number of equations as unknowns always has a unique solution, whereas a system having more equations than unknowns never has a solution. The geometrical interpretation of the problem shows that these statements are not true.

We know that a single linear equation in two unknowns has a line in the plane as solution set. Similarly, a single linear equation in three unknowns has a plane in

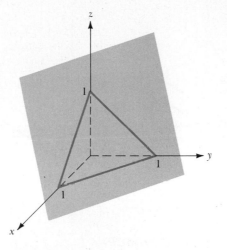

FIGURE 1.1 **The plane $x + y + z = 1$.**

space as solution set. The solution set of the equation $x + y + z = 1$ is the plane sketched in Figure 1.1. This geometric analysis can be extended to an equation that has more than three variables, but it is difficult for us to represent the solution set of such an equation graphically.

Two lines in the plane usually intersect in a single point; here the word *usually* means that, if the lines are selected in some random way, the chance that they are either parallel (have empty intersection) or coincide (have an infinite number of points in their intersection) is very small. Thus we see that a system of two randomly selected equations in two unknowns can be expected to have a unique solution. However, it

---

SYSTEMS OF LINEAR EQUATIONS are found in ancient babylonian and chinese texts dating back well over 2000 years. The problems are generally stated in real-life terms, but it is clear that they are artificial and designed simply to train students in mathematical procedures. As an example of a Babylonian problem, consider the following, which has been slightly modified from the original found on a clay tablet of about 300 B.C.: There are two fields whose total area is 1800 square yards. One produces grain at a rate of $\frac{2}{3}$ bushel per square yard, the other at a rate of $\frac{1}{2}$ bushel per square yard. The total yield of the two fields is 1100 bushels. What is the size of each field? This problem leads to the system

$$x + \quad y = 1800$$
$$\left(\tfrac{2}{3}\right)x + \left(\tfrac{1}{2}\right)y = 1100.$$

A typical Chinese problem, taken from the Han dynasty text *Nine Chapters of the Mathematical Art* (about 200 B.C.), reads as follows: There are three classes of corn, of which three bundles of the first class, two of the second, and one of the third make 39 measures. Two of the first, three of the second, and one of the third make 34 measures. And one of the first, two of the second, and three of the third make 26 measures. How many measures of grain are contained in one bundle of each class? The system of equations here is

$$3x + 2y + \quad z = 39$$
$$2x + 3y + \quad z = 34$$
$$x + 2y + 3z = 26.$$

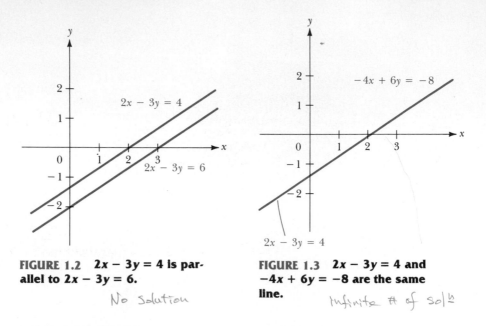

FIGURE 1.2 $2x - 3y = 4$ is parallel to $2x - 3y = 6$.

No Solution

FIGURE 1.3 $2x - 3y = 4$ and $-4x + 6y = -8$ are the same line.

Infinite # of sol^{n}

is *possible* for the system to have no solutions or an infinite number of solutions. For example, the equations

$$2x - 3y = 4$$
$$2x - 3y = 6$$

correspond to distinct parallel lines, as shown in Figure 1.2, and the system consisting of these equations has no solutions. Moreover, the equations

$$2x - 3y = \ \ 4$$
$$-4x + 6y = -8$$

correspond to the same line, as shown in Figure 1.3. All points on this line are solutions of this system of two equations. And because it is possible to have any number of lines in the plane—say, fifty lines—pass through a single point, it is possible for a system of fifty equations in only two unknowns to have a unique solution.

Similar illustrations can be made in space, where a linear equation has as solution set a plane. Three randomly chosen planes can be expected to have a unique point in common. Two of them can be expected to intersect in a line (see Fig. 1.4), which in turn can be expected to meet the third plane in a single point. However, it is possible for three planes to have no point in common, giving rise to a linear system with no solutions. It is also possible for all three planes to contain a common line, in which case the corresponding linear system will have an infinite number of solutions.

Elementary Row Operations

In this subsection we describe operations that can be used to modify the equations of a linear system to obtain a system having the same solutions, but whose solutions are obvious. The most general type of linear system may have m equations in n unknowns.

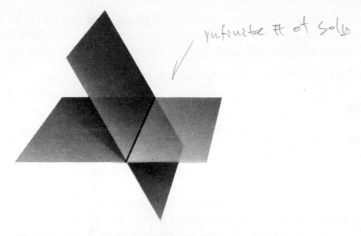

infinite # of solu

FIGURE 1.4 **Two planes intersecting in a line.**

Such a system may be written as

$$
\begin{aligned}
a_{11}x_1 + a_{12}x_2 &+ \cdots + a_{1n}x_n = b_1 \\
a_{21}x_1 + a_{22}x_2 &+ \cdots + a_{2n}x_n = b_2 \\
&\vdots \\
a_{m1}x_1 + a_{m2}x_2 &+ \cdots + a_{mn}x_n = b_m.
\end{aligned}
\tag{1}
$$

System (1) is completely determined by its $m \times n$ **coefficient matrix** $A = [a_{ij}]$ and by the column vector $\mathbf{b}$ with ith entry b_i. The system can be written as the single matrix equation

$$
A\mathbf{x} = \mathbf{b},
\tag{2}
$$

where $\mathbf{x}$ is the column vector with ith entry x_i. Any column vector $\mathbf{s}$ such that $A\mathbf{s} = \mathbf{b}$ is a **solution** of system (1).

The **partitioned matrix** or **augmented matrix**

$$
\left[
\begin{array}{cccc|c}
a_{11} & a_{12} & \cdots & a_{1n} & b_1 \\
a_{21} & a_{22} & \cdots & a_{2n} & b_2 \\
 & & \vdots & & \vdots \\
a_{m1} & a_{m2} & \cdots & a_{mn} & b_m
\end{array}
\right]
\tag{3}
$$

is a shorthand summary of system (1). The coefficient matrix has been *augmented* by the column vector of constants. We denote matrix (3) by $[A \mid \mathbf{b}]$.

We shall see how to determine all solutions of system (1) by manipulating augmented matrix (3) using *elementary row operations*. The elementary row operations correspond to the following familiar operations with equations of system (1):

R1 Interchange two equations in system (1).

R2 Multiply an equation in system (1) by a nonzero constant.

R3 Replace an equation in system (1) with the sum of itself and a multiple of a different equation of the system.

It is clear that operations R1 and R2 do not change the solution sets of the equations they affect. Therefore, they do not change the intersection of the solution sets of the equations; that is, the solution set of the system is unchanged. The fact that R2 does not change the solution set and the familiar algebraic principle, "Equals added to equals yield equals," show that any solution of both the ith and jth equations is also a solution of a new jth equation obtained by adding r times the ith equation to the jth equation. Thus operation R3 yields a system having all the solutions of the original one. Since the original system can be recovered from the new one by multiplying the ith equation by $-r$ and adding it to the new jth equation (an R3 operation), we see that the original system has all the solutions of the new one. Hence R3, too, does not alter the solution set of system (1).

These procedures applied to system (1) correspond to the **elementary row operations** (listed in the box) applied to partitioned matrix (3).

A MATRIX-REDUCTION METHOD of solving a system of linear equations occurs in the ancient Chinese work, *Nine Chapters of the Mathematical Art*. The author presents the following solution to the system

$$3x + 2y + \ z = 39$$
$$2x + 3y + \ z = 34$$
$$x + 2y + 3z = 26.$$

The diagram of the coefficients is to be set up on a "counting board":

$$
\begin{array}{ccc}
1 & 2 & 3 \\
2 & 3 & 2 \\
3 & 1 & 1 \\
26 & 34 & 39
\end{array}
$$

He then instructs the reader to multiply the middle column by 3 and subsequently to subtract the right column "as many times as possible"; the same is to be done to the left column. The new diagrams are then

$$
\begin{array}{ccc}
1 & 0 & 3 \\
2 & 5 & 2 \\
3 & 1 & 1 \\
26 & 24 & 39
\end{array}
\quad \text{and} \quad
\begin{array}{ccc}
0 & 0 & 3 \\
4 & 5 & 2 \\
8 & 1 & 1 \\
39 & 24 & 39
\end{array}
$$

The next instruction is to multiply the left column by 5 and then to subtract the middle column as many times as possible. This gives

$$
\begin{array}{ccc}
0 & 0 & 3 \\
0 & 5 & 2 \\
36 & 1 & 1 \\
99 & 24 & 39
\end{array}
$$

The system has thus been reduced to the system $3x + 2y + z = 39$, $5y + z = 24$, $36z = 99$, from which the complete solution is easily found.

Elementary Row Operations

R1 (*Row interchange*) Interchange two row vectors in a matrix.

R2 (*Row scaling*) Multiply any row vector in the matrix by a nonzero scalar.

R3 (*Row addition*) Replace any row vector in the matrix with the sum of itself and a multiple of a different row vector.

$B \sim A$

If a matrix B can be obtained from a matrix A by means of a sequence of elementary row operations, then B is **row equivalent** to A. Each elementary row operation can be undone by another of the same type. For example, an R2 operation that multiplies a row vector by r can be undone by multiplying the resulting vector by $1/r$, and we saw above that an R3 operation can be undone by another R3 operation. Thus, if B is row equivalent to A, then A is row equivalent to B; and we can simply speak of *row-equivalent matrices A and B*, which we denote by $A \sim B$. We have just seen that the operations on a linear system $A\mathbf{x} = \mathbf{b}$ corresponding to these elementary row operations on the augmented matrix $[A \mid \mathbf{b}]$ do not change the solution set of the system. This gives us at once the following theorem, which is the foundation for the algorithm we will present for solving linear systems.

THEOREM 1.1 Invariance of Solution Sets Under Row Equivalence

If $[A \mid \mathbf{b}]$ and $[H \mid \mathbf{c}]$ are row-equivalent partitioned matrices, then the linear systems $A\mathbf{x} = \mathbf{b}$ and $H\mathbf{x} = \mathbf{c}$ have the same solution sets.

Row Echelon Form

We will solve a linear system $A\mathbf{x} = \mathbf{b}$ by row-reducing the partitioned matrix $[A \mid \mathbf{b}]$ to a partitioned matrix $[H \mid \mathbf{c}]$, where H is a matrix in *row echelon form* (which we now define).

DEFINITION 1.6 Row Echelon Form, Pivot

A matrix is in **row echelon form** if it satisfies two conditions:

1. All rows containing only zeros appear below rows with nonzero entries.
2. The first nonzero entry in any row appears in a column to the right of the first nonzero entry in any preceding row.

For such a matrix, the first nonzero entry in a row is the **pivot** for that row.

EXAMPLE 1 Determine which of the matrices

$$A = \begin{bmatrix} 1 & 3 & 2 \\ 0 & 0 & 0 \\ 0 & 0 & 1 \end{bmatrix}, \quad B = \begin{bmatrix} 2 & 4 & 0 \\ 1 & 3 & 2 \\ 0 & 0 & 0 \end{bmatrix}, \quad C = \begin{bmatrix} 0 & -1 & 2 \\ 0 & 0 & 3 \\ 0 & 0 & 0 \\ 0 & 0 & 0 \end{bmatrix},$$

$$D = \begin{bmatrix} 1 & 3 & 2 & 5 \\ 0 & 0 & 1 & 3 \\ 0 & 0 & 0 & 1 \\ 0 & 0 & 0 & 0 \end{bmatrix}$$

are in row echelon form.

Solution Matrix A is not in row echelon form, since the second row (consisting of all zero entries) is not below the third row (which has a nonzero entry).

Matrix B is not in row echelon form, since the first nonzero entry (1, in the second row) does not appear in a column to the right of the first nonzero entry (2, in the first row).

Matrix C is in row echelon form, since both conditions of Definition 1.6 are satisfied. The pivots are -1 and 3.

Matrix D satisfies both conditions, as well, and is in row echelon form. The pivots are the entries 1. ■

Solutions of $H\mathbf{x} = \mathbf{c}$

We illustrate by examples that, if a linear system $H\mathbf{x} = \mathbf{c}$ has coefficient matrix H in row echelon form, it is easy to determine all solutions of the system. We color the pivots in H in these examples.

EXAMPLE 2 Find all solutions of $H\mathbf{x} = \mathbf{c}$, where

$$[H \mid \mathbf{c}] = \begin{bmatrix} -5 & -1 & 3 & \big| & 3 \\ 0 & 3 & 5 & \big| & 8 \\ 0 & 0 & 2 & \big| & -4 \end{bmatrix}.$$

Solution The equations corresponding to this augmented matrix are

$$-5x_1 - x_2 + 3x_3 = 3$$
$$3x_2 + 5x_3 = 8$$
$$2x_3 = -4.$$

From the last equation, we obtain $x_3 = -2$. Substituting in the second equation, we have

$$3x_2 + 5(-2) = 8, \quad 3x_2 = 18, \quad x_2 = 6.$$

Finally, we substitute these values for x_2 and x_3 into the top equation, obtaining

$$-5x_1 - 6 + 3(-2) = 3, \quad -5x_1 = 15, \quad x_1 = -3.$$

Thus the only solution is

$$\mathbf{x} = \begin{bmatrix} -3 \\ 6 \\ -2 \end{bmatrix},$$

or equivalently, $x_1 = -3$, $x_2 = 6$, $x_3 = -2$. ▪

The procedure for finding the solution of $H\mathbf{x} = \mathbf{c}$ illustrated in Example 2 is called **back substitution**, since the values of the variables are found in backward order, starting with the variable with the largest subscript.

By multiplying each nonzero row in $[H \mid \mathbf{b}]$ by the reciprocal of its pivot, we can assume that each pivot in H is 1. We assume that this has been done in the next two examples.

EXAMPLE 3   Use back substitution to find all solutions of $H\mathbf{x} = \mathbf{c}$, where

$$[H \mid \mathbf{c}] = \begin{bmatrix} 1 & -3 & 5 & \mid & 3 \\ 0 & 1 & 2 & \mid & 2 \\ 0 & 0 & 0 & \mid & -1 \end{bmatrix}.$$

Solution   The equation corresponding to the last row of this partitioned matrix is

$$0x_1 + 0x_2 + 0x_3 = -1.$$

This equation has no solutions, because the left side is 0 for any values of the variables while the right side is -1. ▪

DEFINITION 1.7   Consistent Linear System

A linear system having no solutions is **inconsistent**. If it has one or more solutions, the linear system is said to be **consistent**.

Now we illustrate a many-solutions case.

EXAMPLE 4   Use back substitution to find all solutions of $H\mathbf{x} = \mathbf{c}$, where

$$[H \mid \mathbf{c}] = \begin{bmatrix} 1 & -3 & 0 & 5 & 0 & \mid & 4 \\ 0 & 0 & 1 & 2 & 0 & \mid & -7 \\ 0 & 0 & 0 & 0 & 1 & \mid & 1 \\ 0 & 0 & 0 & 0 & 0 & \mid & 0 \end{bmatrix}.$$

Solution   The reduced linear system corresponding to this partitioned matrix is

$$x_1 - 3x_2 \quad + 5x_4 \quad = \quad 4$$
$$x_3 + 2x_4 \quad = -7$$
$$x_5 = \quad 1.$$

We solve each equation for the variable corresponding to the colored pivot in the matrix. Thus we obtain

$$x_1 = 3x_2 - 5x_4 + 4$$
$$x_3 = -2x_4 - 7$$
$$x_5 = 1. \tag{4}$$

Notice that x_2 and x_4 correspond to columns of H containing no pivot. We can assign any value r we please to x_2 and any value s to x_4, and we can then use system (4) to determine corresponding values for x_1, x_3, and x_5. Thus the system has an infinite number of solutions. We describe all solutions by the vector equation

$$\mathbf{x} = \begin{bmatrix} x_1 \\ x_2 \\ x_3 \\ x_4 \\ x_5 \end{bmatrix} = \begin{bmatrix} 3r - 5s + 4 \\ r \\ -2s - 7 \\ s \\ 1 \end{bmatrix} \quad \text{for any scalars } r \text{ and } s. \tag{5}$$

We call x_2 and x_4 **free variables,** and we refer to Eq. (5) as the **general solution** of the system. We obtain **particular solutions** by setting r and s equal to specific values. For example, we obtain

$$\begin{bmatrix} 2 \\ 1 \\ -9 \\ 1 \\ 1 \end{bmatrix} \text{ for } r = s = 1, \quad \text{and} \quad \begin{bmatrix} 25 \\ 2 \\ -1 \\ -3 \\ 1 \end{bmatrix} \text{ for } r = 2, s = -3. \quad ■$$

Gauss Reduction of $A\mathbf{x} = \mathbf{b}$ *to* $H\mathbf{x} = \mathbf{c}$

We now show how to reduce an augmented matrix $[A \mid \mathbf{b}]$ to the form $[H \mid \mathbf{c}]$, where H is in row echelon form, using a sequence of elementary row operations. Examples 2 through 4 illustrated how to use back substitution afterward to find solutions of the system $H\mathbf{x} = \mathbf{c}$, which are the same as the solutions of $A\mathbf{x} = \mathbf{b}$, by Theorem 1.1. This procedure for solving $A\mathbf{x} = \mathbf{b}$ is known as **Gauss reduction with back substitution**. In the box, we give an outline for reducing a matrix A to row echelon form.

Reducing a Matrix A to Row Echelon Form H

1. If the first column of A contains only zero entries, cross it off mentally. Continue in this fashion until the left column of the remaining matrix has a nonzero entry or until the columns are exhausted.

2. Use row interchange, if necessary, to obtain a nonzero entry (pivot) p in the top row of the first column of the remaining matrix. For each row below that has a nonzero entry r in the first column, add $-r/p$ times the top row to that row, to create a zero in the first column. In

> this fashion, create zeros below p in the entire first column of the remaining matrix.
> 3. Mentally cross off this first column and the first row of the matrix, to obtain a smaller matrix. (See the shaded portion of the third matrix in the solution of Example 5.) Go back to step 1, and repeat the process with this smaller matrix until either no rows or no columns remain.

EXAMPLE 5   Reduce the matrix

$$\begin{bmatrix} 2 & -4 & 2 & -2 \\ 2 & -4 & 3 & -4 \\ 4 & -8 & 3 & -2 \\ 0 & 0 & -1 & 2 \end{bmatrix}$$

to row echelon form, making all pivots 1.

Solution   We follow the boxed outline and color the pivots of 1. Remember that the symbol $\sim$ denotes row-equivalent matrices.

$$\begin{bmatrix} 2 & -4 & 2 & -2 \\ 2 & -4 & 3 & -4 \\ 4 & -8 & 3 & -2 \\ 0 & 0 & -1 & 2 \end{bmatrix}$$ **Multiply the first row by $\frac{1}{2}$, to produce a pivot of 1 in the next matrix.**

$$\sim \begin{bmatrix} 1 & -2 & 1 & -1 \\ 2 & -4 & 3 & -4 \\ 4 & -8 & 3 & -2 \\ 0 & 0 & -1 & 2 \end{bmatrix}$$ **Add -2 times row 1 to row 2, and then add -4 times row 1 to row 3, to obtain the next matrix.**

$$\sim \begin{bmatrix} 1 & -2 & 1 & -1 \\ 0 & 0 & 1 & -2 \\ 0 & 0 & -1 & 2 \\ 0 & 0 & -1 & 2 \end{bmatrix}$$ **Cross off the first shaded column of zeros (mentally), to obtain the next shaded matrix.**

$$\sim \begin{bmatrix} 1 & -2 & 1 & -1 \\ 0 & 0 & 1 & -2 \\ 0 & 0 & -1 & 2 \\ 0 & 0 & -1 & 2 \end{bmatrix}$$ **Add row 2 to rows 3 and 4, to obtain the final matrix.**

THE GAUSS SOLUTION METHOD is so named because Gauss described it in a paper detailing the computations he made to determine the orbit of the asteroid Pallas. The parameters of the orbit had to be determined by observations of the asteroid over a 6-year period from 1803 to 1809. These led to six linear equations in six unknowns with quite complicated coefficients. Gauss showed how to solve these equations by systematically replacing them with a new system in which only the first equation had all six unknowns, the second equation included five unknowns, the third equation only four, and so on, until the sixth equation had but one. This last equation could, of course, be easily solved; the remaining unknowns were then found by back substitution.

$$\sim \begin{bmatrix} 1 & -2 & 1 & -1 \\ 0 & 0 & 1 & -2 \\ 0 & 0 & 0 & 0 \\ 0 & 0 & 0 & 0 \end{bmatrix}.$$

This last matrix is in row echelon form, with both pivots equal to 1. ■

To solve a linear system $A\mathbf{x} = \mathbf{b}$, we form the augmented matrix $[A \mid \mathbf{b}]$ and row-reduce it to $[H \mid \mathbf{c}]$, where H is in row echelon form. We can follow the steps outlined in the box preceding Example 5 for row-reducing A to H. Of course, we always perform the elementary row operations on the full augmented matrix, including the entries in the column to the right of the partition.

EXAMPLE 6   Solve the linear system

$$x_2 - 3x_3 = -5$$
$$2x_1 + 3x_2 - x_3 = 7$$
$$4x_1 + 5x_2 - 2x_3 = 10,$$

using Gauss reduction with back substitution.

Solution   We reduce the corresponding augmented matrix, using elementary row operations. Pivots are colored.

$$\begin{bmatrix} 0 & 1 & -3 & -5 \\ 2 & 3 & -1 & 7 \\ 4 & 5 & -2 & 10 \end{bmatrix} \sim \begin{bmatrix} 2 & 3 & -1 & 7 \\ 0 & 1 & -3 & -5 \\ 4 & 5 & -2 & 10 \end{bmatrix} \quad \text{Interchange row 1 and row 2.}$$

$$\begin{bmatrix} 2 & 3 & -1 & 7 \\ 0 & 1 & -3 & -5 \\ 4 & 5 & -2 & 10 \end{bmatrix} \sim \begin{bmatrix} 2 & 3 & -1 & 7 \\ 0 & 1 & -3 & -5 \\ 0 & -1 & 0 & -4 \end{bmatrix} \quad \text{Add } -2 \text{ times row 1 to row 3.}$$

$$\begin{bmatrix} 2 & 3 & -1 & 7 \\ 0 & 1 & -3 & -5 \\ 0 & -1 & 0 & -4 \end{bmatrix} \sim \begin{bmatrix} 2 & 3 & -1 & 7 \\ 0 & 1 & -3 & -5 \\ 0 & 0 & -3 & -9 \end{bmatrix}. \quad \text{Add row 2 to row 3.}$$

From the last partitioned matrix, we could proceed to write the corresponding equations (as in Example 2) and to solve in succession for x_3, x_2, and x_1 by back substitution. However, it makes sense to keep using our shorthand, without writing out variables, and to do our back substitution in terms of partitioned matrices. Starting with the final partitioned matrix in the preceding set, we obtain

$$\begin{bmatrix} 2 & 3 & -1 & 7 \\ 0 & 1 & -3 & -5 \\ 0 & 0 & -3 & -9 \end{bmatrix} \sim \begin{bmatrix} 2 & 3 & -1 & 7 \\ 0 & 1 & -3 & -5 \\ 0 & 0 & 1 & 3 \end{bmatrix} \quad \begin{array}{l} \text{Multiply row 3 by } -\frac{1}{3}. \text{ This} \\ \text{shows that } x_3 = 3. \end{array}$$

$$\begin{bmatrix} 2 & 3 & -1 & 7 \\ 0 & 1 & -3 & -5 \\ 0 & 0 & 1 & 3 \end{bmatrix} \sim \begin{bmatrix} 2 & 3 & 0 & 10 \\ 0 & 1 & 0 & 4 \\ 0 & 0 & 1 & 3 \end{bmatrix} \quad \begin{array}{l} \text{Add row 3 to row 1; add 3} \\ \text{times row 3 to row 2. This} \\ \text{shows that } x_2 = 4. \end{array}$$

$$\begin{bmatrix} 2 & 3 & 0 & | & 10 \\ 0 & 1 & 0 & | & 4 \\ 0 & 0 & 1 & | & 3 \end{bmatrix} \sim \begin{bmatrix} 2 & 0 & 0 & | & -2 \\ 0 & 1 & 0 & | & 4 \\ 0 & 0 & 1 & | & 3 \end{bmatrix}.$$

Add -3 times row 2 to row 1.
This shows that $x_1 = -1$.

We have found the solution: $x_1 = -1$, $x_2 = 4$, $x_3 = 3$. ■

When working with pencil and paper, we can avoid writing so many matrices by using a modified form of the Gauss method with back substitution, known as the **Gauss–Jordan method**. With the Gauss–Jordan method, we make pivots equal to 1 and create zeros *above* as well as below each pivot, as we reduce the matrix to row echelon form. We show in Chapter 9 that, for a large system, it takes about 50% more time for a computer to use the Gauss–Jordan method than to use Gauss reduction with back substituion. That is, creating the zeros above the pivots requires less computation if we do it *after* the matrix is reduced to row echelon form than if we do it as we go along. However, with pencil and paper work, we are faced with fewer matrices to write out if we use the Gauss–Jordan method. Our final example illustrates the procedure.

EXAMPLE 7 Solve the linear system

$$\begin{aligned} x_1 - 2x_2 + x_3 - x_4 &= 4 \\ 2x_1 - 3x_2 + 2x_3 - 3x_4 &= -1 \\ 3x_1 - 5x_2 + 3x_3 - 4x_4 &= 3 \\ -x_1 + x_2 - x_3 + 2x_4 &= 5, \end{aligned}$$

using the Gauss–Jordan method and making all pivots 1.

Solution We form the corresponding partitioned matrix and reduce it to echelon form, while also creating zeros above the pivots of 1. We do not recopy matrices this time; instead, we fix up an entire column at each step.

$$\begin{bmatrix} 1 & -2 & 1 & -1 & | & 4 \\ 2 & -3 & 2 & -3 & | & -1 \\ 3 & -5 & 3 & -4 & | & 3 \\ -1 & 1 & -1 & 2 & | & 5 \end{bmatrix}$$

Create zeros under the pivot 1 in row 1, column 1, to obtain the next matrix.

THE JORDAN HALF OF THE GAUSS–JORDAN METHOD is essentially a systematic technique of back substitution. In this form it was described by Wilhelm Jordan (1842–1899), a German professor of geodesy, in his *Handbook of Geodesy*. This work was first published in German in 1873 and has since gone through ten editions, as well as translations into several other languages. On the other hand, Jordan himself credits the method to Friedrich Robert Helmert and Peter Andreas Hansen, who developed it a few years earlier.

Wilhelm Jordan was prominent in his field in the late nineteenth century, being involved in several geodetic surveys in Germany and in the first major survey of the Libyan desert. He was also the founding editor of the German geodesy journal. His interest in finding a systematic method of solving large systems of linear equations stems from their frequent appearance in problems of triangulation.

$$\sim \begin{bmatrix} 1 & -2 & 1 & -1 & \bigm| & 4 \\ 0 & 1 & 0 & -1 & \bigm| & -9 \\ 0 & 1 & 0 & -1 & \bigm| & -9 \\ 0 & -1 & 0 & 1 & \bigm| & 9 \end{bmatrix}$$

Create zeros above and below the pivot 1 in row 2, column 2, to obtain the next matrix.

RREF $\longrightarrow$

$$\sim \begin{bmatrix} 1 & 0 & 1 & -3 & \bigm| & -14 \\ 0 & 1 & 0 & -1 & \bigm| & -9 \\ 0 & 0 & 0 & 0 & \bigm| & 0 \\ 0 & 0 & 0 & 0 & \bigm| & 0 \end{bmatrix}.$$

The reduction is complete, since the final two rows are **0**. We see that the system is consistent. Since columns 3 and 4 do not contain pivots, we take x_3 and x_4 as free variables, setting $x_3 = r$ and $x_4 = s$, and solve for x_1 and x_2. We thus obtain the general solution

$$\mathbf{x} = \begin{bmatrix} x_1 \\ x_2 \\ x_3 \\ x_4 \end{bmatrix} = \begin{bmatrix} -r + 3s - 14 \\ s - 9 \\ r \\ s \end{bmatrix}.$$ ■

A matrix in row echelon form with all pivots equal to 1 and with zeros above as well as below each pivot is said to be in **reduced row echleon form**. Thus the Gauss–Jordan method consists of using elementary row operations on an augmented matrix $[A \mid \mathbf{b}]$ to bring the coefficient matrix A into reduced row echelon form. It can be shown that the reduced row echelon form of a matrix A is unique. (See Section 2.9, Exercise 35.)

The examples we have given illustrate the three possibilities for solutions of a linear system—namely, no solutions (inconsistent system), a unique solution, or an infinite number of solutions. We state this formally in a theorem and prove it.

THEOREM 1.2 Solutions of $Ax = b$

Let $A\mathbf{x} = \mathbf{b}$ be a linear system, and let $[A \mid \mathbf{b}] \sim [H \mid \mathbf{c}]$, where H is in row echelon form.

1. The system $A\mathbf{x} = \mathbf{b}$ is inconsistent if and only if the partitioned matrix $[H \mid \mathbf{c}]$ has a row with all entries 0 to the left of the partition and a nonzero entry to the right of the partition.
2. If $A\mathbf{x} = \mathbf{b}$ is consistent and every column of H contains a pivot, the system has a unique solution.
3. If $A\mathbf{x} = \mathbf{b}$ is consistent and some column of H has no pivot, the system has infinitely many solutions, with as many free variables as there are pivot-free columns in H.

PROOF   If $[H \mid \mathbf{c}]$ has an ith row with all entries 0 to the left of the partition and a nonzero entry c_i to the right of the partition, the corresponding ith equation in the system $H\mathbf{x} = \mathbf{c}$ is $0x_1 + 0x_2 + \cdots + 0x_n = c_i$, which has no solutions; therefore, the system $A\mathbf{x} = \mathbf{b}$ has no solutions, by Theorem 1.1. The next paragraph shows that, if H contains no such row, we can find a solution to the system. Thus the system is inconsistent if and only if H contains such a row.

Assume now that $[H \mid \mathbf{c}]$ has no row with all entries 0 to the left of the partition and a nonzero entry to the right. If the ith row of $[H \mid \mathbf{c}]$ is a zero row vector, the corresponding equation $0x_1 + 0x_2 + \cdots + 0x_n = 0$ is satisfied for all values of the variables x_j, and thus it can be deleted from the system $H\mathbf{x} = \mathbf{c}$. Assume that this has been done wherever possible, so that $[H \mid \mathbf{c}]$ has no zero row vectors. For each j such that the jth column has no pivot, we can set x_j equal to any value we please (as in Examples 4 and 7) and then, starting from the last remaining equation of the system and working back to the first, solve in succession for the variables corresponding to the columns containing the pivots. If some column j has no pivot, there are an infinite number of solutions, since x_j can be set equal to any value. On the other hand, if every column has a pivot (as in Examples 2 and 6), the value of each x_j is uniquely determined. ▲

With reference to item (3) of Theorem 1.2, the number of free variables in the solution set of a system $A\mathbf{x} = \mathbf{b}$ depends only on the system, and not on the way in which the matrix A is reduced to row echelon form. This seems intuitively reasonable and will be proved in Chapter 2.

SUMMARY

1. A linear system has an associated partitioned (or augmented) matrix, having the coefficient matrix of the system on the left of the partition and the column vector of constants on the right of the partition.

2. The elementary row operations on a matrix are as follows:
   R1 (*Row interchange*) Interchange of two rows;
   R2 (*Row scaling*) Multiplication of a row by a nonzero scalar;
   R3 (*Row addition*) Addition of a multiple of a row to a different row.

3. Matrices A and B are row equivalent (written $A \sim B$) if A can be transformed into B by a sequence of elementary row operations.

4. If $A\mathbf{x} = \mathbf{b}$ and $H\mathbf{x} = \mathbf{c}$ are systems such that the partitioned matrices $[A \mid \mathbf{b}]$ and $[H \mid \mathbf{c}]$ are row equivalent, the systems $A\mathbf{x} = \mathbf{b}$ and $H\mathbf{x} = \mathbf{c}$ have the same solution set.

5. A matrix is in row echelon form if:

   (i) All rows containing only zero entries are grouped together at the bottom of the matrix.

   (ii) The first nonzero element (the pivot) in any row appears in a column to the right of the first nonzero element in any preceding row.

6. A matrix is in reduced row echelon form if it is in row echelon form and, in addition, each pivot is 1 and is the only nonzero element in its column. Every matrix is row equivalent to a unique matrix in reduced row echelon form.

7. In the Gauss method with back substitution, we solve a linear system by reducing the augmented matrix so that the portion to the left of the partition is in row echelon form. The solution is then found by back substitution.

8. The Gauss–Jordan method is similar to the Gauss method, except that the coefficient matrix of the system is brought into reduced row echelon form.

9. A linear system $A\mathbf{x} = \mathbf{b}$ has no solutions if and only if, after $[A \mid \mathbf{b}]$ is row-reduced so that A is transformed into row echelon form, there exists a row with only zero entries to the left of the partition but with a nonzero entry to the right of the partition. The linear system is then *inconsistent*.

10. If $A\mathbf{x} = \mathbf{b}$ is a consistent linear system and if a row echelon form H of A has at least one column containing no (nonzero) pivot, the system has an infinite number of solutions. The free variables corresponding to the columns containing no pivots can be assigned any values, and the reduced linear system can then be solved for the remaining variables.

EXERCISES

In Exercises 1–6, reduce the matrix to (a) row echelon form, and (b) reduced row echelon form. Answers to (a) are not unique, so your answer may differ from the one at the back of the text.

1. $\begin{bmatrix} 2 & 1 & 4 \\ 1 & 3 & 2 \\ 3 & -1 & 6 \end{bmatrix}$
2. $\begin{bmatrix} 2 & 4 & -2 \\ 4 & 8 & 3 \\ -1 & -3 & 0 \end{bmatrix}$

3. $\begin{bmatrix} 0 & 2 & -1 & 3 \\ -1 & 1 & 2 & 0 \\ 1 & 1 & -3 & 3 \\ 1 & 5 & 5 & 9 \end{bmatrix}$

4. $\begin{bmatrix} 0 & 0 & 3 & -2 \\ 0 & 0 & 1 & 2 \\ 1 & 3 & 2 & -4 \end{bmatrix}$

5. $\begin{bmatrix} -1 & 3 & 0 & 1 & 4 \\ 1 & -3 & 0 & 0 & -1 \\ 2 & -6 & 2 & 4 & 0 \\ 0 & 0 & 1 & 3 & -4 \end{bmatrix}$

6. $\begin{bmatrix} 0 & 0 & 1 & 2 & -1 & 4 \\ 0 & 0 & 0 & 1 & -1 & 3 \\ 2 & 4 & -1 & 3 & 2 & -1 \end{bmatrix}$

In Exercises 7–12, describe all solutions of a linear system whose corresponding partitioned matrix can be row-reduced to the given martrix. If requested, also give the indicated particular solution, if it exists.

7. $\begin{bmatrix} 1 & -1 & 2 & | & 3 \\ 0 & 1 & 4 & | & 2 \end{bmatrix}$, solution with $x_3 = 2$

8. $\begin{bmatrix} 1 & 2 & 3 & | & 3 \\ 0 & 1 & 2 & | & -1 \\ 0 & 0 & 2 & | & 4 \end{bmatrix}$

9. $\begin{bmatrix} 1 & 0 & 2 & 0 & | & 1 \\ 0 & 1 & 1 & 3 & | & -2 \\ 0 & 0 & 0 & 0 & | & 0 \end{bmatrix}$,

solution with $x_3 = 3$, $x_4 = -2$

10. $\begin{bmatrix} 1 & 1 & 0 & 3 & 0 & | & -4 \\ 0 & 0 & 1 & -1 & 0 & | & 0 \\ 0 & 0 & 0 & 0 & 1 & | & -2 \\ 0 & 0 & 0 & 0 & 0 & | & 0 \end{bmatrix}$,

solution with $x_2 = 2$, $x_3 = 1$

11. $\begin{bmatrix} 1 & 0 & 0 & 0 & | & 0 \\ 0 & 1 & 0 & 0 & | & 2 \\ 0 & 0 & 1 & 0 & | & -5 \\ 0 & 0 & 0 & 1 & | & 2 \\ 0 & 0 & 0 & 0 & | & 0 \end{bmatrix}$

12. $\begin{bmatrix} 1 & -1 & 2 & 0 & 3 & | & 1 \\ 0 & 0 & 0 & 1 & 4 & | & 2 \\ 0 & 0 & 0 & 0 & 0 & | & -1 \\ 0 & 0 & 0 & 0 & 0 & | & 0 \end{bmatrix}$

In Exercises 13–20, find all solutions of the given linear system, using the Gauss method with back substitution.

13. $2x - y = 8$
$6x - 5y = 32$

14. $4x_1 - 3x_2 = 10$
$8x_1 - x_2 = 10$

15. $y + z = 6$
$3x - y + z = -7$
$x + y - 3z = -13$

16. $2x + y - 3z = 0$
$6x + 3y - 8z = 0$
$2x - y + 5z = -4$

17. $x_1 - 2x_2 = 3$
$3x_1 - x_2 = 14$
$x_1 - 7x_2 = -2$

18. $x_1 - 3x_2 + x_3 = 2$
$3x_1 - 8x_2 + 2x_3 = 5$

19. $x_1 + 4x_2 - 2x_3 = 4$
$2x_1 + 7x_2 - x_3 = -2$
$2x_1 + 9x_2 - 7x_3 = 1$

20. $x_1 - 3x_2 + 2x_3 - x_4 = 8$
$3x_1 - 7x_2 + x_4 = 0$

In Exercises 21–28, find all solutions of the linear system, using the Gauss–Jordan method.

21. $3x_1 - 2x_2 = -8$
$4x_1 + 5x_2 = -3$

22. $2x_1 + 8x_2 = 16$
$5x_1 - 4x_2 = -8$

23. $x_1 - 2x_3 + x_4 = 6$
$2x_1 - x_2 + x_3 - 3x_4 = 0$
$9x_1 - 3x_2 - x_3 - 7x_4 = 4$

24. $x_1 + 2x_2 - 3x_3 + x_4 = 2$
$3x_1 + 6x_2 - 8x_3 - 2x_4 = 1$

25. $x_2 - 3x_3 = 3$
$2x_1 + 4x_2 - x_3 = 5$
$4x_1 - 2x_2 + 5x_3 = 3$

26. $x_1 - 4x_2 + x_3 = 8$
$3x_1 - 12x_2 + 5x_3 = 26$
$2x_1 - 9x_2 - x_3 = 14$

27. $x_1 + 2x_2 - 3x_3 = 8$
$2x_1 + 5x_2 - 6x_3 = 17$
$-x_1 - 2x_2 + x_3 - x_4 = -8$
$4x_1 + 10x_2 - 9x_3 + x_4 = 33$

28. $x_1 - 3x_2 + x_3 + 2x_4 = 2$
$x_1 - 2x_2 + 2x_3 + 4x_4 = -1$
$2x_1 - 8x_2 - x_3 = 3$
$3x_1 - 9x_2 + 4x_3 = 7$

29. Mark each of the following True or False.

a) Every linear system with the same number of equations as unknowns has a unique solution.

b) Every linear system with the same number of equations as unknowns has at least one solution.

c) A linear system with more equations than unknowns may have an infinite number of solutions.

d) A linear system with fewer equations than unknowns may have no solution.

e) Every matrix is row equivalent to a unique matrix in echelon form.

f) Every matrix is row equivalent to a unique matrix in reduced row echelon form.

g) If $[A \mid \mathbf{b}]$ and $[B \mid \mathbf{c}]$ are row-equivalent partitioned matrices, the linear systems $A\mathbf{x} = \mathbf{b}$ and $B\mathbf{x} = \mathbf{c}$ have the same solution set.

h) A linear system with a square coefficient matrix A has a unique solution if and only if A is row equivalent to the identity matrix.

i) A linear system with coefficient matrix A has an infinite number of solutions if and only if A can be row-reduced to an echelon matrix that includes some column containing no pivot.

j) A consistent linear system with coefficient matrix A has an infinite number of solutions if and only if A can be row-reduced to an echelon matrix that includes some column containing no pivot.

In Exercises 30–37, describe all possible values for the unknowns x_i so that the matrix equation is valid.

30. $2[x_1 \ \ x_2] - [4 \ \ 7] = [-2 \ \ 11]$

31. $4[x_1 \ \ x_2] + 2[x_1 \ \ 3] = [-6 \ \ 18]$

32. $[x_1 \ \ x_2] \begin{bmatrix} 1 \\ -3 \end{bmatrix} = [2]$

33. $[x_1 \ \ x_2] \begin{bmatrix} 3 & -1 \\ 2 & 4 \end{bmatrix} = [0 \ \ -14]$

34. $\begin{bmatrix} 1 & -5 \\ 3 & 2 \end{bmatrix} \begin{bmatrix} x_1 \\ x_2 \end{bmatrix} = \begin{bmatrix} 13 \\ 5 \end{bmatrix}$

35. $\begin{bmatrix} x_1 & x_2 \\ x_3 & x_4 \end{bmatrix} \begin{bmatrix} 1 & 1 \\ 1 & 0 \end{bmatrix} = \begin{bmatrix} 0 & 1 \\ 3 & 1 \end{bmatrix}$

36. $[x_1 \ \ x_2] \begin{bmatrix} 3 & 0 & 4 \\ 2 & 1 & -1 \end{bmatrix} = [3 \ \ \ 3 \ \ -7]$

37. $\begin{bmatrix} x_1 & x_2 \\ x_3 & x_4 \end{bmatrix} \begin{bmatrix} 3 & 5 \\ 2 & 3 \end{bmatrix} = \begin{bmatrix} 1 & 0 \\ 0 & 1 \end{bmatrix}$

38. Determine all values of the b_i that make the linear system

$$x_1 + 2x_2 = b_1$$
$$3x_1 + 6x_2 = b_2$$

consistent.

39. Determine all values of the b_i that make the linear system

$$2x_1 + 5x_2 = b_1$$
$$x_1 - x_2 = b_2$$

consistent.

40. Determine all values of the b_i that make the linear system

$$x_1 + x_2 - x_3 = b_1$$
$$2x_2 + x_3 = b_2$$
$$x_2 - x_3 = b_3$$

consistent.

41. Determine all values of the b_i that make the linear system

$$x_1 + 3x_2 - x_3 = b_1$$
$$x_1 - x_2 + 2x_3 = b_2$$
$$4x_2 - 3x_3 = b_3$$

consistent.

*Matrix A **commutes** with matrix B if AB = BA.*

42. Find all values of r for which

$$\begin{bmatrix} 2 & 0 & 0 \\ 0 & 1 & 0 \\ 0 & 0 & r \end{bmatrix} \text{ commutes with } \begin{bmatrix} 1 & 0 & 1 \\ 0 & 1 & 0 \\ 1 & 0 & 1 \end{bmatrix}.$$

43. Find all values of r for which

$$\begin{bmatrix} 2 & 0 & 0 \\ 0 & r & 0 \\ 0 & 0 & 2 \end{bmatrix} \text{ commutes with } \begin{bmatrix} 1 & 0 & 1 \\ 0 & 1 & 0 \\ 1 & 0 & 1 \end{bmatrix}.$$

44. Find a, b, and c such that the parabola $y = ax^2 + bx + c$ passes through the points $(1, -4)$, $(-1, 0)$, and $(2, 3)$.

45. Find a, b, c, and d such that the quartic curve $y = ax^4 + bx^3 + cx^2 + d$ passes through $(1, 2)$, $(-1, 6)$, $(-2, 38)$, and $(2, 6)$.

46. Let A be an $m \times n$ matrix, and let $\mathbf{c}$ be a column vector such that $A\mathbf{x} = \mathbf{c}$ has a unique solution.

a) Show that $m \geq n$.
b) If $m = n$, must the system $A\mathbf{x} = \mathbf{b}$ be consistent for every choice of $\mathbf{b}$?
c) Answer part (b) for the case where $m > n$.

⏻ *A problem we meet when reducing a matrix with the aid of computer involves determining when a computed entry should be 0. The computer might give an entry as 0.00000001, because of roundoff error, when it should be 0. If the computer uses this entry as a pivot in a future step, the result is chaotic! For this reason, it is common practice to program the computer to replace all sufficiently small computed entries with 0, where the meaning of "sufficiently small" must be specified in terms of the size of the nonzero entries in the original matrix. The software program YUREDUCE in LINTEK provides drill on the steps involved in reducing a matrix without requiring burdensome computation. The program computes the smallest nonzero coefficient magnitude m and asks the user to enter a number r (for ratio); all computed entries of magnitude less than rm produced during reduction of the coefficient matrix will be set equal to zero. In Exercises 47–52, use the program YUREDUCE, specifying r = 0.00001, to solve the linear system.*

47. $\begin{aligned} 3x_1 - x_2 &= -10 \\ 7x_1 + 2x_2 &= 7 \\ 2x_1 - 5x_2 &= -37 \end{aligned}$

48. $\begin{aligned} 5x_1 - 2x_2 &= 11 \\ 8x_1 + x_2 &= 3 \\ 6x_1 - 5x_2 &= -4 \end{aligned}$

49. $\begin{aligned} 7x_1 - 2x_2 + x_3 &= -14 \\ -4x_1 + 5x_2 - 3x_3 &= 17 \\ 5x_1 - x_2 + 2x_3 &= -7 \end{aligned}$

50. $\begin{aligned} -3x_1 + 5x_2 + 2x_3 &= 12 \\ 5x_1 - 7x_2 + 6x_3 &= -16 \\ 11x_1 - 17x_2 + 2x_3 &= -40 \end{aligned}$

51. $\begin{aligned} x_1 - 2x_2 + x_3 - x_4 + 2x_5 &= 1 \\ 2x_1 + x_2 - 4x_3 - x_4 + 5x_5 &= 16 \\ 8x_1 - x_2 + 3x_3 - x_4 - x_5 &= 1 \\ 4x_1 - 2x_2 + 3x_3 - 8x_4 + 2x_5 &= -5 \\ 5x_1 + 3x_2 - 4x_3 + 7x_4 - 6x_5 &= 7 \end{aligned}$

52. $\begin{aligned} x_1 - 2x_2 + x_3 - x_4 + 2x_5 &= 1 \\ 2x_1 + x_2 - 4x_3 - x_4 + 5x_5 &= 10 \\ 8x_1 - x_2 + 3x_3 - x_4 - x_5 &= -5 \\ 4x_1 - 2x_2 + 3x_3 - 8x_4 + 2x_5 &= -3 \\ 5x_1 + 3x_2 - 4x_3 + 7x_4 - 6x_5 &= 1 \end{aligned}$

💾 *The program MATCOMP in LINTEK can also be used to find the solutions of a linear system. MATCOMP will bring the left portion of the augmented matrix to reduced row echelon form and display the result on the screen. The user can then find the solutions. Use MATCOMP in the remaining exercises.*

53. Find the reduced row echelon form of the matrix in Exercise 6, by taking it as a coefficient matrix for zero systems.

54. Solve the linear system in Exercise 49.

55. Solve the linear system in Exercise 50.

56. Solve the linear system in Exercise 51.

<table>
<tr><td>**1.3**</td><td></td></tr>
</table>

MORE ABOUT LINEAR SYSTEMS

In the preceding section, we discussed how to find all solutions of any linear system. Now we consider linear systems in a bit more depth, developing information of practical or theoretical use. We start with a time-saving practical technique.

Systems with the Same Coefficient Matrix

In the next section, where we study matrix inversion, we will have occasion to solve several linear systems $A\mathbf{x} = \mathbf{b}$ having the same coefficient matrix A but different constant column vectors $\mathbf{b}$. Suppose that we want to solve two such systems,

$$A\mathbf{x} = \mathbf{b} \quad \text{and} \quad A\mathbf{y} = \mathbf{b}'.$$

Rather than solving one after the other and row-reducing the same matrix A twice, we can form a single partitioned matrix with the coefficient matrix A to the left of the partition, augmented with two column vectors, $\mathbf{b}$ and $\mathbf{b}'$, to the right of the partition. We then reduce this larger partitioned matrix so that the left part is brought into echelon form, solving both systems at once. Since the entries in the column vectors $\mathbf{b}$ and $\mathbf{b}'$ play no role in the selection of elementary row operations to reduce the matrix A, we are able to avoid doing the same work twice.

In the following example, we solve two systems with the same coefficient matrix, using the Gauss–Jordan method.

EXAMPLE 1 Solve the linear systems

$$
\begin{aligned}
2x_1 - 4x_2 \qquad\qquad &= -10 \\
x_1 - 3x_2 \qquad + x_4 &= -4 \\
x_1 \qquad - x_3 + 2x_4 &= 4 \\
3x_1 - 4x_2 + 3x_3 - x_4 &= -11
\end{aligned}
\quad \text{and} \quad
\begin{aligned}
2y_1 - 4y_2 \qquad\qquad &= -8 \\
y_1 - 3y_2 \qquad + y_4 &= -2 \\
y_1 \qquad - y_3 + 2y_4 &= 9 \\
3y_1 - 4y_2 + 3y_3 - y_4 &= -15,
\end{aligned}
$$

using the Gauss–Jordan method.

Solution We put both columns of constants to the right of the partition, and again color the pivots:

$$\begin{bmatrix} 2 & -4 & 0 & 0 & -10 & -8 \\ 1 & -3 & 0 & 1 & -4 & -2 \\ 1 & 0 & -1 & 2 & 4 & 9 \\ 3 & -4 & 3 & -1 & -11 & -15 \end{bmatrix}$$

Multiply row 1 by $\frac{1}{2}$, to obtain the next matrix.

$$\sim \begin{bmatrix} 1 & -2 & 0 & 0 & -5 & -4 \\ 1 & -3 & 0 & 1 & -4 & -2 \\ 1 & 0 & -1 & 2 & 4 & 9 \\ 3 & -4 & 3 & -1 & -11 & -15 \end{bmatrix}$$

Add -1 times row 1 to rows 2 and 3, and add -3 times row 1 to row 4, to obtain the next matrix.

$$\sim \begin{bmatrix} 1 & -2 & 0 & 0 & -5 & -4 \\ 0 & -1 & 0 & 1 & 1 & 2 \\ 0 & 2 & -1 & 2 & 9 & 13 \\ 0 & 2 & 3 & -1 & 4 & -3 \end{bmatrix}$$

Multiply row 2 by -1, to obtain the next matrix.

$$\sim \begin{bmatrix} 1 & -2 & 0 & 0 & -5 & -4 \\ 0 & 1 & 0 & -1 & -1 & -2 \\ 0 & 2 & -1 & 2 & 9 & 13 \\ 0 & 2 & 3 & -1 & 4 & -3 \end{bmatrix}$$

Add 2 times row 2 to row 1, and add -2 times row 2 to rows 3 and 4, to obtain the next matrix.

$$\sim \begin{bmatrix} 1 & 0 & 0 & -2 & -7 & -8 \\ 0 & 1 & 0 & -1 & -1 & -2 \\ 0 & 0 & -1 & 4 & 11 & 17 \\ 0 & 0 & 3 & 1 & 6 & 1 \end{bmatrix}$$

Multiply row 3 by -1, to obtain the next matrix.

$$\sim \begin{bmatrix} 1 & 0 & 0 & -2 & -7 & -8 \\ 0 & 1 & 0 & -1 & -1 & -2 \\ 0 & 0 & 1 & -4 & -11 & -17 \\ 0 & 0 & 3 & 1 & 6 & 1 \end{bmatrix}$$

Add -3 times row 3 to row 4, to obtain the next matrix.

$$\sim \begin{bmatrix} 1 & 0 & 0 & -2 & -7 & -8 \\ 0 & 1 & 0 & -1 & -1 & -2 \\ 0 & 0 & 1 & -4 & -11 & -17 \\ 0 & 0 & 0 & 13 & 39 & 52 \end{bmatrix}$$

Multiply row 4 by $\frac{1}{13}$, to obtain the next matrix.

$$\sim \begin{bmatrix} 1 & 0 & 0 & -2 & -7 & -8 \\ 0 & 1 & 0 & -1 & -1 & -2 \\ 0 & 0 & 1 & -4 & -11 & -17 \\ 0 & 0 & 0 & 1 & 3 & 4 \end{bmatrix}$$

Add 2 times row 4 to 1, add row 4 to row 2 and add 4 times row 4 to row 3, to obtain the next matrix.

$$\sim \begin{bmatrix} 1 & 0 & 0 & 0 & -1 & 0 \\ 0 & 1 & 0 & 0 & 2 & 2 \\ 0 & 0 & 1 & 0 & 1 & -1 \\ 0 & 0 & 0 & 1 & 3 & 4 \end{bmatrix}.$$

From this final matrix, we see that the solutions are $x_1 = -1$, $x_2 = 2$, $x_3 = 1$, $x_4 = 3$ and $y_1 = 0$, $y_2 = 2$, $y_3 = -1$, $y_4 = 4$. ■

The Unique Solution Case and Square Systems

Consider a consistent linear system $A\mathbf{x} = \mathbf{b}$ of m equations in n unknowns. We have seen that this system has a unique solution if and only if a row echelon form H of A has a pivot in each of its n columns. In this situation, since no two pivots appear in the same row of H, we see that H has at least as many rows as columns; that is, $m \geq n$. Consequently, the reduced row echelon form for A must consist of the identity matrix, followed by $m - n$ zero rows. For example, if $m = 5$ and $n = 3$, the reduced row echelon form for A must be

$$\begin{bmatrix} 1 & 0 & 0 \\ 0 & 1 & 0 \\ 0 & 0 & 1 \\ 0 & 0 & 0 \\ 0 & 0 & 0 \end{bmatrix}.$$

We summarize these observations as a theorem.

THEOREM 1.3 The Unique Solution Case

Let $A\mathbf{x} = \mathbf{b}$ be a consistent system of linear equations, where A is an $m \times n$ matrix. The following are equivalent:

1. The system $A\mathbf{x} = \mathbf{b}$ has a unique solution.
2. The reduced row echelon form of A consists of the $n \times n$ identity matrix followed by $m - n$ rows of zeros.

A linear system having an infinite number of solutions is called **underdetermined**. We now show a corollary of the preceding theorem: that a consistent system is underdetermined if it has fewer equations than unknowns.

COROLLARY 1 The Case $m < n$

If a linear system $A\mathbf{x} = \mathbf{b}$ is consistent and has fewer equations than unknowns, then it has an infinite number of solutions.

PROOF   If $m < n$ in Theorem 1.3, the reduced row echelon form of A cannot contain the $n \times n$ identity matrix, so we cannot be in the unique solution case. Since we are assuming that the system is consistent, there are an infinite number of solutions. ▲

A linear system having the same number n of equations as unknowns is called a **square system**, since the coefficient matrix is a square $n \times n$ matrix. When a square matrix is reduced to echelon form, the result is a square matrix having only zero entries below the *main diagonal*, which runs from the upper left-hand corner to the lower right-hand corner. This follows at once from the fact that the pivot in a nonzero row— say, the ith row—is always in a column j where $j \geq i$. Such a square matrix U with zero entries below the main diagonal is called **upper triangular**. Theorem 1.3 provides an important characterization of the unique solution case for square systems.

COROLLARY 2   The Square Case, $m = n$

A square linear system $A\mathbf{x} = \mathbf{b}$ has a unique solution if and only if A is row equivalent to the identity matrix I.

PROOF   Suppose that A is an $n \times n$ matrix and that the system $A\mathbf{x} = \mathbf{b}$ has a unique solution. Then statement (1) in Theorem 1.3 is true; and consequently A is row equivalent to the $n \times n$ identity matrix, since statement 2 must hold with $m = n$. Conversely, if A is row equivalent to the $n \times n$ identity matrix, the linear system $A\mathbf{x} = \mathbf{b}$ has a unique solution for *every* choice of a vector $\mathbf{b}$ of n components.   ▲

A linear system $A\mathbf{x} = \mathbf{b}$ is **homogeneous** if $\mathbf{b} = \mathbf{0}$. A homogeneous linear system $A\mathbf{x} = \mathbf{0}$ is always consistent, because $\mathbf{x} = \mathbf{0}$, the zero vector, is certainly a solution. This zero vector is called the **trivial solution**. Later we will see that it is of interest to know whether a homogeneous system also has a **nontrivial solution** vector—one that includes some nonzero components. Corollary 1 of Theorem 1.3 shows that this is always the case if the homogeneous system has fewer equations than unknowns. We express this in another corollary.

COROLLARY 3   The Homogeneous Case

1. A homogeneous linear system $A\mathbf{x} = \mathbf{0}$ having fewer equations than unknowns has a nontrivial solution—that is, a solution other than the zero vector.
2. A square homogenous system $A\mathbf{x} = \mathbf{0}$ has a nontrivial solution if and only if A is not row equivalent to the identity matrix of the same size.

The Structure of Solution Sets

Let $A\mathbf{x} = \mathbf{b}$ be a consistent linear system. Let $\mathbf{0}$ be the column vector with the same number of components as $\mathbf{b}$ but with all components zero. The system $A\mathbf{x} = \mathbf{0}$ is called the **homogeneous system corresponding to the system** $A\mathbf{x} = \mathbf{b}$. The solution of this homogeneous system has the same number of free variables as the solution of $A\mathbf{x} = \mathbf{b}$, because the free variables depend only on the row echelon form derived from the coefficient matrix A.

THEOREM 1.4 **Structure of Solution Sets**

Let $Ax = b$ be a linear system.

1. If h and h' solutions of the homogeneous system $Ax = 0$, then so is $rh + sh'$ for any scalars r and s.
2. If p is any *particular* solution of $Ax = b$ and h is a solution of the corresponding homogeneous system $Ax = 0$, then $p + h$ is a solution of $Ax = b$. Moreover, every solution of $Ax = b$ has this form $p + h$, which is called the **general solution** of the system.

PROOF Let h and h' be solutions of $Ax = 0$, so that $Ah = 0$ and $Ah' = 0$. By properties of matrix arithmetic, we have

$$A(rh + sh') = A(rh) + A(sh')$$
$$= r(Ah) + s(Ah')$$
$$= r0 + s0 = 0,$$

for any scalars r and s. This shows that $rh + sh'$ is also a solution of $Ax = 0$.

Turning to part (2), let p be a solution of $Ax = b$, so that $Ap = b$, and let h be a solution of $Ax = 0$, so that $Ah = 0$. Then

$$A(p + h) = Ap + Ah = b + 0 = b.$$

Finally, if p' is any other solution of $Ax = b$, then

$$A(p' - p) = Ap' - Ap = b - b = 0,$$

so $p' - p$ is a solution h of $Ax = 0$. From $p' - p = h$, it follows that $p' = p + h$. This completes the proof of (2). ▲

Notice how easy it was to write down the proof of Theorem 1.4 in matrix notation. What a chore it would have been to write out all the equations and all the components of the vectors!

We now give a specific illustration of Theorem 1.4.

EXAMPLE 2 Illustrate Theorem 1.4 for the linear system $Ax = b$ given by

$$x_1 - 2x_2 + x_3 - x_4 = 4$$
$$2x_1 - 3x_2 + 2x_3 - 3x_4 = -1$$
$$3x_1 - 5x_2 + 3x_3 - 4x_4 = 3$$
$$-x_1 + x_2 - x_3 + 2x_4 = 5.$$

Solution We must solve both $Ax = 0$ and $Ax = b$. We reduce the partitioned matrix $[A \mid 0\ b]$ in order to transform A into reduced row echelon form, although we don't really need to include the column vector 0. We have

$$\begin{bmatrix} 1 & -2 & 1 & -1 & 0 & 4 \\ 2 & -3 & 2 & -3 & 0 & -1 \\ 3 & -5 & 3 & -4 & 0 & 3 \\ -1 & 1 & -1 & 2 & 0 & 5 \end{bmatrix} \sim \begin{bmatrix} 1 & -2 & 1 & -1 & 0 & 4 \\ 0 & 1 & 0 & -1 & 0 & -9 \\ 0 & 1 & 0 & -1 & 0 & -9 \\ 0 & -1 & 0 & 1 & 0 & 9 \end{bmatrix}$$

$$\sim \begin{bmatrix} 1 & 0 & 1 & -3 & 0 & -14 \\ 0 & 1 & 0 & -1 & 0 & -9 \\ 0 & 0 & 0 & 0 & 0 & 0 \\ 0 & 0 & 0 & 0 & 0 & 0 \end{bmatrix}.$$

The reduction is complete. The reason we didn't really need to insert the column vector **0** in the partitioned matrix is that it never changes.

From the reduced matrix, we find that the homogeneous system $A\mathbf{x} = \mathbf{0}$ has the solution set described by

$$\mathbf{x} = \begin{bmatrix} x_1 \\ x_2 \\ x_3 \\ x_4 \end{bmatrix} = \begin{bmatrix} -r + 3s \\ s \\ r \\ s \end{bmatrix} \tag{1}$$

for any scalars r and s. On the other hand, the solution set of $A\mathbf{x} = \mathbf{b}$ can be described by

$$\mathbf{x} = \begin{bmatrix} x_1 \\ x_2 \\ x_3 \\ x_4 \end{bmatrix} = \begin{bmatrix} -r + 3s - 14 \\ s - 9 \\ r \\ s \end{bmatrix} = \begin{bmatrix} -14 \\ -9 \\ 0 \\ 0 \end{bmatrix} + \begin{bmatrix} -r + 3s \\ s \\ r \\ s \end{bmatrix}. \tag{2}$$

Equation (2) expresses the general solution **x** of the system $A\mathbf{x} = \mathbf{b}$ as the sum of a particular solution (found by setting $r = s = 0$) and the general solution (1) of $A\mathbf{x} = \mathbf{0}$. This illustrates part (2) of Theorem 1.4. ■

Elementary Matrices

The elementary row operations we have performed can actually be carried out by means of matrix multiplication. While it is not efficient to row-reduce a matrix by multiplying it by other matrices, representing row reduction as a product of matrices is a useful theoretical tool. For example, we use elementary matrices in the next section to show that, for square matrices A and C, if $AC = I$, then $CA = I$. We use them again in Section 3.2 to demonstrate the multiplicative property of determinants, and again in Section 9.2 to exhibit a factorization of some square matrices A into a product LU of a lower-triangular matrix L and an upper-triangular matrix U.

If we interchange its second and third rows, the 3×3 identity matrix

$$I = \begin{bmatrix} 1 & 0 & 0 \\ 0 & 1 & 0 \\ 0 & 0 & 1 \end{bmatrix} \quad \text{becomes} \quad E = \begin{bmatrix} 1 & 0 & 0 \\ 0 & 0 & 1 \\ 0 & 1 & 0 \end{bmatrix}.$$

If $A = [a_{ij}]$ is a 3×3 matrix, we can compute EA, and we find that

$$EA = \begin{bmatrix} 1 & 0 & 0 \\ 0 & 0 & 1 \\ 0 & 1 & 0 \end{bmatrix} \begin{bmatrix} a_{11} & a_{12} & a_{13} \\ a_{21} & a_{22} & a_{23} \\ a_{31} & a_{32} & a_{33} \end{bmatrix} = \begin{bmatrix} a_{11} & a_{12} & a_{13} \\ a_{31} & a_{32} & a_{33} \\ a_{21} & a_{22} & a_{23} \end{bmatrix}.$$

We have interchanged the second and third rows of A by multiplying A on the left by E.

DEFINITION 1.8 Elementary Matrix

Any matrix that can be obtained from an identity matrix by means of one elementary row operation is an **elementary matrix**.

We leave the proof of the following theorem as Exercises 31 through 33.

THEOREM 1.5 Use of Elementary Matrices

Let A be an $m \times n$ matrix, and let E be an $m \times m$ elementary matrix. Multiplication of A on the left by E effects the same elementary row operation on A that was performed on the identity matrix to obtain E.

Thus row reduction of a matrix to row echelon form can be accomplished by successive multiplication on the left by elementary matrices. In other words, if A can be reduced to H through elementary row operations, there exist elementary matrices E_1, E_2, . . . , E_t such that

$$H = (E_t \cdots E_2 E_1)A.$$

Again, this is by no means an efficient way to execute row reduction, but such an algebraic representation of H in terms of A is sometimes handy in proving theorems.

EXAMPLE 3 Let

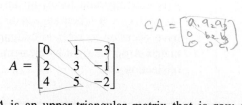

$$A = \begin{bmatrix} 0 & 1 & -3 \\ 2 & 3 & -1 \\ 4 & 5 & -2 \end{bmatrix}.$$

Find a matrix C such that CA is an upper-triangular matrix that is row equivalent to A.

Solution We row reduce A to an upper-triangular matrix U and write down, for each row operation, the elementary matrix obtained by performing the same operation on the 3×3 identity matrix.

Reduction of A	Row Operation	Elementary Matrix
$A = \begin{bmatrix} 0 & 1 & -3 \\ 2 & 3 & -1 \\ 4 & 5 & -2 \end{bmatrix}$	Interchange row 1 and row 2.	$E_1 = \begin{bmatrix} 0 & 1 & 0 \\ 1 & 0 & 0 \\ 0 & 0 & 1 \end{bmatrix}$
$\sim \begin{bmatrix} 2 & 3 & -1 \\ 0 & 1 & -3 \\ 4 & 5 & -2 \end{bmatrix}$	Add -2 times row 1 to row 3.	$E_2 = \begin{bmatrix} 1 & 0 & 0 \\ 0 & 1 & 0 \\ -2 & 0 & 1 \end{bmatrix}$
$\sim \begin{bmatrix} 2 & 3 & -1 \\ 0 & 1 & -3 \\ 0 & -1 & 0 \end{bmatrix}$	Add row 2 to row 3.	$E_3 = \begin{bmatrix} 1 & 0 & 0 \\ 0 & 1 & 0 \\ 0 & 1 & 1 \end{bmatrix}$
$\sim \begin{bmatrix} 2 & 3 & -1 \\ 0 & 1 & -3 \\ 0 & 0 & -3 \end{bmatrix} = U.$		

Thus, we must have $E_3(E_2(E_1A)) = U$; so the desired matrix C is

$$C = E_3E_2E_1 = \begin{bmatrix} 0 & 1 & 0 \\ 1 & 0 & 0 \\ 1 & -2 & 1 \end{bmatrix}.$$

To compute C, we do not actually have to multiply out $E_3E_2E_1$. We know that multiplication of E_1 on the left by E_2 simply adds -2 times row 1 of E_1 to its row 3, and subsequent multiplication on the left by E_3 adds row 2 to row 3 of the matrix E_2E_1. Thus we can find C by executing the same row reduction steps on I that we exectued to change A to U—namely,

$$\begin{bmatrix} 1 & 0 & 0 \\ 0 & 1 & 0 \\ 0 & 0 & 1 \end{bmatrix} \sim \begin{bmatrix} 0 & 1 & 0 \\ 1 & 0 & 0 \\ 0 & 0 & 1 \end{bmatrix} \sim \begin{bmatrix} 0 & 1 & 0 \\ 1 & 0 & 0 \\ 0 & -2 & 1 \end{bmatrix} \sim \begin{bmatrix} 0 & 1 & 0 \\ 1 & 0 & 0 \\ 1 & -2 & 1 \end{bmatrix}.$$

$$I \qquad E_1 \qquad E_2E_1 \qquad C = E_3E_2E_1$$

We can execute analogous *elementary column* operations on a matrix by multiplying the matrix on the *right* by an elementary matrix. Column reduction of a matrix A is not important for us in this chapter, because it does not preserve the solution set of $A\mathbf{x} = \mathbf{b}$ when applied to the augmented matrix $[A \mid \mathbf{b}]$. However, we will have occasion to refer to column reduction later. The effect of multiplication of a matrix A on the right by elementary matrices is explored in Exercises 36 through 38 of Section 1.4. ■

SUMMARY

1. Several linear systems with the same coefficient matrix can all be solved together by augmenting the coefficient matrix with the column vectors of constants from all the systems and proceeding with a row reduction on this partitioned matrix.

2. A consistent linear system $A\mathbf{x} = \mathbf{b}$ of m equations in n unknowns has a unique solution if and only if the reduced row echelon form of A appears as the $n \times n$ identity matrix followed by $m - n$ rows of zeros.

3. A consistent linear system having fewer equations than unknowns is underdetermined; that is, it has an infinite number of solutions.

4. A square linear system has a unique solution if and only if its coefficient matrix is row equivalent to the identity matrix.

5. The solutions of any consistent linear system $A\mathbf{x} = \mathbf{b}$ are precisely the vectors $\mathbf{p} + \mathbf{h}$, where $\mathbf{p}$ is any one particular solution of $A\mathbf{x} = \mathbf{b}$ and $\mathbf{h}$ varies through the solution set of the homogeneous system $A\mathbf{x} = \mathbf{0}$.

6. Sums and scalar multiples of solutions of the *homogeneous* system $A\mathbf{x} = \mathbf{0}$ are also solutions of $A\mathbf{x} = \mathbf{0}$.

7. An elementary matrix E is one obtained by applying a single elementary row operation to an identity matrix I. Multiplication of a matrix A on the left by E effects the same elementary row operation on A.

EXERCISES

In each of Exercises 1–4, two or more linear systems having the same coefficient matrix but different column vectors of constants are given. Find the solutions of each system.

1. $x - 2y = -3,\ 12,\ 3$
   $3x + y = -2,\ \ 1,\ 9$

2. $x_1 - 3x_2 - x_3 = \ \ 3,\ 0,\ -10$
   $2x_1 - 5x_2 + x_3 = \ \ 8,\ 0,\ -11$
   $x_1 + 4x_2 - 2x_3 = -5,\ 0,\ \ \ 9$

3. $2x_1 - x_2 + x_3 = \ \ 1,\ \ 1,\ \ \ 8$
   $-2x_1 + x_2 - 3x_3 = -5,\ -3,\ -14$
   $4x_1 - 3x_2 + x_3 = \ \ 5,\ \ 1,\ \ 14$

4. $x_1 - x_2 + x_3 + x_4 = 1,\ 2,\ 0$
   $2x_1 + x_2 + 5x_3 + 2x_4 = 2,\ 5,\ 0$
   $x_1 + x_2 + 5x_3 \quad\quad = 1,\ 4,\ 0$

In Exercises 5–8, solve the given linear system and express the solution set in a form that illustrates Theorem 1.4.

5. $x_1 - 2x_2 + x_3 + 5x_4 = 7$

6. $2x_1 - x_2 + 3x_3 \quad\quad = -3$
   $4x_1 + 2x_2 \quad\quad - x_4 = \ \ 1$

7. $x_1 - 2x_2 + x_3 + x_4 = 4$
   $2x_1 + x_2 - 3x_3 - x_4 = 6$
   $x_1 - 7x_2 - 6x_3 + 2x_4 = 6$

8. $2x_1 + x_2 + 3x_3 \quad\quad = \ \ 5$
   $x_1 - x_2 + 2x_3 + x_4 = \ \ 0$
   $4x_1 - x_2 + 7x_3 + 2x_4 = \ \ 5$
   $-x_1 - 2x_2 - x_3 + x_4 = -5$

9. Find the value of r such that the homogeneous system

$$2x + ry = 0$$
$$x + y = 0$$

has nontrivial solutions.

10. Find the value of r such that the homogeneous system

$$\begin{bmatrix} 2 - r & 1 & 0 \\ -1 & -r & 1 \\ 1 & 3 & 1 - r \end{bmatrix} \begin{bmatrix} x \\ y \\ z \end{bmatrix} = \begin{bmatrix} 0 \\ 0 \\ 0 \end{bmatrix}$$

has nontrivial solutions.

11. Find an elementary matrix E such that

$$E\begin{bmatrix} 1 & 3 & 1 & 4 \\ 0 & 1 & 2 & 1 \\ 3 & 4 & 5 & 1 \end{bmatrix} = \begin{bmatrix} 1 & 3 & 1 & 4 \\ 0 & 1 & 2 & 1 \\ 0 & -5 & 2 & -11 \end{bmatrix}.$$

12. Find an elementary matrix E such that

$$E\begin{bmatrix} 1 & 3 & 1 & 4 \\ 0 & 1 & 2 & 1 \\ 3 & 4 & 5 & 1 \end{bmatrix} = \begin{bmatrix} 1 & 3 & 1 & 4 \\ 2 & 7 & 4 & 9 \\ 3 & 4 & 5 & 1 \end{bmatrix}.$$

13. Find a matrix C such that

$$C\begin{bmatrix} 1 & 2 \\ 3 & 4 \\ 4 & 2 \end{bmatrix} = \begin{bmatrix} 1 & 2 \\ 0 & -2 \\ 0 & -6 \end{bmatrix}.$$

14. Find a matrix C such that

$$C\begin{bmatrix} 1 & 2 \\ 3 & 4 \\ 4 & 2 \end{bmatrix} = \begin{bmatrix} 3 & 4 \\ 4 & 2 \\ 1 & 2 \end{bmatrix}.$$

In Exercises 15–20, let A be a 4 × 4 matrix. Find a matrix C such that the result of applying the given sequence of elementary row operations to A can also be found by computing the product CA.

15. Interchange row 1 and row 2.

16. Interchange row 1 and row 3; multiply row 3 by 4.

17. Multiply row 1 by 5; interchange rows 2 and 3; add 2 times row 3 to row 4.

18. Add 4 times row 2 to row 4; multiply row 4 by -3; add 5 times row 4 to row 1.

19. Interchange rows 1 and 4; add 6 times row 2 to row 1; add -3 times row 1 to row 3; add -2 times row 4 to row 2.

20. Add 3 times row 2 to row 4; add -2 times row 4 to row 3; add 5 times row 3 to row 1; add -4 times row 1 to row 2.

21. Mark each of the following True or False.

___ a) A linear system with fewer equations than unknowns has an infinite number of solutions.

___ b) A consistent linear system with fewer equations than unknowns has an infinite number of solutions.

___ c) If a square linear system $Ax = b$ has a solution for *every* choice of column vector **b**, then the solution is unique for each **b**.

___ d) If a square system $Ax = 0$ has only the trivial solution, then $Ax = b$ has a unique solution for every column vector **b** with the appropriate number of components.

___ e) If a linear system $Ax = 0$ has only the trivial solution, then $Ax = b$ has a unique solution for every column vector **b** with the appropriate number of components.

___ f) The sum of two solution vectors of any linear system is also a solution vector of the system.

___ g) The sum of two solution vectors of any homogeneous linear system is also a solution vector of the system.

___ h) A scalar multiple of a solution vector of any homogeneous linear system is also a solution vector of the system.

___ i) If an $m \times n$ matrix A is row equivalent to B, then applying the same sequence of elementary row operations to the $m \times m$ identity matrix that reduces A to B yields a matrix C such that $CA = B$.

___ j) The product of two elementary matrices is again an elementary matrix.

22. We have defined a linear system to be *underdetermined* if it has an infinite number of solutions. Explain why this is a reasonable term to use for such a system.

23. A linear system is **overdetermined** if it has more equations than unknowns. Explain why this is a reasonable term to use for such a system.

24. Referring to Exercises 22 and 23, give an example of an overdetermined underdetermined linear system!

25. Is the sum of two $m \times n$ matrices in echelon form again in echelon form? If so, prove it; if not, give a counterexample.

26. Is the sum of two $n \times n$ upper-triangular matrices again upper triangular? If so, prove it; if not, give a counterexample.

27. Repeat Exercise 26 for the product of two $n \times n$ upper-triangular matrices.

28. Use Theorem 1.4 to explain why a homogeneous system of linear equations either has a unique solution or an infinite number of solutions.

29. Use Theorem 1.4 to explain why no system of linear equations can have exactly two solutions.

30. Let A be an $m \times n$ matrix such that the homogeneous system $Ax = 0$ has only the trivial solution.

a) Does it follow that every system $Ax = b$ is consistent?

b) Does it follow that every consistent system $Ax = b$ has a unique solution?

31. Prove Theorem 1.5 for the row-interchange operation.

32. Prove Theorem 1.5 for the row-scaling operation.

33. Prove Theorem 1.5 for the row-addition operation.

In each of Exercises 34 and 35, use either MATCOMP or YUREDUCE to solve the two given systems with a single run of an option. Write down a description of all solutions.

34.
$$
\begin{aligned}
x_1 - 2x_2 + x_3 - x_4 + 2x_5 &= 1,\ 1 \\
2x_1 + x_2 - 4x_3 - x_4 + 5x_5 &= 16,\ 10 \\
8x_1 - x_2 + 3x_3 - x_4 - x_5 &= 1,\ -5 \\
4x_1 - 2x_2 + 3x_3 - 8x_4 + 2x_5 &= -5,\ -3 \\
5x_1 + 3x_2 - 4x_3 + 7x_4 - 6x_5 &= 7,\ 1
\end{aligned}
$$

35.
$$
\begin{aligned}
x_1 + x_2 - x_3 - 2x_4 + 5x_5 + x_6 + 11x_7 &= 9,\ 15 \\
2x_1 + 3x_2 - x_3 + 2x_4 + 5x_5 - 2x_6 + 14x_7 &= -1,\ 2 \\
4x_1 - 3x_2 - 5x_3 - 26x_4 + 25x_5 + x_6 - 5x_7 &= 12,\ 17 \\
-2x_1 + 3x_2 - x_3 + 2x_4 + 5x_5 + 5x_6 + 35x_7 &= -4,\ 1 \\
- 2x_2 + x_3 - 5x_5 - 4x_6 - 26x_7 &= -16,\ 41 \\
2x_1 - x_2 + 3x_3 + 10x_4 - 15x_5 - 12x_7 &= 57,\ 53
\end{aligned}
$$

1.4 **INVERSES OF SQUARE MATRICES**

Matrix Equations and Inverses

A system of n equations in n unknowns $x_1, x_2, \ldots, x_n$ can be expressed in matrix form as

$$A\mathbf{x} = \mathbf{b}, \tag{1}$$

where A is the $n \times n$ coefficient matrix, $\mathbf{x}$ is the $n \times 1$ column vector with ith entry x_i, and $\mathbf{b}$ is an $n \times 1$ column vector with constant entries. The analogous equation using scalars is

$$ax = b \tag{2}$$

for scalars a and b. We usually think of solving Eq. (2) for x by dividing by a, if $a \neq 0$, but we can just as well think of solving it by multiplying by $1/a$. Breaking the solution down into small steps, we have

$(1/a)(ax) = (1/a)b$	**Multiplication by $1/a$**
$[(1/a)a]x = (1/a)b$	**Associativity of multiplication**
$1x = (1/a)b$	**Property of $1/a$**
$x = (1/a)b.$	**Property of 1**

Let us see whether we can solve Eq. (1) similarly. Matrix multiplication is associative, and the $n \times n$ identity matrix I plays the same role for multiplication of $n \times n$ matrices that the number 1 plays for multiplication of numbers. The crucial step is to find an $n \times n$ matrix C such that $CA = I$, so that C plays for matrices the role that $1/a$ does for numbers. If such a matrix C exists, we can obtain from Eq. (1)

Suppose CA = I.

$$C(A\mathbf{x}) = C\mathbf{b} \qquad \text{Multiplication by } C$$

$$(CA)\mathbf{x} = C\mathbf{b} \qquad \text{Associativity of multiplication}$$

$$I\mathbf{x} = C\mathbf{b} \qquad \text{Property of } C$$

$$\mathbf{x} = C\mathbf{b}, \qquad \text{Property of } I$$

which shows that our column vector $\mathbf{x}$ of unknowns must be the column vector $C\mathbf{b}$. The only question is whether there exists an $n \times n$ matrix C such that $CA = I$. The equation

$$\begin{bmatrix} -4 & 9 \\ 1 & -2 \end{bmatrix} \begin{bmatrix} 2 & 9 \\ 1 & 4 \end{bmatrix} = \begin{bmatrix} 1 & 0 \\ 0 & 1 \end{bmatrix}$$

shows that

$$A = \begin{bmatrix} 2 & 9 \\ 1 & 4 \end{bmatrix} \qquad \text{and} \qquad C = \begin{bmatrix} -4 & 9 \\ 1 & -2 \end{bmatrix}$$

satisfy $CA = I$.

Unfortunately, it is not true that, for each $n \times n$ matrix A, we can find an $n \times n$ matrix C such that $CA = I$. For example, if the jth column of A has only zero entries, the jth column of CA also has only zero entries for any matrix C, so $CA \neq I$ for any matrix C. However, for many important $n \times n$ matrices A, there does exist an $n \times n$ matrix C such that $CA = I$. We wonder whether $AC = I$, too. Recall that, in general, matrix multiplication is not commutative. At the end of this section, we will prove the following statement:

For square matrices A and C, we have $AC = I$ if and only if $CA = I$.

This turns out to be a very significant property.

DEFINITION 1.9 Invertible Matrix

An $n \times n$ matrix A is **invertible** if there exists an $n \times n$ matrix C such that $AC = CA = I$, the $n \times n$ identity matrix. If A is not invertible, it is **singular**.

$|A| = 0 \Rightarrow$ not invertible. $\Rightarrow$ singular

If a square matrix A is invertible, then there exists a *unique* matrix C such that $AC = CA = I$, as the following theorem shows. This matrix C is the **inverse of A**.

THEOREM 1.6 Uniqueness Property

Let A be an $n \times n$ matrix. If C and D are matrices such that $AC = DA = I$, then $C = D$.

PROOF Since matrix multiplication is associative, we have

$$D(AC) = (DA)C.$$

But since $AC = I$ and $DA = I$, we find that

$$D(AC) = DI = D \quad \text{and} \quad (DA)C = IC = C.$$

Therefore, $C = D$. ▲

We will denote this unique inverse of A, when it exists, by A^{-1}. Although A^{-1} plays the same role arithmetically as $a^{-1} = 1/a$ (as we showed at the start of this section), we will never write A^{-1} as $1/A$. The powers of an invertible $n \times n$ matrix A are now defined for all integers. That is, for $m > 0$, A^m is the product of m factors A, and A^{-m} is the product of m factors A^{-1}. We consider A^0 to be the $n \times n$ identity matrix I.

Inverses of Elementary Matrices

Let E_1 be an elementary row-interchange matrix, obtained from an identity matrix I by the interchanging of rows i and k. Recall that $E_1 A$ effects the interchange of rows i and k of A for any matrix A such that $E_1 A$ is defined. In particular, taking $A = E_1$, we see that $E_1 E_1$ interchanges rows i and k of E_1, and hence changes E_1 back to I. Thus,

$$E_1 E_1 = I.$$

Consequently, E_1 is an invertible matrix and is its own inverse.

Now let E_2 be an elementary row-scaling matrix, obtained from an identity matrix by the multiplication of row i by a nonzero scalar r. Let E_2' be the matrix obtained from the identity matrix by the multiplication of row i by $1/r$. It is clear that

$$E_2' E_2 = E_2 E_2' = I,$$

so E_2 is invertible, with inverse E_2'.

THE NOTION OF THE INVERSE OF A MATRIX first appears in an 1855 note of Arthur Cayley (1821–1895) and is made more explicit in a 1858 paper entitled "A Memoir on the Theory of Matrices." In that work, Cayley outlines the basic properties of matrices, noting that most of these derive from work with sets of linear equations. In particular, the inverse comes from the idea of solving a system

$$X = a\,x + b\,y + c\,z$$
$$Y = a'x + b'y + c'z$$
$$Z = a''x + b''y + c''z$$

for x, y, z in terms of X, Y, Z. Cayley gives an explicit construction for the inverse in terms of the determinants of the original matrix and of the minors.

In 1842 Arthur Cayley graduated from Trinity College, Cambridge, but could not find a suitable teaching post. So, like Sylvester, he studied law and was called to the bar in 1849. During his 14 years as a lawyer, he wrote about 300 mathematical papers; finally, in 1863 he became a professor at Cambridge, where he remained until his death. It was during his stint as a lawyer that he met Sylvester; their discussions over the next 40 years were extremely fruitful for the progress of algebra. Over his lifetime, Cayley produced close to 1000 papers in pure mathematics, theoretical dynamics, and mathematical astronomy.

Finally, let E_3 be an elementary row-addition matrix, obtained from I by the addition of r times row i to row k. If $E_3{}'$ is obtained from I by the addition of $-r$ times row i to row k, then

$$E_3{}'E_3 = E_3E_3{}' = I.$$

We have established the following fact:

> Every elementary matrix is invertible.

EXAMPLE 1   Find the inverses of the elementary matrices

$$E_1 = \begin{bmatrix} 0 & 1 & 0 \\ 1 & 0 & 0 \\ 0 & 0 & 1 \end{bmatrix}, \qquad E_2 = \begin{bmatrix} 3 & 0 & 0 \\ 0 & 1 & 0 \\ 0 & 0 & 1 \end{bmatrix}, \qquad \text{and} \qquad E_3 = \begin{bmatrix} 1 & 0 & 4 \\ 0 & 1 & 0 \\ 0 & 0 & 1 \end{bmatrix}.$$

Solution   Since E_1 is obtained from

$$I = \begin{bmatrix} 1 & 0 & 0 \\ 0 & 1 & 0 \\ 0 & 0 & 1 \end{bmatrix}$$

by the interchanging of the first and second rows, we see that $E_1{}^{-1} = E_1$.

The matrix E_2 is obtained from I by the multiplication of the first row by 3, so we must multiply the first row of I by $\frac{1}{3}$ to form

$$E_2{}^{-1} = \begin{bmatrix} \frac{1}{3} & 0 & 0 \\ 0 & 1 & 0 \\ 0 & 0 & 1 \end{bmatrix}.$$

Finally, E_3 is obtained from I by the addition of 4 times row 3 to row 1. To form $E_3{}^{-1}$ from I, we add -4 times row 3 to row 1, so

$$E_3{}^{-1} = \begin{bmatrix} 1 & 0 & -4 \\ 0 & 1 & 0 \\ 0 & 0 & 1 \end{bmatrix}. \qquad ■$$

Inverses of Products

The next theorem is fundamental in work with inverses.

THEOREM 1.7   Inverses of Products

Let A and B be invertible $n \times n$ matrices. Then AB is invertible, and
$$(AB)^{-1} = B^{-1}A^{-1}.$$

PROOF By assumption, there exist matrices A^{-1} and B^{-1} such that $AA^{-1} = A^{-1}A = I$ and $BB^{-1} = B^{-1}B = I$. Making use of the associative law for matrix multiplication, we find that

$$(AB)(B^{-1}A^{-1}) = [A(BB^{-1})]A^{-1} = (AI)A^{-1} = AA^{-1} = I.$$

A similar computation shows that $(B^{-1}A^{-1})(AB) = I$. Therefore, the inverse of AB is $B^{-1}A^{-1}$; that is, $(AB)^{-1} = B^{-1}A^{-1}$. ▲

It is instructive to apply Theorem 1.7 to a product $E_t \cdots E_3 E_2 E_1$ of elementary matrices. In the expression

$$(E_t \ldots E_3 E_2 E_1)A,$$

the product $E_t \cdots E_3 E_2 E_1$ performs a sequence of elementary row operations on A. First, E_1 acts on A; then, E_2 acts on $E_1 A$; and so on. To undo this sequence, we must first undo the last elementary row operation, performed by E_t. This is accomplished by using E_t^{-1}. Thus, we should perform the sequence of operations given by

$$E_1^{-1}E_2^{-1}E_3^{-1} \cdots E_t^{-1},$$

in order to effect $(E_t \cdots E_3 E_2 E_1)^{-1}$.

Computation of Inverses

Let $A = [a_{ij}]$ be an $n \times n$ matrix. To find A^{-1}, if it exists, we must find an $n \times n$ matrix $X = [x_{ij}]$ such that $AX = I$—that is, such that

$$\begin{bmatrix} a_{11} & a_{12} & \cdots & a_{1n} \\ a_{21} & a_{22} & \cdots & a_{2n} \\ & \vdots & & \\ a_{n1} & a_{n2} & \cdots & a_{nn} \end{bmatrix} \begin{bmatrix} x_{11} & x_{12} & \cdots & x_{1n} \\ x_{21} & x_{22} & \cdots & x_{2n} \\ & \vdots & & \\ x_{n1} & x_{n2} & \cdots & x_{nn} \end{bmatrix} = \begin{bmatrix} 1 & 0 & \cdots & 0 \\ 0 & 1 & \cdots & 0 \\ & \vdots & & \\ 0 & 0 & \cdots & 1 \end{bmatrix}. \tag{3}$$

Matrix equation (3) corresponds to n^2 linear equations in the n^2 unknowns x_{ij}; there is one linear equation for each of the n^2 positions in an $n \times n$ matrix. For example, equating the entries in the row 2, column 1 position on each side of Eq. (3), we obtain the linear equation

$$a_{21}x_{11} + a_{22}x_{21} + \cdots + a_{2n}x_{n1} = 0.$$

Of these n^2 linear equations, n of them involve the n unknowns x_{i1} for $i = 1, 2, \ldots, n$; and these equations are given by the column-vector equation

$$A \begin{bmatrix} x_{11} \\ x_{21} \\ \vdots \\ x_{n1} \end{bmatrix} = \begin{bmatrix} 1 \\ 0 \\ \vdots \\ 0 \end{bmatrix}, \tag{4}$$

which is a square system of equations. There are also n equations involving the n unknowns x_{i2}, for $i = 1, 2, \ldots, n$; and so on. In addition to solving system (4), we must solve the systems

$$A\begin{bmatrix} x_{12} \\ x_{22} \\ \vdots \\ x_{n2} \end{bmatrix} = \begin{bmatrix} 0 \\ 1 \\ \vdots \\ 0 \end{bmatrix}, \cdots, A\begin{bmatrix} x_{1n} \\ x_{2n} \\ \vdots \\ x_{nn} \end{bmatrix} = \begin{bmatrix} 0 \\ 0 \\ \vdots \\ 1 \end{bmatrix},$$

where each system has the same coefficient matrix A. Our work in Section 1.3 shows that we can solve all of these systems at once by forming the augmented matrix

$$\begin{bmatrix} a_{11} & a_{12} & \cdots & a_{1n} & 1 & 0 & \cdots & 0 \\ a_{21} & a_{22} & \cdots & a_{2n} & 0 & 1 & \cdots & 0 \\ & & \vdots & & & & \vdots & \\ a_{n1} & a_{n2} & \cdots & a_{nn} & 0 & 0 & \cdots & 1 \end{bmatrix}, \tag{5}$$

which we abbreviate by $[A \mid I]$. The matrix A is to the left of the partition, and the identity matrix I is to the right. We then perform a Gauss–Jordan reduction on this augmented matrix. By Corollary 2 of Theorem 1.3, we obtain a unique solution if and only if partitioned matrix (5) can be reduced to

$$\begin{bmatrix} 1 & 0 & \cdots & 0 & c_{11} & c_{12} & \cdots & c_{1n} \\ 0 & 1 & \cdots & 0 & c_{21} & c_{22} & \cdots & c_{2n} \\ & \vdots & & & & \vdots & & \\ 0 & 0 & \cdots & 1 & c_{n1} & c_{n2} & \cdots & c_{nn} \end{bmatrix},$$

where the $n \times n$ identity matrix I is to the left of the partition. The $n \times n$ solution matrix $C = [c_{ij}]$ to the right of the partition then satisfies $AC = I$. We will show at the end of this section that $CA = I$, too, so $A^{-1} = C$. This is an efficient way to compute A^{-1}. We summarize the computation in the following box, and we state the theory in Theorem 1.8.

Computation of A^{-1}

To find A^{-1}, if it exists, proceed as follows:

Step 1 Form the partitioned matrix $[A \mid I]$.

Step 2 Apply the Gauss–Jordan method to attempt to reduce $[A \mid I]$ to $[I \mid C]$. If the reduction can be carried out, then $A^{-1} = C$. Otherwise, A^{-1} does not exist.

EXAMPLE 2 For the matrix $A = \begin{bmatrix} 2 & 9 \\ 1 & 4 \end{bmatrix}$, compute the inverse we exhibited at the start of this section, and use this inverse to solve the linear system

$$2x + 9y = -5$$
$$x + 4y = 7.$$

Solution Reducing the partitioned matrix, we have

$$\begin{bmatrix} 2 & 9 & | & 1 & 0 \\ 1 & 4 & | & 0 & 1 \end{bmatrix} \sim \begin{bmatrix} 1 & 4 & | & 0 & 1 \\ 2 & 9 & | & 1 & 0 \end{bmatrix} \sim \begin{bmatrix} 1 & 4 & | & 0 & 1 \\ 0 & 1 & | & 1 & -2 \end{bmatrix}$$

$$\sim \begin{bmatrix} 1 & 0 & | & -4 & 9 \\ 0 & 1 & | & 1 & -2 \end{bmatrix}.$$

Therefore,

$$A^{-1} = \begin{bmatrix} -4 & 9 \\ 1 & -2 \end{bmatrix}.$$

If A^{-1} exists, the solution of $A\mathbf{x} = \mathbf{b}$ is $\mathbf{x} = A^{-1}\mathbf{b}$. Consequently, the solution of our system

$$\begin{bmatrix} 2 & 9 \\ 1 & 4 \end{bmatrix} \begin{bmatrix} x \\ y \end{bmatrix} = \begin{bmatrix} -5 \\ 7 \end{bmatrix} \quad \text{is} \quad \begin{bmatrix} x \\ y \end{bmatrix} = \begin{bmatrix} -4 & 9 \\ 1 & -2 \end{bmatrix} \begin{bmatrix} -5 \\ 7 \end{bmatrix} = \begin{bmatrix} 83 \\ -19 \end{bmatrix}. \quad ■$$

We emphasize that the computation of the solution of the linear system in Example 2, using the inverse of the coefficient matrix, was for illustration only. When faced with the problem of solving a square system, $A\mathbf{x} = \mathbf{b}$, one should never start by finding the inverse of the coefficient matrix. To do so would involve row reduction of $[A \mid I]$ and subsequent computation of $A^{-1}\mathbf{b}$, whereas the shorter reduction of $[A \mid \mathbf{b}]$ provides the desired solution at once. The inverse of a matrix is primarily useful in symbolic computations. For example, if A is an invertible matrix and we know that $AB = AC$, then we can deduce that $B = C$ by multiplying both sides of $AB = AC$ on the left by A^{-1}.

THEOREM 1.8 Conditions for A^{-1} to Exist

The following conditions for an $n \times n$ matrix A are equivalent:

  (i) A is invertible.

  (ii) A is row equivalent to the identity matrix I.

  (iii) The system $A\mathbf{x} = \mathbf{b}$ has a solution for each n-component column vector $\mathbf{b}$.

  (iv) A can be expressed as a product of elementary matrices.

PROOF Step 2 in the box preceding Example 2 shows that parts (i) and (ii) of Theorem 1.8 are equivalent.

Let us now show the equivalence of parts (i) and (iii). If A is invertible, then $A\mathbf{x} = \mathbf{b}$ has the solution $\mathbf{x} = A^{-1}\mathbf{b}$, so (i) implies (iii). On the other hand, if $A\mathbf{x} = \mathbf{b}$ has a solution for each $\mathbf{b}$, then $AX = I$ has a solution; just regard the jth column vector of X as the solution of $A\mathbf{x} = \mathbf{e}_j$, where $\mathbf{e}_j$ is the jth column vector of the identity matrix. Thus, (iii) implies (i).

Turning to the equivalence of parts (ii) and (iv), we know that the matrix A is row equivalent to I if and only if there is a sequence of elementary matrices E_1, E_2, ..., E_t such that $E_t \cdots E_2 E_1 A = I$; and this is the case if and only if A is expressible as a product $A = E_1^{-1} E_2^{-1} \cdots E_t^{-1}$ of elementary matrices. ▲

EXAMPLE 3 Using Example 2, express $\begin{bmatrix} 2 & 9 \\ 1 & 4 \end{bmatrix}$ as a product of elementary matrices.

Solution The steps we performed in Example 2 can be applied in sequence to the 2×2 identity matrix to generate elementary matrices:

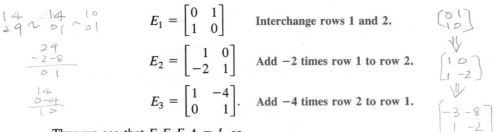

$$E_1 = \begin{bmatrix} 0 & 1 \\ 1 & 0 \end{bmatrix} \qquad \text{Interchange rows 1 and 2.}$$

$$E_2 = \begin{bmatrix} 1 & 0 \\ -2 & 1 \end{bmatrix} \qquad \text{Add } -2 \text{ times row 1 to row 2.}$$

$$E_3 = \begin{bmatrix} 1 & -4 \\ 0 & 1 \end{bmatrix}. \qquad \text{Add } -4 \text{ times row 2 to row 1.}$$

Thus we see that $E_3 E_2 E_1 A = I$, so

$$A = E_1^{-1} E_2^{-1} E_3^{-1} = \begin{bmatrix} 0 & 1 \\ 1 & 0 \end{bmatrix} \begin{bmatrix} 1 & 0 \\ 2 & 1 \end{bmatrix} \begin{bmatrix} 1 & 4 \\ 0 & 1 \end{bmatrix}. \qquad ■$$

Example 3 illustrates the following boxed rule for expressing an invertible matrix A as a product of elementary matrices.

Expressing an Invertible Matrix A as a Product of Elementary Matrices

Write in left-to-right order the inverses of the elementary matrices corresponding to successive row operations that reduce A to I.

EXAMPLE 4 Determine whether the matrix

$$A = \begin{bmatrix} 1 & 3 & -2 \\ 2 & 5 & -3 \\ -3 & 2 & -4 \end{bmatrix}$$

is invertible, and find its inverse if it is.

Solution We have

$$\left[\begin{array}{ccc|ccc} 1 & 3 & -2 & 1 & 0 & 0 \\ 2 & 5 & -3 & 0 & 1 & 0 \\ -3 & 2 & -4 & 0 & 0 & 1 \end{array}\right] \sim \left[\begin{array}{ccc|ccc} 1 & 3 & -2 & 1 & 0 & 0 \\ 0 & -1 & 1 & -2 & 1 & 0 \\ 0 & 11 & -10 & 3 & 0 & 1 \end{array}\right]$$

$$\sim \left[\begin{array}{ccc|ccc} 1 & 0 & 1 & -5 & 3 & 0 \\ 0 & 1 & -1 & 2 & -1 & 0 \\ 0 & 0 & 1 & -19 & 11 & 1 \end{array}\right] \sim \left[\begin{array}{ccc|ccc} 1 & 0 & 0 & 14 & -8 & -1 \\ 0 & 1 & 0 & -17 & 10 & 1 \\ 0 & 0 & 1 & -19 & 11 & 1 \end{array}\right].$$

Therefore, A is an invertible matrix, and

$$A^{-1} = \begin{bmatrix} 14 & -8 & -1 \\ -17 & 10 & 1 \\ -19 & 11 & 1 \end{bmatrix}.$$

▪

EXAMPLE 5   Express the matrix A of Example 4 as a product of elementary matrices.

Solution   In accordance with the box that follows Example 3, we write in left-to-right order the successive inverses of the elementary matrices corresponding to the row reduction of A in Example 4. We obtain

$$A = \begin{bmatrix} 1 & 0 & 0 \\ 2 & 1 & 0 \\ 0 & 0 & 1 \end{bmatrix} \begin{bmatrix} 1 & 0 & 0 \\ 0 & 1 & 0 \\ -3 & 0 & 1 \end{bmatrix} \begin{bmatrix} 1 & 0 & 0 \\ 0 & -1 & 0 \\ 0 & 0 & 1 \end{bmatrix} \begin{bmatrix} 1 & 3 & 0 \\ 0 & 1 & 0 \\ 0 & 0 & 1 \end{bmatrix}$$

$$\times \begin{bmatrix} 1 & 0 & 0 \\ 0 & 1 & 0 \\ 0 & 11 & 1 \end{bmatrix} \begin{bmatrix} 1 & 0 & 2 \\ 0 & 1 & 0 \\ 0 & 0 & 1 \end{bmatrix} \begin{bmatrix} 1 & 0 & 0 \\ 0 & 1 & -1 \\ 0 & 0 & 1 \end{bmatrix}.$$

▪

EXAMPLE 6   Find the inverse of

$$A = \begin{bmatrix} 1 & 3 & 4 \\ -2 & -5 & -3 \\ 1 & 4 & 9 \end{bmatrix},$$

if the inverse exists.

Solution   We have

$$\begin{bmatrix} 1 & 3 & 4 & | & 1 & 0 & 0 \\ -2 & -5 & -3 & | & 0 & 1 & 0 \\ 1 & 4 & 9 & | & 0 & 0 & 1 \end{bmatrix} \sim \begin{bmatrix} 1 & 3 & 4 & | & 1 & 0 & 0 \\ 0 & 1 & 5 & | & 2 & 1 & 0 \\ 0 & 1 & 5 & | & -1 & 0 & 1 \end{bmatrix}$$

$$\sim \begin{bmatrix} 1 & 0 & -11 & | & -5 & -3 & 0 \\ 0 & 1 & 5 & | & 2 & 1 & 0 \\ 0 & 0 & 0 & | & -3 & -1 & 1 \end{bmatrix}.$$

We do not have a pivot in the row 3, column 3 position, so we are not able to complete the reduction. By Theorem 1.8, the matrix is not invertible.   ▪

Our consideration of the inverse of a matrix A arose from the problem of solving a system of equations $A\mathbf{x} = \mathbf{b}$ for a column vector $\mathbf{x}$ of n unknowns. The work involved in finding A^{-1} essentially consists of using the methods described in Section 1.2 for finding the solution of $A\mathbf{x} = \mathbf{b}$, so it is clear that no efficiency results in solving this problem by finding A^{-1}. If we have r systems of equations

$$A\mathbf{x}_1 = \mathbf{b}_1, \qquad A\mathbf{x}_2 = \mathbf{b}_2, \qquad \cdots, \qquad A\mathbf{x}_r = \mathbf{b}_r,$$

all with the same invertible $n \times n$ coefficient matrix A, it might seem to be more efficient (for large r) to solve all the systems by finding A^{-1} and computing the column vectors

$$\mathbf{x}_1 = A^{-1}\mathbf{b}_1, \qquad \mathbf{x}_2 = A^{-1}\mathbf{b}_2, \qquad \cdots, \qquad \mathbf{x}_r = A^{-1}\mathbf{b}_r.$$

Section 9.1 willl show that using the Gauss method with back substitution on the augmented matrix $[A \mid \mathbf{b}_1 \ \mathbf{b}_2 \ \cdots \ \mathbf{b}_r]$ remains more efficient. Thus, inversion of a coefficient matrix is not a good way to solve a linear system. Inverses of matrices are handy, however, for manipulating matrix equations and proving theorems.

Demonstration That $AC = I$ If and Only If $CA = I$

We now prove the boxed result stated prior to Definition 1.9 and used several times already in this section.

THEOREM 1.9 Commutativity with Inverses

> Let A and C be $n \times n$ matrices. Then $AC = I$ if and only if $CA = I$.

PROOF To prove that $AC = I$ if and only if $CA = I$, it suffices to prove that, if $AC = I$, then $CA = I$, since the converse statement is obtained by reversing the roles of A and C.

Suppose now that we do have $AC = I$. We first show that A must be row equivalent to the identity matrix I. The equation $AC = I$ implies that $A\mathbf{x} = \mathbf{b}$ has a solution for every column vector $\mathbf{b}$ of n components; we need only notice that $\mathbf{x} = C\mathbf{b}$ is a solution, since $A(C\mathbf{b}) = (AC)\mathbf{b} = I\mathbf{b} = \mathbf{b}$. Let $E_1, E_2, \ldots, E_t$ be a sequence of elementary matrices such that

$$H = E_t \cdots E_2 E_1 A \tag{6}$$

where H is in reduced row echelon form. Theorem 1.5 assures us that this is possible. If H is not the identity matrix, the bottom row of H is a zero row. Let $\mathbf{e}_n$ be the nth column vector of the $n \times n$ identity matrix. Since the bottom row of H is a zero row, the linear system $H\mathbf{x} = \mathbf{e}_n$ is inconsistent. Then the linear system $A\mathbf{x} = E_1^{-1}E_2^{-1} \cdots E_t^{-1}\mathbf{e}_n$ is also inconsistent, since its row reduction leads to $H\mathbf{x} = \mathbf{e}_n$. But this contradicts our observation that $A\mathbf{x} = \mathbf{b}$ has a solution for every n-component column vector $\mathbf{b}$. Therefore, $H = I$. Then Eq. (6) shows that there is a matrix D (namely, $D = E_t \cdots E_2 E_1$) such that $I = DA$. Since $AC = DA = I$, Theorem 1.6 shows that $D = C$, and so $CA = I$ also. ▲

SUMMARY

1. Let A be a square matrix. A square matrix C such that $CA = AC = I$ is the inverse of A and is denoted by $C = A^{-1}$. If such an inverse of A exists, then A is said to be *invertible*. The inverse of an invertible matrix A is unique. A square matrix that has no inverse is called *singular*.

2. The inverse of a square matrix A exists if and only if A can be reduced to the identity matrix I by means of elementary row operations or (equivalently) if and

only if A is a product of elementary matrices. In this case A is equal to the product, in left-to-right order, of the inverses of the successive elementary matrices corresponding to the sequence of row operations used to reduce A to I.

3. To find A^{-1}, if it exists, form the partitioned matrix $[A \mid I]$ and apply the Gauss–Jordan method to reduce this matrix to $[I \mid C]$. If this can be done, then $A^{-1} = C$. Otherwise, A is not invertible.

4. The inverse of a product of invertible matrices is the product of the inverses in the reverse order.

EXERCISES

In Exercises 1–8, (a) find the inverse of the square matrix, if it exists, and (b) express each invertible matrix as a product of elementary matrices.

1. $\begin{bmatrix} 1 & 1 \\ 0 & 1 \end{bmatrix}$

2. $\begin{bmatrix} 3 & 6 \\ 3 & 8 \end{bmatrix}$

3. $\begin{bmatrix} 3 & 6 \\ 4 & 8 \end{bmatrix}$

4. $\begin{bmatrix} 6 & 7 \\ 8 & 9 \end{bmatrix}$

5. $\begin{bmatrix} 1 & 0 & 1 \\ 0 & 1 & 1 \\ 0 & 0 & -1 \end{bmatrix}$

6. $\begin{bmatrix} 1 & 1 & -1 \\ 2 & 0 & 3 \\ -3 & 1 & -7 \end{bmatrix}$

7. $\begin{bmatrix} 2 & 1 & 4 \\ 3 & 2 & 5 \\ 0 & -1 & 1 \end{bmatrix}$

8. $\begin{bmatrix} -1 & 2 & 1 \\ 2 & -3 & 5 \\ 1 & 0 & 12 \end{bmatrix}$

In Exercises 9–12, find the inverse of the matrix, if it exists.

9. $\begin{bmatrix} 1 & 0 & 1 & -1 \\ 0 & -1 & -3 & 4 \\ 1 & 0 & -1 & 2 \\ -3 & 0 & 0 & -1 \end{bmatrix}$

10. $\begin{bmatrix} 1 & -2 & 1 & 0 \\ -3 & 5 & 0 & 2 \\ 0 & 1 & 2 & -4 \\ -1 & 2 & 4 & -2 \end{bmatrix}$

11. $\begin{bmatrix} 1 & 0 & 0 & 0 & 0 & 0 \\ 0 & -1 & 0 & 0 & 0 & 0 \\ 0 & 0 & 2 & 0 & 0 & 0 \\ 0 & 0 & 0 & 3 & 0 & 0 \\ 0 & 0 & 0 & 0 & 4 & 0 \\ 0 & 0 & 0 & 0 & 0 & 5 \end{bmatrix}$

12. $\begin{bmatrix} 0 & 0 & 0 & 0 & 0 & 6 \\ 0 & 0 & 0 & 0 & 5 & 0 \\ 0 & 0 & 0 & 4 & 0 & 0 \\ 0 & 0 & 3 & 0 & 0 & 0 \\ 0 & 2 & 0 & 0 & 0 & 0 \\ 1 & 0 & 0 & 0 & 0 & 0 \end{bmatrix}$

13. a) Show that the matrix

$$A = \begin{bmatrix} 2 & -3 \\ 5 & -7 \end{bmatrix}$$

is invertible, and find its inverse.

b) Use the result in (a) to find the solution of the system of equations

$$2x_1 - 3x_2 = 4, \qquad 5x_1 - 7x_2 = -3.$$

14. Using the inverse of the matrix in Exercise 7, find the solution of the system of equations

$$2x_1 + x_2 + 4x_3 = 5$$
$$3x_1 + 2x_2 + 5x_3 = 3$$
$$-x_2 + x_3 = 8.$$

15. Find three linear equations that express x, y, z in terms of r, s, t, if

$$2x + y + 4z = r$$
$$3x + 2y + 5z = s$$
$$- y + z = t.$$

[HINT: See Exercise 14.]

16. Let

$$A^{-1} = \begin{bmatrix} 1 & 2 & 1 \\ 0 & 3 & 1 \\ 4 & 1 & 2 \end{bmatrix}.$$

If possible, find a matrix C such that

$$AC = \begin{bmatrix} 1 & 2 \\ 0 & 1 \\ 4 & 1 \end{bmatrix}.$$

17. Let

$$A^{-1} = \begin{bmatrix} 1 & 2 & 1 \\ 0 & 3 & 1 \\ 4 & 1 & 2 \end{bmatrix}.$$

If possible, find a matrix C such that

$$ACA = \begin{bmatrix} 2 & 1 & 3 \\ -1 & 2 & 2 \\ 2 & 1 & 4 \end{bmatrix}.$$

18. Let

$$A = \begin{bmatrix} 4 & 2 & 2 \\ 0 & 3 & 1 \\ 2 & 0 & 1 \end{bmatrix}.$$

If possible, find a matrix B such that $AB = 2A$.

19. Let

$$A = \begin{bmatrix} 1 & 2 & 1 \\ 0 & 1 & 2 \\ 1 & 3 & 2 \end{bmatrix}.$$

$A^{-1}AB = 2A^{-1}A$

$B = 2I$

If possible, find a matrix B such that $AB = A^2 + 2A$.

20. Find all numbers r such that

$$\begin{bmatrix} 2 & 4 & 2 \\ 1 & r & 3 \\ 1 & 2 & 1 \end{bmatrix}$$

is invertible.

21. Find all numbers r such that

$$\begin{bmatrix} 2 & 4 & 2 \\ 1 & r & 3 \\ 1 & 1 & 2 \end{bmatrix}$$

is invertible.

22. Let A and B be two $m \times n$ matrices. Show that A and B are row equivalent if and only if there

exists an invertible $m \times m$ matrix C such that $CA = B$.

23. Mark each of the following True or False. The statements involve matrices A, B, and C, which are assumed to be of appropriate size.

___ a) If $AC = BC$ and C is invertible, then $A = B$.
___ b) If $AB = O$ and B is invertible, then $A = O$.
___ c) If $AB = C$ and two of the matrices are invertible, then so is the third.
___ d) If $AB = C$ and two of the matrices are singular, then so is the third.
___ e) If A^2 is invertible, then A^3 is invertible.
___ f) If A^3 is invertible, then A^2 is invertible.
___ g) Every elementary matrix is invertible.
___ h) Every invertible matrix is an elementary matrix.
___ i) Every invertible matrix is a product of elementary matrices.
___ j) If A and B are invertible, then so is AB, and $(AB)^{-1} = A^{-1}B^{-1}$.

24. Show that, if A is an invertible $n \times n$ matrix, then A^T is invertible. Describe $(A^T)^{-1}$ in terms of A^{-1}.

25. a) If A is invertible, is $A + A^T$ always invertible?
    b) If A is invertible, is $A + A$ always invertible?

26. Let A be a matrix such that A^2 is invertible. Prove that A is invertible.

27. Let A and B be $n \times n$ matrices with A invertible.

    a) Show that $AX = B$ has the unique solution $X = A^{-1}B$.
    b) Show that $X = A^{-1}B$ can be found by the following row reduction:

    $$[A \mid B] \sim [I \mid X].$$

    That is, if the matrix A is reduced to the identity matrix I, then the matrix B will be reduced to $A^{-1}B$.

28. Let A be an $n \times n$ matrix. Show that $Ax = b$ has a solution for each n-component column vector b if and only if $Ax = b$ has a *unique* solution for each n-component column vector b.

29. An $n \times n$ matrix A is **nilpotent** if $A^r = O$ (the $n \times n$ zero matrix) for some positive integer r.

    a) Give an example of a nonzero nilpotent 2×2 matrix.

b) Show that, if A is an invertible $n \times n$ matrix, then A is not nilpotent.

30. A square matrix A is said to be **idempotent** if $A^2 = A$.

   a) Give an example of an idempotent matrix other than O and I.

   b) Show that, if a matrix A is both idempotent and invertible, then $A = I$.

31. Show that

$$\begin{bmatrix} 0 & a_1 & a_2 & a_3 \\ 0 & 0 & b_1 & b_2 \\ 0 & 0 & 0 & c_1 \\ 0 & 0 & 0 & 0 \end{bmatrix}$$

   is nilpotent. (See Exercise 29.)

32. Give an example of a nilpotent 4×4 matrix that is not upper or lower triangular. (See Exercises 29 and 31.)

33. Give an example of two invertible 4×4 matrices whose sum is singular.

34. Give an example of two singular 3×3 matrices whose sum is invertible.

35. Consider the 2×2 matrix

$$A = \begin{bmatrix} a & b \\ c & d \end{bmatrix},$$

   and let $h = ad - bc$.

   a) Show that, if $h \neq 0$, then

$$\begin{bmatrix} d/h & -b/h \\ -c/h & a/h \end{bmatrix}$$

   is the inverse of A.

   b) Show that A is invertible if and only if $h \neq 0$.

Exercises 36–38 develop elementary column operations.

36. For each type of elementary matrix E, explain how E can be obtained from the identity matrix by means of operations on columns.

37. Let A be a square matrix, and let E be an elementary matrix of the same size. Find the effect on A of multiplying A on the right by E. [HINT: Use Exercise 36.]

38. Let A be an invertible square matrix. Recall that $(BA)^{-1} = A^{-1}B^{-1}$, and use Exercise 37 to answer the following questions:

a) If two rows of A are interchanged, how does the inverse of the resulting matrix compare with A^{-1}?

b) Answer the question in part (a) if, instead, a row of A is multiplied by a nonzero scalar r.

c) Answer the question in part (a) if, instead, r times the ith row of A is added to the jth row.

In Exercises 39–42, use the program YURE-DUCE in LINTEK to find the innverse of the matrix, if it exists. If a printer is available, make a copy of the results. Otherwise, copy down the answers to three significant figures.

39. $\begin{bmatrix} 3 & -1 & 2 \\ 1 & 2 & 1 \\ 0 & 3 & -4 \end{bmatrix}$

40. $\begin{bmatrix} -2 & 1 & 4 \\ 3 & 6 & 7 \\ 13 & 15 & -2 \end{bmatrix}$

41. $\begin{bmatrix} 2 & -1 & 3 & 4 \\ -5 & 2 & 0 & 11 \\ 12 & 13 & -6 & 8 \\ 18 & -10 & 3 & 0 \end{bmatrix}$

42. $\begin{bmatrix} 4 & -10 & 3 & 17 \\ 2 & 0 & -3 & 11 \\ 14 & 2 & 12 & -15 \\ 0 & -10 & 9 & -5 \end{bmatrix}$

In Exercises 43–48, follow the instructions for Exercises 39–42, but use the software program MATCOMP in LINTEK. Check to ensure that $AA^{-1} = A^{-1}A = I$ for each matrix A whose inverse is found.

43. The matrix in Exercise 9

44. The matrix in Exercise 10

45. The matrix in Exercise 41

46. The matrix in Exercise 40

47. $\begin{bmatrix} 4 & 1 & -3 & 2 & 6 \\ 0 & 1 & 5 & 2 & 1 \\ 3 & 8 & -11 & 4 & 6 \\ 2 & 1 & -8 & 7 & 2 \\ 1 & 3 & -1 & 4 & 8 \end{bmatrix}$

48. $\begin{bmatrix} 2 & -1 & 0 & 1 & 6 \\ 3 & -1 & 2 & 4 & 6 \\ 0 & 1 & 3 & 4 & 8 \\ -1 & 1 & 1 & 1 & 8 \\ 3 & 1 & 4 & -11 & 10 \end{bmatrix}$

1.5 THE LINEAR TRANSFORMATION $T(\mathbf{x}) = A\mathbf{x}$

Functions are used throughout mathematics to study the structure of sets and relationships between sets. You are familiar with the notation $y = f(x)$, where f is a function that acts on numbers signified by the input variable x and produces numbers signified by the output variable y. In linear algebra, we are interested in functions $\mathbf{y} = f(\mathbf{x})$, where f acts on vectors signified by the input variable $\mathbf{x}$ and produces vectors signified by the output variable $\mathbf{y}$. In this section, we study vector-valued functions T of a vector variable having the form

$$\mathbf{y} = T(\mathbf{x}) = A\mathbf{x}, \tag{1}$$

where A is a matrix. The output of such a function T can be computed by using matrix multiplication, and these functions satisfy certain algebraic properties, mirroring properties of matrix multiplication. Functions having these algebraic properties are called *linear transformations*; the letter T used to denote such a function reflects this name.

After a brief review of the function concept, we present two algebraic properties satisfied by a function $T(\mathbf{x}) = A\mathbf{x}$. In keeping with a practice in modern mathematics, we then proceed to define a linear transformation in terms of these properties. In Section 2.9, we will see examples of linear transformations that are not computed by matrix multiplication.

This section marks the start of the integration of linear transformations into our study of linear algebra. Linear transformations offer a fruitful view of topics in linear algebra, and we will continually take advantage of the insights they offer.

The Function T: $\mathbb{R}^n \to \mathbb{R}^m$ Where $T(\mathbf{x}) = A\mathbf{x}$

A function $f: X \to Y$ is a rule that associates with each x in X an element $y = f(x)$ in Y. We say that f maps the set X into the set Y. The set X is the **domain** of f, and the set Y is often called the **codomain**. To describe a function, we must give its domain and codomain, and then we must specify the action of the function on each element of its domain.

You may be familiar with the set $\mathbb{R}^2$ of all ordered pairs (x, y) in the *Euclidean plane* or the set $\mathbb{R}^3$ of all ordered triples (x, y, z) in *Euclidean 3-space*. In general, $\mathbb{R}^n$ is *Euclidean n-space* and consists of all ordered n-tuples $(x_1, x_2, \ldots, x_n)$ of real numbers. The same ordered sequence of real numbers is also exhibited by the row vector $[x_1 \quad x_2 \quad \cdots \quad x_n]$ and by the column vector

$$\begin{bmatrix} x_1 \\ x_2 \\ \vdots \\ x_n \end{bmatrix}.$$

It is often convenient to view $\mathbb{R}^n$ as the set of all n-component row vectors or n-tuples, or as the set of all n-component column vectors. The context makes clear which viewpoint is intended.

For each positive integer n, let us view $\mathbb{R}^n$ as the set of all column vectors with n real-number components. Let A be an $m \times n$ matrix. If $\mathbf{x} \in \mathbb{R}^n$, then $A\mathbf{x}$ is defined and is a vector in $\mathbb{R}^m$. Thus multiplication of a vector in $\mathbb{R}^n$ on the left by A provides us with a function $T: \mathbb{R}^n \to \mathbb{R}^m$ where $T(\mathbf{x}) = A\mathbf{x}$. We call T the *linear transformation associated with the matrix A*. The set of all vectors of the form $T(\mathbf{x}) = A\mathbf{x}$ is the *range of T*. Figure 1.5 gives a symbolic picture of T.

EXAMPLE 1   Let T be the linear transformation associated with the matrix

$$A = \begin{bmatrix} -1 & 3 & 5 \\ 0 & 4 & 2 \end{bmatrix}.$$

Find the domain and codomain of T, and find formulas for the components of $\mathbf{y} = T(\mathbf{x})$ in terms of the components of $\mathbf{x}$.

Solution   Since A is a 2×3 matrix, the product $A\mathbf{x}$ is defined only for vectors in $\mathbb{R}^3$. Thus, $\mathbb{R}^3$ is the domain of T, and the codomain of T is $\mathbb{R}^2$. Symbolically, $T: \mathbb{R}^3 \to \mathbb{R}^2$. Notice that the order of the superscripts in this symbolic description of T is the opposite of the order of the numbers in the shape description 2×3 of A. Continuing the example, we find that

$$T(\mathbf{x}) = T\left(\begin{bmatrix} x_1 \\ x_2 \\ x_3 \end{bmatrix}\right) = \begin{bmatrix} -1 & 3 & 5 \\ 0 & 4 & 2 \end{bmatrix} \begin{bmatrix} x_1 \\ x_2 \\ x_3 \end{bmatrix} = \begin{bmatrix} -x_1 + 3x_2 + 5x_3 \\ 0x_1 + 4x_2 + 2x_3 \end{bmatrix}.$$

Thus, formulas for the components of $\mathbf{y} = T(\mathbf{x})$ are

$$y_1 = -x_1 + 3x_2 + 5x_3, \qquad y_2 = 4x_2 + 2x_3.$$ ■

Example 1 illustrates that, if T is the linear transformation associated with an $m \times n$ matrix $A = [a_{ij}]$ so that $\mathbf{y} = T(\mathbf{x}) = A\mathbf{x}$, then

$$y_i = a_{i1}x_1 + a_{i2}x_2 + \cdots + a_{in}x_n. \tag{2}$$

Notice that the right-hand side of Eq. (2) is precisely the expression that appears on the left-hand side in the ith equation of a linear system $A\mathbf{x} = \mathbf{b}$. The problem of

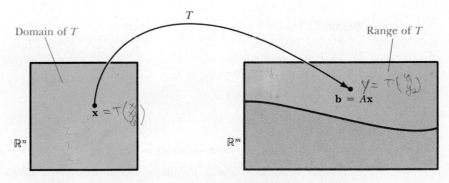

FIGURE 1.5   **The linear transformation $T(\mathbf{x}) = A\mathbf{x}$.**

solving $A\mathbf{x} = \mathbf{b}$ can be phrased as one of determining $\mathbf{x}$ such that $T(\mathbf{x}) = \mathbf{b}$. We know from Section 1.2 that there are three possibilities:

(i) T maps infinitely many $\mathbf{x}$ onto $\mathbf{b}$.　　**Many solutions case**

(ii) T maps just one $\mathbf{x}$ onto $\mathbf{b}$.　　**Unique solution case**

(iii) T maps no vector onto $\mathbf{b}$.　　**Inconsistent case**

If condition (i) holds, then T is not a *one-to-one* function; if condition (ii) holds for each $\mathbf{x}$ in the domain of T, then T is a *one-to-one* function. Condition (iii) reflects the fact that a vector $\mathbf{b}$ is not in the *range* of T. We will develop these ideas more fully in Section 2.9.

Definition of a Linear Transformation T: $\mathbb{R}^n \to \mathbb{R}^m$

We can add any two column vectors in $\mathbb{R}^n$, and we can multiply any column vector in $\mathbb{R}^n$ by a scalar. These two properties of matrix arithmetic,

$$A(\mathbf{u} + \mathbf{v}) = A\mathbf{u} + A\mathbf{v}, \tag{3}$$

$$A(r\mathbf{u}) = r(A\mathbf{u}), \tag{4}$$

played an important role in describing the structure of the solution set of $A\mathbf{x} = \mathbf{b}$ in Section 1.3. These properties exhibit algebraic qualities enjoyed by a function T: $\mathbb{R}^n \to \mathbb{R}^m$, where $T(\mathbf{x}) = A\mathbf{x}$. Written in function notation, the properties become

$$T(\mathbf{u} + \mathbf{v}) = T(\mathbf{u}) + T(\mathbf{v}), \tag{5}$$

$$T(r\mathbf{u}) = rT(\mathbf{u}). \tag{6}$$

These two properties actually characterize the linear transformations associated with matrices. That is, if T: $\mathbb{R}^n \to \mathbb{R}^m$ is a function satisfying these properties for all $\mathbf{u}$, $\mathbf{v} \in \mathbb{R}^n$, then there is an $m \times n$ matrix A such that $T(\mathbf{x}) = A\mathbf{x}$ for all $\mathbf{x} \in \mathbb{R}^n$. We will illustrate this in a moment. It is desirable to define a linear transformation of $\mathbb{R}^n$ into $\mathbb{R}^m$ as a function that has these properties. If you want to know whether some function T: $\mathbb{R}^n \to \mathbb{R}^m$ is a linear transformation, it is often easier to check whether these two properties hold than to try to determine whether there exists an $m \times n$ matrix A such that $T(\mathbf{x}) = A\mathbf{x}$.

DEFINITION 1.10　Linear Transformation

A **linear transformation** T: $\mathbb{R}^n \to \mathbb{R}^m$ is a function satisfying these properties:

1. $T(\mathbf{u} + \mathbf{v}) = T(\mathbf{u}) + T(\mathbf{v})$　　for all $\mathbf{u}, \mathbf{v} \in \mathbb{R}^n$,
2. $T(r\mathbf{u}) = rT(\mathbf{u})$　　for all $\mathbf{u} \in \mathbb{R}^n$ and all scalars r.

From properties 1 and 2, it follows that $T(r\mathbf{u} + s\mathbf{v}) = rT(\mathbf{u}) + sT(\mathbf{v})$ for all $\mathbf{u}$, $\mathbf{v} \in \mathbb{R}^n$ and all scalars r and s. (See Exercise 30.) In fact, this equation can be extended to any number of summands by induction.

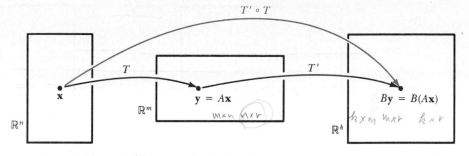

FIGURE 1.6 **The composite map $T' \circ T$.**

Now suppose that A is the $m \times n$ matrix associated with T and that B is the $k \times m$ matrix associated with T'. Then we can compute $T'(T(\mathbf{x}))$ as

$$T'(T(\mathbf{x})) = T'(A\mathbf{x}) = B(A\mathbf{x}).$$

But

$$B(A\mathbf{x}) = (BA)\mathbf{x}, \quad \text{Associativity of matrix multiplication}$$

so $(T' \circ T)(\mathbf{x}) = (BA)\mathbf{x}$. From Theorem 1.10, we see that $T' \circ T$ is again a linear transformation, and that the matrix associated with it is the product of the matrix associated with T' and the matrix associated with T, in that order. Notice how easily this follows from the associativity of matrix multiplication. It really makes us appreciate the power of associativity! We can also show directly from Definition 1.10 that the composite of two linear transformations is again a linear transformation. (See Exercise 29.)

Matrix Multiplication and Composite Transformations

A composition of two linear transformations T and T' yields a linear transformation $T' \circ T$ having as its associated matrix the product of the matrices associated with T' and T, in that order.

Every property of a matrix reflects a corresponding property of the associated linear transformation. Suppose that A is an invertible $n \times n$ matrix, and let $T: \mathbb{R}^n \to \mathbb{R}^n$ be the associated linear transformation, so that $\mathbf{y} = T(\mathbf{x}) = A\mathbf{x}$. There exists a linear transformation of $\mathbb{R}^n$ into $\mathbb{R}^n$ associated with A^{-1}; we denote this by T^{-1}, so that $T^{-1}(\mathbf{y}) = A^{-1}\mathbf{y}$. The matrix of the composite transformation $T^{-1} \circ T$ is the product $A^{-1}A$, as indicated in the preceding box. Since $A^{-1}A = I$ and $I\mathbf{x} = \mathbf{x}$, we see that $(T^{-1} \circ T)(\mathbf{x}) = \mathbf{x}$. That is, $T^{-1} \circ T$ is the _identity transformation,_ leaving all vectors fixed. (See Fig. 1.7.) Since $AA^{-1} = I$ too, we see that $T \circ T^{-1}$ is also the identity transformation on $\mathbb{R}^n$. If $\mathbf{y} = T(\mathbf{x})$, then $\mathbf{x} = T^{-1}(\mathbf{y})$. This transformation T^{-1} is the **inverse transformation of** T, and T is an **invertible** linear transformation.

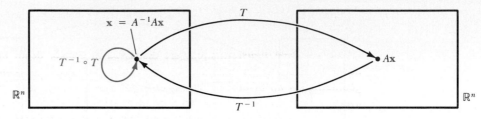

FIGURE 1.7 $T^{-1} \circ T$ **is the identity transformation.**

Invertible Matrices and Inverse Transformations

Let A be an invertible $n \times n$ matrix with associated linear transformation T. The transformation T^{-1} associated with A^{-1} is the inverse transformation of T, and $T \circ T^{-1}$ and $T^{-1} \circ T$ are both the identity transformation on $\mathbb{R}^n$. A linear transformation $T: \mathbb{R}^n \to \mathbb{R}^n$ is invertible if and only if its associated matrix is invertible.

EXAMPLE 5 Show that the linear transformation $T: \mathbb{R}^3 \to \mathbb{R}^3$ defined by $T(x_1, x_2, x_3) = (x_1 - 2x_2 + x_3, x_2 - x_3, 2x_2 - 3x_3)$ is invertible, and find a formula for its inverse.

Solution Using column-vector notation, we see that $T(\mathbf{x}) = A\mathbf{x}$, where

$$A = \begin{bmatrix} 1 & -2 & 1 \\ 0 & 1 & -1 \\ 0 & 2 & -3 \end{bmatrix}.$$

Next, we find the inverse of A:

$$\begin{bmatrix} 1 & -2 & 1 \\ 0 & 1 & -1 \\ 0 & 2 & -3 \end{bmatrix} \begin{array}{|ccc} 1 & 0 & 0 \\ 0 & 1 & 0 \\ 0 & 0 & 1 \end{array} \sim \begin{bmatrix} 1 & 0 & -1 \\ 0 & 1 & -1 \\ 0 & 0 & -1 \end{bmatrix} \begin{array}{|ccc} 1 & 2 & 0 \\ 0 & 1 & 0 \\ 0 & -2 & 1 \end{array}$$

$$\sim \begin{bmatrix} 1 & 0 & 0 \\ 0 & 1 & 0 \\ 0 & 0 & 1 \end{bmatrix} \begin{array}{|ccc} 1 & 4 & -1 \\ 0 & 3 & -1 \\ 0 & 2 & -1 \end{array}.$$

Therefore,

$$T^{-1}(\mathbf{x}) = A^{-1}\mathbf{x} = \begin{bmatrix} 1 & 4 & -1 \\ 0 & 3 & -1 \\ 0 & 2 & -1 \end{bmatrix} \begin{bmatrix} x_1 \\ x_2 \\ x_3 \end{bmatrix} = \begin{bmatrix} x_1 + 4x_2 - x_3 \\ 3x_2 - x_3 \\ 2x_2 - x_3 \end{bmatrix},$$

which we express in row notation as

$$T^{-1}(x_1, x_2, x_3) = (x_1 + 4x_2 - x_3, 3x_2 - x_3, 2x_2 - x_3).$$

In Exercise 28 we ask you to verify that $T^{-1}(T(\mathbf{x})) = \mathbf{x}$, as in Figure 1.7. ■

The Geometry of Invertible Linear Transformations of $\mathbb{R}^2$ into $\mathbb{R}^2$

Invertible linear transformations are discussed in greater detail in Section 2.9. For now, we restrict our attention to $\mathbb{R}^2$ and exploit matrix techniques to describe geometrically all invertible linear transformations of the plane into itself. Recall that ✳ every invertible matrix is a product of elementary matrices. If we can interpret geometrically the effect on the plane of the linear transformations associated with elementary 2×2 matrices, we will gain information about all linear transformations of $\mathbb{R}^2$ into $\mathbb{R}^2$ whose associated matrices are invertible. We change to the familiar x, y-notation for coordinates in $\mathbb{R}^2$.

EXAMPLE 6   Show that flipping the plane $\mathbb{R}^2$ over while keeping the y-axis fixed is a linear transformation of $\mathbb{R}^2$ into itself.

Solution   Figure 1.8 shows that flipping the plane over as indicated carries the point (x, y) to the point $T(x, y) = (-x, y)$. Using column vector notation, we can describe this action with

$$T\left(\begin{bmatrix} x \\ y \end{bmatrix}\right) = \begin{bmatrix} -x \\ y \end{bmatrix} = \begin{bmatrix} -1 & 0 \\ 0 & 1 \end{bmatrix}\begin{bmatrix} x \\ y \end{bmatrix},$$

which is the linear transformation associated with the elementary matrix

$$\begin{bmatrix} -1 & 0 \\ 0 & 1 \end{bmatrix},$$

corresponding to row scaling.   ■

The transformation in Example 6 is the *reflection* of the plane through the y-axis. Similarly, reflection of the plane through the x-axis is a linear transformation with associated matrix

$$T(x, y) = (x, -y) \qquad \begin{bmatrix} 1 & 0 \\ 0 & -1 \end{bmatrix},$$

which is again an elementary matrix corresponding to row scaling. (See Exercise 32.)

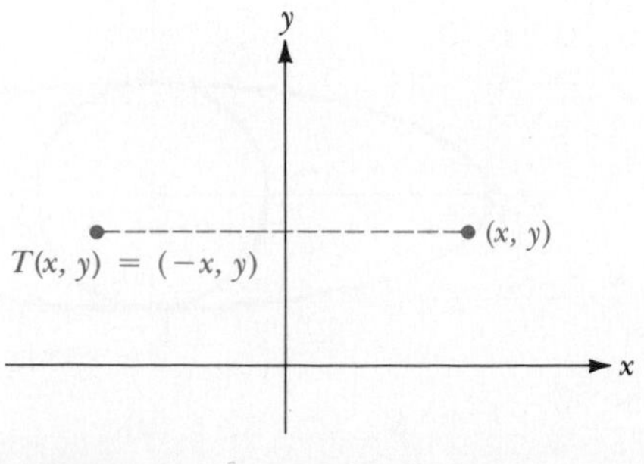

FIGURE 1.8 Reflection of $\mathbb{R}^2$ through the y-axis.

Reflection of the plane through the line $y = x$ is a linear transformation with associated matrix

$$T(x, y) = (y, x) \qquad \begin{bmatrix} 0 & 1 \\ 1 & 0 \end{bmatrix},$$

which is the 2×2 elementary matrix corresponding to row interchange. (See Exercise 33.) Exercise 33 of Section 5.4 shows that reflections in other lines through the origin are also linear transformations. We classify the effect of a linear transformation corresponding to a general 2×2 row-scaling elementary matrix in the next example.

✓ **EXAMPLE 7** Describe geometrically the linear transformation $T\left(\begin{bmatrix} x \\ y \end{bmatrix}\right) = E\begin{bmatrix} x \\ y \end{bmatrix}$, where E is a 2×2 elementary matrix corresponding to <u>row scaling</u>.

Solution The matrix E has the form

$$\begin{bmatrix} r & 0 \\ 0 & 1 \end{bmatrix} \qquad \text{or} \qquad \begin{bmatrix} 1 & 0 \\ 0 & r \end{bmatrix}$$

for some nonzero scalar r. We discuss the first case and leave the second as Exercise 34. The transformation is given by

$$T\left(\begin{bmatrix} x \\ y \end{bmatrix}\right) = \begin{bmatrix} r & 0 \\ 0 & 1 \end{bmatrix}\begin{bmatrix} x \\ y \end{bmatrix} = \begin{bmatrix} rx \\ y \end{bmatrix},$$

or in row notation, $T(x, y) = (rx, y)$. The second component of (x, y) is unchanged. However, the first component is multiplied by the scalar r, resulting in a horizontal expansion if $r > 1$ or in a horizontal contraction if $0 < r < 1$. In Figure 1.9, we illustrate the effect of such a horizontal expansion or contraction on the points of the unit circle. If $r < 0$, we have an expansion or contraction followed by a reflection through the y-axis. For example,

$$\begin{bmatrix} -3 & 0 \\ 0 & 1 \end{bmatrix} = \begin{bmatrix} -1 & 0 \\ 0 & 1 \end{bmatrix}\begin{bmatrix} 3 & 0 \\ 0 & 1 \end{bmatrix},$$

       Reflection **Horizontal expansion**

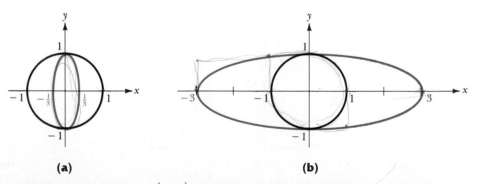

      (a)   **(b)**

FIGURE 1.9   **(a)** $T(x, y) = \left(\frac{1}{3}x, y\right)$ contracts horizontally; **(b)** $T(x, y) = (3x, y)$ expands horizontally.

indicating a horizontal expansion by a factor of 3, followed by a reflection through the y-axis.   ■

EXAMPLE 8   Describe geometrically the linear transformation $T\left(\begin{bmatrix} x \\ y \end{bmatrix}\right) = E\begin{bmatrix} x \\ y \end{bmatrix}$, where E is a 2×2 elementary matrix corresponding to row addition.

Solution   The matrix E has the form

$$\begin{bmatrix} 1 & 0 \\ r & 1 \end{bmatrix} \quad \text{or} \quad \begin{bmatrix} 1 & r \\ 0 & 1 \end{bmatrix}$$

for some nonzero scalar r. We discuss the first case, and leave the second as Exercise 36. The transformation is given by

$$T\left(\begin{bmatrix} x \\ y \end{bmatrix}\right) = \begin{bmatrix} 1 & 0 \\ r & 1 \end{bmatrix}\begin{bmatrix} x \\ y \end{bmatrix} = \begin{bmatrix} x \\ rx + y \end{bmatrix},$$

or in row-vector notation, $T(x, y)$ and $(x, rx + y)$. The first component of the point (x, y) is unchanged. However, the second component is changed by the addition of rx. For example, $(1, 0)$ is carried onto $(1, r)$, and $(1, 1)$ is carried onto $(1, 1 + r)$, while $(0, 0)$ and $(0, 1)$ are carried onto themselves. Notice that every point on the y-axis remains fixed. Figure 1.10 illustrates the effect of this transformation. The squares shaded in black are carried onto the parallelograms shaded in color. This transformation is called a *vertical shear*. Exercise 36 deals with the case of a horizontal shear.   ■

We have noted that a square matrix A is invertible if and only if it is the product of elementary matrices. We also know that a product of matrices corresponds to the

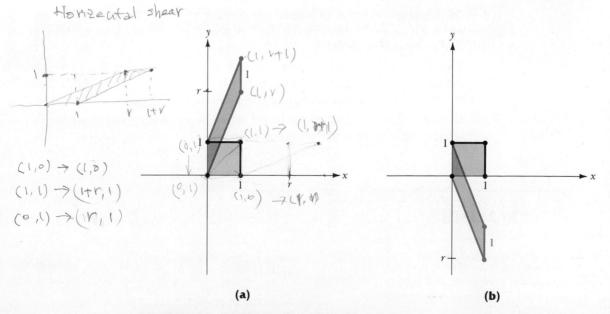

FIGURE 1.10   **(a) The vertical shear $T(x, y) = (x, rx + y)$, $r > 0$; (b) the vertical shear $T(x, y) = (x, rx + y)$, $r < 0$.**

composition of the associated linear transformations, and we have seen the effect of transformations associated with elementary matrices on the plane. Putting all of these ideas together, we obtain the following.

Geometric Description of Invertible Transformations of $\mathbb{R}^2$

A linear transformation T of the plane $\mathbb{R}^2$ into itself is invertible if and only if T consists of a finite sequence of:

Reflections through the x-axis, the y-axis, or the line $y = x$,

Vertical or horizontal expansions or contractions, and

Vertical or horizontal shears.

SUMMARY

1. A function $T: \mathbb{R}^n \to \mathbb{R}^m$ is a linear transformation if $T(\mathbf{u} + \mathbf{v}) = T(\mathbf{u}) + T(\mathbf{v})$ and $T(r\mathbf{u}) = rT(\mathbf{u})$, for all vectors $\mathbf{u}, \mathbf{v} \in \mathbb{R}^n$ and all scalars r.

2. Every linear transformation $T: \mathbb{R}^n \to \mathbb{R}^m$ can be expressed in the form $T(\mathbf{x}) = A\mathbf{x}$, where A is an $m \times n$ matrix and $\mathbf{x}$ is a column vector in $\mathbb{R}^n$. The matrix A is said to be associated with T. Conversely, every function of the form $T(\mathbf{x}) = A\mathbf{x}$ is a linear transformation.

3. The composite $T' \circ T$ of two linear transformations $T(\mathbf{x}) = A\mathbf{x}$ and $T'(\mathbf{y}) = B\mathbf{y}$, where A is an $m \times n$ matrix and B is a $k \times m$ matrix, is the linear transformation with associated matrix BA.

4. If $\mathbf{y} = T(\mathbf{x}) = A\mathbf{x}$, where A is an invertible $n \times n$ matrix, then T is invertible and the transformation T^{-1} defined by $T^{-1}(\mathbf{y}) = A^{-1}\mathbf{y}$ is the inverse of T. Both $T' \circ T$ and $T \circ T'$ are the identity transformation of $\mathbb{R}^n$.

5. An invertible linear transformation of $\mathbb{R}^2$ into itself can be described geometrically as indicated in the box preceding this summary.

EXERCISES

In the statements of many of these exercises, row notation is used to save space. In Exercises 1–6, assume that T is a linear transformation.

1. If $T(1, 0) = (3, -1)$ and $T(0, 1) = (-2, 5)$, find $T(4, -6)$.

2. If $T(-1, 0) = (2, 3)$ and $T(0, 1) = (5, 1)$, find $T(-3, -5)$.

3. If $T(1, 0, 0) = (3, 1, 2)$, $T(0, 1, 0) = (2, -1, 4)$, and $T(0, 0, 1) = (6, 0, 1)$, find $T(2, -5, 1)$.

$T\left(\begin{smallmatrix} x \\ y \\ z \end{smallmatrix}\right) = \left[\quad\right]\left[\begin{smallmatrix} x \\ y \\ z \end{smallmatrix}\right]$ (142)
$V \to V_B \to AV_B = T(V)$
3×3　3×1

4. If $T(1, 0, 0) = (-1, 3)$, $T(0, 1, 0) = (4, -1)$, and $T(0, -1, 1) = (3, -5)$, find $T(-1, 4, 2)$.

5. If $T(-1, 2) = (1, 0, 0)$ and $T(2, 1) = (0, 1, 2)$, find $T(0, 10)$.

6. If $T(-1, 1) = (2, 1, 4)$ and $T(1, 1) = (-6, 3, 2)$, find $T(x, y)$.

In Exercises 7–12, the given formula defines a linear transformation. Give its associated matrix.

7. $T(x, y) = (x + y, x - 3y)$

8. $T(x_1, x_2) = (2x_1 - x_2, x_1 + x_2, x_1 + 3x_2)$

9. $T(x, y, z) = (x + y + z, x + y, x)$

10. $T(x, y, z) = (2x + y + z, x + y + 3z)$

11. $T(x, y, z) = (x - y + 3z, x + y + z, x)$

12. $T(x_1, x_2, x_3) = x_1 + x_2 + x_3$

13. Is $T(x, y, z) = (x, y, z)$ a linear transformation of $\mathbb{R}^3$ into $\mathbb{R}^3$? Why or why not?

14. Is $T(x, y, z) = (0, 0, 0, 0)$ a linear transformation of $\mathbb{R}^3$ into $\mathbb{R}^4$? Why or why not?

15. Is $T(x, y, z) = (1, 1, 1, 1)$ a linear transformation of $\mathbb{R}^3$ into $\mathbb{R}^4$? Why or why not?

16. Is $T(x_1, x_2) = (x_1 - x_2, x_2 + 1, 3x_1 - 2x_2)$ a linear transformation of $\mathbb{R}^2$ into $\mathbb{R}^3$? Why or why not?

17. If $T: \mathbb{R}^2 \to \mathbb{R}^3$ is defined by $T(x, y) = (2x + y, x, x - y)$ and $T': \mathbb{R}^3 \to \mathbb{R}^2$ is defined by $T'(x, y, z) = (x - y + z, x + y)$, find the matrix associated with the linear transformation $T' \circ T$ that carries $\mathbb{R}^2$ into $\mathbb{R}^2$. Find a formula for $(T' \circ T)(x, y)$.

18. Referring to Exercise 17, find the matrix associated with the linear transformation $T \circ T'$ that carries $\mathbb{R}^3$ into $\mathbb{R}^3$. Find a formula for $(T \circ T')(x, y, z)$.

In Exercises 19–26, determine whether the indicated linear transformation T is invertible. If it is, find a formula for $T^{-1}(x)$ in row notation. If it is not, explain why it is not.

19. The transformation T in Exercise 7

20. The transformation T in Exercise 8

21. The transformation T in Exercise 9

22. The transformation T in Exercise 10

23. The transformation T in Exercise 11

24. The transformation T in Exercise 12

25. The transformation $T' \circ T$ in Exercise 17

26. The transformation $T \circ T'$ in Exercise 18

27. Mark each of the following True or False.

____ a) Every linear transformation is a function.

____ b) Every function mapping $\mathbb{R}^n$ into $\mathbb{R}^m$ is a linear transformation.

____ c) Composition of linear transformations corresponds to multiplication of the associated matrices.

____ d) Function composition is associative.

____ e) An invertible linear transformation mapping $\mathbb{R}^n$ into itself has a unique inverse.

____ f) The same matrix may be associated with several different linear transformations.

____ g) The linear transformation associated with an $m \times n$ matrix maps $\mathbb{R}^n$ into $\mathbb{R}^m$.

____ h) The geometric effect of all invertible linear transformations of $\mathbb{R}^3$ into itself can be described in terms of the geometric effect of the linear transformations of $\mathbb{R}^3$ associated with all elementary 3×3 matrices.

____ i) Every linear transformation of the plane into itself can be achieved through a succession of reflections, expansions, contractions, and shears.

____ j) Every invertible linear transformation of the plane into itself can be achieved through a succession of reflections, expansions, contractions, and shears.

28. Verify that $T^{-1}(T(\mathbf{x})) = \mathbf{x}$ for the linear transformation T in Example 5 of the text.

29. Let $T: \mathbb{R}^n \to \mathbb{R}^m$ and $T': \mathbb{R}^m \to \mathbb{R}^k$ be linear transformations. Prove directly from Definition 1.10 that $(T' \circ T): \mathbb{R}^n \to \mathbb{R}^k$ is also a linear transformation.

30. Let $T: \mathbb{R}^n \to \mathbb{R}^m$ be a linear transformation. Show that

　(i) $T(\mathbf{0}) = \mathbf{0}$;

　(ii) $T(r\mathbf{u} + s\mathbf{v}) = rT(\mathbf{u}) + sT(\mathbf{v})$ for all $\mathbf{u}, \mathbf{v}$ in $\mathbb{R}^n$ and for all scalars r and s.

31. Show that the matrix associated with a linear transformation $T: \mathbb{R}^n \to \mathbb{R}^m$ is unique. [HINT: Use Exercise 41 of Section 1.1.]

32. Show that a reflection of the plane through the x-axis corresponds to the linear transformation

$$T\left(\begin{bmatrix} x \\ y \end{bmatrix}\right) = \begin{bmatrix} 1 & 0 \\ 0 & -1 \end{bmatrix}\begin{bmatrix} x \\ y \end{bmatrix}.$$

33. Show that a reflection of the plane through the line $y = x$ corresponds to the linear transformation

$$T\left(\begin{bmatrix} x \\ y \end{bmatrix}\right) = \begin{bmatrix} 0 & 1 \\ 1 & 0 \end{bmatrix}\begin{bmatrix} x \\ y \end{bmatrix}.$$

34. Show that the linear transformation

$$T\left(\begin{bmatrix} x \\ y \end{bmatrix}\right) = \begin{bmatrix} 1 & 0 \\ 0 & r \end{bmatrix}\begin{bmatrix} x \\ y \end{bmatrix}$$

affects the plane $\mathbb{R}^2$ as follows:

(i) A vertical expansion, if $r > 1$;
(ii) A vertical contraction, if $0 < r < 1$;
(iii) A vertical expansion followed by a reflection through the x-axis, if $r < -1$;
(iv) A vertical contraction followed by a reflection through the x-axis, if $-1 < r < 0$.

35. Referring to Exercise 34, explain algebraically why cases (iii) and (iv) can be described by the reflection followed by the expansion or contraction, in that order.

36. Show that the linear transformation

$$T\left(\begin{bmatrix} x \\ y \end{bmatrix}\right) = \begin{bmatrix} 1 & r \\ 0 & 1 \end{bmatrix}\begin{bmatrix} x \\ y \end{bmatrix}$$

corresponds to a horizontal shear of the plane.

In Exercises 37–41, identify the given invertible transformation of $\mathbb{R}^2$ into itself as a product of reflections, expansions, contractions, and shears, by expressing the matrix associated with the transformation as a product of elementary matrices.

37. $T(x, y) = (-y, x)$ (A rotation counterclockwise through 90°)

38. $T(x, y) = (2x, 2y)$ (Expansion away from the origin by a factor of 2)

39. $T(x, y) = (-x, -y)$ (A rotation through 180°)

40. $T(x, y) = (x + y, 2x - y)$

41. $T(x, y) = (x + y, 3x + 5y)$

1.6 **APPLICATIONS TO POPULATION DISTRIBUTION**

Consider situations in which people are split into two or more categories. For example, we might split the citizens of the United States according to income into categories of

poor, middle income, rich.

We might split the inhabitants of North America into categories according to the climate in which they live:

hot, temperate, cold.

In this book, we will speak of a *population* split into *states*. In the two illustrations above, the populations and states are given by the following:

Population	States
Citizens of the United States	poor, middle income, rich
People in North America	hot, temperate, cold

Our populations will often consist of people, but this is not essential. For example, at any moment we can classify the population of cars as operational or not operational.

We are interested in how the distribution of a population between (or among) states may change over a period of time. Matrices and their multiplication can play an important role in such considerations.

Transition Matrices

The tendency of a population to move among n states can sometimes be described using an $n \times n$ matrix. Consider a population distributed among $n = 3$ states, which we call state 1, state 2, and state 3. Suppose that we know the *proportion* t_{ij} of the population of state j that moves to state i over a given fixed time period. Notice that the direction of movement from state j to state i is the right-to-left order of the subscripts in t_{ij}. The matrix $T = [t_{ij}]$ is called a **transition matrix**. (Do not confuse our use of T as a transition matrix in this one section with our use of T as a linear transformation elsewhere in the text.)

EXAMPLE 1 Let the population of a country be classified according to income as

State 1: poor,

State 2: middle income,

State 3: rich.

Suppose that, over each 20-year period (about one generation), we have the following data for people and their offspring:

Of the poor people, 19% become middle income and 1% rich.

Of the middle income people, 15% become poor and 10% rich.

Of the rich people, 5% become poor and 30% middle income.

Give the transition matrix describing these data.

Solution The entry t_{ij} in the transition matrix T represents the proportion of the population moving from state j to state i, not the percentage. Thus the data that 19% of the poor (state 1) will become middle income (state 2) mean that we should take $t_{21} = .19$. Similarly, since 1% of the people in state 1 move to state 3 (rich), we should take $t_{31} = .01$. Now t_{11} represents the proportion of the poor people who remain poor at the end of 20 years. Since this is 80%, we should take $t_{11} = .80$. Continuing in this fashion, starting in state 2 and then in state 3, we obtain the matrix

$$T = \begin{array}{c} \\ \end{array} \begin{array}{ccc} \text{poor} & \text{mid} & \text{rich} \\ \begin{bmatrix} .80 & .15 & .05 \\ .19 & .75 & .30 \\ .01 & .10 & .65 \end{bmatrix} & \begin{array}{l} \text{poor} \\ \text{mid} \\ \text{rich} \end{array} \end{array}.$$

We have labeled the columns and rows with the names of the states. Notice that an entry of the matrix gives the proportion of the population in the state above the entry that moves to the state at the right of the entry during one 20-year period. ■

In Example 1, the sum of the entries in each column of T is 1, since the sum reflects the movement of the entire population for the state listed at the top of the column. Now suppose that the proportions of the entire population in Example 1 that fall into the various states at the start of a time period are given in the column vector

$$\mathbf{p} = \begin{bmatrix} p_1 \\ p_2 \\ p_3 \end{bmatrix}.$$

For example, we would have

$$\mathbf{p} = \begin{bmatrix} \frac{1}{3} \\ \frac{1}{3} \\ \frac{1}{3} \end{bmatrix},$$

if the whole population were initially equally divided among the states. The entries in such a **population distribution vector p** must be nonnegative and must have sum equal to 1.

Let us find the proportion of the entire population that is in state 1 after one time period of 20 years, knowing that initially the proportion in state 1 is p_1. The proportion of the state-1 population that remains in state 1 is t_{11}. This gives a contribution of $t_{11}p_1$ to the proportion of the entire population that will be found in state 1 after the end of 20 years. Of course, we also get contributions to state 1 at the end of 20 years from states 2 and 3. These two states contribute proportions $t_{12}p_2$ and $t_{13}p_3$ of the entire population to state 1. Thus, after 20 years, the proportion in state 1 is

$$t_{11}p_1 + t_{12}p_2 + t_{13}p_3.$$

This is precisely the first entry in the column vector given by the product

$$T\mathbf{p} = \begin{bmatrix} t_{11} & t_{12} & t_{13} \\ t_{21} & t_{22} & t_{23} \\ t_{31} & t_{32} & t_{33} \end{bmatrix} \begin{bmatrix} p_1 \\ p_2 \\ p_3 \end{bmatrix}.$$

In a similar fashion, we find that the second and third components of $T\mathbf{p}$ give the proportions of population in state 2 and in state 3 after one time period.

> For an initial population distribution vector **p** and transition matrix T, the product vector $T\mathbf{p}$ is the population distribution vector after one time period.

Markov Chains

In Example 1, we found the transition matrix governing the flow of a population among three states over a period of 20 years. Suppose that the same transition matrix is valid over the next 20-year period, and for the next 20 years after that, and so on. That is, suppose that there is a sequence or *chain* of 20-year periods over which the transition matrix is valid. Such a situation is called a **Markov chain**. Let us give a formal definition of a transition matrix for a Markov chain.

DEFINITION 1.11 Transition Matrix

A **transition matrix** for an n-state Markov chain is an $n \times n$ matrix in which all entries are nonnegative and in which the sum of the entries in each column is 1.

Markov chains arise naturally in biology, psychology, economics, and many other sciences. Thus they are an important application of linear algebra and of probability. The entry t_{ij} in a transition matrix T is known as the **probability** of moving from state j to state i over one time period.

EXAMPLE 2 Show that the matrix

$$T = \begin{bmatrix} 0 & 0 & 1 \\ 1 & 0 & 0 \\ 0 & 1 & 0 \end{bmatrix}$$

is a transition matrix for a three-state Markov chain, and explain the significance of the zeros and the ones.

Solution The entries are all nonnegative, and the sum of the entries in each column is 1. Thus the matrix is a transition matrix for a Markov chain.

At least for finite populations, a transition probability $t_{ij} = 0$ means that there is no movement from state j to state i over the time period. That is, transition from state j to state i over the time period is impossible. On the other hand, if $t_{ij} = 1$, the entire population of state j moves to state i over the time period. That is, transition from state j to state i in the time period is certain.

For the given matrix, we see that, over one time period, the entire population of state 1 moves to state 2, the entire population of state 2 moves to state 3, and the entire population of state 3 moves to state 1. ■

Let T be the transition matrix over a time period—say, 20 years—in a Markov chain. We can form a new Markov chain by looking at the flow of the population over a time period twice as long—that is, over 40 years. Let us see the relationship of the transition matrix for the 40-year time period to the one for the 20-year time period.

MARKOV CHAINS are named for the Russian mathematician Andrei Andreevich Markov (1856–1922), who first defined them in a paper of 1906 dealing with the Law of Large Numbers and subsequently proved many of the standard results about them. His interest in these sequences stemmed from the needs of probability theory; Markov never dealt with their applications to the sciences. The only real examples he used were from literary texts, where the two possible states were vowels and consonants. To illustrate his results, he did a statistical study of the alternation of vowels and consonants in Pushkin's *Eugene Onegin*.

Andrei Markov taught at St. Petersburg University from 1880 to 1905, when he retired to make room for younger mathematicians. Besides his work in probability, he contributed to such fields as number theory, continued fractions, and approximation theory. He was an active participant in the liberal movement in the pre–World War I era in Russia; on many occasions he made public criticisms of the actions of state authorities. In 1913, when as a member of the Academy of Sciences he was asked to participate in the pompous ceremonies celebrating the 300th anniversary of the Romanov dynasty, he instead organized a celebration of the 200th anniversary of Jacob Bernoulli's publication of the Law of Large Numbers.

We might guess that the transition matrix for 40 years is T^2, for if $\mathbf{p}$ is the initial population distribution vector, then $T\mathbf{p}$ is the population distribution vector after the first time period. Thus $T(T\mathbf{p}) = T^2\mathbf{p}$ is the distribution vector after two periods, which suggests that the two-period transition matrix is T^2. Really, this only shows that $T^2\mathbf{p}$ gives the correct answer for the population distribution vector. We illustrate with an example that T^2 is indeed the correct transition matrix.

EXAMPLE 3 For the Markov chain with time period 20 years described in Example 1, find the transition matrix for a 40-year time period.

Solution Recall that the transition matrix obtained in Example 1 for the 20-year time period is

$$T = \begin{array}{c} \\ \begin{array}{ccc} \text{poor} & \text{mid} & \text{rich} \end{array} \\ \left[\begin{array}{ccc} .80 & .15 & .05 \\ .19 & .75 & .30 \\ .01 & .10 & .65 \end{array}\right] \begin{array}{l} \text{poor} \\ \text{mid} \\ \text{rich} \end{array} \end{array}.$$

Let n_1 be the size of the population in state 1 at the beginning of a time period. We will find the proportion of n_1 in state 2, after two time periods. After the first time period, the distribution of the n_1 members of state 1 among all three states is given by

State 1	State 2	State 3
$(.80)n_1$	$(.19)n_1$	$(.01)n_1$.

During the next time period, 19% of the $(.80)n_1$ in state 1 will make a transition to state 2; 75% of the $(.19)n_1$ in state 2 will remain in state 2; and 30% of the $(0.1)n_1$ in state 3 will move to state 2. Thus, following the fortunes of the original n_1 members of state 1, we find that the number of them who end up in state 2 after two time periods is

$$(.19)(.80)n_1 + (.75)(.19)n_1 + (.30)(.01)n_1 = (.2975)n_1.$$

Thus the entry in the second row and first column of the two-period transition matrix should be

$$(.19)(.80) + (.75)(.19) + (.30)(.01) = .2975.$$

We note that this is precisely the dot product of the second row vector and the first column vector of T, and that consequently it is the entry in the second row and first column of T^2. If we continue in this fashion, we find that the two-period matrix is indeed T^2, and a calculator shows that

$$T^2 = \begin{array}{c} \\ \begin{array}{ccc} \text{poor} & \text{mid} & \text{rich} \end{array} \\ \left[\begin{array}{ccc} .6690 & .2375 & .1175 \\ .2975 & .6210 & .4295 \\ .0335 & .1415 & .4530 \end{array}\right] \begin{array}{l} \text{poor} \\ \text{mid} \\ \text{rich} \end{array} \end{array}.$$

All column sums in T^2 are 1, so T^2 is indeed a transition matrix. ■

If we extend the argument in Example 3, we find that the three-period transition matrix for the Markov chain is T^3, and so on. This exhibits another situation in which matrix multiplication is useful. Although raising even a small matrix like T in Example 3 to a power using pencil and paper (or even a calculator) is tedious, a computer can do it easily. The software program MATCOMP available with this text can be used to compute a power of a matrix.

---

m-Period Transition Matrix

A Markov chain with transition matrix T has T^m as its m-period transition matrix.

EXAMPLE 4   For the transition matrix

$$T = \begin{bmatrix} 0 & 0 & 1 \\ 1 & 0 & 0 \\ 0 & 1 & 0 \end{bmatrix}$$

in Example 2, show that, after three time periods, the distribution of population among the three states is the same as the initial population distribution.

Solution   After three time periods, the population distribution vector is $T^3 \mathbf{p}$, and we easily compute that $T^3 = I$, the 3×3 identity matrix. Thus $T^3 \mathbf{p} = \mathbf{p}$, as asserted. Alternatively, we could note that the entire population of state 1 moves to state 2 in the first time period, then to state 3 in the next time period, and finally back to state 1 in the third time period. Similarly, the populations of the other two states move around and then back to the beginning state over the three periods.   ■

Regular Markov Chains

We now turn to Markov chains where there exists some fixed number m of time periods in which it is possible to get from any state to any other state. This means that the mth power T^m of the transition matrix has no zero entries.

DEFINITION 1.12   Regular Transition Matrix, Regular Chain

A transition matrix T is **regular** if T^m has no zero entries for some integer m. A Markov chain having a regular transition matrix is called a **regular chain**.

EXAMPLE 5   Show that the transition matrix

$$T = \begin{bmatrix} 0 & 0 & 1 \\ 1 & 0 & 0 \\ 0 & 1 & 0 \end{bmatrix}$$

of Example 2 is not regular.

Solution   A computation shows that T^2 still has zero entries. We saw in Example 4 that $T^3 = I$, the 3×3 identity matrix, so we must have $T^4 = T$, $T^5 = T^2$, $T^6 = T^3 = I$, and the powers of T repeat in this fashion. We never eliminate all the zeros. Thus T is not a regular transition matrix. ■

If T^m has no zero entries, then $T^{m+1} = (T^m)T$ has no zero entries, since the entries in any column vector of T are nonnegative with at least one nonzero entry. In determining whether a transition matrix is regular, it is not necessary to compute the entries in powers of the matrix. We need only determine whether or not they are zero.

EXAMPLE 6   If X denotes a nonzero entry, determine whether a transition matrix T with the zero and nonzero configuration given by

$$T = \begin{bmatrix} X & 0 & X & 0 \\ 0 & 0 & X & 0 \\ X & 0 & 0 & X \\ 0 & X & 0 & 0 \end{bmatrix}$$

is regular.

Solution   We compute configurations of high powers of T as rapidly as we can, since once a power has no zero entries, all higher powers must have nonzero entries. We find that

$$T^2 = \begin{bmatrix} X & 0 & X & X \\ X & 0 & 0 & X \\ X & X & X & 0 \\ 0 & 0 & X & 0 \end{bmatrix}, \ T^4 = \begin{bmatrix} X & X & X & X \\ X & 0 & X & X \\ X & X & X & X \\ X & X & X & 0 \end{bmatrix}, \ T^8 = \begin{bmatrix} X & X & X & X \\ X & X & X & X \\ X & X & X & X \\ X & X & X & X \end{bmatrix}$$

so the matrix T is indeed regular. ■

It can be shown that, if a Markov chain is regular, the distribution of population among the states over many time periods approaches a fixed *steady-state distribution vector* **s**. That is, the distribution of population among the states no longer changes significantly as time progresses. This is not to say that there is no longer movement of population between states; the transition matrix T continues to effect changes. But the movement of population out of any state over one time period is balanced by the population moving into that state, so the proportion of the total population in that state remains constant. This is a consequence of the following theorem, whose proof is beyond the scope of this book.

THEOREM 1.11   Achievement of Steady State

Let T be a regular transition matrix. There exists a unique column vector **s** with strictly positive entries whose sum is 1 such that the following hold:

1. As m becomes larger and larger, all columns of T^m approach the column vector **s**.

2. $T\mathbf{s} = \mathbf{s}$, and **s** is the unique column vector with this property and whose components add up to 1.

From Theorem 1.11 we can show that, if **p** is any initial population distribution vector for a regular Markov chain with transition matrix T, the population distribution vector after many time periods approaches the vector **s** described in the theorem. Thus **s** is called a **steady-state distribution vector**. We indicate the argument using a 3×3 matrix T. We know that the population distribution vector after m time periods is $T^m \mathbf{p}$. If we let

$$\mathbf{p} = \begin{bmatrix} p_1 \\ p_2 \\ p_3 \end{bmatrix} \quad \text{and} \quad \mathbf{s} = \begin{bmatrix} s_1 \\ s_2 \\ s_3 \end{bmatrix},$$

then Theorem 1.11 tells us that $T^m \mathbf{p}$ is approximately

$$\begin{bmatrix} s_1 & s_1 & s_1 \\ s_2 & s_2 & s_2 \\ s_3 & s_3 & s_3 \end{bmatrix} \begin{bmatrix} p_1 \\ p_2 \\ p_3 \end{bmatrix} = \begin{bmatrix} s_1 p_1 + s_1 p_2 + s_1 p_3 \\ s_2 p_1 + s_2 p_2 + s_2 p_3 \\ s_3 p_1 + s_3 p_2 + s_3 p_3 \end{bmatrix}.$$

Since $p_1 + p_2 + p_3 = 1$, this vector becomes

$$\begin{bmatrix} s_1 \\ s_2 \\ s_3 \end{bmatrix}.$$

Thus, after many time periods, the population distribution vector is approximately equal to the steady-state vector **s** for any choice of initial population distribution vector **p**.

There are two ways we can attempt to compute the steady-state vector **s** of a regular transition matrix T. If we have a computer handy, we can simply raise T to a sufficiently high power so that all column vectors are the same, as far as the computer can print them. The program MATCOMP in our software can be used to do this. Alternatively, we can use part (2) of Theorem 1.11 and solve for **s** in the equation

$$T\mathbf{s} = \mathbf{s}. \tag{1}$$

In solving Eq. (1), we will be finding our first *eigenvector* in this text. We will have a lot more to say about eigenvectors in Chapter 4.

Using the identity matrix I, we can rewrite Eq. (1) as

$$T\mathbf{s} = I\mathbf{s}$$
$$T\mathbf{s} - I\mathbf{s} = \mathbf{0}$$
$$(T - I)\mathbf{s} = \mathbf{0}.$$

The last equation represents a homogeneous system of linear equations with coefficient matrix $(T - I)$ and column vector **s** of unknowns. From all the solutions of this homogeneous system, choose the solution vector with positive entries that add up to 1. Theorem 1.11 assures us that this solution exists and is unique. Of course, the homogeneous system can be solved easily using a computer. Either of the available programs MATCOMP or YUREDUCE will reduce the augmented matrix to a form from which the solutions can be determined easily. We illustrate both methods with examples.

EXAMPLE 7 Use the program MATCOMP, and raise the transition matrix to powers to find the steady-state distribution vector for the Markov chain in Examples 1 and 3, having states labeled

poor,   middle income,   rich.

Solution Using MATCOMP and experimenting a bit with powers of the matrix T in Examples 1 and 3, we find that

$$T^{60} = \begin{array}{c} \\ \\ \\ \end{array} \overset{\begin{array}{ccc} \text{poor} & \text{mid} & \text{rich} \end{array}}{\begin{bmatrix} .3872054 & .3872054 & .3872054 \\ .4680135 & .4680135 & .4680135 \\ .1447811 & .1447811 & .1447811 \end{bmatrix}} \begin{array}{c} \text{poor} \\ \text{mid} \\ \text{rich} \end{array}.$$

Thus eventually about 38.7% of the population is poor, about 46.8% is middle income, and about 14.5% is rich, and these percentages no longer change as time progresses further over 20-year periods. ▪

EXAMPLE 8 The inhabitants of a vegetarian-prone community agree on the following rules:

1. No one will eat meat two days in a row.
2. A person who eats no meat one day will flip a fair coin and eat meat on the next day if and only if a head appears.

Determine whether this Markov-chain situation is regular; and if so, find the steady-state distribution vector for the proportions of the population eating no meat and eating meat.

Solution The transition matrix T is

$$T = \overset{\begin{array}{cc} \text{no meat} & \text{meat} \end{array}}{\begin{bmatrix} \frac{1}{2} & 1 \\ \frac{1}{2} & 0 \end{bmatrix}} \begin{array}{c} \text{no meat} \\ \text{meat} \end{array}$$

Since T^2 has no zero entries, the Markov chain is regular. We solve

$$(T - I)\mathbf{s} = \mathbf{0}, \quad \text{or} \quad \begin{bmatrix} -\frac{1}{2} & 1 \\ \frac{1}{2} & -1 \end{bmatrix} \begin{bmatrix} s_1 \\ s_2 \end{bmatrix} = \begin{bmatrix} 0 \\ 0 \end{bmatrix}.$$

We reduce the augmented matrix:

$$\begin{bmatrix} -\frac{1}{2} & 1 & \bigg| & 0 \\ \frac{1}{2} & -1 & \bigg| & 0 \end{bmatrix} \sim \begin{bmatrix} 1 & -2 & \bigg| & 0 \\ -\frac{1}{2} & 1 & \bigg| & 0 \end{bmatrix} \sim \begin{bmatrix} 1 & -2 & \bigg| & 0 \\ 0 & 0 & \bigg| & 0 \end{bmatrix}.$$

Thus we have

$$\begin{bmatrix} s_1 \\ s_2 \end{bmatrix} = \begin{bmatrix} 2r \\ r \end{bmatrix} \quad \text{for some scalar } r.$$

But we must have $s_1 + s_2 = 1$, so $2r + r = 1$ and $r = \frac{1}{3}$. Consequently, the steady-state population distribution is given by the vector $\begin{bmatrix} \frac{2}{3} \\ \frac{1}{3} \end{bmatrix}$. We see that, eventually, on each day about $\frac{2}{3}$ of the people eat no meat and the other $\frac{1}{3}$ eat meat. This is independent of the initial distribution of population between the states. All might eat meat the first day, or all might meat no meat; the steady-state vector remains $\begin{bmatrix} \frac{2}{3} \\ \frac{1}{3} \end{bmatrix}$ in either case.

■

If we were to solve Example 8 by reducing an augmented matrix with a computer, we should add a new row corresponding to the condition $s_1 + s_2 = 1$ for the desired steady-state vector. That way, the unique solution can be seen at once from the reduction of the augmented matrix. This can be done using pencil-and-paper computations just as well. If we insert this as the first condition on **s** and rework Example 8, our work appears as follows:

$$\left[\begin{array}{cc|c} 1 & 1 & 1 \\ -\frac{1}{2} & 1 & 0 \\ \frac{1}{2} & -1 & 0 \end{array}\right] \sim \left[\begin{array}{cc|c} 1 & 1 & 1 \\ 0 & \frac{3}{2} & \frac{1}{2} \\ 0 & -\frac{3}{2} & -\frac{1}{2} \end{array}\right] \sim \left[\begin{array}{cc|c} 1 & 0 & \frac{2}{3} \\ 0 & 1 & \frac{1}{3} \\ 0 & 0 & 0 \end{array}\right].$$

Again, we obtain the steady-state vector $\begin{bmatrix} \frac{2}{3} \\ \frac{1}{3} \end{bmatrix}$.

SUMMARY

1. A transition matrix for a Markov chain is a square matrix with nonnegative entries such that the sum of the entries in each column is 1.

2. The entry in the ith row and jth column of a transition matrix is the proportion of the population in state j that moves to state i during one time period of the chain.

3. If the column vector **p** is the initial population distribution vector between states in a Markov chain with transition matrix T, the population distribution vector after one time period of the chain is $T\mathbf{p}$.

4. If T is the transition matrix for one time period of a Markov chain, then T^m is the transition matrix for m time periods.

5. A Markov chain and its associated transition matrix T are called *regular* if there exists an integer m such that T^m has no zero entries.

6. If T is a regular transition matrix for a Markov chain,
   a. The columns of T^m all approach the same probability distribution vector **s** as m becomes large;

b. s is the unique probability distribution vector satisfying $T\mathbf{s} = \mathbf{s}$; and

c. As the number of time periods increases, the population distribution vectors approach s regardless of the initial population distribution vector **p**. Thus s is the steady-state population distribution vector.

EXERCISES

In Exercises 1–8, determine whether the given matrix is a transition matrix. If it is, determine whether it is regular.

1. $\begin{bmatrix} \frac{1}{2} & \frac{1}{2} \\ \frac{1}{3} & \frac{2}{3} \end{bmatrix}$

2. $\begin{bmatrix} 0 & \frac{1}{4} \\ 1 & \frac{3}{4} \end{bmatrix}$

3. $\begin{bmatrix} .2 & .1 & .3 \\ .4 & .5 & -.1 \\ .4 & .4 & .8 \end{bmatrix}$

4. $\begin{bmatrix} .1 & .2 \\ .3 & .4 \\ .6 & .4 \end{bmatrix}$

5. $\begin{bmatrix} .5 & 0 & 0 & .5 \\ .5 & .5 & 0 & 0 \\ 0 & .5 & .5 & 0 \\ 0 & 0 & .5 & .5 \end{bmatrix}$

6. $\begin{bmatrix} .3 & .2 & 0 & .5 \\ .4 & .2 & 0 & .1 \\ .1 & .2 & 1 & .2 \\ .2 & .4 & 0 & .2 \end{bmatrix}$

7. $\begin{bmatrix} 0 & .5 & .2 & .2 & .1 \\ .3 & 0 & .1 & .8 & .5 \\ 0 & 0 & .4 & 0 & .1 \\ .7 & .5 & .1 & 0 & .1 \\ 0 & 0 & .2 & 0 & .2 \end{bmatrix}$

8. $\begin{bmatrix} 0 & .1 & 0 & 0 & .9 \\ 0 & .2 & 0 & 0 & .1 \\ 0 & .3 & 0 & 1 & 0 \\ 1 & .1 & 0 & 0 & 0 \\ 0 & .3 & 1 & 0 & 0 \end{bmatrix}$

In Exercises 9–12, let $T = \begin{bmatrix} .3 & .7 & .4 \\ .4 & .2 & .1 \\ .3 & .1 & .5 \end{bmatrix}$ be the transition matrix for a Markov chain, and let

$P = \begin{bmatrix} .3 \\ .2 \\ .5 \end{bmatrix}$ *be the initial population distribution vector.*

9. Find the proportion of the state 2 population that is in state 3 after two time periods.

10. Find the proportion of the state 3 population that is in state 2 after two time periods.

11. Find the proportion of the total population that is in state 3 after two time periods.

12. Find the population distribution vector after two time periods.

In Exercises 13–18, determine whether the given transition matrix with the indicated distribution of zero entries and nonzero X entries is regular.

13. $\begin{bmatrix} X & X & X \\ 0 & X & X \\ 0 & X & X \end{bmatrix}$

14. $\begin{bmatrix} X & 0 & X \\ 0 & 0 & X \\ X & X & 0 \end{bmatrix}$

15. $\begin{bmatrix} 0 & X & 0 \\ X & X & X \\ 0 & X & 0 \end{bmatrix}$

16. $\begin{bmatrix} X & 0 & X & X \\ X & X & X & X \\ X & 0 & X & X \\ X & 0 & X & X \end{bmatrix}$

17. $\begin{bmatrix} 0 & X & X & X \\ X & 0 & X & X \\ 0 & 0 & X & X \\ 0 & 0 & X & X \end{bmatrix}$

18. $\begin{bmatrix} 0 & 0 & X & X \\ X & 0 & 0 & X \\ 0 & X & 0 & X \\ 0 & 0 & 0 & X \end{bmatrix}$

In Exercises 19–24, find the steady-state distribution vector for the given transition matrix of a Markov chain.

19. $\begin{bmatrix} \frac{2}{3} & 1 \\ \frac{1}{3} & 0 \end{bmatrix}$

20. $\begin{bmatrix} \frac{3}{4} & \frac{1}{2} \\ \frac{1}{4} & \frac{1}{2} \end{bmatrix}$

21. $\begin{bmatrix} 0 & \frac{1}{3} & \frac{1}{3} \\ 0 & \frac{2}{3} & \frac{1}{3} \\ 1 & 0 & \frac{1}{3} \end{bmatrix}$

22. $\begin{bmatrix} 1 & \frac{1}{4} & \frac{1}{2} \\ 0 & \frac{1}{4} & 0 \\ 0 & \frac{1}{2} & \frac{1}{2} \end{bmatrix}$

23. $\begin{bmatrix} \frac{1}{4} & \frac{1}{2} & \frac{1}{3} \\ \frac{1}{4} & 0 & \frac{1}{3} \\ \frac{1}{2} & \frac{1}{2} & \frac{1}{3} \end{bmatrix}$

24. $\begin{bmatrix} \frac{1}{5} & \frac{1}{5} & \frac{1}{3} \\ \frac{2}{5} & \frac{1}{5} & \frac{1}{3} \\ \frac{2}{5} & \frac{3}{5} & \frac{1}{3} \end{bmatrix}$

25. Mark each of the following True or False.

____ a) All entries in a transition matrix are non-negative.

____ b) Every matrix whose entries are all non-negative is a transition matrix.

____ c) The sum of all the entries in an $n \times n$ transition matrix is 1.

____ d) The sum of all the entries in an $n \times n$ transition matrix is n.

____ e) If a transition matrix contains no zero entries, it is regular.

____ f) If a transition matrix is regular, it contains no nonzero entries.

____ g) Every power of a transition matrix is again a transition matrix.

____ h) If a transition matrix is regular, its square has equal column vectors.

____ i) If a transition matrix T is regular, there exists a unique vector **s** such that $T\mathbf{s} = \mathbf{s}$.

____ j) If a transition matrix T is regular, there exists a unique population distribution vector **s** such that $T\mathbf{s} = \mathbf{s}$.

26. Estimate A^{100}, if A is the matrix in Exercise 20.

27. Estimate A^{100}, if A is the matrix in Exercise 23.

Exercises 28–33 deal with the following Markov chain. We classify the women in a country according as to whether they live in an urban (U), suburban (S), or rural (R) area. Suppose that each woman has just one daughter, who in turn has just one daughter, and so on. Suppose further that the following are true:

For urban women, 10% of the daughters settle in rural areas, and 50% in suburban areas.

For suburban women, 20% of the daughters settle in rural areas, and 30% in urban areas.

For rural women, 20% of the daughters settle in the suburbs, and 70% in rural areas.

Let this Markov chain have as its period the time required to produce the next generation.

28. Give the transition matrix for this Markov chain, taking states in the order U, S, R.

29. Find the proportion of urban women whose granddaughters are suburban women.

30. Find the proportion of rural women whose granddaughters are also rural women.

31. If the initial population distribution vector for all the women is $\begin{bmatrix} .4 \\ .5 \\ .1 \end{bmatrix}$, find the population distribution vector for the next generation.

32. Repeat Exercise 31, but find the population distribution vector for the following (third) generation.

33. Show that this Markov chain is regular, and find the steady-state probability distribution vector.

Exercises 34–39 deal with a simple genetic model involving just two types of genes, G and g. Suppose that a physical trait, such as eye color, is controlled by a pair of these genes, one inherited from each parent. A person may be classified as being in one of three states:

> *Dominant (type GG),   Hybrid (type Gg), Recessive (type gg).*

We assume that the gene inherited from a parent is a random choice from the parent's two genes; that is, the gene inherited is just as likely to be one of the parent's two genes as to be the other. We form a Markov chain by starting with a population and always crossing with hybrids to produce offspring. We take the time required to produce a subsequent generation as the time period for the chain.

34. Give an intuitive argument in support of the idea that the transition matrix for this Markov chain is

$$T = \begin{array}{c} \\ \\ \\ \\ \end{array} \begin{matrix} \mathbf{D} & \mathbf{H} & \mathbf{R} \\ \begin{bmatrix} \frac{1}{2} & \frac{1}{4} & 0 \\ \frac{1}{2} & \frac{1}{2} & \frac{1}{2} \\ 0 & \frac{1}{4} & \frac{1}{2} \end{bmatrix} & \begin{matrix} \mathbf{D} \\ \mathbf{H}. \\ \mathbf{R} \end{matrix} \end{matrix}$$

35. What proportion of the third-generation offspring (after two time periods) of the recessive (gg) population is dominant (GG)?

36. What proportion of the third-generation offspring (after two time periods) of the hybrid (Gg) population is not hybrid?

37. If initially the entire population is hybrid, find the population distribution vector in the next generation.

38. If initially the population is evenly divided among the three states, find the population distribution vector in the third generation (after two time periods).

39. Show that this Markov chain is regular, and find the steady-state population distribution vector.

40. A state in a Markov chain is called **absorbing** if it is impossible to leave that state over the next time period. What characterizes the transition matrix of a Markov chain with an absorbing state? Can a Markov chain with an absorbing state be regular?

41. Consider the genetic model for Exercises 34–39. Suppose that, instead of always crossing with hydrids to produce offspring, we always cross with recessives. Give the transition matrix for this Markov chain, and show that there is an absorbing state. (See Exercise 40.)

42. A Markov chain is termed **absorbing** if it contains at least one absorbing state (see Exercise 40) and if it is possible to get from any state to an absorbing state in some number of time periods.

a) Give an example of a transition matrix for a three-state absorbing Markov chain.

b) Give an example of a transition matrix for a three-state Markov chain that is not absorbing but has an absorbing state.

43. With reference to Exercise 42, consider an absorbing Markov chain with transition matrix T and a single absorbing state. Argue that, for any initial distribution vector **p**, the vectors $T^n\mathbf{p}$ for large n approach the vector containing 1 in the component that corresponds to the absorbing state and zeros elsewhere. [SUGGESTION: Let m be such that it is possible to reach the absorbing state from any state in m time periods, and let q be the smallest entry in the row of T^m corresponding to the absorbing state. Form a new chain with just two states, Absorbing (A) and Free (F), which has as time period m time periods of the original chain, and with probability q of moving from state F to state A in that time period. Argue that, by starting in any state in the original chain, you are more likely to reach an absorbing state in mr time periods than you are by starting from state F in the new chain and going for r time periods. Using the fact that large powers of a positive number less than 1 are almost 0, show that for the two-state chain, the population distribution vector approaches $\begin{bmatrix} 1 \\ 0 \end{bmatrix}$ as the number of time periods increases, regardless of the initial population distribution vector.]

44. Let T be an $n \times n$ transition matrix. Show that, if every row and every column have fewer than $n/2$ zero entries, the matrix is regular.

In Exercises 45–54, find the steady-state population distribution vector for the given transition matrix. Use MATCOMP, YUREDUCE, or similar software for Exercises 50–54. See the comment following Example 8.

45. $\begin{bmatrix} 0 & \frac{1}{4} \\ 1 & \frac{3}{4} \end{bmatrix}$

46. $\begin{bmatrix} \frac{1}{2} & \frac{1}{4} \\ \frac{1}{2} & \frac{3}{4} \end{bmatrix}$

47. $\begin{bmatrix} \frac{1}{5} & 1 \\ \frac{4}{5} & 0 \end{bmatrix}$

48. $\begin{bmatrix} 0 & \frac{1}{4} & \frac{1}{2} \\ \frac{1}{2} & 0 & \frac{1}{2} \\ \frac{1}{2} & \frac{3}{4} & 0 \end{bmatrix}$

49. $\begin{bmatrix} \frac{1}{5} & \frac{3}{4} & \frac{1}{8} \\ \frac{4}{5} & 0 & \frac{3}{8} \\ 0 & \frac{1}{4} & \frac{1}{2} \end{bmatrix}$

50. $\begin{bmatrix} .1 & .3 & .4 \\ .2 & 0 & .2 \\ .7 & .7 & .4 \end{bmatrix}$

51. $\begin{bmatrix} .3 & .3 & .1 \\ .1 & .3 & .5 \\ .6 & .4 & .4 \end{bmatrix}$

52. $\begin{bmatrix} .1 & 0 & .2 & .5 \\ .4 & 0 & .2 & .5 \\ .2 & .5 & .4 & 0 \\ .3 & .5 & .2 & 0 \end{bmatrix}$

53. $\begin{bmatrix} .5 & 0 & 0 & 0 & .9 \\ .5 & .4 & 0 & 0 & 0 \\ 0 & .6 & .3 & 0 & 0 \\ 0 & 0 & .7 & .2 & 0 \\ 0 & 0 & 0 & .8 & .1 \end{bmatrix}$

54. The matrix in Exercise 8

1.7 SYSTEMS OF LINEAR INEQUALITIES

The preceding sections of Chapter 1 have been concerned with linear equalities. To close this chapter, we consider systems of linear inequalities, such as

$$2x + 3y \geq 6$$
$$2x + y \geq 2$$
$$x - y \leq 3. \tag{1}$$

A great many practical problems lead directly to a system of linear inequalities, although many of these problems were not formulated and studied until the 1940s. With the advent of the computer, it is possible to deal with very large systems. Consequently, work with these systems has mushroomed.

The linear programming problem we will discuss arose partly from modern technology in medicine, transportation, forestry, agriculture, fuels, and so on. Linear programming has contributed to the preservation of environmental resources. It has also saved industries (and eventually the consumer) millions of dollars. New applications arise constantly. George Dantzig developed the simplex method for solving linear programming problems in 1947.

Real-world linear programming problems may involve hundreds or even thousands of variables. However, we will discuss only the two-variable case of linear programming at this time, and our treatment will be very geometric and intuitive. Chapter 10 considers linear programming problems with more variables.

Statement of a Linear Programming Problem

To illustrate the type of problem we wish to study, we state a simple linear programming problem in two variables, and then analyze it mathematically.

PROBLEM A lumber company owns two mills that produce hardwood veneer sheets of plywood. Each mill produces the same three types of plywood. Table 1.1 shows the daily production and the daily cost of operation of each mill. The rightmost column shows the amount of plywood required of the lumber company for a period of 6 months. Find the number of days that each mill should operate during the 6 months to supply the required sheets in the most economical way.

Analysis If mill 1 operates for x days and mill 2 operates for y days, the cost is $3000x + 2000y$. In order to fulfill the company's requirements, these inequalities must be satisfied simultaneously:

$$100x + 20y \geq 2000$$

$$40x + 80y \geq 3200$$

$$60x + 60y \geq 3600$$

$$x \geq 0, \qquad y \geq 0.$$

Formulated mathematically, our problem is to minimize the linear function $3000x + 2000y$, subject to these inequalities. ■

TABLE 1.1

Plywood Type	Mill 1 per Day	Mill 2 per Day	Six-month Demand
A	100 sheets	20 sheets	2000 sheets
B	40 sheets	80 sheets	3200 sheets
C	60 sheets	60 sheets	3600 sheets
Daily costs	$3000	$2000	

The Solution Set of a System of Linear Inequalities

Every linear programming problem is concerned with maximizing or minimizing a linear function on the solution set of a system composed of linear equations or inequalities. To solve a problem like our lumber mill example, we consider the nature of such solution sets.

We know that a single linear equation in x and y has a line in the plane as its graph. The equation $y = mx + b$ describes a line of slope m and y-intercept b, as shown in Figure 1.11.

The inequality $y < mx_0 + b$ describes the points (x_0, y) on $x = x_0$ lying under the line $y = mx + b$. (See Fig. 1.12.) Since x_0 can be any value, we see that $y \leq mx + b$ describes the half-plane consisting of the points on or below the line $y = mx + b$. This half-plane is shaded in Figure 1.12. Similarly, the inequality $y \geq mx + b$ describes all points on or above the line $y = mx + b$, which gives the half-plane shaded in Figure 1.13.

As the preceding arguments indicate, any linear inequality $ax + by \leq c$, where not both a and b are zero, has as its solution set a half-plane consisting of the line $ax + by = c$ together with the planar region on one side of the line.

Given a system of linear inequalities in x and y such as system (1), we want to find all (x, y) for which all the inequalities hold simultaneously. Then (x, y) must lie in all the half-planes defined by the inequalities—that is, in the intersection of these half-planes.

> The solution set of a system of linear inequalities consists of an intersection of half-planes.

EXAMPLE 1   In the plane, shade the solution set of system (1), given at the beginning of this section:

$$2x + 3y \geq 6$$
$$2x + y \geq 2$$
$$x - y \leq 3.$$

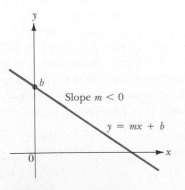

FIGURE 1.11   The line $y = mx + b$.

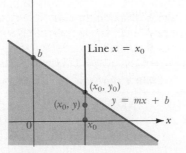

FIGURE 1.12   The half-plane $y \leq mx + b$.

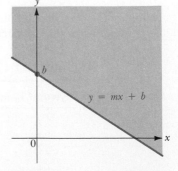

FIGURE 1.13   The half-plane $y \geq mx + b$.

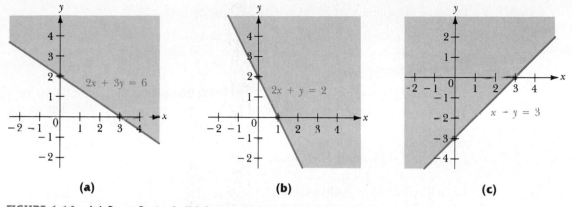

FIGURE 1.14   (a) $2x + 3y \geq 6$; (b) $2x + y \geq 2$; (c) $x - y \leq 3$.

Solution   The three half-planes corresponding to these inequalities are shown in Figure 1.14. Once the line $2x + 3y = 6$ has been drawn, as in Figure 1.14(a), an easy way to determine which side of the line to shade is to pick a point on one side and see whether it satisfies the inequality. For example, if we pick $(0, 0)$, we see that it does not satisfy $2x + 3y \geq 6$, so we shade the half-plane on the opposite side of the line in Figure 1.14(a).

In Figure 1.15, the intersection of the half-planes in Figure 1.14 is shaded. Points in this shaded region are precisely those in the solution set of system (1).   ■

EXAMPLE 2   Shade the solution set of the system $3x + 4y \leq 12$, $x \geq 0$, $y \geq 0$.

Solution   The solution set is the shaded region in Figure 1.16.   ■

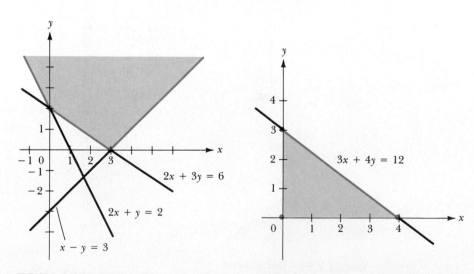

FIGURE 1.15   The set $2x + 3y \geq 6$, $2x + y \geq 2$, $x - y \leq 3$.

FIGURE 1.16   The set $3x + 4y \leq 12$, $x \geq 0$, $y \geq 0$.

EXAMPLE 3 Describe the solution set of the system $3x + 4y \leq -12$, $x \geq 0$, $y \geq 0$.

Solution The half-plane described by $3x + 4y \leq -12$ is shaded in Figure 1.17. Since none of the points in that half-plane satisfies both $x \geq 0$ and $y \geq 0$, the solution set is empty. ■

From the preceding examples, it is apparent that a nonempty solution set of a system of linear inequalities is bounded by lines or line segments, and that all the corners point outward. We call such a set a *convex polygonal set*. Let us define the notion of convexity, and then prove the convexity of such a solution set as a theorem. Figure 1.18 illustrates the definition.

DEFINITION 1.13 Convex Set

A subset of the plane is **convex** if, for each two points P and Q in the set, the line segment joining them also lies in the set.

THEOREM 1.12 Convexity of Solution Sets

Any nonempty intersection of half-planes is a convex set.

PROOF Let P and Q be points in the intersection. Then P and Q lie in each of the half-planes. A half-plane is convex, so the line segment joining P and Q lies in each half-plane. Therefore, the line segment lies in the intersection of the half-planes. This completes the proof. ▲

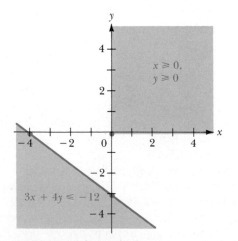

FIGURE 1.17 The shaded regions do not intersect.

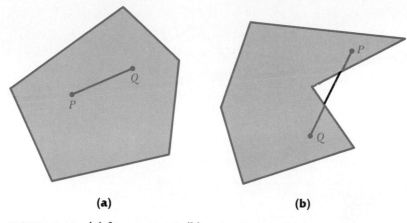

(a)   **(b)**

FIGURE 1.18   **(a) A convex set; (b) not a convex set.**

Maximizing or Minimizing $ax + by$ on an Intersection of Half-planes

Consider the linear programming problem of maximizing or minimizing a linear function $ax + by$ on the solution set of a system of linear inequalities. The inequalities are called the **constraints**, and the function $ax + by$ is the **objective function**. With economic applications in mind, the objective function is called the **profit function** if it is to be maximized or the **cost function** if it is to be minimized. In applied problems, two of the constraints are always taken to be $x \geq 0$ and $y \geq 0$, but we will not assume this for the moment. The convex solution set of the constraints is called the **feasible set**, and any point in this set is a **feasible solution**. Our task is thus to determine from among all the feasible solutions those that maximize or minimize $ax + by$.

Suppose that, for a feasible solution (x_0, y_0) and cost function $ax + by$, we have $ax_0 + by_0 = c_0$. Then every feasible solution lying on the line $ax + by = c_0$ also gives the same value c_0 for this cost function. For a feasible solution (x_1, y_1) not on this line, we obtain some cost c_1, where $c_1 = ax_1 + by_1$. Again, all feasible solutions (x, y) on the line $ax + by = c_1$ give this same value c_1. Now the lines $ax + by = c_0$ and $ax + by = c_1$ are parallel, because the slope of a line is completely determined by the coefficients of x and y. This is illustrated in Figure 1.19. If we hold a and b fixed and increase c, the parallel lines $ax + by = c$ move upward or downward in the plane, depending on the sign of b. They move upward if $b > 0$ and downward if $b < 0$, as we see by considering the y-intercepts c/b of the lines. Of course, if $b = 0$, the lines $ax = c$ for fixed a move either left or right as c increases.

Suppose that the feasible set is as shown in Figure 1.20(a). Then if $b > 0$, the maximum value c_{max} is the value of $ax + by$ at the last point where the lines $ax + by = c$ touch the feasible set as c increases and the lines move upward. The minimum value c_{min} is assumed at the last point where the lines touch the set as c decreases and the lines move downward. The case $b < 0$ where the lines move

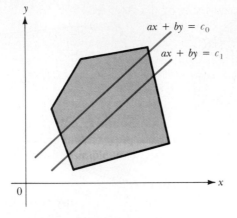

FIGURE 1.19 Different values of c give parallel lines.

downward as c increases is shown in Figure 1.20(b). The case for vertical lines where $b = 0$ is handled in a similar way.

It may be that the final position of a line $ax + by = c$ before leaving a feasible set falls on a whole boundary segment of the set. Figure 1.20(b) shows the case where $c_{\max}$ is assumed at every point on such a line segment.

If the feasible set is unbounded, as in Figure 1.15, there may be no maximum or minimum attained by the objective function. For example, if $b > 0$, then $ax + by$ attains no maximum value on the set in Figure 1.15. Similarly, if $b < 0$, no minimum value is attained.

From considering all these cases, we obtain a theorem.

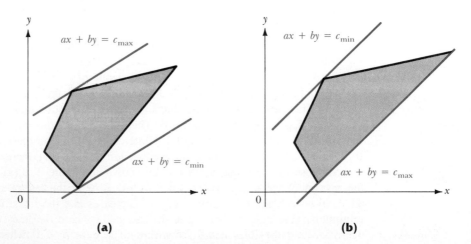

(a)   **(b)**

FIGURE 1.20 (a) The case $b > 0$; (b) the case $b < 0$.

THEOREM 1.13 Assumption of Optimal Cost or Profit

If a linear objective function $ax + by$ assumes a maximum value on a set defined by linear constraints, this maximum value is assumed at a corner point of the set. The same is true if an objective function assumes a minimum value on the feasible set.

Theorem 1.13 and the discussion preceding it suggest the following procedure for finding the maximum or minimum attained by an objective function $ax + by$.

Finding the Maximum (Minimum) Attained by $ax + by$ Subject to a System of Linear Constraints

Step 1 Sketch the feasible set defined by the constraints.

Step 2 Determine whether a maximum (minimum) is attainable.

Step 3 If so, find the coordinates of all corner points of the set.

Step 4 Compute the value of $ax + by$ at each corner point of the set. The largest (smallest) value obtained is the maximum (minimum) of the objective function on the feasible set.

Concerning step 2, it is usually clear in a paractical problem that the desired maximum or minimum is assumed. If the feasible set is bounded, both a maximum and a minimum are assumed. If the set is unbounded, one or both may fail to be assumed, as shown in Example 5.

EXAMPLE 4 Find the maximum and minimum values attained by $x + 2y$, subject to the constraints $3x + 4y \leq 12$, $x \geq 0$, $y \geq 0$ given in Example 2.

Solution Referring to Figure 1.16, we see that the feasible set is bounded, so both a maximum and a minimum are attained by $x + 2y$. The figure shows that the corner points are $(0, 0)$, $(4, 0)$, and $(0, 3)$. Thus we have the following:

Corner Point	Value of $x + 2y$
$(0, 0)$	0
$(0, 3)$	6
$(4, 0)$	4

Therefore, the maximum value is 6, attained at $(0, 3)$, and the minimum value is 0, attained at $(0, 0)$. ■

EXAMPLE 5 Discuss the values of a and b for which $ax + by$ attains a maximum or minimum, subject to the constraints $2x + 3y \geq 6$, $2x + y \geq 2$, $x - y \leq 3$ of Example 1.

Solution The feasible set was found in Example 1, and is shown again in Figure 1.21. In order for $ax + by$ to attain either a maximum or a minimum value on this set, the parallel lines $ax + by = c$ must eventually fail to intersect the feasible set as c increases or decreases. Referring to Figure 1.21, we see that the rightmost boundary line of the feasible set has slope 1, whereas the leftmost boundary line has slope -2. As illustrated in Figure 1.21(a), every line of slope greater than 1 intersects the feasible set, so an objective function $ax + by$ attains no maximum or minimum on this set if the line $ax + by = c$ has slope greater than 1. A similar argument shows that this is also true if the line is steeper than the left boundary line—that is, if it has slope less than -2. However, as shown in Figure 1.21(b), if $ax + by = c$ has slope m, where $-2 \leq m \leq 1$, then the cost function $ax + by$ attains a minimum value if $b > 0$ and a maximum value if $b < 0$. Thus $ax + by$ attains a maximum or a minimum if $-2 \leq -a/b \leq 1$. ■

EXAMPLE 6 As a continuation of Example 5, find the maximum or minimum attained by $x + 3y$ and by $x - 2y$ on the feasible set of Figure 1.21.

Solution The slope of $x + 3y = c$ is $-\frac{1}{3}$, whereas the slope of $x - 2y = c$ is $\frac{1}{2}$, both of which are between -2 and 1. Example 5 shows that a maximum or minimum will be attained in each case. We see from the figure that the corner points are $(0, 2)$ and $(3, 0)$.

The value of $x + 3y$ is 6 at $(0, 2)$ and 3 at $(3, 0)$. Clearly, $x + 3y$ takes on very large values in this unbounded feasible set. Thus 3 is the minimum value attained.

The value of $x - 2y$ is -4 at $(0, 2)$ and 3 at $(3, 0)$. Clearly, $x - 2y$ attains negative values of great magnitude at some points $(0, y)$ in the feasible set, so 3 is the maximum value attained by $x - 2y$. ■

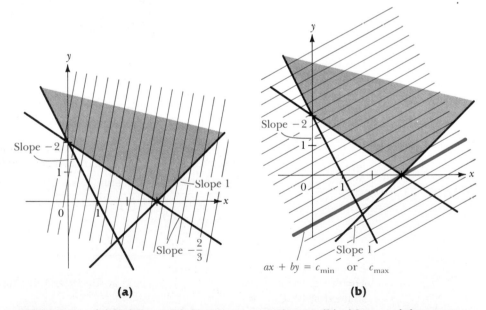

(a)   **(b)**

FIGURE 1.21   (a) Neither a minimum nor a maximum; (b) either a minimum or a maximum.

TABLE 1.2

Plywood Type	Mill 1 per Day	Mill 2 per Day	Six-month Demand
A	100 sheets	20 sheets	2000 sheets
B	40 sheets	80 shcets	3200 sheets
C	60 sheets	60 sheets	3600 sheets
Daily costs	$3000	$2000	

The Solution to Our Opening Problem

We restate and solve the problem posed at the beginning of this section.

EXAMPLE 7   Each of two mills produces the same three types of plywood. Table 1.2 gives the production, demand, and cost data. Find the number of days that each mill should operate during the 6 months in order to supply the required sheets in the most economical way.

Solution   If mill 1 operates x days and mill 2 operates y days, the cost is $3000x + 2000y$. In order to fulfill the company's requirements, we must have the following constraints:

$$100x + 20y \geq 2000$$

$$40x + 80y \geq 3200$$

$$60x + 60y \geq 3600$$

$$x \geq 0, \quad y \geq 0.$$

The feasible set is sketched in Figure 1.22. For this practical problem, a minimum cost must surely exist, so we need only find the corner points and compute $3000x + 2000y$ there.

We can see from the figure that two corner points are $(0, 100)$ and $(80, 0)$. To find another corner point, we solve the equations for two consecutive boundary lines of the feasible set simultaneously. We have

$$
\begin{array}{ll}
100x + 20y = 2000 & \text{Multiply by 3 and subtract.} \\
\underline{60x + 60y = 3600} & \\
240x \phantom{{}+60y} = 2400, & x = 10, \quad y = 50.
\end{array}
$$

For our final corner point, we solve:

$$
\begin{array}{ll}
40x + 80y = 3200 & \text{Multiply by 3.} \\
\underline{60x + 60y = 3600} & \text{Multiply by 2 and subtract.} \\
\phantom{40x +{}} 120y = 2400, & y = 20, \quad x = 40.
\end{array}
$$

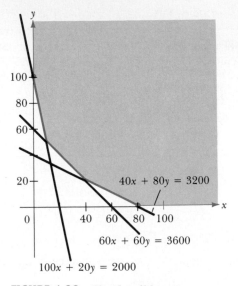

FIGURE 1.22 The feasible set.

Thus we have the following:

Corner Point (x, y)	$3000x + 2000y$
(0, 100)	$200,000
(80, 0)	$240,000
(10, 50)	$130,000
(40, 20)	$160,000

We see that it is best to run mill 1 for 10 days and mill 2 for 50 days. This will produce the necessary 2000 sheets of type A plywood and the 3600 sheets of type C. Notice that it will actually produce 4400 sheets of type B, for a surplus of 1200 sheets over the 3200 required of that type! ▪

Problems in linear inequalities involving just two variables serve to illustrate the concepts. Chapter 10 presents the *simplex method*, developed by Dantzig in 1947, which is presently the main algebraic algorithm for solving similar problems involving more than two variables. It is simply not feasible to graph and find all corner points in spaces of more than two dimensions. The simplex method has been implemented on computers to solve a wide variety of important applications.

At the present time, there is great interest in another algorithm as an alternative to the simplex method, one that was developed by Narendra Karmarkar at AT&T Bell Labs. Dantzig's simplex method finds a corner point of the convex set and then travels along edges to adjacent corner points until an optimal solution is found. Karmarkar's method attempts to take shortcuts through the interior of the convex set. The AT&T report to stockholders for the quarter ending December 31, 1984, included a paragraph on Karmarkar's work. The report noted that finding the most efficient way to route an overseas call to the Pacific area might involve 20,000 variables. People at Bell Labs think it possible that a computer could solve the problem using Karmarkar's algorithm in less than 1 minute. Clearly, we can't give problems involving 20,000 variables in this text! We have tried instead to present a concept and to provide a little insight into its importance.

SUMMARY

1. The solution set of a linear inequality $ax + by + c \leq 0$ is a half-plane.

2. A subset of the plane is convex if, for every two points in the set, the line segment joining them is in the set.

3. The solution set of a system of linear inequalities in two variables is either empty or a convex polygonal set in the plane. The solution set may or may not be bounded.

4. A linear programming problem consists of maximizing or minimizing a linear function on the solution set of a system containing linear equations and/or inequalities. The conditions in the system are called *constraints*, their solution set is called the *feasible set*, and the linear function to be optimized is called the *objective function*.

5. The box on page 85 lists a four-step procedure for solving linear programming problems in two variables.

EXERCISES

In Exercises 1–6, shade the solution set in the plane of the system of linear inequalities.

1. $x \geq 0, y \leq 0$
2. $x + y \geq -3$
3. $x - y \leq 5, x + y \geq 2$
4. $x \geq 2, y \leq 4, x + y \leq 8$
5. $x \geq 0, y \geq 0, x + 3y \geq 12, y + 2x \geq 6$
6. $x \geq 0, y \geq 0, 4x + 3y \geq 24, 5 + 2y \geq 20,$ $6x + y \geq 12$

In Exercises 7–12, find the maximum and minimum values (if they exist) of the objective functions on the feasible set defined by the constraints.

7. Constraints: $x \geq 2, y \geq 3$
   Objective functions: (a) $2x + y$; (b) $y - 2x$; (c) $-2x - 3y$

8. Constraints: $x \geq 0, y \geq 0, x + y \geq 1$
   Objective functions: (a) $x - 2y$; (b) $4x + 5y$; (c) $3x + 2y$

9. Constraints: $x - y \leq 0$, $x + y \geq 0$
   Objective functions: (a) $2x - 3y$; (b) $3y + x$;
   (c) $y + 2x$

10. Same as Exercise 9, but with the additional constraint $x + 9y \leq 80$.

11. Constraints: $x \geq 0$, $y \geq 0$, $x + 2y \geq 5$,
    $2x + y \geq 4$, $x + 4y \geq 7$
    Objective functions: (a) $2x + 9y$; (b) $x + 3y$;
    (c) $x + y$

12. Same as Exercise 11, but with the additional constraint $7x + 4y \leq 28$.

13. Referring to Example 7 in the text, suppose that production and requirements remain the same, but daily costs are subject to change. Find the condition on the ratio

    $$\frac{\text{(Daily cost for mill 1)}}{\text{(Daily cost for mill 2)}}$$

    in order that:

    a) it is best to run only mill 1;
    b) it is best to run only mill 2.

14. Solve the problem in Example 7 if the data for the two mills are changed to those given in Table 1.3.

15. A food packaging house receives daily 1500 lb of type C coffee and 2000 lb of type K coffee. It can sell blend A, which consists of 2 parts C to 1 part K, at a profit of 10 cents per pound; and it can sell blend B, which consists of 2 parts K to 1 part C, at a profit of 12 cents per pound. Find the amount of each blend that the house should prepare daily to maximize the profit for sales of just these two blends.

16. Referring to Exercise 15, find the amount of each blend necessary for a maximum profit if the house is also able to sell the leftover types C and K for a profit of:

    a) 6 cents per pound each;
    b) 9 cents per pound each;
    c) 12 cents per pound each.

17. A dietician wants to use food 1 and food 2 to supply certain minimum amounts of vitamins A, B, and C. The data for the amounts of the vitamins, the requirements, and the costs are given in Table 1.4. How much of each food should be used to meet the vitamin requirements in the most economical way?

TABLE 1.3

Plywood Type	Mill 1 per Day	Mill 2 per Day	Six-month Demand
A	200 sheets	100 sheets	5000 sheets
B	40 sheets	10 sheets	1100 sheets
C	20 sheets	80 sheets	2800 sheets
Daily costs	$4000	$2000	

TABLE 1.4

Vitamin	Food 1	Food 2	Requirements
A	30 units/oz	20 units/oz	120 units
B	40 units/oz	10 units/oz	80 units
C	20 units/oz	40 units/oz	100 units
Cost	10 units/oz	15 units/oz	

18. Answer Exercise 17 if other data remain the same, but:

a) Food 1 costs 5 cents/oz and food 2 costs 15 cents/oz;

b) Food 1 costs 25 cents/oz and food 2 costs 15 cents/oz;

c) Food 1 costs 25 cents/oz and food 2 costs 5 cents/oz.

19. A gardener has two bags of fertilizer, each weighing 80 lb. Bag 1 contains 10% nitrogen, 5% phosphate, and 10% potash, while bag 2 contains 5% nitrogen, 10% phosphate, and 10% potash. Bag 1 costs $8 and bag 2 costs $10. The gardener wishes to put at least 5 lb of nitrogen, 4 lb of phosphate, and 7 lb of potash on his garden. How many pounds from each bag should he mix together to fertilize most economically?

20. Answer Exercise 19 if the cost of the fertilizer is:

a) $10 for bag 1 and $8 for bag 2;

b) $4 for bag 1 and $10 for bag 2.

21. Mark each of the following True or False.

____ a) Every intersection of a finite number of half-planes is a convex set.

____ b) Every convex set in the plane is the intersection of a finite number of half-planes.

____ c) If a set in the plane contains two points and the line segment joining them, it is a convex set.

____ d) The solution set of every system of linear inequalities in x and y is a nonempty set.

____ e) Every linear function achieves a maximum value on a convex set.

____ f) Every linear function achieves a minimum value on a bounded convex set.

____ g) A linear function can achieve a maximum value at at most one point of a convex set.

____ h) A linear function can achieve a maximum value at at most two points of a convex set.

____ i) It is conceivable that a linear function may achieve a maximum value at every point of a nonempty convex set in the plane.

____ j) It is conceivable that a linear function may achieve a maximum value at every point of a convex set of area greater than zero in the plane.

▣ *The program LINPROG in the software available to schools using this text can be used to graph the feasible set of a system of linear inequalities in two variables, where the system includes $x \geq 0$ and $y \geq 0$. The user may then specify an objective function, and estimate graphically the maximum and minimum values of this function on the feasible set, if they exist. Information on using the program is given at its beginning. In the remaining exercises, use this program, or similar software, to find as accurately as you can from the graph the maximum and minimum of the given objective function on the feasible set defined by the constraints.*

22. Constraints: $x \geq 0$, $y \geq 0$, $x + y \geq 10$, $x \leq 15$, $x - y \leq 10$, $y \leq 7$
Objective function: (a) $x + 2y$; (b) $2x + y$

23. Constraints: $x \geq 0$, $y \geq 0$, $x + y \geq 4$, $2x + y \leq 8$, $5x + 8y \leq 50$
Objective function: (a) $2x + 3y$; (b) $3x + 2y$; (c) $3x + y$; (d) $x - 2y$

24. Same as Exercise 23, but with the additional constraint $2x - y \leq 4$.

25. Constraints: $x \geq 0$, $y \geq 0$, $x \leq 8$, $5x + y \geq 10$, $3x - 2y \leq 15$
Objective function: (a) $x + 6y$; (b) $2x - y$; (c) $x + 4y$; (d) $y - x$

26. Same as Exercise 25, but with the additional constraint $6x + 5y \leq 60$.

2

VECTOR SPACES

For the sake of efficiency, mathematicians often study objects just in terms of their mathematical structure, deemphasizing such things as particular symbols used, names of things, and applications. Any properties derived exclusively from mathematical structure will hold for all objects having that structure. Organizing mathematics in this way avoids repeating the same arguments in different contexts. Viewed from this perspective, linear algebra is the study of all objects that have a *vector-space structure*. The well-known model $\mathbb{R}^n$ described in Section 1.5 serves as our guide.

Recall that, for each positive integer n, we let $\mathbb{R}^n$ be the set of all n-tuples $(x_1, x_2, \ldots, x_n)$ of real numbers; we also refer to n-tuples as vectors $\mathbf{x}$ with n components. These Euclidean spaces $\mathbb{R}^n$ are the most familiar vector spaces. We worked with them algebraically in Chapter 1. Section 2.1 discusses geometry in $\mathbb{R}^n$. We should always take advantage of a chance to visualize a structure geometrically when undertaking its study.

Section 2.2 defines the general notion of a *vector space*, motivated by the familiar algebraic structure of the spaces $\mathbb{R}^n$. Many examples other than $\mathbb{R}^n$ are given there. We have an intuitive idea that the space $\mathbb{R}$ of real numbers is one-dimensional, the space $\mathbb{R}^2$ is two-dimensional, the space $\mathbb{R}^3$ is three-dimensional, and so on. In advanced work, it can be shown that there is an analogous concept of *dimension* for every vector space. Sections 2.3, 2.4, and 2.5 introduce the notions of independent vectors, bases, and finitely generated vector spaces.

Section 2.6 defines the *dimension* of a finitely generated vector space. The *rank* (dimensionality as opposed to size) of a matrix is then discussed. Section 2.7 shows that every finite-dimensional (real) vector space can be *coordinatized* to become algebraically indistinguishable from one of the spaces $\mathbb{R}^n$. The study of finite-dimensional vector spaces is essentially the study of the vector spaces $\mathbb{R}^n$.

Section 2.8 uses the ideas already developed to continue the study of geometry in $\mathbb{R}^n$. Section 2.9 discusses linear transformations for vector spaces. Section 2.10 on inner product spaces completes the chapter; the section describes how we try to access such geometric notions as length and angle even in infinite-dimensional vectors spaces.

Our text is primarily concerned with the vector spaces $\mathbb{R}^n$—that is, with finite-dimensional real vector spaces. However, when a proof of a property of $\mathbb{R}^n$ amounts to a proof of the property for a general vector space, we phrase the property and proof in terms of general vector spaces.

2.1 THE GEOMETRY OF $\mathbb{R}^n$

The *vector space* $\mathbb{R}^n$ can be regarded as either a space of row vectors or a space of column vectors. The appropriate row versus column viewpoint is indicated by the context. Even when column-vector notation is appropriate, we sometimes use row notation just to save space.

Geometric Representation of Vectors

We are already familiar with the Euclidean two-space $\mathbb{R}^2$ and the usual manner in which it is represented as the x,y-plane using a rectangular coordinate system. We will generalize our work in $\mathbb{R}^2$ to $\mathbb{R}^n$; consequently, we find it convenient to use (x_1, x_2) as well as (x, y). Thus each vector $\mathbf{v} = (v_1, v_2)$ in $\mathbb{R}^2$ has associated with it a unique point in the plane, as shown in Figure 2.1. The *vector* $\mathbf{v}$ can be represented geometrically by the arrow that begins at the *origin* $\mathbf{0} = (0, 0)$ and terminates at the *point* (v_1, v_2). Analytically, there is no distinction between the vector $\mathbf{v}$ and the point (v_1, v_2), because each consists of the ordered pair (v_1, v_2). In physical applications, we think of a vector as having length and direction—concepts that can be represented geometrically by an arrow. Thus it is customary to regard the vector $\mathbf{v} = (v_1, v_2)$ geometrically as an arrow, and the point (v_1, v_2) as the tip of the arrow.

The **length** or **magnitude** $\|\mathbf{v}\|$ of $\mathbf{v} = (v_1, v_2)$ is defined to be the length of the arrow in Figure 2.1. Using the Pythagorean theorem, we have

$$\|\mathbf{v}\| = \sqrt{v_1{}^2 + v_2{}^2}. \tag{1}$$

THE IDEA OF AN n-DIMENSIONAL SPACE FOR $n > 3$ reached acceptance gradually during the nineteenth century; it is thus difficult to pinpoint a first "invention" of this concept. Among the various early uses of this notion are its appearances in a work on the divergence theorem by the Russian mathematician Mikhail Ostrogradskii (1801–1862) in 1836, in the geometrical tracts of Hermann Grassmann (1809–1877) in the early 1840s, and in a brief paper of Arthur Cayley in 1846. Unfortunately, the first two authors were virtually ignored in their lifetimes. In particular, the work of Grassmann was quite philosophical and extremely difficult to read. Cayley's note merely stated that one can generalize certain results to dimensions greater than three "without recourse to any metaphysical notion with regard to the possibility of a space of four dimensions." Sir William Rowan Hamilton (1805–1865), in an 1841 letter, also noted that "it *must* be possible, in *some* way or other, to introduce not only triplets but *polyplets*, so as in some sense to satisfy the symbolical equation

$$a = (a_1, a_2, \ldots, a_n);$$

a being here one symbol, as indicative of one (complex) thought; and $a_1, a_2, \ldots, a_n$ denoting n real numbers, positive or negative."

Hamilton, whose work on quaternions will be mentioned later, and who spent much of his professional life as the Royal Astronomer of Ireland, is most famous for his work in dynamics. As Erwin Shrödinger wrote, "the Hamiltonian principle has become the cornerstone of modern physics, the thing with which a physicist expects *every* physical phenomenon to be in conformity."

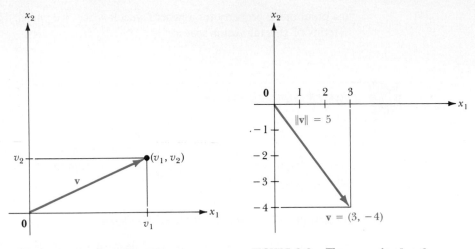

FIGURE 2.1 Geometric represen-tation of a vector in $\mathbb{R}^2$.

FIGURE 2.2 The magnitude of v.

EXAMPLE 1 Represent the vector $\mathbf{v} = (3, -4)$ geometrically, and find its magnitude.

Solution The vector $\mathbf{v} = (3, -4)$ has magnitude

$$\|\mathbf{v}\| = \sqrt{3^2 + (-4)^2} = \sqrt{25} = 5$$

and is shown in Figure 2.2. ■

To represent $\mathbb{R}^3$ geometrically, we choose three perpendicular lines through the origin $\mathbf{0} = (0, 0, 0)$ as coordinate axes, as shown in Figure 2.3. The coordinate system in this figure is called a *right-hand system* because, when the fingers of the right hand are curved in the direction required to rotate the positive x_1-axis toward the positive x_2-axis, the right thumb points up the x_3-axis.

Each vector $\mathbf{v} = (v_1, v_2, v_3)$ in $\mathbb{R}^3$ corresponds to a unique point in space, as shown in Figure 2.3. The vector $\mathbf{v}$ is again represented by an arrow from the origin $\mathbf{0} = (0, 0, 0)$ to the point (v_1, v_2, v_3). The **magnitude** $\|\mathbf{v}\|$ of $\mathbf{v}$ is the length of this arrow. This magnitude appears in Figure 2.3 as the hypotenuse of a right triangle whose altitude is $|v_3|$ and whose base in the x_1,x_2-plane has length $\sqrt{v_1^2 + v_2^2}$. Using the Pythagorean theorem, we obtain

$$\|\mathbf{v}\| = \sqrt{v_1^2 + v_2^2 + v_3^2}. \tag{2}$$

EXAMPLE 2 Represent the vector $\mathbf{v} = (2, 3, 4)$ geometrically, and find its magnitude.

Solution The vector $\mathbf{v} = (2, 3, 4)$ has magnitude $\|\mathbf{v}\| = \sqrt{2^2 + 3^2 + 4^2} = \sqrt{29}$ and is represented in Figure 2.4. ■

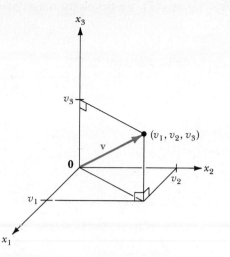

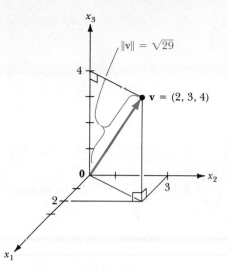

FIGURE 2.3 Geometric representation of a vector in $\mathbb{R}^3$.

FIGURE 2.4 The magnitude of v.

By analogy, we consider any vector $\mathbf{v} = (v_1, v_2, \ldots, v_n)$ in $\mathbb{R}^n$ to be represented by an arrow from the origin $\mathbf{0} = (0, 0, \ldots, 0)$ to the point $\mathbf{v}$, as shown in Figure 2.5. Notice that this picture is less cluttered than Figures 2.3 and 2.4, because we do not attempt to represent n mutually perpendicular coordinate axes for the variables x_1, $x_2, \ldots, x_n$. Indeed, we will leave out the coordinate axes whenever doing so will make our diagrams cleaner and more general.

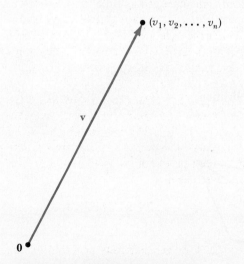

FIGURE 2.5 Geometric representation of an n-component vector.

As suggested by Eqs. (1) and (2), we define the magnitude $\|\mathbf{v}\|$ of the vector $\mathbf{v}$ as follows.

DEFINITION 2.1 Magnitude of a vector in $\mathbb{R}^n$

Let $\mathbf{v} = (v_1, v_2, \ldots, v_n)$ be a vector in $\mathbb{R}^n$. The **magnitude** of $\mathbf{v}$ is

$$\|\mathbf{v}\| = \sqrt{v_1^2 + v_2^2 + \cdots + v_n^2}. \tag{3}$$

EXAMPLE 3 Find the magnitude of the vector

$$\mathbf{v} = (-2, 1, 3, -1, 4, 2, 1).$$

Solution We have

$$\|\mathbf{v}\| = \sqrt{(-2)^2 + 1^2 + 3^2 + (-1)^2 + 4^2 + 2^2 + 1^2} = \sqrt{36} = 6. \qquad ■$$

Addition, Subtraction, and Scalar Multiplication

We already know how to add and subtract matrices of the same size and how to multiply a matrix by a scalar. Since the vectors in $\mathbb{R}^n$ can be regarded as $1 \times n$ matrices, the following definition is a review.

DEFINITION 2.2 Vector-space Operations in $\mathbb{R}^n$

Let $\mathbf{v} = (v_1, v_2, \ldots, v_n)$ and $\mathbf{w} = (w_1, w_2, \ldots, w_n)$ be vectors in $\mathbb{R}^n$. The vectors are added and subtracted as follows:

Vector addition: $\mathbf{v} + \mathbf{w} = (v_1 + w_1, v_2 + w_2, \ldots, v_n + w_n)$

Vector subtraction: $\mathbf{v} - \mathbf{w} = (v_1 - w_1, v_2 - w_2, \ldots, v_n - w_n)$

If r is any scalar, the vector $\mathbf{v}$ is multiplied by r as follows:

Scalar multiplication: $r\mathbf{v} = (rv_1, rv_2, \ldots, rv_n)$

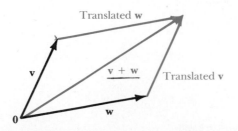

FIGURE 2.6 Geometric representation of v + w in $\mathbb{R}^n$.

The sum of the vectors **v** and **w** can be represented geometrically as shown in Figure 2.6. The tip of the vector **v** + **w** can be located by drawing the vector **w** as an arrow emanating from the tip of **v**. This is nothing more than a translation of axes, using the point at the tip of **v** as a local origin. Equivalently, we can translate the axes to the point at the tip of **w** and draw the vector **v** as an arrow emanating from this point. When both of these translated vectors are drawn, we have *completed the parallelogram* shown in Figure 2.6 and **v** + **w** is the indicated diagonal.

Application to Force Vectors

A practical physical model of vector addition is given by force vectors. We illustrate for force vectors in the plane. Suppose that we are moving some object in the plane by pushing it. The motion of the object is influenced by the direction in which we are pushing and by how hard we are pushing. Thus it is natural to represent the force with which we are pushing by an arrow pointing in the direction in which we are pushing, and having a length that represents how hard we are pushing. Such an arrow represents the physicist's notion of a *force vector*. If we choose coordinate axes in the plane at the point where the force is applied to the object, the force can be represented by an arrow emanating from the origin.

Suppose that two forces, represented by vectors $\mathbf{v} = (v_1, v_2)$ and $\mathbf{w} = (w_1, w_2)$, are acting on a body at $\mathbf{0} = (0, 0)$ in the plane, as shown in Figure 2.7. It can be shown that the effect the two forces have on the body is the same as the effect of a single force represented by the diagonal of the parallelogram determined by the given vectors, as shown in the figure. We see from the figure that the vector along the diagonal is $(v_1 + w_1, v_2 + w_2)$; this vector is the sum of the two given vectors.

If $\mathbf{v} = (v_1, v_2)$ is a force vector, the vector corresponding to double the force in the same direction is $2\mathbf{v} = (2v_1, 2v_2)$, while the vector corresponding to one-third the force in the opposite direction is $-\frac{1}{3}\mathbf{v} = (-v_1/3, -v_2/3)$, as shown in Figure 2.8. Thus this physical model also illustrates scalar multiplication.

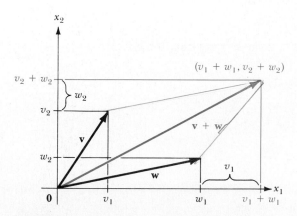

FIGURE 2.7 Geometric representation of v + w in $\mathbb{R}^2$.

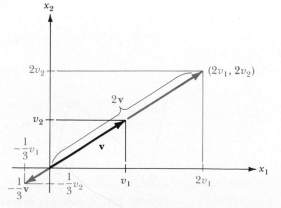

FIGURE 2.8 Geometric representation of scalar multiplication in $\mathbb{R}^2$.

The Distance from v to w

The difference $\mathbf{v} - \mathbf{w}$ of two vectors in $\mathbb{R}^n$ can be represented geometrically as the arrow from the tip of $\mathbf{w}$ to the tip of $\mathbf{v}$, as shown in Figure 2.9. Here $\mathbf{v} - \mathbf{w}$ is the vector that, when added to $\mathbf{w}$, yields $\mathbf{v}$. The dashed arrow in Figure 2.9 shows $\mathbf{v} - \mathbf{w}$ in *standard position*—that is, emanating from the origin. The **distance** from $\mathbf{v}$ to $\mathbf{w}$ is $\|\mathbf{v} - \mathbf{w}\|$.

EXAMPLE 4   Find the distance from $\mathbf{v} = (1, 2, 5, 1)$ to $\mathbf{w} = (2, 1, 0, 3)$ in $\mathbb{R}^4$.

Solution   We see that the required distance is

$$\|\mathbf{v} - \mathbf{w}\| = \|(-1, 1, 5, -2)\| = \sqrt{(-1)^2 + 1^2 + 5^2 + (-2)^2} = \sqrt{31}. \qquad ■$$

Application to Velocity Vectors and Navigation

The following two examples are concerned with another physical model for vector algebra. A vector is the **velocity vector** of a moving object at an instant if it points in the direction of the motion and if its magnitude is the speed of the object at that instant. Common navigation problems can be solved using velocity vectors.

EXAMPLE 5   Suppose that a sloop is sailing at 8 knots, following a course of 010° (that is, 10° east of north), on a bay that has a 2-knot current setting in the direction 070° (that is, 70° east of north). Find the course and speed made good. (The expression *made good* is standard navigation terminology for the actual course and speed of a vessel over the bottom.)

Solution   The velocity vectors $\mathbf{s}$ for the sloop and $\mathbf{c}$ for the current are shown in Figure 2.10, in which the vertical axis points due north. We find $\mathbf{s}$ and $\mathbf{c}$ by using a calculator and computing

$$\mathbf{s} = (8 \cos 80°, 8 \sin 80°) \approx (1.39, 7.88)$$

and

$$\mathbf{c} = (2 \cos 20°, 2 \sin 20°) \approx (1.88, 0.684).$$

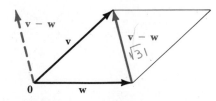

FIGURE 2.9   Geometric representation of $\mathbf{v} - \mathbf{w}$ in $\mathbb{R}^n$.

By adding **s** and **c**, we find the vector **v** representing the course and speed of the sloop over the bottom—that is, the course and speed made good. Thus we have **v** = **s** + **c** ≈ (3.27, 8.56). Therefore, the speed of the sloop is

$$\|\mathbf{v}\| \approx \sqrt{(3.27)^2 + (8.56)^2} \approx 9.16 \text{ knots,}$$

and the course made good is given by

$$90° - \arctan(8.56/3.27) \approx 90° - 69° = 21°.$$

That is, the course is 021°. ■

EXAMPLE 6   Suppose the captain of our sloop realizes the importance of keeping track of the current. He wishes to sail in 5 hours to a harbor that bears 120° and is 35 nautical miles away. That is, he wishes to make good the course 120° and the speed 7 knots. He knows from a tide and current table that the current is setting due south at 2 knots. What should be his course and speed through the water?

Solution   In a vector diagram (see Fig. 2.11), we again represent the course and speed to be made good by a vector **v** and the velocity of the current by **c**. The correct course and speed to follow is represented by the vector **s** and is obtained by computing

$$\mathbf{s} = \mathbf{v} - \mathbf{c}$$
$$= (7 \cos 30°, -7 \sin 30°) - (0, -2)$$
$$\approx (6.06, -3.5) - (0, -2) = (6.06, -1.5).$$

Thus the captain should steer course $90° - \arctan(-1.5/6.06) \approx 90° + 13.9° = 103.9°$ and should proceed at

$$\|\mathbf{s}\| \approx \sqrt{(6.06)^2 + (-1.5)^2} \approx 6.24 \text{ knots.}$$ ■

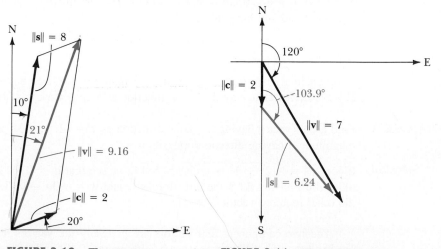

FIGURE 2.10   The vector v = s + c.

FIGURE 2.11   The vector s = v - c.

Parallel Vectors

We turn to the geometric representation of scalar multiplication. Our work with force vectors suggests that, for vector $\mathbf{v}$ in $\mathbb{R}^n$ and scalar r, the vector $r\mathbf{v}$ has the same direction as $\mathbf{v}$ if $r > 0$ and has the opposite direction if $r < 0$. The zero vector is considered to have all directions. The magnitude of $r\mathbf{v}$ is

$$\|r\mathbf{v}\| = |r|\|\mathbf{v}\|. \tag{4}$$

This was illustrated in Figure 2.8. We can verify Eq. (4) by computing

$$\|r\mathbf{v}\| = \|(rv_1, rv_2, \ldots, rv_n)\| = \sqrt{(rv_1)^2 + (rv_2)^2 + \cdots + (rv_n)^2}$$
$$= |r|\sqrt{v_1^2 + v_2^2 + \cdots + v_n^2} = |r|\|\mathbf{v}\|.$$

With this discussion as motivation, we define the general notion of parallel vectors.

DEFINITION 2.3 Parallel Vectors

Two nonzero vectors $\mathbf{v}$ and $\mathbf{w}$ in $\mathbb{R}^n$ are **parallel**, and we write $\mathbf{v} \parallel \mathbf{w}$, if one is a scalar multiple of the other. If $\mathbf{v} = r\mathbf{w}$ with $r > 0$, then $\mathbf{v}$ and $\mathbf{w}$ have the **same direction**; if $r < 0$, then $\mathbf{v}$ and $\mathbf{w}$ have **opposite direction**.

EXAMPLE 7   Determine whether the vectors $\mathbf{v} = (2, 1, 3, -4)$ and $\mathbf{w} = (6, 3, 9, -12)$ are parallel.

Solution   We put $\mathbf{v} = r\mathbf{w}$ and try to solve for r. This gives rise to four component equations:

$$2 = 6r, \qquad 1 = 3r, \qquad 3 = 9r, \qquad -4 = -12r.$$

Since $r = \frac{1}{3} > 0$ is a common solution to the four equations, we conclude that $\mathbf{v}$ and $\mathbf{w}$ are parallel and have the same direction.   ■

Unit Vectors

A vector in $\mathbb{R}^n$ is a **unit vector** if it has magnitude 1. Given any nonzero vector $\mathbf{v}$ in $\mathbb{R}^n$, a unit vector having the same direction as $\mathbf{v}$ is given by $(1/\|\mathbf{v}\|)\mathbf{v}$.

EXAMPLE 8   Find a unit vector having the same direction as $\mathbf{v} = (2, 1, -3)$, and find a vector of magnitude 3 having direction opposite to $\mathbf{v}$.

Solution   Since $\|\mathbf{v}\| = \sqrt{2^2 + 1^2 + (-3)^2} = \sqrt{14}$, we see that $\mathbf{u} = (1/\sqrt{14})(2, 1, -3)$ is the unit vector having the same direction as $\mathbf{v}$, and $\mathbf{w} = -3\mathbf{u} = (-3/\sqrt{14})(2, 1, -3)$ is the other required vector.   ■

The two-component unit vectors are precisely the vectors that extend from the origin to the unit circle $x^2 + y^2 = 1$ with center $(0, 0)$ and radius 1 in $\mathbb{R}^2$. (See Fig. 2.12(a)). The three-component unit vectors extend from $(0, 0, 0)$ to the unit sphere in $\mathbb{R}^3$, as illustrated in Figure 2.12(b).

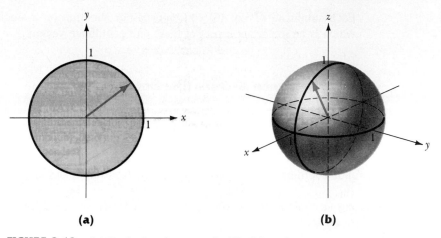

FIGURE 2.12 (a) **Typical unit vector in** $\mathbb{R}^2$; (b) **typical unit vector in** $\mathbb{R}^3$.

The vectors $\mathbf{e}_r = (0, 0, \ldots, 1, 0, \ldots, 0)$ in $\mathbb{R}^n$ with rth component 1 and zeros elsewhere are known as the **unit coordinate vectors** in $\mathbb{R}^n$. (See Fig. 2.13.) The unit coordinate vectors in $\mathbb{R}^2$ and $\mathbb{R}^3$ are often written as

$$\mathbf{i} = \mathbf{e}_1, \qquad \mathbf{j} = \mathbf{e}_2, \qquad \text{and} \qquad \mathbf{k} = \mathbf{e}_3.$$

Each vector $\mathbf{v}$ in $\mathbb{R}^n$ can be expressed in a unique way as a sum of multiples of the unit coordinate vectors. Namely, for $\mathbf{v} = (v_1, v_2, \ldots, v_n)$, we have

$$\mathbf{v} = v_1\mathbf{e}_1 + v_2\mathbf{e}_2 + \cdots + v_n\mathbf{e}_n. \tag{5}$$

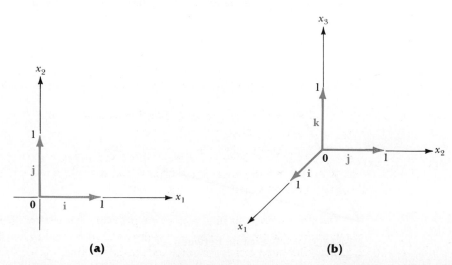

FIGURE 2.13 (a) **Unit coordinate vectors in** $\mathbb{R}^2$; (b) **unit coordinate vectors in** $\mathbb{R}^3$.

Each summand of Eq. (5) concentrates our attention on a single component of **v**, which is an important aspect of these unit coordinate vectors.

Angle Between Vectors; The Dot Product

We often describe a direction to someone by pointing. Our extended arm then represents a vector whose direction is clear. We describe directions in $\mathbb{R}^n$ by using vectors. The direction of a vector **v** in $\mathbb{R}^n$ is determined by its coordinates. We can specify this direction further by indicating the angle that the vector **v** makes with each unit coordinate vector. More generally, let us find the angle θ that a vector $\mathbf{v} = (v_1, v_2, \ldots, v_n)$ makes with a vector $\mathbf{w} = (w_1, w_2, \ldots, w_n)$ in $\mathbb{R}^n$. The law of cosines is just what we need to do this. If we refer to Figure 2.14, the law of cosines tells us that

$$\|\mathbf{v}\|^2 + \|\mathbf{w}\|^2 = \|\mathbf{v} - \mathbf{w}\|^2 + 2\|\mathbf{v}\|\|\mathbf{w}\|(\cos \theta)$$

or

$$v_1^2 + \cdots + v_n^2 + w_1^2 + \cdots + w_n^2$$
$$= (v_1 - w_1)^2 + \cdots + (v_n - w_n)^2 + 2\|\mathbf{v}\|\|\mathbf{w}\|(\cos \theta). \quad (6)$$

After computing the squares on the right-hand side of Eq. (6) and simplifying, we obtain

$$\|\mathbf{v}\|\|\mathbf{w}\|(\cos \theta) = v_1 w_1 + \cdots + v_n w_n. \quad (7)$$

We recognize the expression on the right-hand side of Eq. (7) as the dot product $\mathbf{v} \cdot \mathbf{w}$ of **v** and **w**, which was defined in Section 1.2. Thus, for nonzero vectors **v** and **w** in $\mathbb{R}^n$, we can attempt to define the angle θ between **v** and **w** by the following.

The **angle** between nonzero vectors **v** and **w** is
$$\arccos\left(\frac{\mathbf{v} \cdot \mathbf{w}}{\|\mathbf{v}\|\|\mathbf{w}\|}\right). \quad (8)$$

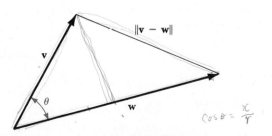

FIGURE 2.14 **The angle between v and w.**

Expression (8) makes sense, provided that

$$-1 \le \frac{\mathbf{v} \cdot \mathbf{w}}{\|\mathbf{v}\|\|\mathbf{w}\|} \le 1, \tag{9}$$

which allows us to compute the arccosine of this quantity. Condition (9) can be rewritten as

$$|\mathbf{v} \cdot \mathbf{w}| \le \|\mathbf{v}\|\|\mathbf{w}\|. \tag{10}$$

Relation (10) does hold for all $\mathbf{v}$ and $\mathbf{w}$ in $\mathbb{R}^n$. It is known as the *Schwarz inequality*, and we will prove it in Section 2.10.

EXAMPLE 9   Find the angle θ between the vectors $(1, 2, 0, 2)$ and $(-3, 1, 1, 5)$ in $\mathbb{R}^4$.

Solution   We have

$$\cos \theta = \frac{(1, 2, 0, 2) \cdot (-3, 1, 1, 5)}{\sqrt{1^2 + 2^2 + 0^2 + 2^2} \sqrt{(-3)^2 + 1^2 + 1^2 + 5^2}} = \frac{9}{(3)(6)} = \frac{1}{2}.$$

Thus, $\theta = 60°$.   ■

We see from Eq. (7) that the dot product $\mathbf{v} \cdot \mathbf{w}$ of two vectors in $\mathbb{R}^n$ is described geometrically by

$$\mathbf{v} \cdot \mathbf{w} = \|\mathbf{v}\|\|\mathbf{w}\|(\cos \theta), \tag{11}$$

where θ is the angle between $\mathbf{v}$ and $\mathbf{w}$.

Perpendicular Vectors

Two vectors $\mathbf{v}$ and $\mathbf{w}$ in $\mathbb{R}^n$ are perpendicular if the angle between them is 90°. This is the case precisely when the numerator in Eq. (8) is zero.

DEFINITION 2.4   Perpendicular or Orthogonal Vectors

Two vectors $\mathbf{v}$ and $\mathbf{w}$ in $\mathbb{R}^n$ are **perpendicular** or **orthogonal**, and we write $\mathbf{v} \perp \mathbf{w}$, if their dot product is zero (that is, if $\mathbf{v} \cdot \mathbf{w} = 0$).

EXAMPLE 10   Determine whether the vectors $\mathbf{v} = (4, 1, -2, 1)$ and $\mathbf{w} = (3, -4, 2, -4)$ are perpendicular.

Solution   We have

$$\mathbf{v} \cdot \mathbf{w} = (4)(3) + (1)(-4) + (-2)(2) + (1)(-4) = 0.$$

Thus, $\mathbf{v} \perp \mathbf{w}$.   ■

SUMMARY

Let $\mathbf{v} = (v_1, v_2, \ldots, v_n)$ and $\mathbf{w} = (w_1, w_2, \ldots, w_n)$ be vectors in $\mathbb{R}^n$.

1. The vectors $\mathbf{v}$ and $\mathbf{w}$ can be added and subtracted, and each can be multiplied by any scalar r. The result in each case is a vector in $\mathbb{R}^n$. These operations are special cases of the matrix operations defined in Section 1.1.

2. The angle θ between the vectors $\mathbf{v}$ and $\mathbf{w}$ can be found by using the relation $\mathbf{v} \cdot \mathbf{w} = \|\mathbf{v}\|\|\mathbf{w}\|(\cos\theta)$, where $\mathbf{v} \cdot \mathbf{w}$ is the dot product of $\mathbf{v}$ and $\mathbf{w}$.

3. Nonzero vectors $\mathbf{v}$ and $\mathbf{w}$ are parallel if one is a scalar multiple of the other. In that case, the vectors have the same direction if the scalar is positive, and they have opposite directions if the scalar is negative.

4. The vectors $\mathbf{v}$ and $\mathbf{w}$ are perpendicular or orthogonal if their dot product is zero.

EXERCISES

In Exercises 1–4, find the indicated distance.

1. The distance from $(-1, 4, 2)$ to $(0, 8, 1)$ in $\mathbb{R}^3$

2. The distance from $(2, -1, 3)$ to $(4, 1, -2)$ in $\mathbb{R}^3$

3. The distance from $(3, 1, 2, 4)$ to $(-1, 2, 1, 2)$ in $\mathbb{R}^4$

4. The distance from $(-1, 2, 1, 4, 7, -3)$ to $(2, 1, -3, 5, 4, 5)$ in $\mathbb{R}^6$

In Exercises 5–22, let $\mathbf{u} = (-1, 3, 4)$, $\mathbf{v} = (2, 1, -1)$, *and* $\mathbf{w} = (-2, -1, 3)$. *Find the indicated quantity.*

5. $-\mathbf{u}$

6. $\|\mathbf{v}\|$

7. $\mathbf{u} + \mathbf{v}$

8. $\mathbf{v} - 2\mathbf{u}$

9. $3\mathbf{u} - \mathbf{v} + 2\mathbf{w}$

10. $\mathbf{v} - \mathbf{u}$

11. $\left(\frac{4}{5}\right)\mathbf{w}$

12. The unit vector parallel to $\mathbf{u}$, having the same direction

13. The unit vector parallel to $\mathbf{w}$, having the opposite direction

14. $\mathbf{u} \cdot \mathbf{v}$

15. $\mathbf{u} \cdot (\mathbf{v} + \mathbf{w})$

16. $(\mathbf{u} + \mathbf{v}) \cdot \mathbf{w}$

17. The angle between $\mathbf{u}$ and $\mathbf{v}$

18. The angle between $\mathbf{u}$ and $\mathbf{w}$

19. The value of x such that $(x, -3, 5)$ is perpendicular to $\mathbf{u}$

20. The value of y such that $(-3, y, 10)$ is perpendicular to $\mathbf{u}$

21. A nonzero vector perpendicular to both $\mathbf{u}$ and $\mathbf{v}$

22. A nonzero vector perpendicular to both $\mathbf{u}$ and $\mathbf{w}$

23. Find the angle between $\mathbf{v} = (1, -1, 2, 3, 0, 4)$ and $\mathbf{w} = (7, 0, 1, 3, 2, 4)$ in $\mathbb{R}^6$.

24. Show that $(2, 0, 4)$, $(4, 1, -1)$, and $(6, 7, 7)$ are vertices of a right triangle in $\mathbb{R}^3$.

In Exercises 25–30, classify the vectors as parallel, perpendicular, or neither. If they are parallel, state whether they have the same direction or opposite directions.

25. $(-1, 4)$ and $(8, 2)$

26. $(-2, -1)$ and $(5, 2)$

27. $(3, 2, 1)$ and $(-9, -6, -3)$

28. $(2, 1, 4, -1)$ and $(0, 1, 2, 4)$

29. $(10, 4, -1, 8)$ and $(-5, -2, 3, -4)$

30. $(4, 1, 2, 1, 6)$ and $(8, 2, 4, 2, 3)$

31. The captain of a barge wishes to get to a point directly across a straight river that runs from north to south. If the current flows directly downstream at 5 knots and if the barge steams at 13 knots, in what direction should the captain steer his barge?

32. A 100-lb weight is suspended by a rope passed through an eyelet on top of the weight and making angles of 30° with the vertical, as shown in Figure 2.15. Find the tension (magnitude of the force vector) along the rope. [HINT: The sum of the force vectors along the two halves of the rope at the eyelet must be an upward vertical vector of magnitude 100.]

33. a) Answer Exercise 32 if each half of the rope makes an angle of θ with the vertical at the eyelet.

   b) Find the tension in the rope if both halves are vertical ($\theta = 0$).

   c) What happens if an attempt is made to stretch the rope out straight (horizontal) while the 100-lb weight hangs on it?

34. Suppose that a weight of 100 lb is suspended by two different ropes tied at an eyelet on top of the weight, as shown in Figure 2.16. Let the angles the ropes make with the vertical be θ_1 and θ_2, as shown in the figure. Let the tensions in the ropes be T_1 for the right-hand rope and T_2 for the left-hand rope.

   a) Show that the force vector $\mathbf{F}_1$ shown in Figure 2.16 is $T_1(\sin \theta_1)\mathbf{i} + T_1(\cos \theta_1)\mathbf{j}$.

   b) Find the corresponding expression for $\mathbf{F}_2$ in terms of T_2 and θ_2.

   c) If the system is in equilibrium, $\mathbf{F}_1 + \mathbf{F}_2 = 100\mathbf{j}$, so $\mathbf{F}_1 + \mathbf{F}_2$ must have $\mathbf{i}$-component 0 and $\mathbf{j}$-component 100. Write two equations reflecting this fact, using the answers to parts (a) and (b).

   d) Find T_1 and T_2 if $\theta_1 = 45°$ and $\theta_2 = 30°$.

35. Mark each of the following True or False.

   ___ **a)** Every nonzero vector in $\mathbb{R}^n$ has nonzero magnitude.

   ___ **b)** Every vector of nonzero magnitude in $\mathbb{R}^n$ is nonzero.

   ___ **c)** The magnitude of $\mathbf{v} + \mathbf{w}$ must be at least as large as the magnitude of either $\mathbf{v}$ or $\mathbf{w}$ in $\mathbb{R}^n$.

   ___ **d)** Every nonzero vector $\mathbf{v}$ in $\mathbb{R}^n$ has exactly one unit vector parallel to it.

   ___ **e)** There are exactly two unit vectors parallel to any given nonzero vector in $\mathbb{R}^n$.

   ___ **f)** There are exactly two unit vectors perpendicular to any given nonzero vector in $\mathbb{R}^n$.

   ___ **g)** The angle between two nonzero vectors in $\mathbb{R}^n$ is less than 90° if and only if the dot product of the vectors is positive.

   ___ **h)** The dot product of a vector with itself yields the magnitude of the vector.

   ___ **i)** For a vector $\mathbf{v}$ in $\mathbb{R}^n$, the magnitude of r times $\mathbf{v}$ is r times the magnitude of $\mathbf{v}$.

   ___ **j)** If $\mathbf{v}$ and $\mathbf{w}$ are vectors in $\mathbb{R}^n$ of the same magnitude, then the magnitude of $\mathbf{v} - \mathbf{w}$ is 0.

36. The software available to schools using this text includes a program called VECTGRPH, which is designed to strengthen students' geometric understanding of vector concepts. The program uses graphics for vectors in the plane only. Run the program, and work with the first item involving vector addition and subtraction until you consistently achieve a score of 80% or better.

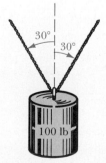

FIGURE 2.15 Both halves of the rope make an angle of 30° with the vertical.

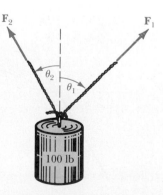

FIGURE 2.16 Two ropes tied at the eyelet and making angles θ_1 and θ_2 with the vertical.

2.2 VECTOR SPACES

We suggest that you now reread the first paragraph of the introduction to this chapter, so that you understand the why and wherefore of this section.

The Vector-Space Operations

In each Euclidean space $\mathbb{R}^n$, we know how to add two vectors and how to perform scalar multiplication of a vector by a real number (scalar). These are the two *vector-space operations*. The first requirement a set V must satisfy in order to be a vector space is that it have two well-defined algebraic operations:

Addition of two elements of V, and

Scalar multiplication of an element of V.

For example, we know how to add the two functions x^2 and $\sin x$, and we know how to multiply $\tan x$ by a real number. We require that, whenever addition and/or scalar multiplication is performed with V, *the answers obtained lie again in V.* We express this by saying that V must be **closed under addition** and **closed under scalar multiplication**. We give two examples illustrating these **closure properties**.

EXAMPLE 1   Determine whether the set S of vectors in the first quadrant of the plane $\mathbb{R}^2$ is closed under (a) addition and (b) scalar multiplication.

Solution   If two vectors in the first quadrant are added, their sum is also a vector in the first quadrant, as illustrated in Figure 2.17. However, if a vector in the first quadrant is

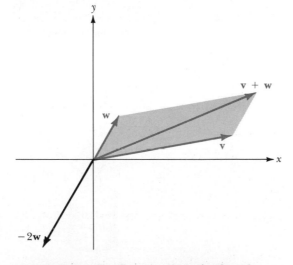

FIGURE 2.17   The first quadrant is closed under addition, but it is not closed under scalar multiplication.

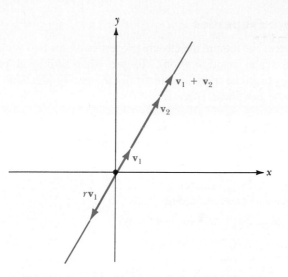

FIGURE 2.18 **The line $y = 2x$ is closed under addition and scalar multiplication.**

multiplied by a negative scalar, the result is a vector in the third quadrant. Thus S is closed under addition, but it is not closed under scalar multiplication. ∎

EXAMPLE 2   Determine whether the set of vectors on the line $y = 2x$ in the plane $\mathbb{R}^2$ is closed under addition and scalar multiplication.

Solution   The line may be described as the set $L = \{\mathbf{v} \mid \mathbf{v} = (x, 2x)$ and $x \in \mathbb{R}\}$.* If we take two vectors $\mathbf{v}_1 = (x_1, 2x_1)$ and $\mathbf{v}_2 = (x_2, 2x_2)$ in L and add them, we obtain

$$\mathbf{v}_1 + \mathbf{v}_2 = (x_1 + x_2, 2x_1 + 2x_2),$$

which may be written in the form

$$(\underset{x}{x_1 + x_2}, \underset{2x}{2(x_1 + x_2)}).$$

Thus the sum is indeed in L.

Similarly, for any scalar r, we have

$$r\mathbf{v}_1 = r(x_1, 2x_1) = (rx_1, 2rx_1) = (\underset{x}{rx_1}, \underset{2x}{2(rx_1)}),$$

so $r\mathbf{v}_1$ is again in L.

Thus L is closed under both addition and scalar multiplication. (See Fig. 2.18.)

*Recall that the set $\{x \mid P(x)\}$ is read as, "the set of all x such that the property $P(x)$ is true," and $x \in \mathbb{R}$ is read as "x is a member of $\mathbb{R}$."

Vector Space Properties

In a vector space, we require the closure properties of our two vector-space operations (addition and scalar multiplication). To see what additional properties might be required, let us examine more systematically the purely algebraic properties of these operations in the Euclidean space $\mathbb{R}^n$.

THEOREM 2.1 Vector-space Properties of $\mathbb{R}^n$

Let $\mathbf{u}$, $\mathbf{v}$, and $\mathbf{w}$ be any vectors in $\mathbb{R}^n$, and let r and s be any scalars in $\mathbb{R}$.

Properties of Vector Addition

A1	$(\mathbf{u} + \mathbf{v}) + \mathbf{w} = \mathbf{u} + (\mathbf{v} + \mathbf{w})$	Associative law
A2	$\mathbf{v} + \mathbf{w} = \mathbf{w} + \mathbf{v}$	Commutative law
A3	$\mathbf{0} + \mathbf{v} = \mathbf{v}$	Nature of the zero vector
A4	$\mathbf{v} + (-\mathbf{v}) = \mathbf{0}$	$-\mathbf{v}$ as additive inverse of $\mathbf{v}$

Properties Involving Scalar Multiplication

S1	$r(\mathbf{v} + \mathbf{w}) = r\mathbf{v} + r\mathbf{w}$	A distributive law
S2	$(r + s)\mathbf{v} = r\mathbf{v} + s\mathbf{v}$	A distributive law
S3	$r(s\mathbf{v}) = (rs)\mathbf{v}$	Associative law
S4	$1\mathbf{v} = \mathbf{v}$	Preservation of scale

The eight properties given in Theorem 2.1 are quite easy to prove, and we leave most of them as exercises. The proofs in Examples 3 and 4 are typical.

EXAMPLE 3 Prove property A2 of Theorem 2.1.

Solution Writing

$$\mathbf{v} = (v_1, v_2, \ldots, v_n) \quad \text{and} \quad \mathbf{w} = (w_1, w_2, \ldots, w_n),$$

we have

$$\mathbf{v} + \mathbf{w} = (v_1 + w_1, v_2 + w_2, \ldots, v_n + w_n)$$

and

$$\mathbf{w} + \mathbf{v} = (w_1 + v_1, w_2 + v_2, \ldots, w_n + v_n).$$

These two vectors are equal because $v_i + w_i = w_i + v_i$ for each i. Thus, the commutative law of vector addition follows directly from the commutative law of addition of numbers. ■

EXAMPLE 4 Prove property S2 of Theorem 2.1.

Solution   Writing $\mathbf{v} = (v_1, v_2, \ldots, v_n)$, we have

$$
\begin{aligned}
(r + s)\mathbf{v} &= (r + s)(v_1, v_2, \ldots, v_n) \\
&= ((r + s)v_1, (r + s)v_2, \ldots, (r + s)v_n) \\
&= (rv_1 + sv_1, rv_2 + sv_2, \ldots, rv_n + sv_n) \\
&= (rv_1, rv_2, \ldots, rv_n) + (sv_1, sv_2, \ldots, sv_n) \\
&= r\mathbf{v} + s\mathbf{v}.
\end{aligned}
$$

Thus the property $(r + s)\mathbf{v} = r\mathbf{v} + s\mathbf{v}$ involving vectors follows from the analogous property $(r + s)a_i = ra_i + sa_i$ for numbers.   ■

The vector space $\mathbb{R}^n$ serves as a model for the general definition of a real vector space.

DEFINITION 2.5 Vector Space

A **(real) vector space** is a set V of objects called **vectors**, together with a rule for adding any two vectors $\mathbf{v}$ and $\mathbf{w}$ to produce a vector $\mathbf{v} + \mathbf{w}$ in V and a rule for multiplying any vector $\mathbf{v}$ in V by any scalar r in $\mathbb{R}$ to produce a vector $r\mathbf{v}$ in V. Moreover, there must exist a vector $\mathbf{0}$ in V, and for each $\mathbf{v}$ in V there must exist a vector $-\mathbf{v}$ in V such that properties A1 through A4 and S1 through S4 of Theorem 2.1 are satisfied for all choices of vectors and scalars.

The vector $\mathbf{0}$ in V is called the **zero vector**, while $-\mathbf{v}$ is called the **additive inverse of v** and is usually read "minus $\mathbf{v}$." We write $\mathbf{v} - \mathbf{w}$ for $\mathbf{v} + (-\mathbf{w})$.

EXAMPLE 5   Show that the set $M_{m,n}$ of all $m \times n$ matrices is a vector space, using as vector addition and scalar multiplication the usual addition of matrices and multiplication of a matrix by a scalar.

THOUGH VECTORS IN $\mathbb{R}^n$ were dealt with by mathematicians and physicists throughout the second half of the nineteenth century (as were other objects that are today considered vectors), it was not until the appearance of Hermann Weyl's (1885–1955) treatise *Space–Time–Matter* in 1918 that an abstract, axiomatic definition of a vector space appeared in print. Weyl wrote this book as an introduction to Einstein's general theory of relativity; in Chapter 1 he discussed the nature of Euclidean space, and as a part of that discussion he formulated what are now the standard axioms of a vector space (A1–A4, S1–S4). Since he was dealing only with spaces V of finite dimension, he included the "axiom of dimensionality"—that for some whole number h, there are h linearly independent vectors in V, but every set of $h + 1$ vectors is linearly dependent (see Section 2.4).

The vector-space axioms themselves had been known for years, but they were generally proved as consequences of other definitions of vectors. For example, these appear in Giuseppe Peano's (1858–1932) brief Italian text, *Geometric Calculus* (1888), in which he explains the work of Hermann Grassmann (see note on page 127).

Solution We have seen that addition of $m \times n$ matrices and multiplication of an $m \times n$ matrix by a scalar again yield an $m \times n$ matrix. Thus, $M_{m,n}$ is closed under vector addition and scalar multiplication. We take as zero vector in $M_{m,n}$ the usual zero matrix, all of whose entries are zero. For any matrix A in $M_{m,n}$ we consider $-A$ to be the matrix $(-1)A$. The properties of matrix arithmetic on page 11 show that all eight properties A1–A4 and S1–S4 required of a vector space are satisfied. ■

The preceding example introduced the notation $M_{m,n}$ for the vector space of all $m \times n$ matrices. We use M_n for the vector space of all square $n \times n$ matrices.

EXAMPLE 6 Show that the set P of all polynomials with coefficients in $\mathbb{R}$ is a vector space, using for vector addition and scalar multiplication the usual addition of polynomials and multiplication of a polynomial by a scalar.

Solution Let p and q be polynomials

$$p = a_0 + a_1 x + a_2 x^2 + \cdots + a_n x^n$$

and

$$q = b_0 + b_1 x + b_2 x^2 + \cdots + b_m x^m.$$

If $m \geq n$, the **sum** of p and q is given by

$$p + q = (a_0 + b_0) + (a_1 + b_1)x + \cdots + (a_n + b_n)x^n$$
$$+ b_{n+1}x^{n+1} + \cdots + b_m x^m.$$

For example, if $p = 1 + 2x + 3x^2$ and $q = x + x^3$, then $p + q = 1 + 3x + 3x^2 + x^3$. A similar definition is made if $m < n$. The **product** of p by a scalar r is given by

$$rp = ra_0 + ra_1 x + ra_2 x^2 + \cdots + ra_n x^n.$$

Taking the usual notions of the zero polynomial and of $-p$, we recognize that the eight properties A1–A4 and S1–S4 required of a vector space are familiar properties for these polynomial operations. Thus, P is a vector space. ■

the set of all polynomials with coefficients in $\mathbb{R}$

EXAMPLE 7 Let F be the set of all real-valued functions of a real variable; that is, let F be the set of all functions mapping $\mathbb{R}$ into $\mathbb{R}$. The vector **sum** $f + g$ of two functions f and g in F is defined in the usual way to be the function whose value at any x in $\mathbb{R}$ is $f(x) + g(x)$; that is,

$$(f + g)(x) = f(x) + g(x).$$

For any scalar r in $\mathbb{R}$ and function f in F, the **product** rf is the function whose value at x is $rf(x)$, so that

$$(rf)(x) = rf(x).$$

Show that F with these operations is a vector space.

Solution   We observe that, for f and g in F, both $f + g$ and rf are functions mapping $\mathbb{R}$ into $\mathbb{R}$, so $f + g$ and rf are in F. Thus, F is closed under vector addition and under scalar multiplication. We take as zero vector in F the constant function whose value at each x in $\mathbb{R}$ is 0. For each function f in F, we take as $-f$ the function $(-1)f$ in F.

There are four vector-addition properties to verify, and they are all easy. We illustrate by verifying condition A4. For f in F, the function $f + (-f) = f + (-1)f$ has as value at x in $\mathbb{R}$ the number $f(x) + (-1)f(x)$, which is 0. Consequently, $f + (-f)$ is the zero function, and A4 is verified.

The scalar multiplicative properties are just as easy to verify. For example, to verify S4, we must compute $1f$ at any x in $\mathbb{R}$ and compare the result with $f(x)$. We obtain $(1f)(x) = 1f(x) = f(x)$, so $1f = f$.   ■

EXAMPLE 8   Let $\mathbb{R}^2$ have the usual operation of addition, but define scalar multiplication $r(x, y)$ by $r(x, y) = (0, 0)$. Determine whether $\mathbb{R}^2$ with these operations is a vector space.

Solution   Since conditions A1–A4 of Definition 2.1 do not involve scalar multiplication, and since addition is the usual operation, we need only check conditions S1–S4. We see that all of these hold, except for condition S4: since $1(x, y) = (0, 0)$, the scale is not preserved. Thus, $\mathbb{R}^2$ is not a vector space with these particular two operations.   ■

EXAMPLE 9   Let $\mathbb{R}^2$ have the usual scalar multiplication, but let addition $\curlyvee$ be defined on $\mathbb{R}^2$ by the formula

$$(x, y) \curlyvee (r, s) = (x + r, 2y + s).$$

Determine whether $\mathbb{R}^2$ with these operations is a vector space. (We use the symbol $\curlyvee$ for the warped addition of vectors in $\mathbb{R}^2$, to distinguish it from the usual addition.)

Solution   We check the associative law for $\curlyvee$:

$$[(x, y) \curlyvee (r, s)] \curlyvee (a, b) = (x + r, 2y + s) \curlyvee (a, b)$$
$$= ((x + r) + a, 2(2y + s) + b)$$
$$= (x + r + a, 4y + 2s + b),$$

while

$$(x, y) \curlyvee [(r, s) \curlyvee (a, b)] = (x, y) \curlyvee (r + a, 2s + b)$$
$$= (x + (r + a), 2y + (2s + b))$$
$$= (x + r + a, 2y + 2s + b).$$

Since the two colored scalars are not equal, we expect that $\curlyvee$ is not associative. We can find a specific violation of the associative law by choosing $y \neq 0$; for example,

$$[(0, 1) \curlyvee (0, 0)] \curlyvee (0, 0) = (0, 4),$$

while

$$(0, 1) \curlyvee [(0, 0) \curlyvee (0, 0)] = (0, 2).$$

Therefore, $\mathbb{R}^2$ is not a vector space with these two operations.   ■

We now indicate that vector addition and scalar multiplication possess still more of the properties we are accustomed to expect.

THEOREM 2.2 Elementary Properties of Vector Spaces

Every vector space V has the following properties:

(i) The vector $\mathbf{0}$ is the unique vector $\mathbf{x}$ satisfying the equation $\mathbf{x} + \mathbf{v} = \mathbf{v}$ for all vectors $\mathbf{v}$ in V.

(ii) For each vector $\mathbf{v}$ in V, the vector $-\mathbf{v}$ is the unique vector $\mathbf{y}$ satisfying $\mathbf{v} + \mathbf{y} = \mathbf{0}$.

(iii) If $\mathbf{u} + \mathbf{v} = \mathbf{u} + \mathbf{w}$ for vectors $\mathbf{u}$, $\mathbf{v}$, and $\mathbf{w}$ in V, then $\mathbf{v} = \mathbf{w}$.

(iv) $0\mathbf{v} = \mathbf{0}$ for all vectors $\mathbf{v}$ in V.

(v) $r\mathbf{0} = \mathbf{0}$ for all scalars r in $\mathbb{R}$.

(vi) $(-r)\mathbf{v} = r(-\mathbf{v}) = -(r\mathbf{v})$ for all scalars r in $\mathbb{R}$ and vectors $\mathbf{v}$ in V.

PROOF We prove only property (i), leaving proofs of the remaining properties as Exercises 30 through 34.

Suppose that vectors $\mathbf{0}$ and $\mathbf{0}'$ both satisfy the equation $\mathbf{x} + \mathbf{v} = \mathbf{v}$ for all $\mathbf{v}$ in V. We then obtain

$$\mathbf{0} + \mathbf{0}' = \mathbf{0}' \quad \text{and} \quad \mathbf{0}' + \mathbf{0} = \mathbf{0}.$$

By the commutative property A2, we know that $\mathbf{0} + \mathbf{0}' = \mathbf{0}' + \mathbf{0}$, and we conclude that $\mathbf{0} = \mathbf{0}'$. ▲

SUMMARY

1. A vector space is a <u>nonempty</u> set V of objects called *vectors*, together with rules for adding any two vectors $\mathbf{v}$ and $\mathbf{w}$ in V and for multiplying any vector $\mathbf{v}$ in V by any scalar r in $\mathbb{R}$. Furthermore, V must be closed under this vector addition and scalar multiplication so that $\mathbf{v} + \mathbf{w}$ and $r\mathbf{v}$ are both in V. Moreover, the following axioms must be satisfied for all vectors $\mathbf{u}$, $\mathbf{v}$, and $\mathbf{w}$ in V and all scalars r and s in $\mathbb{R}$:

A1 $(\mathbf{u} + \mathbf{v}) + \mathbf{w} = \mathbf{u} + (\mathbf{v} + \mathbf{w})$.

A2 $\mathbf{v} + \mathbf{w} = \mathbf{w} + \mathbf{v}$.

A3 There exists a zero vector $\mathbf{0}$ in V such that $\mathbf{0} + \mathbf{v} = \mathbf{v}$ for all vectors $\mathbf{v}$.

A4 Each vector $\mathbf{v}$ has an additive inverse $-\mathbf{v}$ in V such that $\mathbf{v} + (-\mathbf{v}) = \mathbf{0}$.

$$S1 \quad r(\mathbf{v} + \mathbf{w}) = r\mathbf{v} + r\mathbf{w}.$$
$$S2 \quad (r + s)\mathbf{v} = r\mathbf{v} + s\mathbf{v}.$$
$$S3 \quad r(s\mathbf{v}) = (rs)\mathbf{v}.$$
$$S4 \quad 1\mathbf{v} = \mathbf{v}.$$

2. $\mathbb{R}^n$ is a vector space for each positive integer n.

EXERCISES

1. Prove the indicated property of vector addition in $\mathbb{R}^n$, stated in Theorem 2.1.

a) The property A1
b) The property A3
c) The property A4

2. Prove the indicated property of scalar multiplication in $\mathbb{R}^n$, stated in Theorem 2.1.

a) The property S1
b) The property S3
c) The property S4

In Exercises 3–6, let $\mathbf{v} = (2, 1, 4)$ *and* $\mathbf{w} = (1, -3, 2)$. *Compute the indicated quantities in at least two ways.*

3. $4\mathbf{v} - 4\mathbf{w}$ **4.** $2\mathbf{v} - 5\mathbf{v}$

5. $[3\mathbf{v} - 2(\mathbf{v} + \mathbf{w})] + 3(\mathbf{w} - \mathbf{v})$

6. $\mathbf{v} + \mathbf{w} + \mathbf{v} + \mathbf{w} + \mathbf{v}$

In Exercises 7–13, determine whether the given set is closed under (a) addition and (b) scalar multiplication.

7. The line $y = 3x + 2$ in the plane $\mathbb{R}^2$

8. The line $y = 5x$ in the plane $\mathbb{R}^2$

9. The first and third quadrants in the plane $\mathbb{R}^2$, that is, all (x, y) in $\mathbb{R}^2$ such that $xy \geq 0$

10. The set of all upper-triangular 4×4 matrices

11. The set of all invertible 4×4 matrices

12. The set of all symmetric 4×4 matrices

13. The set of all vectors in $\mathbb{R}^3$ with middle component equal to zero

In Exercises 14–21, decide whether or not the given set, together with the indicated operations of addition and scalar multiplication, is a (real) vector space.

14. The set $\mathbb{R}^2$, with the usual addition but with scalar multiplication defined by $r(x, y) = (ry, rx)$.

15. The set $\mathbb{R}^2$, with the usual scalar multiplication but with addition defined by $(x, y) + (r, s) = (y + s, x + r)$.

16. The set $\mathbb{R}^2$, with addition defined by $(x, y) + (a, b) = (x + a + 1, y + b)$ and with scalar multiplication defined by $r(x, y) = (rx + r - 1, ry)$.

17. The set of all 2×2 matrices, with the usual scalar multiplication but with addition defined by $A + B = O$, the 2×2 zero matarix.

18. The set of all 2×2 matrices, with the usual addition but with scalar multiplication defined by $rA = O$, the 2×2 zero matrix.

19. The sct F of all functions mapping $\mathbb{R}$ into $\mathbb{R}$, with scalar multiplication defined as in Example 7 but with addition defined by $(f + g)(x) = \max\{f(x), g(x)\}$.

20. The set F of all functions mapping $\mathbb{R}$ into $\mathbb{R}$, with scalar multiplication defined as in Example 7 but with addition defined by $(f + g)(x) = f(x) + 2g(x)$.

21. The set F of all functions mapping $\mathbb{R}$ into $\mathbb{R}$, with scalar multiplication defined as in Example 7 but with addition defined by $(f + g)(x) = 2f(x) + 2g(x)$.

In Exercises 22–28, determine whether the given set is closed under the usual operations of addition and scalar multiplication, and is a (real) vector space.

22. The set of all upper-triangular $n \times n$ matrices.

23. The set of all 2×2 matrices of the form

$$\begin{bmatrix} X & 1 \\ 1 & X \end{bmatrix},$$

where each X may be any scalar.

24. The set of all diagonal $n \times n$ matrices.

25. The set of all 3×3 matrices of the form

$$\begin{bmatrix} X & 0 & X \\ 0 & X & 0 \\ X & 0 & X \end{bmatrix},$$

where each X may be any scalar.

26. The set $\{0\}$ consisting only of the number zero.

27. The set $\mathbb{Q}$ of all rational numbers.

28. The set $\mathbb{C}$ of complex numbers; that is

$$\mathbb{C} = \{a + b\sqrt{-1} \mid a, b \text{ in } \mathbb{R}\},$$

with the usual addition of complex numbers and with scalar multiplication defined in the usual way by $r(a + b\sqrt{-1}) = ra + rb\sqrt{-1}$ for any numbers a, b, and r in $\mathbb{R}$.

29. Mark each of the following True or False.

____ a) Matrix multiplication is a vector-space operation on the set $M_{m \times n}$ of all $m \times n$ matrices.

____ b) Matrix multiplication is a vector-space operation on the set M_n of all square $n \times n$ matrices.

____ c) Multiplication of any vector by the zero scalar always yields the zero vector.

____ d) Multiplication of a nonzero vector by a nonzero scalar never yields the zero vector.

____ e) No vector is its own additive inverse.

____ f) The zero vector is the only vector that is its own additive inverse.

____ g) Multiplication of two scalars is of no concern in the definition of a vector space.

____ h) One of the axioms for a vector space relates addition of scalars, multiplication of a vector by scalars, and addition of vectors.

____ i) Every vector space has at least two vectors.

____ j) Every vector space has at least one vector.

30. Prove property (ii) of Theorem 2.2.

31. Prove property (iii) of Theorem 2.2.

32. Prove property (iv) of Theorem 2.2.

33. Prove property (v) of Theorem 2.2.

34. Prove property (vi) of Theorem 2.2.

35. Your answer to Exercise 16 should be that $\mathbb{R}^2$ with the given operations is a vector space.

a) Describe the "zero vector" in this vector space.

b) Explain why the relations $r(0, 0) = (r - 1, 0) \neq (0, 0)$ do not violate part (v) of Theorem 2.2.

36. Let V be a vector space. Show that, if $\mathbf{v}$ is in V and if r is a scalar and if $r\mathbf{v} = \mathbf{0}$, then either $r = 0$ or $\mathbf{v} = \mathbf{0}$.

37. Let V be a vector space, and let $\mathbf{v}$ and $\mathbf{w}$ be vectors in V. Prove that there is a *unique* vector $\mathbf{x}$ in V such that $\mathbf{x} + \mathbf{v} = \mathbf{w}$.

2.3 LINEAR COMBINATIONS AND SUBSPACES

We have seen that $\mathbb{R}^n$ is a vector space, where vector addition consists of adding corresponding components of column (or row) vectors, and where scalar multiplication is achieved by multiplying each component by the scalar. Any subset of $\mathbb{R}^n$ that has the same operations and is also a vector space is called a *subspace* of $\mathbb{R}^n$. For example, the x_1, x_2-plane in $\mathbb{R}^3$ consisting of all vectors having zero as third entry is a subspace of $\mathbb{R}^3$. However, the subset $\{(m, n, p) \mid m, n, p \text{ any integers}\}$ of $\mathbb{R}^3$ consisting of all vectors in $\mathbb{R}^3$ with integer components is not a subspace; although this subset is closed under vector addition, it is not closed under scalar multiplication. For example, $0.5(1, 2, 5) = (0.5, 1, 2.5)$ is not in the subset.

DEFINITION 2.6   Subspace

A subset W of a vector space V is a **subspace** of V if W itself fulfills the requirements of a vector space, where addition and scalar multiplication of vectors in W produce the same vectors as these operations did in V.

In order for a nonempty subset W of a vector space V to be a subspace, the subset (together with the operations of vector addition and scalar multiplication) must form a self-contained system. That is, any addition or scalar multiplication using vectors in the subset W must always yield a vector that lies again in W. Then taking any $\mathbf{v}$ in W, we see that $0\mathbf{v} = \mathbf{0}$ and $(-1)\mathbf{v} = -\mathbf{v}$ are also in W. The eight properties A1–A4 and S1–S4 required of a vector space in Definition 2.5 are sure to be true for the subset, since they hold in all of V. That is, if W is nonempty and is closed under addition and scalar multiplication, it is sure to be a vector space in its own right. We have arrived at an efficient test for determining whether a subset is a subspace of a vector space.

THEOREM 2.3   Test for a Subspace

A nonempty subset W of a vector space V is a subspace of V if and only if it satisfies the following two conditions:

 (i) If $\mathbf{v}$ and $\mathbf{w}$ are in W, then $\mathbf{v} + \mathbf{w}$ is in W.

Closure under vector addition

 (ii) If r is any scalar in $\mathbb{R}$ and $\mathbf{v}$ is in W, then $r\mathbf{v}$ is in W.

Closure under scalar multiplication

As observed earlier, condition (ii) of the definition with $r = 0$ shows that the zero vector lies in every subspace. Consequently, a subspace of $\mathbb{R}^n$ always contains the origin.

The entire vector space V satisfies the conditions of Theorem 2.3. That is, V is a subspace of itself. Other subspaces of V are called **proper subspaces**. One such subspace is the subset $\{\mathbf{0}\}$, consisting of only the zero vector. We call $\{\mathbf{0}\}$ the **zero subspace** of V.

EXAMPLE 1   Verify that the x_1,x_2-plane in $\mathbb{R}^n$ is a subspace of $\mathbb{R}^n$ for $n \geq 2$.

Solution   The x_1,x_2-plane in $\mathbb{R}^n$ can be described as the set

$$W = \{\mathbf{x} \in \mathbb{R}^n \mid x_3 = x_4 = \cdots = x_n = 0\}.$$

Of course, W is nonempty. To prove that W is a subspace of $\mathbb{R}^n$, we choose $\mathbf{v}$, $\mathbf{w}$ in W, writing

$$\mathbf{v} = (v_1, v_2, 0, \ldots, 0) \quad \text{and} \quad \mathbf{w} = (w_1, w_2, 0, \ldots, 0).$$

For any scalar r, both

$$r\mathbf{v} = (rv_1, rv_2, 0, \ldots, 0)$$

and

$$\mathbf{v} + \mathbf{w} = (v_1 + w_1, v_2 + w_2, 0, \ldots, 0)$$

have the required zero entries and are therefore in W. Thus, W is indeed a subspace. ■

EXAMPLE 2   Verify that the line $y = 2x$ is a subspace of $\mathbb{R}^2$.

Solution   The line $y = 2x$ can be described as the set

$$L = \{(x, y) \mid y = 2x \text{ for any scalar } x\}$$

or more simply as $L = \{(x, 2x) \mid x \in \mathbb{R}\}$. We can see that L is nonempty. We choose two vectors $(a, 2a)$ and $(b, 2b)$ in L and proceed to compute $(a, 2a) + (b, 2b) = (a + b, 2(a + b))$, which has the form $(x, 2x)$ and consequently is in L. Similarly, $r(a, 2a) = (ra, 2ra)$ is in L for any scalar r. Thus L is a subspace.

The subspace L is shown in Figure 2.19. Notice that it is a line through the origin. Similarly, every line through the origin is a subspace of $\mathbb{R}^2$. ■

Let A be an $m \times n$ matrix, and regard $\mathbb{R}^n$ as a space of column vectors. The solution set of the homogeneous system $A\mathbf{x} = \mathbf{0}$ is a subspace of $\mathbb{R}^n$, by Theorem 1.4. It is called the **nullspace** of the matrix A, and is denoted by N_A. The nullspace of A is one of several very important spaces associated with a matrix A. We box its description for easy reference.

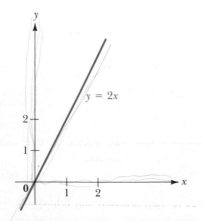

FIGURE 2.19   The subspace of $\mathbb{R}^2$ where $y = 2x$.

The Nullspace of a Matrix

Let A be an $m \times n$ matrix. The nullspace N_A of A is the solution space of the homogeneous system $A\mathbf{x} = \mathbf{0}$. That is,

$$N_A = \{\mathbf{x} \in \mathbb{R}^n \mid A\mathbf{x} = \mathbf{0}\}.$$

For example, the line in Example 2 is the nullspace of the 1×2 matrix $[-2 \quad 1]$, since the equation $y = 2x$ can be expressed as

$$[-2 \quad 1]\begin{bmatrix} x \\ y \end{bmatrix} = \mathbf{0}.$$

EXAMPLE 3 Show that, for each integer $n \geq 0$, the set P_n of all polynomials of degree at most n, together with the zero polynomial, is a subspace of the vector space P of Example 6 in Section 2.2.

Solution The set P_n is nonempty since the zero polynomial is in P_n. The sum $p + q$ of two polynomials p and q of degrees $\leq n$ is again a polynomial of degree $\leq n$, and the product rp of a polynomial p in P_n times a scalar r is a polynomial of degree at most n. Thus P_n is closed under addition and scalar multiplication, so it is a subspace of P. ▪

EXAMPLE 4 (*Calculus*) Show that the set W of all differentiable functions mapping $\mathbb{R}$ into $\mathbb{R}$ is a subspace of the space F of all functions mapping $\mathbb{R}$ into $\mathbb{R}$ described in Example 7 of Section 2.2.

Solution We know that W is not empty. If f and g are two functions in W, then f and g are differentiable. We know that $f + g$ is also differentiable; in fact, its derivative is $(f + g)' = f' + g'$. Therefore, $f + g$ is in W. Similarly, if f is in W, so is any scalar multiple rf; the derivative of rf is $(rf)' = rf'$. Thus, W is a subspace of F. ▪

The Subspace Spanned (or Generated) by Vectors

Examples 1 and 2 might suggest that any line or plane in $\mathbb{R}^3$ is a subspace. This cannot be the case, since a subspace of $\mathbb{R}^3$ must contain the zero vector $\mathbf{0}$. However, we will show that any line or plane containing $\mathbf{0}$ is a subspace of $\mathbb{R}^3$.

Geometrically, we see that a plane in $\mathbb{R}^3$ through the origin is completely determined by any two nonzero and nonparallel vectors $\mathbf{v}_1$ and $\mathbf{v}_2$ that lie in it, as illustrated in Figure 2.20. The figure indicates that any vector in this plane has the form $r_1\mathbf{v}_1 + r_2\mathbf{v}_2$. This expression is a **linear combination** of the vectors $\mathbf{v}_1$ and $\mathbf{v}_2$. For example, $(8, -7, 7) = 2(1, 1, 2) - 3(-2, 3, -1)$ is a linear combination of $(1, 1, 2)$ and $(-2, 3, -1)$.

More generally, let $\mathbf{v}_1$ and $\mathbf{v}_2$ be two nonzero and nonparallel vectors in $\mathbb{R}^n$, and consider the set of all linear combinations of $\mathbf{v}_1$ and $\mathbf{v}_2$. We write this set as

$$\mathrm{sp}(\mathbf{v}_1, \mathbf{v}_2) = \{r_1\mathbf{v}_1 + r_2\mathbf{v}_2 \mid r_1, r_2 \in \mathbb{R}\},$$

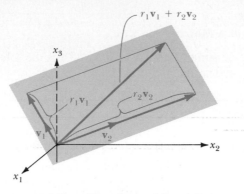

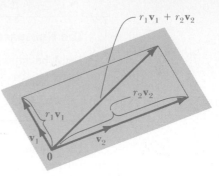

FIGURE 2.20 The plane sp(v_1, v_2) in $\mathbb{R}^3$.

FIGURE 2.21 The plane sp(v_1, v_2) in $\mathbb{R}^n$.

called the **span** of $\{v_1, v_2\}$ or the **span** of v_1 and v_2. Referring to Figure 2.21, we see that every linear combination of v_1 and v_2 corresponds to some point in the plane determined by the vectors v_1 and v_2.

We claim that $S = \text{sp}(v_1, v_2)$ is a subspace of $\mathbb{R}^n$. Since S is evidently nonempty, we simply check the closure conditions of Theorem 2.3:

1. $(r_1v_1 + r_2v_2) + (s_1v_1 + s_2v_2) = (r_1 + s_1)v_1 + (r_2 + s_2)v_2$, which shows that $\text{sp}(v_1, v_2)$ is closed under addition; and

2. $s(r_1v_1 + r_2v_2) = (sr_1)v_1 + (sr_2)v_2$, which shows that $\text{sp}(v_1, v_2)$ is closed under scalar multiplication.

The preceding argument is valid for any vectors v_1 and v_2 in $\mathbb{R}^n$. In particular, if v_1 and v_2 are parallel, so that $v_1 = sv_2$ for some scalar s in $\mathbb{R}$, then $\text{sp}(v_1, v_2)$ in Figure 2.21 reduces to a line that contains both vectors v_1 and v_2. In this case, $\text{sp}(v_1, v_2) = \text{sp}(v_1)$.

Since a subspace of $\mathbb{R}^n$ must be closed under addition and scalar multiplication, every subspace containing vectors v_1 and v_2 must contain $\text{sp}(v_1, v_2)$. It is clear that $\text{sp}(v_1, v_2)$ is the *smallest* subspace of $\mathbb{R}^n$ containing both v_1 and v_2. This subspace is *generated* or *spanned* by these two vectors.

EXAMPLE 5   Describe geometrically the subspace of $\mathbb{R}^4$ generated by the vector $v = (2, 3, -1, 4)$.

Solution   The subspace sp(v) is a *line* in $\mathbb{R}^4$ containing the origin. Each point on this line is a scalar multiple rv of v; that is, each point has the form (x_1, x_2, x_3, x_4), where $x_1 = 2r$, $x_2 = 3r$, $x_3 = -r$, $x_4 = 4r$ for some scalar r.   ■

The computations preceding Example 5 that show sp(v_1, v_2) to be a subspace of $\mathbb{R}^n$ can be carried out just as easily with linear combinations of more than two vectors in a general vector space, proving the following theorem.

THEOREM 2.4   Subspace Generated or Spanned by Vectors

Let $\mathbf{v}_1, \mathbf{v}_2, \ldots, \mathbf{v}_k$ be vectors in a vector space V. The set

$$sp(\mathbf{v}_1, \mathbf{v}_2, \ldots, \mathbf{v}_k) = \{r_1\mathbf{v}_1 + r_2\mathbf{v}_2 + \cdots + r_k\mathbf{v}_k \mid r_i \in \mathbb{R}\}$$

of all linear combinations of these vectors $\mathbf{v}_i$ is a subspace of V.

the subspace of V
spanned by $v_1, \ldots, v_k$.

As the title of Theorem 2.4 indicates, the subspace $W - sp(\mathbf{v}_1, \mathbf{v}_2, \ldots, \mathbf{v}_k)$ of V is referred to as the subspace of V **generated** or **spanned** by the vectors $\mathbf{v}_1, \mathbf{v}_2, \ldots, \mathbf{v}_k$. The set $\{\mathbf{v}_1, \mathbf{v}_2, \ldots, \mathbf{v}_k\}$ is a **spanning** or **generating set** for W.

Any subset, finite or infinite, of a vector space V is a **spanning** or **generating set** for V if each vector in V is a linear combination of a *finite* number of vectors in the set. If V has a finite generating set, then V is **finitely generated**. Thus a vector space V is finitely generated if and only if it has the form $V = sp(\mathbf{v}_1, \mathbf{v}_2, \ldots, \mathbf{v}_n)$ for some finite number n of vectors in V.

EXAMPLE 6   Describe geometrically the subspace of $\mathbb{R}^3$ spanned by $\mathbf{v}_1 = (1, 2, 3)$ and $\mathbf{v}_2 = (2, 3, -1)$ in $\mathbb{R}^3$.

Solution   The subspace $sp(\mathbf{v}_1, \mathbf{v}_2)$ is a plane in $\mathbb{R}^3$, part of which is shown in Figure 2.22. Each point in this plane has the form

$$(x_1, x_2, x_3) = r(1, 2, 3) + s(2, 3, -1),$$

so $x_1 = r + 2s$, $x_2 = 2r + 3s$, and $x_3 = 3r - s$ for any scalars r and s.   ■

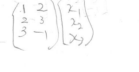

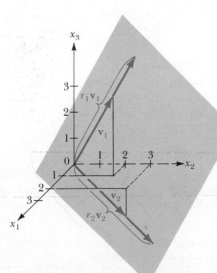

FIGURE 2.22   The plane $sp(\mathbf{v}_1, \mathbf{v}_2)$.

EXAMPLE 7   Describe the subspace $sp(1, x, x^2)$ of the vector space P of all polynomials with coefficents in $\mathbb{R}$.

Solution   Since $sp(1, x, x^2) = \{a + bx + cx^2 \mid a, b, c \in \mathbb{R}\}$, the space $sp(1, x, x^2)$ is the space P_2 of all polynomials of degree ≤ 2.   ■

As illustrated in Example 7, the subspace $sp(1, x, x^2, \ldots, x^n)$ of P is the space P_n of all polynomials of degree $\leq n$. The vector space P is an example of a vector space that is not finitely generated. No finite set of polynomials can possibly generate all of P, since P contains polynomials of arbitrarily high degree. The space P is generated by the infinite set $\{1, x, x^2, x^3, \ldots\}$.

EXAMPLE 8   Find a set of vectors in $\mathbb{R}^4$ that spans the solution space of the homogeneous system

$$
\begin{aligned}
x_1 + x_2 + 3x_3 + x_4 &= 0 \\
2x_1 + 3x_2 + x_3 + x_4 &= 0 \\
x_1 \qquad\quad + 8x_3 + 2x_4 &= 0.
\end{aligned}
$$

Solution   We solve the system of equations by reducing its coefficient matrix, using the Gauss-Jordan method. We have

$$
\begin{bmatrix} 1 & 1 & 3 & 1 \\ 2 & 3 & 1 & 1 \\ 1 & 0 & 8 & 2 \end{bmatrix} \sim \begin{bmatrix} 1 & 1 & 3 & 1 \\ 0 & 1 & -5 & -1 \\ 0 & -1 & 5 & 1 \end{bmatrix} \sim \begin{bmatrix} 1 & 0 & 8 & 2 \\ 0 & 1 & -5 & -1 \\ 0 & 0 & 0 & 0 \end{bmatrix}.
$$

The system has a solution space described by

$$
\mathbf{x} = \begin{bmatrix} x_1 \\ x_2 \\ x_3 \\ x_4 \end{bmatrix} = \begin{bmatrix} -8r - 2s \\ 5r + s \\ r \\ s \end{bmatrix} = r\begin{bmatrix} -8 \\ 5 \\ 1 \\ 0 \end{bmatrix} + s\begin{bmatrix} -2 \\ 1 \\ 0 \\ 1 \end{bmatrix} \tag{1}
$$

for any scalars r and s. The vectors

$$
\mathbf{v}_1 = \begin{bmatrix} -8 \\ 5 \\ 1 \\ 0 \end{bmatrix} \quad \text{and} \quad \mathbf{v}_2 = \begin{bmatrix} -2 \\ 1 \\ 0 \\ 1 \end{bmatrix}
$$

are obtained from the column vector after the second equal sign in Eq. (1) if we choose $r = 1$, $s = 0$ for $\mathbf{v}_1$ and $r = 0$, $s = 1$ for $\mathbf{v}_2$. Thus the solution space for the given homogeneous system is the subspace $sp(\mathbf{v}_1, \mathbf{v}_2)$ of $\mathbb{R}^4$ spanned by $\mathbf{v}_1$ and $\mathbf{v}_2$. This is also the nullspace of the coefficient matrix with which we started.   ■

The Column Space and the Row Space of a Matrix

In addition to discussing the nullspace of an $m \times n$ matrix A, we present two other subspaces related to the matrix. If we regard $\mathbb{R}^n$ as a space of row vectors, the subspace of $\mathbb{R}^n$ generated by the m rows of A is the **row space** of A. If we regard $\mathbb{R}^m$ as a space of column vectors, the subspace of $\mathbb{R}^m$ generated by the n columns of A is the

column space of A. The column space of A holds the key to the existence of a solution of a linear system $A\mathbf{x} = \mathbf{b}$. We can write $A\mathbf{x} = \mathbf{b}$ in the form

$$x_1 \begin{bmatrix} a_{11} \\ a_{21} \\ \vdots \\ a_{m1} \end{bmatrix} + x_2 \begin{bmatrix} a_{12} \\ a_{22} \\ \vdots \\ a_{m2} \end{bmatrix} + \cdots + x_n \begin{bmatrix} a_{1n} \\ a_{2n} \\ \vdots \\ a_{mn} \end{bmatrix} = \begin{bmatrix} b_1 \\ b_2 \\ \vdots \\ b_m \end{bmatrix}. \tag{2}$$

As $x_1, x_2, \ldots, x_n$ vary over all scalars in $\mathbb{R}$, the left-hand side of Eq. (2) runs through the entire column space of A.

The Column Space of a Matrix

Let A be an $m \times n$ matrix. The column space of A is the set of all vectors $\mathbf{b}$ in $\mathbb{R}^m$ of the form given in Eq. (2). That is, the column space of A is

$$\{A\mathbf{x} \mid \mathbf{x} \in \mathbb{R}^n\}.$$

We obtain the following criterion at once for the existence of a solution of $A\mathbf{x} = \mathbf{b}$.

THEOREM 2.5   Existence of Solutions

A linear system $A\mathbf{x} = \mathbf{b}$ has a solution if and only if $\mathbf{b}$ lies in the column space of A.

EXAMPLE 9   Determine whether $\mathbf{b} = (7, 6, 1)$ lies in $\mathrm{sp}(\mathbf{v}_1, \mathbf{v}_2, \mathbf{v}_3)$, where $\mathbf{v}_1 = (1, 2, 3)$, $\mathbf{v}_2 = (-2, -5, 2)$, and $\mathbf{v}_3 = (1, 2, -1)$.

Solution   If we use column-vector notation, we see that we must try to solve the system of equations

$$x_1 \begin{bmatrix} 1 \\ 2 \\ 3 \end{bmatrix} + x_2 \begin{bmatrix} -2 \\ -5 \\ 2 \end{bmatrix} + x_3 \begin{bmatrix} 1 \\ 2 \\ -1 \end{bmatrix} = \begin{bmatrix} 7 \\ 6 \\ 1 \end{bmatrix}. \tag{3}$$

Reducing the partitioned matrix for this system, we obtain

$$\begin{bmatrix} 1 & -2 & 1 & | & 7 \\ 2 & -5 & 2 & | & 6 \\ 3 & 2 & -1 & | & 1 \end{bmatrix} \sim \begin{bmatrix} 1 & -2 & 1 & | & 7 \\ 0 & -1 & 0 & | & -8 \\ 0 & 8 & -4 & | & -20 \end{bmatrix} \sim \begin{bmatrix} 1 & -2 & 1 & | & 7 \\ 0 & -1 & 0 & | & -8 \\ 0 & 0 & -4 & | & -84 \end{bmatrix}.$$

Since the linear system is consistent, $\mathbf{b}$ is in $\mathrm{sp}(\mathbf{v}_1, \mathbf{v}_2, \mathbf{v}_3)$. ▪

An important problem involves determining whether n vectors $\mathbf{v}_1, \mathbf{v}_2, \ldots, \mathbf{v}_n$ generate all of $\mathbb{R}^n$. To illustrate, we claim that the vectors $\mathbf{v}_1, \mathbf{v}_2$, and $\mathbf{v}_3$ in Example 9 generate all of $\mathbb{R}^3$. To verify this, we solve system (3) but using a general vector

$\mathbf{b} = (b_1, b_2, b_3)$ in $\mathbb{R}^3$ in place of $(7, 6, 1)$. We know that we will obtain a solution again, since the coefficient matrix has not changed. We need not find the solution explicitly to conclude that $\text{sp}(\mathbf{v}_1, \mathbf{v}_2, \mathbf{v}_3) = \mathbb{R}^3$.

To test whether n vectors $\mathbf{v}_1, \mathbf{v}_2, \ldots, \mathbf{v}_n$ in $\mathbb{R}^n$ generate $\mathbb{R}^n$, we form the matrix A whose jth column vector is $\mathbf{v}_j$. By Theorem 1.8, we know that $A\mathbf{x} = \mathbf{b}$ has a solution for every choice of $\mathbf{b}$ in $\mathbb{R}^n$ if and only if A is invertible. Thus every $\mathbf{b}$ in $\mathbb{R}^n$ lies in the column space of A if and only if A is invertible. We rephrase this result in the following box.

Criterion for n Vectors to Generate $\mathbb{R}^n$

Vectors $\mathbf{v}_1, \mathbf{v}_2, \ldots, \mathbf{v}_n$ in $\mathbb{R}^n$ generate all of $\mathbb{R}^n$ if and only if the $n \times n$ matrix A having $\mathbf{v}_1, \mathbf{v}_2, \ldots, \mathbf{v}_n$ as column vectors is invertible.

EXAMPLE 10   Test whether the vectors $(2, 1, 3)$, $(-1, 2, 0)$, and $(1, 8, 6)$ generate $\mathbb{R}^3$.

Solution   Following the preceding criterion, we form the matrix A having these vectors as column vectors, and then we row-reduce A to see if it is invertible. We obtain

$$A = \begin{bmatrix} 2 & -1 & 1 \\ 1 & 2 & 8 \\ 3 & 0 & 6 \end{bmatrix} \sim \begin{bmatrix} 1 & 2 & 8 \\ 2 & -1 & 1 \\ 3 & 0 & 6 \end{bmatrix} \sim \begin{bmatrix} 1 & 2 & 8 \\ 0 & -5 & -15 \\ 0 & -6 & -18 \end{bmatrix}$$

$$\sim \begin{bmatrix} 1 & 2 & 8 \\ 0 & 1 & 3 \\ 0 & 0 & 0 \end{bmatrix} = U.$$

Thus, A is not invertible, so the given vectors do not generate all of $\mathbb{R}^3$.   ■

This criterion shows at once that a set of fewer than n vectors does not generate $\mathbb{R}^n$. For if $\mathbf{v}_1, \mathbf{v}_2, \ldots, \mathbf{v}_k$ generated $\mathbb{R}^n$ and if $k < n$, then the $n \times n$ matrix having column vectors $\mathbf{v}_1, \mathbf{v}_2, \ldots, \mathbf{v}_k, \mathbf{0}, \ldots, \mathbf{0}$, using $n - k$ zero vectors, would be invertible, which is not the case.

On the other hand, for $k > n$ we can determine whether a set $\{\mathbf{v}_1, \mathbf{v}_2, \ldots, \mathbf{v}_k\}$ of vectors in $\mathbb{R}^n$ generates $\mathbb{R}^n$, by forming the rectangular matrix A having all of these vectors as columns and then reducing the matrix to row echelon form H. If the row echelon form contains n pivots, the columns of A corresponding to the pivot columns of H must generate $\mathbb{R}^n$, by the boxed criterion preceding Example 10. If there are fewer than n pivots in H, the bottom row of H consists of zeros and the system $H\mathbf{x} = \mathbf{c}$ will be inconsistent for any vector $\mathbf{c}$ in $\mathbb{R}^n$ whose last entry is nonzero. The partitioned matrix $[H \mid \mathbf{c}]$ is row equivalent to a partitioned matrix $[A \mid \mathbf{b}]$, which then represents an inconsistent system $A\mathbf{x} = \mathbf{b}$. Thus, $\mathbf{b}$ is not in the column space of A; that is, the column vectors of A do not generate $\mathbb{R}^n$.

We have obtained the following criterion.

> **Criterion for *k* Vectors to Generate $\mathbb{R}^n$**
>
> Vectors $\mathbf{v}_1, \mathbf{v}_2, \ldots, \mathbf{v}_k$ in $\mathbb{R}^n$ generate $\mathbb{R}^n$ if and only if the $n \times k$ matrix A having $\mathbf{v}_1, \mathbf{v}_2, \ldots, \mathbf{v}_k$ as column vectors is row-equivalent to a matrix in row echelon form containing n pivots.

EXAMPLE 11 Test whether the vectors $(1, 2, 0)$, $(1, 1, 1)$, $(1, 4, -2)$, and $(2, 3, 2)$ generate $\mathbb{R}^3$.

Solution We form the matrix A having these vectors as column vectors, and then we reduce it to row echelon form, obtaining

$$A = \begin{bmatrix} 1 & 1 & 1 & 2 \\ 2 & 1 & 4 & 3 \\ 0 & 1 & -2 & 2 \end{bmatrix} \sim \begin{bmatrix} 1 & 1 & 1 & 2 \\ 0 & -1 & 2 & -1 \\ 0 & 1 & -2 & 2 \end{bmatrix} \sim \begin{bmatrix} 1 & 1 & 1 & 2 \\ 0 & 1 & -2 & 1 \\ 0 & 0 & 0 & 1 \end{bmatrix}.$$

Since in row echelon form the matrix contains three pivots, the given list of vectors generates $\mathbb{R}^3$. Note that the third vector in the list is not needed to generate $\mathbb{R}^3$ since no pivot appears in the third column of the final matrix. ■

SUMMARY

1. A subset W of a vector space V is a subspace of V if and only if it is nonempty and satisfies the two closure properties:

   $\mathbf{v} + \mathbf{w}$ is contained in W for all vectors $\mathbf{v}$ and $\mathbf{w}$ in W; and

   $r\mathbf{v}$ is contained in W for all vectors $\mathbf{v}$ in W and all scalars r.

2. Let $\mathbf{v}_1, \mathbf{v}_2, \ldots, \mathbf{v}_k$ be vectors in a vector space V. The set $\mathrm{sp}(\mathbf{v}_1, \mathbf{v}_2, \ldots, \mathbf{v}_k)$ of all linear combinations of the $\mathbf{v}_i$ is a subspace of V called the *span* of the $\mathbf{v}_i$ or the subspace *generated* by the $\mathbf{v}_i$. It is the smallest subspace of V containing all the vectors $\mathbf{v}_i$.

3. A vector space V is finitely generated if $V = \mathrm{sp}(\mathbf{v}_1, \mathbf{v}_2, \ldots, \mathbf{v}_n)$ for some vectors $\mathbf{v}_i \in V$.

4. The nullspace of an $m \times n$ matrix A is the subspace of $\mathbb{R}^n$ consisting of all solutions of $A\mathbf{x} = \mathbf{0}$.

5. Let A be an $m \times n$ matrix. The row space of A is the subspace of $\mathbb{R}^n$ spanned by the m row vectors of A, while the column space is the subspace of $\mathbb{R}^m$ spanned by the n column vectors of A.

6. The linear system $A\mathbf{x} = \mathbf{b}$ has a solution if and only if $\mathbf{b}$ lies in the column space of A.

7. Vectors $\mathbf{v}_1, \mathbf{v}_2, \ldots, \mathbf{v}_n$ in $\mathbb{R}^n$ span all of $\mathbb{R}^n$ if and only if the matrix A having these vectors as column vectors is invertible.

EXERCISES

In Exercises 1–18, determine whether the indicated subset is a subspace of the given vector space.

1. $\{(r, -r) \mid r \in \mathbb{R}\}$ in $\mathbb{R}^2$

2. $\{(x, x + 1) \mid x \in \mathbb{R}\}$ in $\mathbb{R}^2$

3. $\{(n, m) \mid n \text{ and } m \text{ are integers}\}$ in $\mathbb{R}^2$

4. $\{(x, y) \mid x, y \in \mathbb{R} \text{ and } x, y \geq 0\}$ (the first quadrant of $\mathbb{R}^2$)

5. $\{(x, y) \mid x, y \in \mathbb{R}, \text{ and } x, y \geq 0 \text{ or } x, y \leq 0\}$ (the first and third quadrants of $\mathbb{R}^2$)

6. $\{(x, y, z) \mid x, y, z \in \mathbb{R} \text{ and } z = 3x + 2\}$ in $\mathbb{R}^3$

7. The set of all polynomials of degree greater than 3 in the vector space P of all polynomials with coefficients in $\mathbb{R}$

8. $\{(x, y, z) \mid x, y, z \in \mathbb{R} \text{ and } x = 2y + z\}$ in $\mathbb{R}^3$

9. $\{(x, y, z) \mid x, y, z \in \mathbb{R} \text{ and } z = 1, y = 2x\}$ in $\mathbb{R}^3$

10. $\{(2x, x + y, y) \mid x, y \in \mathbb{R}\}$ in $\mathbb{R}^3$

11. $\{(2x_1, 3x_2, 4x_3, 5x_4) \mid x_i \in \mathbb{R}\}$ in $\mathbb{R}^4$

12. The set of all upper-triangular matrices in the space M_n of all $n \times n$ matrices

13. The set of all invertible matrices in the space M_n of all $n \times n$ matrices

14. $\{(x_1, x_2, \ldots, x_n) \mid x_i \in \mathbb{R}, x_2 = 0\}$ in $\mathbb{R}^n$

15. The set of all functions f such that $f(0) = 1$ in the vector space F of all functions mapping $\mathbb{R}$ into $\mathbb{R}$

16. The set of all functions f such that $f(1) = 0$ in the vector space F of all functions mapping $\mathbb{R}$ into $\mathbb{R}$

17. (*Calculus*) The set of all functions f in the vector space W of differentiable functions mapping $\mathbb{R}$ into $\mathbb{R}$ (see Example 4) such that $f'(2) = 0$

18. (*Calculus*) The set of all functions f in the vector space W of differentiable functions mapping $\mathbb{R}$ into $\mathbb{R}$ (see Example 4) such that f has derivatives of all orders

In Exercises 19–21, let $\mathbf{v}_1 = (1, 3, 4)$, $\mathbf{v}_2 = (2, 7, 2)$, and $\mathbf{v}_3 = (-1, 2, 1)$.

19. Determine whether $(-3, 1, -2)$ is in $\text{sp}(\mathbf{v}_1, \mathbf{v}_2, \mathbf{v}_3)$.

20. Determine whether $(2, 1, 1)$ is in $\text{sp}(\mathbf{v}_1, \mathbf{v}_2, \mathbf{v}_3)$.

21. Determine whether $\text{sp}(\mathbf{v}_1, \mathbf{v}_2, \mathbf{v}_3) = \mathbb{R}^3$.

22. Determine whether $\text{sp}((1, 2, 1), (2, 1, 0)) = \mathbb{R}^3$.

23. Determine whether the vectors $(1, 2, 1)$, $(2, 1, 3)$, $(3, 3, 4)$, and $(-1, 2, 0)$ generate $\mathbb{R}^3$.

24. Determine whether the vectors $(1, 2, 1)$, $(2, 1, 3)$, and $(3, 3, 4)$ generate $\mathbb{R}^3$.

25. Let F be the vector space of functions mapping $\mathbb{R}$ into $\mathbb{R}$. Show that

   a) $\text{sp}(\sin^2 x, \cos^2 x)$ contains all constant functions,

   b) $\text{sp}(\sin^2 x, \cos^2 x)$ contains the function $\cos 2x$,

   c) $\text{sp}(7, \sin^2 2x)$ contains the function $8 \cos 4x$.

In Exercises 26–29, express the given vector $\mathbf{b}$ as a linear combination of the other vectors $\mathbf{v}_i$ in the given space $\mathbb{R}^n$, if possible.

26. $\mathbf{b} = (1, 2)$, $\mathbf{v}_1 = (3, 1)$, $\mathbf{v}_2 = (1, 3)$ in $\mathbb{R}^2$

27. $\mathbf{b} = (1, 2, 1)$, $\mathbf{v}_1 = (1, 1, 1)$, $\mathbf{v}_2 = (1, 1, 2)$, $\mathbf{v}_3 = (0, 1, 1)$ in $\mathbb{R}^3$

28. $\mathbf{b} = (1, 3, 1)$, $\mathbf{v}_1 = (2, 1, 1)$, $\mathbf{v}_2 = (1, 1, 2)$, $\mathbf{v}_3 = (3, 1, 0)$ in $\mathbb{R}^3$

29. $\mathbf{b} = (1, 2, 3)$, $\mathbf{v}_1 = (2, 2, 2)$, $\mathbf{v}_2 = (1, 0, 1)$ in $\mathbb{R}^3$

30. Let P be the vector space of polynomials. Show that $\text{sp}(1, x) = \text{sp}(1 + 2x, x)$. [HINT: Show that each of these subspaces is a subset of the other.]

31. Let V be a vector space, and let $\mathbf{v}_1$ and $\mathbf{v}_2$ be vectors in V. Follow the hint of Exercise 30 to show that

   a) $\text{sp}(\mathbf{v}_1, \mathbf{v}_2) = \text{sp}(\mathbf{v}_1, 2\mathbf{v}_1 + \mathbf{v}_2)$,

   b) $\text{sp}(\mathbf{v}_1, \mathbf{v}_2) = \text{sp}(\mathbf{v}_1 + \mathbf{v}_2, \mathbf{v}_1 - \mathbf{v}_2)$.

32. Let $\mathbf{v}_1, \mathbf{v}_2, \ldots, \mathbf{v}_k$ and $\mathbf{w}_1, \mathbf{w}_2, \ldots, \mathbf{w}_m$ be vectors in a vector space V. Give a necessary and sufficient condition, involving linear combinations, for

$$\text{sp}(\mathbf{v}_1, \mathbf{v}_2, \ldots, \mathbf{v}_k) = \text{sp}(\mathbf{w}_1, \mathbf{w}_2, \ldots, \mathbf{w}_m).$$

33. Consider the homogeneous system of equations

$$4x_1 + 9x_2 - x_3 + 9x_4 = 0$$
$$x_1 + 2x_2 - x_3 + 3x_4 = 0$$
$$2x_1 + 5x_2 + x_3 + 3x_4 = 0$$
$$x_1 + x_2 - 4x_3 + 6x_4 = 0,$$

which has solutions $\mathbf{v} = (7, -3, 1, 0)$ and $\mathbf{w} = (-9, 3, 0, 1)$. Show that each linear combination of $\mathbf{v}$ and $\mathbf{w}$ is also a solution to the system.

34. Referring to Exercise 33, verify that the solution space of the given system of equations is $\mathrm{sp}(\mathbf{v}, \mathbf{w})$, using a Gauss–Jordan reduction.

35. Mark each of the following True or False.

___ a) The set consisting of the zero vector is a subspace of every vector space.

___ b) Every vector space has at least two distinct subspaces.

___ c) Every vector space with a nonzero vector has at least two distinct subspaces.

___ d) If $\{\mathbf{v}_1, \mathbf{v}_2, \ldots, \mathbf{v}_n\}$ is a subset of a vector space V, then $\mathbf{v}_i$ is in $\mathrm{sp}(\mathbf{v}_1, \mathbf{v}_2, \ldots, \mathbf{v}_n)$ for $i = 1, 2, \ldots, n$.

___ e) If $\{\mathbf{v}_1, \mathbf{v}_2, \ldots, \mathbf{v}_n\}$ is a subset of a vector space V, then the sum $\mathbf{v}_i + \mathbf{v}_j$ is in $\mathrm{sp}(\mathbf{v}_1, \mathbf{v}_2, \ldots, \mathbf{v}_n)$ for all choices of i and j from 1 to n.

___ f) If $\mathbf{u} + \mathbf{v}$ lies in a subspace W of a vector space V, then both $\mathbf{u}$ and $\mathbf{v}$ lie in W.

___ g) Two subspaces of a vector space V may have empty intersection.

___ h) Every line in $\mathbb{R}^2$ is a subspace of $\mathbb{R}^2$ generated by a single vector.

___ i) Every line through the origin in $\mathbb{R}^2$ is a subspace of $\mathbb{R}^2$ generated by a single vector.

___ j) Every generating set for $\mathbb{R}^2$ must contain at least two vectors.

36. Let W_1 and W_2 be subspaces of a vector space V. Show that the intersection $W_1 \cap W_2$ is also a subspace of V.

37. Let $W_1 = \mathrm{sp}((1, 2, 3), (2, 1, 1))$ and $W_2 = \mathrm{sp}((1, 0, 1), (3, 0, -1))$ in $\mathbb{R}^3$. Find a set of generating vectors for $W_1 \cap W_2$.

In Exercises 38–45, find as small a set of vectors as you can that generates the solution space of the given homogeneous system of equations.

38. $x - y = 0$
$2x - 2y = 0$

39. $2x + 5y = 0$
$3x + y = 0$

40. $3x_1 - 2x_2 + x_3 = 0$
$2x_1 + x_2 - 5x_3 = 0$
$x_1 + x_2 - 6x_3 = 0$

41. $3x_1 + x_2 + x_3 = 0$
$6x_1 + 2x_2 + 2x_3 = 0$
$-9x_1 - 3x_2 - 3x_3 = 0$

42. $x_1 - x_2 + x_3 - x_4 = 0$
$x_2 + x_3 = 0$
$x_1 + 2x_2 - x_3 + 3x_4 = 0$

43. $2x_1 + x_2 + x_3 + x_4 = 0$
$x_1 - 6x_2 + x_3 = 0$
$3x_1 - 5x_2 + 2x_3 + x_4 = 0$
$5x_1 - 4x_2 + 3x_3 + 2x_4 = 0$

44. $2x_1 + x_2 + x_3 + x_4 = 0$
$3x_1 + x_2 - x_3 + 2x_4 = 0$
$x_1 + x_2 + 3x_3 = 0$
$x_1 - x_2 - 7x_3 + 2x_4 = 0$

45. $x_1 - x_2 + 6x_3 + x_4 - x_5 = 0$
$3x_1 + 2x_2 - 3x_3 + 2x_4 + 5x_5 = 0$
$4x_1 + 2x_2 - x_3 + 3x_4 - x_5 = 0$
$3x_1 - 2x_2 + 14x_3 + x_4 - 8x_5 = 0$
$2x_1 - x_2 + 8x_3 + 2x_4 - 7x_5 = 0$

46. Use the second option with the program VECTGRPH for graphic drill on linear combinations until you regularly attain a score of at least 80%.

In Exercises 47–50, use YUREDUCE or MATCOMP or similar software.

47. Determine whether the vectors $(1, 3, 2, 1, 4)$, $(0, 2, 1, 2, 2)$, $(0, 0, 3, 1, 1)$, $(0, 0, 0, 4, 1)$, and $(0, 0, 0, 0, 2)$ generate $\mathbb{R}^5$.

48. Determine whether the vectors $(2, 1, 3, -4, 0)$, $(1, 1, 1, -1, 1)$, $(4, -2, -3, 1, 7)$, $(4, -3, -2, -1, 5)$, $(6, -3, 1, 2, 4)$, and $(8, -8, -3, 3, 7)$ generate $\mathbb{R}^5$.

49. Express $(2, -1, 1, 1)$ as a linear combination of $(0, 1, 1, 1)$, $(1, 0, 1, 1)$, $(1, 1, 0, 1)$, and $(1, 1, 1, 0)$, if possible.

50. Express $(1, 3, 2, -1)$ as a linear combination of $(1, 1, 0, 0)$, $(1, 0, 1, 0)$, $(0, 0, 1, 1)$, and $(0, 1, 0, 1)$, if possible.

2.4 INDEPENDENCE

We now turn to the question of producing an efficient generating set for a finitely generated vector space. Consider a vector space $V = \text{sp}(\mathbf{v}_1, \mathbf{v}_2, \ldots, \mathbf{v}_n)$, where the generating vectors listed are all distinct. We are interested in seeing whether a shorter list consisting of some of these same vectors $\mathbf{v}_i$ can generate V as well; the smaller the number of generating vectors is, the simpler will be the description of the vector space V. Assuming $V \neq \{\mathbf{0}\}$, if some $\mathbf{v}_i = \mathbf{0}$, then that $\mathbf{v}_i$ can be omitted. Furthermore, if $\mathbf{v}_2 = r_1\mathbf{v}_1$ for some scalar r_1 in $\mathbb{R}$, then $\mathbf{v}_2$ can be omitted from the list. More generally, if some $\mathbf{v}_j$ can be written in terms of its predecessors as

$$\mathbf{v}_j = r_1\mathbf{v}_1 + r_2\mathbf{v}_2 + \cdots + r_{j-1}\mathbf{v}_{j-1}, \tag{1}$$

then each linear combination

$$\mathbf{v} = s_1\mathbf{v}_1 + s_2\mathbf{v}_2 + \cdots + \boxed{s_j\mathbf{v}_j} + \cdots + s_n\mathbf{v}_n$$

in V can be written without using $\mathbf{v}_j$—namely, $\quad s_j\,(r_1 v_1 + r_2 v_2 + \cdots + r_{j-1} v_{j-1})$

$$\mathbf{v} = (s_1 + s_jr_1)\mathbf{v}_1 + (s_2 + s_jr_2)\mathbf{v}_2 + \cdots + (s_{j-1} + s_jr_{j-1})\mathbf{v}_{j-1}$$
$$+ s_{j+1}\mathbf{v}_{j+1} + s_{j+2}\mathbf{v}_{j+2} + \cdots + s_n\mathbf{v}_n.$$

Therefore, in this case,

$$V = \text{sp}(\mathbf{v}_1, \mathbf{v}_2, \ldots, \mathbf{v}_{j-1}, \mathbf{v}_{j+1}, \ldots, \mathbf{v}_n),$$

and we have succeeded in shortening our list of generating vectors.

The preceding discussion suggests that we study relations between vectors of the form given in Eq. (1).

Dependent and Independent Sets of Vectors

By moving all terms to one side of the equation, we can write relation (1) in the form

$$r_1\mathbf{v}_1 + r_2\mathbf{v}_2 + \cdots + r_n\mathbf{v}_n = \mathbf{0},$$

where $r_j = -1$ and $r_{j+1} = r_{j+2} = \cdots = r_n = 0$. This is an example of a *dependence relation*.

DEFINITION 2.7 Linearly Dependent Set of Vectors

A set $\{\mathbf{v}_1, \mathbf{v}_2, \ldots, \mathbf{v}_k\}$ of vectors in a vector space V is **linearly dependent** if there exists a **dependence relation**

$$r_1\mathbf{v}_1 + r_2\mathbf{v}_2 + \cdots + r_k\mathbf{v}_k = \mathbf{0}, \qquad \text{some } r_j \neq 0. \tag{2}$$

For convenience, we will often drop the word *linearly* from the term *linearly dependent* and just speak of a *dependent set of vectors*. We will sometimes drop the words *set of* and refer to *dependent vectors* $\mathbf{v}_1, \mathbf{v}_2, \ldots, \mathbf{v}_k$.

Any set of vectors that includes the zero vector is a dependent set, since, if $\mathbf{v}_j = \mathbf{0}$, we may take $r_j = 1$ and all other $r_i = 0$ and obtain a dependence relation (2). In practice, we are seldom concerned with a dependent set of vectors that contains the zero vector.

If the set $\{\mathbf{v}_1, \mathbf{v}_2, \ldots, \mathbf{v}_k\}$ does not contain the zero vector and is dependent, we can deduce that a dependence relation (2) must contain at least two nonzero coefficients r_i. Suppose that we let r_j be the nonzero coefficient with maximum subscript. Since the coefficients with higher subscripts are zero, dependence relation (2) can be rewritten as

$$\mathbf{v}_j = (-r_1/r_j)\mathbf{v}_1 + (-r_2/r_j)\mathbf{v}_2 + \cdots + (-r_{j-1}/r_j)\mathbf{v}_{j-1}.$$

This is a relation of the form (1), expressing $\mathbf{v}_j$ as a linear combination of its predecessors. We immediately obtain the criterion shown in the following box for dependence of a finite set of nonzero vectors.

Dependence of a Set of Nonzero Vectors

A finite list of nonzero vectors in a vector space V is linearly dependent if and only if some vector in the list is equal to a linear combination of its predecessors.

For example, two nonzero vectors are dependent if and only if one is a multiple of the other. As another example, a set of the form $\{\mathbf{v}_1, \mathbf{v}_2, 2\mathbf{v}_1 + 3\mathbf{v}_2\}$ is dependent in any vector space V. This is illustrated in Figure 2.23(a) for the vector space $\mathbb{R}^n$. Notice that the third vector lies in the plane generated by the first two vectors.

A finite set of vectors that is not dependent is said to be *independent*. For example, three nonzero vectors are independent if no one of them is in the subspace generated by the other two. (See Fig. 2.23(b).) Because independent sets of vectors will play an important role for us, we write out the negations of Definition 2.7 and the subsequent boxed statement.

A COORDINATE-FREE TREATMENT of vector-space concepts appeared in 1862 in the second version of Hermann Grassmann's *Ausdehnungslehre* (*The Calculus of Extension*). In this version he was able to suppress somewhat the philosophical bias that had made his earlier work so unreadable and to concentrate on his new mathematical ideas. These included the basic ideas of the theory of n-dimensional vector spaces, including linear combinations, linear independence, and the notions of a subspace and a basis. He developed the idea of the dimension of a subspace as the maximal number of linearly independent vectors and proved the fundamental relation for two subspaces V and W that dim $(V + W)$ = dim V + dim W − dim $(V \cap W)$.

Grassmann's notions derived from the attempt to translate geometric ideas about n-dimensional space into the language of algebra without dealing with coordinates, as is done in ordinary analytic geometry. He was the first to produce a complete system in which such concepts as points, line segments, planes, and their analogues in higher dimensions are represented as single elements. Though his ideas were initially difficult to understand, ultimately they entered the mathematical mainstream in such fields as vector analysis and the exterior algebra. Grassmann himself, unfortunately, never attained his goal of becoming a German university professor, spending most of his professional life as a mathematics teacher at a gymnasium (high school) in Stettin. In the final decades of his life, he turned away from mathematics and established himself as an expert in linguistics.

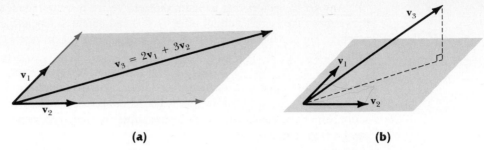

FIGURE 2.23 (a) Vectors v_1, v_2, and v_3 in $\mathbb{R}^n$ are dependent; (b) vectors v_1, v_2, and v_3 in $\mathbb{R}^n$ are independent.

DEFINITION 2.8 Independent Set of Vectors

A set $\{v_1, v_2, \ldots, v_k\}$ of vectors in a vector space V is **linearly independent** if no dependence relation of form (2) exists, so that, if $r_1v_1 + r_2v_2 + \cdots + r_kv_k = 0$, then all the scalars r_i are zero.

The negation of the condition for dependence of a set of nonzero vectors takes the following form.

Independence of a Set of Nonzero Vectors

A finite list of nonzero vectors in a vector space V is linearly independent if and only if no vector in the list is equal to a linear combination of its predecessors.

EXAMPLE 1 Determine whether the vectors $v_1 = (1, 2, 3, 1)$, $v_2 = (2, 2, 1, 3)$, and $v_3 = (-1, 2, 7, -3)$ in $\mathbb{R}^4$ are independent.

Solution We must determine whether the vector equation

$$r_1v_1 + r_2v_2 + r_3v_3 = 0$$

has a nontrivial solution for r_1, r_2, and r_3. Writing vectors as column vectors, we may express the previous equation in the form

$$r_1\begin{bmatrix} 1 \\ 2 \\ 3 \\ 1 \end{bmatrix} + r_2\begin{bmatrix} 2 \\ 2 \\ 1 \\ 3 \end{bmatrix} + r_3\begin{bmatrix} -1 \\ 2 \\ 7 \\ -3 \end{bmatrix} = \begin{bmatrix} 0 \\ 0 \\ 0 \\ 0 \end{bmatrix}. \tag{3}$$

This gives rise to a homogeneous system of equations with partitioned matrix

$$\left[\begin{array}{rrr|r} 1 & 2 & -1 & 0 \\ 2 & 2 & 2 & 0 \\ 3 & 1 & 7 & 0 \\ 1 & 3 & -3 & 0 \end{array}\right].$$

As usual, we leave off the column of zeros on the right-hand side of the partition and reduce the coefficient matrix A:

$$A = \begin{bmatrix} 1 & 2 & -1 \\ 2 & 2 & 2 \\ 3 & 1 & 7 \\ 1 & 3 & -3 \end{bmatrix} \sim \begin{bmatrix} 1 & 2 & -1 \\ 0 & -2 & 4 \\ 0 & -5 & 10 \\ 0 & 1 & -2 \end{bmatrix} \sim \begin{bmatrix} 1 & 0 & 3 \\ 0 & 1 & -2 \\ 0 & 0 & 0 \\ 0 & 0 & 0 \end{bmatrix}.$$

Thus there are an infinite number of nontrivial solutions of Eq. (3). For example, the solution $r_3 = 1$, $r_2 = 2$, $r_1 = -3$ shows that $-3\mathbf{v}_1 + 2\mathbf{v}_2 + \mathbf{v}_3 = \mathbf{0}$. The vectors $\mathbf{v}_1$, $\mathbf{v}_2$, and $\mathbf{v}_3$ are dependent. Notice that $\mathbf{v}_3$ is a linear combination of $\mathbf{v}_1$ and $\mathbf{v}_2$. ■

$$V_3 = 3V_1 - 2V_2$$
$$= (3,6,9,3) - (4,4,2,6)$$
$$= (-1,2,2,-3)$$

EXAMPLE 2 Show that $\{x^{1/3}, x^2\}$ is an independent set of functions in the vector space F of all functions mapping $\mathbb{R}$ into $\mathbb{R}$.

Solution Suppose that we have a relation of the form

$$r_1 x^{1/3} + r_2 x^2 = \mathbf{0},$$

where we regard $\mathbf{0}$ as the zero function. Evaluating this equation when $x = 1$ and when $x = -1$, we obtain

$$r_1 + r_2 = 0 \qquad \text{Taking } x = 1$$
$$-r_1 + r_2 = 0. \qquad \text{Taking } x = -1$$

We find that these two equations in the two unknowns r_1 and r_2 have only the trivial solution $r_1 = r_2 = 0$. It follows from Definition 2.8 that $x^{1/3}$ and x^2 are independent functions in the vector space F. ■

The technique used in Example 2 is a standard one for testing the independence of n functions in the vector space F. We consider a dependence relation and evaluate it at n different values for x. This gives rise to n linear equations with the n coefficients in the dependence relation as unknowns. If the system has only the trivial solution, the functions are independent. If the system has a nontrivial solution, we have to work harder. Perhaps a different choice of values for x would yield only the trivial solution, or perhaps the functions actually are dependent. Dependence of functions can be difficult to establish, unless we are able to spot a dependence relation by inspection.

EXAMPLE 3 Show that $\{1, \sin^2 x, \cos^2 x\}$ is a dependent set of functions in the vector space F of all functions mapping $\mathbb{R}$ into $\mathbb{R}$.

Solution From the familiar trigonometric identity $\sin^2 x + \cos^2 x = 1$, we obtain the dependence relation

$$(-1)1 + (1)\sin^2 x + (1)\cos^2 x = \mathbf{0}. \qquad \blacksquare$$

Definition 2.8 can be stated for column vectors in $\mathbb{R}^n$ in another way: A set $\{\mathbf{v}_1, \mathbf{v}_2, \ldots, \mathbf{v}_k\}$ of vectors in $\mathbb{R}^n$ is independent if and only if the homogeneous $n \times k$ system of equations

$$x_1\mathbf{v}_1 + x_2\mathbf{v}_2 + \cdots + x_k\mathbf{v}_k = \mathbf{0} \qquad (4)$$

has no nontrivial solution—that is, if and only if it has only the zero solution $\mathbf{x} = \mathbf{0}$. We state this as a theorem.

THEOREM 2.6   Application to Linear Systems

> Let A be an $n \times k$ matrix. The homogeneous system $A\mathbf{x} = \mathbf{0}$ has no nontrivial solution if and only if the column vectors of A are independent. Equivalently, the homogeneous system $A\mathbf{x} = \mathbf{0}$ has a nontrivial solution if and only if the column vectors of A are dependent.

We saw in Corollary 3 of Theorem 1.3 on page 34 that a system such as (4) will have a nontrivial solution if the number of unknowns exceeds the number of equations—that is, if $k > n$ in Eq. (4). In other words, any set $\{\mathbf{v}_1, \mathbf{v}_2, \ldots, \mathbf{v}_k\}$ of vectors in $\mathbb{R}^n$ is dependent if $k > n$.

> Any set of more than n vectors in $\mathbb{R}^n$ is linearly dependent.

EXAMPLE 4 Determine whether the vectors $(1, -2, 1)$, $(3, -5, 2)$, $(2, -3, 6)$, and $(1, 2, 1)$ in $\mathbb{R}^3$ are independent.

Solution Since the number of vectors, 4, is greater than the number of components, 3, the vectors are dependent. ■

In the case $k = n$, system (4) has no nontrivial solution if and only if the square matrix A containing the vectors $\mathbf{v}_j$ as columns is invertible. Invertibility of A also serves as a test for the $\mathbf{v}_j$ to span $\mathbb{R}^n$. (See the box on page 122.) We obtain three equivalent conditions, which we state as a theorem.

THEOREM 2.7 Equivalent Conditions for *n* Vectors in $\mathbb{R}^n$

Let $\mathbf{v}_1, \mathbf{v}_2, \ldots, \mathbf{v}_n$ be n vectors in $\mathbb{R}^n$. The following conditions are equivalent:

1. The vectors are independent.
2. The vectors generate all of $\mathbb{R}^n$.
3. The matrix A having these vectors as column vectors is invertible.

EXAMPLE 5 Determine whether the vectors $(1, 0, 0)$, $(1, 1, 0)$, and $(1, 1, 1)$ generate $\mathbb{R}^3$.

Solution These vectors are independent, so they must generate $\mathbb{R}^3$, according to Theorem 2.7. Alternatively, we can see that the matrix

$$A = \begin{bmatrix} 1 & 1 & 1 \\ 0 & 1 & 1 \\ 0 & 0 & 1 \end{bmatrix}$$

is invertible. Theorem 2.7 shows again that the vectors must generate $\mathbb{R}^3$. ■

Independence of an Infinite Set of Vectors

Let V be a vector space, and let S be any subset (finite or infinite) of V. We can form linear combinations of any finite list $\mathbf{v}_1, \mathbf{v}_2, \ldots, \mathbf{v}_k$ of vectors in S. We have defined the notion of a *dependence relation* for such a finite list of vectors. If such a dependence relation exists in S, then S is a **dependent set** of vectors. If no such dependence relation exists for vectors in S, then S is an **independent set** of vectors. This is a very natural extension of our development previously for finite subsets S of V. However, even in the general case, we are concerned only with taking linear combinations of a finite number of vectors. We have no notion of adding an infinite number of vectors in a general vector space.

EXAMPLE 6 Find an independent set of vectors generating the vector space P of all polynomials with real coefficients.

Solution We observed in Section 2.3 that the infinite set

$$\{1, x, x^2, x^3, \ldots\}$$

of monomials is a generating set for P. Clearly no linear combination of a finite number of these monomials can yield the zero polynomial unless all of the coefficients are zero. In other words, every finite subset of S is independent. Thus, S is also independent. ■

SUMMARY

Let V be a vector space.

1. A set of vectors $\{v_1, v_2, \ldots, v_k\}$ in V is linearly dependent if there exists a dependence relation

$$r_1v_1 + r_2v_2 + \cdots + r_kv_k = 0, \qquad \text{at least one } r_j \neq 0.$$

The set is linearly independent if no such dependence relation exists, so a linear combination of the v_i is the zero vector only if all scalars are zero.

2. A finite list of nonzero vectors in V forms a linearly independent set if and only if no vector in the list can be expressed as a linear combination of its predecessors.

3. The following statements are equivalent for n vectors in $\mathbb{R}^n$:
   a. The vectors are linearly independent.
   b. The vectors generate $\mathbb{R}^n$.
   c. A matrix having the vectors as columns is invertible.

4. An infinite set S of vectors in a vector space V is independent if there is no dependence relation involving a finite number of vectors in S.

EXERCISES

1. Give a geometric criterion for a set of two distinct nonzero vectors in $\mathbb{R}^2$ to be dependent.

2. Argue geometrically that any set of three distinct vectors in $\mathbb{R}^2$ is dependent.

3. Give a geometric criterion for a set of two distinct nonzero vectors in $\mathbb{R}^3$ to be dependent.

4. Give a geometric description of the subspace of $\mathbb{R}^3$ generated by an independent set of two vectors.

5. Give a geometric criterion for a set of three distinct nonzero vectors in $\mathbb{R}^3$ to be dependent.

6. Argue geometrically that every set of four distinct vectors in $\mathbb{R}^3$ is dependent.

In Exercises 7–19, determine whether the given set of vectors is dependent or independent.

7. $\{(1, 3), (-2, -6)\}$ in $\mathbb{R}^2$

8. $\{(1, 3), (2, -4)\}$ in $\mathbb{R}^2$

9. $\{(-3, 1), (6, 4)\}$ in $\mathbb{R}^2$

10. $\{(-3, 1), (9, -3)\}$ in $\mathbb{R}^2$

11. $\{(2, 1), (-6, -3), (1, 4)\}$ in $\mathbb{R}^2$

12. $\{(-1, 2, 1), (2, -4, 3)\}$ in $\mathbb{R}^3$

13. $\{(1, -3, 2), (2, -5, 3), (4, 0, 1)\}$ in $\mathbb{R}^3$

14. $\{(1, -4, 3), (3, -11, 2), (1, -3, -4)\}$ in $\mathbb{R}^3$

15. $\{(1, 4, -1, 3), (-1, 5, 6, 2), (1, 13, 4, 7)\}$ in $\mathbb{R}^4$

16. $\{(-2, 3, 1), (3, -1, 2), (1, 2, 3), (-1, 5, 4)\}$ in $\mathbb{R}^3$

17. $\{x^2 - 1, x^2 + 1, 4x, 2x - 3\}$ in P

18. $\{1, 4x + 3, 3x - 4, x^2 + 2, x - x^2\}$ in P

19. $\{1, \sin^2 x, \cos 2x, \cos^2 x\}$ in F

In Exercises 20–23, use the technique discussed following Example 2 to determine whether the given set of functions in the vector space F is independent or dependent.

20. $\{\sin x, \cos x\}$ **21.** $\{1, x, x^2\}$

22. $\{\sin x, \sin 2x, \sin 3x\}$

23. $\{\sin x, \sin(-x)\}$

24. Let $\mathbf{v}_1, \mathbf{v}_2, \mathbf{v}_3$ be independent vectors in a vector space V. Show that $\mathbf{w}_1 = 3\mathbf{v}_1$, $\mathbf{w}_2 = 2\mathbf{v}_1 - \mathbf{v}_2$, $\mathbf{w}_3 = \mathbf{v}_1 + \mathbf{v}_3$ are also independent.

25. Let $\mathbf{v}_1, \mathbf{v}_2, \mathbf{v}_3$ be any vectors in a vector space V. Show that $\mathbf{w}_1 = 2\mathbf{v}_1 + 3\mathbf{v}_2$, $\mathbf{w}_2 = \mathbf{v}_2 - 2\mathbf{v}_3$, $\mathbf{w}_3 = -\mathbf{v}_1 - 3\mathbf{v}_3$ are dependent.

26. Find all scalars s, if any exist, such that $(1, 0, 1), (2, s, 3), (2, 3, 1)$ are independent.

27. Find all scalars s, if any exist, such that $(1, 0, 1), (2, s, 3), (1, -s, 0)$ are independent.

28. Let $\mathbf{v}$ and $\mathbf{w}$ be independent column vectors in $\mathbb{R}^3$, and let A be an invertible 3×3 matrix. Show that the vectors $A\mathbf{v}$ and $A\mathbf{w}$ are independent.

29. Give an example to show that the conclusion of the preceding exercise need not hold if A is nonzero but singular. Can you also find specific independent vectors $\mathbf{v}$ and $\mathbf{w}$ and a singular matrix A such that $A\mathbf{v}$ and $A\mathbf{w}$ are still independent?

30. Let $\mathbf{v}$ and $\mathbf{w}$ be column vectors in $\mathbb{R}^n$, and let A be an $n \times n$ matrix. Show that, if $A\mathbf{v}$ and $A\mathbf{w}$ are independent, $\mathbf{v}$ and $\mathbf{w}$ are independent.

31. Show that, if $\{\mathbf{v}_1, \mathbf{v}_2, \ldots, \mathbf{v}_k\}$ is an independent subset of a vector space V, each element of $\text{sp}(\mathbf{v}_1, \mathbf{v}_2, \ldots, \mathbf{v}_k)$ can be expressed *uniquely* as a linear combination of the $\mathbf{v}_i$.

32. Show the converse of Exercise 31—namely, that, if each vector in $\text{sp}(\mathbf{v}_1, \mathbf{v}_2, \ldots, \mathbf{v}_k)$ can be expressed uniquely as a linear combination of the

$\mathbf{v}_i$, then $\{\mathbf{v}_1, \mathbf{v}_2, \ldots, \mathbf{v}_k\}$ is an independent set of vectors.

33. Let $S = \{\mathbf{v}_1, \mathbf{v}_2, \ldots, \mathbf{v}_k\}$ be a set of vectors in a vector space V. Mark each of the following True or False.

____ a) A subset of $\mathbb{R}^n$ containing two nonzero distinct parallel vectors is dependent.

____ b) If a set of nonzero vectors in $\mathbb{R}^n$ is dependent, any two vectors in the set are parallel.

____ c) Every subset $\{\mathbf{v}\}$ of V, where $\mathbf{v} \neq \mathbf{0}$, is independent.

____ d) If S is independent, each vector in V can be expressed uniquely as a linear combination of vectors in S.

____ e) If S is independent and generates V, each vector in V can be expressed uniquely as a linear combination of vectors in S.

____ f) If each vector in V can be expressed uniquely as a linear combination of vectors in S, then S is an independent set.

____ g) The subset S of V is independent if and only if each vector in $\text{sp}(\mathbf{v}_1, \mathbf{v}_2, \ldots, \mathbf{v}_k)$ has a unique expression as a linear combination of vectors in S.

____ h) Every subset of three vectors in $\mathbb{R}^2$ is dependent.

____ i) Every subset of two vectors in $\mathbb{R}^2$ is independent.

____ j) If a subset of two vectors in $\mathbb{R}^2$ generates $\mathbb{R}^2$, the subset is independent.

34. Generalizing Exercise 28, let $\mathbf{v}_1, \mathbf{v}_2, \ldots, \mathbf{v}_k$ be independent column vectors in $\mathbb{R}^n$, and let C be an invertible $n \times n$ matrix. Show that the vectors $C\mathbf{v}_1, C\mathbf{v}_2, \ldots, C\mathbf{v}_k$ are independent.

2.5   BASES

We return to our discussion of a vector space V generated by vectors $\mathbf{v}_1, \mathbf{v}_2, \ldots, \mathbf{v}_n$. We desire to have this list of generating vectors be as short as possible. We assume that the vectors $\mathbf{v}_i$ are all nonzero. In the previous section, we saw that, if we delete from this list of vectors a vector that is a linear combination of its predecessors, we obtain a shorter list of vectors that still generates V. We can continue this process until no vector remaining in the shortened list is a linear combination of its predecessors. From the boxed statement following Definition 2.8, we know that the vectors then remaining in the list form an independent set of vectors generating V.

We will refer to the process of deleting vectors that can be expressed as a linear combination of predecessors from a list as the *casting-out technique*. It is best to perform this technique systematically—say, from left to right in the list.

General Casting-out Technique

Let $\mathbf{v}_1, \mathbf{v}_2, \ldots, \mathbf{v}_n$ be a list of nonzero vectors generating a vector space V. Start with $\mathbf{v}_2$ in the list, and work to the right, casting out any vector that is equal to a linear combination of its remaining predecessors. After all vectors have been tested in this manner, the vectors remaining form an independent generating set for V.

Independent generating sets of vectors for a vector space are very important in our subsequent work. We start with the following definition.

DEFINITION 2.9   Basis for a Vector Space

Let V be a vector space. A set of vectors in V is a **basis** for V if the following conditions are met:

1. The set of vectors generates V; and
2. The set of vectors is linearly independent.

The set $\{\mathbf{e}_1, \mathbf{e}_2, \ldots, \mathbf{e}_n\}$ of unit coordinate vectors in $\mathbb{R}^n$ is a basis for $\mathbb{R}^n$. This basis is called the **standard basis**. The set $\{1, x, x^2, x^3, \ldots, x^n\}$ is linearly independent in the vector space P of all polynomials with coefficients in $\mathbb{R}$. This set of monomials is a basis for the subspace P_n of all polynomials of degree $\leq n$. The set $\{1, x, x^2, x^3, \ldots\}$ is a basis for the space P of all polynomials. We think of a basis for a vector space as being a set of *reference vectors* for the space. Our next theorem underscores this viewpoint.

THEOREM 2.8   Unique Representation of Vectors

Let V be a vector space with basis $B = \{\mathbf{b}_1, \mathbf{b}_2, \ldots, \mathbf{b}_n\}$. Each vector $\mathbf{v}$ in V can be uniquely expressed in the form

$$\mathbf{v} = r_1\mathbf{b}_1 + r_2\mathbf{b}_2 + \cdots + r_n\mathbf{b}_n. \tag{1}$$

That is, there is only one choice for each of the scalars $r_1, r_2, \ldots, r_n$.

PROOF   Condition 1 of Definition 2.9 tells us that a vector $\mathbf{v}$ in V can be expressed in form (1). We prove uniqueness by using condition 2. The usual way to prove that something is unique is to suppose that there are two of the things, and then to show that they have to be the same. Accordingly, let us suppose that $\mathbf{v}$ has form (1) and can also be written as

$$\mathbf{v} = s_1\mathbf{b}_1 + s_2\mathbf{b}_2 + \cdots + s_n\mathbf{b}_n.$$

Equating the two expressions for $\mathbf{v}$, we have

$$r_1\mathbf{b}_1 + r_2\mathbf{b}_2 + \cdots + r_n\mathbf{b}_n = s_1\mathbf{b}_1 + s_2\mathbf{b}_2 + \cdots + s_n\mathbf{b}_n$$

or

$$(r_1 - s_1)\mathbf{b}_1 + (r_2 - s_2)\mathbf{b}_2 + \cdots + (r_n - s_n)\mathbf{b}_n - \mathbf{0}.$$

Condition 2 in the definition of a *basis*—namely, that the vectors $\mathbf{b}_i$ are independent—leads us to conclude that each $r_i - s_i = 0$; that is, each $r_i = s_i$. The proof is complete. ▲

We will make essential use of Theorem 2.8 in the next section.

EXAMPLE 1   Determine whether the vectors

$$(1, 2, -1, 0), \qquad (0, 1, 0, 1), \qquad (-1, -5, 2, 0), \qquad \text{and} \qquad (2, 3, -2, 7)$$

form a basis for $\mathbb{R}^4$.

Solution   We form the matrix A having these vectors as columns, and obtain

$$\begin{bmatrix} 1 & 0 & -1 & 2 \\ 2 & 1 & -5 & 3 \\ -1 & 0 & 2 & -2 \\ 0 & 1 & 0 & 7 \end{bmatrix} \sim \begin{bmatrix} 1 & 0 & -1 & 2 \\ 0 & 1 & -3 & -1 \\ 0 & 0 & 1 & 0 \\ 0 & 1 & 0 & 7 \end{bmatrix} \sim \begin{bmatrix} 1 & 0 & -1 & 2 \\ 0 & 1 & -3 & -1 \\ 0 & 0 & 1 & 0 \\ 0 & 0 & 3 & 8 \end{bmatrix}$$

$$\sim \begin{bmatrix} 1 & 0 & -1 & 2 \\ 0 & 1 & -3 & -1 \\ 0 & 0 & 1 & 0 \\ 0 & 0 & 0 & 8 \end{bmatrix},$$

which shows A to be invertible. Therefore, the given vectors form a basis for $\mathbb{R}^4$, by Theorem 2.7. ▪

EXAMPLE 2   Find a basis for the nullspace of the matrix

$$A = \begin{bmatrix} 3 & 1 & 9 \\ 1 & 2 & -2 \\ 2 & 1 & 5 \end{bmatrix}.$$

Solution   The nullspace N_A of the matrix A is the set of solutions of the system $A\mathbf{x} = \mathbf{0}$. Using a Gauss–Jordan reduction, we obtain

$$\left[\begin{array}{ccc|c} 3 & 1 & 9 & 0 \\ 1 & 2 & -2 & 0 \\ 2 & 1 & 5 & 0 \end{array}\right] \sim \left[\begin{array}{ccc|c} 1 & 2 & -2 & 0 \\ 0 & -5 & 15 & 0 \\ 0 & -3 & 9 & 0 \end{array}\right] \sim \left[\begin{array}{ccc|c} 1 & 0 & 4 & 0 \\ 0 & 1 & -3 & 0 \\ 0 & 0 & 0 & 0 \end{array}\right].$$

Therefore,

$$N_A = \left\{ \begin{bmatrix} x_1 \\ x_2 \\ x_3 \end{bmatrix} \, \middle| \, x_2 = 3x_3, \, x_1 = -4x_3 \text{ for } x_3 \in \mathbb{R}^n \right\}$$

$$= \left\{ x_3 \begin{bmatrix} -4 \\ 3 \\ 1 \end{bmatrix} \, \middle| \, x_3 \in \mathbb{R} \right\} = \text{sp}\left(\begin{bmatrix} -4 \\ 3 \\ 1 \end{bmatrix} \right).$$

Thus, one basis for N_A contains the single vector

$$\begin{bmatrix} -4 \\ 3 \\ 1 \end{bmatrix}.$$

▪

EXAMPLE 3 Find a basis for the subspace

$$W = \{(x_1, x_2, x_3, x_4, x_5) \mid x_2 = 3x_1, \, x_4 = 2x_1, \, x_5 = x_3 - x_1\}$$

of $\mathbb{R}^5$.

Solution We can write

$$W = \{(x_1, 3x_1, x_3, 2x_1, x_3 - x_1) \mid x_1, x_3 \in \mathbb{R}\}$$
$$= \{x_1(1, 3, 0, 2, -1) + x_3(0, 0, 1, 0, 1) \mid x_1, x_3 \in \mathbb{R}\}$$
$$= \text{sp}(\mathbf{v}_1, \mathbf{v}_2),$$

where $\mathbf{v}_1 = (1, 3, 0, 2, -1)$ is obtained from $(x_1, 3x_1, x_3, 2x_1, x_3 - x_1)$ by taking $x_1 = 1$ and $x_3 = 0$, and $\mathbf{v}_2 = (0, 0, 1, 0, 1)$ is obtained by taking $x_1 = 0$ and $x_3 = 1$. Since $\mathbf{v}_2$ is not a multiple of $\mathbf{v}_1$, we conclude that $\{\mathbf{v}_1, \mathbf{v}_2\}$ is a basis for W. ▪

The casting-out technique described earlier gives us the following theorem.

THEOREM 2.9 Reducing a Generating Set to a Basis

Let V be a vector space generated by a set $\{\mathbf{v}_1, \mathbf{v}_2, \ldots, \mathbf{v}_n\}$ of nonzero vectors. Then V can be generated by a possibly smaller subset of these vectors, which is a basis for V. Such a basis can be found by casting out from $\{\mathbf{v}_1, \mathbf{v}_2, \ldots, \mathbf{v}_n\}$ the vectors that are linear combinations of predecessors.

Restated, Theorem 2.9 tells us that every nonzero finitely generated vector space has a basis. This is also true of vector spaces that are not finitely generated, but the demonstration is beyond the scope of our text; it is often difficult to exhibit a basis for such a vector space. Notice, however, that we have exhibited a basis for the space P of all polynomials—namely, the basis $\{1, x, x^2, x^3, \ldots\}$. It is convenient to define the basis of the vector space $\{\mathbf{0}\}$ to be the null set. Thus, every vector space has a basis.

EXAMPLE 4   Consider the subspace W of $\mathbb{R}^3$ given by

$$W = \text{sp}((2, 1, 3), (-1, 2, 0), (1, 8, 6)).$$

Use the casting-out technique to find a basis for W.

Solution   We apply the casting-out technique systematically, starting with the second vector and working to the right. Since $(-1, 2, 0)$ is not a multiple of $(2, 1, 3)$, we keep $(-1, 2, 0)$ in the list. We proceed to the vector $(1, 8, 6)$ and check whether it can be expressed in the form $(1, 8, 6) = r_1(2, 1, 3) + r_2(-1, 2, 0)$ for scalars r_1 and r_2. Rewriting this equation in the form

$$r_1\begin{bmatrix} 2 \\ 1 \\ 3 \end{bmatrix} + r_2\begin{bmatrix} -1 \\ 2 \\ 0 \end{bmatrix} = \begin{bmatrix} 1 \\ 8 \\ 6 \end{bmatrix},$$

we recognize that it leads to a system of equations in the unknowns r_1 and r_2. Next, we reduce the partitioned matrix of the system:

$$\left[\begin{array}{rr|r} 2 & -1 & 1 \\ 1 & 2 & 8 \\ 3 & 0 & 6 \end{array}\right] \sim \left[\begin{array}{rr|r} 1 & 2 & 8 \\ 2 & -1 & 1 \\ 3 & 0 & 6 \end{array}\right] \sim \left[\begin{array}{rr|r} 1 & 2 & 8 \\ 0 & -5 & -15 \\ 0 & -6 & -18 \end{array}\right] \sim \left[\begin{array}{rr|r} 1 & 0 & 2 \\ 0 & 1 & 3 \\ 0 & 0 & 0 \end{array}\right].$$

We obtain the solution $r_2 = 3$ and $r_1 = 2$, so we conclude that $(1, 8, 6)$ is in $\text{sp}((2, 1, 3), (-1, 2, 0))$. Consequently, we cast out $(1, 8, 6)$ and arrive at the basis

$$\{(2, 1, 3), (-1, 2, 0)\}$$

for W.   ■

EXAMPLE 5   Use the casting-out technique to find a basis for the subspace

$$W = \text{sp}(x^2 - 1, x^2 + 1, 3, 2x - 1, x)$$

of the vector space P of polynomials.

Solution   We see that $x^2 + 1 \neq r(x^2 - 1)$ for any scalar r, so we retain $x^2 + 1$ in the casting-out technique. We do, however, have a dependence relation

$$3 = -\tfrac{3}{2}(x^2 - 1) + \tfrac{3}{2}(x^2 + 1),$$

so we cast out 3 and work with the list $x^2 - 1, x^2 + 1, 2x - 1, x$. It is impossible to express $2x - 1$ as a linear combination of $x^2 - 1$ and $x^2 + 1$, since no first power of x appears in these two predecessors, so $2x - 1$ is retained in the list. Finally, let us see if scalars c_1, c_2, and c_3 can be found such that

$$x = c_1(x^2 - 1) + c_2(x^2 + 1) + c_3(2x - 1).$$

Rearranging terms on the right-hand side, this equation can be written as

$$x = (c_1 + c_2)x^2 + 2c_3x + (-c_1 + c_2 - c_3).$$

Equating coefficients of like powers of x we obtain the system

$$
\begin{array}{rl}
c_1 + c_2 & = 0 \\
2c_3 & = 1 \\
-c_1 + c_2 - c_3 & = 0
\end{array}
\qquad \text{whose solution is} \qquad
\mathbf{c} = \begin{bmatrix} -\frac{1}{4} \\ \frac{1}{4} \\ \frac{1}{2} \end{bmatrix}.
$$

Consequently,

$$
x = -\tfrac{1}{4}(x^2 - 1) + \tfrac{1}{4}(x^2 + 1) + \tfrac{1}{2}(2x - 1),
$$

and we should cast out x. Thus we arrive at the basis $\{x^2 - 1, \, x^2 + 1, \, 2x - 1\}$ for W. ■

Executing the Casting-out Technique in $\mathbb{R}^n$

As we have presented it up to now, the casting-out technique seems potentially laborious. For example, we might expect that, to apply it to a list $\mathbf{v}_1, \mathbf{v}_2, \ldots, \mathbf{v}_6$, we should test for solutions of five vector equations:

$$
\mathbf{v}_2 = r_1 \mathbf{v}_1, \qquad \mathbf{v}_3 = r_1 \mathbf{v}_1 + r_2 \mathbf{v}_2, \qquad \text{and so on.}
$$

In the vector space $\mathbb{R}^n$, each of these equations would lead to a linear system, as illustrated in Example 4. The following box outlines a more efficient method for solving this problem.

The Casting-out Technique in $\mathbb{R}^n$

Elimination of nonzero vectors that are linear combinations of predecessors in a list $\mathbf{v}_1, \mathbf{v}_2, \ldots, \mathbf{v}_k$ can be accomplished as follows:

1. Form the matrix A, with $\mathbf{v}_j$ as the jth column vector.
2. Reduce A to row echelon form, as in the Gauss method.
3. The vectors in the columns of A that give rise to pivots should be retained; they will generate the entire column space of A. That is, <u>cast out from the list all vectors in the columns of A that do not give rise to pivots.</u>

To see why this simple procedure works, suppose that the matrix

$$
A = \begin{bmatrix} | & | & & | \\ \mathbf{v}_1 & \mathbf{v}_2 & \cdots & \mathbf{v}_k \\ | & | & & | \end{bmatrix}
$$

is reduced to a row echelon form H. Recall from page 121 that $A\mathbf{x}$ is a linear combination of the column vectors of A, namely $x_1 \mathbf{v}_1 + x_2 \mathbf{v}_2 + \cdots + x_k \mathbf{v}_k$. Any dependence relation among the column vectors of A is thus reflected by a nontrivial solution of the homogeneous system $A\mathbf{x} = \mathbf{0}$. From Theorem 1.1, we know that the solutions of

$Ax = 0$ are precisely the solutions of $Hx = 0$. This means that a dependence relation among the column vectors of A corresponds to a dependence relation among the column vectors of H, and vice versa, using the *same* scalars.

Let us take H to be the *reduced* row echelon form of A. For example, we might have

$$H = \begin{bmatrix} 1 & 0 & X & 0 & X & 0 \\ 0 & 1 & X & 0 & X & 0 \\ 0 & 0 & 0 & 1 & X & 0 \\ 0 & 0 & 0 & 0 & 0 & 1 \\ 0 & 0 & 0 & 0 & 0 & 0 \end{bmatrix},$$

where each X may be any real number. Since the columns of H that have pivots are distinct unit coordinate vectors, these pivot column vectors are independent. Moreover, any column of H that lacks a pivot is not only dependent on the pivot columns, but is in fact a linear combination of the pivot columns preceding it, as the row echelon form shows. Therefore, the columns of A that are linear combinations of predecessors and should be cast out are precisely the columns that do not yield pivots in the reduced row echelon form H. We should retain the columns of A that do yield pivots in H.

In practice, we save time by applying the boxed procedure using any echelon form, rather than going all the way to the reduced echelon form.

EXAMPLE 6   Let W be the subspace of $\mathbb{R}^5$ generated by

$$\mathbf{v}_1 = (1, -1, 0, 2, 1), \qquad \mathbf{v}_2 = (2, 1, -2, 0, 0),$$
$$\mathbf{v}_3 = (0, -3, 2, 4, 2), \qquad \mathbf{v}_4 = (3, 3, -4, -2, -1),$$
$$\mathbf{v}_5 = (2, 4, 1, 0, 1), \qquad \mathbf{v}_6 = (5, 7, -3, -2, 0).$$

Use the casting-out technique to attempt to shorten the list $\mathbf{v}_1, \mathbf{v}_2, \mathbf{v}_3, \mathbf{v}_4, \mathbf{v}_5, \mathbf{v}_6$ of generators of W to a basis for W.

Solution   We reduce the matrix that has $\mathbf{v}_j$ as jth column vector, and we obtain an echelon form:

$$\begin{bmatrix} 1 & 2 & 0 & 3 & 2 & 5 \\ -1 & 1 & -3 & 3 & 4 & 7 \\ 0 & -2 & 2 & -4 & 1 & -3 \\ 2 & 0 & 4 & -2 & 0 & -2 \\ 1 & 0 & 2 & -1 & 1 & 0 \end{bmatrix} \sim \begin{bmatrix} 1 & 2 & 0 & 3 & 2 & 5 \\ 0 & 3 & -3 & 6 & 6 & 12 \\ 0 & -2 & 2 & -4 & 1 & -3 \\ 0 & -4 & 4 & -8 & -4 & -12 \\ 0 & -2 & 2 & -4 & -1 & -5 \end{bmatrix}$$

$$\sim \begin{bmatrix} 1 & 2 & 0 & 3 & 2 & 5 \\ 0 & 1 & -1 & 2 & 2 & 4 \\ 0 & 0 & 0 & 0 & 5 & 5 \\ 0 & 0 & 0 & 0 & 4 & 4 \\ 0 & 0 & 0 & 0 & 3 & 3 \end{bmatrix} \sim \begin{bmatrix} 1 & 2 & 0 & 3 & 2 & 5 \\ 0 & 1 & -1 & 2 & 2 & 4 \\ 0 & 0 & 0 & 0 & 1 & 1 \\ 0 & 0 & 0 & 0 & 0 & 0 \\ 0 & 0 & 0 & 0 & 0 & 0 \end{bmatrix}.$$

Since there are pivots in columns 1, 2, and 5 of the echelon form, the vectors $\mathbf{v}_1$, $\mathbf{v}_2$, and $\mathbf{v}_5$ are retained and are independent. The vectors $\mathbf{v}_3$, $\mathbf{v}_4$, and $\mathbf{v}_6$ are cast out, and we obtain $\{\mathbf{v}_1, \mathbf{v}_2, \mathbf{v}_5\}$ as a basis of W.   ■

We emphasize that the vectors retained are in columns of the matrix A formed at the start of the reduction process. A common error is to take instead the actual column vectors containing pivots in the echelon form H. There is no reason even to expect that the column vectors of H lie in the column space of A. For example,

$$A = \begin{bmatrix} 1 & 1 \\ 1 & 1 \end{bmatrix} \sim \begin{bmatrix} 1 & 1 \\ 0 & 0 \end{bmatrix} = H,$$

and certainly $\begin{bmatrix} 1 \\ 0 \end{bmatrix}$ is not in the column space of A.

EXAMPLE 7 Find a basis for the column space of the matrix

$$A = \begin{bmatrix} 1 & -2 & 2 & -1 \\ -3 & 6 & 1 & 10 \\ 1 & -2 & -4 & -7 \end{bmatrix}.$$

Solution We reduce the matrix A to an echelon form, obtaining

$$\begin{bmatrix} 1 & -2 & 2 & -1 \\ -3 & 6 & 1 & 10 \\ 1 & -2 & -4 & -7 \end{bmatrix} \sim \begin{bmatrix} 1 & -2 & 2 & -1 \\ 0 & 0 & 7 & 7 \\ 0 & 0 & -6 & -6 \end{bmatrix} \sim \begin{bmatrix} 1 & -2 & 2 & -1 \\ 0 & 0 & 1 & 1 \\ 0 & 0 & 0 & 0 \end{bmatrix}.$$

Thus the column vectors

$$\begin{bmatrix} 1 \\ -3 \\ 1 \end{bmatrix} \quad \text{and} \quad \begin{bmatrix} 2 \\ 1 \\ -4 \end{bmatrix}$$

form a basis for the column space of A. ■

Reductions used in the casting-out technique can be executed easily on a computer. You can use either of the available programs YUREDUCE and MATCOMP, or similar software.

SUMMARY

Let V be a finitely generated vector space.

1. A set of vectors is a basis for V if it is both a generating set for V and an independent set of vectors.

2. If $B = \{\mathbf{b}_1, \mathbf{b}_2, \ldots, \mathbf{b}_n\}$ is a basis for V, each vector in V can be expressed in the form

$$\mathbf{v} = r_1\mathbf{b}_1 + r_2\mathbf{b}_2 + \cdots + r_n\mathbf{b}_n$$

for unique scalars $r_1, r_2, \ldots, r_n$.

3. Any finite generating set for V can be reduced (if necessary) to a basis for V by deleting zero vectors, and then casting out those that are linear combinations of predecessors.

4. The casting-out technique in $\mathbb{R}^n$ can be executed efficiently, as explained in the box on page 138.

EXERCISES

In Exercises 1–7, determine whether or not the given set of vectors is a basis for the indicated vector space.

1. $\{(-1, 1), (1, 2)\}$ for $\mathbb{R}^2$

2. $\{(-1, 3, 1), (2, 1, 4)\}$ for $\mathbb{R}^3$

3. $\{(-1, 3, 4), (1, 5, -1), (1, 13, 2)\}$ for $\mathbb{R}^3$

4. $\{(2, 1, -3), (4, 0, 2), (2, -1, 3)\}$ for $\mathbb{R}^3$

5. $\{(2, 1, 0, 2), (2, -3, 1, 0), (3, 2, 0, 0), (5, 0, 0, 0)\}$ for $\mathbb{R}^4$

6. $\{x, x^2 + 1, (x - 1)^2\}$ for P_2

7. $\{x, (x + 1)^2, (x - 1)^2\}$ for P_2

8. Argue geometrically that, if two vectors $\mathbf{v}_1$ and $\mathbf{v}_2$ are chosen at random from $\mathbb{R}^2$, it is very likely that $\{\mathbf{v}_1, \mathbf{v}_2\}$ is a basis for $\mathbb{R}^2$. (The probability of obtaining a basis by such a random selection is actually 1.)

9. Repeat Exercise 8 for a random selection of vectors $\mathbf{v}_1$, $\mathbf{v}_2$, and $\mathbf{v}_3$ in $\mathbb{R}^3$.

In Exercises 10–17, use the casting-out technique to find a basis for the given subspace of the vector space.

10. $\text{sp}((-3, 1), (6, 4))$ in $\mathbb{R}^2$

11. $\text{sp}((-3, 1), (9, -3))$ in $\mathbb{R}^2$

12. $\text{sp}((2, 1), (-6, -3), (1, 4))$ in $\mathbb{R}^2$

13. $\text{sp}((-2, 3, 1), (3, -1, 2), (1, 2, 3), (-1, 5, 4))$ in $\mathbb{R}^3$

14. $\text{sp}((1, 2, 1, 2), (2, 1, 0, -1), (-1, 4, 3, 8), (0, 3, 2, 5))$ in $\mathbb{R}^4$

15. $\text{sp}(x^2 - 1, x^2 + 1, 4, 2x - 3)$ in P

16. $\text{sp}(1, 4x + 3, 3x - 4, x^2 + 2, x - x^2)$ in P

17. $\text{sp}(1, \sin^2 x, \cos 2x, \cos^2 x)$ in F

18. Find a basis for the column space of the matrix

$$A = \begin{bmatrix} 2 & 3 & 1 \\ 5 & 2 & 1 \\ 1 & 7 & 2 \\ 6 & -2 & 0 \end{bmatrix}.$$

19. We describe two methods for finding a basis for the row space of a matrix A:

   (i) Use the casting-out technique, reducing A^T to row echelon form. The column vectors of A^T that are not cast out form a basis for the row space of A.

   (ii) Reduce A to row echelon form; the nonzero rows in this echelon form comprise a basis for the row space of A.

   a) Method (i) is valid since the row space of A is the column space of A^T. Explain why method (ii) is valid.

   b) The bases obtained by using methods (i) and (ii) are usually different. What is a feature of the basis obtained by using method (i) that is not in general true of the basis obtained by using method (ii)?

   c) If you want to find bases for both the column space of A and the row space of A, how should you proceed in order to minimize the amount of computation?

20. Find a basis for the row space of the matrix A in Exercise 18.

21. Find a basis for the nullspace of the matrix A in Exercise 18.

22. Find a basis for the row space of the matrix

$$A = \begin{bmatrix} 1 & 3 & 5 & 7 \\ 2 & 0 & 4 & 2 \\ 3 & 2 & 8 & 7 \end{bmatrix}.$$

23. Find a basis for the column space of the matrix A in Exercise 22.

24. Find a basis for the nullspace of the matrix A in Exercise 22.

25. Find a basis for the space $M_{2,3}$ of all 2×3 matrices.

26. Find a basis for the plane in $\mathbb{R}^3$ that passes through the origin (and hence constitutes a subspace) and is given by the equation

$$2x - 3y + 4z = 0.$$

27. Let V be a vector space. Mark each of the following True or False.

____ a) Every generating subset of V is a basis for V.

____ b) Every independent subset of V is a basis for V.

____ c) Every independent set of vectors in V is a basis for the subspace the vectors span.

____ d) If $\{v_1, v_2, \ldots, v_n\}$ generates V, then each $v \in V$ is a linear combination of the vectors in this set.

____ e) If $\{v_1, v_2, \ldots, v_n\}$ generates V, then each $v \in V$ is a unique linear combination of the vectors in this set.

____ f) If $\{v_1, v_2, \ldots, v_n\}$ generates V and is independent, then each $v \in V$ is a unique linear combination of the vectors in this set.

____ g) If $\{v_1, v_2, \ldots, v_n\}$ generates V, then this set of vectors is independent.

____ h) If each vector in V is a unique linear combination of the vectors in the set $\{v_1, v_2, \ldots, v_n\}$, then this set is independent.

____ i) If each vector in V is a unique linear combination of the vectors in the set $\{v_1, v_2, \ldots, v_n\}$, then this set is a basis for V.

____ j) All vector spaces having a basis are finitely generated.

28. Let V be a vector space with basis $\{v_1, v_2, v_3\}$. Show that $\{v_1, v_1 + v_2, v_1 + v_2 + v_3\}$ is also a basis for V.

29. Let V be a vector space with basis $\{v_1, v_2, \ldots, v_n\}$, and let $W = \text{sp}(v_3, v_4, \ldots, v_n)$. If $w = r_1 v_1 + r_2 v_2$ is in W, show that $w = 0$.

30. Let $\{v_1, v_2, v_3\}$ be a basis for a vector space V. Show that the vectors $w_1 = v_1 + v_2$, $w_2 = v_2 + v_3$, $w_3 = v_1 - v_3$ do not generate V.

31. Let $\{v_1, v_2, v_3\}$ be a basis for a vector space V.

Show that, if w is not in $\text{sp}(v_1, v_2)$, then $\{v_1, v_2, w\}$ is also a basis for V.

32. Let $\{v_1, v_2, \ldots, v_n\}$ be a basis for a vector-space V, and let $w = t_1 v_1 + t_2 v_2 + \cdots + t_k v_k$, with $t_k \neq 0$. Show that

$$\{v_1, v_2, \ldots, v_{k-1}, w, v_{k+1}, \ldots, v_n\}$$

is a basis for V.

33. Let $V = \text{sp}(v_1, v_2, \ldots, v_n)$, and suppose that each vector in V is a unique linear combination of the v_i. Show that $\{v_1, v_2, \ldots, v_n\}$ is a basis for V.

34. Let W and U be subspaces of a vector space V, and let $W \cap U = \{0\}$. Let $\{w_1, w_2, \ldots, w_k\}$ be a basis for W, and let $\{u_1, u_2, \ldots, u_m\}$ be a basis for U. Show that, if each vector v in V is expressible in the form $v = w + u$ for $w \in W$ and $u \in U$, then

$$\{w_1, w_2, \ldots, w_k, u_1, u_2, \ldots, u_m\}$$

is a basis for V.

35. Illustrate Exercise 34 with nontrivial subspaces W and U of $\mathbb{R}^5$.

▣ *In Exercises 36–39, use the program YUREDUCE or MATCOMP or similar software to cast out vectors in the given list that are linear combinations of predecessors. (See the description of the proper technique on page 138.) List by number the vectors retained.*

36. $v_1 = (5, 4, 3)$, $\quad v_2 = (2, 1, 6)$,
$v_3 = (4, 5, -12)$, $\quad v_4 = (6, 1, 4)$,
$v_5 = (1, 1, 1)$

37. $v_1 = (0, 1, 1, 2)$, $\quad v_2 = (-3, -2, 4, 5)$,
$v_3 = (1, 2, 0, 1)$, $\quad v_4 = (-1, 4, 6, 11)$,
$v_5 = (1, 1, 1, 3)$, $\quad v_6 = (3, 7, 3, 9)$

38. $v_1 = (3, 1, 2, 4, 1)$,
$v_2 = (3, -2, 6, 7, -3)$,
$v_3 = (3, 4, -2, 1, 5)$,
$v_4 = (1, 2, 3, 2, 1)$,
$v_5 = (7, 1, 11, 13, -1)$,
$v_6 = (2, -1, 2, 3, 1)$

39. $v_1 = (2, -1, 3, 4, 1, 2)$,
$v_2 = (-2, 5, 3, -2, 1, -4)$,
$v_3 = (2, 4, 6, 5, 2, 1)$,
$v_4 = (1, -1, 1, -1, 2, 2)$,
$v_5 = (1, 8, 10, 2, 5, -1)$,
$v_6 = (3, 0, 0, 2, 1, 5)$,
$v_7 = (2, 4, 5, 0, 3, 2)$

2.6 **DIMENSION AND RANK**

The Dimension of a Vector Space

We think of $\mathbb{R}$ as being one-dimensional, of $\mathbb{R}^2$ as being two-dimensional, of $\mathbb{R}^3$ as being three-dimensional, and so on. We are finally in a position to give an *algebraic* explanation of this. Indeed, we will establish a notion of dimension for any vector space that can be generated by a finite number of vectors.

Let V be a finitely generated vector space. Although V may have several bases, we will show that they all must have the same number of elements. We demonstrate a key step in the argument.

THEOREM 2.10 Relative Size of Generating and Independent Sets

> Let V be a finitely generated vector space. Any finite generating set for V contains at least as many vectors as exist in any independent set of vectors in V.

PROOF If $V = \{0\}$, the result is trivial. Let $\{\mathbf{v}_1, \mathbf{v}_2, \ldots, \mathbf{v}_k\}$ be an independent set of vectors in V, and let $\{\mathbf{w}_1, \mathbf{w}_2, \ldots, \mathbf{w}_m\}$ be a generating set for V. We will show that $m \geq k$. We may as well assume $k > 1$, since we already have $m \geq 1$.

Consider the set of vectors

$$\{\mathbf{v}_1, \mathbf{w}_1, \mathbf{w}_2, \ldots, \mathbf{w}_m\}.$$

This set generates V, because the $\mathbf{w}_j$ generate V. Moreover, a dependence relation exists among these vectors, since $\mathbf{v}_1$ can be expressed as a linear combination of the $\mathbf{w}_j$. Using the casting-out technique described on page 138, we obtain, after casting out at least one $\mathbf{w}_j$ and renumbering the rest of the $\mathbf{w}_j$ (if necessary), a new set

$$B_1 = \{\mathbf{v}_1, \mathbf{w}_1, \ldots, \mathbf{w}_{m_1}\},$$

which is a basis for V, where $m_1 \leq m - 1$. Since $k > 1$ and since the $\mathbf{v}_i$ are independent, the set $\{\mathbf{v}_1\}$ is not a basis for V, so not all of the $\mathbf{w}_j$ were cast out. We now consider the set of vectors

$$\{\mathbf{v}_1, \mathbf{v}_2, \mathbf{w}_1, \ldots, \mathbf{w}_{m_1}\}$$

and repeat the technique, casting out at least one $\mathbf{w}_j$ to obtain, after renumbering (if necessary), a basis

$$B_2 = \{\mathbf{v}_1, \mathbf{v}_2, \mathbf{w}_1, \ldots, \mathbf{w}_{m_2}\}$$

for V, where $m_2 \leq m - 2$. If $k > 2$, then, since $\mathbf{v}_3$ is not in $\mathrm{sp}(\mathbf{v}_1, \mathbf{v}_2)$, some $\mathbf{w}_j$ remain in B_2; therefore, we consider

$$\{\mathbf{v}_1, \mathbf{v}_2, \mathbf{v}_3, \mathbf{w}_1, \ldots, \mathbf{w}_{m_2}\}$$

and repeat the process. No $\mathbf{v}_i$ is ever cast out, since the set $\{\mathbf{v}_1, \mathbf{v}_2, \ldots, \mathbf{v}_k\}$ is independent, and we must have had at least one $\mathbf{w}_j$ to cast out for each $\mathbf{v}_i$. That is, $m \geq k$, which is precisely what we wanted to show. ▲

Theorem 2.10 shows that a finitely generated vector space cannot have arbitrarily long lists of independent vectors. Suppose that W is a nonzero subspace of a finitely generated vector space V. We can find a basis for W in the following way. First, choose a nonzero vector in W. Next, if possible, choose another vector in W that is not a multiple of the first vector chosen. Continue this process until it is no longer possible to choose another vector in W that is not a linear combination of those already chosen. This process must terminate because V is finitely generated. Thus we must eventually obtain a basis for W. In particular, this shows the following:

> Every subspace of a finitely generated vector space must itself be finitely generated.

We are now ready to show that the number of vectors in a basis for a finitely generated vector space V is an *invariant* of V. That is, the number of elements in a basis is the same for all choices of bases for V. This invariance will enable us to define the *dimension* of V as the number of elements in any basis for V.

THEOREM 2.11   Invariance of Dimension

Any two bases of a finitely generated vector space V have the same number of elements.

PROOF   Let

$$B = \{\mathbf{b}_1, \mathbf{b}_2, \ldots, \mathbf{b}_k\} \quad \text{and} \quad B' = \{\mathbf{b}_1', \mathbf{b}_2', \ldots, \mathbf{b}_m'\}$$

be any two bases for V. Viewing B as an independent set in V, and viewing B' as a generating set for V, we have $k \leq m$. Now viewing B as a generating set for V, and viewing B' as an independent set in V, we obtain $m \leq k$. Therefore, $m = k$.   ▲

DEFINITION 2.10   Dimension

The **dimension** of a finitely generated vector space V is the number of elements in any basis for V, denoted by $\dim(V)$. A vector space V is **finite-dimensional** if it is finitely generated; otherwise, it is **infinite-dimensional**.

Since the unit coordinate vectors $\mathbf{e}_1, \mathbf{e}_2, \ldots, \mathbf{e}_n$ form a basis for $\mathbb{R}^n$, we see that $\dim(\mathbb{R}^n) = n$. Since a basis for the vector space P_n of polynomials of degree $\leq n$ is $\{1, x, x^2, x^3, \ldots, x_n\}$, we see that P_n has dimension $n + 1$. Notice that the vector space $\{\mathbf{0}\}$ has dimension 0. We defined the basis of $\{\mathbf{0}\}$ to be the null set so that this would be true.

EXAMPLE 1   Find the dimension of the subspace

$$W = \text{sp}((1, -3, 1), (-2, 6, -2), (2, 1, -4), (-1, 10, -7))$$

of $\mathbb{R}^3$.

Solution   Example 7 in Section 2.5 showed that

$$\{(1, -3, 1), (2, 1, -4)\}$$

is a basis for W. Consequently, $\dim(W) = 2$.   ▪

We now show how any independent set $\{\mathbf{v}_1, \mathbf{v}_2, \ldots, \mathbf{v}_k\}$ in a finite-dimensional vector space V can be expanded, if necessary, into a basis for V. We select a basis $\{\mathbf{b}_1, \mathbf{b}_2, \ldots, \mathbf{b}_n\}$ for V and form the list of vectors

$$\mathbf{v}_1, \mathbf{v}_2, \ldots, \mathbf{v}_k, \mathbf{b}_1, \mathbf{b}_2, \ldots, \mathbf{b}_n.$$

Since these vectors clearly generate V, we use the casting-out technique to arrive at a basis for V. (See Theorem 2.9.) The basis will contain the independent set $\{\mathbf{v}_1, \mathbf{v}_2, \ldots, \mathbf{v}_k\}$, because no $\mathbf{v}_i$ can be cast out. We summarize this result in an "expanding" theorem, to complement the "reducing" Theorem 2.9.

THEOREM 2.12   Expanding an Independent Set to Become a Basis

Let V be a finite-dimensional vector space. Any independent subset of V can be expanded, if necessary, to become a basis for V.

independent ≠ basis. (but can be "=" set by expansion)

EXAMPLE 2   Expand the independent set $\{(1, 1, 0), (-1, 2, 0)\}$ to a basis for $\mathbb{R}^3$.

Solution   Since $\{(1, 0, 0), (0, 1, 0), (0, 0, 1)\}$ is a basis for $\mathbb{R}^3$, we apply the casting-out technique to the list of vectors

$$(1, 1, 0), (-1, 2, 0), (1, 0, 0), (0, 1, 0), (0, 0, 1),$$

and we arrive at the basis

$$\{(1, 1, 0), (-1, 2, 0), (0, 0, 1)\}$$

for $\mathbb{R}^3$.   ▪

Let V be a finite-dimensional vector space. In order for a subset B of V to be a basis for V, two conditions are required:

1. B must generate V; and
2. B must be independent.

If the dimension of V is known and if the number of vectors in B matches the dimension of V, then either one of the two conditions for a basis is enough to ensure that the other holds. If B consists of n independent vectors, Theorem 2.12 shows that B can be expanded, if necessary, to become a basis for V. But if $\dim(V) = n$, Theorem 2.11 tells us that B cannot be expanded further to become a basis. Consequently B must already be a basis and must generate V. On the other hand, Theorem 2.9 shows that, if B generates V, then B can be reduced to a basis for V. But if the number of elements in B matches $\dim(V)$, we again conclude that B must already be a basis for V, so B is an independent set. We summarize this analysis in a theorem that is similar to Theorem 2.7.

THEOREM 2.13   A Test for dim(V) Vectors to Form a Basis for V

Let V be a finite-dimensional vector space with $\dim(V) = n$. For a set B of n vectors from V, the following statements are equivalent:

1. B is linearly independent.

2. B generates V.

In either case, B is a basis for V.

EXAMPLE 3   Use Theorem 2.13 to show that $\{1, x - 1, (x - 1)^2\}$ is a basis for the polynomial space P_2.

Solution   We check that the given set is independent in P_2. If

$$c_1 + c_2(x - 1) + c_3(x - 1)^2 = 0,$$

then

$$(c_1 - c_2 + c_3) + (c_2 - 2c_3)x + c_3x^2 = 0.$$

Equating the coefficients to zero, we find that $c_3 = c_2 = c_1 = 0$, so the vectors $1, x - 1$, and $(x - 1)^2$ are independent. Since $\dim(P_2) = 3$ and our set has three independent vectors, Theorem 2.13 shows that the set is a basis for P_2.   ■

The Rank of a Matrix

Three important numbers are associated with an $m \times n$ matrix A, in addition to m and n:

1. The dimension of the column space of A, which is called the **column rank** of A and is written as colrank(A).

2. The dimension of the row space of A, which is called the **row rank** of A and is written as rowrank(A).

3. The dimension of the nullspace of A, which is called the **nullity** of A and is written as nullity(A).

The relationship among these three numbers can be seen by row-reducing the matrix A to echelon form H. For example, suppose that a matrix A has been reduced to the echelon form

to find dim(A)
of m×n matrix.

$$H = \begin{bmatrix} p_1 & X & X & X & X & X & X \\ 0 & 0 & p_3 & X & X & X & X \\ 0 & 0 & 0 & p_4 & X & X & X \\ 0 & 0 & 0 & 0 & 0 & 0 & p_7 \\ 0 & 0 & 0 & 0 & 0 & 0 & 0 \\ 0 & 0 & 0 & 0 & 0 & 0 & 0 \end{bmatrix}, \qquad (1)$$

where the entries p_i are the pivots. As explained in the description of the casting-out technique on page 138, the column vectors of A that correspond to the columns containing the pivots in Eq. (1) form a basis for the column space of A. Consequently, the column rank of A is the number of pivots. On the other hand, we find the dimension of the nullspace of A by counting the columns in Eq. (1) that do not contain pivots; they give rise to free variables in the solution of the homogeneous system $Ax = 0$, as explained in Section 1.2. In our illustration, we have free variables x_2, x_5, and x_6 in the system

process to
find dim(A)
of nullspace

$$\begin{bmatrix} p_1 & X & X & X & X & X & X \\ 0 & 0 & p_3 & X & X & X & X \\ 0 & 0 & 0 & p_4 & X & X & X \\ 0 & 0 & 0 & 0 & 0 & 0 & p_7 \\ 0 & 0 & 0 & 0 & 0 & 0 & 0 \\ 0 & 0 & 0 & 0 & 0 & 0 & 0 \end{bmatrix} \begin{bmatrix} x_1 \\ x_2 \\ x_3 \\ x_4 \\ x_5 \\ x_6 \\ x_7 \end{bmatrix} = \begin{bmatrix} 0 \\ 0 \\ 0 \\ 0 \\ 0 \\ 0 \\ 0 \end{bmatrix}. \qquad (2)$$

The bound variables x_1, x_3, x_4, and x_7 can be solved for in terms of the free variables. The technique of back substitution provides numbers a_1, a_2, b_1, b_2, c_1, c_2, and c_3 such that

$$\begin{bmatrix} x_1 \\ x_2 \\ x_3 \\ x_4 \\ x_5 \\ x_6 \\ x_7 \end{bmatrix} = \begin{bmatrix} c_1 x_2 + c_2 x_5 + c_3 x_6 \\ x_2 \\ b_1 x_5 + b_2 x_6 \\ a_1 x_5 + a_2 x_6 \\ x_5 \\ x_6 \\ 0 \end{bmatrix} = x_2 \begin{bmatrix} c_1 \\ 1 \\ 0 \\ 0 \\ 0 \\ 0 \\ 0 \end{bmatrix} + x_5 \begin{bmatrix} c_2 \\ 0 \\ b_1 \\ a_1 \\ 1 \\ 0 \\ 0 \end{bmatrix} + x_6 \begin{bmatrix} c_3 \\ 0 \\ b_2 \\ a_2 \\ 0 \\ 1 \\ 0 \end{bmatrix}.$$

$$\qquad\qquad\qquad \mathbf{v}_1 \qquad\quad \mathbf{v}_2 \qquad\quad \mathbf{v}_3$$

In this way the nullspace of A is seen to be $\mathrm{sp}(\mathbf{v}_1, \mathbf{v}_2, \mathbf{v}_3)$. The vectors $\mathbf{v}_1$, $\mathbf{v}_2$, $\mathbf{v}_3$ are independent, since, in coordinate positions 2, 5, and 6, the color components are the same as those of the independent vectors $\mathbf{e}_1$, $\mathbf{e}_2$, $\mathbf{e}_3$ in the standard basis for $\mathbb{R}^3$. Thus, nullity(A) is the number of columns that contain no pivots of the echelon matrix H.

Finally, since every column of a row-echelon matrix H either contains a pivot or does not contain a pivot, we see that the sum of the column rank of A and the nullity of A must be the number of columns in A. We summarize our discussion in a theorem; its proof follows precisely the preceding line of reasoning and so is omitted.

THEOREM 2.14 Rank Equation

Let A be an $m \times n$ matrix with row echelon form H. Then:

1. nullity(A) = (Number of free variables in the solution space of $A\mathbf{x} = \mathbf{0}$) = (Number of pivot-free columns in H);
2. colrank(A) = (Number of pivots in H);
3. (*Rank equation*): colrank(A) + nullity(A) = (Number of columns of A).

Since nullity(A) is defined as the number of vectors in a basis of the nullspace of A, <u>the invariance of dimension shows that the number of free variables obtained in</u> <u>the solution of a linear system $A\mathbf{x} = \mathbf{b}$ is independent of the steps in the row reduction</u> <u>to echelon form</u>, as we asserted in Section 1.2.

EXAMPLE 4 Find bases for the column space of A and for the nullspace of A if

$$A = \begin{bmatrix} 1 & 2 & 0 & -1 & 1 \\ 1 & 3 & 1 & 1 & -1 \\ 2 & 5 & 1 & 0 & 0 \\ 3 & 6 & 0 & 0 & -6 \\ 1 & 5 & 3 & 5 & -5 \end{bmatrix}.$$

Solution We find an echelon form of A:

$$A = \begin{bmatrix} 1 & 2 & 0 & -1 & 1 \\ 1 & 3 & 1 & 1 & -1 \\ 2 & 5 & 1 & 0 & 0 \\ 3 & 6 & 0 & 0 & -6 \\ 1 & 5 & 3 & 5 & -5 \end{bmatrix} \sim \begin{bmatrix} 1 & 2 & 0 & -1 & 1 \\ 0 & 1 & 1 & 2 & -2 \\ 0 & 1 & 1 & 2 & -2 \\ 0 & 0 & 0 & 3 & -9 \\ 0 & 3 & 3 & 6 & -6 \end{bmatrix}$$

$$\sim \begin{bmatrix} 1 & 2 & 0 & -1 & 1 \\ 0 & 1 & 1 & 2 & -2 \\ 0 & 0 & 0 & 0 & 0 \\ 0 & 0 & 0 & 3 & -9 \\ 0 & 0 & 0 & 0 & 0 \end{bmatrix} \sim \begin{bmatrix} 1 & 2 & 0 & -1 & 1 \\ 0 & 1 & 1 & 2 & -2 \\ 0 & 0 & 0 & 3 & -9 \\ 0 & 0 & 0 & 0 & 0 \\ 0 & 0 & 0 & 0 & 0 \end{bmatrix} = H.$$

THE RANK OF A MATRIX was defined in 1879 by Georg Frobenius (1849–1917) as follows: if all determinants of the $(r + 1)$st degree vanish, but not all those of the rth degree, then r is the rank of the matrix. He used this concept to deal with the questions of canonical forms for certain matrices of integers and with the solutions of certain systems of linear congruences.

On the other hand, the nullity was defined by James Sylvester in 1884 for square matrices as follows: the nullity of an $n \times n$ matrix is i if every minor (determinant) of order $n - i + 1$ (and therefore of every higher order) equals 0, and i is the largest such number for which this is true. Sylvester was interested here, as in much of his mathematical research, in discovering *invariants*—properties of particular mathematical objects that do not change under specified types of transformation. He proceeded to prove what he called one of the cardinal laws in the theory of matrices: the nullity of the product of two matrices is not less than the nullity of any factor nor greater than the sum of the nullities of the factors.

Since we have pivots in columns 1, 2, and 4 of H, the corresponding column vectors

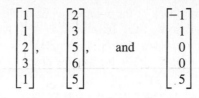

of A form a basis for the column space of A, and colrank$(A) = 3$. Turning to the nullspace of A, we see that columns 3 and 5 of H contain no pivot. This gives rise to free variables x_3 and x_5 in the solution set of the homogeneous system $A\mathbf{x} = \mathbf{0}$. Looking at the matrix H and using back substitution, we can describe all solutions of this system by the equations

$$\begin{bmatrix} x_1 \\ x_2 \\ x_3 \\ x_4 \\ x_5 \end{bmatrix} = \begin{bmatrix} 2x_3 + 10x_5 \\ -x_3 - 4x_5 \\ x_3 \\ 3x_5 \\ x_5 \end{bmatrix} = x_3 \begin{bmatrix} 2 \\ -1 \\ 1 \\ 0 \\ 0 \end{bmatrix} + x_5 \begin{bmatrix} 10 \\ -4 \\ 0 \\ 3 \\ 1 \end{bmatrix}.$$

The two explicit solutions in this last equation form a basis for the nullspace of A and are obtained as follows:

Basis for $\mathbb{R}^2$	x_3	x_5	**Basis for the Nullspace**
$\mathbf{e}_1$	1	0	$\mathbf{v}_1 = (2,\ -1,\ 1,\ 0,\ 0)$
$\mathbf{e}_2$	0	1	$\mathbf{v}_2 = (10,\ -4,\ 0,\ 3,\ 1)$

Thus, $\{\mathbf{v}_1, \mathbf{v}_2\}$ is a basis for the nullspace of A, and nullity$(A) = 2$. Notice that colrank(A) + nullity$(A) = 3 + 2 = 5 = $ (Number of columns of A). ■

We turn to the row rank of a matrix A. The elementary row operations do not alter the row space of A (see Exercise 20); thus the row rank of A is the same as the row rank of its echelon form. The row rank of a matrix H in row echelon form is the number of nonzero rows, which is also the number of pivots. For example, the first three rows of the matrix H in the previous example form a basis for the row space of H. In view of Theorem 2.14, we conclude that the row rank of a matrix is the same as its column rank. That is:

$$\text{rowrank}(A) = \text{colrank}(A).$$

We often refer simply to the **rank** of a matrix, which is this common value, and we write rank(A).

EXAMPLE 5   Find the rank of the matrix A in Example 4, and find a basis for the row space of A.

Solution   Looking back at the solution of Example 4, we see three pivots in the echelon form H of A, so we must have rank$(A) = 3$. A basis for the row space of A consists of the first three rows (that is, all nonzero rows) of H. Notice that the first three rows of A *do not* form a basis for the row space of A. ■

Let us summarize the way to find bases for three vector spaces related to a matrix A from a row echelon form H of A.

Finding Bases for Spaces Associated with a Matrix

Let A be an $m \times n$ matrix with row echelon form H.

1. For a basis of the row space of A, use the nonzero rows of H.
2. For a basis of the column space of A, use the columns of A corresponding to the columns of H containing pivots.
3. For a basis of the nullspace of A, use H and back substitution to solve $H\mathbf{x} = \mathbf{0}$ in the usual way (see Example 4).

Our work has given us still another criterion for the invertibility of a square matrix.

THEOREM 2.15   An Invertibility Criterion

An $n \times n$ matrix A is invertible if and only if rank$(A) = n$.

We close this chapter with a result that will be needed in Chapter 5. Recall that the transpose A^T of an $m \times n$ matrix A is an $n \times m$ matrix. The product $(A^T)A$ is then an $n \times n$ symmetric matrix, as we ask you to show in Exercise 25.

THEOREM 2.16   The Rank of $(A^T)A$

Let A be an $m \times n$ matrix of rank r. Then the $n \times n$ symmetric matrix $(A^T)A$ also has rank r.

PROOF   We will work with nullspaces. If $\mathbf{v}$ is any solution vector to the system $A\mathbf{x} = \mathbf{0}$, so that $A\mathbf{v} = \mathbf{0}$, then upon multiplying the sides of this last equation on the left by A^T, we see that $\mathbf{v}$ is also a solution to the system $(A^T)A\mathbf{x} = \mathbf{0}$. Conversely, assume that $(A^T)A\mathbf{w} = \mathbf{0}$ for an $n \times 1$ vector $\mathbf{w}$. Then

$$0 = \mathbf{w}^T[(A^T)A\mathbf{w}] = (A\mathbf{w})^T(A\mathbf{w}),$$

which may be written $\|A\mathbf{w}\|^2 = 0$. That is, $A\mathbf{w}$ is a vector of magnitude 0, and consequently $A\mathbf{w} = \mathbf{0}$. It follows that A and $(A^T)A$ have the same nullspace. Since they have the same number of columns, we apply Theorem 2.14 and conclude that rank(A) = rank((A^T)A). ▲

SUMMARY

1. Let V be a finite-dimensional vector space. All bases of V have the same number of vectors. This number dim(V) is called the *dimension* of V.

2. The dimensions of the row space, column space, and nullspace of an $m \times n$ matrix A are called the *row rank*, *column rank*, and *nullity* of A, respectively.

3. The row rank and column rank of any matrix A are equal and are referred to as the *rank* of A. The rank of A is equal to the number of (nonzero) pivots in an echelon reduction of A.

4. The nullity of A plus the rank of A equals the number of columns of A.

5. An $n \times n$ matrix A is invertible if and only if rank(A) = n.

EXERCISES

In Exercises 1–4, find the dimension of the given subspace of $\mathbb{R}^n$.

1. The subspace $\{(x_1, x_2, x_3) \mid x_1 + x_2 + x_3 = 0\}$ in $\mathbb{R}^3$

   Homogeneous system

2. The subspace sp$((1, 2, 0), (2, 1, 1), (0, -3, 1), (3, 0, 2))$ in $\mathbb{R}^3$

3. The nullspace of the matrix
$$\begin{bmatrix} 2 & 0 & 1 \\ -1 & 1 & 0 \\ 3 & 1 & 2 \end{bmatrix}$$

4. The subspace
$$\{(x_1, x_2, x_3, x_4, x_5) \mid x_1 + x_2 = 0, x_2 = x_4,$$
$$x_5 = x_1 + 3x_3\}$$
   of $\mathbb{R}^5$ *Homo. System.*

In Exercises 5–8, expand the given independent set to become a basis of the entire space.

5. $\{(1, 2, 1)\}$ in $\mathbb{R}^3$

6. $\{(2, 1, 1, 1), (1, 0, 1, 1)\}$ in $\mathbb{R}^4$

7. $\{x^2 + 1\}$ in P_3 8. $\{1 + x, (1 + x)^2\}$ in P_3

For the matrices in Exercises 9–14, find (a) the rank of the matrix, (b) a basis for the row space, (c) a basis for the column space, and (d) a basis for the nullspace.

9. $\begin{bmatrix} 2 & 0 & -3 & 1 \\ 3 & 4 & 2 & 2 \end{bmatrix}$

10. $\begin{bmatrix} 5 & -1 & 0 & 2 \\ 1 & 2 & 1 & 0 \\ 3 & 1 & -2 & 4 \\ 0 & 4 & -1 & 2 \end{bmatrix}$

11. $\begin{bmatrix} 0 & 6 & 6 & 3 \\ 1 & 2 & 1 & 1 \\ 4 & 1 & -3 & 4 \\ 1 & 3 & 2 & 0 \end{bmatrix}$

12. $\begin{bmatrix} 3 & 1 & 4 & 2 \\ -1 & 0 & -1 & 0 \\ 2 & 1 & 0 & 1 \\ 1 & 0 & -1 & 1 \end{bmatrix}$

13. $\begin{bmatrix} 0 & 1 & 2 & 1 \\ 2 & 1 & 0 & 2 \\ 0 & 2 & 1 & 1 \end{bmatrix}$

14. $\begin{bmatrix} 0 & 2 & 3 & 1 \\ -4 & 4 & 1 & 4 \\ 3 & 3 & 2 & 0 \\ -4 & 0 & 1 & 2 \end{bmatrix}$

Whenever there exists the fraction, make it integer.

In Exercises 15–18, determine whether the given matrix is invertible, by finding its rank.

15. $\begin{bmatrix} 0 & -9 & -9 & 2 \\ 1 & 2 & 1 & 1 \\ 4 & 1 & -3 & 4 \\ 1 & 3 & 2 & 0 \end{bmatrix}$

16. $\begin{bmatrix} 2 & 3 & 1 \\ 4 & -1 & 2 \\ 1 & 0 & 1 \end{bmatrix}$

17. $\begin{bmatrix} 2 & 0 & 1 \\ 0 & 0 & 4 \\ 2 & 4 & 0 \end{bmatrix}$ **18.** $\begin{bmatrix} 3 & 0 & -1 & 2 \\ 4 & 2 & 1 & 8 \\ 1 & 4 & 0 & 1 \\ 2 & 6 & -3 & 1 \end{bmatrix}$

19. Mark each of the following True or False.

____ a) The dimension of the space P_n of polynomials of degree $\leq n$ is n.

____ b) The rank of an $m \times n$ matrix can be any number from 0 to the maximum of m and n.

____ c) The rank of an $m \times n$ matrix can be any number from 0 to the minimum of m and n.

____ d) The nullity of an $m \times n$ matrix can be any number from 0 to n.

____ e) The nullity of an $m \times n$ matrix can be any number from 0 to m.

____ f) The rank of an invertible matrix must equal the number of rows of the matrix.

____ g) If the rank of an $m \times n$ matrix is as large as possible for a matrix of that size, either AA^T or A^TA is an invertible matrix.

____ h) Every independent subset of a vector space V is a subset of every basis for V.

____ i) Every independent subset of a vector space V is a part of some basis for V.

____ j) Any two bases in a finite-dimensional vector space V have the same number of elements.

20. Show that the row space and the row rank of a matrix are not affected by the three elementary row operations.

21. Show that the column space and the column rank of a matrix are not affected by the three elementary column operations. [HINT: Consider the transpose.]

22. Prove that, if W is a subspace of an n-dimensional vector space V and $\dim(W) = n$, then $W = V$.

23. Show that, if A is a square matrix, the nullity of A is the same as the nullity of A^T.

24. Let A be an $m \times n$ matrix, and let $\mathbf{b}$ be an $n \times 1$ vector. Prove that the system of equations $A\mathbf{x} = \mathbf{b}$ has a solution for $\mathbf{x}$ if and only if $\text{rank}(A) = \text{rank}(A \mid \mathbf{b})$, where $\text{rank}(A \mid \mathbf{b})$ represents the rank of the associated partitioned matrix $[A \mid \mathbf{b}]$ of the system.

25. Prove that, for any $m \times n$ matrix A, both $(A^T)A$ and $A(A^T)$ are symmetric matrices.

In Exercises 26–30, let A and C be matrices such that the product AC is defined.

26. Show that the column space of AC is contained in the column space of A. [HINT: See the box on page 121.]

27. Show that $\text{rank}(AC) \leq \text{rank}(A)$.

28. Give an example where $\text{rank}(AC) < \text{rank}(A)$.

29. Is it true that the column space of AC is contained in the column space of C? Explain.

30. Is it true that $\text{rank}(AC) \leq \text{rank}(C)$? Explain.

31. Let A be an $m \times n$ matrix. Prove that $\text{rank}(A(A^T)) = \text{rank}(A)$. (See Theorem 2.16.)

32. If $\mathbf{a}$ is an $n \times 1$ vector and $\mathbf{b}$ is a $1 \times m$ vector, show that $\mathbf{ab}$ is an $n \times m$ matrix of rank at most one.

2.7   COORDINATIZATION OF VECTORS

Most of the work in this text is phrased in terms of the vector spaces $\mathbb{R}^n$ for $n = 1$, $2, 3, \ldots$. In this section, we show that, for finite-dimensional vector spaces, no loss of generality results from restricting ourselves to the spaces $\mathbb{R}^n$. Specifically, we will see that, if a vector space V has a finite basis of n vectors, V can be *coordinatized* so that it will look just like $\mathbb{R}^n$. We can then work with these coordinates, utilizing the matrix techniques we have developed for the spaces $\mathbb{R}^n$.

Ordered Bases

The vector $(2, 5)$ in $\mathbb{R}^2$ can be expressed in terms of the standard basis as $2\mathbf{e}_1 + 5\mathbf{e}_2$. The components of $(2, 5)$ are precisely the coefficients of these standard basis vectors. The vector $(2, 5)$ is different from the vector $(5, 2)$. We regard the standard basis vectors as having a natural order, $\mathbf{e}_1 = (1, 0)$ and $\mathbf{e}_2 = (0, 1)$. In a nonzero vector space V with a basis $B = \{\mathbf{b}_1, \mathbf{b}_2, \ldots, \mathbf{b}_n\}$, there is usually no natural order for the basis vectors. For example, the vectors $\mathbf{b}_1 = (-1, 5)$ and $\mathbf{b}_2 = (3, 2)$ form a basis for $\mathbb{R}^2$, but there is no natural order for these vectors. If we want the vectors to have an order, we must specify their order. By convention, set notation does not denote order; for example, $\{\mathbf{b}_1, \mathbf{b}_2\} = \{\mathbf{b}_2, \mathbf{b}_1\}$. To describe order, we use parentheses, ( ), in place of set brackets, { }; we are used to paying attention to order in the notation $(\mathbf{b}_1, \mathbf{b}_2)$. We denote an *ordered basis* of n vectors in V by $B = (\mathbf{b}_1, \mathbf{b}_2, \ldots, \mathbf{b}_n)$. For example, the standard basis $\{\mathbf{e}_1, \mathbf{e}_2, \mathbf{e}_3\}$ of $\mathbb{R}^3$ gives rise to six different ordered bases—namely,

$$(\mathbf{e}_1, \mathbf{e}_2, \mathbf{e}_3) \qquad (\mathbf{e}_2, \mathbf{e}_1, \mathbf{e}_3) \qquad (\mathbf{e}_3, \mathbf{e}_1, \mathbf{e}_2)$$
$$(\mathbf{e}_1, \mathbf{e}_3, \mathbf{e}_2) \qquad (\mathbf{e}_2, \mathbf{e}_3, \mathbf{e}_1) \qquad (\mathbf{e}_3, \mathbf{e}_2, \mathbf{e}_1).$$

These correspond to the six possible orders for the unit coordinate vectors. The ordered basis $(\mathbf{e}_1, \mathbf{e}_2, \mathbf{e}_3)$ is the *standard ordered basis* for $\mathbb{R}^3$, and in general, the basis $E = (\mathbf{e}_1, \mathbf{e}_2, \ldots, \mathbf{e}_n)$ is the *standard ordered basis* for $\mathbb{R}^n$.

Coordinatization of Vectors

Let V be a finite-dimensional vector space, and let $B = (\mathbf{b}_1, \mathbf{b}_2, \ldots, \mathbf{b}_n)$ be a basis for V. By Theorem 2.8, every vector $\mathbf{v}$ in V can be expressed in the form

$$\mathbf{v} = r_1\mathbf{b}_1 + r_2\mathbf{b}_2 + \cdots + r_n\mathbf{b}_n$$

for unique scalars $r_1, r_2, \ldots, r_n$. We associate the vector $(r_1, r_2, \ldots, r_n)$ in $\mathbb{R}^n$ with $\mathbf{v}$. This gives us a way of coordinatizing V.

DEFINITION 2.11 Coordinate Vector Relative to an Ordered Basis

Let $B = (\mathbf{b}_1, \mathbf{b}_2, \ldots, \mathbf{b}_n)$ be an ordered basis for a finite-dimensional vector space V, and let

$$\mathbf{v} = r_1\mathbf{b}_1 + r_2\mathbf{b}_2 + \cdots + r_n\mathbf{b}_n.$$

The vector $(r_1, r_2, \ldots, r_n)$ is the **coordinate vector of v relative to the ordered basis** B, and is denoted by $\mathbf{v}_B$.

EXAMPLE 1 Find the coordinate vectors of $(1, -1)$ and of $(-1, -8)$ relative to the ordered basis $B = ((1, -1), (1, 2))$ of $\mathbb{R}^2$.

Solution   We see that $(1, -1)_B = (1, 0)$, since

$$(1, -1) = 1(1, -1) + 0(1, 2).$$

To find $(-1, -8)_B$, we must find r_1 and r_2 such that $(-1, -8) = r_1(1, -1) + r_2(1, 2)$. Equating components of this vector equation, we obtain the linear system

$$r_1 + r_2 = -1$$
$$-r_1 + 2r_2 = -8.$$

The solution of this system is $r_1 = 2$, $r_2 = -3$, so we have $(-1, -8)_B = (2, -3)$. Figure 2.24 indicates the geometric meaning of these coordinates.   ■

EXAMPLE 2   Find the coordinate vector of $(1, 2, -2)$ relative to the ordered basis $B = ((1, 1, 1), (1, 2, 0), (1, 0, 1))$ in $\mathbb{R}^3$.

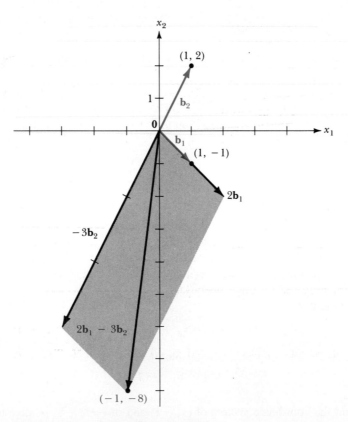

FIGURE 2.24 $(-1, -8)_B = (2, -3)$.

Solution   We must express $(1, 2, -2)$ as a linear combination of the basis vectors in B. Working with column vectors, we must solve the equation

$$r_1 \begin{bmatrix} 1 \\ 1 \\ 1 \end{bmatrix} + r_2 \begin{bmatrix} 1 \\ 2 \\ 0 \end{bmatrix} + r_3 \begin{bmatrix} 1 \\ 0 \\ 1 \end{bmatrix} = \begin{bmatrix} 1 \\ 2 \\ -2 \end{bmatrix}$$

for r_1, r_2, and r_3. We find the unique solution by a Gauss–Jordan reduction:

$$\begin{bmatrix} 1 & 1 & 1 & | & 1 \\ 1 & 2 & 0 & | & 2 \\ 1 & 0 & 1 & | & -2 \end{bmatrix} \sim \begin{bmatrix} 1 & 1 & 1 & | & 1 \\ 0 & 1 & -1 & | & 1 \\ 0 & -1 & 0 & | & -3 \end{bmatrix}$$

$$\sim \begin{bmatrix} 1 & 0 & 2 & | & 0 \\ 0 & 1 & -1 & | & 1 \\ 0 & 0 & -1 & | & -2 \end{bmatrix} \sim \begin{bmatrix} 1 & 0 & 0 & | & -4 \\ 0 & 1 & 0 & | & 3 \\ 0 & 0 & 1 & | & 2 \end{bmatrix}.$$

Therefore, $(1, 2, -2)_B = (-4, 3, 2)$. ■

We now box the procedure illustrated by Example 2.

Finding the Coordinate Vector of v in $\mathbb{R}^n$ Relative to an Ordered Basis $B = (\mathbf{b}_1, \mathbf{b}_2, \ldots, \mathbf{b}_n)$

Step 1   Writing vectors as column vectors, form the partitioned matrix $[\mathbf{b}_1, \mathbf{b}_2, \ldots, \mathbf{b}_n \mid \mathbf{v}]$.

Step 2   Use a Gauss–Jordan reduction to obtain the partitioned matrix $[I \mid \mathbf{v}_B]$, where I is the $n \times n$ identity matrix and $\mathbf{v}_B$ is the desired coordinate vector.

Coordinatization of a Finite-dimensional Vector Space

We can coordinatize a finite-dimensional vector space V by selecting an ordered basis $B = (\mathbf{b}_1, \mathbf{b}_2, \ldots, \mathbf{b}_n)$ and associating with each vector in V its coordinate vector relative to B. To show that we may now work in $\mathbb{R}^n$ rather than in V, we have to show that the vector-space operations of vector addition and scalar multiplication in V are mirrored by those operations on coordinate vectors in $\mathbb{R}^n$. That is, we must show that

$$(\mathbf{v} + \mathbf{w})_B = \mathbf{v}_B + \mathbf{w}_B \quad \text{and} \quad (t\mathbf{v})_B = t\mathbf{v}_B \tag{1}$$

for all vectors $\mathbf{v}$ and $\mathbf{w}$ in V and for all scalars t in $\mathbb{R}$. To do this, suppose that

$$\mathbf{v} = r_1\mathbf{b}_1 + r_2\mathbf{b}_2 + \cdots + r_n\mathbf{b}_n$$

and

$$\mathbf{w} = s_1\mathbf{b}_1 + s_2\mathbf{b}_2 + \cdots + s_n\mathbf{b}_n.$$

Since

$$\mathbf{v} + \mathbf{w} = (r_1 + s_1)\mathbf{b}_1 + (r_2 + s_2)\mathbf{b}_2 + \cdots + (r_n + s_n)\mathbf{b}_n,$$

we see that the coordinate vector of $\mathbf{v} + \mathbf{w}$ is

$$
\begin{aligned}
(\mathbf{v} + \mathbf{w})_B &= (r_1 + s_1, r_2 + s_2, \ldots, r_n + s_n) \\
&= (r_1, r_2, \ldots, r_n) + (s_1, s_2, \ldots, s_n) \\
&= \mathbf{v}_B + \mathbf{w}_B,
\end{aligned}
$$

which is the sum of the coordinate vectors of $\mathbf{v}$ and of $\mathbf{w}$. Similarly, for any scalar t, we have

$$
\begin{aligned}
t\mathbf{v} &= t(r_1\mathbf{b}_1 + r_2\mathbf{b}_2 + \cdots + r_n\mathbf{b}_n) \\
&= (tr_1)\mathbf{b}_1 + (tr_2)\mathbf{b}_2 + \cdots + (tr_n)\mathbf{b}_n,
\end{aligned}
$$

so the coordinate vector of $t\mathbf{v}$ is

$$
\begin{aligned}
(t\mathbf{v})_B &= (tr_1, tr_2, \ldots, tr_n) \\
&= t(r_1, r_2, \ldots, r_n) = t\mathbf{v}_B.
\end{aligned}
$$

This completes the demonstration of relations (1). These relations tell us that, when we rename the vectors in V by coordinates relative to B, the resulting vector space of coordinates, namely $\mathbb{R}^n$, has the same vector-space structure as V. Whenever the vectors in a vector space V can be renamed to make V appear structurally identical to a vector space W, we say that V and W are *isomorphic vector spaces*. Relations (1) show that every real vector space having a basis of n vectors is isomorphic to $\mathbb{R}^n$. For example, the space P_n of all polynomials of degree at most n is isomorphic to $\mathbb{R}^{n+1}$, because P_n has an ordered basis $B = (x^n, x^{n-1}, \ldots, x^2, x, 1)$ of $n + 1$ vectors. Each polynomial

$$a_n x^n + a_{n-1} x^{n-1} + \cdots + a_1 x + a_0$$

can be renamed by its coordinate vector

$$(a_n, a_{n-1}, \ldots, a_1, a_0)$$

relative to B. The adjective *isomorphic* is used throughout algebra to signify that two algebraic structures are identical except in the names of their elements.

For a vector space V isomorphic to a vector space W, all of the algebraic properties of vectors in V that can be derived solely from the axioms of a vector space correspond to identical properties in W. However, we cannot expect other features—such as whether the vectors are functions, matrices, or n-tuples—to be carried over from one space to the other. But a generating set of vectors or an independent set of vectors in one space corresponds to a set with the same property in the other space. Here is an example showing how we can simplify computations in a finitely generated vector space V, by working instead in $\mathbb{R}^n$.

EXAMPLE 3 Determine whether $x^2 - 3x + 2$, $3x^2 + 5x - 4$, and $7x^2 + 21x - 16$ are independent in the vector space P_2 of polynomials of degree at most 2.

Solution   We take $B = (x^2, x, 1)$ as an ordered basis for P_2. The coordinate vectors relative to B of the given polynomials are

$$(x^2 - 3x + 2)_B = (1, -3, 2),$$

$$(3x^2 + 5x - 4)_B = (3, 5, -4),$$

$$(7x^2 + 21x - 16)_B = (7, 21, -16).$$

We can determine whether the polynomials are independent by determining whether the corresponding coordinate vectors in $\mathbb{R}^3$ are independent. We set up the usual matrix, with these vectors as column vectors, and then we row-reduce it, obtaining

$$\begin{bmatrix} 1 & 3 & 7 \\ -3 & 5 & 21 \\ 2 & -4 & -16 \end{bmatrix} \sim \begin{bmatrix} 1 & 3 & 7 \\ 0 & 14 & 42 \\ 0 & -10 & -30 \end{bmatrix} \sim \begin{bmatrix} 1 & 3 & 7 \\ 0 & 1 & 3 \\ 0 & 0 & 0 \end{bmatrix}. \sim \begin{matrix} 1 & 0 & -2 \\ 0 & 1 & 3 \\ 0 & 0 & 2 \end{matrix}$$

Since the third column in the echelon form has no pivot, these three coordinate vectors in $\mathbb{R}^3$ are dependent, so the three polynomials are dependent. ▪

check: $-2(1,-3,2)+3(3,5,-4)$

$= (-2+9, 6+15, -4-12) = (7, 21, -16)$

Change of Basis

standard ordered basis.

Let $B = (\mathbf{b}_1, \mathbf{b}_2, \ldots, \mathbf{b}_n)$ be an ordered basis in $\mathbb{R}^n$, and let $E = (\mathbf{e}_1, \mathbf{e}_2, \ldots, \mathbf{e}_n)$ be the standard ordered basis. If $\mathbf{v}$ is a vector in $\mathbb{R}^n$ with column coordinate vector

$$\mathbf{v}_B = \begin{bmatrix} r_1 \\ r_2 \\ \vdots \\ r_n \end{bmatrix} \qquad (2)$$

relative to B, then

$$r_1\mathbf{b}_1 + r_2\mathbf{b}_2 + \cdots + r_n\mathbf{b}_n = \mathbf{v}. \qquad (3)$$

Let M_B be the matrix having the vectors in the ordered basis B as column vectors; this is the **basis matrix** for B, which we display as

$$M_B = \begin{bmatrix} | & | & & | \\ \mathbf{b}_1 & \mathbf{b}_2 & \cdots & \mathbf{b}_n \\ | & | & & | \end{bmatrix} \qquad (4)$$

Equation (3), which expresses $\mathbf{v}$ as a vector in the column space of the matrix M_B, can be written in the form

ex. $\{(1,2,3) \ (1,0,2), (2,30)\}$
ordered basis.

$$M_B\mathbf{v}_B = \mathbf{v}$$

$M_B = \begin{matrix} 1 & 1 & 2 \\ 2 & 0 & 3 \\ 3 & 2 & 0 \end{matrix}$.

or, equivalently,

$$M_B\mathbf{v}_B = \boxed{\mathbf{v}_E}. \qquad (5)$$

$V_B = \begin{bmatrix} r_1 \\ r_2 \\ r_3 \end{bmatrix}$

Multiplication on the left by M_B thus converts the column coordinate vector of **v** relative to B into the column coordinate vector of **v** relative to the standard ordered basis E. If B' is another ordered basis for $\mathbb{R}^n$, we can similarly obtain

$$M_{B'}\mathbf{v}_{B'} = \mathbf{v}_E. \tag{6}$$

Equations 5 and 6 together yield

$$M_{B'}\mathbf{v}_{B'} = M_B\mathbf{v}_B. \tag{7}$$

Equation (7) is easy to remember, and it turns out to be very useful. Both M_B and $M_{B'}$ are invertible, since they are square matrices and since their column vectors are independent. Thus, Eq. (7) yields

$$\mathbf{v}_{B'} = M_{B'}^{-1}M_B\mathbf{v}_B. \tag{8}$$

Equation (8) shows that, given any two ordered bases B and B' of $\mathbb{R}^n$, there exists an invertible matrix C, namely,

$$C = M_{B'}^{-1}M_B, \tag{9}$$

such that

$$\mathbf{v}_{B'} = C\mathbf{v}_B. \tag{10}$$

If we know this matrix C, we can conveniently convert coordinates relative to B into coordinates relative to B'. The matrix C in Eq. (9) is the unique matrix satisfying Eq. (10). This can be seen by assuming that D is also a matrix satisfying Eq. (10), so that $\mathbf{v}_{B'} = D\mathbf{v}_B$; then $C\mathbf{v}_B = D\mathbf{v}_B$ for all vectors **v** in $\mathbb{R}^n$. Exercise 41 of Section 1.1 shows that we must have $C = D$.

In view of our earlier work in this section, we see that all our results here are valid for coordinates with respect to ordered bases B and B' in any finite-dimensional vector space. We phrase the definition that follows in these terms.

DEFINITION 2.12 Change-of-Coordinates Matrix

Let B and B' be ordered bases for a finite-dimensional vector space V. The **change-of-coordinates matrix** from B to B' is the unique matrix C such that $C\mathbf{v}_B = \mathbf{v}_{B'}$ for all vectors **v** in V. It is also denoted by $C_{B,B'}$.

The term *transition matrix* is used in some texts in place of *change-of-coordinates matrix.* But we used *transition matrix* with reference to Markov chains in Section 1.6, so we avoid duplicate terminology here by using the more descriptive term *change-of-coordinates matrix*.

Definition 2.12 and Eq. (5) show that M_B of Eq. (4) is the change-of-coordinates matrix from B to E.

Equation (10), written in the form $C^{-1}\mathbf{v}_{B'} = \mathbf{v}_B$, shows that the inverse of the change-of-coordinates matrix from B to B' is the change-of-coordinates matrix from B' to B.

Equation (9) shows us exactly what the change-of-coordinates matrix must be for any two ordered bases B and B' in $\mathbb{R}^n$. A direct way to compute this product $M_{B'}^{-1}M_B$, if M_B^{-1} is not already available, is to form the partitioned matrix

$$[M_{B'} \mid M_B]$$

and reduce it (by using elementary row operations) to the form $[I \mid C]$. Exercise 27 in Section 1.4 shows that the matrix C that results to the right of the partition will indeed be $M_{B'}^{-1}M_B$. We now have a convenient procedure for finding a change-of-coordinates matrix.

Finding the Change-of-Coordinates Matrix from B to B' in $\mathbb{R}^n$

Let $B = (\mathbf{b}_1, \mathbf{b}_2, \ldots, \mathbf{b}_n)$ and $B' = (\mathbf{b}_1', \mathbf{b}_2', \ldots, \mathbf{b}_n')$ be ordered bases of $\mathbb{R}^n$. The change-of-coordinates matrix from B to B' is the matrix $C_{B,B'}$ obtained by the row reduction

$$\left[\begin{array}{ccc|ccc} \mathbf{b}_1' & \mathbf{b}_2' \cdots \mathbf{b}_n' & \mathbf{b}_1 & \mathbf{b}_2 \cdots \mathbf{b}_n \end{array}\right] \sim [I \mid C_{B,B'}].$$

New basis matrix   **Old basis matrix**

As an aid in remembering the form of this partitioned matrix, notice that it represents n different systems of equations $A\mathbf{x} = \mathbf{b}_i$ with the same coefficient matrix $A = M_{B'}$. Each system expresses a basis vector from B as a linear combination of the basis vectors from B'; thus it expresses the change in coordinates from basis B to basis B'.

EXAMPLE 4   Let $B = ((1, 1, 0), (2, 0, 1), (1, -1, 0))$. Find the change-of-coordinates matrix from E to B, and use it to find the coordinate vector of

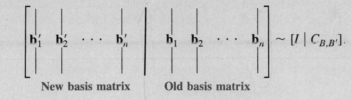

$$\mathbf{v} = \begin{bmatrix} 2 \\ 3 \\ 4 \end{bmatrix}$$

relative to B.

Solution   Following the boxed procedure, we place the "new" basis matrix to the left and the "old" basis matrix to the right in a partitioned matrix, and then we proceed with the row reduction:

$$\begin{bmatrix} 1 & 2 & 1 & 1 & 0 & 0 \\ 1 & 0 & -1 & 0 & 1 & 0 \\ 0 & 1 & 0 & 0 & 0 & 1 \end{bmatrix} \sim \begin{bmatrix} 1 & 2 & 1 & 1 & 0 & 0 \\ 0 & -2 & -2 & -1 & 1 & 0 \\ 0 & 1 & 0 & 0 & 0 & 1 \end{bmatrix}$$

$\qquad\quad M_B \qquad\qquad M_E$

$$\sim \begin{bmatrix} 1 & 0 & -1 & 0 & 1 & 0 \\ 0 & 1 & 1 & \frac{1}{2} & -\frac{1}{2} & 0 \\ 0 & 0 & -1 & -\frac{1}{2} & \frac{1}{2} & 1 \end{bmatrix} \sim \begin{bmatrix} 1 & 0 & 0 & \frac{1}{2} & \frac{1}{2} & -1 \\ 0 & 1 & 0 & 0 & 0 & 1 \\ 0 & 0 & 1 & \frac{1}{2} & -\frac{1}{2} & -1 \end{bmatrix}.$$

Thus the change-of-coordinates matrix from E to B is

$$C = \begin{bmatrix} \frac{1}{2} & \frac{1}{2} & -1 \\ 0 & 0 & 1 \\ \frac{1}{2} & -\frac{1}{2} & -1 \end{bmatrix}.$$

The coordinate vector of $\mathbf{v} = \begin{bmatrix} 2 \\ 3 \\ 4 \end{bmatrix}$ relative to B is

$$\mathbf{v}_B = C\mathbf{v} = \begin{bmatrix} \frac{1}{2} & \frac{1}{2} & -1 \\ 0 & 0 & 1 \\ \frac{1}{2} & -\frac{1}{2} & -1 \end{bmatrix} \begin{bmatrix} 2 \\ 3 \\ 4 \end{bmatrix} = \begin{bmatrix} -\frac{3}{2} \\ 4 \\ -\frac{9}{2} \end{bmatrix}.$$

Notice that C is the inverse of the matrix appearing on the left-hand side of the original partitioned matrix; that is, C is the inverse of the change-of-coordinates matrix from B to E. ■

Let us now compute a change-of-coordinates matrix for the vector space P_2 of polynomials of degree at most 2, showing how the work can be carried out with coordinate vectors in $\mathbb{R}^3$.

EXAMPLE 5 Let $B = (x^2, x, 1)$ and $B' = (x^2 - x, 2x^2 - 2x + 1, x^2 - 2x)$ be ordered bases of the vector space P_2. Find the change-of-coordinates matrix from B to B', and use it to find the coordinate vector of $2x^2 + 3x - 1$ relative to B'.

Solution Identifying each polynomial $a_2x^2 + a_1x + a_0$ with its coordinate vector (a_2, a_1, a_0) relative to the basis B, we obtain the following correspondence from P_2 to $\mathbb{R}^3$:

	Polynomial Basis in P_2	Coordinate Basis in $\mathbb{R}^3$
Old	$B = (x^2, x, 1)$	$((1, 0, 0), (0, 1, 0), (0, 0, 1))$
New	$B' = (x^2 - x, 2x^2 - 2x + 1, x^2 - 2x)$	$((1, -1, 0), (2, -2, 1), (1, -2, 0))$.

Working with coordinate vectors in $\mathbb{R}^3$, we compute the desired change-of-coordinates matrix, as described in the box preceding Example 4.

$$\begin{bmatrix} 1 & 2 & 1 \\ -1 & -2 & -2 \\ 0 & 1 & 0 \end{bmatrix} \begin{array}{|ccc} 1 & 0 & 0 \\ 0 & 1 & 0 \\ 0 & 0 & 1 \end{array} \sim \begin{bmatrix} 1 & 2 & 1 \\ 0 & 0 & -1 \\ 0 & 1 & 0 \end{bmatrix} \begin{array}{|ccc} 1 & 0 & 0 \\ 1 & 1 & 0 \\ 0 & 0 & 1 \end{array}$$

$$\underset{\text{New basis}}{} \qquad \underset{\text{Old basis}}{}$$

$$\sim \begin{bmatrix} 1 & 0 & 1 \\ 0 & 1 & 0 \\ 0 & 0 & -1 \end{bmatrix} \begin{array}{|ccc} 1 & 0 & -2 \\ 0 & 0 & 1 \\ 1 & 1 & 0 \end{array} \sim \begin{bmatrix} 1 & 0 & 0 \\ 0 & 1 & 0 \\ 0 & 0 & 1 \end{bmatrix} \begin{array}{|ccc} 2 & 1 & -2 \\ 0 & 0 & 1 \\ -1 & -1 & 0 \end{array}.$$

The change-of-coordinates matrix is

$$C = \begin{bmatrix} 2 & 1 & -2 \\ 0 & 0 & 1 \\ -1 & -1 & 0 \end{bmatrix}.$$

We compute the coordinate vector of $\mathbf{v} = 2x^2 + 3x - 1$ relative to B', using $C_{B,B'}$ and the coordinate vector of $\mathbf{v}$ relative to B, as follows:

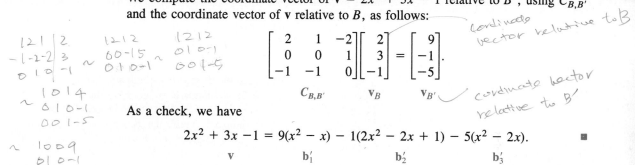

$$\underbrace{\begin{bmatrix} 2 & 1 & -2 \\ 0 & 0 & 1 \\ -1 & -1 & 0 \end{bmatrix}}_{C_{B,B'}} \underbrace{\begin{bmatrix} 2 \\ 3 \\ -1 \end{bmatrix}}_{\mathbf{v}_B} = \underbrace{\begin{bmatrix} 9 \\ -1 \\ -5 \end{bmatrix}}_{\mathbf{v}_{B'}}.$$

coordinate vector relative to B

coordinate vector relative to B'

As a check, we have

$$\underset{\mathbf{v}}{2x^2 + 3x - 1} = 9\underset{\mathbf{b}'_1}{(x^2 - x)} - 1\underset{\mathbf{b}'_2}{(2x^2 - 2x + 1)} - 5\underset{\mathbf{b}'_3}{(x^2 - 2x)}. \qquad ■$$

SUMMARY

Let V be a vector space with basis $\{\mathbf{b}_1, \mathbf{b}_2, \ldots, \mathbf{b}_n\}$.

1. $B = (\mathbf{b}_1, \mathbf{b}_2, \ldots, \mathbf{b}_n)$ is an ordered basis; the vectors are regarded as being in a specified order in this n-tuple notation.

2. Each vector $\mathbf{v}$ in V has a unique expression as a linear combination:

   $$\mathbf{v} = r_1\mathbf{b}_1 + r_2\mathbf{b}_2 + \cdots + r_n\mathbf{b}_n.$$

3. The vector $\mathbf{v}_B = (r_1, r_2, \ldots, r_n)$ for the uniquely determined scalars r_i in the preceding equation (summary item 2) is the coordinate vector of $\mathbf{v}$ relative to B.

4. The vector space V can be coordinatized, using summary item 3, so that V is *isomorphic* to $\mathbb{R}^n$.

5. Let B and B' be ordered bases of $\mathbb{R}^n$. There exists a unique matrix $C = C_{B,B'}$ such that $C\mathbf{v}_B = \mathbf{v}_{B'}$ for all vectors $\mathbf{v}$ in $\mathbb{R}^n$. This is the *change-of-coordinates* matrix from B to B'. It can be computed by means of a row reduction of $[B' \mid B]$ to $[I \mid C]$, as illustrated in the box on page 159.

EXERCISES

In Exercises 1–10, find the coordinate vector of the given vector relative to the indicated ordered basis.

1. $(-1, 1)$ in $\mathbb{R}^2$ relative to $((0, 1), (1, 0))$

2. $(-2, 4)$ in $\mathbb{R}^2$ relative to $((0, -2), (-\frac{1}{2}, 0))$

3. $(4, 6, 2)$ in $\mathbb{R}^3$ relative to $((2, 0, 0), (0, 1, 1), (0, 0, 1))$

4. $(4, -2, 1)$ in $\mathbb{R}^3$ relative to $((0, 1, 1), (2, 0, 0), (0, 3, 0))$

5. $(3, 13, -1)$ in $\mathbb{R}^3$ relative to $((1, 3, -2),$
$(4, 1, 3), (-1, 2, 0))$

6. $(9, 6, 11, 0)$ in $\mathbb{R}^4$ relative to $((1, 0, 1, 0),$
$(2, 1, 1, -1), (0, 1, 1, -1), (2, 1, 3, 1))$

7. $x^3 + x^2 - 2x + 4$ in P_3 relative to $(1, x^2, x, x^3)$

8. $x^3 + 3x^2 - 4x + 2$ in P_3 relative to
$(x, x^2 - 1, x^3, 2x^2)$

9. $x + x^4$ in P_4 relative to
$(1, 2x - 1, x^3 + x^4, 2x^3, x^2 + 2)$

10. $\begin{bmatrix} 1 & -2 \\ 3 & 4 \end{bmatrix}$ in M_2 relative to

$$\left(\begin{bmatrix} 0 & 1 \\ 1 & 0 \end{bmatrix}, \begin{bmatrix} 0 & -1 \\ 0 & 0 \end{bmatrix}, \begin{bmatrix} 1 & -1 \\ 0 & 3 \end{bmatrix}, \begin{bmatrix} 0 & 1 \\ 0 & 1 \end{bmatrix} \right)$$

11. Let V be a nonzero finite-dimensional vector space. Mark each of the following True or False.

____ a) The vector space V is isomorphic to $\mathbb{R}^n$ for some positive integer n.

____ b) There is a unique coordinate vector associated with each vector $\mathbf{v} \in V$.

____ c) There is a unique coordinate vector associated with each vector $\mathbf{v} \in V$ relative to a basis for V.

____ d) There is a unique coordinate vector associated with each vector $\mathbf{v} \in V$ relative to an ordered basis for V.

____ e) Distinct vectors in V have distinct coordinate vectors relative to an ordered basis B for V.

____ f) The same vector in V cannot have the same coordinate vector relative to different ordered bases for V.

____ g) There are six possible ordered bases for $\mathbb{R}^3$.

____ h) There are six possible ordered bases for $\mathbb{R}^3$, consisting of the standard unit coordinate vectors in $\mathbb{R}^3$.

____ i) A reordering of elements in an ordered basis for V corresponds to a similar reordering of components in coordinate vectors with respect to the basis.

____ j) Addition and multiplication by scalars in V can be computed in terms of coordinate vectors with respect to any fixed ordered basis for V.

In Exercises 12–15, find the change-of-coordinates matrix (a) from B to B', and (b) from B' to B. Verify that these matrices are inverses of each other.

12. $B = ((1, 1), (1, 0))$ and $B' = ((0, 1), (1, 1))$ in $\mathbb{R}^2$

13. $B = ((2, 3, 1), (1, 2, 0), (2, 0, 3))$ and $B' = ((1, 0, 0), (0, 1, 0), (0, 0, 1))$ in $\mathbb{R}^3$

14. $B = ((1, 0, 1), (1, 1, 0), (0, 1, 1))$ and $B' = ((0, 1, 1), (1, 1, 0), (1, 0, 1))$ in $\mathbb{R}^3$

15. $B = ((1, 1, 1, 1), (1, 1, 1, 0), (1, 1, 0, 0), (1, 0, 0, 0))$ and the standard ordered basis $B' = E$ for $\mathbb{R}^4$

16. Find the change-of-coordinates matrix from B' to B for the bases $B = (x^2, x, 1)$ and $B' = (x^2 - x, 2x^2 - 2x + 1, x^2 - 2x)$ of P_2 in Example 5. Verify that this matrix is the inverse of the change-of-coordinates matrix from B to B' found in that example.

17. Proceeding as in Example 5, find the change-of-coordinates matrix from $B = (x^3, x^2, x, 1)$ to $B' = (x^3 - x^2, x^2 - x, x - 1, x^3 + 1)$ in the vector space P_3 of polynomials of degree at most 3.

18. Find the change-of-coordinates matrix from B' to B for the vector space and ordered bases given in Exercise 17. Show that this matrix is the inverse of the matrix found in Exercise 17.

19. For ordered bases B and B' in $\mathbb{R}^n$, explain how the change-of-coordinates matrix from B to B' is related to the change-of-coordinates matrices from B to E and from E to B'.

20. Let B and B' be ordered bases of a vector space V. Show that the change-of-coordinates matrix C from B to B' can be expressed as

$$C = \begin{bmatrix} | & | & & | \\ (\mathbf{b}_1)_{B'} & (\mathbf{b}_2)_{B'} & \cdots & (\mathbf{b}_n)_{B'} \\ | & | & & | \end{bmatrix}.$$

In Exercises 21–25, use the result of Exercise 20 to find the change-of-coordinates matrix from B to B' for the given vector space V with ordered bases B and B'.

21. $V = P_3$, B is the basis in Exercise 8, $B' = (x^3, x^2, x, 1)$

22. $V = P_3$, $B = (x^3 + x^2 + 1, 2x^3 - x^2 + x + 1,$
    $x^3 - x + 1, x^2 + x + 1)$, $B' = (x^3, x^2, x, 1)$

23. V is the space M_2 of all 2×2 matrices, B is the basis in Exercise 10,

$$B' = \left(\begin{bmatrix} 1 & 0 \\ 0 & 0 \end{bmatrix}, \begin{bmatrix} 0 & 1 \\ 0 & 0 \end{bmatrix}, \begin{bmatrix} 0 & 0 \\ 1 & 0 \end{bmatrix}, \begin{bmatrix} 0 & 0 \\ 0 & 1 \end{bmatrix} \right)$$

24. V is the subspace $\mathrm{sp}(\sin^2 x, \cos^2 x)$ of the vector space F of all real-valued functions with domain $\mathbb{R}$, $B = (\cos 2x, 1)$, $B' = (\sin^2 x, \cos^2 x)$

25. Find the change-of-coordinates matrix from B' to B for the vector space V and ordered bases B and B' of Exercise 24.

26. Let B be an ordered basis for $\mathbb{R}^3$. If

$$C_{E,B} = \begin{bmatrix} 3 & 1 & 2 \\ 4 & 1 & 2 \\ -1 & 2 & 1 \end{bmatrix} \quad \text{and} \quad \mathbf{v} = \begin{bmatrix} 2 \\ 5 \\ -1 \end{bmatrix},$$

find the coordinate vector $\mathbf{v}_B$.

27. Let V be a vector space with ordered bases B and B'. If

$$C_{B,B'} = \begin{bmatrix} 1 & 2 & 0 \\ 0 & 1 & -2 \\ -1 & 0 & 1 \end{bmatrix} \quad \text{and}$$

$$\mathbf{v} = 3\mathbf{b}_1 - 2\mathbf{b}_2 + \mathbf{b}_3,$$

find the coordinate vector $\mathbf{v}_{B'}$.

28. Find the polynomial in P_2 whose coordinate vector relative to the ordered basis $B = (x + x^2, x - x^2, 1 + x)$ is $(3, 1, 2)$.

29. Let B, B', and B'' be ordered bases for $\mathbb{R}^n$. Let C be the change-of-coordinates matrix from B to B', and let C' be the change-of-coordinates matrix from B' to B''. Find the change-of-coordinates matrix from B to B'' in terms of C and C'. [HINT: For a vector $\mathbf{v}$ in $\mathbb{R}^n$, what matrix times $\mathbf{v}_B$ gives $\mathbf{v}_{B''}$?]

2.8   LINES AND OTHER TRANSLATES

In this section, we return for the moment to geometry. Our work in the preceding sections of this chapter enables us to describe geometrically the solution set of any consistent linear system.

In discussing the subspace of $\mathbb{R}^n$ spanned by nonzero vectors $\mathbf{d}_1, \mathbf{d}_2, \dots, \mathbf{d}_k$, we noted in Section 2.3 that $\mathrm{sp}(\mathbf{d}_1)$ is a line containing $\mathbf{0}$ and that $\mathrm{sp}(\mathbf{d}_1, \mathbf{d}_2)$ is a plane containing $\mathbf{0}$ in $\mathbb{R}^n$ if $\mathbf{d}_1$ and $\mathbf{d}_2$ are not parallel. Lines and planes in general need not contain the origin, and so need not be subspaces.

Lines in $\mathbb{R}^n$

Let $\mathbf{a}$ be a point and $\mathbf{d}$ be a vector in $\mathbb{R}^n$. We wish to translate the line $\mathrm{sp}(\mathbf{d})$ to obtain a parallel line through the point $\mathbf{a}$, as indicated in Figure 2.25. Geometrically, translation of a set by a vector $\mathbf{a}$ is accomplished by sliding the set in the direction determined by $\mathbf{a}$ through a distance $\|\mathbf{a}\|$. Analytically, translating by $\mathbf{a}$ corresponds to adding $\mathbf{a}$ to every vector in the set.

For the line L in Figure 2.25, the vector $\mathbf{d}$ parallel to the line determines the direction of L, while the point $\mathbf{a}$ is a specific point lying on the line. That is, we think of the line as being determined by point $\mathbf{a}$ and parallel vector $\mathbf{d}$. Every point $\mathbf{x}$ on the line can be expressed in the form

$$\mathbf{x} = \mathbf{a} + t\mathbf{d}$$

as indicated in Figure 2.25.

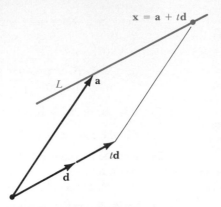

$$\mathbf{x} = \mathbf{a} + t\mathbf{d}$$

L $\mathbf{a}$

$t\mathbf{d}$

$\mathbf{d}$

FIGURE 2.25 **The line L through the point a with parallel vector d.**

DEFINITION 2.13   A Line in $\mathbb{R}^n$

The **line in $\mathbb{R}^n$ through $\mathbf{a} = (a_1, a_2, \ldots, a_n)$ with nonzero parallel vector $\mathbf{d} = (d_1, d_2, \ldots, d_n)$** is the set of points $\mathbf{x} = (x_1, x_2, \ldots, x_n)$ satisfying

$$\mathbf{x} = \mathbf{a} + t\mathbf{d} \qquad \text{for some scalar } t. \qquad (1)$$

Equation (1) is the **vector equation of the line**, and **parametric equations** are the corresponding component equations

$$x_i = a_i + d_i t \qquad \text{for } i = 1, 2, \ldots, n. \qquad (2)$$

EXAMPLE 1   Find paramatric equations of the line in $\mathbb{R}^2$ through $(2, 1)$ having parallel vector $(3, 4)$.

Solution   The vector equation of the line is

$$(x_1, x_2) = (2, 1) + t(3, 4),$$

and parametric equations are

$$x_1 = 2 + 3t, \qquad x_2 = 1 + 4t.$$

EXAMPLE 2   Find parametric equations of the line in $\mathbb{R}^3$ that passes through the points $\mathbf{a} = (1, 2, -1)$ and $\mathbf{b} = (3, 1, 4)$.

Solution   In order to find a suitable parallel vector for the line, we compute $\mathbf{d} = \mathbf{b} - \mathbf{a} = (2, -1, 5)$, which is the vector that starts at $\mathbf{a}$ and terminates at $\mathbf{b}$. (See Fig. 2.26.) Therefore, a vector equation of the line is

$$(x_1, x_2, x_3) = (1, 2, -1) + t(2, -1, 5),$$

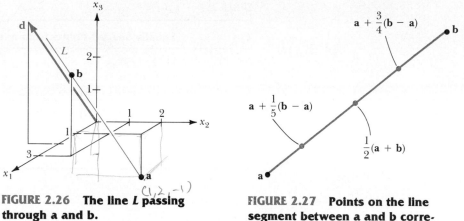

FIGURE 2.26 The line *L* passing through a and b.

FIGURE 2.27 Points on the line segment between a and b corresponding to $t = \frac{1}{5}, \frac{1}{2},$ and $\frac{3}{4}$.

and parametric equations are

$$x_1 = 1 + 2t, \qquad x_2 = 2 - t, \qquad x_3 = -1 + 5t. \qquad ■$$

Line Segments

Example 2 illustrates that a suitable parallel vector for the line passing through two points **a** and **b** in $\mathbb{R}^n$ is $\mathbf{d} = \mathbf{b} - \mathbf{a}$, and is given by the following vector equation.

Line Passing Through a and b in $\mathbb{R}^n$

$$\mathbf{x} = \mathbf{a} + t(\mathbf{b} - \mathbf{a}) \qquad \text{for scalars } t. \qquad (3)$$

Equations (1) and (3) in effect present the line as a *t*-axis whose origin is at **a** and for which one unit on the *t*-axis yields $\|\mathbf{d}\|$ units in $\mathbb{R}^n$.

We see that each point in $\mathbb{R}^n$ on the **line segment** between **a** and **b** consists of all points **x** in Eq. (3) obtained for some value *t* from 0 to 1. By choosing *t* appropriately, we can locate any point on the line (3) that we please, as illustrated in Figure 2.27. In particular, the **midpoint of the line segment** between **a** and **b** is given by

$$\mathbf{a} + \tfrac{1}{2}(\mathbf{b} - \mathbf{a}) = \tfrac{1}{2}(\mathbf{a} + \mathbf{b}).$$

EXAMPLE 3   Find the points that divide the line segment between $\mathbf{a} = (1, 2, 1, 3)$ and $\mathbf{b} = (2, 1, 4, 2)$ in $\mathbb{R}^4$ into five equal parts.

Solution   The line segment between **a** and **b** is given by Eq. (3) for values of *t* from 0 to 1. Equation (3) becomes

$$(x_1, x_2, x_3, x_4) = (1, 2, 1, 3) + t(1, -1, 3, -1), \qquad 0 \le t \le 1.$$

TABLE 2.1

t	Equally Spaced Points from a to b
0	(1, 2, 1, 3)
$\frac{1}{5}$	(1.2, 1.8, 1.6, 2.8)
$\frac{2}{5}$	(1.4, 1.6, 2.2, 2.6)
$\frac{3}{5}$	(1.6, 1.4, 2.8, 2.4)
$\frac{4}{5}$	(1.8, 1.2, 3.4, 2.2)
1	(2, 1, 4, 2)

By choosing $t = 0, \frac{1}{5}, \frac{2}{5}, \frac{3}{5}, \frac{4}{5}$, and 1, we obtain the division required, as shown in Table 2.1. ■

Flats in $\mathbb{R}^n$

Just as a line is a translate of a one-dimensional subspace in $\mathbb{R}^n$, a plane in $\mathbb{R}^n$ is a translate of a two-dimensional subspace $\text{sp}(\mathbf{a}_1, \mathbf{a}_2)$, where $\mathbf{a}_1$ and $\mathbf{a}_2$ are nonzero and nonparallel vectors in $\mathbb{R}^n$. A plane appears as a flat piece of $\mathbb{R}^n$ (see Fig. 2.28). More generally, a k-dimensional subspace of $\mathbb{R}^n$ appears flat if $k < n$; and translation of such a subspace then produces a flat piece of $\mathbb{R}^n$, not necessarily containing the origin. Let us give a formal definition.

THE EQUATION OF A PLANE IN $\mathbb{R}^3$ appears as early as 1732 in a paper of Jacob Hermann (1678–1733). He was able to determine the plane's position by using intercepts, and he also noted that the sine of the angle between the plane and the one coordinate plane he dealt with (what we call the x_1,x_2-plane) was

$$\frac{\sqrt{d_1{}^2 + d_2{}^2}}{\sqrt{d_1{}^2 + d_2{}^2 + d_3{}^2}}.$$

In his 1748 *Introduction to Infinitesimal Analysis*, Leonhard Euler (1707–1783) used, instead, the cosine of this angle, $d_3/\sqrt{d_1{}^2 + d_2{}^2 + d_3{}^2}$.

At the end of the eighteenth century, Gaspard Monge (1746–1818), in his notes for a course on solid analytic geometry at the Ecole Polytechnique, related the equation of a plane to all three coordinate planes and gave the cosines of the angles the plane made with each of these (the so-called direction cosines). He also presented many of the standard problems of solid analytic geometry, examples of which appear in the exercises. For example, he showed how to find the plane passing through three given points, the line passing through a point perpendicular to a plane, the distance between two parallel planes, and the angle between a line and a plane.

Known as "the greatest geometer of the eighteenth century," Monge developed new graphical geometric techniques as a student and later as a professor at a military school. The first problem he solved involved a procedure enabling soldiers to make quickly a fortification capable of shielding a position from both the view and the firepower of the enemy. Monge served the French revolutionary government as minister of the navy and later served Napoleon in various scientific offices. Ultimately, he was appointed senator for life by the emperor.

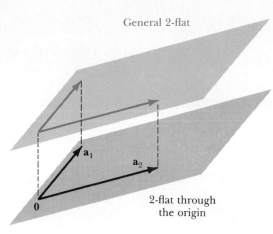

General 2-flat

a_1

a_2

0

2-flat through
the origin

FIGURE 2.28 Planes or 2-flats in $\mathbb{R}^n$.

DEFINITION 2.14 A *k*-flat in $\mathbb{R}^n$

Let $\{a_1, a_2, \ldots, a_k\}$ be a basis for a subspace W of $\mathbb{R}^n$ and let **b** be a point in $\mathbb{R}^n$. The **translate b + W**, consisting of all vectors **x** of the form

$$x = b + t_1 a_1 + t_2 a_2 + \cdots + t_k a_k \qquad (4)$$

for scalars t_i in $\mathbb{R}$, is called a **k-flat** containing **b** in $\mathbb{R}^n$. A 1-flat is a **line**, a 2-flat is a **plane**, and an $(n-1)$-flat is a **hyperplane** in $\mathbb{R}^n$.

Eq. (4) is the **vector equation** of the k-flat. Equating corresponding components yields **parametric equations**.

EXAMPLE 4   Find parametric equations of the plane in $\mathbb{R}^4$ passing through the points **a** = $(1, 1, 1, 1)$, **b** = $(2, 1, 1, 0)$, and **c** = $(3, 2, 1, 0)$.

Solution   If we translate the given plane by the vector $-\mathbf{a}$, we obtain a plane containing

$$\mathbf{a} - \mathbf{a} = \mathbf{0}, \qquad \mathbf{b} - \mathbf{a} = (1, 0, 0, -1), \qquad \text{and} \qquad \mathbf{c} - \mathbf{a} = (2, 1, 0, -1).$$

(See Fig. 2.29.) Thus the translated plane through the origin is sp(**b** − **a**, **c** − **a**), and the original desired plane has vector equation

$$x = a + s(b - a) + t(c - a).$$

Parametric equations obtained by equating components are

$$x_1 = 1 + s + 2t$$
$$x_2 = 1 \qquad + t$$
$$x_3 = 1$$
$$x_4 = 1 - s - t.$$

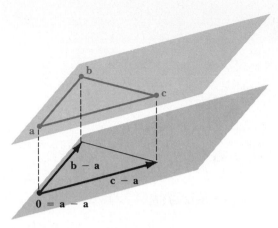

FIGURE 2.29 The plane passing through a, b, and c.

The Geometry of Linear Systems

Let $A\mathbf{x} = \mathbf{b}$ be any system of m equations in n unknowns that has at least one solution. (See Theorem 1.4 on page 35.) The solution set of the system is a k-flat in $\mathbb{R}^n$, where k is the dimension of the nullspace of A. If a system of equations has a unique solution, its solution set is a *zero-flat*.

EXAMPLE 5   Show that the linear equation $c_1 x_1 + c_2 x_2 + c_3 x_3 = d$, where not all of c_1, c_2, c_3 are zero, represents a plane in $\mathbb{R}^3$.

Solution   Let us assume that $c_1 \neq 0$. A particular solution of the given equation is $\mathbf{b} = (d/c, 0, 0)$, The corresponding homogeneous equation has a solution space generated by $\mathbf{a}_1 = (c_3, 0, -c_1)$ and $\mathbf{a}_2 = (c_2, -c_1, 0)$. Thus the solution set of the linear equation is a 2-flat in $\mathbb{R}^3$ with equation $\mathbf{x} = \mathbf{b} + t_1 \mathbf{a}_1 + t_2 \mathbf{a}_2$, which is a plane in $\mathbb{R}^3$. ◾

EXAMPLE 6   Solve the system of equations

$$x_1 + 2x_2 - 2x_3 + x_4 + 3x_5 = 1$$
$$2x_1 + 5x_2 - 3x_3 - x_4 + 2x_5 = 2$$
$$-3x_1 - 8x_2 + 6x_3 - x_4 - 5x_5 = 1,$$

and write the solution set as a k-flat.

Solution Reducing the corresponding partitioned matrix, we have

$$\left[\begin{array}{ccccc|c} 1 & 2 & -2 & 1 & 3 & 1 \\ 2 & 5 & -3 & -1 & 2 & 2 \\ -3 & -8 & 6 & -1 & -5 & 1 \end{array}\right] \sim \left[\begin{array}{ccccc|c} 1 & 2 & -2 & 1 & 3 & 1 \\ 0 & 1 & 1 & -3 & -4 & 0 \\ 0 & -2 & 0 & 2 & 4 & 4 \end{array}\right]$$

$$\sim \left[\begin{array}{ccccc|c} 1 & 0 & -4 & 7 & 11 & 1 \\ 0 & 1 & 1 & -3 & -4 & 0 \\ 0 & 0 & 2 & -4 & -4 & 4 \end{array}\right] \sim \left[\begin{array}{ccccc|c} 1 & 0 & 0 & -1 & 3 & 9 \\ 0 & 1 & 0 & -1 & -2 & -2 \\ 0 & 0 & 1 & -2 & -2 & 2 \end{array}\right].$$

Thus, $\mathbf{b} = (9, -2, 2, 0, 0)$ is a particular solution to the given system, and $\mathbf{a}_1 = (1, 1, 2, 1, 0)$ and $\mathbf{a}_2 = (-3, 2, 2, 0, 1)$ form a basis for the solution space of the corresponding homgeneous system. The solution set of the given system is the 2-flat in $\mathbb{R}^5$ with vector equation $\mathbf{x} = \mathbf{b} + t_1\mathbf{a}_1 + t_2\mathbf{a}_2$, which can be written in the form

$$\begin{bmatrix} x_1 \\ x_2 \\ x_3 \\ x_4 \\ x_5 \end{bmatrix} = \begin{bmatrix} 9 \\ -2 \\ 2 \\ 0 \\ 0 \end{bmatrix} + t_1 \begin{bmatrix} 1 \\ 1 \\ 2 \\ 1 \\ 0 \end{bmatrix} + t_2 \begin{bmatrix} -3 \\ 2 \\ 2 \\ 0 \\ 1 \end{bmatrix}.$$

$$\begin{pmatrix} 9 + t_1 - 3t_2 \\ -2 + t_1 + 2t_2 \\ 2 + 2t_1 + 2t_2 \\ t_1 \\ t_2 \end{pmatrix}$$

$$x_3 - 2x_4 - 2x_5 = 2$$
$$x_3 = 2 + 2x_4 + 2x_5$$
$$x_2 = x_4 + 2x_5 - 2$$
$$x_1 = 9 + x_4 - 3x_5$$

In the preceding example, we described the solution set of a system of equations as a 2-flat in $\mathbb{R}^5$. Notice that the original form of the system represents an intersection of three hyperplanes in $\mathbb{R}^5$—one for each equation. We generalize this example to the solution set of any consistent linear system.

Consider a system of m equations in n unknowns. Let the rank of the coefficient matrix be r so that, when the matrix is reduced to row echelon form, there are r nonzero rows. According to the rank equation, the number of free variables is then $n - r$; the corresponding homogeneous system has as its solution set an $(n - r)$-dimensional subspace—that is, an $(n - r)$-flat through the origin. The solution set of the original nonhomogeneous system is a translate of this subspace and is an $(n - r)$-flat in $\mathbb{R}^n$. In particular, a single consistent linear equation has as its solution set an $(n - 1)$-flat (that is, a hyperplane) in $\mathbb{R}^n$. In general, if we adjoin an additional linear equation to a given linear system, we expect the dimension of the solution flat to be reduced by 1. This is the case precisely when the new system is still consistent and when the new equation is independent of the others (in the sense that it yields a new nonzero row when the augmented matrix is row-reduced to echelon form).

We have shown that a system $A\mathbf{x} = \mathbf{b}$ of m equations in n unknowns has as its solution set an $(n - r)$-flat, where r is the rank of A. Conversely, it can be shown that a k-flat in $\mathbb{R}^n$ is the solution set of some system of $n - k$ linear equations in n unknowns. That is, a k-flat in $\mathbb{R}^n$ is the intersection of $n - k$ hyperplanes. Thus there are two ways to view a k-flat in $\mathbb{R}^n$:

1. As a translate of a k-dimensional subspace of $\mathbb{R}^n$, described using parametric equations, and

2. As an intersection of $n - k$ hyperplanes, described with a system of linear equations.

EXAMPLE 6   Describe the line (1-flat) in $\mathbb{R}^3$ that passes through $(1, 2, -1)$ and $(3, 1, 4)$, in terms of

(1) parametric equations, and

$$\begin{bmatrix} x_1 \\ x_2 \\ x_3 \end{bmatrix} = \begin{bmatrix} 1 \\ 2 \\ -1 \end{bmatrix} + t \begin{bmatrix} 2 \\ -1 \\ 5 \end{bmatrix}$$

(2) a system of linear equations.

Solution   (1) In Example 2, we found the parametric equations for the line:

$$x_1 = 1 + 2t, \qquad x_2 = 2 - t, \qquad x_3 = -1 + 5t. \tag{5}$$

(2) In order to describe the line with a system of linear equations, we eliminate the parameter t from Eqs. (5):

$$\begin{array}{ll} x_1 + 2x_2 \qquad = 5 & \text{Add two times the second equation to the first.} \\ \qquad 5x_2 + x_3 = 9 & \text{Add five times the second equation to the third.} \end{array} \tag{6}$$

$x_1 = 1 + 2t$

$\begin{array}{r} \tau \\ \hline \end{array} \; 2x_2 = 4 - 2t$

$\overline{x_1 + 2x_2 = 5}$

intersection.

$\begin{array}{cccc} 1 & 2 & 0 & 5 \\ 0 & 5 & 1 & 9 \end{array}$

This system describes the line as an intersection of two planes. The line can be represented as the intersection of any two distinct planes, each containing the line. This is illustrated by the equivalent systems we have at the various stages in the Gauss reduction of system (6) to obtain solution (5). ■

$5x_2 = 9 - x_3$

$x_2 = \frac{9}{5} - \frac{1}{5}x_3 \qquad x_1 + 2\left(\frac{9}{5} - \frac{1}{5}x_3\right).$

SUMMARY

1. A k-flat in $\mathbb{R}^n$ is a translate of a k-dimensional subspace and has the form $\mathbf{b} + \text{sp}(\mathbf{a}_1, \mathbf{a}_2, \ldots, \mathbf{a}_k)$, where $\mathbf{b}$ is a vector in $\mathbb{R}^n$ and $\mathbf{a}_1, \mathbf{a}_2, \ldots, \mathbf{a}_k$ are independent vectors in $\mathbb{R}^n$. The vector equation of the k-flat is $\mathbf{x} = \mathbf{b} + t_1\mathbf{a}_1 + t_2\mathbf{a}_2 + \cdots + t_k\mathbf{a}_k$ for scalars t_i in $\mathbb{R}$.

2. A line in $\mathbb{R}^n$ is a 1-flat. The line passing through the point $\mathbf{b}$ with parallel vector $\mathbf{d}$ is given by $\mathbf{x} = \mathbf{b} + t\mathbf{d}$, where t runs through all scalars. Parametric equations of the line are the component equations $x_i = b_i + d_i t$ for $i = 1, 2, \ldots, n$.

3. The line segment from point $\mathbf{a}$ to point $\mathbf{b}$ in $\mathbb{R}^n$ is the set of points $\mathbf{x}$ such that $\mathbf{x} = \mathbf{a} + t(\mathbf{b} - \mathbf{a})$, $0 \le t \le 1$.

4. A plane in $\mathbb{R}^n$ is a 2-flat; a hyperplane in $\mathbb{R}^n$ is an $(n - 1)$-flat.

5. The solution set of a consistent linear system in n variables with coefficient matrix of rank r is an $(n - r)$ flat in $\mathbb{R}^n$.

6. Every k-flat in $\mathbb{R}^n$ can be viewed both as a translate of a k-dimensional subspace and as the intersection of $n - k$ hyperplanes.

EXERCISES

1. Give parametric equations for the line in $\mathbb{R}^2$ through $(3, -3)$ with parallel vector $\mathbf{d} = (-8, 4)$. Sketch the line in an appropriate figure.

2. Give parametric equations for the line in $\mathbb{R}^3$ through $(-1, 3, 0)$ with parallel vector $\mathbf{d} = (-2, -1, 4)$. Sketch the line in an appropriate figure.

3. Consider the line in $\mathbb{R}^2$ that is given by the equation $d_1x_1 + d_2x_2 = c$ for numbers d_1, d_2, and c in $\mathbb{R}$, where d_1 and d_2 are not both zero. Find parametric equations of the line.

4. Find parametric equations for the line in $\mathbb{R}^2$ through $(5, -1)$ and orthogonal to the line with parametric equations $x_1 = 4 - 2t$, $x_2 = 7 + t$.

5. For each pair of points, find parametric equations of the line containing them.

a) $(-2, 4)$ and $(3, -1)$ in $\mathbb{R}^2$
b) $(3, -1, 6)$ and $(0, -3, -1)$ in $\mathbb{R}^3$
c) $(2, 0, 4)$ and $(-1, 5, -8)$ in $\mathbb{R}^3$

6. For each of the given pairs of lines in $\mathbb{R}^3$, determine whether the lines intersect. If they do intersect, find the point of intersection, and determine whether the lines are orthogonal.

a) $x_1 = 4 + t$, $x_2 = 2 - 3t$,
   $x_3 = -3 + 5t$

and

   $x_1 = 11 + 3s$, $x_2 = -9 - 4s$,
   $x_3 = -4 - 3s$
b) $x_1 = 11 + 3t$, $x_2 = -3 - t$,
   $x_3 = 4 + 3t$

and

   $x_1 = 6 - 2s$, $x_2 = -2 + s$,
   $x_3 = -15 + 7s$

7. Find all points in common to the lines in $\mathbb{R}^2$ given by $x_1 = 5 - 3t$, $x_2 = -1 + t$, and $x_1 = -7 + 6s$, $x_2 = 3 - 2s$.

8. Find parametric equations for the line in $\mathbb{R}^3$ through $(-1, 2, 3)$ that is orthogonal to each of the two lines having parametric equations $x_1 = -2 + 3t$, $x_2 = 4$, $x_3 = 1 - t$ and $x_1 = 7 - t$, $x_2 = 2 + 3t$, $x_3 = 4 + t$.

9. Find the midpoint of the line segment joining each pair of points.

a) $(-2, 4)$ and $(3, -1)$ in $\mathbb{R}^2$
b) $(3, -1, 6)$ and $(0, -3, -1)$ in $\mathbb{R}^3$
c) $(0, 4, 8)$ and $(-4, 5, 9)$ in $\mathbb{R}^3$

10. Find the point in $\mathbb{R}^2$ on the line segment joining $(-1, 3)$ and $(2, 5)$ that is twice as close to $(-1, 3)$ as to $(2, 5)$.

11. Find the point in $\mathbb{R}^3$ on the line segment joining $(-2, 1, 3)$ and $(0, -5, 6)$ that is one-fourth of the way from $(-2, 1, 3)$ to $(0, -5, 6)$.

12. Find the points that divide the line segment between $(2, 1, 3, 4)$ and $(-1, 2, 1, 3)$ in $\mathbb{R}^4$ into three equal parts.

13. Find the midpoint of the line segment between $(2, 1, 3, 4, 0)$ and $(1, 2, -1, 3, -1)$ in $\mathbb{R}^5$.

14. Find the intersection in $\mathbb{R}^3$ of the line given by

$$x_1 = 5 + t, \qquad x_2 = -3t, \qquad x_3 = -2 + 4t$$

and the plane with equation $x_1 - 3x_2 + 2x_3 = -25$.

15. Find the intersection in $\mathbb{R}^3$ of the line given by

$$x_1 = 2, \qquad x_2 = 5 - t, \qquad x_3 = 2t$$

and the plane with equation $x_1 + 2x_3 = 10$.

16. Find parametric equations of the plane that passes through the unit coordinate points $(1, 0, 0)$, $(0, 1, 0)$, and $(0, 0, 1)$ in $\mathbb{R}^3$.

17. Find a single linear equation in three variables whose solution set is the plane in Exercise 16. [HINT: We suggest two general methods of attack: (1) eliminate the parameters from your answer to Exercise 16; or (2) solve an appropriate linear system. Actually, this particular answer can be found by inspection.]

18. Find parametric equations of the plane in $\mathbb{R}^3$ that passes through $(1, 0, 0)$, $(0, 1, -1)$, and $(1, 1, 1)$.

19. Find a single linear equation in three variables whose solution set is the plane in Exercise 18. [See the hint for Exercise 17.]

20. Find a vector equation of the plane that passes through the points $(1, 2, 1)$, $(-1, 2, 3)$, and $(2, 1, 4)$ in $\mathbb{R}^3$.

21. Find a single linear equation in three variables whose solution set is the plane in Exercise 20. [See the hint for Exercise 17.]

22. Find a vector equation for the plane in $\mathbb{R}^4$ that passes through the points $(1, 2, 1, 3)$, $(4, 1, 2, 1)$, and $(3, 1, 2, 0)$.

23. Find a linear system with two equations in four variables whose solution set is the plane in Exercise 22. [See the hint for Exercise 17.]

24. Find a vector equation of the hyperplane that passes through the points $(1, 2, 1, 2, 3)$, $(0, 1, 2, 1, 3)$, $(0, 0, 3, 1, 2)$, $(0, 0, 0, 1, 4)$, and $(0, 0, 0, 0, 2)$ in $\mathbb{R}^5$.

25. Find a single linear equation in five variables whose solution set is the hyperplane in Exercise 24. [See the hint for Exercise 17.]

26. Find a vector equation of the hyperplane in $\mathbb{R}^6$ passing through the points $\mathbf{e}_1, \mathbf{e}_2, \ldots, \mathbf{e}_6$.

27. Find a single linear equation in six variables whose solution set is the hyperplane in Exercise 26. [See the hint for Exercise 17.]

In Exercises 28–35, solve the given system of linear equations and write the solution set as a k-flat.

28. $\begin{aligned} x_1 - 2x_2 &= 3 \\ 3x_1 - x_2 &= 14 \end{aligned}$

29. $\begin{aligned} x_1 + 2x_2 - x_3 &= -3 \\ 3x_1 + 7x_2 + 2x_3 &= 1 \\ 4x_1 - 2x_2 + x_3 &= -2 \end{aligned}$

30. $\begin{aligned} x_1 + 4x_2 - 2x_3 &= 4 \\ 2x_1 + 7x_2 - x_3 &= -2 \\ x_1 + 3x_2 + x_3 &= -6 \end{aligned}$

31. $\begin{aligned} x_1 - 3x_2 + x_3 &= 2 \\ 3x_1 - 8x_2 + 2x_3 &= 5 \\ 3x_1 - 7x_2 + x_3 &= 4 \end{aligned}$

32. $\begin{aligned} x_1 - 3x_2 + 2x_3 - x_4 &= 8 \\ 3x_1 - 7x_2 + x_4 &= 0 \end{aligned}$

33. $\begin{aligned} x_1 - 2x_3 + x_4 &= 6 \\ 2x_1 - x_2 + x_3 - 3x_4 &= 0 \\ 9x_1 - 3x_2 - x_3 - 7x_4 &= 4 \end{aligned}$

34. $\begin{aligned} x_1 + 2x_2 - 3x_3 + x_4 &= 2 \\ 3x_1 + 6x_2 - 8x_3 - 2x_4 &= 1 \end{aligned}$

35. $\begin{aligned} x_1 - 3x_2 + x_3 + 2x_4 &= 2 \\ x_1 - 2x_2 + 2x_3 + 4x_4 &= -1 \\ 2x_1 - 8x_2 - x_3 &= 3 \\ 3x_1 - 9x_2 + 4x_3 &= 7 \end{aligned}$

36. $2x_1 - 5x_2 + x_3 - 10x_4 + 15x_5 = 60$

37. Mark each of the following True or False.

___ a) The solution set of a linear equation in x_1 and x_2 can be regarded as a hyperplane in $\mathbb{R}^2$.

___ b) Every line and hyperplane in $\mathbb{R}^n$ intersect in a single point.

___ c) The intersection of two distinct hyperplanes in $\mathbb{R}^5$ is a line, if the intersection is nonempty.

___ d) The Euclidean space $\mathbb{R}^5$ has no physical existence, but exists only in our minds.

___ e) The Euclidean spaces $\mathbb{R}$, $\mathbb{R}^2$, and $\mathbb{R}^3$ have no physical existence, but exist only in our minds.

___ f) The mathematical existence of Euclidean 5-space is as substantial as the mathematical existence of Euclidean 3-space.

___ g) Every plane in $\mathbb{R}^n$ is a two-dimensional subspace of $\mathbb{R}^n$.

___ h) Every plane through the origin in $\mathbb{R}^n$ is a two-dimensional subspace of $\mathbb{R}^n$.

___ i) Every k-flat in $\mathbb{R}^n$ contains the origin.

___ j) Every k-flat in $\mathbb{R}^n$ is a translate of a k-dimensional subspace.

2.9 LINEAR TRANSFORMATIONS

Linear transformations mapping $\mathbb{R}^n$ into $\mathbb{R}^m$ were defined in Section 1.5. Now that we have considered vectors in more general spaces than $\mathbb{R}^k$, it is natural to extend the notion to linear transformations of other vector spaces, not necessarily finite-dimensional. In this section, we introduce linear transformations in a general setting. Recall that a linear transformation $T: \mathbb{R}^n \rightarrow \mathbb{R}^m$ is a function that satisfies

$$T(\mathbf{u} + \mathbf{v}) = T(\mathbf{u}) + T(\mathbf{v}) \tag{1}$$

and

$$T(r\mathbf{u}) = rT(\mathbf{u}) \tag{2}$$

for all vectors $\mathbf{u}$ and $\mathbf{v}$ in $\mathbb{R}^n$ and for all scalars r. In Section 1.5, we indicated without proof that, for each such transformation T, there is an associated $m \times n$ matrix A such that $T(\mathbf{x}) = A\mathbf{x}$ for all $\mathbf{x}$ in $\mathbb{R}^n$. Using the notion of bases for vector spaces, we demonstrate this association rigorously in this section. In Chapter 6, we will carry this correspondence between matrices and linear transformations even further, working with general finite-dimensional vector spaces and their bases. It is our policy to intersperse ideas involving linear transformations throughout our text so that they become familiar to you and so that you become comfortable viewing linear algebra in terms of these mappings.

Linear Transformations $T: V \to V'$

The definition of <u>a linear transformation of a vector space V into a vector space V' is practically identical to the definition for the Euclidean vector spaces in Section 1.5. We need only replace $\mathbb{R}^n$ by V and $\mathbb{R}^m$ by V'.</u> *[handwritten: $T: \mathbb{R}^n \to \mathbb{R}^m$ is same as]*

DEFINITION 2.15 Linear Transformation   *[handwritten: $T: V \to V'$]*

> A function T that maps a vector space V into a vector space V' is a **linear transformation** if it satisfies two criteria:
>
> 1. $T(\mathbf{u} + \mathbf{v}) = T(\mathbf{u}) + T(\mathbf{v})$,   **Preservation of addition**
> 2. $T(r\mathbf{u}) = rT(\mathbf{u})$,   **Preservation of scalar multiplication**
>
> for all vectors $\mathbf{u}$ and $\mathbf{v}$ in V and for all scalars r in $\mathbb{R}$.

Exercise 26 shows that the two conditions of Definition 2.15 may be combined into the single condition

$$T(r\mathbf{u} + s\mathbf{v}) = rT(\mathbf{u}) + sT(\mathbf{v}) \tag{3}$$

for all vectors $\mathbf{u}$ and $\mathbf{v}$ in V and for all scalars r and s in $\mathbb{R}$. Mathematical induction can be used to verify the analogous relation for n summands:

$$T(r_1\mathbf{v}_1 + r_2\mathbf{v}_2 + \cdots + r_n\mathbf{v}_n) = r_1T(\mathbf{v}_1) + r_2T(\mathbf{v}_2) + \cdots + r_nT(\mathbf{v}_n). \tag{4}$$

For a linear transformation $T: V \to V'$, the set V is the **domain** of T, and the set V' is the **codomain** of T. <u>The set of all vectors $T(\mathbf{v})$, as $\mathbf{v}$ varies in V, is the **range** of T.</u> The range of T need not be all of V'.

Let V, V', and V'' be vector spaces, and let $T: V \to V'$ and $T': V' \to V''$ be linear transformations. The **composite transformation** $T' \circ T: V \to V''$ is defined by $(T' \circ T)(\mathbf{v}) = T'(T(\mathbf{v}))$ for $\mathbf{v}$ in V. Exercise 27 shows that $T' \circ T$ is again a linear transformation.

Recall that, if A is an $m \times n$ matrix, the function $T: \mathbb{R}^n \to \mathbb{R}^m$ defined by $T(\mathbf{x}) = A\mathbf{x}$ is a linear transformation. Our first two examples use this fact.

EXAMPLE 1   Show that the function $T: \mathbb{R}^3 \to \mathbb{R}^2$ defined by $T(x_1, x_2, x_3) = (x_1 - 2x_2, x_2 + 3x_3)$ is a linear transformation.

Solution Using column vectors, we see that

$$T\left(\begin{bmatrix} x_1 \\ x_2 \\ x_3 \end{bmatrix}\right) = \begin{bmatrix} 1 & -2 & 0 \\ 0 & 1 & 3 \end{bmatrix}\begin{bmatrix} x_1 \\ x_2 \\ x_3 \end{bmatrix},$$

which is of the form $T(\mathbf{x}) = A\mathbf{x}$. Thus, T is a linear transformation. ■

In Section 1.5, we examined invertible linear transformations of the plane $\mathbb{R}^2$ into itself from a geometric point of view. Following is another such example.

EXAMPLE 2 (*Rotation of the Plane*) Show that the function T that maps $\mathbb{R}^2$ into $\mathbb{R}^2$ by rotating the plane counterclockwise through a positive angle α is a linear transformation.

Solution We view $\mathbb{R}^2$ as a space of column vectors. Referring to Figure 2.30, we compute a formula for $T(\mathbf{v})$, where $\mathbf{v}$ is any vector in $\mathbb{R}^2$. If $\mathbf{v}$ makes an angle of θ with the x-axis and if $r = \|\mathbf{v}\|$, then

$$\mathbf{v} = \begin{bmatrix} r \cos \theta \\ r \sin \theta \end{bmatrix}$$

and

$$T(\mathbf{v}) = \begin{bmatrix} r \cos(\alpha + \theta) \\ r \sin(\alpha + \theta) \end{bmatrix} = \begin{bmatrix} r(\cos \alpha \cos \theta - \sin \alpha \sin \theta) \\ r(\sin \alpha \cos \theta + \cos \alpha \sin \theta) \end{bmatrix}.$$

This last equation can be written as

$$T(\mathbf{v}) = \underbrace{\begin{bmatrix} \cos \alpha & -\sin \alpha \\ \sin \alpha & \cos \alpha \end{bmatrix}}_{A} \underbrace{\begin{bmatrix} r \cos \theta \\ r \sin \theta \end{bmatrix}}_{\mathbf{v}}. \tag{5}$$

Thus the function T has the form $T(\mathbf{v}) = A\mathbf{v}$, where A is the 2×2 matrix in Eq. (5); and therefore, T is a linear transformation. ■

THE CONCEPT OF A LINEAR SUBSTITUTION dates back to the eighteenth century. But it was only after physicists became used to dealing with vectors that the idea of a function of vectors became explicit. One of the founders of vector analysis, Oliver Heaviside (1850–1925), introduced the idea of a linear vector operator in one of his works on electromagnetism in 1885. He defined it using coordinates: $\vec{B}$ comes from $\vec{H}$ by a linear vector operator if, when $\vec{B}$ has components B_1, B_2, B_3 and $\vec{H}$ has components H_1, H_2, H_3, there are numbers μ_{ij} for $i, j = 1, 2, 3$, where

$$B_1 = \mu_{11}H_1 + \mu_{12}H_2 + \mu_{13}H_3$$
$$B_2 = \mu_{21}H_1 + \mu_{22}H_2 + \mu_{23}H_3$$
$$B_3 = \mu_{31}H_1 + \mu_{32}H_2 + \mu_{33}H_3.$$

In his lectures at Yale, which were published in 1901, J. Willard Gibbs called this same tranformation a *linear vector function*. But he also defined this more abstractly as a continuous function f such that $f(\vec{v} + \vec{w}) = f(\vec{v}) + f(\vec{w})$. A fully abstract definition, exactly like Definition 2.15, was given by Hermann Weyl in *Space–Time–Matter* (1918).

Oliver Heaviside was a self-taught expert on mathematical physics who played an important role in the development of electromagnetic theory and especially its practical applications. In 1901 he predicted the existence of a reflecting ionized region surrounding the earth; the existence of this layer, now called the *ionosphere*, was soon confirmed.

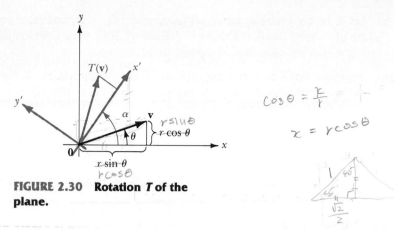

FIGURE 2.30 **Rotation *T* of the plane.**

Our next three examples illustrate the use of the two conditions in Definition 2.15 in determining whether a given function on a vector space is a linear transformation.

EXAMPLE 3   Show from Definition 2.15 that the function $T: \mathbb{R} \to \mathbb{R}$ defined by $T(x) = \sin x$ is not a linear transformation.

Solution   We know that

$$\sin\left(\tfrac{\pi}{4} + \tfrac{\pi}{4}\right) \neq \sin\left(\tfrac{\pi}{4}\right) + \sin\left(\tfrac{\pi}{4}\right),$$

because $\sin(\pi/4 + \pi/4) = \sin(\pi/2) = 1$, whereas $\sin(\pi/4) + \sin(\pi/4) = 1/\sqrt{2} + 1/\sqrt{2} = 2/\sqrt{2}$. Thus, $\sin x$ is not a linear transformation, because it does not preserve addition. ■

Here is a favorite example from calculus, involving infinite-dimensional vector spaces.

EXAMPLE 4   (*Calculus*) Let F be the vector space of all functions $f: \mathbb{R} \to \mathbb{R}$, and let V be its subspace of all differentiable functions. Show that differentiation is a linear transformation of V into F.

Solution   Let $T: V \to F$ be defined by $T(f) = f'$, the derivative of f. Using the familiar rules

$$(f + g)' = f' + g' \quad \text{and} \quad (rf)' = r(f')$$

for differentiation from calculus, we see that

$$T(f + g) = (f + g)' = f' + g' = T(f) + T(g)$$

and

$$T(rf) = (rf)' = r(f') = rT(f)$$

for all functions f and g in V. In other words, these two rules for differentiating a sum of functions and for differentiating a constant times a function constitute precisely the assertion that differentiation is a linear transformation. ■

EXAMPLE 5   Let F be the vector space of all functions $f: \mathbb{R} \to \mathbb{R}$, and let c be in $\mathbb{R}$. Show that the evaluation function $T: F \to \mathbb{R}$ defined by $T(f) = f(c)$, which maps each function f in F into its value at c, is a linear transformation.

Solution   We show that T preserves addition and scalar multiplication. If f and g are functions in the vector space F, then, evaluating $f + g$ at c, we obtain

$$(f + g)(c) = f(c) + g(c).$$

Therefore,

$$
\begin{aligned}
T(f + g) &= (f + g)(c) && \text{Definition of } T \\
&= f(c) + g(c) && \text{Definition of } f + g \text{ in } F \\
&= T(f) + T(g). && \text{Definition of } T
\end{aligned}
$$

This shows that T preserves addition. In a similar manner, the compuation

$$
\begin{aligned}
T(rf) &= (rf)(c) && \text{Definition of } T \\
&= r(f(c)) && \text{Definition of } rf \text{ in } F \\
&= r(T(f)) && \text{Definition of } T
\end{aligned}
$$

shows that T preserves scalar multiplication.   ■

Properties of Linear Transformations

The two properties of linear transformations in our next theorem are useful and easy to prove.

THEOREM 2.17   Preservation of Zero and Subtraction

Let V and V' be vector spaces, and let $T: V \to V'$ be a linear transformation. Then

(i) $T(\mathbf{0}) = \mathbf{0}'$, and   Preservation of zero

(ii) $T(\mathbf{v}_1 - \mathbf{v}_2) = T(\mathbf{v}_1) - T(\mathbf{v}_2)$ Preservation of subtraction

for any vectors $\mathbf{v}_1$ and $\mathbf{v}_2$ in V.

PROOF   We establish preservation of zero by taking $r = 0$ and $\mathbf{v} = \mathbf{0}$ in condition 2 of Definition 2.15 for a linear transformation. Condition 2 and the property $0\mathbf{v} = \mathbf{0}$ (see Theorem 2.2) yield

$$T(\mathbf{0}) = T(0\mathbf{0}) = 0T(\mathbf{0}) = \boxed{\mathbf{0}'}.$$

Preservation of substraction follows from Eq. (3), as follows:

$$
\begin{aligned}
T(\mathbf{v}_1 - \mathbf{v}_2) &= T(\mathbf{v}_1 + (-1)\mathbf{v}_2) \\
&= T(\mathbf{v}_1) + (-1)T(\mathbf{v}_2) \\
&= T(\mathbf{v}_1) - T(\mathbf{v}_2).
\end{aligned}
$$

▲

EXAMPLE 6 Determine whether the function $T: \mathbb{R}^2 \to \mathbb{R}^2$ defined by $T(x_1, x_2) = (x_1 + x_2, x_1 + 1)$ is a linear transformation.

Solution Since $T(0, 0) = (0, 1)$, we see that T does not preserve zero, so the function cannot be a linear transformation. ■

Example 6 illustrates our assertion in Section 1.5 that the component functions involved in a linear transformation of $\mathbb{R}^n$ into $\mathbb{R}^m$ cannot contain a nonzero constant summand. We see that $T(\mathbf{0}) \neq \mathbf{0}'$ whenever such a nonzero constant summand is present, as in the second component function in Example 6.

Section 1.5 shows that, for an $m \times n$ matrix A, the function $T: \mathbb{R}^n \to \mathbb{R}^m$ given by $T(\mathbf{x}) = A\mathbf{x}$ is a linear transformation. Let us see why every function $T: \mathbb{R}^n \to \mathbb{R}^m$ that is a linear transformation in the sense of Definition 2.15 is given by $T(\mathbf{x}) = A\mathbf{x}$ for some $m \times n$ matrix A. The next theorem is a first step.

THEOREM 2.18 Bases and Linear Transformations

Let $T: V \to V'$ be a linear transformation, and let $B = \{\mathbf{b}_1, \mathbf{b}_2, \ldots, \mathbf{b}_n\}$ be a basis for V. For any vector $\mathbf{v}$ in V, the vector $T(\mathbf{v})$ is uniquely determined by the vectors $T(\mathbf{b}_1), T(\mathbf{b}_2), \ldots, T(\mathbf{b}_n)$. In other words, if two linear transformations have the same value at each basis vector $\mathbf{b}_i$, the two transformations have the same value at each vector in V; that is, they are the same transformation.

PROOF Let $\mathbf{v}$ be any vector in V. We know from Theorem 2.8 that there exist unique scalars $r_1, r_2, \ldots, r_n$ such that

$$\mathbf{v} = r_1\mathbf{b}_1 + r_2\mathbf{b}_2 + \cdots + r_n\mathbf{b}_n.$$

Using Eq. (4), we see that

$$T(\mathbf{v}) = T(r_1\mathbf{b}_1 + r_2\mathbf{b}_2 + \cdots + r_n\mathbf{b}_n)$$
$$= r_1T(\mathbf{b}_1) + r_2T(\mathbf{b}_2) + \cdots + r_nT(\mathbf{b}_n).$$

Since the coefficients r_i are uniquely determined by $\mathbf{v}$, it follows that $T(\mathbf{v})$ is completely determined by the vectors $T(\mathbf{b}_i)$ for $i = 1, 2, , \ldots, n$. ▲

As an illustration of Theorem 2.18 for $T: \mathbb{R}^2 \to \mathbb{R}^2$, we can find the transformation $T(x_1, x_2) = (ax_1 + bx_2, cx_1 + dx_2)$ if we know $T(1, 0)$ and $T(0, 1)$. For example, if $T(1, 0) = (3, -5)$ and $T(0, 1) = (-7, 6)$, then we have $(a, c) = (3, -5)$ and $(b, d) = (-7, 6)$. Thus the transformation is given by

$$T(x_1, x_2) = (3x_1 - 7x_2, -5x_1 + 6x_2).$$

If we write vectors as column vectors, then

$$T\left(\begin{bmatrix} x_1 \\ x_2 \end{bmatrix}\right) = \begin{bmatrix} 3 & -7 \\ -5 & 6 \end{bmatrix}\begin{bmatrix} x_1 \\ x_2 \end{bmatrix},$$

so this transformation has the associated matrix

$$A = \begin{bmatrix} 3 & -7 \\ -5 & 6 \end{bmatrix}.$$

Notice that the columns of A are precisely

$$T\left(\begin{bmatrix} 1 \\ 0 \end{bmatrix}\right) \quad \text{and} \quad T\left(\begin{bmatrix} 0 \\ 1 \end{bmatrix}\right).$$

Now we can show that a linear transformation $T: \mathbb{R}^n \to \mathbb{R}^m$ is indeed given by an equation $T(\mathbf{x}) = A\mathbf{x}$ for some $m \times n$ matrix A. Let $(\mathbf{e}_1, \mathbf{e}_2, \ldots, \mathbf{e}_n)$ be the standard ordered basis for $\mathbb{R}^n$. We form the matrix

$$A = \begin{bmatrix} | & | & & | \\ T(\mathbf{e}_1) & T(\mathbf{e}_2) & \cdots & T(\mathbf{e}_n) \\ | & | & & | \end{bmatrix}, \tag{6}$$

whose jth column vector is $T(\mathbf{e}_j)$. For each column vector $\mathbf{e}_j$ in our standard basis, we have

$$A\mathbf{e}_j = T(\mathbf{e}_j),$$

since $A\mathbf{e}_j$ gives the jth column vector of A. This shows that the transformation given by $A\mathbf{x}$ and the transformation $T(\mathbf{x})$ have the same values on the standard basis vectors in $\mathbb{R}^n$. By Theorem 2.18, they must be the same transformation.

We refer to matrix (6) as the **standard matrix representation** of the transformation $T: \mathbb{R}^n \to \mathbb{R}^m$, rather than as "the matrix associated with T"—the wording used in Chapter 1. As we shall see in Chapter 6, a linear transformation T has many different matrix representations, depending on the choice of bases of the vector spaces.

Finding the standard matrix representation of a linear transformation is quite straightforward, as illustrated by Example 1. The form of matrix (6) indicates another way to find the standard matrix representation of a linear transformation of $\mathbb{R}^n$ into $\mathbb{R}^m$.

EXAMPLE 7 Let $T: \mathbb{R}^2 \to \mathbb{R}^3$ be defined by

$$T(x_1, x_2) = (3x_1 - x_2, 4x_1 + 5x_2, -7x_1 + 2x_2).$$

Use formula (6) to find the standard matrix representation A of T.

Solution We see that $T(1, 0) = (3, 4, -7)$ and $T(0, 1) = (-1, 5, 2)$. To form matrix (6), we take these as column vectors of the matrix A, and we obtain

$$A = \begin{bmatrix} 3 & -1 \\ 4 & 5 \\ -7 & 2 \end{bmatrix}. \qquad ■$$

Invertible Transformations

In Section 1.5, we saw that, if A is an invertible $n \times n$ matrix, the linear transformation $T: \mathbb{R}^n \to \mathbb{R}^n$ defined by $T(\mathbf{x}) = A\mathbf{x}$ has an inverse $T^{-1}: \mathbb{R}^n \to \mathbb{R}^n$, where $T^{-1}(\mathbf{y}) = A^{-1}\mathbf{y}$, and that both composite transformations $T^{-1} \circ T$ and $T \circ T^{-1}$ are the identity transformation of $\mathbb{R}^n$. We now extend this idea to linear transformations of vector spaces in general.

DEFINITION 2.16 Invertible Transformation

> Let V and V' be vector spaces. A linear transformation $T: V \to V'$ is **invertible** if there exists a linear transformation $T^{-1}: V' \to V$ such that $T^{-1} \circ T$ is the identity transformation on V and $T \circ T^{-1}$ is the identity transformation on V'.

EXAMPLE 8   Determine whether the evaluation transformation $T: F \to \mathbb{R}$ of Example 5, where $T(f) = f(c)$ for some selected c in $\mathbb{R}$, is invertible.

Solution   Consider the polynomial functions f and g, where $f(x) = 2x + c$ and $g(x) = 4x - c$. Then $T(f) = T(g) = 3c$. If T were invertible, there would have to be a linear transformation $T^{-1}: \mathbb{R} \to F$ such that $T^{-1}(3c)$ is both f and g. But this is impossible. ■

Example 8 illustrates that, if $T: V \to V'$ is an invertible linear transformation, T must satisfy the following property:

$$\text{if} \quad \mathbf{v}_1 \neq \mathbf{v}_2, \quad \text{then} \quad T(\mathbf{v}_1) \neq T(\mathbf{v}_2). \quad \textbf{One-to-one} \qquad (7)$$

As in the argument of Example 8, if $T(\mathbf{v}_1) = T(\mathbf{v}_2) = \mathbf{v}'$ and T is invertible, there would have to be a linear transformation $T^{-1}: V' \to V$ such that $T^{-1}(\mathbf{v}')$ is simultaneously $\mathbf{v}_1$ and $\mathbf{v}_2$, which is impossible when $\mathbf{v}_1 \neq \mathbf{v}_2$. A linear transformation satisfying property (7) is **one-to-one**.

An invertible linear transformation $T: V \to V'$ must also satisfy another property:

$$\text{if } \mathbf{v}' \text{ is in } V', \quad T(\mathbf{v}) = \mathbf{v}' \quad \text{for some } \mathbf{v} \text{ in } V. \quad \textbf{Onto} \qquad (8)$$

This follows at once from the fact that, for T^{-1} with the properties in Definition 2.16 and for $\mathbf{v}'$ in V', we have $T^{-1}(\mathbf{v}') = \mathbf{v}$ for some $\mathbf{v}$ in V. But then $\mathbf{v}' = (T \circ T^{-1})(\mathbf{v}') = T(T^{-1}(\mathbf{v}')) = T(\mathbf{v})$. A transformation $T: V \to V'$ satisfying property (8) is **onto** V'; in this case the range of T is all of V'. We have thus proved half of the following theorem.

THEOREM 2.19 Invertibility of T

> A linear transformation $T: V \to V'$ is invertible if and only if it is one-to-one and onto V'.

PROOF We have just shown that, if T is invertible, it must be one-to-one and onto V'.

Suppose now that T is one-to-one and onto V'. Since T is onto V', for each $\mathbf{v}'$ in V' we can find $\mathbf{v}$ in V such that $T(\mathbf{v}) = \mathbf{v}'$. Since T is one-to-one, this vector $\mathbf{v}$ in V is *unique*. Let $T^{-1}: V' \to V$ be defined by $T^{-1}(\mathbf{v}') = \mathbf{v}$, where $\mathbf{v}$ is the unique vector in V such that $T(\mathbf{v}) = \mathbf{v}'$. Then

$$(T \circ T^{-1})(\mathbf{v}') = T(T^{-1}(\mathbf{v}')) = T(\mathbf{v}) = \mathbf{v}'$$

and

$$(T^{-1} \circ T)(\mathbf{v}) = T^{-1}(T(\mathbf{v})) = T^{-1}(\mathbf{v}') = \mathbf{v},$$

which shows that $T \circ T^{-1}$ is the identity map of V' and that $T^{-1} \circ T$ is the identity map of V. It only remains to be shown that T^{-1} is indeed a linear transformation— that is, that

$$T^{-1}(\mathbf{v}_1' + \mathbf{v}_2') = T^{-1}(\mathbf{v}_1') + T^{-1}(\mathbf{v}_2') \qquad \text{and} \qquad T^{-1}(r\mathbf{v}_1') = rT^{-1}(\mathbf{v}_1')$$

for all $\mathbf{v}_1'$ and $\mathbf{v}_2'$ in V' and for all scalars r. Let $\mathbf{v}_1$ and $\mathbf{v}_2$ be the unique vectors in V such that $T(\mathbf{v}_1) = \mathbf{v}_1'$ and $T(\mathbf{v}_2) = \mathbf{v}_2'$. Remembering that $T^{-1} \circ T$ is the identity map, we have

$$T^{-1}(\mathbf{v}_1' + \mathbf{v}_2') = T^{-1}(T(\mathbf{v}_1) + T(\mathbf{v}_2)) = T^{-1}(T(\mathbf{v}_1 + \mathbf{v}_2))$$
$$= (T^{-1} \circ T)(\mathbf{v}_1 + \mathbf{v}_2) = \mathbf{v}_1 + \mathbf{v}_2 = T^{-1}(\mathbf{v}_1') + T^{-1}(\mathbf{v}_2')$$

Similarly,

$$T^{-1}(r\mathbf{v}_1') = T^{-1}(rT(\mathbf{v}_1)) = T^{-1}(T(r\mathbf{v}_1)) = r\mathbf{v}_1 = rT^{-1}(\mathbf{v}_1'). \qquad \blacktriangle$$

The proof of Theorem 2.19 shows that, if $T: V \to V'$ is invertible, the linear transformation $T^{-1}: V' \to V$ described in Definition 2.16 is unique. This transformation T^{-1} is the **inverse transformation** of T.

EXAMPLE 9 Show that the rotation of the plane counterclockwise through a positive angle α, leaving the origin fixed as described in Example 2, is an invertible transformation.

Solution We can proceed in a number of ways. Because such a rotation is one-to-one and onto $\mathbb{R}^2$, it is invertible. Alternatively, $T^{-1}: \mathbb{R}^2 \to \mathbb{R}^2$, defined as rotation clockwise through α, is a linear transformation such that both $T \circ T^{-1}$ and $T^{-1} \circ T$ are the identity transformation of $\mathbb{R}^2$. For still another demonstration, we could show that the standard matrix representation A of T found in Example 2 is invertible, finding that

$$A^{-1} = \begin{bmatrix} \cos \alpha & \sin \alpha \\ -\sin \alpha & \cos \alpha \end{bmatrix},$$

and use our work in Section 1.5. ■

Isomorphism

An **isomorphism** is a linear transformation $T: V \to V'$ that is one-to-one and onto V'. Theorem 2.19 shows that isomorphisms are precisely the invertible linear transformations $T: V \to V'$. If an isomorphism T exists, then it is invertible and its inverse is also an isomorphism. The vector spaces V and V' are said to be **isomorphic** in this case. We view isomorphic vector spaces V and V' as being structurally identical in the following sense. Let $T: V \to V'$ be an isomorphism. Rename each $\mathbf{v}$ in V by the $\mathbf{v}' = T(\mathbf{v})$ in V'. Since T is one-to-one, no two different elements of V get the same name from V', and since T is onto V', all names in V' are used. The renamed V and V' then appear identical as sets. But they also have the same algebraic structure as vector spaces, as Figure 2.31 illustrates. We discussed a special case of this concept at the end of Section 2.7, indicating informally that every finite-dimensional vector space is structurally the same as $\mathbb{R}^n$ for some n. We are now in a position to state this formally.

THEOREM 2.20 Coordinatization of Finite-dimensional Spaces

Let V be a finite-dimensional vector space with ordered basis $B = (\mathbf{b}_1, \mathbf{b}_2, \ldots, \mathbf{b}_n)$. The map $T: V \to \mathbb{R}^n$ defined by $T(\mathbf{v}) = \mathbf{v}_B$, the coordinate vector of $\mathbf{v}$ relative to B, is an isomorphism.

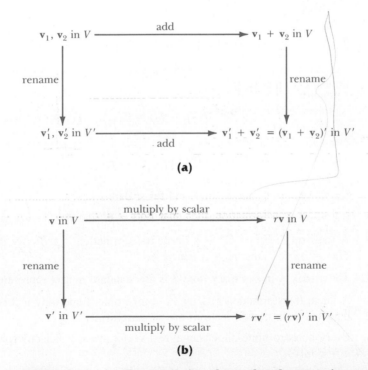

FIGURE 2.31 (a) The equal sign shows that the renaming preserves vector addition. (b) The equal sign shows that the renaming preserves scalar multiplication.

PROOF Equation 1 of Section 2.7 shows that T preserves addition and scalar multiplication. Moreover, T is one-to-one, because the coordinate vector $\mathbf{v}_B$ of $\mathbf{v}$ uniquely determines $\mathbf{v}$, and the range of T is all of $\mathbb{R}^n$. Therefore, T is an isomorphism. ▲

The isomorphism of V with $\mathbb{R}^n$, described in Theorem 2.20, is by no means unique. There is one such isomorphism for each choice of an ordered basis B of V.

Let V and V' be vector spaces of dimensions n and m, respectively. By Theorem 2.20, we can choose ordered bases B for V and B' for V', and essentially convert V into $\mathbb{R}^n$ and V' into $\mathbb{R}^m$ by renaming vectors by their coordinate vectors relative to these bases. Each linear transformation $T: V \rightarrow V'$ then corresponds to a linear transformation $\bar{T}: \mathbb{R}^n \rightarrow \mathbb{R}^m$ in a natural way, and we can answer questions about T by studying $\bar{T}$ instead. But we can study $\bar{T}$ in turn by studying its standard matrix representation A. This matrix changes if we change the ordered bases B or B'. A good deal of the remainder of our text is devoted to studying how to choose B and B' when $m = n$ so that the square matrix A has a simple form that illuminates the structure of the transformation T. This is the thrust of Chapters 4 and 6.

SUMMARY

Let V and V' be vector spaces, and let T be a function mapping V into V'.

1. The function T is a linear transformation if it preserves addition and scalar multiplication—that is, if

$$T(\mathbf{v}_1 + \mathbf{v}_2) = T(\mathbf{v}_1) + T(\mathbf{v}_2)$$

and

$$T(r\mathbf{v}_1) = rT(\mathbf{v}_1)$$

for all vectors $\mathbf{v}_1$ and $\mathbf{v}_2$ in V and for scalars r in $\mathbb{R}$.

2. If T is a linear transformation, then $T(\mathbf{0}) = \mathbf{0}'$ and $T(\mathbf{v}_1 - \mathbf{v}_2) = T(\mathbf{v}_1) - T(\mathbf{v}_2)$.

3. A function $T: \mathbb{R}^n \rightarrow \mathbb{R}^m$ is a linear transformation if and only if T has the form $T(\mathbf{x}) = A\mathbf{x}$ for some $m \times n$ matrix A.

4. The matrix A in summary item 3 is the standard matrix representation of T.

5. A linear transformation $T: V \rightarrow V'$ is invertible if and only if it is one-to-one and onto V'. Such transformations are isomorphisms.

6. Every nonzero finite-dimensional real vector space V is isomorphic to $\mathbb{R}^n$, where $n = \dim(V)$.

P124

12.   upper triangular.
       addition.

P124   # 35.

P152   # (9 8)

P171   # 8 (compare to Q in Test

√ P284   # 25   2) ⟷ j)

* P183   #25(e) (b)

Check the exam location and
time.

USD — Price changes in 32nds

TeriCoupon		Maturity	Settle	Price	Yield	DOWN	
						−0.01	−0.02
2	4.250	31–Oct–94	24–Nov–92	99.08	4.659	4.676	4.693
3	5.125	15–Nov–95	24–Nov–92	99.22	5.239	5.251	5.262
5	5.750	31–Oct–97	24–Nov–92	98.18	6.091	6.098	6.106
7	6.000	15–Oct–99	24–Nov–92	97.09	6.494	6.500	6.506
10	6.375	15–Aug–2002	24–Nov–92	96.18	6.863	6.868	6.872
						UP	
						0.01	0.02

EXERCISES

In Exercises 1–12, determine whether the given function T is a linear transformation. If it is a linear transformation, determine whether it is invertible. In Exercises 8–12 let F be the vector space of all functions $f: \mathbb{R} \to \mathbb{R}$.

1. $T: \mathbb{R}^2 \to \mathbb{R}^2$ defined by $T(x, y) = (x + y, 2y)$

2. $T: \mathbb{R}^2 \to \mathbb{R}^2$ defined by $T(x, y) = (x + 4, x - y)$

3. $T: \mathbb{R}^2 \to \mathbb{R}^3$ defined by $T(x, y) =$ $(2x, 3x + y, x + y)$

4. $T: \mathbb{R}^3 \to \mathbb{R}^3$ defined by $T(x_1, x_2, x_3) =$ $(x_3, 0, x_1 + x_2)$

5. $T: \mathbb{R}^3 \to \mathbb{R}^2$ defined by $T(x_1, x_2, x_3) =$ $(x_1 + x_2 + x_3, 2x_1)$

6. $T: \mathbb{R}^3 \to \mathbb{R}^4$ defined by $T(x_1, x_2, x_3) =$ $(x_1 + x_2, x_2 + x_3, x_1 + x_3, x_1 - x_3)$

7. $T: \mathbb{R}^4 \to \mathbb{R}^4$ defined by $T(x_1, x_2, x_3, x_4) =$ $(x_1, 1, x_3, 1)$

8. $T: F \to \mathbb{R}$ defined by $T(f) = f(-4)$

9. $T: F \to \mathbb{R}$ defined by $T(f) = f(5)^2$

10. $T: F \to F$ defined by $T(f) = f + f$

11. $T: F \to F$ defined by $T(f) = f + 3$, where 3 is the constant function with value 3 for all x in $\mathbb{R}$

12. $T: F \to F$ defined by $T(f) = -f$

In Exercises 13–18, use Eq. (6) to find the standard matrix representation of the given linear transformation.

13. $T: \mathbb{R}^2 \to \mathbb{R}^2$ defined by $T(x, y) =$ $(2x - 3y, x + y)$

14. $T: \mathbb{R}^2 \to \mathbb{R}^3$ defined by $T(x, y) =$ $(2x + y, x + 2y, x - 3y)$

15. $T: \mathbb{R}^3 \to \mathbb{R}^3$ defined by $T(x_1, x_2, x_3) =$ $(x_1 + x_2 + x_3, x_1 - x_2 - x_3, -x_1 - x_2 + x_3)$

16. $T: \mathbb{R}^3 \to \mathbb{R}^3$ defined by $T(x_1, x_2, x_3) =$ $(x_1 + 2x_2 + x_3, x_1, x_2 + x_3)$

17. $T: \mathbb{R}^3 \to \mathbb{R}^4$ defined by $T(x_1, x_2, x_3) =$ $(x_1 + x_2, x_2 + x_3, x_3 + x_1, x_1 + x_2 + x_3)$

18. $T: \mathbb{R}^3 \to \mathbb{R}^2$ defined by $T(x_1, x_2, x_3) =$ $(x_1 + 2x_2 + 3x_3, x_1)$

19. Argue geometrically that, if the line L in $\mathbb{R}^2$ contains the origin, reflection through L is a linear transformation of $\mathbb{R}^2$ into itself.

20. Find the standard matrix representation of the reflection of $\mathbb{R}^2$ through the line $y = x$.

21. Express rotation of the plane counterclockwise through an angle of 60° as a sequence of one or more reflections, expansions or contractions, and vertical or horizontal shears, as described in the box at the end of Section 1.5.

22. Argue geometrically that reflection through a plane containing the origin is a linear transformation of $\mathbb{R}^3$ into $\mathbb{R}^3$.

23. Find the standard matrix representation of the reflection of $\mathbb{R}^3$ through
    a) the plane $x_2 = 0$,
    b) the plane $x_3 = 0$.

24. Find the standard matrix representation of the reflection of $\mathbb{R}^3$ through
    a) the plane $x_2 = x_1$,
    b) the plane $x_1 = x_3$.

25. Let V and V' be vector spaces. Mark each of the following True or False.
    ___ a) A linear transformation of vector spaces preserves the vector-space operations.
    ___ b) Every function mapping V into V' relates the algebraic structure of V to that of V'.
    ___ c) A linear transformation $T: V \to V'$ carries the zero vector of V into the zero vector of V'.
    ___ d) A linear transformation $T: V \to V'$ carries a pair $\mathbf{v}, -\mathbf{v}$ in V into a pair $\mathbf{v}', -\mathbf{v}'$ in V'.
    ___ e) For every vector $\mathbf{b}'$ in V', the function $T_{\mathbf{b}'}: V \to V'$ defined by $T_{\mathbf{b}'}(\mathbf{v}) = \mathbf{b}'$ for all $\mathbf{v}$ in V is a linear transformation.
    ___ f) The function $T_{\mathbf{0}'}: V \to V'$ defined by $T_{\mathbf{0}'}(\mathbf{v}) = \mathbf{0}'$, the zero vector of V', for all $\mathbf{v}$ in V is a linear transformation.
    ___ g) The vector space P_{10} of polynomials of degree ≤ 10 is isomorphic to $\mathbb{R}^{10}$.
    ___ h) There is exactly one isomorphism $T: P_{10} \to \mathbb{R}^{11}$.

i) Let V and V' be vector spaces of dimensions n and m, respectively. A linear transformation $T: V \rightarrow V'$ is invertible if and only if $m = n$.

j) If T in part (i) is an invertible transformation, then $m = n$.

26. Prove that the two conditions in Definition 2.15 for a linear transformation are equivalent to the single condition in Eq. (3).

27. Let V, V', and V'' be vector spaces, and let $T: V \rightarrow V'$ and $T': V' \rightarrow V''$ be linear transformations. Show that the composite function $(T' \circ T): V \rightarrow V''$ defined by $(T' \circ T)(\mathbf{v}) = T'(T(\mathbf{v}))$ for each $\mathbf{v}$ in V is again a linear transformation.

28. Show that, if T and T' are invertible linear transformations of vector spaces such that $T' \circ T$ is defined, $T' \circ T$ is also invertible.

29. State conditions for an $m \times n$ matrix A that are equivalent to the condition that the linear transformation $T(\mathbf{x}) = A\mathbf{x}$ for $\mathbf{x}$ in $\mathbb{R}^n$ is an isomorphism.

30. Let $\mathbf{v}$ and $\mathbf{w}$ be independent vectors in V, and let $T: V \rightarrow V'$ be a one-to-one linear transformation of V into V'. Show that $T(\mathbf{v})$ and $T(\mathbf{w})$ are independent vectors in V'.

31. Let $T: V \rightarrow V'$ be a linear transformation. Show that the set of all $\mathbf{v}$ in V such that $T(\mathbf{v}) = \mathbf{0}'$, the zero vector in V', is a subspace of V. (This subspace is called the **kernel** or **nullspace** of T.)

32. Referring to Exercise 31, show that the kernel of a linear transformation $T: V \rightarrow V'$ is $\{\mathbf{0}\}$ if and only if T is one-to-one.

33. Let $T: V \rightarrow V'$ be a linear transformation, and let W be a subspace of V. Show that $T[W] = \{T(\mathbf{w}) \mid \mathbf{w} \text{ in } V\}$ is a subspace of V'.

34. Let $T: V \rightarrow V'$ be a linear transformation, and let W' be a subspace of V'. Show that $T^{-1}[W'] = \{\mathbf{v} \text{ in } V \mid T(\mathbf{v}) \text{ is in } W\}$ is a subspace of V.

Exercise 35 shows that the reduced row echelon form of a matrix is unique.

35. Let A be an $m \times n$ matrix with row echelon form H, and let V be the row space of A (and hence of H). Let $W_k = \mathrm{sp}(\mathbf{e}_1, \mathbf{e}_2, \ldots, \mathbf{e}_k)$ be the subspace of $\mathbb{R}^n$ generated by the first k rows of the $n \times n$ identity matrix. Consider $T_k: V \rightarrow W_k$ defined by

$$T_k(x_1, x_2, \ldots, x_n)$$
$$= (x_1, x_2, \ldots, x_k, 0, \ldots, 0).$$

a) Show that T_k is a linear transformation of V into W_k and that $T_k[V] = \{T_k(\mathbf{v}) \mid \mathbf{v} \text{ in } V\}$ is a subspace of W_k. [HINT: See Exercise 33.]

b) If $T_k[V]$ has dimension d_k, show that, for each $j < n$, we have either $d_{j+1} = d_j$ or $d_{j+1} = d_j + 1$.

c) Assume that A has four columns. Referring to part (b), suppose that $d_1 = d_2 = 1$ and $d_3 = d_4 = 2$. Find the number of pivots in H, and give the location of each.

d) Repeat part (c) for the case where A has six columns and $d_1 = 1$, $d_2 = d_3 = d_4 = 2$, and $d_5 = d_6 = 3$.

e) Argue that, for any matrix A, the number of pivots and the location of each pivot in any row echelon form of A is always the same.

f) Show that the reduced row echelon form of a matrix A is unique. [HINT: Consider the nature of the basis for the row space of A given by the nonzero rows of H.]

Exercises 36–38 are concerned with the algebra of all linear transformations of a vector space V into a vector space V'. The exercises show that this collection of linear transformations has a natural vector-space structure. We let $L(V, V')$ be the collection of all linear transformations of V into V'.

36. Let T_1 and T_2 be in $L(V, V')$, and let $(T_1 + T_2): V \rightarrow V'$ be defined by

$$(T_1 + T_2)(\mathbf{v}) = T_1(\mathbf{v}) + T_2(\mathbf{v})$$

for each vector $\mathbf{v}$ in V. Show that $T_1 + T_2$ is again a linear transformation of V into V'.

37. Let T be in $L(V, V')$, let r be any scalar in $\mathbb{R}$, and let $rT: V \rightarrow V'$ be defined by

$$(rT)(\mathbf{v}) = r(T(\mathbf{v}))$$

for each vector $\mathbf{v}$ in V. Show that rT is again a linear transformation of V into V'.

38. Verify the properties required for addition and scalar multiplication in a vector space for the addition and scalar multiplication defined on $L(V, V')$ in Exercises 36 and 37. Careful proofs of all the properties need not be written out, but sufficient thought should be given to determining why they are true. What is the zero vector in $L(V, V')$?

39. Work with Topic 4 of the available program VECTGRPH until you can consistently achieve a score of at least 80%.

40. Work with Topic 5 of the available program VECTGRPH until you can regularly attain a score of at least 82%.

2.10 INNER-PRODUCT SPACES

In Section 2.1, we introduced the concept of the length of a vector and the concept of the angle between vectors in $\mathbb{R}^n$. *Length* and *angle* are defined and computed in $\mathbb{R}^n$ using the dot product of vectors. In this section, we discuss these notions for more general vector spaces. We start by listing the properties of the dot product in $\mathbb{R}^n$.

THEOREM 2.21 Properties of the Dot Product in $\mathbb{R}^n$

Let **u**, **v**, and **w** be vectors in $\mathbb{R}^n$ and let r be any scalar in $\mathbb{R}$. The following properties hold:

D1 $\mathbf{v} \cdot \mathbf{w} = \mathbf{w} \cdot \mathbf{v}$,   Commutative law

D2 $\mathbf{u} \cdot (\mathbf{v} + \mathbf{w}) = \mathbf{u} \cdot \mathbf{v} + \mathbf{u} \cdot \mathbf{w}$,   Distributive law

D3 $r(\mathbf{v} \cdot \mathbf{w}) = (r\mathbf{v}) \cdot \mathbf{w} = \mathbf{v} \cdot (r\mathbf{w})$,   Homogeneity

D4 $\mathbf{v} \cdot \mathbf{v} \geq 0$, and $\mathbf{v} \cdot \mathbf{v} = 0$ if and only if $\mathbf{v} = \mathbf{0}$.   Positivity

All of the properties in Theorem 2.21 are easy to verify, as is illustrated in the following example.

EXAMPLE 1   Verify D4 of Theorem 2.21.

Solution   We let $\mathbf{v} = (v_1, v_2, \ldots, v_n)$, and we find that

$$\mathbf{v} \cdot \mathbf{v} = v_1^2 + v_2^2 + \cdots + v_n^2.$$

A sum of squares is nonnegative and can be zero if and only if each summand is zero. But a summand v_i^2 is itself a square, and will be zero if and only if $v_i = 0$. This completes the demonstration. ■

There are many vector spaces for which we can define useful dot products satisfying the conditions of Theorem 2.21. In phrasing a general definition for such vector spaces, it is customary to speak of an *inner product* rather than of a dot product, and to use the notation $\langle \mathbf{v}, \mathbf{w} \rangle$ in place of $\mathbf{v} \cdot \mathbf{w}$. For example, $\langle (2, 3), (4, -1) \rangle = 2(4) + 3(-1) = 5$.

DEFINITION 2.17 Inner-product Space

An **inner product** on a vector space V is a function that associates with each ordered pair of vectors $\mathbf{v}, \mathbf{w}$ in V a real number, written $\langle \mathbf{v}, \mathbf{w} \rangle$, satisfying the following properties for all $\mathbf{u}$, $\mathbf{v}$, and $\mathbf{w}$ in V and for all scalars r:

P1 $\langle \mathbf{v}, \mathbf{w} \rangle = \langle \mathbf{w}, \mathbf{v} \rangle$,   Symmetry

P2 $\langle \mathbf{u}, \mathbf{v} + \mathbf{w} \rangle = \langle \mathbf{u}, \mathbf{v} \rangle + \langle \mathbf{u}, \mathbf{w} \rangle$,   Additivity

P3 $r\langle \mathbf{v}, \mathbf{w} \rangle = \langle r\mathbf{v}, \mathbf{w} \rangle = \langle \mathbf{v}, r\mathbf{w} \rangle$,   Homogeneity

P4 $\langle \mathbf{v}, \mathbf{v} \rangle \geq 0$, and $\langle \mathbf{v}, \mathbf{v} \rangle = 0$ if and only if $\mathbf{v} = \mathbf{0}$.   Positivity

An **inner-product space** is a vector space V together with an inner product on V.

EXAMPLE 2   Verify that $\mathbb{R}^n$ is an inner-product space if we let $\langle \mathbf{v}, \mathbf{w} \rangle = \mathbf{v} \cdot \mathbf{w}$ for all vectors $\mathbf{v}$ and $\mathbf{w}$ in $\mathbb{R}$.

Solution   This example is just a restatement of Theorem 2.21. ■

The inner product for $\mathbb{R}^n$ using the usual dot product is referred to as the **standard inner product** for $\mathbb{R}^n$.

EXAMPLE 3   Determine whether $\mathbb{R}^2$ is an inner-product space if, for $\mathbf{v} = (v_1, v_2)$ and $\mathbf{w} = (w_1, w_2)$, we define

$$\langle \mathbf{v}, \mathbf{w} \rangle = 2v_1 w_1 + 5v_2 w_2.$$

Solution   We check each of the four conditions in Definition 2.17.

P1: Since $\langle \mathbf{v}, \mathbf{w} \rangle = 2v_1 w_1 + 5v_2 w_2$ and since $\langle \mathbf{w}, \mathbf{v} \rangle = 2w_1 v_1 + 5w_2 v_2$, the first condition holds.

P2: We compute

$$\langle \mathbf{u}, \mathbf{v} + \mathbf{w} \rangle = 2u_1(v_1 + w_1) + 5u_2(v_2 + w_2),$$

$$\langle \mathbf{u}, \mathbf{v} \rangle = 2u_1 v_1 + 5u_2 v_2,$$

$$\langle \mathbf{u}, \mathbf{w} \rangle = 2u_1 w_1 + 5u_2 w_2.$$

The sum of the right-hand sides of the last two equations equals the right-hand side of the first equation. This establishes condition P2.

P3: We compute

$$r\langle \mathbf{v}, \mathbf{w} \rangle = r(2v_1 w_1 + 5v_2 w_2),$$

$$\langle r\mathbf{v}, \mathbf{w} \rangle = 2(rv_1)w_1 + 5(rv_2)w_2,$$

$$\langle \mathbf{v}, r\mathbf{w} \rangle = 2v_1(r_1 w_1) + 5v_2(rw_2).$$

Since the three right-hand expressions are equal, property P3 holds.

P4: We see that $\langle \mathbf{v}, \mathbf{v} \rangle = 2v_1{}^2 + 5v_2{}^2 \geq 0$. Since both terms of the sum are nonnegative, the sum is zero if and only if $v_1 = v_2 = 0$—that is, if and only if $\mathbf{v} = \mathbf{0}$. Therefore, we have an inner product space. ■

EXAMPLE 4   Determine whether $\mathbb{R}^2$ is an inner-product space when, for $\mathbf{v} = (v_1, v_2)$ and $\mathbf{w} = (w_1, w_2)$, we define

$$\langle \mathbf{v}, \mathbf{w} \rangle = 2v_1w_1 - 5v_2w_2.$$

Solution   A solution similar to the one for Example 3 goes along smoothly until we check condition P4, which fails: $(V, V) = 2 V_1 V_1 - 5 V_2 V_2 = 2V_1^2 - 5 V_2^2$

$$\langle (1, 1), (1, 1) \rangle = 2 - 5 = -3 < 0. \quad \text{if } V = (1, 1) \text{ then}$$
$$\text{it fails.}$$

Therefore, $\mathbb{R}^2$ with this definition of $\langle \ , \ \rangle$ <u>is not an inner-product space</u>. ■

EXAMPLE 5   (*Calculus*) Determine whether the space P of all polynomial functions with real coefficients is an inner-product space if for p and q in P we define

$$\langle p, q \rangle = \int_0^1 p(x)q(x) \, dx.$$

Solution   We check the conditions for an inner product.

P1: Clearly $\langle p, q \rangle = \langle q, p \rangle$ because

$$\int_0^1 p(x)q(x) \, dx = \int_0^1 q(x)p(x) \, dx.$$

P2: For polynomial functions p, q, and h, we have

$$\langle p, q + h \rangle = \int_0^1 p(x)(q(x) + h(x)) \, dx$$
$$= \int_0^1 p(x)q(x) \, dx + \int_0^1 p(x)h(x) \, dx$$
$$= \langle p, q \rangle + \langle p, h \rangle.$$

P3: We have

$$r\int_0^1 p(x)q(x) \, dx = \int_0^1 r(p(x))q(x) \, dx$$
$$= \int_0^1 p(x)r(q(x)) \, dx,$$

so P3 holds.

P4: Since $\langle p, p \rangle = \int_0^1 p(x)^2 \, dx$ and since $p(x)^2 \geq 0$ for all x, we have $\langle p, p \rangle = \int_0^1 p(x)^2 \, dx \geq 0$. Now $p(x)^2$ is a continuous nonnegative polynomial function and can

square of the
polynomial fn
is ≥ 0

be zero only at a finite number of points unless $p(x)$ is the zero polynomial. It follows that

$$\langle p, p \rangle = \int_0^1 p(x)^2 \, dx > 0,$$

unless $p(x)$ is the zero polynomial. This establishes P4.

Since all four conditions of the definition hold, the space P is an inner-product space with the given inner product. ■

We mention that Example 5 is a very famous inner product, and the same definition

works for all continuous real-valued fu

$$\langle f, g \rangle = \int_0^1 f(x)g(x) \, dx$$

gives an inner product on the space of all continuous real-valued functions whose domains contain the interval [0, 1]. The hypothesis of continuity is essential for the demonstration of P4, as is shown in advanced calculus. Of course, there is nothing unique about the interval [0, 1]. Any interval [a, b] can be used in its place. The choice of interval in an application depends on where you are working with the functions.

Magnitude

The condition $\langle \mathbf{v}, \mathbf{v} \rangle \geq 0$ in the definition of an inner-product space allows us to define the magnitude of a vector, just as we did in Section 2.1 using the dot product.

DEFINITION 2.18 Magnitude or Norm of a Vector

Let V be an inner-product space. The **magnitude** or **norm** of a vector $\mathbf{v}$ in V is $\|\mathbf{v}\| = \sqrt{\langle \mathbf{v}, \mathbf{v} \rangle}$.

This definition of *norm* reduces to the usual definition of magnitude of vectors in $\mathbb{R}^n$ when the dot product is used. That is, $\|\mathbf{v}\| = \sqrt{\mathbf{v} \cdot \mathbf{v}}$.

Since $\|\mathbf{a} - \mathbf{b}\|$ gives the distance from $\mathbf{a}$ to $\mathbf{b}$ in $\mathbb{R}^n$, we define the **distance** from $\mathbf{v}$ to $\mathbf{w}$ in an inner-product space V to be $d(\mathbf{v}, \mathbf{w}) = \|\mathbf{v} - \mathbf{w}\|$.

EXAMPLE 6 (*Calculus*) In the inner-product space P of all polynomial functions with real coefficients and with inner product defined by

$$\langle p, q \rangle = \int_0^1 p(x)q(x) \, dx,$$

(a) find the magnitude of the polynomial $p(x) = x + 1$, and (b) compute the distance $d(x^2, x)$ from x^2 to x.

Solution   For part (a), we have

$$\|x + 1\|^2 = \langle x + 1, x + 1 \rangle = \int_0^1 (x + 1)^2 \, dx$$

$$= \int_0^1 (x^2 + 2x + 1) \, dx = \left(\frac{x^3}{3} + x^2 + x \right) \Big|_0^1 = \frac{7}{3}.$$

Therefore, $\|x + 1\| = \sqrt{7/3}$.

For part (b), we have $d(x^2, x) = \|x^2 - x\|$. We compute

$$\|x^2 - x\|^2 = \langle x^2 - x, x^2 - x \rangle = \int_0^1 (x^2 - x)^2 \, dx$$

$$= \int_0^1 (x^4 - 2x^3 + x^2) \, dx = \frac{1}{5} - \frac{1}{2} + \frac{1}{3} = \frac{1}{30}.$$

Therefore, $d(x^2, x) = 1/\sqrt{30}$.   ■

The inner product used in Example 6 was not contrived just for this illustration. Another measure of the distance between the functions x and x^2 over $[0, 1]$ is the maximum vertical distance between their graphs over $[0, 1]$, which calculus easily shows to be $\frac{1}{4}$ where $x = \frac{1}{2}$. In contrast, the inner product in this example uses an *integral* to measure the distance between the functions, not only at one point of the interval, but over the interval as a whole. Notice that the distance $1/\sqrt{30}$ we obtained is less than the maximum distance $\frac{1}{4}$ between the graphs, reflecting the fact that the graphs are less than $\frac{1}{4}$ unit apart over much of the interval. The functions are shown in Figure 2.32. The notion (used in Example 6) of distance between functions over an interval is very important in advanced mathematics, where it is used in approximating

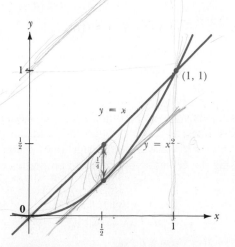

FIGURE 2.32   Graphs of x and x^2 over $[0, 1]$.

a complicated function over an interval as closely as possible by a function that is easier to handle.

It is often best, when working with the norm of a vector $\mathbf{v}$, to work with $\|\mathbf{v}\|^2 = \langle \mathbf{v}, \mathbf{v} \rangle$ and to introduce the radical in the final stages of the computation.

EXAMPLE 7   Let V be an inner-product space. Verify that

$$\|r\mathbf{v}\| = |r| \|\mathbf{v}\|$$

for any vector $\mathbf{v}$ in V and for any scalar r.

Solution   We have

$$\|r\mathbf{v}\|^2 = \langle r\mathbf{v}, r\mathbf{v} \rangle \; = r\langle v, rv \rangle = r^2 \langle v, v \rangle$$
$$= r^2 \langle \mathbf{v}, \mathbf{v} \rangle \quad \textbf{Applying P3 twice}$$
$$= r^2 \|\mathbf{v}\|^2.$$

On taking square roots, we have $\|r\mathbf{v}\| = |r| \|\mathbf{v}\|$.   ■

The property of norms in the preceding example can be useful in computing the magnitude of a vector and in establishing relations between magnitudes, as illustrated in the following two examples.

EXAMPLE 8   Using the standard inner product in $\mathbb{R}^n$, find the magnitude of the vector

$$\mathbf{v} = (-6, -12, 6, 18, -6).$$

Solution   Since $\mathbf{v} = -6(1, 2, -1, -3, 1)$, the property of norms in Example 7 tells us that

$$\|\mathbf{v}\| = |-6| \|(1, 2, -1, -3, 1)\| = 6\sqrt{16} = 24.$$ ■

$1 + 4 + 1 + 9 + 1 = 16$

EXAMPLE 9   Show that the sum of the squares of the lengths of the diagonals of a parallelogram in $\mathbb{R}^n$ is equal to the sum of the squares of the lengths of the sides. (This is the *parallelogram relation*).

Solution   We take our parallelogram with vertex at the origin and with vectors $\mathbf{v}$ and $\mathbf{w}$ emanating from the origin to form two sides, as shown in Figure 2.33. The lengths of the diagonals are then $\|\mathbf{v} + \mathbf{w}\|$ and $\|\mathbf{v} - \mathbf{w}\|$. We will do the algebraic computation for a general inner-product space. Using the definition of *norm* and properties of an inner product, we have

$$\|\mathbf{v} + \mathbf{w}\|^2 + \|\mathbf{v} - \mathbf{w}\|^2 = \langle (\mathbf{v} + \mathbf{w}), (\mathbf{v} + \mathbf{w}) \rangle + \langle (\mathbf{v} - \mathbf{w}), (\mathbf{v} - \mathbf{w}) \rangle$$
$$= \langle \mathbf{v}, \mathbf{v} \rangle + 2\langle \mathbf{v}, \mathbf{w} \rangle + \langle \mathbf{w}, \mathbf{w} \rangle + \langle \mathbf{v}, \mathbf{v} \rangle - 2\langle \mathbf{v}, \mathbf{w} \rangle + \langle \mathbf{w}, \mathbf{w} \rangle$$
$$= 2\langle \mathbf{v}, \mathbf{v} \rangle + 2\langle \mathbf{w}, \mathbf{w} \rangle$$
$$= 2\|\mathbf{v}\|^2 + 2\|\mathbf{w}\|^2,$$

which is what we wished to prove.   ■

geometrical expression??

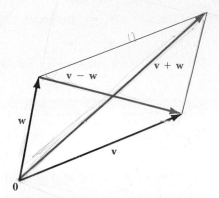

FIGURE 2.33 The parallelogram has v + w and v − w as vector diagonals.

The Schwarz and Triangle Inequalities

In Section 2.1, we defined the angle θ between two nonzero vectors $\mathbf{v}$ and $\mathbf{w}$ in $\mathbb{R}^n$ to be

$$\theta = \arccos \frac{\mathbf{v} \cdot \mathbf{w}}{\|\mathbf{v}\|\|\mathbf{w}\|},$$

provided that

$$-1 \leq \frac{\mathbf{v} \cdot \mathbf{w}}{\|\mathbf{v}\|\|\mathbf{w}\|} \leq 1,$$

that is, provided that $|\mathbf{v} \cdot \mathbf{w}| \leq \|\mathbf{v}\|\|\mathbf{w}\|$. We now prove that this inequality is true for all choices of $\mathbf{v}$ and $\mathbf{w}$ in $\mathbb{R}^n$. It is no more difficult to work in a general inner-product space than in $\mathbb{R}^n$.

THEOREM 2.22 Schwarz Inequality

Let V be an inner-product space, and let $\mathbf{v}$ and $\mathbf{w}$ be vectors in V. Then $|\langle \mathbf{v}, \mathbf{w} \rangle| \leq \|\mathbf{v}\|\|\mathbf{w}\|$.

PROOF We shall make repeated use of Definitions 2.17 and 2.18. For any scalars r and s, we have

$$\|r\mathbf{v} + s\mathbf{w}\|^2 \geq 0. \tag{1}$$

Now

$$\begin{aligned}
\|r\mathbf{v} + s\mathbf{w}\|^2 &= \langle r\mathbf{v} + s\mathbf{w}, r\mathbf{v} + s\mathbf{w} \rangle \\
&= r^2\langle \mathbf{v}, \mathbf{v} \rangle + 2rs\langle \mathbf{v}, \mathbf{w} \rangle + s^2\langle \mathbf{w}, \mathbf{w} \rangle.
\end{aligned} \tag{2}$$

In particular, setting $r = \langle \mathbf{w}, \mathbf{w} \rangle$ and $s = -\langle \mathbf{v}, \mathbf{w} \rangle$, we obtain from Eqs. (1) and (2)

$$\langle \mathbf{w}, \mathbf{w} \rangle^2 \langle \mathbf{v}, \mathbf{v} \rangle - 2\langle \mathbf{w}, \mathbf{w} \rangle \langle \mathbf{v}, \mathbf{w} \rangle \langle \mathbf{v}, \mathbf{w} \rangle + \langle \mathbf{v}, \mathbf{w} \rangle^2 \langle \mathbf{w}, \mathbf{w} \rangle$$
$$= \langle \mathbf{w}, \mathbf{w} \rangle^2 \langle \mathbf{v}, \mathbf{v} \rangle - \langle \mathbf{w}, \mathbf{w} \rangle \langle \mathbf{v}, \mathbf{w} \rangle^2 \geq 0 \quad (3)$$

or

$$\langle \mathbf{w}, \mathbf{w} \rangle (\langle \mathbf{v}, \mathbf{v} \rangle \langle \mathbf{w}, \mathbf{w} \rangle - \langle \mathbf{v}, \mathbf{w} \rangle^2) \geq 0. \quad (4)$$

Now if $\langle \mathbf{w}, \mathbf{w} \rangle = 0$, then $\mathbf{w} = \mathbf{0}$, and our result follows. If $\langle \mathbf{w}, \mathbf{w} \rangle \neq 0$, then $\langle \mathbf{w}, \mathbf{w} \rangle > 0$, so Eq. (4) yields

$$\langle \mathbf{v}, \mathbf{v} \rangle \langle \mathbf{w}, \mathbf{w} \rangle - \langle \mathbf{v}, \mathbf{w} \rangle^2 \geq 0$$

or

$$\langle \mathbf{v}, \mathbf{w} \rangle^2 \leq \langle \mathbf{v}, \mathbf{v} \rangle \langle \mathbf{w}, \mathbf{w} \rangle = \|\mathbf{v}\|^2 \|\mathbf{w}\|^2.$$

On taking square roots, we obtain the desired result. ▲

As we observed, the Schwarz inequality permits us to define the angle between two vectors in a general inner-product space. In particular, we define vectors $\mathbf{v}$ and $\mathbf{w}$ in an inner-product space to be **orthogonal** (or **perpendicular**) if $\langle \mathbf{v}, \mathbf{w} \rangle = 0$.

Another important inequality that follows from the Schwarz inequality is this: for vectors $\mathbf{v}$ and $\mathbf{w}$ in an inner-product space,

$$\|\mathbf{v} + \mathbf{w}\| \leq \|\mathbf{v}\| + \|\mathbf{w}\|. \quad \textbf{Triangle inequality} \quad (5)$$

THE SCHWARZ INEQUALITY is due independently to Augustin-Louis Cauchy (see note on page 214), Hermann Amandus Schwarz (1843–1921), and Viktor Yakovlevich Bunyakovsky (1804–1889).

It was first stated as a theorem about coordiantes in an appendix to Cauchy's 1821 text for his course on analysis at the Ecole Polytechnique, as follows:

$$|a\alpha + a'\alpha' + a''\alpha'' + \cdots| \leq \sqrt{a^2 + a'^2 + a''^2 + \cdots} \sqrt{\alpha^2 + \alpha'^2 + \alpha''^2 + \cdots}.$$

Cauchy's proof follows from the algebraic identity

$$(a\alpha + a'\alpha' + a''\alpha'' + \cdots)^2 + (a\alpha' - a'\alpha)^2 + (a\alpha'' - a''\alpha)^2 + \cdots + (u'u'' - u''u')^2 + \cdots$$
$$= (a^2 + a'^2 + a''^2 + \cdots)(\alpha^2 + \alpha'^2 + \alpha''^2 + \cdots).$$

Bunyakovsky proved the inequality for functions in 1859; that is, he stated the result

$$\left[\int_a^b f(x)g(x) \, dx \right]^2 \leq \int_a^b f^2(x) \, dx \cdot \int_a^b g^2(x) \, dx,$$

where we can consider $\int_a^b f(x)g(x) \, dx$ to be the inner product of the functions $f(x)$, $g(x)$ in the vector space of continuous functions on $[a, b]$. Bunyakovsky served as vice-president of the St. Petersburg Academy of Sciences from 1864 until his death. In 1875, the Academy established a mathematics prize in his name in recognition of his 50 years of teaching and research.

Schwarz stated the inequality in 1884. In his case, the vectors were functions ϕ, X of two variables in a region T in the plane, and the inner product of these functions was given by $\iint_T \phi X \, dx \, dy$, where this integral is assumed to exist. The inequality then states that

$$\left| \iint_T \phi X \, dx \, dy \right| \leq \sqrt{\iint_T \phi^2 \, dx \, dy} \cdot \sqrt{\iint_T X^2 \, dx \, dy}.$$

Schwarz's proof is similar to the one given in the text (page 191). Schwarz was the leading mathematician in Berlin around the turn of the century; the work in which the inequality appears is devoted to a question about minimal surfaces.

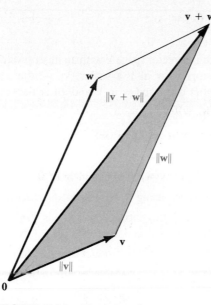

FIGURE 2.34 The triangle inequality.

Figure 2.34 indicates the origin of the name *triangle inequality* for Eq. (5). Viewed geometrically in $\mathbb{R}^n$, the inequality asserts that the sum of the lengths of two sides of a triangle is at least as great as the length of the third side. We give an algebraic proof that applies to any inner-product space.

THEOREM 2.23 **The Triangle Inequality**

Let V be an inner-product space, and let $\mathbf{v}$ and $\mathbf{w}$ be vectors in V. Then

$$\|\mathbf{v} + \mathbf{w}\| \le \|\mathbf{v}\| + \|\mathbf{w}\|.$$

PROOF Using the properties of the inner product, as well as the Schwarz inequality, we have

$$\langle v, w \rangle \le \|v\| \|w\|$$

$$\begin{aligned}
\|\mathbf{v} + \mathbf{w}\|^2 &= \langle \mathbf{v} + \mathbf{w}, \mathbf{v} + \mathbf{w} \rangle \\
&= \langle \mathbf{v}, \mathbf{v} \rangle + 2\langle \mathbf{v}, \mathbf{w} \rangle + \langle \mathbf{w}, \mathbf{w} \rangle \\
&\le \langle \mathbf{v}, \mathbf{v} \rangle + 2\|\mathbf{v}\|\|\mathbf{w}\| + \langle \mathbf{w}, \mathbf{w} \rangle \\
&= \|\mathbf{v}\|^2 + 2\|\mathbf{v}\|\|\mathbf{w}\| + \|\mathbf{w}\|^2 \\
&= (\|\mathbf{v}\| + \|\mathbf{w}\|)^2.
\end{aligned}$$

The desired relation follows at once, by taking square roots. ▲

SUMMARY

1. An inner-product space is a vector space V with an inner product $\langle\ ,\ \rangle$ that associates with each pair of vectors $\mathbf{v}$, $\mathbf{w}$ in V a scalar $\langle \mathbf{v}, \mathbf{w} \rangle$ that satisfies the following conditions for all vectors $\mathbf{u}$, $\mathbf{v}$, and $\mathbf{w}$ in V and all scalars r:

   P1 $\langle \mathbf{v}, \mathbf{w} \rangle = \langle \mathbf{w}, \mathbf{v} \rangle$,

   P2 $\langle \mathbf{u}, \mathbf{v} + \mathbf{w} \rangle = \langle \mathbf{u}, \mathbf{v} \rangle + \langle \mathbf{u}, \mathbf{w} \rangle$,

   P3 $r\langle \mathbf{v}, \mathbf{w} \rangle = \langle r\mathbf{v}, \mathbf{w} \rangle = \langle \mathbf{v}, r\mathbf{w} \rangle$,

   P4 $\langle \mathbf{v}, \mathbf{v} \rangle \geq 0$, and $\langle \mathbf{v}, \mathbf{v} \rangle = 0$ if and only if $\mathbf{v} = \mathbf{0}$.

2. $\mathbb{R}^n$ is an inner-product space using the usual dot product as inner product.

3. In an inner-product space V, the norm of a vector is $\|\mathbf{v}\| = \sqrt{\langle \mathbf{v}, \mathbf{v} \rangle}$ and satisfies $\|r\mathbf{v}\| = |r| \|\mathbf{v}\|$. The distance between vectors $\mathbf{v}$ and $\mathbf{w}$ is $\|\mathbf{v} - \mathbf{w}\|$.

4. For all vectors $\mathbf{v}$ and $\mathbf{w}$ in an inner-product space, we have the following two inequalities:

   Schwarz inequality: $|\langle \mathbf{v}, \mathbf{w} \rangle| \leq \|\mathbf{v}\| \|\mathbf{w}\|$,

   Triangle inequality: $\|\mathbf{v} + \mathbf{w}\| \leq \|\mathbf{v}\| + \|\mathbf{w}\|$.

5. Vectors $\mathbf{v}$ and $\mathbf{w}$ in an inner-product space are orthogonal (perpendicular) if and only if $\langle \mathbf{v}, \mathbf{w} \rangle = 0$.

EXERCISES

1. Prove the indicated property of the dot product, stated in Theorem 2.21.

   a) The property D1
   b) The property D2
   c) The property D3

In Exercises 2–5, use properties of the dot product and norm to compute the indicated quantities mentally, without pencil or paper.

2. $\|(42, 14)\|$

3. $\|(10, 20, 25, -15)\|$

4. $(14, 21, 28) \cdot (4, 8, 20)$

5. $(12, -36, 24) \cdot (25, 30, 10)$

6. For vectors $\mathbf{v}$ and $\mathbf{w}$ in $\mathbb{R}^n$, show that $\mathbf{v} - \mathbf{w}$ and $\mathbf{v} + \mathbf{w}$ are perpendicular if and only if $\|\mathbf{v}\| = \|\mathbf{w}\|$.

7. For vectors $\mathbf{u}$, $\mathbf{v}$, and $\mathbf{w}$ in $\mathbb{R}^n$ and for scalars r and s, show that, if $\mathbf{w}$ is perpendicular to both $\mathbf{u}$ and $\mathbf{v}$, then $\mathbf{w}$ is perpendicular to $r\mathbf{u} + s\mathbf{v}$.

8. Use vector methods to show that the diagonals of a rhombus (parallelogram with equal sides) are perpendicular. [HINT: Use a figure like Figure 2.31 and one of the preceding exercises.]

9. Use vector methods to show that the midpoint of the hypotenuse of a right triangle is equidistant from the three vertices. [HINT: See Figure 2.35. Show that

$$\|(\mathbf{v} + \mathbf{w})/2\| = \|(\mathbf{v} - \mathbf{w})/2\|.]$$

In Exercises 10–18, determine whether or not the indicated product satisfies the conditions for an inner product in the given vector space.

10. In $\mathbb{R}^2$, let $\langle (x_1, x_2), (y_1, y_2) \rangle = x_1 y_1 - x_2 y_2$

11. In $\mathbb{R}^2$, let $\langle (x_1, x_2), (y_1, y_2) \rangle = x_1 x_2 + y_1 y_2$

12. In $\mathbb{R}^2$, let $\langle (x_1, x_2), (y_1, y_2) \rangle = x_1^2 y_1 + x_2 y_2$

13. In $\mathbb{R}^2$, let $\langle (x_1, x_2), (y_1, y_2) \rangle = x_1 y_2$

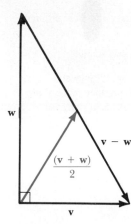

FIGURE 2.35 The vector $\frac{1}{2}(\mathbf{v} + \mathbf{w})$ to the midpoint of the hypotenuse.

14. In $\mathbb{R}^3$, let $\langle (x_1, x_2, x_3), (y_1, y_2, y_3) \rangle = x_1 y_1$.

15. In $\mathbb{R}^3$, let $\langle (x_1, x_2, x_3), (y_1, y_2, y_3) \rangle = x_1 + y_1$.

16. In the vector space M_2 of all 2×2 matrices, let

$$\left\langle \begin{bmatrix} a_1 & a_2 \\ a_3 & a_4 \end{bmatrix}, \begin{bmatrix} b_1 & b_2 \\ b_3 & b_4 \end{bmatrix} \right\rangle$$
$$= a_1 b_1 + a_2 b_2 + a_3 b_3 + a_4 b_4.$$

17. (*Calculus*) Let C be the vector space of all continuous functions mapping $\mathbb{R}$ into $\mathbb{R}$, and let $\langle f, g \rangle = \int_{-1}^{1} f(x) g(x) \, dx$.

18. (*Calculus*) Let C be as in Exercise 17, and let $\langle f, g \rangle = f(0) g(0)$.

19. (*Calculus*) Let $C[a, b]$ be the vector space of all continuous real-valued functions with domain $[a, b]$. Show that $\langle \, , \, \rangle$, defined in $C[a, b]$ by $\langle f, g \rangle = \int_a^b f(x) g(x) \, dx$, is an inner product in $C[a, b]$.

20. (*Calculus*) Let $C[0, 1]$ be the vector space of all continuous real-valued functions with domain $[0, 1]$. Let $\langle \, , \, \rangle$ be defined in $C[0, 1]$ by $\langle f, g \rangle = \int_0^1 f(x) g(x) \, dx$.
   a) Find $\langle (x + 1), x \rangle$.   b) Find $\|x\|$.
   c) Find $\|x^2 - x\|$.   d) Find $\|\sin \pi x\|$.

21. (*Calculus*) Show that $\sin x$ and $\cos x$ are orthogonal functions in the vector space $C[0, \pi]$ of Exercise 19, with the inner product defined there.

22. (*Calculus*) Let $\langle \, , \, \rangle$ be defined in $C[0, 1]$ as in Exercise 20. Find a set of two independent functions in $C[0, 1]$, each of which is orthogonal to the constant function 1.

23. Let $\mathbf{u}$ and $\mathbf{v}$ be vectors in an inner-product space, and suppose that $\|\mathbf{u}\| = 3$ and $\|\mathbf{v}\| = 5$. Find $\langle \mathbf{u} + 2\mathbf{v}, \mathbf{u} - 2\mathbf{v} \rangle$.

24. Suppose that the vectors $\mathbf{u}$ and $\mathbf{v}$ of Exercise 23 are perpendicular. Find $\langle \mathbf{u} + 2\mathbf{v}, 3\mathbf{u} + \mathbf{v} \rangle$.

25. Let V be an inner-product space. Mark each of the following True or False.
   ___ a) The norm of every vector in V is a positive real number.
   ___ b) The norm of every nonzero vector in V is a positive real number.
   ___ c) We have $\|r\mathbf{v}\| = r\|\mathbf{v}\|$ for every scalar r and vector $\mathbf{v}$ in V.
   ___ d) We have $\|\mathbf{u} + \mathbf{v}\|^2 = \|\mathbf{u}\|^2 + \|\mathbf{v}\|^2$ for all vectors $\mathbf{u}$ and $\mathbf{v}$ in V.
   ___ e) Two nonzero orthogonal vectors in V are independent.
   ___ f) If $\|\mathbf{u} + \mathbf{v}\|^2 = \|\mathbf{u}\|^2 + \|\mathbf{v}\|^2$ for two nonzero vectors $\mathbf{u}$ and $\mathbf{v}$ in V, then $\mathbf{u}$ and $\mathbf{v}$ are orthogonal.
   ___ g) An inner product can be defined on every finite-dimensional vector space.
   ___ h) Let r be any real scalar. Then $\langle \, , \, \rangle'$, defined by $\langle \mathbf{u}, \mathbf{v} \rangle' = r\langle \mathbf{u}, \mathbf{v} \rangle$ for vectors $\mathbf{u}$ and $\mathbf{v}$ in V, is also an inner product on V.
   ___ i) $\langle \, , \, \rangle'$, defined in part (h), is an inner product on V if r is nonzero.
   ___ j) The distance between two vectors $\mathbf{u}$ and $\mathbf{v}$ in V is given by $|\langle \mathbf{u} - \mathbf{v}, \mathbf{u} - \mathbf{v} \rangle|$.

26. Let S be a subset of nonzero vectors in an inner-product space V, and suppose that any two different vectors in S are orthogonal. Show that S is an independent set.

27. Let V be an inner-product space, and let S be a subset of V. Show that

$$S^\perp = \{\mathbf{v} \in V \mid \mathbf{v} \text{ is orthogonal to each vector in } S\}$$

is a subspace of V.

28. Referring to Exercise 27, show that $S \subseteq (S^\perp)^\perp$.

29. Give an example of an inner-product space V for which there exists a subspace W such that $(W^\perp)^\perp \neq W$.

30. (*Pythagorean theorem*) Let $\mathbf{u}$ and $\mathbf{v}$ be orthogonal vectors in an inner-product space V. Prove that $\|\mathbf{u} + \mathbf{v}\|^2 = \|\mathbf{u}\|^2 + \|\mathbf{v}\|^2$.

31. Use the triangle inequality to prove that

$$\|\mathbf{v} - \mathbf{w}\| \leq \|\mathbf{v}\| + \|\mathbf{w}\|$$

for any vectors $\mathbf{v}$ and $\mathbf{w}$ in an inner-product space V.

check!!

32. Show that, for any vectors $\mathbf{v}$ and $\mathbf{w}$ in an inner-product space V, we have

$$\|\mathbf{v} - \mathbf{w}\| \geq \|\mathbf{v}\| - \|\mathbf{w}\|.$$

33. Show that the vectors $\|\mathbf{v}\|\mathbf{w} + \|\mathbf{w}\|\mathbf{v}$ and $\|\mathbf{v}\|\mathbf{w} - \|\mathbf{w}\|\mathbf{v}$ in an inner-product space V are perpendicular.

3

DETERMINANTS

Each square matrix has associated with it a number called the *determinant* of the matrix. We begin this chapter with an introduction to determinants of 2×2 and 3×3 matrices, motivated by computations of area and volume. In Section 3.2, we discuss determinants of order n and their properties. An efficient way to compute determinants is presented in Section 3.3. In this section we also give a general formula for the *volume* of an m-dimensional box in $\mathbb{R}^n$ for $m \leq n$. This fabulous formula lies at the heart of much of integral calculus. By itself, it justifies the study of determinants. Section 3.4 extends this interpretation of determinants and presents the volume-change factor of a linear transformation.

Cramer's rule for solving a square linear system with a unique solution is presented in Section 3.5. This section is of mostly theoretical interest. Since references and formulas involving Cramer's rule do appear in advanced calculus and other fields, we believe that students should at least read the statement of Cramer's rule and look at an illustration. However, this rule must be one of the most inefficient methods known for solving a square linear system of large size.

3.1 AREAS, VOLUMES, AND CROSS PRODUCTS

We introduce determinants by discussing one of their most important applications: finding areas and volumes. We will find areas and volumes of very simple boxlike regions. In calculus, one finds areas and volumes of regions having more general shapes, using formulas that involve determinants.

The Area of a Parallelogram

The parallelogram determined by two nonzero and nonparallel vectors $\mathbf{a} = (a_1, a_2)$ and $\mathbf{b} = (b_1, b_2)$ in $\mathbb{R}^2$ is shown in Figure 3.1. This parallelogram has a vertex at the origin, and we regard the arrows representing $\mathbf{a}$ and $\mathbf{b}$ as forming the two sides of the parallelogram having the origin as a common vertex.

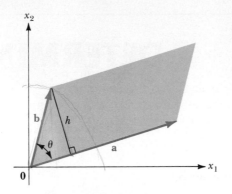

FIGURE 3.1 The parallelogram determined by a and b.

$\sin \theta = \dfrac{?}{r}$

We can find the area of this parallelogram by multiplying the length $\|\mathbf{a}\|$ of its base by the altitude h, obtaining

$$\text{Area} = \|\mathbf{a}\|\, h = \|\mathbf{a}\|\|\mathbf{b}\|(\sin\theta) = \|\mathbf{a}\|\|\mathbf{b}\|\sqrt{1 - \cos^2\theta}.$$

Recall from page 103 of Section 2.1 that $\mathbf{a} \cdot \mathbf{b} = \|\mathbf{a}\|\|\mathbf{b}\|(\cos\theta)$. Squaring our area equation, we have

$$\|\mathbf{a}\|\|\mathbf{b}\|\dfrac{x}{r}$$

$$
\begin{aligned}
(\text{Area})^2 &= \|\mathbf{a}\|^2\|\mathbf{b}\|^2 - \|\mathbf{a}\|^2\|\mathbf{b}\|^2\cos^2\theta \\
&= \|\mathbf{a}\|^2\|\mathbf{b}\|^2 - (\mathbf{a}\cdot\mathbf{b})^2 \\
&= (a_1{}^2 + a_2{}^2)(b_1{}^2 + b_2{}^2) - (a_1b_1 + a_2b_2)^2 \\
&= (a_1b_2 - a_2b_1)^2.
\end{aligned}
\tag{1}
$$

The last equality should be checked using pencil and paper. On taking square roots, we obtain

$$\text{Area} = |a_1b_2 - a_2b_1|.$$

The number within the absolute value symbol is known as the **determinant** of the matrix

$$A = \begin{bmatrix} a_1 & a_2 \\ b_1 & b_2 \end{bmatrix}$$

and is denoted by $|A|$ or $\det(A)$, so that

$$\det(A) = \begin{vmatrix} a_1 & a_2 \\ b_1 & b_2 \end{vmatrix}.$$

That is, if

$$A = \begin{bmatrix} a_1 & a_2 \\ b_1 & b_2 \end{bmatrix},$$

then

$$\det(A) = \begin{vmatrix} a_1 & a_2 \\ b_1 & b_2 \end{vmatrix} = a_1 b_2 - a_2 b_1 \tag{2}$$

We can remember this formula for the determinant by taking the product of the black entries on the main diagonal of the matrix, minus the product of the colored entries on the other diagonal.

EXAMPLE 1   Find the determinant of the matrix

$$\begin{bmatrix} 2 & 3 \\ 1 & 4 \end{bmatrix}.$$

Solution   We have

$$\begin{vmatrix} 2 & 3 \\ 1 & 4 \end{vmatrix} = (2)(4) - (3)(1) = 5. \qquad ▪$$

EXAMPLE 2   Find the area of the parallelogram in $\mathbb{R}^2$ with vertices $(1, 1)$, $(2, 3)$, $(2, 1)$, $(3, 3)$.

Solution   The parallelogram is sketched in Figure 3.2. The sides having $(1, 1)$ as common vertex can be regarded as the vectors

$$\mathbf{a} = (2, 1) - (1, 1) = (1, 0)$$

and

$$\mathbf{b} = (2, 3) - (1, 1) = (1, 2),$$

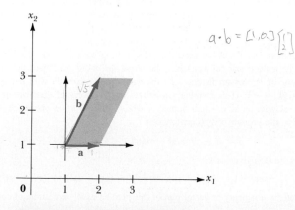

FIGURE 3.2   The parallelogram determined by a = (1, 0) and b = (1, 2).

as shown in the figure. Therefore, the area of the parallelogram is given by the determinant

$$\begin{vmatrix} 1 & 0 \\ 1 & 2 \end{vmatrix} = (1)(2) - (0)(1) = 2.$$ ■

The Cross Product

Equation (2) defines a **second-order determinant**, associated with a 2×2 matrix. Another application of these second-order determinants appears when we find a vector in $\mathbb{R}^3$ that is perpendicular to each of two given independent vectors $\mathbf{b} = (b_1, b_2, b_3)$ and $\mathbf{c} = (c_1, c_2, c_3)$. Recall that the unit coordinate vectors in $\mathbb{R}^3$ are $\mathbf{i} = (1, 0, 0)$, $\mathbf{j} = (0, 1, 0)$, and $\mathbf{k} = (0, 0, 1)$. We leave as an exercise the verification that

$$\mathbf{p} = \begin{vmatrix} b_2 & b_3 \\ c_2 & c_3 \end{vmatrix} \mathbf{i} - \begin{vmatrix} b_1 & b_3 \\ c_1 & c_3 \end{vmatrix} \mathbf{j} + \begin{vmatrix} b_1 & b_2 \\ c_1 & c_2 \end{vmatrix} \mathbf{k} \tag{3}$$

is a vector perpendicular to both $\mathbf{b}$ and $\mathbf{c}$. (See Exercise 5.) This can be seen by computing $\mathbf{p} \cdot \mathbf{b} = \mathbf{p} \cdot \mathbf{c} = 0$. The vector $\mathbf{p}$ in formula (3) is known as the **cross product** of $\mathbf{b}$ and $\mathbf{c}$, and is denoted $\mathbf{p} = \mathbf{b} \times \mathbf{c}$.

There is a very easy way to remember formula (3) for the cross product $\mathbf{b} \times \mathbf{c}$. Form the 3×3 *symbolic matrix*

$$\begin{bmatrix} \mathbf{i} & \mathbf{j} & \mathbf{k} \\ b_1 & b_2 & b_3 \\ c_1 & c_2 & c_3 \end{bmatrix}.$$

THE NOTION OF A CROSS PRODUCT grew out of Sir William Rowan Hamilton's attempt to develop a multiplication for "triplets," that is, vectors in $\mathbb{R}^3$. In a paper published in 1837 but written four years earlier, he developed the theory of complex numbers as pairs (a, b) of real numbers with the addition $(a, b) + (a', b') = (a + a', b + b')$ and the multiplication $(a, b)(a', b') = (aa' - bb', ab' + a'b)$. Over a period of many years, he attempted to generalize this to triples. On October 16, 1843 after many unsuccessful attempts, he finally succeeded in discovering an analogous result, not for triples but for quadruples. As he walked that day in Dublin, he wrote, he could not "resist the impulse . . . to cut with a knife on a stone of Brougham Bridge . . . the fundamental formula with the symbols i, j, k; namely $i^2 = j^2 = k^2 = ijk = -1$." This formula symbolizes his discovery of *quaternions*, elements of the form $w + xi + yj + zk$ with w, x, y, z real numbers, whose multiplication obeys the laws just given. Therefore, in particular, the product of two quaternions

$$\alpha = xi + yj + zk \quad \text{and} \quad \alpha' = x'i + y'j + z'k$$

whose real parts are 0 is given as

$$-xx' - yy' - zz' + (yz' - zy')i + (zx' - xz')j + (xy' - yx')k.$$

Hamilton denoted the imaginary or vector part of this product as $V \cdot \alpha\alpha'$. It is, of course, our modern cross product.

The first appearance of our current symbolism for the cross product is in the brief text *Elements of Vector Analysis* (1881) by the American Josiah Willard Gibbs (1839–1903), professor of mathematical physics at Yale, who wrote it for use in his courses in electricity and magnetism at Yale and in mechanics at Johns Hopkins.

Formula (3) can be obtained from this matrix in a simple way. Multiply the vector $\mathbf{i}$ by the determinant of the 2×2 matrix obtained by crossing out the row and column containing $\mathbf{i}$, as in

$$\begin{bmatrix} \mathbf{i} & \mathbf{j} & \mathbf{k} \\ b_1 & b_2 & b_3 \\ c_1 & c_2 & c_3 \end{bmatrix}.$$

Similarly, multiply $(-\mathbf{j})$ by the determinant of the matrix obtained by crossing out the row and column in which $\mathbf{j}$ appears. Finally, multiply $\mathbf{k}$ by the determinant of the matrix obtained by the crossing out the row and column containing $\mathbf{k}$, and add these multiples of $\mathbf{i}$, $\mathbf{j}$, and $\mathbf{k}$ to obtain formula (3).

EXAMPLE 3 Find a vector perpendicular to both $(2, 1, 1)$ and $(1, 2, 3)$ in $\mathbb{R}^3$.

Solution We form the symbolic matrix

$$\begin{bmatrix} \mathbf{i} & \mathbf{j} & \mathbf{k} \\ 2 & 1 & 1 \\ 1 & 2 & 3 \end{bmatrix}$$

and find that

$$(2, 1, 1) \times (1, 2, 3) = \begin{vmatrix} 1 & 1 \\ 2 & 3 \end{vmatrix} \mathbf{i} - \begin{vmatrix} 2 & 1 \\ 1 & 3 \end{vmatrix} \mathbf{j} + \begin{vmatrix} 2 & 1 \\ 1 & 2 \end{vmatrix} \mathbf{k}$$

$$= \mathbf{i} - 5\mathbf{j} + 3\mathbf{k} = (1, -5, 3). \qquad ■$$

The cross product $\mathbf{p} = \mathbf{b} \times \mathbf{c}$ as defined in Eq. (3) not only is perpendicular to both $\mathbf{b}$ and $\mathbf{c}$ but points in the direction determined by the familiar *right-hand rule*: when the fingers of the right hand curve in the direction from $\mathbf{b}$ to $\mathbf{c}$, the thumb points in the direction of $\mathbf{b} \times \mathbf{c}$. (See Fig. 3.3.) We do not attempt to prove this.

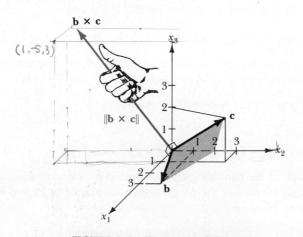

FIGURE 3.3 The area of the parallelogram determined by b and c is $\|\mathbf{b} \times \mathbf{c}\|$.

The magnitude of the vector $\mathbf{p} = \mathbf{b} \times \mathbf{c}$ in formula (3) is of interest as well: it is the area of the parallelogram with a vertex at the origin in $\mathbb{R}^3$ and edges at that vertex given by the vectors $\mathbf{b}$ and $\mathbf{c}$. To see this, we refer to a diagram like the one in Figure 3.1, but with $\mathbf{a}$ replaced by $\mathbf{c}$, and we repeat the computation for area. This time, Eq. (1) takes the form

$$
\begin{aligned}
(\text{Area})^2 &= \|\mathbf{c}\|^2 \|\mathbf{b}\|^2 - (\mathbf{c} \cdot \mathbf{b})^2 \\
&= (c_1^2 + c_2^2 + c_3^2)(b_1^2 + b_2^2 + b_3^2) - (c_1 b_1 + c_2 b_2 + c_3 b_3)^2 \\
&= \begin{vmatrix} b_2 & b_3 \\ c_2 & c_3 \end{vmatrix}^2 + \begin{vmatrix} b_1 & b_3 \\ c_1 & c_3 \end{vmatrix}^2 + \begin{vmatrix} b_1 & b_2 \\ c_1 & c_2 \end{vmatrix}^2 .
\end{aligned}
$$

Again pencil and paper are needed to check this last equality. Taking square roots, we obtain

$$\|\mathbf{b} \times \mathbf{c}\| = \text{Area of the parallelogram in } \mathbb{R}^3 \text{ determined by } \mathbf{b} \text{ and } \mathbf{c}.$$

EXAMPLE 4 Find the area of the parallelogram in $\mathbb{R}^3$ determined by the vectors $\mathbf{b} = (3, 1, 0)$ and $\mathbf{c} = (1, 3, 2)$.

Solution From the symbolic matrix

$$\begin{bmatrix} \mathbf{i} & \mathbf{j} & \mathbf{k} \\ 3 & 1 & 0 \\ 1 & 3 & 2 \end{bmatrix},$$

we find that

$$
\begin{aligned}
\mathbf{b} \times \mathbf{c} &= \begin{vmatrix} 1 & 0 \\ 3 & 2 \end{vmatrix} \mathbf{i} - \begin{vmatrix} 3 & 0 \\ 1 & 2 \end{vmatrix} \mathbf{j} + \begin{vmatrix} 3 & 1 \\ 1 & 3 \end{vmatrix} \mathbf{k} \\
&= (2, -6, 8).
\end{aligned}
$$

Therefore, the area of the parallelogram shown in Figure 3.3 is

$$\|\mathbf{b} \times \mathbf{c}\| = 2\sqrt{1 + 9 + 16} = 2\sqrt{26}. \qquad ■$$

EXAMPLE 5 Find the area of the triangle in $\mathbb{R}^3$ with vertices $(-1, 2, 0)$, $(2, 1, 3)$, and $(1, 1, -1)$.

Solution We think of $(-1, 2, 0)$ as a local origin, and we take the vectors corresponding to arrows starting there and reaching to $(2, 1, 3)$ and to $(1, 1, -1)$—namely,

$$\mathbf{a} = (2, 1, 3) - (-1, 2, 0) = (3, -1, 3)$$

and

$$\mathbf{b} = (1, 1, -1) - (-1, 2, 0) = (2, -1, -1).$$

Now $\|\mathbf{a} \times \mathbf{b}\|$ is the area of the parallelogram determined by these vectors, and the area of the triangle is half the area of the parallelogram, as shown in Figure 3.4. We form the symbolic matrix

$$\begin{bmatrix} \mathbf{i} & \mathbf{j} & \mathbf{k} \\ 3 & -1 & 3 \\ 2 & -1 & -1 \end{bmatrix},$$

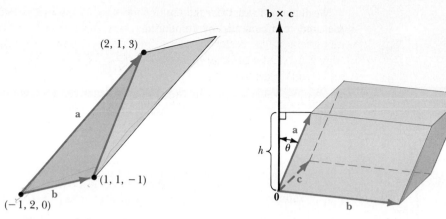

FIGURE 3.4 **The triangle constitutes half the parallelogram.**

FIGURE 3.5 **The box in $\mathbb{R}^3$ determined by a, b, and c.**

and we find that $\mathbf{a} \times \mathbf{b} = 4\mathbf{i} + 9\mathbf{j} - \mathbf{k}$. Thus,

$$\|\mathbf{a} \times \mathbf{b}\| = \sqrt{16 + 81 + 1} = \sqrt{98} = 7\sqrt{2},$$

so the area of the triangle is $7\sqrt{2}/2$. ■

The Volume of a Box

The cross product is useful in finding the volume of the box, or parallelepiped, determined by the three nonzero vectors $\mathbf{a} = (a_1, a_2, a_3)$, $\mathbf{b} = (b_1, b_2, b_3)$, and $\mathbf{c} = (c_1, c_2, c_3)$ in $\mathbb{R}^3$, as shown in Figure 3.5. The volume of the box can be computed by multiplying the area of the base by the altitude h.

THE VOLUME INTERPRETATION of a determinant first appeared in a 1773 paper on mechanics by Joseph Louis Lagrange (1736–1813). He noted that if the points M, M', M'' have coordinates (x, y, z), (x', y', z'), (x'', y'', z''), respectively, then the tetrahedron with vertices at the origin and at those three points will have volume

$$\left(\tfrac{1}{6}\right)[z(x'y'' - y'x'') + z'(yx'' - xy'') + z''(xy' - yx')];$$

that is,

$$\left(\tfrac{1}{6}\right) \det \begin{bmatrix} x & y & z \\ x' & y' & z' \\ x'' & y'' & z'' \end{bmatrix}.$$

Lagrange was born in Turin, but spent most of his mathematical career in Berlin and in Paris. He contributed important results to such varied fields as the calculus of variation, celestial mechanics, number theory, and the theory of equations. Among his most famous works are the *Treatise on Analytical Mechanics* (1788), in which he presented the various principles of mechanics from a single point of view, and the *Theory of Analytic Functions* (1797), in which he attempted to base the differential calculus on the theory of power series.

We have just seen that the area of the base of the box is equal to $\|\mathbf{b} \times \mathbf{c}\|$, and the altitude can be found by computing

$$h = \|\mathbf{a}\||\cos \theta| = \frac{\|\mathbf{b} \times \mathbf{c}\|\|\mathbf{a}\||\cos \theta|}{\|\mathbf{b} \times \mathbf{c}\|} = \frac{|(\mathbf{b} \times \mathbf{c}) \cdot \mathbf{a}|}{\|\mathbf{b} \times \mathbf{c}\|}.$$

The absolute value is used in case $\cos \theta$ is negative. This would be the case if the direction of $\mathbf{b} \times \mathbf{c}$ were opposite to that shown in Figure 3.5. Thus,

$$\text{Volume} = (\text{Area of base})h = \|\mathbf{b} \times \mathbf{c}\|\frac{|(\mathbf{b} \times \mathbf{c}) \cdot \mathbf{a}|}{\|\mathbf{b} \times \mathbf{c}\|} = |(\mathbf{b} \times \mathbf{c}) \cdot \mathbf{a}|.$$

That is, referring to Formula (3), which defines $\mathbf{b} \times \mathbf{c}$, we see that

$$\text{Volume} = |a_1(b_2c_3 - b_3c_2) - a_2(b_1c_3 - b_3c_1) + a_3(b_1c_2 - b_2c_1)|. \tag{4}$$

The number within the absolute value symbol is known as a **third-order determinant**. It is the **determinant of the matrix**

$$A = \begin{bmatrix} a_1 & a_2 & a_3 \\ b_1 & b_2 & b_3 \\ c_1 & c_2 & c_3 \end{bmatrix}$$

and is denoted by

$$\det(A) = \begin{vmatrix} a_1 & a_2 & a_3 \\ b_1 & b_2 & b_3 \\ c_1 & c_2 & c_3 \end{vmatrix}.$$

It can be computed as

$$\det(A) = a_1 \begin{vmatrix} b_2 & b_3 \\ c_2 & c_3 \end{vmatrix} - a_2 \begin{vmatrix} b_1 & b_3 \\ c_1 & c_3 \end{vmatrix} + a_3 \begin{vmatrix} b_1 & b_2 \\ c_1 & c_2 \end{vmatrix}. \tag{5}$$

Notice the similarity of formula (5) to our computation of the cross product $\mathbf{b} \times \mathbf{c}$ in formula (3). We simply replace $\mathbf{i}$, $\mathbf{j}$, and $\mathbf{k}$ by a_1, a_2, and a_3, respectively.

EXAMPLE 6   Find the determinant of the matrix

$$A = \begin{bmatrix} 2 & 1 & 3 \\ 4 & 1 & 2 \\ 1 & 2 & -3 \end{bmatrix}.$$

Solution   Using Formula (5), we have

$$\begin{vmatrix} 2 & 1 & 3 \\ 4 & 1 & 2 \\ 1 & 2 & -3 \end{vmatrix} = 2 \begin{vmatrix} 1 & 2 \\ 2 & -3 \end{vmatrix} - 1 \begin{vmatrix} 4 & 2 \\ 1 & -3 \end{vmatrix} + 3 \begin{vmatrix} 4 & 1 \\ 1 & 2 \end{vmatrix}$$

$$= 2(-7) - (-14) + 3(7) = 21. \qquad ■$$

EXAMPLE 7   Find the volume of the box with vertex at the origin determined by the vectors $\mathbf{a} = (4, 1, 1)$, $\mathbf{b} = (2, 1, 0)$ and $\mathbf{c} = (0, 2, 3)$, and sketch the box in a figure.

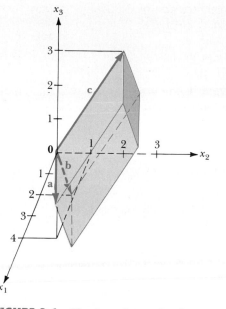

FIGURE 3.6 **The box determined by a, b, and c.**

Solution The box is shown in Figure 3.6. Its volume is given by the absolute value of the determinant

$$
\begin{vmatrix} 4 & 1 & 1 \\ 2 & 1 & 0 \\ 0 & 2 & 3 \end{vmatrix} = 4\begin{vmatrix} 1 & 0 \\ 2 & 3 \end{vmatrix} - \begin{vmatrix} 2 & 0 \\ 0 & 3 \end{vmatrix} + \begin{vmatrix} 2 & 1 \\ 0 & 2 \end{vmatrix}
$$

$$
= 10.
$$
■

The computation of $\det(A)$ in Eq. (5) is referred to as an *expansion of the determinant on the first row*. It is a particular case of a more general procedure for computing $\det(A)$, which is described in the next section.

We list the results of our work with the cross product and some of its algebraic properties in one place as a theorem. The algebraic properties can be checked through computation with the components of the vectors. Example 8 gives an illustration.

THEOREM 3.1 **Properties of the Cross Product**

Let **a**, **b**, and **c** be vectors in $\mathbb{R}^3$.

1. $\mathbf{b} \times \mathbf{c} = -(\mathbf{c} \times \mathbf{b})$.   Anticommutativity

2. $\mathbf{a} \times (\mathbf{b} \times \mathbf{c})$ is generally different from   Nonassociativity of $\times$
   $(\mathbf{a} \times \mathbf{b}) \times \mathbf{c}$.

3. $\mathbf{a} \times (\mathbf{b} + \mathbf{c}) = (\mathbf{a} \times \mathbf{b}) + (\mathbf{a} \times \mathbf{c})$ Distributive properties
   $(\mathbf{a} + \mathbf{b}) \times \mathbf{c} = (\mathbf{a} \times \mathbf{c}) + (\mathbf{b} \times \mathbf{c})$.

4. $\mathbf{b} \cdot (\mathbf{b} \times \mathbf{c}) = (\mathbf{b} \times \mathbf{c}) \cdot \mathbf{c} = 0.$ Perpendicularity of $\mathbf{b} \times \mathbf{c}$ to both $\mathbf{b}$ and $\mathbf{c}$

5. $\|\mathbf{b} \times \mathbf{c}\| =$ Area of the parallelogram determined by $\mathbf{b}$ and $\mathbf{c}$.   Area property

6. $\mathbf{a} \cdot (\mathbf{b} \times \mathbf{c}) = (\mathbf{a} \times \mathbf{b}) \cdot \mathbf{c} =$ $\pm$ Volume of the box determined by $\mathbf{a}$, $\mathbf{b}$, and $\mathbf{c}$.   Volume property

7. $\mathbf{a} \times (\mathbf{b} \times \mathbf{c}) = (\mathbf{a} \cdot \mathbf{c})\mathbf{b} - (\mathbf{a} \cdot \mathbf{b})\mathbf{c}.$ Formula for computation of $\mathbf{a} \times (\mathbf{b} \times \mathbf{c})$

EXAMPLE 8   Show that $\mathbf{b} \times \mathbf{c} = -(\mathbf{c} \times \mathbf{b})$ for any vectors $\mathbf{b}$ and $\mathbf{c}$ in $\mathbb{R}^3$.

Solution   We compute

$$\mathbf{c} \times \mathbf{b} = \begin{vmatrix} \mathbf{i} & \mathbf{j} & \mathbf{k} \\ c_1 & c_2 & c_3 \\ b_1 & b_2 & b_3 \end{vmatrix}$$

$$= \begin{vmatrix} c_2 & c_3 \\ b_2 & b_3 \end{vmatrix} \mathbf{i} - \begin{vmatrix} c_1 & c_3 \\ b_1 & b_3 \end{vmatrix} \mathbf{j} + \begin{vmatrix} c_1 & c_2 \\ b_1 & b_2 \end{vmatrix} \mathbf{k}.$$

A simple computation shows that interchanging the rows of a 2×2 matrix having a determinant d gives a matrix with determinant $-d$ (see Exercise 11). Comparison of the preceding formula for $\mathbf{c} \times \mathbf{b}$ with the formula for $\mathbf{b} \times \mathbf{c}$ in Eq. (3) then shows that $\mathbf{b} \times \mathbf{c} = -(\mathbf{c} \times \mathbf{b})$. ■

SUMMARY

1. A second-order determinant is defined by

$$\begin{vmatrix} a & b \\ c & d \end{vmatrix} = ad - bc.$$

A third-order determinant is defined by Eq. (5).

2. The area of the parallelogram with vertex at the origin determined by nonzero vectors $\mathbf{a}$ and $\mathbf{b}$ in $\mathbb{R}^2$ is the absolute value of the determinant of the matrix having row vectors $\mathbf{a}$ and $\mathbf{b}$.

3. The cross product of vectors $\mathbf{b}$ and $\mathbf{c}$ in $\mathbb{R}^3$ can be computed by using the symbolic determinant

$$\begin{vmatrix} \mathbf{i} & \mathbf{j} & \mathbf{k} \\ b_1 & b_2 & b_3 \\ c_1 & c_2 & c_3 \end{vmatrix}.$$

This cross product $\mathbf{b} \times \mathbf{c}$ is perpendicular to both $\mathbf{b}$ and $\mathbf{c}$.

4. The area of the parallelogram determined by nonzero vectors **b** and **c** in $\mathbb{R}^3$ is $\|\mathbf{b} \times \mathbf{c}\|$.

5. The volume of the box determined by nonzero vectors **a**, **b**, and **c** in $\mathbb{R}^3$ is the absolute value of the determinant of the matrix having row vectors **a**, **b**, and **c**. This determinant is also equal to $\mathbf{a} \cdot (\mathbf{b} \times \mathbf{c})$.

EXERCISES

In Exercises 1–4, find the indicated determinant.

1. $\begin{vmatrix} -1 & 3 \\ 5 & 0 \end{vmatrix}$

2. $\begin{vmatrix} -1 & 0 \\ 0 & 7 \end{vmatrix}$

3. $\begin{vmatrix} 0 & -3 \\ 5 & 0 \end{vmatrix}$

4. $\begin{vmatrix} 21 & -4 \\ 10 & 7 \end{vmatrix}$

5. Show that the vector $\mathbf{p} = \mathbf{b} \times \mathbf{c}$ given in Eq. (3) is indeed perpendicular to both **b** and **c**.

In Exercises 6–9, find the indicated determinant.

6. $\begin{vmatrix} 1 & 4 & -2 \\ 5 & 13 & 0 \\ 2 & -1 & 3 \end{vmatrix}$

7. $\begin{vmatrix} 2 & -5 & 3 \\ 1 & 3 & 4 \\ -2 & 3 & 7 \end{vmatrix}$

8. $\begin{vmatrix} 1 & -2 & 7 \\ 0 & 1 & 4 \\ 1 & 0 & 3 \end{vmatrix}$

9. $\begin{vmatrix} 2 & -1 & 1 \\ -1 & 0 & 3 \\ 2 & 1 & -4 \end{vmatrix}$

10. Show by direct computation that:

a) $\begin{vmatrix} a_1 & a_2 & a_3 \\ a_1 & a_2 & a_3 \\ c_1 & c_2 & c_3 \end{vmatrix} = 0;$

b) $\begin{vmatrix} a_1 & a_2 & a_3 \\ b_1 & b_2 & b_3 \\ a_1 & a_2 & a_3 \end{vmatrix} = 0.$

11. Show by direct computation that

$$\begin{vmatrix} a_1 & a_2 \\ b_1 & b_2 \end{vmatrix} = -\begin{vmatrix} b_1 & b_2 \\ a_1 & a_2 \end{vmatrix}.$$

12. Show by direct computation that

$$\begin{vmatrix} a_1 & a_2 & a_3 \\ b_1 & b_2 & b_3 \\ c_1 & c_2 & c_3 \end{vmatrix} = -\begin{vmatrix} a_1 & a_2 & a_3 \\ c_1 & c_2 & c_3 \\ b_1 & b_2 & b_3 \end{vmatrix}.$$

In Exercises 13–18, Find $\mathbf{a} \times \mathbf{b}$.

13. $\mathbf{a} = 2\mathbf{i} - \mathbf{j} + 3\mathbf{k}$, $\mathbf{b} = \mathbf{i} + 2\mathbf{j}$

14. $\mathbf{a} = -5\mathbf{i} + \mathbf{j} + 4\mathbf{k}$, $\mathbf{b} = 2\mathbf{i} + \mathbf{j} - 3\mathbf{k}$

15. $\mathbf{a} = -\mathbf{i} + 2\mathbf{j} + 4\mathbf{k}$, $\mathbf{b} = 2\mathbf{i} - 4\mathbf{j} - 8\mathbf{k}$

16. $\mathbf{a} = \mathbf{i} - \mathbf{j} + \mathbf{k}$, $\mathbf{b} = 3\mathbf{i} - 2\mathbf{j} + 7\mathbf{k}$

17. $\mathbf{a} = 2\mathbf{i} - 3\mathbf{j} + 5\mathbf{k}$, $\mathbf{b} = 4\mathbf{i} - 5\mathbf{j} + \mathbf{k}$

18. $\mathbf{a} = -2\mathbf{i} + 3\mathbf{j} - \mathbf{k}$, $\mathbf{b} = 4\mathbf{i} - 6\mathbf{j} + \mathbf{k}$

19. Mark each of the following True or False.

____ a) The determinant of a 2 × 2 matrix is a vector.

____ b) If two rows of a 3 × 3 matrix are interchanged, the sign of the determinant is changed.

____ c) The determinant of a 3 × 3 matrix is zero if two rows of the matrix are parallel vectors in $\mathbb{R}^3$.

____ d) In order for the determinant of a 3 × 3 matrix to be zero, two rows of the matrix must be parallel vectors in $\mathbb{R}^3$.

____ e) The determinant of a 3 × 3 matrix is zero if the points in $\mathbb{R}^3$ given by the rows of the matrix lie in a plane.

____ f) The determinant of a 3 × 3 matrix is zero if the points in $\mathbb{R}^3$ given by the rows of the matrix lie in a plane through the origin.

____ g) The parallelogram in $\mathbb{R}^2$ determined by nonzero vectors **a** and **b** is a square if and only if $\mathbf{a} \cdot \mathbf{b} = 0$.

____ h) The box in $\mathbb{R}^3$ determined by vectors **a**, **b**, and **c** is a cube if and only if $\mathbf{a} \cdot \mathbf{b} = \mathbf{a} \cdot \mathbf{c} = \mathbf{b} \cdot \mathbf{c} = 0$ and $\mathbf{a} \cdot \mathbf{a} = \mathbf{b} \cdot \mathbf{b} = \mathbf{c} \cdot \mathbf{c}$.

____ i) If the angle between vectors **a** and **b** in $\mathbb{R}^3$ is $\pi/4$, then $\|\mathbf{a} \times \mathbf{b}\| = |\mathbf{a} \cdot \mathbf{b}|$.

____ j) For any vector **a** in $\mathbb{R}^3$, we have $\|\mathbf{a} \times \mathbf{a}\| = \|\mathbf{a}\|^2$.

In Exercises 20–24, find the area of the parallelogram with vertex at the origin and with the given vectors as edges.

20. $-\mathbf{i} + 4\mathbf{j}$ and $2\mathbf{i} + 3\mathbf{j}$

21. $-5\mathbf{i} + 3\mathbf{j}$ and $\mathbf{i} + 7\mathbf{j}$

22. $\mathbf{i} + 3\mathbf{j} - 5\mathbf{k}$ and $2\mathbf{i} + 4\mathbf{j} - \mathbf{k}$

23. $2\mathbf{i} - \mathbf{j} + \mathbf{k}$ and $\mathbf{i} + 3\mathbf{j} - \mathbf{k}$

24. $\mathbf{i} - 4\mathbf{j} + \mathbf{k}$ and $2\mathbf{i} + 3\mathbf{j} - 2\mathbf{k}$

In Exercises 25–32, find the area of the given geometric configuration.

25. The triangle with vertices $(-1, 2)$, $(3, -1)$, and $(4, 3)$

26. The triangle with vertices $(3, -4)$, $(1, 1)$ and $(5, 7)$

27. The triangle with vertices $(2, 1, -3)$, $(3, 0, 4)$, and $(1, 0, 5)$

28. The triangle with vertices $(3, 1, -2)$, $(1, 4, 5)$, and $(2, 1, -4)$

29. The triangle in the plane $\mathbb{R}^2$ bounded by the lines $y = x$, $y = -3x + 8$, and $3y + 5x = 0$

30. The parallelogram with vertices $(1, 3)$, $(-2, 6)$, $(1, 11)$, and $(4, 8)$

31. The parallelogram with vertices $(1, 0, 1)$, $(3, 1, 4)$, $(0, 2, 9)$, and $(-2, 1, 6)$

32. The parallelogram in the plane $\mathbb{R}^2$ bounded by the lines $x - 2y = 3$, $x - 2y = 10$, $2x + 3y = -1$, and $2x + 3y = -8$

In Exercises 33–36, find $\mathbf{a} \cdot (\mathbf{b} \times \mathbf{c})$ and $\mathbf{a} \times (\mathbf{b} \times \mathbf{c})$.

33. $\mathbf{a} = \mathbf{i} + 2\mathbf{j} - 3\mathbf{k}$, $\mathbf{b} = 4\mathbf{i} - \mathbf{j} + 2\mathbf{k}$, $\mathbf{c} = 3\mathbf{i} + \mathbf{k}$

34. $\mathbf{a} = -\mathbf{i} + \mathbf{j} + 2\mathbf{k}$, $\mathbf{b} = \mathbf{i} + \mathbf{k}$, $\mathbf{c} = 3\mathbf{i} - 2\mathbf{j} + 5\mathbf{k}$

35. $\mathbf{a} = \mathbf{i} - 3\mathbf{k}$, $\mathbf{b} = -\mathbf{i} + 4\mathbf{j}$, $\mathbf{c} = \mathbf{i} + \mathbf{j} + \mathbf{k}$

36. $\mathbf{a} = 4\mathbf{i} - \mathbf{j} + 2\mathbf{k}$, $\mathbf{b} = 3\mathbf{i} + 5\mathbf{j} - 2\mathbf{k}$, $\mathbf{c} = \mathbf{i} - 3\mathbf{j} + \mathbf{k}$

In Exercises 37–40, find the volume of the box having the given vectors as adjacent edges.

37. $-\mathbf{i} + 4\mathbf{j} + 7\mathbf{k}$, $3\mathbf{i} - 2\mathbf{j} - \mathbf{k}$, $4\mathbf{i} + 2\mathbf{k}$

38. $2\mathbf{i} + \mathbf{j} - 4\mathbf{k}$, $3\mathbf{i} - \mathbf{j} + 2\mathbf{k}$, $\mathbf{i} + 3\mathbf{j} - 8\mathbf{k}$

39. $-2\mathbf{i} + \mathbf{j}$, $3\mathbf{i} - 4\mathbf{j} + \mathbf{k}$, $\mathbf{i} - 2\mathbf{k}$

40. $3\mathbf{i} - \mathbf{j} + 4\mathbf{k}$, $\mathbf{i} - 2\mathbf{j} + 7\mathbf{k}$, $5\mathbf{i} - 3\mathbf{j} + 10\mathbf{k}$

In Exercises 41–44, find the volume of the tetrahedron having the given vertices. (Consider how the volume of a tetrahedron having three vectors from one point as edges is related to the volume of the box having the same three vectors as adjacent edges.)

41. $(-3, 0, 1)$, $(4, 2, 1)$, $(0, 1, 7)$, $(1, 1, 1)$

42. $(0, 1, 1)$, $(8, 2, -7)$, $(3, 1, 6)$, $(-4, -2, 0)$

43. $(-1, 1, 2)$, $(3, 1, 4)$, $(-1, 6, 0)$, $(2, -1, 5)$

44. $(-1, 2, 4)$, $(2, -3, 0)$, $(-4, 2, -1)$, $(0, 3, -2)$

*In Exercises 45–48, use a determinant to ascertain whether the given points lie on a line in $\mathbb{R}^2$. [*HINT: *What is the area of a "parallelogram" with collinear vertices?]*

45. $(0, 0)$, $(3, 5)$, $(6, 9)$

46. $(0, 0)$, $(4, 2)$, $(-6, -3)$

47. $(1, 5)$, $(3, 7)$, $(-3, 1)$

48. $(2, 3)$, $(1, -4)$, $(6, 2)$

*In Exercises 49–52, use a determinant to ascertain whether the given points lie in a plane in $\mathbb{R}^3$. [*HINT: *What is the "volume" of a box with coplanar vertices?]*

49. $(0, 0, 0)$, $(1, 4, 3)$, $(2, 5, 8)$, $(-1, 2, -5)$

50. $(0, 0, 0)$, $(2, 1, 1)$, $(3, -2, 1)$, $(-1, 2, 3)$

51. $(1, -1, 3)$, $(4, 2, 3)$, $(3, 1, -2)$, $(5, 5, -5)$

52. $(1, 2, 1)$, $(3, 3, 4)$, $(2, 2, 2)$, $(4, 3, 5)$

Let $\mathbf{a}$, $\mathbf{b}$, and $\mathbf{c}$ be any vectors in $\mathbb{R}^3$. In Exercises 53–56, simplify the given expression.

53. $\mathbf{a} \cdot (\mathbf{a} \times \mathbf{b})$

54. $(\mathbf{b} \times \mathbf{c}) - (\mathbf{c} \times \mathbf{b})$

55. $\|\mathbf{a} \times \mathbf{b}\|^2 + |\mathbf{a} \cdot \mathbf{b}|^2$

56. $\mathbf{a} \times (\mathbf{b} \times \mathbf{c}) + \mathbf{b} \times (\mathbf{c} \times \mathbf{a}) + \mathbf{c} \times (\mathbf{a} \times \mathbf{b})$

57. Prove property (2) of Theorem 3.1.

58. Prove property (3) of Theorem 3.1.

59. Prove property (6) of Theorem 3.1.

60. Option 7 of the available program VECTGRPH provides drill on the determinant of a 2×2 matrix as the area of the parallelogram determined by its row vectors, with an associated plus or minus sign. Run this option until you can regularly achieve a score of 80% or better.

THE DETERMINANT OF A SQUARE MATRIX

The Definition

We defined a third-order determinant in terms of second-order determinants in Eq. (5) on page 204. A second-order determinant can be defined in terms of first-order determinants if we interpret the determinant of a 1×1 matrix to be its sole entry. We define an nth-order determinant in terms of determinants of order $n - 1$. In order to facilitate this, we introduce the **minor matrix** A_{ij} of an $n \times n$ matrix $A = [a_{ij}]$; it is the $(n - 1) \times (n - 1)$ matrix obtained by crossing out the ith row and jth column of A. The minor matrix is the portion shown in color shading in the matrix

$$A_{ij} = \begin{bmatrix} a_{11} \cdots a_{1j} \cdots a_{1n} \\ \vdots \quad\quad \vdots \quad\quad \vdots \\ a_{i1} \cdots a_{ij} \cdots a_{in} \\ \vdots \quad\quad \vdots \quad\quad \vdots \\ a_{n1} \cdots a_{nj} \cdots a_{nn} \end{bmatrix} i\text{th row.} \tag{1}$$

jth column

Using $|A_{ij}|$ as notation for the determinant of the minor matrix A_{ij}, we can express the determinant of a 3×3 matrix A as

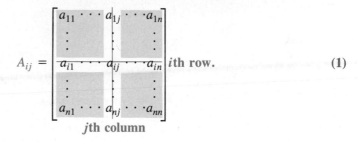

Cofactors

$$\begin{vmatrix} a_{11} & a_{12} & a_{13} \\ a_{21} & a_{22} & a_{23} \\ a_{31} & a_{32} & a_{33} \end{vmatrix} = a_{11}|A_{11}| - a_{12}|A_{12}| + a_{13}|A_{13}|.$$

The numbers $a'_{11} = |A_{11}|$, $a'_{12} = -|A_{12}|$, and $a'_{13} = |A_{13}|$ are appropriately called the **cofactors** of a_{11}, a_{12}, and a_{13}. We now proceed to define the *determinant of any square matrix*, using mathematical induction. (See Appendix A for a discussion of mathematical induction.)

THE FIRST APPEARANCE OF THE DETERMINANT OF A SQUARE MATRIX in Western Europe occurred in a 1683 letter from Gottfried von Leibniz to the Marquis de L'Hôpital. Leibniz wrote a system of three equations in two unknowns with abstract "numerical" coefficients, as follows:

$$10 + 11x + 12y = 0$$
$$20 + 21x + 22y = 0$$
$$30 + 31x + 32y = 0,$$

in which he noted that each coefficient number has "two characters, the first marking in which equation it occurs, the second marking which letter it belongs to." He then proceeded to eliminate first y and then x to show that the criterion for the system of equations to have a solution is that

$$10 \cdot 21 \cdot 32 + 11 \cdot 22 \cdot 30 + 12 \cdot 20 \cdot 31 = 10 \cdot 22 \cdot 31 + 11 \cdot 20 \cdot 32 + 12 \cdot 21 \cdot 30.$$

This is, of course, equivalent to our condition that the determinant of the matrix of coefficients must be zero. Unfortunately, the letter was not published until 1850 and therefore had no influence on subsequent work.

Determinants also appeared in the contemporaneous work of the Japanese mathematician Takakazu Seki (1642–1708).

DEFINITION 3.1 Cofactors and Determinants

The **determinant** of a 1×1 matrix is its sole entry; it is a **first-order** determinant. Let $n > 1$, and assume that determinants of order less than n have been defined. Let $A = [a_{ij}]$ be an $n \times n$ matrix. The **cofactor** of a_{ij} in A is

$$a'_{ij} = (-1)^{i+j} \det(A_{ij}), \qquad (2)$$

where A_{ij} is the minor matrix of A given in Eq. (1). The **determinant** of A is

$$\det(A) = \begin{vmatrix} a_{11} & a_{12} & \cdots & a_{1n} \\ a_{21} & a_{22} & \cdots & a_{2n} \\ \vdots & \vdots & & \vdots \\ a_{n1} & a_{n2} & \cdots & a_{nn} \end{vmatrix}$$

$$= a_{11}a'_{11} + a_{12}a'_{12} + \cdots + a_{1n}a'_{1n}, \qquad (3)$$

and is an ***n*th-order** determinant.

In addition to the notation $\det(A)$ for the determinant of A, we will sometimes use $|A|$ when determinants appear in equations, to make the equations easier to read.

EXAMPLE 1   Find the cofactor of the entry 3 in the matrix

$$A = \begin{bmatrix} 2 & 1 & 0 & 1 \\ 3 & 2 & 1 & 2 \\ 4 & 0 & 1 & 4 \\ 1 & 0 & 2 & 1 \end{bmatrix}. \qquad -(1-8)=7$$

Solution   Since 3 is in the row 2, column 1 position of A, we cross out the second row and first column of A and obtain

$$3' = a'_{21} = (-1)^{2+1} \begin{vmatrix} 1 & 0 & 1 \\ 0 & 1 & 4 \\ 0 & 2 & 1 \end{vmatrix} = -\begin{vmatrix} 1 & 4 \\ 2 & 1 \end{vmatrix} - \begin{vmatrix} 0 & 1 \\ 0 & 2 \end{vmatrix}$$

$$= -(-7 - 0) = 7. \qquad ■$$

EXAMPLE 2   Use Eq. (3) in Definition 3.1 to find the determinant of the matrix

$$A = \begin{bmatrix} 5 & -2 & 4 & -1 \\ 0 & 1 & 5 & 2 \\ 1 & 2 & 0 & 1 \\ -3 & 1 & -1 & 1 \end{bmatrix}.$$

Solution We have

$$\det(A) = \begin{vmatrix} 5 & -2 & 4 & -1 \\ 0 & 1 & 5 & 2 \\ 1 & 2 & 0 & 1 \\ -3 & 1 & -1 & 1 \end{vmatrix}$$

$$= 5(-1)^2 \begin{vmatrix} 1 & 5 & 2 \\ 2 & 0 & 1 \\ 1 & -1 & 1 \end{vmatrix} + (-2)(-1)^3 \begin{vmatrix} 0 & 5 & 2 \\ 1 & 0 & 1 \\ -3 & -1 & 1 \end{vmatrix}$$

$$+ 4(-1)^4 \begin{vmatrix} 0 & 1 & 2 \\ 1 & 2 & 1 \\ -3 & 1 & 1 \end{vmatrix} + (-1)(-1)^5 \begin{vmatrix} 0 & 1 & 5 \\ 1 & 2 & 0 \\ -3 & 1 & -1 \end{vmatrix}.$$

Computing the third-order determinants, we have

$$\begin{vmatrix} 1 & 5 & 2 \\ 2 & 0 & 1 \\ 1 & -1 & 1 \end{vmatrix} - 1 \begin{vmatrix} 0 & 1 \\ -1 & 1 \end{vmatrix} - 5 \begin{vmatrix} 2 & 1 \\ 1 & 1 \end{vmatrix} + 2 \begin{vmatrix} 2 & 0 \\ 1 & -1 \end{vmatrix}$$

$$= 1(1) - 5(1) + 2(-2) = -8;$$

$$\begin{vmatrix} 0 & 5 & 2 \\ 1 & 0 & 1 \\ -3 & -1 & 1 \end{vmatrix} = 0 \begin{vmatrix} 0 & 1 \\ -1 & 1 \end{vmatrix} - 5 \begin{vmatrix} 1 & 1 \\ -3 & 1 \end{vmatrix} + 2 \begin{vmatrix} 1 & 0 \\ -3 & -1 \end{vmatrix}$$

$$= 0(1) - 5(4) + 2(-1) = -22;$$

$$\begin{vmatrix} 0 & 1 & 2 \\ 1 & 2 & 1 \\ -3 & 1 & 1 \end{vmatrix} = 0 \begin{vmatrix} 2 & 1 \\ 1 & 1 \end{vmatrix} - 1 \begin{vmatrix} 1 & 1 \\ -3 & 1 \end{vmatrix} + 2 \begin{vmatrix} 1 & 2 \\ -3 & 1 \end{vmatrix}$$

$$= 0(1) - 1(4) + 2(7) = 10;$$

$$\begin{vmatrix} 0 & 1 & 5 \\ 1 & 2 & 0 \\ -3 & 1 & -1 \end{vmatrix} = 0 \begin{vmatrix} 2 & 0 \\ 1 & -1 \end{vmatrix} - 1 \begin{vmatrix} 1 & 0 \\ -3 & -1 \end{vmatrix} + 5 \begin{vmatrix} 1 & 2 \\ -3 & 1 \end{vmatrix}$$

$$= 0(-2) - 1(-1) + 5(7) = 36.$$

Therefore, $\det(A) = 5(-8) + 2(-22) + 4(10) + 1(36) = -8.$ ■

The preceding example makes one thing plain:

Computation of determinants of matrices of even moderate size using only
Definition 3.1 is a tremendous chore.

According to modern astronomical theory, our solar system would be dead long before
a present-day computer could find the determinant of a 50×50 matrix using just the
inductive Definition 3.1. (See Exercise 37.) Section 3.3 gives an alternative efficient
method for computing determinants.

Although it is difficult to explain their importance without using notions from calculus for functions of several variables, determinants are important. Let us just say that in some cases where primary values depend on some secondary values, the single number that best measures the rate at which the primary values change as the secondary values change is given by a determinant. This is closely connected with the geometric interpretation of a determinant as a *volume*. In Section 3.1, we motivated the determinant of a 2 × 2 matrix by using area, and we motivated the determinant of a 3 × 3 matrix by using volume. In the next section we show how an nth-order determinant can be interpreted as a "volume."

It is desirable to have an efficient way to compute a determinant. We will spend the remainder of this section developing properties of determinants that will enable us to find a good method for their computation. The computation of det(A) using Eq. (3) is called **expansion by minors** on the first row. Appendix B gives a proof by mathematical induction that det(A) can be obtained by using an expansion by minors *on any row or on any column*. We state this more precisely in a theorem.

THEOREM 3.2   General Expansion by Minors

Let A be an $n \times n$ matrix, and let r and s be any selections from the list of numbers 1, 2, . . . , n. Then

$$\det(A) = a_{r1}a'_{r1} + a_{r2}a'_{r2} + \cdots + a_{rn}a'_{rn}, \qquad (4)$$

and also

$$\det(A) = a_{1s}a'_{1s} + a_{2s}a'_{2s} + \cdots + a_{ns}a'_{ns}, \qquad (5)$$

where a'_{ij} is the cofactor of a_{ij} given in Definition 3.1.

Equation (4) is the **expansion of** det(A) **by minors on the rth row** of A, and Eq. (5) is the **expansion of** det(A) **by minors on the sth column** of A. Theorem 3.2 thus says that det(Λ) can be found by expanding by minors on any row or on any column of A.

EXAMPLE 3   Find the determinant of the matrix

$$A = \begin{bmatrix} 3 & 2 & 0 & 1 & 3 \\ -2 & 4 & 1 & 2 & 1 \\ 0 & -1 & 0 & 1 & -5 \\ -1 & 2 & 0 & -1 & 2 \\ 0 & 0 & 0 & 0 & 2 \end{bmatrix}.$$

Solution   An inductive computation like the one in Definition 3.1 is still the only way we have of computing det(A) at the moment, but we can expedite the computation if we expand by minors at each step on the row or column containing the most zeros. We have

$$\det(A) = 2(-1)^{5+5} \begin{vmatrix} 3 & 2 & 0 & 1 \\ -2 & 4 & 1 & 2 \\ 0 & -1 & 0 & 1 \\ -1 & 2 & 0 & -1 \end{vmatrix} \qquad \text{Expanding on row 5}$$

$$= (2)(1)(-1)^{2+3}\begin{vmatrix} 3 & 2 & 1 \\ 0 & -1 & 1 \\ -1 & 2 & -1 \end{vmatrix}$$ **Expanding on column 3**

$$= -2\left(3(-1)^{1+1}\begin{vmatrix} -1 & 1 \\ 2 & -1 \end{vmatrix} - 1(-1)^{3+1}\begin{vmatrix} 2 & 1 \\ -1 & 1 \end{vmatrix}\right)$$ **Expanding on column 1**

$$= -2(3(1 - 2) - 1(2 + 1)) = -2(-3 - 3) = 12.$$ ■

EXAMPLE 4 Show that the determinant of an upper- or lower-triangular square matrix is the product of its diagonal elements.

Solution We work with an upper-triangle matrix, the other case being analogous. If

$$U = \begin{bmatrix} u_{11} & u_{12} & \cdot & \cdot & \cdot & u_{1n} \\ & u_{22} & \cdot & \cdot & \cdot & u_{2n} \\ & & \cdot & & & \cdot \\ & \mathbf{0} & & \cdot & & \cdot \\ & & & & \cdot & \cdot \\ & & & & & u_{nn} \end{bmatrix}$$

is an upper-triangular matrix, then by expanding on first columns each time, we have

$$\det(U) = u_{11}\begin{vmatrix} u_{22} & u_{23} & \cdots & u_{2n} \\ 0 & u_{33} & \cdots & u_{3n} \\ \vdots & \vdots & & \vdots \\ 0 & 0 & \cdots & u_{nn} \end{vmatrix} = u_{11}u_{22}\begin{vmatrix} u_{33} & \cdots & u_{3n} \\ 0 & \cdots & u_{4n} \\ \vdots & & \vdots \\ 0 & \cdots & u_{nn} \end{vmatrix}$$

$$= \cdots = u_{11}u_{22}\cdots u_{nn}.$$ ■

Properties of the Determinant

Using Theorem 3.2, we can establish several properties of the determinant that will be of tremendous help in its computation. Since Defintion 3.1 was an inductive one, we use mathematical induction as we start to prove properties of the determinant. (Again, mathematical induction is reviewed in Appendix A.) We will always consider A to be an $n \times n$ matrix.

PROPERTY 1   The Transpose Property

For any square matrix A, we have $\det(A) = \det(A^T)$.

PROOF Verification of this property is trivial for determinants of orders 1 or 2. Let $n > 2$, and assume that the property holds for square matrices of size smaller than $n \times n$. We proceed to prove Property 1 for an $n \times n$ matrix A. We have

$$\det(A) = a_{11}|A_{11}| - a_{12}|A_{12}| + \cdots + (-1)^{n+1}a_{1n}|A_{1n}|.$$ **Expanding on row 1 of A**

Writing $B = A^T$, we have

$$\det(B) = b_{11}|B_{11}| - b_{21}|B_{21}| + \cdots + (-1)^{n+1}b_{n1}|B_{n1}|.$$ **Expanding on column 1 of B**

However, $a_{1j} = b_{j1}$ and $B_{j1} = A_{1j}{}^T$, because $B = A^T$. Applying our induction hypothesis to the $(n - 1)$st-order determinant $|A_{1j}|$, we have $|A_{1j}| = |B_{j1}|$. We conclude that $\det(A) = \det(B) = \det(A^T)$. ▲

This transpose property has a very useful consequence. It guarantees that any property of the determinant involving rows of a matrix is equally valid if we replace *rows* by *columns* in the statement of that property. For example, the next property has an analogue for columns.

PROPERTY 2 The Row-interchange Property

> If two different rows of a square matrix A are interchanged, the determinant of the resulting matrix is $-\det(A)$.

PROOF Again we find that the proof is trivial for the case $n = 2$. Assume that $n > 2$, and that this row-interchange property holds for matrices of size smaller than $n \times n$. Let A be an $n \times n$ matrix, and let B be the matrix obtained from A by interchanging the ith and rth rows, leaving the other rows unchanged. Since $n > 2$, we can choose a kth row for expansion by minors, where k is different from both r and i. Consider the cofactors

$$(-1)^{k+j}|A_{kj}| \quad \text{and} \quad (-1)^{k+j}|B_{kj}|.$$

These numbers must have opposite signs, by our induction hypothesis, since the minor matrices A_{kj} and B_{kj} have size $(n - 1) \times (n - 1)$, and B_{kj} can be obtained from A_{kj} by interchanging two rows. That is, $|B_{kj}| = -|A_{kj}|$. Expanding by minors on the kth row to find $\det(A)$ and $\det(B)$, we see that $\det(A) = -\det(B)$. ▲

THE THEORY OF DETERMINANTS grew from the efforts of many mathematicians of the late eighteenth and early nineteenth centuries. Besides Gabriel Cramer, whose work we will discuss in the note on page 239, Etienne Bezout (1739–1783) in 1764 and Alexandre-Theophile Vandermonde (1735–1796) in 1771 gave various methods for computing determinants. In a work on integral calculus, Pierre Simon Laplace (1749–1827) had to deal with systems of linear equations. He repeated the work of Cramer, but he also stated and proved the rule that interchanging two adjacent columns of the determinant changes the sign and showed that a determinant with two equal columns will be 0.

The most complete of the early works on determinants is that of Augustin-Louis Cauchy (1789–1857) in 1812. In this work, Cauchy introduced the name *determinant* to replace several older terms, used our current double-subscript notation for a square array of numbers, defined the array of adjoints (or minors) to a given array, and showed that one can calculate the determinant by expanding on any row or column. In addition, Cauchy re-proved many of the standard theorems on determinants that had been more or less known for the past 50 years.

Cauchy was the most prolific mathematician of the nineteenth century, contributing to such areas as complex analysis, calculus, differential equations, and mechanics. In particular, he wrote the first calculus text using our modern ε, δ-approach to continuity. Politically he was a conservative; when the July Revolution of 1830 replaced the Bourbon king Charles X with the Orleans king Louis-Philippe, Cauchy refused to take the oath of allegiance, thereby forfeiting his chairs at the Ecole Polytechnique and the Collège de France and going into exile in Turin and Prague.

PROPERTY 3 The Equal-rows Property

If two rows of a square matrix A are equal, then $\det(A) = 0$.

PROOF Let B be the matrix obtained from A by interchanging the two equal rows of A. By the row-interchange property, we have $\det(B) = -\det(A)$. On the other hand, $B = A$, so $\det(A) = -\det(A)$. Therefore, $\det(A) = 0$. ▲

PROPERTY 4 The Scalar-multiplication Property

If a single row of a square matrix A is multiplied by a scalar r, the determinant of the resulting matrix is $r \cdot \det(A)$.

PROOF Let r be any scalar, and let B be the matrix obtained from A by replacing the kth row $(a_{k1}, a_{k2}, \ldots, a_{kn})$ of A by $(ra_{k1}, ra_{k2}, \ldots, ra_{kn})$. Since the rows of B are equal to those of A except possibly for the kth row, it follows that the minor matrices A_{kj} and B_{kj} are equal for each j. Therefore, $a'_{kj} = b'_{kj}$, and computing $\det(B)$ by expanding by minors on the kth row, we have

$$\begin{aligned}
\det(B) &= b_{k1}b'_{k1} + b_{k2}b'_{k2} + \cdots + b_{kn}b'_{kn} \\
&= r \cdot a_{k1}a'_{k1} + r \cdot a_{k2}a'_{k2} + \cdots + r \cdot a_{kn}a'_{kn} \\
&= r \cdot \det(A).
\end{aligned}$$

▲

EXAMPLE 5 Find the determinant of the matrix

$$A = \begin{bmatrix} 2 & 1 & 3 & 4 & 2 \\ 6 & 2 & 1 & 4 & 1 \\ 6 & 3 & 9 & 12 & 6 \\ 2 & 1 & 3 & 4 & 1 \\ 1 & 4 & 2 & 1 & 1 \end{bmatrix}.$$

Solution We note that the third row of A is three times the first row. Therefore, we have

$$\det(A) = 3 \begin{vmatrix} 2 & 1 & 3 & 4 & 2 \\ 6 & 2 & 1 & 4 & 1 \\ 2 & 1 & 3 & 4 & 2 \\ 2 & 1 & 3 & 4 & 1 \\ 1 & 4 & 2 & 1 & 1 \end{vmatrix} \qquad \text{Property 4}$$

$$= 3(0) = 0. \qquad \text{Property 3} \qquad ■$$

The equal-rows property and the scalar-multiplication property indicate how the determinant of a matrix changes when two of the three elementary row operations are used. The next property deals with the most complicated of the elementary row operations, and lies at the heart of the efficient computation of determinants given in the next section.

PROPERTY 5 The Row-addition Property

> If the product of one row of a square matrix A by a scalar is added to a different row of A, the determinant of the resulting matrix is the same as $\det(A)$.

PROOF   Let $\mathbf{a}_i = (a_{i1}, a_{i2}, \ldots, a_{in})$ be the ith row of A. Suppose that $r\mathbf{a}_i$ is added to the kth row $\mathbf{a}_k$ of A, where r is any scalar and $k \neq i$. We obtain a matrix B whose rows are the same as the rows of A except possibly for the kth row, which is

$$\mathbf{b}_k = (ra_{i1} + a_{k1}, ra_{i2} + a_{k2}, \ldots, ra_{in} + a_{kn}).$$

Clearly the minor matrices A_{kj} and B_{kj} are equal for each j. Therefore, $a'_{kj} = b'_{kj}$, and computing $\det(B)$ by expanding by minors on the kth row, we have

$$\begin{aligned}
\det(B) &= b_{k1}b'_{k1} + b_{k2}b'_{k2} + \cdots + b_{kn}b'_{kn} \\
&= (ra_{i1} + a_{k1})a'_{k1} + (ra_{i2} + a_{k2})a'_{k2} + \cdots + (ra_{in} + a_{kn})a'_{kn} \\
&= (ra_{i1}a'_{k1} + ra_{i2}a'_{k2} + \cdots + ra_{in}a'_{kn}) \\
&\quad + (a_{k1}a'_{k1} + a_{k2}a'_{k2} + \cdots + a_{kn}a'_{kn}) \\
&= r \cdot \det(C) + \det(A),
\end{aligned}$$

where C is the matrix obtained from A by replacing the kth row of A with the ith row of A. Since C has two equal rows, its determinant is zero, so $\det(B) = \det(A)$. ▲

We now know how the three types of elementary row operations affect the determinant of a matrix A. In particular, if we reduce A to an echelon form H and avoid the use of row scaling, then $\det(A) = \pm\det(H)$, and $\det(H)$ is the product of its diagonal entries. (See Example 4.) We know that an echelon form of A has only nonzero entries on its main diagonal if and only if A is invertible. Thus, $\det(A) \neq 0$ if and only if A is invertible. We state this new condition for invertibility as a theorem.

THEOREM 3.3 Determinant Criterion for Invertibility

> A square matrix A is invertible if and only if $\det(A) \neq 0$.

We conclude with a multiplicative property of determinants. Section 3.3 indicates that this property has important geometric significance. Rather than labeling it "Property 6," we emphasize its increased level of importance over Properties 1 through 5 by stating it as a theorem.

THEOREM 3.4 The Multiplicative Property

> If A and B are $n \times n$ matrices, then $\det(AB) = \det(A) \cdot \det(B)$.

PROOF First we note that, if A is a diagonal matrix, the result follows easily, since the product

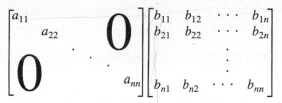

has its ith row equal to a_{ii} times the ith row of B. Using the scalar-multiplication property in each of these rows, we obtain

$$\det(AB) = (a_{11}a_{22} \cdots a_{nn}) \cdot \det(B) = \det(A) \cdot \det(B).$$

To deal with the nondiagonal case, we begin by reducing the problem to the case where A is invertible. For if A is singular, then so is AB (see Exercise 30); so both A and AB have a zero determinant, by Theorem 3.3, and $\det(A) \cdot \det(B) = 0$, too.

If we assume that A is invertible, it can be row-reduced through row-interchange and row-addition operations to an upper-triangular matrix with nonzero entries on the diagonal. We continue such row reduction analogous to the Gauss–Jordan method but without making pivots 1, and finally we reduce A to a diagonal matrix D with nonzero diagonal entries. We can write $D = EA$, where E is the product of elementary matrices corresponding to the row interchanges and row additions used to reduce A to D. By the properties of determinants, we have $\det(A) = (-1)^r \cdot \det(D)$, where r is the number of row interchanges. The same sequence of steps will reduce the matrix AB to the matrix $E(AB) = (EA)B = DB$, so $\det(AB) = (-1)^r \cdot \det(DB)$. Therefore,

$$\det(AB) = (-1)^r \cdot \det(DB) = (-1)^r \cdot \det(D) \cdot \det(B) = \det(A) \cdot \det(B),$$

and the proof is complete. ▲

EXAMPLE 6 Find $\det(A)$ if

$$A = \begin{bmatrix} 2 & 0 & 0 \\ 1 & 3 & 0 \\ 4 & 2 & 1 \end{bmatrix} \begin{bmatrix} 1 & 2 & 3 \\ 0 & 1 & 2 \\ 0 & 0 & 2 \end{bmatrix}.$$

Solution Since the determinant of an upper- or lower-triangular matrix is the product of the diagonal elements (see Example 4), Theorem 3.4 shows that

$$\det(A) = \begin{vmatrix} 2 & 0 & 0 \\ 1 & 3 & 0 \\ 4 & 2 & 1 \end{vmatrix} \begin{vmatrix} 1 & 2 & 3 \\ 0 & 1 & 2 \\ 0 & 0 & 2 \end{vmatrix} = (6)(2) = 12. \qquad ■$$

EXAMPLE 7 If $\det(A) = 3$, find $\det(A^5)$.

Solution Applying Theorem 3.4 several times, we have

$$\det(A^5) = [\det(A)]^5 = 3^5 = 243. \qquad ■$$

SUMMARY

1. The cofactor of an element a_{ij} in a square matrix A is $(-1)^{i+j}|A_{ij}|$, where A_{ij} is the matrix obtained from A by deleting the ith row and the jth column.

2. The *determinant* of an $n \times n$ matrix may be defined inductively by expansion by minors on the first row. The determinant can be computed by expansion by minors on any row or on any column; it is the sum of the products of the entries in that row or column with the cofactors of the entries. For large matrices, such a computation is hopelessly long.

3. The elementary row operations have the following effect on the determinant of a square matrix A.
    a. If two different rows of A are interchanged, the sign of the determinant is changed.
    b. If a single row of A is multiplied by a scalar, the determinant is multiplied by the scalar.
    c. If a multiple of one row is added to a different row, the determinant is not changed.

4. We have $\det(A) = \det(A^T)$. As a consequence, the properties just listed for elementary row operations are also true for elementary column operations.

5. If two rows or two columns of a matrix are the same, its determinant is zero.

6. The determinant of an upper-triangular matrix or of a lower-triangular matrix is the product of the diagonal entries.

7. An $n \times n$ matrix A is invertible if and only if $\det(A) \neq 0$.

8. If A and B are $n \times n$ matrices, then $\det(AB) = \det(A) \cdot \det(B)$.

EXERCISES

In Exercises 1–10, find the determinant of the given matrix.

1. $\begin{bmatrix} 5 & 2 & 1 \\ 1 & -1 & 4 \\ 3 & 0 & 2 \end{bmatrix}$

2. $\begin{bmatrix} 1 & 0 & 6 \\ 4 & 1 & -1 \\ 5 & 0 & 1 \end{bmatrix}$

3. $\begin{bmatrix} 3 & 2 & 4 \\ 0 & 1 & 2 \\ 1 & 4 & 1 \end{bmatrix}$

4. $\begin{bmatrix} 4 & -1 & 2 \\ 3 & 1 & 0 \\ -1 & 2 & 1 \end{bmatrix}$

5. $\begin{bmatrix} 0 & 1 & 4 \\ 2 & 3 & 1 \\ 1 & 4 & 1 \end{bmatrix}$

6. $\begin{bmatrix} 6 & 2 & 1 \\ 0 & 4 & 1 \\ 0 & 0 & 5 \end{bmatrix}$

7. $\begin{bmatrix} 2 & 3 & 4 & 6 \\ 2 & 0 & -9 & 6 \\ 4 & 1 & 0 & 2 \\ 0 & 1 & -1 & 0 \end{bmatrix}$

8. $\begin{bmatrix} 2 & 0 & -1 & 7 \\ 6 & 1 & 0 & 4 \\ 8 & -2 & 1 & 0 \\ 4 & 1 & 0 & 2 \end{bmatrix}$

9. $\begin{bmatrix} 1 & 2 & 0 & -1 & 2 & 4 \\ 6 & 2 & 8 & 1 & -1 & 1 \\ 4 & 2 & 1 & 2 & 2 & -5 \\ 4 & 5 & 4 & 5 & 1 & 2 \\ 1 & 2 & 0 & -1 & 2 & 4 \\ 1 & 0 & 1 & 8 & 1 & 5 \end{bmatrix}$

10. $\begin{bmatrix} 1 & 0 & 1 & 2 \\ 3 & 4 & 1 & 2 \\ 6 & 1 & 0 & 0 \\ 0 & 1 & 2 & 1 \end{bmatrix}$

11. Find the cofactor of 5 for the matrix in Exercise 2.

12. Find the cofactor of 3 for the matrix in Exercise 4.

13. Find the cofactor of 7 for the matrix in Exercise 8.

14. Find the cofactor of -5 for the matrix in Exercise 9.

In Exercises 15–20, let A be a 3 × 3 matrix with $\det(A) = 2$.

15. Find $\det(A^2)$.

16. Find $\det(A^k)$.

17. Find $\det(3A)$.

18. Find $\det(A + A)$.

19. Find $\det(A^{-1})$.

20. Find $\det(A^T)$.

21. Mark each of the following True or False.

____ a) The determinant $\det(A)$ is defined for any matrix A.

____ b) The determinant $\det(A)$ is defined for each square matrix A.

____ c) The determinant of a square matrix is a scalar.

____ d) If a matrix A is multiplied by a scalar c, the determinant of the resulting matrix is $c \cdot \det(A)$.

____ e) If an $n \times n$ matrix A is multiplied by a scalar c, the determinant of the resulting matrix is $c^n \cdot \det(A)$.

____ f) For every square matrix A, we have $\det(AA^T) = \det(A^TA) = [\det(A)]^2$.

____ g) If two rows and also two columns of a square matrix A are interchanged, the determinant changes sign.

____ h) The determinant of an elementary matrix is nonzero.

____ i) If $\det(A) = 2$ and $\det(B) = 3$, then $\det(A + B) = 5$.

____ j) If $\det(A) = 2$ and $\det(B) = 3$, then $\det(AB) = 6$.

In Exercises 22–25, let A be a 3 × 3 matrix with row vectors **a**, **b**, **c** *and with determinant equal to 3. Find the determinant of the matrix having the indicated row vectors.*

22. $\mathbf{a} + \mathbf{a}, \mathbf{a} + \mathbf{b}, \mathbf{a} + \mathbf{c}$

23. $\mathbf{a}, \mathbf{b}, 2\mathbf{a} + 3\mathbf{b}$

24. $\mathbf{a}, \mathbf{b}, 2\mathbf{a} + 3\mathbf{b} + 2\mathbf{c}$

25. $\mathbf{a} + \mathbf{b}, \mathbf{b} + \mathbf{c}, \mathbf{c} + \mathbf{a}$

In Exercises 26–29, find the values of λ for which the given matrix is singular.

26. $\begin{bmatrix} 1 - \lambda & 2 \\ 3 & 2 - \lambda \end{bmatrix}$

27. $\begin{bmatrix} -\lambda & 5 \\ 2 & 3 - \lambda \end{bmatrix}$

28. $\begin{bmatrix} 2 - \lambda & 0 & 0 \\ 0 & 1 - \lambda & 4 \\ 0 & 1 & 1 - \lambda \end{bmatrix}$

29. $\begin{bmatrix} 1 - \lambda & 0 & 2 \\ 0 & 4 - \lambda & 3 \\ 0 & 4 & -\lambda \end{bmatrix}$

30. If A and B are $n \times n$ matrices and if A is singular, prove (without using Theorem 3.4) that AB is also singular. [HINT: Assume that AB is invertible, and derive a contradiction.]

31. If A is invertible, find $\det(A^{-1})$ in terms of $\det(A)$.

32. If A and C are $n \times n$ matrices, with C invertible, show that $\det(A) = \det(C^{-1}AC)$.

33. Without using the multiplicative property of determinants (Theorem 3.4), prove that $\det(AB) = \det(A) \cdot \det(B)$ for the case where B is a diagonal matrix.

34. Continuing Exercise 33, find two other types of matrices B for which it is easy to show that $\det(AB) = \det(A) \cdot \det(B)$.

35. Prove that, if three $n \times n$ matrices A, B, and C are identical except for the kth rows $\mathbf{a}_k$, $\mathbf{b}_k$, and $\mathbf{c}_k$, respectively, which are related by $\mathbf{a}_k = \mathbf{b}_k + \mathbf{c}_k$, then

$$\det(A) = \det(B) + \det(C).$$

36. Notice that

$$\begin{vmatrix} a_{11} & a_{12} \\ a_{21} & a_{22} \end{vmatrix} = (a_{11}a_{22}) + (-a_{12}a_{21})$$

is a sum of signed products, where each product contains precisely one factor from each row and one factor from each column of the corresponding matrix. Prove by induction that this is true for an $n \times n$ matrix $A = [a_{ij}]$.

37. This exercise is for the reader who is skeptical of our assertion that the solar system would be dead long before a present-day computer could find the determinant of a 50 × 50 matrix using just Definition 3.1 with expansion by minors.

a) Recall that $n! = n(n - 1) \cdots (3)(2)(1)$. Show by induction that expansion of an $n \times n$ matrix by minors requires at least $n!$ multiplications for $n > 1$.

b) Run the available program TIMING, and find the number of seconds required for 3000 multiplications by your computer. Choose either interpretive or compiled BASIC. Then run the available program EBYMTIME and find the time required to perform $n!$ multiplications

for $n = 2, 4, 8, 12, 16, 20, 25, 30, 40, 50, 70,$ and 100.

38. MATCOMP can be used to compute determinants. Check Example 2 and the answers to some of the exercises, using MATCOMP.

3.3 COMPUTATION OF DETERMINANTS AND VOLUMES OF BOXES

We have seen that computation of determinants of high order is an unreasonable task if it is done directly from Definition 3.1, relying entirely upon repeated expansion by minors. In the special case where a square matrix is triangular, Example 4 of Section 3.2 shows that the determinant is simply the product of the diagonal entries. We know that a matrix can be reduced to row echelon form by means of elementary row operations, and row echelon form for a square matrix is always triangular. The discussion leading to Theorem 3.3 in the last section actually shows how the determinant of a matrix can be computed by a row reduction to echelon form. We rephrase part of this discussion in a box as an algorithm that a computer might follow to find a determinant.

Computation of a Determinant

The determinant of an $n \times n$ matrix A can be computed as follows:

1. Reduce A to an echelon form, using only row addition and row interchanges.
2. If any of the matrices appearing in the reduction contains a row of zeros, then $\det(A) = 0$.
3. Otherwise,

$$\det(A) = (-1)^r \cdot (\text{Product of pivots}),$$

where r is the number of row interchanges performed.

When doing a computation with pencil and paper rather than with a computer, we often use row scaling to make pivots 1, in order to ease calculations. As you study the following example, notice how the pivots accumulate as factors when the scalar-multiplication property of determinants is repeatedly used.

EXAMPLE 1 Find the determinant of the following matrix by reducing it to row echelon form.

$$A = \begin{bmatrix} 2 & 2 & 0 & 4 \\ 3 & 3 & 2 & 2 \\ 0 & 1 & 3 & 2 \\ 2 & 0 & 2 & 1 \end{bmatrix},$$

Solution We find that

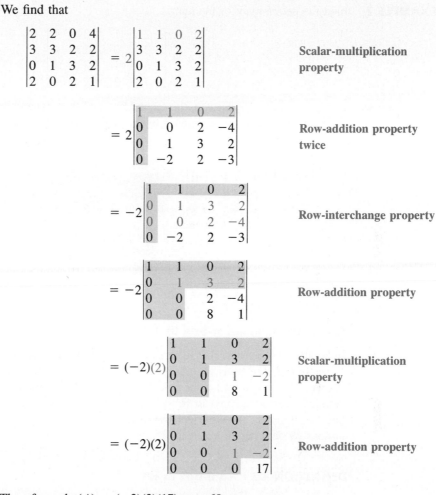

$$\begin{vmatrix} 2 & 2 & 0 & 4 \\ 3 & 3 & 2 & 2 \\ 0 & 1 & 3 & 2 \\ 2 & 0 & 2 & 1 \end{vmatrix} = 2 \begin{vmatrix} 1 & 1 & 0 & 2 \\ 3 & 3 & 2 & 2 \\ 0 & 1 & 3 & 2 \\ 2 & 0 & 2 & 1 \end{vmatrix}$$ Scalar-multiplication property

$$= 2 \begin{vmatrix} 1 & 1 & 0 & 2 \\ 0 & 0 & 2 & -4 \\ 0 & 1 & 3 & 2 \\ 0 & -2 & 2 & -3 \end{vmatrix}$$ Row-addition property twice

$$= -2 \begin{vmatrix} 1 & 1 & 0 & 2 \\ 0 & 1 & 3 & 2 \\ 0 & 0 & 2 & -4 \\ 0 & -2 & 2 & -3 \end{vmatrix}$$ Row-interchange property

$$= -2 \begin{vmatrix} 1 & 1 & 0 & 2 \\ 0 & 1 & 3 & 2 \\ 0 & 0 & 2 & -4 \\ 0 & 0 & 8 & 1 \end{vmatrix}$$ Row-addition property

$$= (-2)(2) \begin{vmatrix} 1 & 1 & 0 & 2 \\ 0 & 1 & 3 & 2 \\ 0 & 0 & 1 & -2 \\ 0 & 0 & 8 & 1 \end{vmatrix}$$ Scalar-multiplication property

$$= (-2)(2) \begin{vmatrix} 1 & 1 & 0 & 2 \\ 0 & 1 & 3 & 2 \\ 0 & 0 & 1 & -2 \\ 0 & 0 & 0 & 17 \end{vmatrix}.$$ Row-addition property

Therefore, $\det(A) = (-2)(2)(17) = -68$. ■

In our written work, we usually omit the shaded portion of the computation in the preceding example.

Row reduction offers an efficient way to program a computer to compute a determinant. If we are using pencil and paper, a further modification is more practical. We can use elementary row or column operations and the properties of determinants to reduce the computation to the determinant of a matrix having some row or column with a sole nonzero entry. A computer program generally modifies the matrix so that the first column has a single nonzero entry, but we can look at the matrix and choose the row or column where this can be achieved most easily. Expanding by minors on that row or column reduces the computation to a determinant of order one less, and we can continue the process until we are left with the computation of a determinant of a 2 × 2 matrix. Here is an illustration.

EXAMPLE 2   Find the determinant of the matrix

$$A = \begin{bmatrix} 2 & -1 & 3 & 5 \\ 2 & 0 & 1 & 0 \\ 6 & 1 & 3 & 4 \\ -7 & 3 & -2 & 8 \end{bmatrix}.$$

Solution   It is easiest to create zeros in the second row and then expand by minors on that row. We start by multiplying column 3 by -2 and adding the result to column 1, and we continue in a similar way:

$$\begin{vmatrix} 2 & -1 & 3 & 5 \\ 2 & 0 & 1 & 0 \\ 6 & 1 & 3 & 4 \\ -7 & 3 & -2 & 8 \end{vmatrix} = \begin{vmatrix} -4 & -1 & 3 & 5 \\ 0 & 0 & 1 & 0 \\ 0 & 1 & 3 & 4 \\ -3 & 3 & -2 & 8 \end{vmatrix} = -\begin{vmatrix} -4 & -1 & 5 \\ 0 & 1 & 4 \\ -3 & 3 & 8 \end{vmatrix}$$

$$= -\begin{vmatrix} -4 & -1 & 9 \\ 0 & 1 & 0 \\ -3 & 3 & -4 \end{vmatrix} = -\begin{vmatrix} -4 & 9 \\ -3 & -4 \end{vmatrix}$$

$$= -(16 + 27) = -43. \qquad ■$$

The Volume of an *n*-box in $\mathbb{R}^m$

In Section 3.1, we saw that the area of the parallelogram (or *2-box*) in $\mathbb{R}^2$ determined by vectors $\mathbf{a}_1$ and $\mathbf{a}_2$ is the absolute value $|\det(A)|$ of the determinant of the 2×2 matrix A having $\mathbf{a}_1$ and $\mathbf{a}_2$ as column vectors.* We also saw that the volume of the parallelepiped (or *3-box*) in $\mathbb{R}^3$ determined by vectors $\mathbf{a}_1$, $\mathbf{a}_2$, and $\mathbf{a}_3$ is $|\det(A)|$ for the 3×3 matrix A whose column vectors are $\mathbf{a}_1$, $\mathbf{a}_2$, $\mathbf{a}_3$. We wish to extend these notions by defining an *n-box* in $\mathbb{R}^m$ for $m \geq n$ and finding its "volume."

DEFINITION 3.2   An *n*-box in $\mathbb{R}^m$

Let $\mathbf{a}_1, \mathbf{a}_2, \ldots, \mathbf{a}_n$ be n independent vectors in $\mathbb{R}^m$ for $n \leq m$. The **n-box** in $\mathbb{R}^m$ determined by these vectors is the set of all vectors $\mathbf{x}$ satisfying

$$\mathbf{x} = t_1\mathbf{a}_1 + t_2\mathbf{a}_2 + \cdots + t_n\mathbf{a}_n$$

for $0 \leq t_i \leq 1$ and $i = 1, 2, \ldots, n$.

If the vectors $\mathbf{a}_1, \mathbf{a}_2, \ldots, \mathbf{a}_n$ in Definition 3.2 are dependent, the set described is a *degenerate n-box*.

EXAMPLE 3   Describe geometrically the 1-box determined by the "vector" 2 in $\mathbb{R}$ and the 1-box determined by a nonzero vector $\mathbf{a}$ in $\mathbb{R}^m$.

---

*Anticipating Section 3.4, we choose to arrange the vectors $\mathbf{a}_i$ as columns rather than as rows of A.

Solution The 1-box determined by the "vector" 2 in $\mathbb{R}$ consists of all numbers $t(2)$ for $0 \leq t \leq 1$, which is simply the closed interval $0 \leq x \leq 2$. Similarly, the 1-box in $\mathbb{R}^m$ determined by a nonzero vector $\mathbf{a}$ is the line segment joining the origin to the point $\mathbf{a}$. ▪

EXAMPLE 4 Draw symbolic sketches of a 2-box in $\mathbb{R}^m$ and a 3-box in $\mathbb{R}^m$.

Solution A 2-box in $\mathbb{R}^m$ is a parallelogram with a vertex at the origin, as shown in Figure 3.7. Similarly, a 3-box in $\mathbb{R}^m$ is a parallelepiped with a vertex at the origin, as illustrated in Figure 3.8. ▪

Notice that our boxes need not be rectangular boxes; perhaps we should have used the term *skew box* to make this clear.

We are accustomed to speaking of the *length* of a line segment, of the *area* of a piece of the plane, and of the *volume* of a piece of space. To avoid introducing new terms when discussing a general n-box, we will use the three-dimensional term *volume* when speaking of its size. Notice that we already used the three-dimensional term *box* as the name for the object! Thus, by the *volume* of a 1-box, we simply mean its length; the *volume* of a 2-box is its area, and so on.

The volume of the 1-box determined by $\mathbf{a}_1$ in $\mathbb{R}^m$ is its length $\|\mathbf{a}_1\|$. Since the determinant of a 1×1 matrix is its sole entry, Exercise 31 shows that this length can be written as

$$\text{Length} = \|\mathbf{a}_1\| = \sqrt{\det[\mathbf{a}_1 \cdot \mathbf{a}_1]}. \tag{1}$$

Let us turn to a 2-box in $\mathbb{R}^m$ determined by nonzero and nonparallel vectors $\mathbf{a}_1$ and $\mathbf{a}_2$. We repeat an argument that we made in Section 3.1 for a 2-box in $\mathbb{R}^2$, using a slightly different notation for this general m. From Figure 3.9, we see that the area of this parallelogram is given by

$$\text{Area} = \|\mathbf{b}\|\|\mathbf{a}_2\|,$$

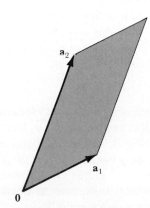

FIGURE 3.7 A 2-box in $\mathbb{R}^m$.

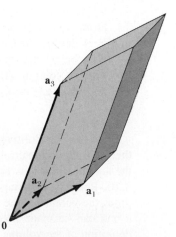

FIGURE 3.8 A 3-box in $\mathbb{R}^m$.

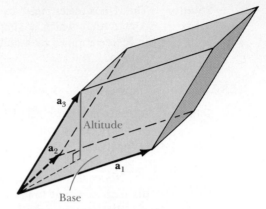

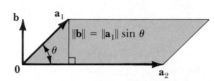

FIGURE 3.9 The volume of a 2-box.

FIGURE 3.10 The volume of a 3-box in $\mathbb{R}^m$ is (area of the base) × (altitude).

where, for the angle θ between $\mathbf{a}_1$ and $\mathbf{a}_2$, we have $\|\mathbf{b}\| = \|\mathbf{a}_1\| \sin \theta$. We then have

$$
\begin{aligned}
(\text{Area})^2 &= \|\mathbf{a}_1\|^2 \|\mathbf{a}_2\|^2 \sin^2\theta \\
&= \|\mathbf{a}_1\|^2 \|\mathbf{a}_2\|^2 (1 - \cos^2\theta) \\
&= (\mathbf{a}_1 \cdot \mathbf{a}_1)(\mathbf{a}_2 \cdot \mathbf{a}_2)\left(1 - \frac{(\mathbf{a}_1 \cdot \mathbf{a}_2)^2}{(\mathbf{a}_1 \cdot \mathbf{a}_1)(\mathbf{a}_2 \cdot \mathbf{a}_2)}\right) \\
&= (\mathbf{a}_1 \cdot \mathbf{a}_1)(\mathbf{a}_2 \cdot \mathbf{a}_2) - (\mathbf{a}_1 \cdot \mathbf{a}_2)(\mathbf{a}_2 \cdot \mathbf{a}_1) \\
&= \begin{vmatrix} \mathbf{a}_1 \cdot \mathbf{a}_1 & \mathbf{a}_1 \cdot \mathbf{a}_2 \\ \mathbf{a}_2 \cdot \mathbf{a}_1 & \mathbf{a}_2 \cdot \mathbf{a}_2 \end{vmatrix} = \det([\mathbf{a}_i \cdot \mathbf{a}_j]).
\end{aligned}
\tag{2}
$$

From Eqs. (1) and (2), we might guess that the square of the volume of an n-box in $\mathbb{R}^m$ is $\det([\mathbf{a}_i \cdot \mathbf{a}_j])$. Of course, we must define what we mean by the volume of such a box, but with the natural definition, this conjecture is true. If A is the matrix with jth column vector $\mathbf{a}_j$, then A^T is the matrix with ith row vector $\mathbf{a}_i$, and the $n \times n$ matrix $[\mathbf{a}_i \cdot \mathbf{a}_j]$ is A^TA, so Eq. (2) can be written as

$$
(\text{Area}) = \sqrt{\det(A^TA)}.
$$

We have an intuitive idea of the volume of an n-box in $\mathbb{R}^m$ for $n \leq m$. For example, the 3-box in $\mathbb{R}^m$ determined by independent vectors $\mathbf{a}_1$, $\mathbf{a}_2$, $\mathbf{a}_3$ has a volume equal to the altitude of the box times the volume (that is, area) of the base, as shown in Figure 3.10. Roughly speaking, the volume of an n-box is equal to the altitude of the box times the volume of the base, which is an $(n - 1)$ box. This notion of the *altitude* of a box can be made precise after we develop projections in Chapter 5. The formal definition of the volume of an n-box appears in Appendix B, as does the proof of our main result on volumes (Theorem 3.5). For the remainder of this section, we are content to proceed with our intuitive notion of volume.

THEOREM 3.5 Volume of a Box

The volume of the n-box in $\mathbb{R}^m$ determined by independent vectors $\mathbf{a}_1$, $\mathbf{a}_2$, . . . , $\mathbf{a}_n$ is given by

$$\text{Volume} = \sqrt{\det(A^T A)},$$

where A is the $m \times n$ matrix with $\mathbf{a}_j$ as jth column vector.

The volume of an n-box in $\mathbb{R}^n$ is of such importance that we restate this special case as a corollary.

COROLLARY Volume of an *n*-box in $\mathbb{R}^n$

If A is an $n \times n$ matrix with independent column vectors $\mathbf{a}_1$, $\mathbf{a}_2$, . . . , $\mathbf{a}_n$, then $|\det(A)|$ is the volume of the n-box in $\mathbb{R}^n$ determined by these n vectors.

PROOF By Theorem 3.5, the square of the volume of the n-box is $\det(A^T A)$. But since A is an $n \times n$ matrix, we have

$$\det(A^T A) = \det(A^T) \cdot \det(A) = (\det(A))^2.$$

The conclusion of the corollary then follows at once. ▲

EXAMPLE 5 Find the area of the parallelogram in $\mathbb{R}^4$ determined by the vectors $(2, 1, -1, 3)$ and $(0, 2, 4, -1)$.

Solution If

$$A = \begin{bmatrix} 2 & 0 \\ 1 & 2 \\ -1 & 4 \\ 3 & -1 \end{bmatrix},$$

then

$$A^T A = \begin{bmatrix} 2 & 1 & -1 & 3 \\ 0 & 2 & 4 & -1 \end{bmatrix} \begin{bmatrix} 2 & 0 \\ 1 & 2 \\ -1 & 4 \\ 3 & -1 \end{bmatrix} = \begin{bmatrix} 15 & -5 \\ -5 & 21 \end{bmatrix}.$$

By Theorem 3.5, we have

$$(\text{Area})^2 = \begin{vmatrix} 15 & -5 \\ -5 & 21 \end{vmatrix} = 290.$$

Thus the area of the parallelogram is $\sqrt{290}$. ∎

EXAMPLE 6 Find the volume of the parallelepiped in $\mathbb{R}^3$ determined by the vectors $(1, 0, -1)$, $(-1, 1, 3)$, and $(2, 4, 1)$.

Solution We compute the determinant

$$\begin{vmatrix} 1 & -1 & 2 \\ 0 & 1 & 4 \\ -1 & 3 & 1 \end{vmatrix} = \begin{vmatrix} 1 & -1 & 2 \\ 0 & 1 & 4 \\ 0 & 2 & 3 \end{vmatrix} = \begin{vmatrix} 1 & 4 \\ 2 & 3 \end{vmatrix} = -5.$$

By the corollary of Theorem 3.5, the volume of the parallelepiped is therefore 5. ▪

Comparing Theorem 3.5 and its corollary, we see that the formula for the volume of an n-box in a space of larger dimension m involves a square root, whereas the formula for the volume of a box in a space of its own dimension does not involve a square root. The student of calculus discovers that the calculus formulas used to find the length of a curve (which is one-dimensional) in the plane or in space involve a square root. The same is true of the formulas used to find the area of a surface (two-dimensional) in space. However, the calculus formulas for finding the area of part of the plane or the volume of some part of space do not involve square roots. Theorem 3.5 and its corollary lie at the heart of this difference in the calculus formulas.

SUMMARY

1. A computationally feasible algorithm for finding the determinant of a matrix is to reduce the matrix to echelon form, using just row-addition and row-interchange operations. If a row of zeros is formed during the process, the determinant is zero. Otherwise, the determinant of the original matrix is found by computing $(-1)^r \cdot$ (Product of pivots) in the echelon form, where r is the number of row interchanges performed. This is one way to program a computer to find a determinant.

2. The determinant of a matrix can be found by row or column reduction of the matrix to a matrix having a sole nonzero entry in some column or row. One then expands by minors on that column or row, and continues this process. If a matrix having a zero row or column is encountered, the determinant is zero. Otherwise, one continues until the computation is reduced to the determinant of a 2×2 matrix. This is a good way to find a determinant when working with pencil and paper.

3. An n-box in $\mathbb{R}^m$, where $m \geq n$, is determined by n independent vectors $\mathbf{a}_1, \mathbf{a}_2, \dots, \mathbf{a}_n$ and consists of all vectors $\mathbf{x}$ in $\mathbb{R}^m$ such that

$$\mathbf{x} = t_1\mathbf{a}_1 + t_2\mathbf{a}_2 + \cdots + t_n\mathbf{a}_n,$$

where $0 \leq t_i \leq 1$ for $i = 1, 2, \dots, n$.

4. A 1-box in $\mathbb{R}^m$ is a line segment, and its "volume" is its length.

5. A 2-box in $\mathbb{R}^m$ is a parallelogram determined by two independent vectors, and the "volume" of the 2-box is the area of the parallelogram.

6. A 3-box in $\mathbb{R}^m$ is a skewed box (parallelepiped) in the usual sense, and its volume is the usual volume.

7. Let $\mathbf{a}_1, \mathbf{a}_2, \ldots, \mathbf{a}_n$ be independent vectors in $\mathbb{R}^m$ for $m \geq n$, and let A be the $m \times n$ matrix with jth column vector $\mathbf{a}_j$. The volume of the n-box in $\mathbb{R}^m$ determined by the n vectors is $\sqrt{\det(A^T A)}$.

8. For the case of an n-box in the space $\mathbb{R}^n$ of the same dimension, the formula for its volume given in summary item 7 reduces to $|\det(A)|$.

EXERCISES

In Exercises 1–10, find the determinant of the given matrix.

1. $\begin{bmatrix} 2 & 3 & -1 \\ 5 & -7 & 1 \\ -3 & 2 & -1 \end{bmatrix}$

2. $\begin{bmatrix} 4 & -3 & 2 \\ -1 & -1 & 1 \\ -5 & 5 & 7 \end{bmatrix}$

3. $\begin{bmatrix} 5 & 2 & 4 & 0 \\ 2 & -3 & -1 & 2 \\ 3 & -4 & 3 & 7 \\ 1 & -1 & 0 & 1 \end{bmatrix}$

4. $\begin{bmatrix} 3 & -5 & -1 & 7 \\ 0 & 3 & 1 & -6 \\ 2 & -5 & -1 & 8 \\ -8 & 8 & 2 & -9 \end{bmatrix}$

5. $\begin{bmatrix} 2 & 1 & 0 & 0 & 0 \\ 3 & -1 & 2 & 0 & 0 \\ 0 & 4 & 1 & -1 & 2 \\ 0 & 0 & -3 & 2 & 4 \\ 0 & 0 & 0 & -1 & 3 \end{bmatrix}$

6. $\begin{bmatrix} 3 & 2 & 0 & 0 & 0 \\ -1 & 4 & 1 & 0 & 0 \\ 0 & -3 & 5 & 2 & 0 \\ 0 & 0 & 0 & 1 & 4 \\ 0 & 0 & 0 & -1 & 2 \end{bmatrix}$

7. $\begin{bmatrix} 0 & 0 & 0 & 3 & 1 \\ 0 & 0 & 2 & 0 & -3 \\ 0 & -2 & 1 & 0 & 0 \\ 5 & -3 & 2 & 0 & 0 \\ -3 & 4 & 0 & 0 & 0 \end{bmatrix}$

8. $\begin{bmatrix} 2 & -1 & 0 & 0 \\ 4 & 5 & 0 & 0 \\ 0 & 0 & 3 & 6 \\ 0 & 0 & -4 & 2 \end{bmatrix}$

9. $\begin{bmatrix} 2 & -1 & 3 & 0 & 0 \\ 0 & 1 & 4 & 0 & 0 \\ -5 & 2 & 6 & 0 & 0 \\ 0 & 0 & 0 & 1 & 4 \\ 0 & 0 & 0 & -2 & 8 \end{bmatrix}$

10. $\begin{bmatrix} 0 & 0 & 0 & 3 & -4 \\ 0 & 0 & 0 & 2 & 1 \\ -1 & 2 & 4 & 0 & 0 \\ 3 & 1 & -2 & 0 & 0 \\ 5 & 1 & 5 & 0 & 0 \end{bmatrix}$

11. The matrices in Exercises 8 and 9 have zero entries except for entries in an $r \times r$ submatrix R and a separate $s \times s$ submatrix S whose main diagonals lie on the main diagonal of the whole $n \times n$ matrix, and where $r + s = n$. Prove that, if A is such a matrix with submatrices R and S, then $\det(A) = \det(R) \cdot \det(S)$.

12. The matrix A in Exercise 10 has a structure similar to that discussed in Exercise 11, except that the square submatrices R and S lie along the other diagonal. State a result similar to that in Excricse 11 for such a matrix.

13. State a generalization of the result in Exercise 11, when the matrix A has zero entries except for entries in k submatrices positioned along the diagon l.

14. Find the area of the parallelogram in $\mathbb{R}^3$ determined by the vectors $(0, 1, 4)$ and $(-1, 3, -2)$.

15. Find the area of the parallelogram in $\mathbb{R}^5$ determined by the vectors $(1, 0, 1, 2, -1)$ and $(0, 1, -1, 1, 3)$.

16. Find the volume of the 3-box in $\mathbb{R}^4$ determined by the vectors $(-1, 2, 0, 1)$, $(0, 1, 3, 0)$, and $(0, 0, 2, -1)$.

17. Find the volume of the 4-box in $\mathbb{R}^5$ determined by the vectors $(1, 0, 1, 0, 1)$, $(0, 1, 1, 0, 0)$, $(3, 0, 1, 0, 0)$, and $(1, -1, 0, 0, 1)$.

In Exercises 18–23, find the volume of the n-box determined by the given vectors in $\mathbb{R}^n$.

18. $(-1, 4)$, $(2, 3)$ in $\mathbb{R}^2$

19. $(-5, 3)$, $(1, 7)$ in $\mathbb{R}^2$

20. $(1, 3, -5)$, $(2, 4, -1)$, $(3, 1, 2)$ in $\mathbb{R}^3$

21. $(-1, 4, 7)$, $(3, -2, -1)$, $(4, 0, 2)$ in $\mathbb{R}^3$

22. $(1, 0, 0, 1)$, $(2, -1, 3, 0)$, $(0, 1, 3, 4)$, $(-1, 1, -2, 1)$ in $\mathbb{R}^4$

23. $(1, -1, 0, 1)$, $(2, -1, 3, 1)$, $(-1, 4, 2, -1)$, $(0, 1, 0, 2)$ in $\mathbb{R}^4$

24. Find the area of the triangle in $\mathbb{R}^3$ with vertices $(-1, 2, 3)$, $(0, 1, 4)$, and $(2, 1, 5)$. [HINT: Think of vectors emanating from $(-1, 2, 3)$. The triangle may be viewed as half a parallelogram.]

25. Find the volume of the tetrahedron in $\mathbb{R}^3$ with vertices $(1, 0, 3)$, $(-1, 2, 4)$, $(3, -1, 2)$, and $(2, 0, -1)$. [HINT: Think of vectors emanating from $(1, 0, 3)$.]

26. Find the volume of the tetrahedron in $\mathbb{R}^4$ with vertices $(1, 0, 0, 1)$, $(-1, 2, 0, 1)$, $(3, 0, 1, 1)$, and $(-1, 4, 0, 1)$. [HINT: See the hint for Exercise 25.]

27. Give a geometric interpretation of the fact that an $n \times n$ matrix with two equal rows has determinant zero.

28. Using the results of this section, give a criterion that four points **a**, **b**, **c**, and **d** in $\mathbb{R}^n$ lie in a plane.

29. Determine whether the points $(1, 0, 1, 0)$, $(-1, 1, 0, 1)$, $(0, 1, -1, 1)$, and $(1, -1, 4, -1)$ lie in a plane in $\mathbb{R}^4$. (See Exercise 28.)

30. Determine whether the points $(2, 0, 1, 3)$, $(3, 1, 0, 1)$, $(-1, 2, 0, 4)$, and $(3, 1, 2, 4)$ lie in a plane in $\mathbb{R}^4$. (See Exercise 28.)

31. Prove Eq. (1); that is, prove that the square of the length of the line segment determined by $\mathbf{a}_1$ in $\mathbb{R}^n$ is $\|\mathbf{a}_1\|^2 = \det([\mathbf{a}_1 \cdot \mathbf{a}_1])$.

32. a) If one attempts to define an n-box in $\mathbb{R}^m$ for $n > m$, what will its volume as an n-box be?

b) Let A be an $m \times n$ matrix with $n > m$. Find $\det(A^T A)$.

33. Mark each of the following True or False.

___ a) The determinant of a square matrix is the product of the entries on its main diagonal.

___ b) The determinant of an upper-triangular square matrix is the product of the entries on its main diagonal.

___ c) The determinant of a lower-triangular square matrix is the product of the entries on its main diagonal.

___ d) A square matrix is nonsingular if and only if its determinant is positive.

___ e) The column vectors of an $n \times n$ matrix are independent if and only if the determinant of the matrix is nonzero.

___ f) A homogeneous square linear system has a nontrivial solution if and only if the determinant of its coefficient matrix is zero.

___ g) The determinant of every square matrix can be viewed as representing the "volume" of some "box."

___ h) The absolute value of the determinant of every nonsingular square matrix can be viewed as representing the "volume" of a "box."

___ i) If k vectors in $\mathbb{R}^n$ have integers for all components, the volume of the k-box they determine is also an integer.

___ j) If n vectors in $\mathbb{R}^n$ have integers for all components, the volume of the n-box they determine is also an integer.

The available software program YUREDUCE has a menu option D that will compute and display the product of the diagonal elements of a square matrix. The program MATCOMP has a menu option D to compute a determinant. Use YUREDUCE, or similar software, to compute the determinant of the matrices in Exercise 34–36. Write down your results. After this is done, use MATCOMP, or similar software, to compute the determinants of the same matrices again. Compare the answers.

34.
$$\begin{bmatrix} 11 & -9 & 28 \\ 32 & -24 & 21 \\ 10 & 13 & -19 \end{bmatrix}$$

35.
$$\begin{bmatrix} 13 & -15 & 33 \\ -15 & 25 & 40 \\ 12 & -33 & 27 \end{bmatrix}$$

36.
$$\begin{bmatrix} 7.6 & 2.8 & -3.9 & 19.3 & 25.0 \\ -33.2 & 11.4 & 13.2 & 22.4 & 18.3 \\ 21.4 & -32.1 & 45.7 & -8.9 & 12.5 \\ 17.4 & 11.0 & -6.8 & 20.3 & -35.1 \\ 22.7 & 11.9 & 33.2 & 2.5 & 7.8 \end{bmatrix}$$

37. MATCOMP computes determinants in essentially the way described in this section. The matrix

$$A = \begin{bmatrix} 3 & 4 \\ 2 & 3 \end{bmatrix}$$

has determinant 1, so every power of it should have determinant 1. Use MATCOMP with single-precision printing and with the default roundoff control ratio r. Start computing determinants of powers of A. Find the smallest positive integer m such that $\det(A^m) \neq 1$, according to MATCOMP. How bad is the error? What does MATCOMP give for $\det(A^{20})$? At what integer exponent does the break occur between the incorrect value 0 and incorrect values of large magnitude?

Repeat the above, taking zero for roundoff control ratio r. Try to explain why the results are different and what is happening in each case.

3.4 VOLUME-CHANGE FACTOR OF A LINEAR TRANSFORMATION

We continue our program of exhibiting the relationship between matrices and linear transformations. Associated with an $m \times n$ matrix A is the linear transformation T mapping $\mathbb{R}^n$ into $\mathbb{R}^m$, where $T(\mathbf{x}) = A\mathbf{x}$ for $\mathbf{x}$ in $\mathbb{R}^n$. If $n = m$, so that A is a square matrix, then $\det(A)$ is defined. But what is the meaning of the number $\det(A)$ for the transformation T? We now tackle this question, and the answer is so important that it merits a section all to itself. The notion of the determinant associated with a linear transformation T mapping $\mathbb{R}^n$ into $\mathbb{R}^n$ lies at the heart of variable substitution in integral calculus. We present an informal and intuitive explanation of this.

Recall the multiplicative property of determinants: $\det(AB) = \det(A) \cdot \det(B)$, where A and B are $n \times n$ matrices. Suppose that both A and B have rank n. Then the column vectors $\mathbf{b}_1, \mathbf{b}_2, \ldots, \mathbf{b}_n$ of B determine an n-box having volume $|\det(B)| \neq 0$. From the definition of matrix multiplication, the jth column vector of AB is $A\mathbf{b}_j$, and we write

$$AB = \begin{bmatrix} | & | & & | \\ A\mathbf{b}_1 & A\mathbf{b}_2 & \cdots & A\mathbf{b}_n \\ | & | & & | \end{bmatrix}.$$

The linear transformation T defined by $T(\mathbf{x}) = A\mathbf{x}$ thus carries the original n-box determined by the column vectors of B into a new n-box determined by the column vectors of AB. The new n-box has volume $|\det(AB)| = |\det(A)| \cdot |\det(B)|$. That is, the volume of the new n-box, or *image box*, is equal to $|\det(A)|$ times the volume of the original box. Thus $|\det(A)|$ is referred to as the **volume-change factor** for the linear transformation T. We are interested in this concept only when $\det(A) \neq 0$, a requirement that ensures that A is invertible, that the n vectors $A\mathbf{b}_i$ are independent, and that T is an invertible transformation.

To illustrate this idea of the volume-change factor, consider the n-cube in $\mathbb{R}^n$ determined by the vectors $c\mathbf{e}_1, c\mathbf{e}_2, \ldots, c\mathbf{e}_n$ for $c > 0$. This n-cube with edges of length c has volume c^n. It is carried by T into an n-box having volume $|\det(A)| \cdot c^n$; the image n-box need not be a cube, nor even rectangular. We illustrate the image n-box in Figure 3.11 for the case $n = 1$ and in Figure 3.12 for the case $n = 3$.

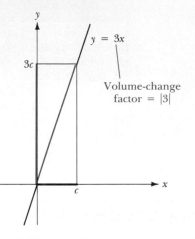

FIGURE 3.11 The volume-change factor of $T(x) = 3x$.

EXAMPLE 1 Consider the linear transformation $T: \mathbb{R}^3 \to \mathbb{R}^3$ defined by

$$T(x_1, x_2, x_3) = (x_1 + x_3, 2x_2, 2x_1 + 5x_3).$$

Find the volume of the image box when T acts on the cube determined by the vectors $c\mathbf{e}_1$, $c\mathbf{e}_2$, and $c\mathbf{e}_3$ for $c > 0$.

Solution The image box is determined by the vectors

$$T(c\mathbf{e}_1) = cT(\mathbf{e}_1) = c(1, 0, 2)$$

$$T(c\mathbf{e}_2) = cT(\mathbf{e}_2) = c(0, 2, 0)$$

$$T(c\mathbf{e}_3) = cT(\mathbf{e}_3) = c(1, 0, 5).$$

The standard matrix representation of T is

$$A = \begin{bmatrix} 1 & 0 & 1 \\ 0 & 2 & 0 \\ 2 & 0 & 5 \end{bmatrix},$$

and the volume-change factor of T is $|\det(A)| = 6$. Therefore, the image box has volume $6c^3$. This volume can also be computed by taking the determinant of the matrix having as column vectors $T(c\mathbf{e}_1)$, $T(c\mathbf{e}_2)$, and $T(c\mathbf{e}_3)$. This matrix is cA, and has determinant $c^3 \cdot \det(A) = 6c^3$. ■

The sign of the determinant associated with an invertible linear transformation $T: \mathbb{R}^n \to \mathbb{R}^n$ depends on whether T preserves orientation. In the plane where $n = 2$, orientation is preserved by T if $T(\mathbf{e}_1)$ can be rotated counterclockwise through an angle of less than $180°$ to lie along $T(\mathbf{e}_2)$. It can be shown that this is the case if and only if $\det(A) > 0$, where A is the standard matrix representation of T. In general, a linear transformation $T: \mathbb{R}^n \to \mathbb{R}^n$ is said to *preserve orientation* if the determinant of its standard matrix representation A is positive. Thus the linear transformation in

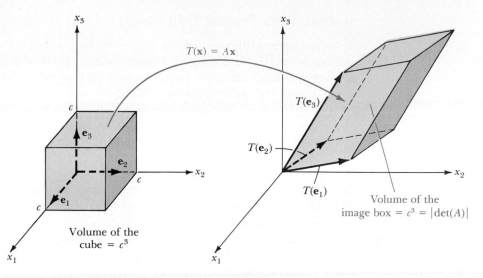

FIGURE 3.12 **The volume-change factor of $T(x) = Ax$.**

Example 1 preserves orientation since $\det(A) = 6$. Since this topic is more suitable for a course in analysis than for one in linear algebra, we do not pursue it further.

Application to Calculus

We can get an intuitive feel for the connection between the volume-change factor of T and integral calculus. The definition of an integral involves summing products of the form

$$\text{(Function value at some point of a box)(Volume of the box).} \qquad \textbf{(1)}$$
$$\phantom{\text{(Function value at some point}}f(\mathbf{x})\phantom{\text{of a box)(Volume of the box).}}\mathbf{dx}$$

Under a change of variables—say, from **x**-variables to **t**-variables—the boxes in the **dx**-space are replaced by boxes in **dt**-space via an invertible linear transformation— namely, the *differential* of the variable substitution function. Thus the second factor in product (1) must be expressed in terms of volumes of boxes in the **dt**-space. The determinant of the differential transformation must play a role, since the volume of the box in **dx**-space is the volume of the corresponding box in **dt**-space multiplied by the absolute value of the determinant.

Let us look at a one-dimensional example. In making the substitution $x = \sin t$ in an integral in single-variable calculus, we associate with each t-value t_0 the linear transformation of dt-space into dx-space given by the equation $dx = (\cos t_0)dt$. The determinant of this linear transformation is $\cos t_0$. A little 1-box of volume (length) dt and containing the point t_0 is carried by this linear transformation into a little 1-box in the dx-space of volume (length) $|\cos t_0|dt$. Having conveyed a rough idea of this topic and its importance, we leave its further development to an upper-level course in analysis.

Regions More General Than Boxes

In the remainder of this section, we will be working with the volume of *sufficiently nice regions* in $\mathbb{R}^n$. To define such a notion carefully would require an excursion into calculus of several variables. We will simply assume that we all have an intuitive notion of such regions having a well-defined volume.

Let $T: \mathbb{R}^n \to \mathbb{R}^n$ be a linear transformation of rank n, and let A be its standard matrix representation. Recall that the jth column vector of A is $T(\mathbf{e}_j)$ (see p. 178). We will show that, if a region G in $\mathbb{R}^n$ has volume V, the image of G under T has volume

$$|\det(A)| \cdot V.$$

That is, the volume of a region is multiplied by $|\det(A)|$ when the region is transformed by T. This result has important applications to integral calculus.

The volume of a sufficiently nice region G in $\mathbb{R}^n$ may be approximated by adding the volumes of small n-cubes contained in G and having edges of length c parallel to the coordinate axes. Figure 3.13(a) illustrates this situation for a plane region in $\mathbb{R}^2$, where a grid of small squares (2-cubes) is placed on the region. As the length c of the edges of the squares approaches zero, the sum of the areas of the colored squares inside the region approaches the area of the region. These squares inside G are mapped by T into parallelograms of area $c^2|\det(A)|$ inside the image of G under T. (See the colored parallelograms in Figure 3.13(b).) As c approaches zero, the sum of the areas of these parallelograms approaches the area of the image G under T, which thus must be the area of G multiplied by $|\det(A)|$. A similar construction can be made with a grid of n-cubes for a region G in $\mathbb{R}^n$. Each such cube is mapped by T into an n-box of volume $c^n|\det(A)|$. Adding the volumes of these n-boxes and taking the limiting value of the sum as c approaches zero, we see that the volume of the image under T of the region G is given by

$$\text{Volume of image of } G = |\det(A)| \cdot (\text{Volume of } G). \qquad (2)$$

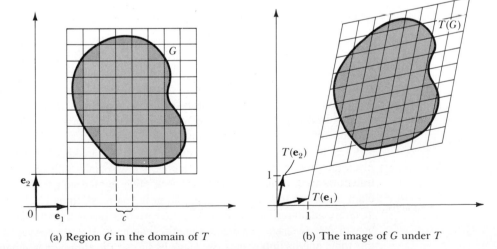

(a) Region G in the domain of T (b) The image of G under T

FIGURE 3.13

We summarize this work in a theorem.

THEOREM 3.6 Volume-change Factor for *T*: $\mathbb{R}^n \to \mathbb{R}^n$

Let G be a region in $\mathbb{R}^n$ of volume V, and let $T: \mathbb{R}^n \to \mathbb{R}^n$ be a linear transformation of rank n with standard matrix representation A. Then the volume of the image of G under T is $|\det(A)| \cdot V$.

EXAMPLE 2   Let $T: \mathbb{R}^2 \to \mathbb{R}^2$ be the linear transformation given by $T(x, y) = (2x - y, x + 3y)$. Find the area of the image under T of the disk $x^2 + y^2 \leq 4$ in the domain of T.

Solution   The disk $x^2 + y^2 \leq 4$ has radius 2 and area 4π. It is mapped by T into a region (actually bounded by an ellipse) of area

$$|\det(A)| \cdot (4\pi) = \begin{vmatrix} 2 & -1 \\ 1 & 3 \end{vmatrix} \cdot (4\pi) = (6 + 1)(4\pi) = 28\pi. \qquad \blacksquare$$

We can generalize Theorem 3.6 to a linear transformation $T: \mathbb{R}^n \to \mathbb{R}^m$, where $m \geq n$ and T has rank n. This time, the standard matrix representation A is an $m \times n$ matrix. The image under T of the unit n-cube in $\mathbb{R}^n$ outlined by $\mathbf{e}_1$, $\mathbf{e}_2$, . . . , $\mathbf{e}_n$ is the n-box in $\mathbb{R}^m$ outlined by $T(\mathbf{e}_1)$, $T(\mathbf{e}_2)$, . . . , $T(\mathbf{e}_n)$. According to Theorem 3.5, the volume of this box in $\mathbb{R}^m$ is

$$\sqrt{\det(A^T A)}.$$

The same grid argument used earlier and illustrated in Figure 3.13 shows that a region G in $\mathbb{R}^n$ of volume V is mapped by T into a region of $\mathbb{R}^m$ of volume

$$\sqrt{\det(A^T A)} \cdot V.$$

We summarize this generalization in a theorem.

THEOREM 3.7 Volume-change Factor for *T*: $\mathbb{R}^n \to \mathbb{R}^m$

Let G be a region in $\mathbb{R}^n$ of volume V. Let $m \geq n$ and let $T: \mathbb{R}^n \to \mathbb{R}^m$ be a linear transformation of rank n. Let A be the standard matrix representation of T. Then the volume of the image of G in $\mathbb{R}^m$ under T is

$$\sqrt{\det(A^T A)} \cdot V.$$

EXAMPLE 3   Let $T: \mathbb{R}^2 \to \mathbb{R}^3$ be given by $T(x, y) = (2x + 3y, x - y, 2y)$. Find the area of the image in $\mathbb{R}^3$ under T of the disk $x^2 + y^2 \leq 4$ in $\mathbb{R}^2$.

Solution   The standard matrix representation A of T is

$$A = \begin{bmatrix} 2 & 3 \\ 1 & -1 \\ 0 & 2 \end{bmatrix}$$

and

$$A^TA = \begin{bmatrix} 2 & 1 & 0 \\ 3 & -1 & 2 \end{bmatrix} \begin{bmatrix} 2 & 3 \\ 1 & -1 \\ 0 & 2 \end{bmatrix} = \begin{bmatrix} 5 & 5 \\ 5 & 14 \end{bmatrix}.$$

Thus,

$$\sqrt{\det(A^TA)} = \sqrt{70 - 25} = \sqrt{45} = 3\sqrt{5}.$$

A region G of $\mathbb{R}^2$ having area V is mapped by T into a plane region of area $3\sqrt{5} \cdot V$ in $\mathbb{R}^3$. Thus the disk $x^2 + y^2 \leq 4$ of area 4π is mapped into a plane region in $\mathbb{R}^3$ of area

$$(3\sqrt{5})(4\pi) = 12\pi\sqrt{5}. \qquad ■$$

SUMMARY

1. If $T: \mathbb{R}^n \to \mathbb{R}^n$ is a linear transformation of rank n with standard matrix representation A, then T maps a region in its domain of volume V into a region of volume $|\det(A)|V$.

2. If $T: \mathbb{R}^n \to \mathbb{R}^m$ is a linear transformation of rank n with standard matrix representation A, then T maps a region in its domain of volume V into a region of $\mathbb{R}^m$ of volume $\sqrt{\det(A^TA)} \cdot V$.

EXERCISES

In Exercises 1–4, let $T: \mathbb{R}^2 \to \mathbb{R}^2$ be the linear transformation defined by $T(x, y) = (4x - 2y, 2x + 3y)$. Find the area of the image under T of each of the given regions in $\mathbb{R}^2$.

1. The square $0 \leq x \leq 1, 0 \leq y \leq 1$

2. The rectangle $-1 \leq x \leq 1, 1 \leq y \leq 2$

3. The parallelogram determined by $2\mathbf{e}_1 + 3\mathbf{e}_2$ and $4\mathbf{e}_1 - \mathbf{e}_2$

4. The disk $(x - 1)^2 + (y + 2)^2 \leq 9$

In Exercises 5–8, let $T: \mathbb{R}^3 \to \mathbb{R}^3$ be defined by $T(x, y, z) = (x - 2y, 3x + z, 4x + 3y)$. Find the volume of the image under T of each of the given regions in $\mathbb{R}^3$.

5. The cube $0 \leq x \leq 1, 0 \leq y \leq 1, 0 \leq z \leq 1$

6. The box $0 \leq x \leq 2, -1 \leq y \leq 3, 2 \leq z \leq 5$

7. The box determined by $2\mathbf{e}_1 + 3\mathbf{e}_2 - \mathbf{e}_3$,  $4\mathbf{e}_1 - 2\mathbf{e}_3$, and $\mathbf{e}_1 - \mathbf{e}_2 + 2\mathbf{e}_3$

8. The ball $x^2 + (y - 3)^2 + (z + 2)^2 \leq 16$

In Exercises 9–12, let $T: \mathbb{R}^2 \to \mathbb{R}^3$ be the linear transformation defined by $T(x, y) = (y, x, x + y)$. Find the area of the image under T of each of the following regions in $\mathbb{R}^2$.

9. The square $0 \leq x \leq 1, 0 \leq y \leq 1$

10. The rectangle $2 \leq x \leq 3, -1 \leq y \leq 4$

11. The triangle with vertices $(0, 0), (6, 0), (0, 3)$

12. The disk $x^2 + y^2 \leq 25$

In Exercises 13–15, let $T: \mathbb{R}^2 \to \mathbb{R}^4$ be defined by $T(x, y) = (x - y, x, -y, 2x + y)$. Find the area of the image under T of each of the following regions in $\mathbb{R}^2$.

13. The square $0 \leq x \leq 1, 0 \leq y \leq 1$

14. The square $-1 \le x \le 3$, $-1 \le y \le 3$

15. The disk $x^2 + y^2 \le 9$

16. We have seen that, for $n \times n$ matrices A and B, we have $\det(AB) = \det(A) \cdot \det(B)$, but the proof was not intuitive. Give an intuitive geometric argument showing that at least $|\det(AB)| = |\det(A)| \cdot |\det(B)|$. [HINT: Use the fact that, if A is the standard matrix representation of $T: \mathbb{R}^n \to \mathbb{R}^n$ and B is the standard matrix representation of $T': \mathbb{R}^n \to \mathbb{R}^n$, then AB is the standard matrix representation $T \circ T'$.]

17. Let $T: \mathbb{R}^n \to \mathbb{R}^n$ be a linear transformation of rank n with standard matrix representation A. Mark each of the following True or False.

____ a) The image under T of a box in $\mathbb{R}^n$ is again a box in $\mathbb{R}^n$.

____ b) The image under T of an n-box in $\mathbb{R}^n$ of volume V is a box in $\mathbb{R}^n$ of volume $\det(A) \cdot V$.

____ c) The image under T of an n-box in $\mathbb{R}^n$ of volume > 0 is a box in $\mathbb{R}^n$ of volume > 0.

____ d) If the image under T of an n-box B in $\mathbb{R}^n$ has volume 12, the box B has volume $|\det(A)| \cdot 12$.

____ e) If the image under T of an n-box B in $\mathbb{R}^n$ has volume 12, the box B has volume $12/|\det(A)|$.

____ f) If $n = 2$, the image under T of the unit disk $x^2 + y^2 \le 1$ has area $|\det(A)|$.

____ g) The linear transformation T is an isomorphism.

____ h) The image under $T \circ T$ of an n-box in $\mathbb{R}^n$ of volume V is a box in $\mathbb{R}^n$ of volume $\det(A^2) \cdot V$.

____ i) The image under $T \circ T \circ T$ of an n-box in $\mathbb{R}^n$ of volume V is a box in $\mathbb{R}^n$ of volume $\det(A^3) \cdot V$.

____ j) The image under T of a nondegenerate 1-box is again nondegenerate.

3.5 ## Cramer's Rule

In this section we describe two classical applications of determinants. One gives a formula for the inverse A^{-1} of an invertible matrix A. The other gives a formula for the components of the solution vector of a square linear system $A\mathbf{x} = \mathbf{b}$ if A is invertible. Both of these formulas are computationally very inefficient compared with the methods we have developed to compute A^{-1} and the solution of $A\mathbf{x} = \mathbf{b}$. However, the structure of the formulas proves useful in theoretical considerations.

The Adjoint Matrix

We begin by finding a formula in terms of determinants for the inverse of an invertible $n \times n$ matrix $A = [a_{ij}]$. Recall the definition of the cofactor a'_{ij} from Eq. (2) of Section 3.2. Let $A_{i \to j}$ be the matrix obtained from A by replacing the jth row of A by the ith row. That is,

$$A_{i \to j} = \begin{bmatrix} a_{11} & a_{12} & \cdots & a_{1n} \\ \vdots & \vdots & & \vdots \\ a_{i1} & a_{i2} & \cdots & a_{in} \\ \vdots & \vdots & & \vdots \\ a_{i1} & a_{i2} & \cdots & a_{in} \\ \vdots & \vdots & & \vdots \\ a_{n1} & a_{n2} & \cdots & a_{nn} \end{bmatrix} \begin{matrix} \\ \\ i\text{th row} \\ \\ j\text{th row} \\ \\ \end{matrix}.$$

Then

$$\det(A_{i \to j}) = \begin{cases} \det(A) & \text{if } i = j, \\ 0 & \text{if } i \neq j. \end{cases}$$

If we expand $\det(A_{i \to j})$ by minors on the jth row, we have

$$\det(A_{i \to j}) = \sum_{s=1}^{n} a_{is} a_{js}',$$

and we obtain the important relation

$$\sum_{s=1}^{n} a_{is} a_{js}' = \begin{cases} \det(A) & \text{if } i = j, \\ 0 & \text{if } i \neq j. \end{cases} \tag{1}$$

The term on the left-hand side in Eq. (1) is the entry in the ith row and jth column in the product $A(A')^T$, where $A' = [a_{ij}']$ is the matrix whose entries are the cofactors of the entries of A. Thus Eq. (1) can be written in matrix form as

$$A(A')^T = (\det(A))I,$$

where I is the $n \times n$ identity matrix. Similarly, replacing the jth column of A by the ith column, we have

$$\sum_{r=1}^{n} a_{ri}' a_{rj} = \begin{cases} \det(A) & \text{if } i = j, \\ 0 & \text{if } i \neq j. \end{cases} \tag{2}$$

Relation (2) yields $(A')^T A = (\det(A))I$.

The matrix $(A')^T$ is called the **adjoint** of A and is denoted by $\text{adj}(A)$. We have established an important relationship between a matrix and its adjoint.

THEOREM 3.8 Property of the Adjoint

Let A be an $n \times n$ matrix. The adjoint $\text{adj}(A) = (A')^T$ of A satisfies

$$(\text{adj}(A))A = A(\text{adj}(A)) = \underline{(\det(A))I},$$

where I is the $n \times n$ identity matrix.

Theorem 3.8 provides a formula for the inverse of an invertible matrix, which we present as a corollary.

COROLLARY A Formula for the Inverse of an Invertible Matrix

Let $A = [a_{ij}]$ be an $n \times n$ matrix with $\det(A) \neq 0$. Then A is invertible, and

$$A^{-1} = \frac{1}{\det(A)} \text{adj}(A),$$

where $\text{adj}(A) = [a_{ij}']^T$ is the transposed matrix of cofactors.

EXAMPLE 1   Find the inverse of

$$(2-6) + 8 = 4.$$

$$A = \begin{bmatrix} 4 & 0 & 1 \\ 2 & 2 & 0 \\ 3 & 1 & 1 \end{bmatrix}$$

if the matrix is invertible, using the corollary of Theorem 3.8.

Solution   We find that $\det(A) = 4$, so A is invertible. The cofactors a'_{ij} are

$$a'_{11} = (-1)^2 \begin{vmatrix} 2 & 0 \\ 1 & 1 \end{vmatrix} = 2, \qquad a'_{12} = (-1)^3 \begin{vmatrix} 2 & 0 \\ 3 & 1 \end{vmatrix} = -2,$$

$$a'_{13} = (-1)^4 \begin{vmatrix} 2 & 2 \\ 3 & 1 \end{vmatrix} = -4, \qquad a'_{21} = (-1)^3 \begin{vmatrix} 0 & 1 \\ 1 & 1 \end{vmatrix} = 1,$$

$$a'_{22} = (-1)^4 \begin{vmatrix} 4 & 1 \\ 3 & 1 \end{vmatrix} = 1, \qquad a'_{23} = (-1)^5 \begin{vmatrix} 4 & 0 \\ 3 & 1 \end{vmatrix} = -4,$$

$$a'_{31} = (-1)^4 \begin{vmatrix} 0 & 1 \\ 2 & 0 \end{vmatrix} = -2, \qquad a'_{32} = (-1)^5 \begin{vmatrix} 4 & 1 \\ 2 & 0 \end{vmatrix} = 2,$$

$$a'_{33} = (-1)^6 \begin{vmatrix} 4 & 0 \\ 2 & 2 \end{vmatrix} = 8.$$

Hence,

$$\det(\text{adj}(A)) = 16.$$
$$\det(A)^{3-1} = 16 \qquad \det(A) = \sqrt{16} = 4$$

$$A' = [a'_{ij}] = \begin{bmatrix} 2 & -2 & -4 \\ 1 & 1 & -4 \\ -2 & 2 & 8 \end{bmatrix}, \quad \text{so} \quad \text{adj}(A) = \begin{bmatrix} 2 & 1 & -2 \\ -2 & 1 & 2 \\ -4 & -4 & 8 \end{bmatrix}$$

and

$$A^{-1} = \frac{1}{\det(A)} \text{adj}(A) = \frac{1}{4} \begin{bmatrix} 2 & 1 & -2 \\ -2 & 1 & 2 \\ -4 & -4 & 8 \end{bmatrix} = \begin{bmatrix} \frac{1}{2} & \frac{1}{4} & -\frac{1}{2} \\ -\frac{1}{2} & \frac{1}{4} & \frac{1}{2} \\ -1 & -1 & 2 \end{bmatrix}. \qquad ■$$

The method described in Section 1.4 for finding the inverse of an invertible matrix is more efficient than the method illustrated in the preceding example, especially if the matrix is large. The corollary is often used to find the inverse of a 2 × 2 matrix. We see that if $ad - bc \neq 0$, then

$$\begin{bmatrix} a & b \\ c & d \end{bmatrix}^{-1} = \frac{1}{ad - bc} \begin{bmatrix} d & -b \\ -c & a \end{bmatrix}.$$

Cramer's Rule

We turn to the problem of finding formulas in terms of determinants for the components in the solution vector of a square linear system $A\mathbf{x} = \mathbf{b}$, where A is an invertible matrix. We will prove the following theorem.

THEOREM 3.9 Cramer's Rule

Consider the linear system $A\mathbf{x} = \mathbf{b}$, where $A = [a_{ij}]$ is an $n \times n$ invertible matrix,

$$\mathbf{x} = \begin{bmatrix} x_1 \\ \vdots \\ x_n \end{bmatrix}, \quad \text{and} \quad \mathbf{b} = \begin{bmatrix} b_1 \\ \vdots \\ b_n \end{bmatrix}.$$

The system has a unique solution given by

$$x_k = \frac{\det(B_k)}{\det(A)} \quad \text{for} \quad k = 1, \ldots, n, \tag{3}$$

where B_k is the matrix obtained from A by replacing the kth column of A by the column vector $\mathbf{b}$.

PROOF Since A is invertible, the unique solution of the system $A\mathbf{x} = \mathbf{b}$ can be expressed as

$$\mathbf{x} = A^{-1}\mathbf{b} = \frac{1}{\det(A)}(A')^T\mathbf{b}, \tag{4}$$

where $A' = [a'_{ij}]$ is the matrix of cofactors. Comparing Eq. (4) with the desired result in Eq. (3), we see that we need only show that the kth component of the column vector $(A')^T\mathbf{b}$ is given by the determinant of the matrix

$$B_k = \begin{bmatrix} a_{11} & \cdots & b_1 & \cdots & a_{1n} \\ a_{21} & \cdots & b_2 & \cdots & u_{2n} \\ \vdots & & \vdots & & \vdots \\ a_{n1} & \cdots & b_n & \cdots & a_{nn} \end{bmatrix},$$

obtained from A by replacing the kth column by $\mathbf{b}$, that is, by replacing a_{jk} by b_j. Expanding $\det(B_k)$ by minors on the kth column, we have

$$\det(B_k) = \sum_{j=1}^{n} a'_{jk}b_j. \tag{5}$$

But this is indeed just the kth component of the column vector

$$(A')^T\mathbf{b} = \begin{bmatrix} a'_{11} & a'_{21} & \cdots & a'_{n1} \\ \vdots & \vdots & & \vdots \\ a'_{1k} & a'_{2k} & \cdots & a'_{nk} \\ \vdots & \vdots & & \vdots \\ a'_{1n} & a'_{2n} & \cdots & a'_{nn} \end{bmatrix} \begin{bmatrix} b_1 \\ b_2 \\ \vdots \\ b_n \end{bmatrix}.$$

This completes our proof. ▲

EXAMPLE 2   Solve the linear system

$$5x_1 - 2x_2 + x_3 = 1$$
$$3x_1 + 2x_2 \quad\quad = 3$$
$$x_1 + x_2 - x_3 = 0,$$

using Cramer's rule.

Solution   Using the notation in Theorem 3.9, we find that

$$\det(A) = \begin{vmatrix} 5 & -2 & 1 \\ 3 & 2 & 0 \\ 1 & 1 & -1 \end{vmatrix} = -15, \quad \det(B_1) = \begin{vmatrix} 1 & -2 & 1 \\ 3 & 2 & 0 \\ 0 & 1 & -1 \end{vmatrix} = -5,$$

$$\det(B_2) = \begin{vmatrix} 5 & 1 & 1 \\ 3 & 3 & 0 \\ 1 & 0 & -1 \end{vmatrix} = -15, \quad \det(B_3) = \begin{vmatrix} 5 & -2 & 1 \\ 3 & 2 & 3 \\ 1 & 1 & 0 \end{vmatrix} = -20.$$

Hence,

$$x_1 = (-5)/(-15) = \tfrac{1}{3},$$
$$x_2 = (-15)/(-15) = 1,$$
$$x_3 = (-20)/(-15) = \tfrac{4}{3}.$$
■

The most efficient way we have presented for computing a determinant is to row-reduce a matrix to triangular form. This is also the way we solve a square linear system. If A is a 10×10 invertible matrix, solving $A\mathbf{x} = \mathbf{b}$ using Cramer's rule involves row-reducing eleven 10×10 matrices $A, B_1, B_2, \ldots, B_{10}$ to triangular form. Solving the linear system by the method of Section 1.3 requires row-reducing just one 10×11 matrix so that the first 10 columns are in upper-triangular form. This illustrates the folly of using Cramer's rule to solve linear systems. However, the

CRAMER'S RULE appeared for the first time in full generality in a work of Gabriel Cramer (1704–1752) entitled *Introduction to the Analysis of Algebraic Curves* (1750). The problem in which Cramer was interested was that of determining the equation of a plane curve of given degree passing through a certain number of given points. He stated the theorem that a curve of nth degree is determined when $\left(\tfrac{1}{2}\right)n(n + 3)$ points of the curve are known. For example, a second-degree curve, which he wrote as

$$A + By + Cx + Dy^2 + Exy + x^2 = 0, \tag{*}$$

is determined by five points. The question then is how to determine A, B, C, D, and E, given the five points. The obvious method is to substitute into equation (*) the coordinates of each of the five points in turn. This gives us five equations for the five unknown coefficients. Cramer then referred to the appendix of the tract, in which he gave his general rule: "One finds the value of each unknown by forming n fractions of which the common denominator has as many terms as there are permutations of n things." He went on to explain exactly how one calculates these terms as products of certain coefficients of the n equations, how one determines the appropriate sign for each term, and how one determines the n numerators of the fractions by replacing certain coefficients in this calculation by the constant terms of the system.

Cramer was a Swiss mathematician who taught at the Académie de Calvin in Geneva from 1724 until his death. He was an excellent scholar, but one whose work was overshadowed by such brilliant contemporaries as Euler and the Bernoullis.

structure of the components of the solution vector, as given by the Cramer's rule formula $x_k = \det(B_k)/\det(A)$, is of interest in the study of advanced calculus, for example.

SUMMARY

1. Let A be an $n \times n$ matrix, and let A' be its matrix of cofactors. The adjoint adj(A) is the matrix $(A')^T$ and satisfies $(\mathrm{adj}(A))A = A(\mathrm{adj}(A)) = (\det(A))I$, where I is the $n \times n$ identity matrix.

2. The inverse of an invertible matrix A is given by the explicit formula

$$A^{-1} = \frac{1}{\det(A)}\mathrm{adj}(A).$$

3. If A is invertible, a linear system $A\mathbf{x} = \mathbf{b}$ has the unique solution $\mathbf{x}$ whose kth component is given explicitly by the formula

$$x_k = \frac{\det(B_k)}{\det(A)},$$

where the matrix B_k is obtained from A by replacing the kth column of A by $\mathbf{b}$.

4. The methods of Chapter 1 are far more efficient than those described in this section for actual computation of both the inverse of A and the solution of the system $A\mathbf{x} = \mathbf{b}$.

EXERCISES

In Exercises 1–6, use the corollary to Theorem 3.8 to find A^{-1} if A is invertible.

1. $A = \begin{bmatrix} 2 & 0 \\ 1 & -1 \end{bmatrix}$

2. $A = \begin{bmatrix} 4 & 1 \\ 2 & 1 \end{bmatrix}$

3. $A = \begin{bmatrix} 2 & 1 & 1 \\ 0 & 1 & 1 \\ -2 & 1 & 1 \end{bmatrix}$

4. $A = \begin{bmatrix} 3 & 0 & 4 \\ -2 & 1 & 1 \\ 3 & 1 & 2 \end{bmatrix}$

5. $A = \begin{bmatrix} 3 & 0 & 3 \\ 4 & 1 & -2 \\ -5 & 1 & 4 \end{bmatrix}$

6. $A = \begin{bmatrix} 2 & 1 & 3 \\ 0 & 1 & 4 \\ 1 & 2 & 1 \end{bmatrix}$

7. Find the adjoint of the matrix $\begin{bmatrix} 4 & 5 \\ -3 & 6 \end{bmatrix}$.

8. Find the adjoint of the matrix $\begin{bmatrix} 2 & 1 & 0 \\ 3 & 1 & 4 \\ 0 & 2 & 1 \end{bmatrix}$.

9. Given that $A^{-1} = \begin{bmatrix} a & b \\ c & d \end{bmatrix}$ and $\det(A^{-1}) = 3$, find the matrix A.

10. If A is a matrix with integer entries and if $\det(A) = \pm 1$, show that A^{-1} also has the same properties.

In Exercises 11–18, solve the given system of linear equations by Cramer's rule wherever it is possible.

11. $\begin{array}{l} x_1 - 2x_2 = 1 \\ 3x_1 + 4x_2 = 3 \end{array}$

12. $\begin{array}{l} 2x_1 - 3x_2 = 1 \\ -4x_1 + 6x_2 = -2 \end{array}$

13. $\begin{array}{l} 3x_1 + x_2 = 5 \\ 2x_1 + x_2 = 0 \end{array}$

14. $\begin{array}{l} x_1 + x_2 = 1 \\ x_2 + 2x_2 = 2 \end{array}$

15. $\begin{array}{l} 5x_1 - 2x_2 + x_3 = 1 \\ x_2 + x_3 = 0 \\ x_1 + 6x_2 - x_3 = 4 \end{array}$

16. $\begin{aligned} x_1 + 2x_2 - x_3 &= -2 \\ 2x_1 + x_2 + x_3 &= 0 \\ 3x_1 - x_2 + 5x_3 &= 1 \end{aligned}$

17. $\begin{aligned} x_1 - x_2 + x_3 &= 0 \\ x_1 + 2x_2 - x_3 &= 1 \\ x_1 - x_2 + 2x_3 &= 0 \end{aligned}$

18. $\begin{aligned} 3x_1 + 2x_2 - x_3 &= 1 \\ x_1 - 4x_2 + x_3 &= -2 \\ 5x_1 + 2x_2 &= 1 \end{aligned}$

In Exercises 19 and 20, find the component x_2 of the solution vector for the given linear system.

19. $\begin{aligned} x_1 + x_2 - 3x_3 + x_4 &= 1 \\ 2x_1 + x_2 + 2x_4 &= 0 \\ x_2 - 6x_3 - x_4 &= 5 \\ 3x_1 + x_2 + x_4 &= 1 \end{aligned}$

20. $\begin{aligned} 6x_1 + x_2 - x_3 &= 4 \\ x_1 - x_2 + 5x_4 &= -2 \\ -x_1 + 3x_2 + x_3 &= 2 \\ x_1 + x_2 - x_3 + 2x_4 &= 0 \end{aligned}$

21. Find the unique solution (assuming that it exists) of the system of equations represented by the partitioned matrix

$$\left[\begin{array}{cccc|c} a_1 & b_1 & c_1 & d_1 & 3b_1 \\ a_2 & b_2 & c_2 & d_2 & 3b_2 \\ a_3 & b_3 & c_3 & d_3 & 3b_3 \\ a_4 & b_4 & c_4 & d_4 & 3b_4 \end{array}\right].$$

22. Prove that the inverse of a nonsingular upper-triangular matrix is upper triangular.

23. Let A be a square matrix. Mark each of the following True or False.

____ a) The product of a square matrix and its adjoint is the identity matrix.

____ b) The product of a square matrix and its adjoint is equal to some scalar times the identity matrix.

____ c) The transpose of the adjoint of A is the matrix of cofactors of A.

____ d) The formula $A^{-1} = (1/\det(A))\text{adj}(A)$ is of practical use in computing the inverse of a large nonsingular matrix.

____ e) If A has only integer entries, its adjoint matrix has only integer entries.

____ f) The matrix A has only integer entries if and only if its adjoint matrix has only integer entries.

____ g) Cramer's rule, in theory, can be used to find all solutions of any linear system.

____ h) Cramer's rule, in theory, can be used to find all solutions of any square linear system.

____ i) Cramer's rule, in theory, can be used to find all solutions of any square linear system having a unique solution.

____ j) Cramer's rule is a practical way to find a solution of a large linear system having a unique solution.

24. Prove that a square matrix is invertible if and only if its adjoint is an invertible matrix.

25. Let A be an $n \times n$ matrix. Show that $\det(\text{adj}(A)) = \det(A)^{n-1}$.

26. Let A be an invertible $n \times n$ matrix with $n > 1$. Using Exercises 24 and 25 show that $\text{adj}(\text{adj}(A)) = (\det(A))^{n-2}A$.

In Exercises 27–29, use MATCOMP or similar software and the corollary of Theorem 3.8 to find the matrix of cofactors of the given matrix.

27. $\begin{bmatrix} 1 & 2 & -3 \\ 2 & 3 & 0 \\ 3 & 1 & 4 \end{bmatrix}$

28. $\begin{bmatrix} -52 & 31 & 47 \\ 21 & -11 & 28 \\ 43 & -71 & 87 \end{bmatrix}$

29. $\begin{bmatrix} 6 & -3 & 2 & 14 \\ -3 & 7 & 8 & 1 \\ 4 & 9 & -5 & 3 \\ -8 & -40 & 47 & 29 \end{bmatrix}$

[HINT: Entries in the matrix of cofactors are integers. The cofactors of a matrix are continuous functions of its entries; that is, changing an entry by a very slight amount will change a cofactor only slightly. Change some entry just a bit to make the determinant nonzero.]

4

EIGENVALUES AND EIGENVECTORS

This chapter introduces the important topic of eigenvalues and eigenvectors of a square matrix and the associated linear transformation. Determinants enable us to give illustrative computations and applications involving matrices of very small size. Eigenvectors and eigenvalues continue to appear at intervals throughout much of the rest of the text. In Section 7.4 we discuss some other methods for computing them.

4.1 EIGENVALUES AND EIGENVECTORS

Encounters with $A^k\mathbf{x}$

In Section 1.6 we studied Markov chains dealing with the distribution of a population among states, measured over evenly spaced time intervals. An $n \times n$ transition matrix T describes the movement of the population among the states during one time interval. The matrix T has the property that all entries are nonnegative and the sum of the entries in any column is 1. Suppose that $\mathbf{p}$ is the initial population distribution vector—that is, the column vector whose ith component is the proportion of the population in the ith state at the start of the process. Then $T\mathbf{p}$ is the corresponding population distribution vector after one time interval. Similarly, $T^2\mathbf{p}$ is the population distribution vector after two time intervals, and in general, $T^k\mathbf{p}$ gives the distribution of population among the states after k time intervals.

Markov chains provide one example in which we are interested in computing $A^k\mathbf{x}$ for an $n \times n$ matrix A and a column vector $\mathbf{x}$ of n components. We give a famous classical problem that provides another illustration.

EXAMPLE 1   (*Fibonacci's rabbits*) Suppose that newly born pairs of rabbits produce no offspring during the first month of their lives, but each pair produces one new pair each subsequent month. Starting with $F_1 = 1$ newly born pair in the first month, find the number F_k of pairs in the kth month, assuming that no rabbit dies.

Solution   In the kth month, the number of pairs of rabbits is

$$F_k = \text{(Number of pairs alive the preceding month)}$$

$$+ \text{(Number of newly born pairs for the } k\text{th month)}.$$

Since our rabbits do not produce offspring during the first month of their lives, we see that the number of newly born pairs for the kth month is the number F_{k-2} of pairs alive two months before. Thus we can write the equation above as

$$F_k = F_{k-1} + F_{k-2}. \quad \textbf{Fibonacci's relation} \tag{1}$$

It is convenient to set $F_0 = 0$, denoting 0 pairs for month 0 before the arrival of the first newly born pair, which is presumably a gift. Thus the sequence

$$F_0, F_1, F_2, \ldots, F_k, \ldots$$

for the number of pairs of rabbits becomes the **Fibonacci sequence**

$$0, 1, 1, 2, 3, 5, 8, 13, 21, 34, \ldots, \tag{2}$$

where each term starting with $F_2 = 0 + 1 = 1$ is the sum of the two preceding terms. For any particular k, we could compute F_k by writing out the sequence far enough. ■

Fibonacci published this problem early in the thirteenth century. The Fibonacci sequence (2) occurs naturally in a surprising number of places. For example, leaves appear in a spiral pattern along a branch. Some trees have 5 growths of leaves for every 2 turns, others have 8 growths for every 3 turns, and still others have 13 growths for every 5 turns; note the appearance of these numbers in the sequence (2). A mathematical journal, the *Fibonacci Quarterly*, has published many papers dealing with the Fibonacci sequence.

We said in Example 1 that F_k can be found by simply writing out enough terms of the sequence (2). That can be a tedious task, even if we want to compute only F_{30}. Linear algebra gives us another approach to this problem. The Fibonacci relation (1) can be expressed in matrix form. We see that

$$\begin{bmatrix} F_k \\ F_{k-1} \end{bmatrix} = \begin{bmatrix} 1 & 1 \\ 1 & 0 \end{bmatrix} \begin{bmatrix} F_{k-1} \\ F_{k-2} \end{bmatrix}.$$

Thus, if we set

$$\mathbf{x}_k = \begin{bmatrix} F_k \\ F_{k-1} \end{bmatrix} \quad \text{and} \quad A = \begin{bmatrix} 1 & 1 \\ 1 & 0 \end{bmatrix},$$

we find that

$$\mathbf{x}_k = A\mathbf{x}_{k-1}. \tag{3}$$

Applying Eq. (3) repeatedly, we see that

$$\mathbf{x}_2 = A\mathbf{x}_1, \quad \mathbf{x}_3 = A\mathbf{x}_2 = A^2\mathbf{x}_1, \quad \mathbf{x}_4 = A\mathbf{x}_3 = A^3\mathbf{x}_1,$$

and in general

$$\mathbf{x}_k = A^{k-1}\mathbf{x}_1 = \begin{bmatrix} 1 & 1 \\ 1 & 0 \end{bmatrix}^{k-1}\begin{bmatrix} 1 \\ 0 \end{bmatrix}. \tag{4}$$

Thus we can compute the kth Fibonacci number F_k by finding A^{k-1} and multiplying it on the right by the column vector $\mathbf{x}_1$. Raising a matrix to a power is also a bit of a job, but the program MATCOMP can easily find F_{30} for us. (See Exercise 41.)

Both Markov chains and the Fibonacci sequence lead us to computations of the form $A^k\mathbf{x}$. Other examples leading to $A^k\mathbf{x}$ abound in the physical and social sciences.

> Computations of $A^k\mathbf{x}$ arise in any process in which information given by a column vector gives rise to analogous information at a later time by multiplying the vector by a matrix A.

Suppose that A is an $n \times n$ matrix and $\mathbf{v}$ is a column vector with n components such that

$$A\mathbf{v} = \lambda\mathbf{v} \tag{5}$$

for some scalar λ. Then

$$A^2\mathbf{v} = A(A\mathbf{v}) = A(\lambda\mathbf{v}) = \lambda(A\mathbf{v}) = \lambda(\lambda\mathbf{v}) = \lambda^2\mathbf{v}.$$

It is easy to show, in general, that $A^k\mathbf{v} = \lambda^k\mathbf{v}$. (See Exercise 25.) Thus, $A^k\mathbf{x}$ is easily computed if $\mathbf{x}$ is equal to *this vector* $\mathbf{v}$. For many matrices A, the compuation of $A^k\mathbf{x}$ for a *general vector* $\mathbf{x}$ is greatly simplified by finding first all nonzero vectors $\mathbf{v}$ and scalars λ satisfying Eq. (5). In Section 4.3 we will illustrate how this works.

Geometrically, Eq. (5) asserts that $A\mathbf{v}$ is a vector parallel to $\mathbf{v}$. We turn our attention to finding such vectors $\mathbf{v}$ and scalars λ.

Eigenvalues and Eigenvectors

DEFINITION 4.1 Eigenvalues and Eigenvectors

Let A be an $n \times n$ matrix. A scalar λ is an **eigenvalue** of A if there is a *nonzero* column vector $\mathbf{v}$ in $\mathbb{R}^n$ such that $A\mathbf{v} = \lambda\mathbf{v}$. The vector $\mathbf{v}$ is then an **eigenvector** of A corresponding to λ. (The terms **characteristic vector** and **characteristic value** or **proper vector** and **proper value** are also used in place of *eigenvector* and *eigenvalue*, respectively.)

For example, the computations

$$\begin{bmatrix} 2 & 2 \\ 3 & 1 \end{bmatrix}\begin{bmatrix} 1 \\ 1 \end{bmatrix} = \begin{bmatrix} 4 \\ 4 \end{bmatrix} = 4\begin{bmatrix} 1 \\ 1 \end{bmatrix}$$

show that the vector $\begin{bmatrix} 1 \\ 1 \end{bmatrix}$ is an eigenvector of the matrix $\begin{bmatrix} 2 & 2 \\ 3 & 1 \end{bmatrix}$ corresponding to the eigenvalue 4.

In this section we show how a determinant can be used to find eigenvalues; the technique is practical only for relatively small matrices. Some further computational techniques for finding eigenvalues are described in Section 7.4.

We write the equation $A\mathbf{v} = \lambda\mathbf{v}$ as $A\mathbf{v} - \lambda\mathbf{v} = \mathbf{0}$, or as $A\mathbf{v} - \lambda I\mathbf{v} = \mathbf{0}$, where I is the $n \times n$ identity matrix. This last equation can be written as $(A - \lambda I)\mathbf{v} = \mathbf{0}$, so $\mathbf{v}$ must be a solution of the homogeneous linear system

$$(A - \lambda I)\mathbf{x} = \mathbf{0}. \qquad (6)$$

non-zero

An eigenvalue of A is thus a scalar λ for which system (6) has a nontrivial solution $\mathbf{v}$. (Recall that an eigenvector is nonzero by definition.) We know that system (6) has a nontrivial solution precisely when the determinant of the coefficient matrix is zero— that is, if and only if

$$\det(A - \lambda I) = 0. \qquad (7)$$

If $A = [a_{ij}]$, then the previous equation can be written

$$\begin{vmatrix} a_{11} - \lambda & a_{12} & \cdots & a_{1n} \\ a_{21} & a_{22} - \lambda & \cdots & a_{2n} \\ \vdots & \vdots & & \vdots \\ a_{n1} & a_{n2} & \cdots & a_{nn} - \lambda \end{vmatrix} = 0. \qquad (8)$$

If we expand the determinant in Eq. (8), we obtain a polynomial expression $p(\lambda)$ of degree n with coefficients involving the a_{ij}. That is,

$$\det(A - \lambda I) = p(\lambda).$$

THE FIRST APPEARANCE OF EIGENVALUES occurred in connection with their use in solving differential equations (see Section 4.3). In 1743 Leonhard Euler first introduced the standard method of solving an nth-order differential equation with constant coefficients

$$y^{(n)} + a_{n-1}y^{(n-1)} + \cdots + a_1 y' + a_0 y = 0,$$

by using functions of the form $y = e^{\lambda t}$, where λ is a root of the characteristic equation

$$z^n + a_{n-1}z^{n-1} + \cdots + a_1 z + a_0 = 0.$$

This is the same equation one gets by making the substitutions $y_1 = y$, $y_2 = y'$, $y_3 = y''$, ..., $y_n = y^{(n-1)}$, replacing the single nth-order equation by a system of n first-order equations

$$\begin{aligned} y_1' &= y_2 \\ y_2' &= y_3 \\ &\cdots \\ y_n' &= -a_0 y_1 - a_1 y_2 - \cdots - a_{n-1} y_n, \end{aligned}$$

and calculating the characteristic equation of the matrix of coefficients of this system.

About 20 years later, Lagrange gave a more explicit version of this same idea when he found the solution of a system of differential equations by finding the roots of what amounted to the characteristic equation of the matrix of coefficients. The particular system of differential equations came from examining the "infinitesimal movements" of a mechanical system in the neighborhood of its position of equilibrium. Lagrange solved a similar problem in celestial mechanics by using the same technique in 1774.

The polynomial $p(\lambda)$ is the **characteristic polynomial** of the matrix A. The eigenvalues of A are precisely the solutions of the **characteristic equation** $p(\lambda) = 0$.

EXAMPLE 2   Find the eigenvalues of the matrix

$$A = \begin{bmatrix} 3 & 2 \\ 2 & 0 \end{bmatrix}.$$

Solution   The characteristic polynomial of A is

$$\det(A - \lambda I) = \begin{vmatrix} 3 - \lambda & 2 \\ 2 & -\lambda \end{vmatrix} = \lambda^2 - 3\lambda - 4.$$

The characteristic equation is

$$\lambda^2 - 3\lambda - 4 = 0,$$

and we obtain $(\lambda - 4)(\lambda + 1) = 0$; therefore $\lambda_1 = -1$ and $\lambda_2 = 4$ are eigenvalues of A.   ■

EXAMPLE 3   Show that $\lambda_1 = 1$ is an eigenvalue of the transition matrix for any Markov chain.

Solution   Let T be an $n \times n$ transition matrix for a Markov chain; that is, all entries in T are nonnegative and the sum of the entries in each column of T is 1. We easily see that the sum of the entries in any column of $T - I$ must be zero. Thus the sum of the row vectors in T is the zero vector, so the rows of $T - I$ are linearly dependent. Consequently, the rank of $T - I$ is less than n, and the equation $(T - I)\mathbf{x} = 0$ has a nontrivial solution, so $\lambda_1 = 1$ is an eigenvalue of T.   ■

The characteristic equation of an $n \times n$ matrix is a polynomial equation of degree n. This equation has n solutions if we allow both real and complex numbers and if we count the possible multiplicities greater than 1 of some solutions. Linear algebra can be done using complex scalars as well as real scalars. Introducing complex scalars makes the theory simpler and illuminates behavior in the real-scalar case. However, pencil-and-paper computations involving complex numbers can be very laborious. Exercises 44 and 49 through 52 deal with some computations involving complex eigenvalues.

With the exception of the exercises just mentioned, we leave the discussion of complex scalars and complex eigenvalues to Chapter 8. Throughout this section, we will always mean *real eigenvalues* when we refer to *eigenvalues*. In fact, all the questions we study concerning matrices in this section will be considered only for real scalars. We will suppress the word *real* most of the time, for convenience. Later we will discuss similarity and diagonalization of matrices, and until we reach Chapter 8, we will always have in mind similarity and diagonalization using only real numbers as scalars.

The characteristic polynomial of a matrix A may have multiple roots; perhaps it has -2 as a root of multiplicity 1, and 5 as a root of multiplicity 2, corresponding to factors $(\lambda + 2)(\lambda - 5)^2$ of the characteristic polynomial. Suppose that these are the only roots. We will say that the eigenvalues are $\lambda_1 = -2$ and $\lambda_2 = \lambda_3 = 5$. That is, if there are a total of k roots of the characteristic polynomial, it is convenient for us to denote them by $\lambda_1, \lambda_2, \ldots, \lambda_k$ even if not all these roots are distinct. Our next example illustrates this.

EXAMPLE 4 Find the eigenvalues of the matrix

$$A = \begin{bmatrix} 2 & 1 & 0 \\ -1 & 0 & 1 \\ 1 & 3 & 1 \end{bmatrix}.$$

Solution The characteristic polynomial is

$$p(\lambda) = \begin{vmatrix} 2 - \lambda & 1 & 0 \\ -1 & -\lambda & 1 \\ 1 & 3 & 1 - \lambda \end{vmatrix} = (2 - \lambda) \begin{vmatrix} -\lambda & 1 \\ 3 & 1 - \lambda \end{vmatrix} - 1 \begin{vmatrix} -1 & 1 \\ 1 & 1 - \lambda \end{vmatrix}$$

$$= (2 - \lambda)(\lambda^2 - \lambda - 3) - (\lambda - 2) = -(\lambda - 2)(\lambda^2 - \lambda - 2)$$

$$= -(\lambda - 2)(\lambda - 2)(\lambda + 1).$$

Hence, the eigenvalues of A are $\lambda_1 = -1$ and $\lambda_2 = \lambda_3 = 2$. ▪

Computation of Eigenvectors

We turn to the computation of the eigenvectors corresponding to an eigenvalue λ of a matrix A. Having found the eigenvalue, we substitute it in homogeneous system (6) and solve to find the nontrivial solutions of the system. We will obtain an infinite number of nontrivial solutions, each of which is an eigenvector corresponding to the eigenvalue λ.

EXAMPLE 5 Find the eigenvectors corresponding to each eigenvalue found in Example 4 for the matrix

$$A = \begin{bmatrix} 2 & 1 & 0 \\ -1 & 0 & 1 \\ 1 & 3 & 1 \end{bmatrix}.$$

Solution The eigenvalues of A were found to be $\lambda_1 = -1$ and $\lambda_2 = \lambda_3 = 2$. We substitute each of these values in the homogeneous system (6). The eigenvectors are obtained by reducing the coefficient matrix $A - \lambda I$ in the augmented matrix for the system. For $\lambda_1 = -1$, we obtain

$$[A - \lambda_1 I \mid \mathbf{0}] = [A + I \mid \mathbf{0}]$$

$$= \begin{bmatrix} 3 & 1 & 0 & \big| & 0 \\ -1 & 1 & 1 & \big| & 0 \\ 1 & 3 & 2 & \big| & 0 \end{bmatrix} \sim \begin{bmatrix} 1 & 3 & 2 & \big| & 0 \\ 0 & 4 & 3 & \big| & 0 \\ 0 & -8 & -6 & \big| & 0 \end{bmatrix}$$

$$\sim \begin{bmatrix} 1 & 3 & 2 & \big| & 0 \\ 0 & 4 & 3 & \big| & 0 \\ 0 & 0 & 0 & \big| & 0 \end{bmatrix}.$$

The solution of the homogeneous system is given by

$$\begin{bmatrix} r/4 \\ -3r/4 \\ r \end{bmatrix} \quad \text{for any scalar } r.$$

Therefore,

$$\mathbf{v}_1 = \begin{bmatrix} r/4 \\ -3r/4 \\ r \end{bmatrix} = r\begin{bmatrix} \frac{1}{4} \\ -\frac{3}{4} \\ 1 \end{bmatrix} \quad \text{for any } \textit{nonzero} \text{ scalar } r$$

is an eigenvector corresponding to the eigenvalue $\lambda_1 = -1$. Replacing r by $4r$, we can express this result without fractions as

$$\mathbf{v}_1 = \begin{bmatrix} r \\ -3r \\ 4r \end{bmatrix} = r\begin{bmatrix} 1 \\ -3 \\ 4 \end{bmatrix} \quad \text{for any nonzero scalar } r.$$

For $\lambda_2 = 2$, we obtain

$$[A - \lambda_2 I \mid \mathbf{0}] = [A - 2I \mid \mathbf{0}]$$

$$= \begin{bmatrix} 0 & 1 & 0 & | & 0 \\ -1 & -2 & 1 & | & 0 \\ 1 & 3 & -1 & | & 0 \end{bmatrix} \sim \begin{bmatrix} 1 & 3 & -1 & | & 0 \\ 0 & 1 & 0 & | & 0 \\ 0 & 0 & 0 & | & 0 \end{bmatrix}.$$

This time we find that

$$\mathbf{v}_2 = \begin{bmatrix} s \\ 0 \\ s \end{bmatrix} = s\begin{bmatrix} 1 \\ 0 \\ 1 \end{bmatrix} \quad \text{for any nonzero scalar } s$$

is an eigenvector. As a check, we could compute $A\mathbf{v}_1$ and $A\mathbf{v}_2$. For example, we have

$$A\mathbf{v}_2 = \begin{bmatrix} 2 & 1 & 0 \\ -1 & 0 & 1 \\ 1 & 3 & 1 \end{bmatrix}\begin{bmatrix} s \\ 0 \\ s \end{bmatrix} = \begin{bmatrix} 2s \\ 0 \\ 2s \end{bmatrix} = 2\begin{bmatrix} s \\ 0 \\ s \end{bmatrix} = 2\mathbf{v}_2 = \lambda_2\mathbf{v}_2. \qquad \blacksquare$$

EXAMPLE 6 Find the eigenvalues and eigenvectors of the matrix

$$A = \begin{bmatrix} 1 & 0 & 0 \\ -8 & 4 & -6 \\ 8 & 1 & 9 \end{bmatrix}.$$

Solution The characteristic polynomial of A is

$$p(\lambda) = |A - \lambda I| = \begin{vmatrix} 1-\lambda & 0 & 0 \\ -8 & 4-\lambda & -6 \\ 8 & 1 & 9-\lambda \end{vmatrix} = (1-\lambda)\begin{vmatrix} 4-\lambda & -6 \\ 1 & 9-\lambda \end{vmatrix}$$

$$= (1-\lambda)(\lambda^2 - 13\lambda + 42) = (1-\lambda)(\lambda - 6)(\lambda - 7).$$

The eigenvalues of A are $\lambda_1 = 1$, $\lambda_2 = 6$, and $\lambda_3 = 7$. We will drop the augmentation of $A - \lambda I$ by the column of zeros as we compute the eigenvectors for each of these eigenvalues. For $\lambda_1 = 1$, we have

$$A - \lambda_1 I = A - I = \begin{bmatrix} 0 & 0 & 0 \\ -8 & 3 & -6 \\ 8 & 1 & 8 \end{bmatrix} \sim \begin{bmatrix} 0 & 0 & 0 \\ 0 & 4 & 2 \\ 8 & 1 & 8 \end{bmatrix} \sim \begin{bmatrix} 8 & 1 & 8 \\ 0 & 2 & 1 \\ 0 & 0 & 0 \end{bmatrix},$$

so

$$\mathbf{v}_1 = \begin{bmatrix} -15r/16 \\ -r/2 \\ r \end{bmatrix} = r \begin{bmatrix} -\frac{15}{16} \\ -\frac{1}{2} \\ 1 \end{bmatrix} \quad \text{for any nonzero scalar } r$$

is an eigenvector. Replacing r by $-16r$, we can express this result without the fractions as

$$\mathbf{v}_1 = \begin{bmatrix} 15r \\ 8r \\ -16r \end{bmatrix} = r \begin{bmatrix} 15 \\ 8 \\ -16 \end{bmatrix} \quad \text{for any nonzero scalar } r.$$

For $\lambda_2 = 6$, we have

$$A - \lambda_2 I = A - 6I = \begin{bmatrix} -5 & 0 & 0 \\ -8 & -2 & -6 \\ 8 & 1 & 3 \end{bmatrix} \sim \begin{bmatrix} 1 & 0 & 0 \\ 0 & -2 & -6 \\ 0 & 1 & 3 \end{bmatrix} \sim \begin{bmatrix} 1 & 0 & 0 \\ 0 & 1 & 3 \\ 0 & 0 & 0 \end{bmatrix},$$

so

$$\mathbf{v}_2 = \begin{bmatrix} 0 \\ -3s \\ s \end{bmatrix} = s \begin{bmatrix} 0 \\ -3 \\ 1 \end{bmatrix} \quad \text{for any nonzero scalar } s$$

is an eigenvector. Finally, for $\lambda_3 = 7$, we have

$$A - \lambda_3 I = A - 7I = \begin{bmatrix} -6 & 0 & 0 \\ -8 & -3 & -6 \\ 8 & 1 & 2 \end{bmatrix} \sim \begin{bmatrix} 1 & 0 & 0 \\ 0 & -3 & -6 \\ 0 & 1 & 2 \end{bmatrix} \sim \begin{bmatrix} 1 & 0 & 0 \\ 0 & 1 & 2 \\ 0 & 0 & 0 \end{bmatrix},$$

so

$$\mathbf{v}_3 = \begin{bmatrix} 0 \\ -2t \\ t \end{bmatrix} = t \begin{bmatrix} 0 \\ -2 \\ 1 \end{bmatrix} \quad \text{for any nonzero scalar } t$$

is an eigenvector. ■

Properties of Eigenvalues and Eigenvectors

We turn now to algebraic properties of eigenvalues and eigenvectors. The properties given in Theorem 4.1 are so easy to prove that we leave the proofs as Exercises 25–27.

THEOREM 4.1   Properties of Eigenvalues and Eigenvectors

Let A be an $n \times n$ matrix.

1. If λ is an eigenvalue of A with $\mathbf{v}$ as a corresponding eigenvector, then λ^k is an eigenvalue of A^k, again with $\mathbf{v}$ as a corresponding eigenvector, for any positive integer k.

2. If λ is an eigenvalue of an invertible matrix A with $\mathbf{v}$ as a corresponding eigenvector, then $1/\lambda$ is an eigenvalue of A^{-1}, again with $\mathbf{v}$ as a corresponding eigenvector.

3. If λ is an eigenvalue of A, then the set E_λ consisting of the zero vector together with all eigenvectors of A for this eigenvalue λ is a subspace of $\mathbb{R}^n$, the **eigenspace** of λ.

Eigenvalues and Transformations

Let us consider the significance of an eigenvalue λ and corresponding eigenvector $\mathbf{v}$ of an $n \times n$ matrix A for the associated linear transformation $T(\mathbf{x}) = A\mathbf{x}$. The equation

$$A\mathbf{v} = \lambda\mathbf{v}$$

takes the form

$$T(\mathbf{v}) = \lambda\mathbf{v}.$$

Thus, the linear transformation T maps the vector $\mathbf{v}$ onto a vector that is parallel to $\mathbf{v}$. (See Fig. 4.1.) We present a definition of eigenvalues and eigenvectors for linear transformations that is more general than for matrices, in that we need not restrict ourselves to a finite-dimensional vector space.

DEFINITION 4.2   Eigenvalues and Eigenvectors

Let T be a linear transformation of a vector space V into itself. A scalar λ is an **eigenvalue** of T if there is a *nonzero* vector $\mathbf{v}$ in V such that $T(\mathbf{v}) = \lambda\mathbf{v}$. The vector $\mathbf{v}$ is then an **eigenvector** of T corresponding to λ.

It is significant that we can define eigenvalues and eigenvectors for a linear transformation $T: V \rightarrow V$ without any reference to a matrix representation and without even assuming that V is finite-dimensional. Example 8 will discuss the eigenvalues and eigenvectors of a linear transformation that lies at the heart of calculus and deals with an infinite-dimensional vector space. Students who have studied exponential growth problems in calculus will recognize the importance of this example.

Not every linear transformation has eigenvectors. For example, rotation of the plane counterclockwise through a positive angle α is a linear transformation (see page 174). If $0 < \alpha < 180°$, then no vector is mapped onto one parallel to it, that is, no

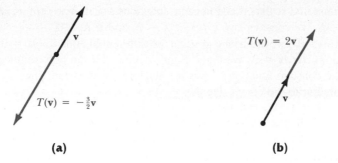

FIGURE 4.1 (a) **v** has eigenvalue $\lambda = -\frac{3}{2}$; (b) **v** has eigenvalue $\lambda = 2$.

vector is an eigenvector. If $\alpha = 180°$, then every nonzero vector is an eigenvector, and they all have the same associated eigenvalue $\lambda_1 = -1$.

For a linear transformation $T: \mathbb{R}^n \to \mathbb{R}^n$, we can find any of the transformation's eigenvalues and eigenvectors by finding those of its standard matrix representation. The following example illustrates how this works.

EXAMPLE 7 Find the eigenvalues λ and eigenvectors **v** of the linear transformation $T: \mathbb{R}^3 \to \mathbb{R}^3$ defined by $T(x_1, x_2, x_3) = (x_1, -8x_1 + 4x_2 - 6x_3, 8x_1 + x_2 + 9x_3)$. Illustrate the equation $T(\mathbf{v}) = \lambda\mathbf{v}$ for each eigenvalue.

$$A = \begin{bmatrix} 1 & 0 & 0 \\ -8 & 4 & -6 \\ 8 & 1 & 9 \end{bmatrix}$$

Solution Writing vectors as column vectors, we may express the linear transformation T as $T(\mathbf{x}) = A\mathbf{x}$, where A is the matrix given in Example 6. The rest of the solution proceeds precisely as in that example. Let us return to row notation and illustrate the action of T on the basic eigenvectors obtained by taking $r = s = t = 1$ in the expressions for $\mathbf{v}_1$, $\mathbf{v}_2$, and $\mathbf{v}_3$ in Example 6. We have

$$\lambda_1 = 1$$

$$T(15, 8, -16) = (15, -8(15) + 4(8) - 6(-16), 8(15) + 8 + 9(-16))$$
$$= (15, 8, -16),$$

$$\lambda_2 = 6$$

$$T(0, -3, 1) = (0, -8(0) + 4(-3) - 6(1), 8(0) - 3 + 9(1))$$
$$= (0, -18, 6) = 6(0, -3, 1),$$

$$\lambda_3 = 7$$

$$T(0, -2, 1) = (0, -8(0) + 4(-2) - 6(1), 8(0) - 2 + 9(1))$$
$$= (0, -14, 7) = 7(0, -2, 1). \qquad ▪$$

$$\begin{bmatrix} 1 & 0 & 0 \\ -8 & 4 & -6 \\ 8 & 1 & 9 \end{bmatrix} \begin{bmatrix} 15 \\ 8 \\ -16 \end{bmatrix} = \begin{bmatrix} 15 \\ 8 \\ -16 \end{bmatrix}$$

We now give an example from calculus, involving a vector space that is not finite-dimensional.

EXAMPLE 8 (*Calculus*) Let W be the vector space of all functions mapping $\mathbb{R}$ into $\mathbb{R}$ and having derivatives of all orders. Let $T: W \to W$ be the differentiation map, so that $T(f) = f'$, the derivative of f. Describe all eigenvalues and eigenvectors of T. (We have seen in Section 2.9 that differentiation does give a linear transformation.)

Solution   We must find scalars λ and nonzero functions f such that $T(f) = \lambda f$, that is, such that $f' = \lambda f$. We consider two cases: if $\lambda = 0$, and if $\lambda \neq 0$.

If $\lambda = 0$, we are trying to solve the differential equation $f' = 0$, or to use Leibniz notation, $dy/dx = 0$. We know from calculus that the only solutions of this equation are the constant functions. Thus the nonzero constant functions are eigenvectors corresponding to the eigenvalue 0.

If $\lambda \neq 0$, the differential equation becomes $f' = \lambda f$ or $dy/dx = \lambda y$. It can be confirmed that $y = e^{\lambda x}$ is a solution of this equation, so $f(x) = ke^{\lambda x}$ is an eigenvector for every nonzero scalar k. To see that these are the only solutions, you can solve the differential equation by separating variables, which yields the equation

$$\frac{dy}{y} = \lambda \, dx.$$

Integrating both sides of the equation yields

$$\ln|y| = \lambda x + c.$$

Solving for y, we obtain $y = \pm e^c e^{\lambda x} = ke^{\lambda x}$, so the only solutions of the differential equation are indeed of the form $y = ke^{\lambda x}$. ■

SUMMARY

Let A be an $n \times n$ matrix.

1. If $A\mathbf{v} = \lambda\mathbf{v}$, where $\mathbf{v}$ is a nonzero column vector and λ is a scalar, then λ is an eigenvalue of A and $\mathbf{v}$ is an eigenvector of A corresponding to λ.

2. The characteristic polynomial $p(\lambda)$ of A is obtained by expanding the determinant $|A - \lambda I|$, where I is the $n \times n$ identity matrix.

3. The eigenvalues λ of A can be found by solving the characteristic equation $p(\lambda) = |A - \lambda I| = 0$. There are at most n real solutions λ of this equation.

4. The eigenvectors of A corresponding to λ are the nontrivial solutions of the homogeneous system $(A - \lambda I)\mathbf{x} = \mathbf{0}$, as illustrated in Examples 5 and 6.

5. Let k be a positive integer. If λ is an eigenvalue of A having $\mathbf{v}$ as eigenvector, then λ^k is an eigenvalue of A^k with $\mathbf{v}$ as eigenvector. If A is invertible, this statement is also true for $k = -1$.

6. Let λ be an eigenvalue of A. The set E_λ in $\mathbb{R}^n$ consisting of the zero vector and all eigenvectors for λ is a subspace of $\mathbb{R}^n$.

7. Let T be a linear transformation of a vector space V into itself. A nonzero $\mathbf{v}$ in V is an eigenvector of T if $T(\mathbf{v}) = \lambda\mathbf{v}$ for some scalar λ, which is called the eigenvalue of T corresponding to $\mathbf{v}$.

8. If T is a linear transformation of the vector space $\mathbb{R}^n$ into itself, the eigenvalues of T are the eigenvalues of the standard matrix representation of T.

EXERCISES

In Exercises 1–16, find the characteristic polynomial, the real eigenvalues, and the corresponding eigenvectors of the given matrix.

1. $\begin{bmatrix} 1 & 0 \\ 1 & 2 \end{bmatrix}$ 2. $\begin{bmatrix} 7 & 5 \\ -10 & -8 \end{bmatrix}$

3. $\begin{bmatrix} -1 & -2 \\ 4 & 5 \end{bmatrix}$ 4. $\begin{bmatrix} -7 & -5 \\ 16 & 17 \end{bmatrix}$

5. $\begin{bmatrix} 0 & -1 \\ 1 & 0 \end{bmatrix}$ 6. $\begin{bmatrix} 1 & -2 \\ 1 & 2 \end{bmatrix}$

7. $\begin{bmatrix} 2 & 0 & 0 \\ 1 & -1 & -2 \\ -1 & 0 & 1 \end{bmatrix}$ 8. $\begin{bmatrix} -1 & 0 & 0 \\ -4 & 2 & -1 \\ 4 & 0 & 3 \end{bmatrix}$

9. $\begin{bmatrix} 8 & 0 & 0 \\ 7 & -1 & -2 \\ -7 & 0 & 1 \end{bmatrix}$ 10. $\begin{bmatrix} 1 & 0 & 0 \\ -8 & 4 & -5 \\ 8 & 0 & 9 \end{bmatrix}$

11. $\begin{bmatrix} -2 & 0 & 0 \\ -5 & -2 & -5 \\ 5 & 0 & 3 \end{bmatrix}$ 12. $\begin{bmatrix} -4 & 0 & 0 \\ -7 & 2 & -1 \\ 7 & 0 & 3 \end{bmatrix}$

13. $\begin{bmatrix} -1 & 0 & 1 \\ -7 & 2 & 5 \\ 3 & 0 & 1 \end{bmatrix}$ 14. $\begin{bmatrix} 4 & 0 & 0 \\ 8 & 4 & 8 \\ 0 & 0 & 4 \end{bmatrix}$

15. $\begin{bmatrix} 0 & 0 & 1 \\ -2 & -2 & 1 \\ 2 & 0 & -1 \end{bmatrix}$ 16. $\begin{bmatrix} 2 & 0 & 1 \\ 6 & 4 & -3 \\ 2 & 0 & 3 \end{bmatrix}$

In Exercises 17–22, find the eigenvalues λ_i and the corresponding eigenvectors $\mathbf{v}_i$ of the linear transformation T.

17. T defined on $\mathbb{R}^2$ by $T(x, y) = (2x - 3y, -3x + 2y)$

18. T defined on $\mathbb{R}^2$ by $T(x, y) = (x - y, -x + y)$

19. T defined on $\mathbb{R}^3$ by $T(x_1, x_2, x_3) = (x_1 + x_3, x_2, x_1 + x_3)$

20. T defined on $\mathbb{R}^3$ by $T(x_1, x_2, x_3) = (x_1, 4x_2 + 7x_3, 2x_2 - x_3)$

21. T defined on $\mathbb{R}^3$ by $T(x_1, x_2, x_3) = (x_1, -5x_1 + 3x_2 - 5x_3, -3x_1 - 2x_2)$

22. T defined on $\mathbb{R}^3$ by $T(x_1, x_2, x_3) = (3x_1 - x_2 + x_3, -2x_1 + 2x_2 - x_3, 2x_1 + x_2 + 4x_3)$

23. Mark each of the following True or False.
____ a) Every square matrix has real eigenvalues.
____ b) Every $n \times n$ matrix has n distinct (possibly complex) eigenvalues.
____ c) Every $n \times n$ matrix has n not necessarily distinct and possibly complex eigenvalues.
____ d) There can be only one eigenvalue associated with an eigenvector of a linear transformation.
____ e) There can be only one eigenvector associated with an eigenvalue of a linear transformation.
____ f) If $\mathbf{v}$ is an eigenvector of a matrix A, then $\mathbf{v}$ is an eigenvector of $A + cI$ for all scalars c.
____ g) If λ is an eigenvalue of a matrix A, then λ is an eigenvalue of $A + cI$ for all scalars c.
____ h) If $\mathbf{v}$ is an eigenvector of an invertible matrix A, then $c\mathbf{v}$ is an eigenvector of A^{-1} for all nonzero scalars c.
____ i) Every vector in a vector space V is an eigenvector of the identity transformation of V into V.
____ j) Every nonzero vector in a vector space V is an eigenvector of the identity transformation of V into V.

24. (*Calculus*) Following the idea in Example 8, show that the functions e^{ax}, e^{-ax}, $\sin ax$, and $\cos ax$ are eigenvectors for the linear transformation $T: W \to W$ defined by $T(f) = f^{(4)}$, the fourth derivative of f. Indicate the eigenvalue in each case.

25. Prove part (1) of Theorem 4.1.

26. Prove part (2) of Theorem 4.1.

27. Prove part (3) of Theorem 4.1.

28. Show that a square matrix is invertible if and only if no eigenvalue is zero.

29. Find the eigenvalues and eigenvectors of the matrix

$$A = \begin{bmatrix} 1 & 1 \\ 1 & 0 \end{bmatrix}$$

used to generate the Fibonacci sequence (2).

30. Let A be an $n \times n$ matrix and let I be the $n \times n$ identity matrix. Compare the eigenvectors and eigenvalues of A with those of $A + rI$ for a scalar r.

31. Let a square matrix A with real eigenvalues have a unique eigenvalue of greatest magnitude. Numerical computation of this eigenvalue by the *power method* (discussed in Section 7.4) can be difficult if there is another eigenvalue of almost equal magnitude, so the ratio of these magnitudes is close to 1.
    a) Suppose we know that a 4×4 matrix A has eigenvalues of approximately 20, 2, -3, and -19.5. Using Exercise 30, how might we modify A so that the aforementioned ratio is not so close to 1?
    b) Repeat part (a), given that the eigenvalues are known to be approximately 19.5, 2, -3, and -20.

32. Let A be an $n \times n$ matrix. An eigenvector $\mathbf{w}$ and a corresponding eigenvalue α of A^T are also called a *left eigenvector and eigenvalue* of A. Explain the reason for this name.

33. (*Principle of biorthogonality*) Let A be an $n \times n$ matrix. Let $\mathbf{v}$ be an eigenvector of A with corresponding eigenvalue λ, and let $\mathbf{w}$ be an eigenvector of A^T with corresponding eigenvalue α. Show that if $\lambda \neq \alpha$, then $\mathbf{v}$ and $\mathbf{w}$ are perpendicular vectors. [HINT: Refer to Exercise 32, and compute $\mathbf{w}^T A \mathbf{v}$ in two ways, using associativity of matrix multiplication.]

34. a) Show that the eigenvalues of an $n \times n$ matrix A are the same as the eigenvalues of A^T.
    b) With reference to part (a), show by an example that an eigenvector of A need not be an eigenvector of A^T.

35. The **trace** of an $n \times n$ matrix A is defined by
$$\text{tr}(A) = a_{11} + a_{22} + \cdots + a_{nn}.$$
Let the characteristic polynomial $p(\lambda)$ factor into linear factors, so that A has n (not necessarily distinct) eigenvalues $\lambda_1, \lambda_2, \ldots, \lambda_n$. Show that
$$\text{tr}(A) = (-1)^{n-1}(\text{Coefficient of } \lambda^{n-1} \text{ in } p(\lambda))$$
$$= \lambda_1 + \lambda_2 + \cdots + \lambda_n.$$

36. Let A be an $n \times n$ matrix, and let C be an invertible $n \times n$ matrix. Show that the eigenvalues of A and of $C^{-1}AC$ are the same. [HINT: Show that

the characteristic polynomials of the two matrices are the same.]

37. The Cayley–Hamilton theorem states that every square matrix A satisfies its characteristic equation. That is, if the characteristic equation is
$$p_n\lambda^n + p_{n-1}\lambda^{n-1} + \cdots + p_1\lambda + p_0 = 0, \text{ then}$$
$$p_n A^n + p_{n-1}A^{n-1} + \cdots + p_1A + p_0I = O,$$
the zero matrix. Illustrate the Cayley–Hamilton theorem for the matrix $\begin{bmatrix} 2 & -1 \\ 1 & 3 \end{bmatrix}$.

38. Let $\mathbf{v}_1$ and $\mathbf{v}_2$ be eigenvectors of a linear transformation $T: V \to V$ with corresponding eigenvalues λ_1 and λ_2, respectively. Show that, if $\lambda_1 \neq \lambda_2$, then $\mathbf{v}_1$ and $\mathbf{v}_2$ are independent vectors.

39. The analogue of Exercise 38 for a list of r eigenvectors in V having distinct eigenvalues is also true; that is, the vectors are independent. See if you can prove it. [HINT: Suppose that the vectors are dependent; consider the first vector in the list that is a linear combination of its predecessors, and apply T.]

40. State the result for matrices corresponding to Exercise 39. Explain why successful completion of Exercise 39 gives a proof of this statement for matrices.

41. Use MATCOMP, or similar software, and relation (4) to find the following terms of the Fibonacci sequence (2) as accurately as possible. (Use double-precision printing if possible.)
    a) F_8 (Note that $F_8 = 21$; this part is to check procedure.)
    b) F_{30}
    c) F_{50}
    d) F_{77}
    e) F_{150}

42. The first two terms of a sequence are $a_0 = 0$ and $a_1 = 1$. Subsequent terms are generated using the relation
$$a_k = 2a_{k-1} + a_{k-2} \quad \text{for} \quad k \geq 2.$$
    a) Write the terms of the sequence through a_8.
    b) Find a matrix that can be used to generate the sequence, as the matrix A in Exercise 29 can be used to generate the Fibonacci sequence.
    c) Use MATCOMP or similar software to find a_{30}.

43. Repeat Exercise 42 for a sequence where $a_0 = 0$, $a_1 = 1$, $a_2 = 2$, and $a_k = 2a_{k-1} - 3a_{k-2} + a_{k-3}$ for $k \geq 3$.

44. Let

$$A = \begin{bmatrix} 0 & -1 \\ 1 & 0 \end{bmatrix}$$

It can be shown that, if $\mathbf{x}$ is any column vector in $\mathbb{R}^2$, then $A\mathbf{x}$ can be obtained geometrically from $\mathbf{x}$ by rotating $\mathbf{x}$ counterclockwise through an angle of 90°. For example, we find that $A\mathbf{e}_1 = \mathbf{e}_2$ and $A\mathbf{e}_2 = -\mathbf{e}_1$.

a) Argue geometrically that A has no real eigenvalues.
b) Find the complex eigenvalues and eigenvectors of A.
c) Argue geometrically that A^2 should have real eigenvalues _____ . (Fill in the blank.)
d) Use part (b) and Theorem 4.1 to find the eigenvalues of A^2, and compare with the answer obtained for part (c). What are the real eigenvectors of A^2? the complex eigenvectors of A^2?
e) Find eigenvalues and eigenvectors for A^3.
f) Find eigenvalues and eigenvectors for A^4.

■ *In Exercises 45–48, use the program MATCOMP, or similar software, to find the real eigenvalues and corresponding eigenvectors of the given matrix.*

45. $\begin{bmatrix} 7 & 10 & 6 \\ 2 & -1 & -6 \\ -2 & -5 & 0 \end{bmatrix}$

46. $\begin{bmatrix} -1 & 0 & 0 & 0 \\ 3 & 2 & 0 & 0 \\ -3 & 0 & 2 & 0 \\ -3 & 1 & 0 & 3 \end{bmatrix}$

47. $\begin{bmatrix} 0 & 0 & -2 & 2 \\ -3 & 4 & -3 & 3 \\ -4 & 0 & 2 & 4 \\ -2 & 0 & 2 & 4 \end{bmatrix}$

48. $\begin{bmatrix} 4 & 0 & 0 & 0 \\ -6 & 16 & -6 & 6 \\ -16 & 0 & 20 & 16 \\ -16 & 0 & 16 & 20 \end{bmatrix}$

■ *The program ALLROOTS can be used to find both real and complex solutions of a polynomial equation. The program uses Newton's method. The program is designed so that the user can watch approximations approach a solution. Of course, a program designed for research would simply spit out the answers. In Exercises 49–52, first use MATCOMP to find the characteristic equation of the given matrix. Copy down the equation, and then use ALLROOTS to find both the real and the complex eigenvalues of the matrix.*

49. $\begin{bmatrix} -1 & 4 & 6 \\ 2 & 7 & 9 \\ -3 & 11 & 13 \end{bmatrix}$

50. $\begin{bmatrix} 10 & -13 & 8 \\ 3 & -20 & 5 \\ -11 & 7 & -6 \end{bmatrix}$

51. $\begin{bmatrix} -7 & 11 & -7 & 10 \\ 5 & 8 & -13 & 3 \\ -15 & 8 & -9 & 2 \\ 3 & -4 & 20 & -6 \end{bmatrix}$

52. $\begin{bmatrix} 21 & -8 & 0 & 32 \\ -14 & 17 & -6 & 9 \\ 15 & 11 & -13 & 16 \\ -18 & 30 & 43 & 31 \end{bmatrix}$

■ **53.** Use MATCOMP or similar software to illustrate the Cayley–Hamilton theorem for the matrix

$$\begin{bmatrix} -2 & 4 & 6 & -1 \\ 5 & -8 & 3 & 2 \\ 11 & -3 & 7 & 1 \\ 0 & -5 & 9 & 10 \end{bmatrix}.$$

(See Exercise 37.)

4.2 DIAGONALIZATION

A square matrix is called **diagonal** if all entries not on the main diagonal are zero. In the preceding section we indicated the importance of being able to compute $A^k\mathbf{x}$ for an $n \times n$ matrix A and a column vector $\mathbf{x}$ in $\mathbb{R}^n$. In this section, we show that, if A

has n distinct eigenvalues, then computation of A^k can be essentially replaced by computation of D^k, where D is a diagonal matrix with the eigenvalues of A as diagonal entries. Notice that D^k is the diagonal matrix obtained from D by raising each diagonal entry to the power k. For example,

$$\begin{bmatrix} 2 & 0 & 0 \\ 0 & -1 & 0 \\ 0 & 0 & -2 \end{bmatrix}^3 = \begin{bmatrix} 8 & 0 & 0 \\ 0 & -1 & 0 \\ 0 & 0 & -8 \end{bmatrix}.$$

The theorem that follows shows how to summarize the action of a square matrix on its eigenvectors in a single matrix equation. This theorem is the first step toward our goal of *diagonalizing* a matrix.

THEOREM 4.2   Matrix Summary of Eigenvalues of A

Let A be an $n \times n$ matrix having eigenvalues $\lambda_1, \lambda_2, \ldots, \lambda_n$. Let C be the $n \times n$ matrix having $\mathbf{v}_j$ as jth column vector, and let

$$D = \begin{bmatrix} \lambda_1 & & & \\ & \lambda_2 & & \mathbf{0} \\ & & \ddots & \\ \mathbf{0} & & & \lambda_n \end{bmatrix}.$$

Then $AC = CD$ if and only if $\mathbf{v}_j$ is an eigenvector of A corresponding to λ_j for $j = 1, 2, \ldots, n$.

PROOF   We have

$$CD = \begin{bmatrix} | & | & & | \\ \mathbf{v}_1 & \mathbf{v}_2 & \cdots & \mathbf{v}_n \\ | & | & & | \end{bmatrix} \begin{bmatrix} \lambda_1 & & & \\ & \lambda_2 & & \mathbf{0} \\ & & \ddots & \\ \mathbf{0} & & & \lambda_n \end{bmatrix}$$

$$= \begin{bmatrix} | & | & & | \\ \lambda_1\mathbf{v}_1 & \lambda_2\mathbf{v}_2 & \cdots & \lambda_n\mathbf{v}_n \\ | & | & & | \end{bmatrix}.$$

On the other hand,

$$AC = A \begin{bmatrix} | & | & & | \\ \mathbf{v}_1 & \mathbf{v}_2 & \cdots & \mathbf{v}_n \\ | & | & & | \end{bmatrix} = \begin{bmatrix} | & | & & | \\ A\mathbf{v}_1 & A\mathbf{v}_2 & \cdots & A\mathbf{v}_n \\ | & | & & | \end{bmatrix}.$$

Thus, $AC = CD$ if and only if $A\mathbf{v}_j = \lambda_j\mathbf{v}_j$. ▲

In Theorem 4.2, the matrix C whose column vectors are eigenvectors of A is invertible if and only if $\text{rank}(C) = n$, that is, if and only if n eigenvectors can be found that are linearly independent. In this case, the conclusion of Theorem 4.2 can be written as $D = C^{-1}AC$ and also as $A = CDC^{-1}$. From $A = CDC^{-1}$, we obtain

$$A^k = \underbrace{(CDC^{-1})(CDC^{-1})(CDC^{-1}) \cdots (CDC^{-1})}_{k \text{ factors}} \tag{1}$$

The adjacent terms $C^{-1}C$ cancel in Eq. (1) to give $A^k = CD^kC^{-1}$. Thus, the computation of A^k is essentially reduced to the computation of D^k. We summarize these observations as a corollary of Theorem 4.2.

COROLLARY 1   Computation of A^k

Let an $n \times n$ matrix A have n eigenvectors and eigenvalues, giving rise to the matrices C and D described in Theorem 4.2. If the n eigenvectors are independent, then C is an invertible matrix and $C^{-1}AC = D$. Under these conditions, we have

$$A^k = CD^kC^{-1}$$

DEFINITION 4.3   Diagonalizable Matrix

An $n \times n$ matrix A is **diagonalizable** if there exists an invertible matrix C such that $C^{-1}AC = D$, a diagonal matrix. The matrix C is said to **diagonalize** the matrix A.

Theorem 4.2 and the discussion following it give us another corollary.

COROLLARY 2   A Criterion for Diagonalization

An $n \times n$ matrix A is diagonalizable if and only if $\mathbb{R}^n$ has a basis consisting of eigenvectors of A.

Diagonalization of matrices plays a very important role in linear algebra. As we show in the next theorem, diagonalization of an $n \times n$ matrix A can always be achieved if the characteristic polynomial has n distinct roots.

THEOREM 4.3   Independence of Eigenvectors

Let A be an $n \times n$ matrix. If $\mathbf{v}_1, \mathbf{v}_2, \ldots, \mathbf{v}_n$ are eigenvectors of A corresponding to *distinct* eigenvalues $\lambda_1, \lambda_2, \ldots, \lambda_n$, respectively, the set $\{\mathbf{v}_1, \mathbf{v}_2, \ldots, \mathbf{v}_n\}$ is linearly independent and A is diagonalizable.

PROOF Suppose that the conclusion is false, so the eigenvectors $\mathbf{v}_1$, $\mathbf{v}_2$, . . . , $\mathbf{v}_n$ are linearly dependent. Then one of them is a linear combination of its predecessors. Let $\mathbf{v}_k$ be the first such vector, so that

$$\mathbf{v}_k = d_1\mathbf{v}_1 + d_2\mathbf{v}_2 + \cdots + d_{k-1}\mathbf{v}_{k-1} \tag{2}$$

and $\{\mathbf{v}_1, \mathbf{v}_2, . . . , \mathbf{v}_{k-1}\}$ is independent. Multiplying Eq. (2) by λ_k, we obtain

$$\lambda_k\mathbf{v}_k = d_1\lambda_k\mathbf{v}_1 + d_2\lambda_k\mathbf{v}_2 + \cdots + d_{k-1}\lambda_k\mathbf{v}_{k-1}. \tag{3}$$

On the other hand, multiplying both sides of Eq. (2) on the left by the matrix A yields

$$\lambda_k\mathbf{v}_k = d_1\lambda_1\mathbf{v}_1 + d_2\lambda_2\mathbf{v}_2 + \cdots + d_{k-1}\lambda_{k-1}\mathbf{v}_{k-1}, \tag{4}$$

since $A\mathbf{v}_i = \lambda_i\mathbf{v}_i$. Subtracting Eq. (4) from Eq. (3), we see that

$$\mathbf{0} = d_1(\lambda_k - \lambda_1)\mathbf{v}_1 + d_2(\lambda_k - \lambda_2)\mathbf{v}_2 + \cdots + d_{k-1}(\lambda_k - \lambda_{k-1})\mathbf{v}_{k-1}.$$

This last equation is a dependence relation because not all the coefficients are zero. (Not all d_i are zero because of Eq. (2) and since the λ_i are distinct.) But this contradicts the linear independence of the set $\{\mathbf{v}_1, \mathbf{v}_2, . . . , \mathbf{v}_{k-1}\}$. We conclude that $\{\mathbf{v}_1, \mathbf{v}_2, . . . , \mathbf{v}_n\}$ is independent. That A is diagonalizable follows at once from Corollary 2 of Theorem 4.2. ▲

EXAMPLE 1 Diagonalize the matrix $A = \begin{bmatrix} -3 & 5 \\ -2 & 4 \end{bmatrix}$, and compute A^k for each positive integer k.

Solution We compute

$$\det(A - \lambda I) = \begin{vmatrix} -3 - \lambda & 5 \\ -2 & 4 - \lambda \end{vmatrix} = \lambda^2 - \lambda - 2 = (\lambda - 2)(\lambda + 1).$$

The eigenvalues of A are $\lambda_1 = 2$ and $\lambda_2 = -1$. For $\lambda_1 = 2$, we have

$$A - 2I = \begin{bmatrix} -5 & 5 \\ -2 & 2 \end{bmatrix} \sim \begin{bmatrix} 1 & -1 \\ 0 & 0 \end{bmatrix},$$

which yields an eigenvector

$$\mathbf{v}_1 = \begin{bmatrix} 1 \\ 1 \end{bmatrix}.$$

For $\lambda_2 = -1$, we have

$$A + I = \begin{bmatrix} -2 & 5 \\ -2 & 5 \end{bmatrix} \sim \begin{bmatrix} 2 & -5 \\ 0 & 0 \end{bmatrix},$$

which yields an eigenvector

$$\mathbf{v}_2 = \begin{bmatrix} 5 \\ 2 \end{bmatrix}.$$

A diagonalization of A is given by

$$A = CDC^{-1} = \begin{bmatrix} 1 & 5 \\ 1 & 2 \end{bmatrix}\begin{bmatrix} 2 & 0 \\ 0 & -1 \end{bmatrix}\begin{bmatrix} -\frac{2}{3} & \frac{5}{3} \\ \frac{1}{3} & -\frac{1}{3} \end{bmatrix}.$$

(We omit the computation of C^{-1}.) Thus,

$$A^k = \begin{bmatrix} 1 & 5 \\ 1 & 2 \end{bmatrix} \begin{bmatrix} 2^k & 0 \\ 0 & (-1)^k \end{bmatrix} \begin{bmatrix} -\frac{2}{3} & \frac{5}{3} \\ \frac{1}{3} & -\frac{1}{3} \end{bmatrix} = \frac{1}{3} \begin{bmatrix} -2^{k+1} \pm 5 & 5(2^k) \mp 5 \\ -2^{k+1} \pm 2 & 5(2^k) \mp 2 \end{bmatrix},$$

the colored sign being used only when k is odd. ■

The preceding matrices A and D provide examples of *similar matrices*, a term we now define.

DEFINITION 4.4   Similar Matrices

An $n \times n$ matrix P is **similar** to an $n \times n$ matrix Q if there exists an invertible $n \times n$ matrix C such that $C^{-1}PC = Q$.

The relationship "P is similar to Q" satisfies the properties required for an *equivalence relation* (see Exercise 16). In particular, if P is similar to Q, then Q is also similar to P. This means that similarity need not be stated in a directional way; we can simply say that P and Q are similar matrices.

EXAMPLE 2   Find a diagonal matrix similar to the matrix

$$A = \begin{bmatrix} 1 & 0 & 0 \\ -8 & 4 & -6 \\ 8 & 1 & 9 \end{bmatrix}$$

of Example 6 in Section 4.1.

Solution   Taking $r = s = t = 1$ in Example 6 of Section 4.1, we see that eigenvalues and corresponding eigenvectors of A are given by

$$\lambda_1 = 1, \qquad \lambda_2 = 6, \qquad \lambda_3 = 7,$$

$$\mathbf{v}_1 = \begin{bmatrix} 15 \\ 8 \\ -16 \end{bmatrix}, \qquad \mathbf{v}_2 = \begin{bmatrix} 0 \\ -3 \\ 1 \end{bmatrix}, \qquad \mathbf{v}_3 = \begin{bmatrix} 0 \\ -2 \\ 1 \end{bmatrix}.$$

THE IDEA OF SIMILARITY, like many matrix notions, appears without a definition in works from as early as the 1820s. In fact, in his 1826 work on quadratric forms (see note on page 366), Cauchy showed that if two quadratic forms are related by a change of variables—that is, if their matrices are similar—then their characteristic equations are the same. But like the concept of orthogonality, that of similiarity was first formally defined and discussed by Georg Frobenius in 1878. Frobenius began by discussing the general case: he called two matrices A, D *equivalent* if there existed invertible matrices P, Q such that $D = PAQ$. The latter matrices were called the *substitutions* through which A was transformed into D.

Frobenius then dealt with the special cases where $P = Q^T$ (the two matrices were then called *congruent*) and where $P = Q^{-1}$ (the similarity case of this section). Frobenius went on to prove many results on similarity, including the useful theorem that, if A is similar to D, then $f(A)$ is similar to $f(D)$, where f is any polynomial matrix function.

If we let

$$C = \begin{bmatrix} 15 & 0 & 0 \\ 8 & -3 & -2 \\ -16 & 1 & 1 \end{bmatrix},$$

then Theorem 4.3 tells us that C is invertible. Theorem 4.2 then shows that

$$C^{-1}AC = D = \begin{bmatrix} 1 & 0 & 0 \\ 0 & 6 & 0 \\ 0 & 0 & 7 \end{bmatrix}.$$

$AC = CD$

Thus D is similar to A. We are not eager to check that $C^{-1}AC = D$; however, it is easy to check the equivalent statement:

$$AC = CD = \begin{bmatrix} 15 & 0 & 0 \\ 8 & -18 & -14 \\ -16 & 6 & 7 \end{bmatrix}. \qquad ■$$

It is not always essential that a matrix have distinct eigenvalues in order to be diagonalizable. As long as the $n \times n$ matrix A has n *independent* eigenvectors to form the column vectors of an invertible C, we will have $C^{-1}AC = D$, the diagonal matrix of the eigenvalues.

EXAMPLE 3   Diagonalize the matrix

$$A = \begin{bmatrix} 1 & -3 & 3 \\ 0 & -5 & 6 \\ 0 & -3 & 4 \end{bmatrix}.$$

Solution   We find that the characteristic equation of A is

$$(1 - \lambda)((-5 - \lambda)(4 - \lambda) + 18) = (1 - \lambda)(\lambda^2 + \lambda - 2)$$
$$= (1 - \lambda)(\lambda + 2)(\lambda - 1) = 0.$$

Thus, the eigenvalues of A are $\lambda_1 = 1$, $\lambda_2 = 1$, and $\lambda_3 = -2$. Notice that 1 is a root of multiplicity 2 of the characteristic equation; we say that the eigenvalue 1 has *algebraic multiplicity* 2.

Reducing $A - I$, we obtain

$$A - I = \begin{bmatrix} 0 & -3 & 3 \\ 0 & -6 & 6 \\ 0 & -3 & 3 \end{bmatrix} \sim \begin{bmatrix} 0 & 1 & -1 \\ 0 & 0 & 0 \\ 0 & 0 & 0 \end{bmatrix}.$$

We see that the eigenspace E_1 (that is, the nullspace of $A - I$) has dimension 2 and consists of vectors of the form

$$\begin{bmatrix} s \\ r \\ r \end{bmatrix} \qquad \text{for any scalars } r \text{ and } s.$$

Taking $s = 1$ and $r = 0$, and then taking $s = 0$ and $r = 1$, we obtain the independent eigenvectors

$$\mathbf{v}_1 = \begin{bmatrix} 1 \\ 0 \\ 0 \end{bmatrix} \quad \text{and} \quad \mathbf{v}_2 = \begin{bmatrix} 0 \\ 1 \\ 1 \end{bmatrix}$$

corresponding to the eigenvalues $\lambda_1 = \lambda_2 = 1$.

Reducing $A + 2I$, we find that

$$A + 2I = \begin{bmatrix} 3 & -3 & 3 \\ 0 & -3 & 6 \\ 0 & -3 & 6 \end{bmatrix} \sim \begin{bmatrix} 3 & 0 & -3 \\ 0 & 1 & -2 \\ 0 & 0 & 0 \end{bmatrix}.$$

Thus an eigenvector corresponding to $\lambda_3 = -2$ is

$$\mathbf{v}_3 = \begin{bmatrix} 1 \\ 2 \\ 1 \end{bmatrix}.$$

Therefore, if we take

$$C = \begin{bmatrix} 1 & 0 & 1 \\ 0 & 1 & 2 \\ 0 & 1 & 1 \end{bmatrix},$$

we should have

$$C^{-1}AC = D = \begin{bmatrix} 1 & 0 & 0 \\ 0 & 1 & 0 \\ 0 & 0 & -2 \end{bmatrix}.$$

A check shows that indeed

$$AC = CD = \begin{bmatrix} 1 & 0 & -2 \\ 0 & 1 & -4 \\ 0 & 1 & -2 \end{bmatrix}. \qquad ■$$

As we indicated in Example 3, the **algebraic multiplicity** of an eigenvalue λ_i of A is its multiplicity as a root of the characteristic equation of A. Its **geometric multiplicity** is the dimension of the eigenspace E_{λ_i}. Of course, the geometric multiplicity of each eigenvalue must be at least 1, since there always exists a nonzero eigenvector in the eigenspace. However, it is possible for the algebraic multiplicity to be greater than the geometric multiplicity.

EXAMPLE 4 Referring back to Examples 4 and 5 in Section 4.1, find the algebraic and geometric multiplicities of the eigenvalue 2 of the matrix

$$A = \begin{bmatrix} 2 & 1 & 0 \\ -1 & 0 & 1 \\ 1 & 3 & 1 \end{bmatrix}.$$

Solution   Example 4 on page 247 shows that the characteristic equation of A is $-(\lambda - 2)^2(\lambda + 1) = 0$, so 2 is an eigenvalue of algebraic multiplicity 2. Example 5 on page 247 shows that the reduced form of $A - 2I$ is

$$\begin{bmatrix} 1 & 3 & -1 \\ 0 & 1 & 0 \\ 0 & 0 & 0 \end{bmatrix}.$$

Thus the eigenspace E_2, which is the set of vectors of the form

$$\begin{bmatrix} r \\ 0 \\ r \end{bmatrix} \quad \text{for } r \in \mathbb{R},$$

has dimension 1, so the eigenvalue 2 has geometric multiplicity 1.   ■

We state a relationship between the algebraic multiplicity and the geometric multiplicity of a (possibly complex) eigenvalue. (See Exercise 33 of Section 8.4.)

> The geometric multiplicity of an eigenvalue of a matrix A is less than or equal to its algebraic multiplicity.

Let $\lambda_1, \lambda_2, \ldots, \lambda_m$ be the distinct (possibly complex) eigenvalues of an $n \times n$ matrix A. Let B_i be a basis for the eigenspace of λ_i for $i = 1, 2, \ldots, m$. It can be shown by an argument similar to the proof of Theorem 4.3 that the union of these bases B_i is an independent set of vectors in $\mathbb{R}^n$ (see Exercise 21). Corollary 2 on page 257 shows that the matrix A is diagonalizable if this union of the B_i is a basis for $\mathbb{R}^n$. This will occur precisely when the geometric multiplicity of each eigenvalue is equal to its algebraic multiplicity. (See the boxed statement.) Conversely, it can be shown that, <u>if A is diagonalizable, the algebraic multiplicity of each eigenvalue is the</u> <u>same as its geometric multiplicity.</u> We summarize this in a theorem.

THEOREM 4.4   A Criterion for Diagonalization

> An $n \times n$ matrix A is diagonalizable if and only if the algebraic multiplicity of each (possibly complex) eigenvalue is equal to its geometric multiplicity.

Thus the 3×3 matrix in Example 3 is diagonalizable, because its eigenvalue 1 has algebraic and geometric multiplicity 2, and its eigenvalue -2 has algebraic and geometric multiplicity 1. However, the matrix in Example 4 is not diagonalizable, because the eigenvalue 2 has algebraic multiplicity 2 but geometric multiplicity 1.

In Section 8.4, we show that every square matrix A is similar to a matrix J, its *Jordan canonical form*. If A is diagonalizable, then J is a diagonal matrix, found precisely as in the preceding examples. If A is not diagonalizable, then J again has the eigenvalues of A on its main diagonal, but it also has entries 1 immediately above some of the diagonal entries. The remaining entries are all zero. For example, the matrix

$$A = \begin{bmatrix} 2 & 1 & 0 \\ -1 & 0 & 1 \\ 1 & 3 & 1 \end{bmatrix}$$

of Example 4 has Jordan canonical form

$$J = \begin{bmatrix} -1 & 0 & 0 \\ 0 & 2 & 1 \\ 0 & 0 & 2 \end{bmatrix}.$$

Section 8.4 describes a technique for finding J. This Jordan canonical form is as close to a diagonalization of A as we can come. The Jordan canonical form has applications to the solution of systems of differential equations.

To conclude this discussion, we state a result whose proof requires an excursion into complex numbers. The proof is given in Chapter 8. Thus far, we have seen nothing to indicate that symmetric matrices play any significant role in linear algebra. The following theorem immediately elevates them into a position of prominence.

THEOREM 4.5 **Diagonalization of Real Symmetric Matrices**

Let A be an $n \times n$ (real) symmetric matrix. Then every root of the characteristic polynomial is a real number. The algebraic multiplicity of each eigenvalue of A equals its geometric multiplicity, so A is diagonalizable.

A diagonalizing matrix C for a symmetric matrix A can be chosen to have some very nice properties, as we will show for real symmetric matrices in Chapter 5.

SUMMARY

Let A be an $n \times n$ matrix.

1. If A has n distinct eigenvalues $\lambda_1, \lambda_2, \ldots, \lambda_n$, and C is an $n \times n$ matrix having as jth column vector an eigenvector corresponding to λ_j, then $C^{-1}AC$ is the diagonal matrix having λ_j on the main diagonal in the jth column.

2. If $C^{-1}AC = D$, then $A^k = CD^kC^{-1}$.

3. A matrix P is similar to a matrix Q if there exists an invertible matrix C such that $C^{-1}PC = Q$.

4. The algebraic multiplicity of an eigenvalue λ of A is its multiplicity as a root of the characteristic equation; its geometric multiplicity is the dimension of the corresponding eigenspace E_λ.

5. Any eigenvalue's geometric multiplicity is less than or equal to its algebraic multiplicity.

6. The matrix A is diagonalizable if and only if the geometric multiplicity of each of its eigenvalues is the same as the algebraic multiplicity.

7. Every symmetric matrix is diagonalizable. All eigenvalues of a real symmetric matrix are real numbers.

EXERCISES

1. Consider the matrices

$$A_1 = \begin{bmatrix} 1 & -1 & -1 \\ -1 & 1 & -1 \\ -1 & -1 & 1 \end{bmatrix}, A_2 = \begin{bmatrix} 3 & -2 & 0 \\ -2 & 3 & 0 \\ 0 & 0 & 5 \end{bmatrix}$$

and the vectors

$$\mathbf{v}_1 = \begin{bmatrix} 1 \\ 1 \\ 1 \end{bmatrix}, \mathbf{v}_2 = \begin{bmatrix} 0 \\ 0 \\ 1 \end{bmatrix}, \mathbf{v}_3 = \begin{bmatrix} -1 \\ 1 \\ 0 \end{bmatrix},$$

$$\mathbf{v}_4 = \begin{bmatrix} 1 \\ 1 \\ 0 \end{bmatrix}, \mathbf{v}_5 = \begin{bmatrix} -1 \\ 0 \\ 1 \end{bmatrix}.$$

List the vectors that are eigenvectors of A_1 and the ones that are eigenvectors of A_2. Give the eigenvalue in each case.

In Exercises 2–8, find the eigenvalues λ_i and the corresponding eigenvectors $\mathbf{v}_i$ of the given matrix A, and also find an invertible matrix C and a diagonal matrix D such that $D = C^{-1}AC$.

2. $A = \begin{bmatrix} 3 & 2 \\ 1 & 4 \end{bmatrix}$ 3. $A = \begin{bmatrix} 7 & 8 \\ -4 & -5 \end{bmatrix}$

4. $A = \begin{bmatrix} 6 & 3 & -3 \\ -2 & -1 & 2 \\ 16 & 8 & -7 \end{bmatrix}$ 5. $A = \begin{bmatrix} -3 & 10 & -6 \\ 0 & 7 & -6 \\ 0 & 0 & 1 \end{bmatrix}$

6. $A = \begin{bmatrix} -3 & 5 & -20 \\ 2 & 0 & 8 \\ 2 & 1 & 7 \end{bmatrix}$ 7. $A = \begin{bmatrix} -2 & 0 & -1 \\ 0 & 2 & 0 \\ 3 & 0 & 2 \end{bmatrix}$

8. $A = \begin{bmatrix} -4 & 6 & -12 \\ 3 & -1 & 6 \\ 3 & -3 & 8 \end{bmatrix}$

In Exercises 9–12, determine whether the given matrix is diagonalizable.

9. $\begin{bmatrix} 1 & 2 & 6 \\ 2 & 0 & -4 \\ 6 & -4 & 3 \end{bmatrix}$ 10. $\begin{bmatrix} 3 & 1 & 0 \\ 0 & 3 & 1 \\ 0 & 0 & 3 \end{bmatrix}$

11. $\begin{bmatrix} -1 & 4 & 2 & -7 \\ 0 & 5 & -3 & 6 \\ 0 & 0 & -5 & 1 \\ 0 & 0 & 0 & 11 \end{bmatrix}$

12. $\begin{bmatrix} 3 & 2 & 5 & 1 \\ 2 & 0 & 2 & 6 \\ 5 & 2 & 7 & -1 \\ 1 & 6 & -1 & 3 \end{bmatrix}$

13. Mark each of the following True or False.

____ a) Every $n \times n$ matrix is diagonalizable.

____ b) If an $n \times n$ matrix has n distinct real eigenvalues, it is diagonalizable.

____ c) Every $n \times n$ symmetric matrix is diagonalizable.

____ d) An $n \times n$ matrix is diagonalizable if and only if it has n distinct eigenvalues.

____ e) An $n \times n$ matrix is diagonalizable if and only if the algebraic multiplicity of each of its eigenvalues equals the geometric multiplicity.

____ f) Every invertible matrix is diagonalizable.

____ g) Every triangular matrix is diagonalizable.

____ h) If A and B are similar square matrices and A is diagonalizable, then B is also diagonalizable.

____ i) If an $n \times n$ matrix A is diagonalizable, there is a unique diagonal matrix D that is similar to A.

____ j) If A and B are similar square matrices, then $\det(A) = \det(B)$.

14. Give two different diagonal matrices that are similar to the matrix $\begin{bmatrix} 1 & 4 \\ 0 & -3 \end{bmatrix}$.

15. Show that, if a matrix is diagonalizable, so is its transpose.

16. Let P, Q, and R be $n \times n$ matrices. Recall that P is similar to Q if there exists an invertible $n \times n$ matrix C such that $C^{-1}PC = Q$. This exercise shows that similarity is an *equivalence relation*.
    a) (*Reflexive.*) Show that P is similar to itself.
    b) (*Symmetric.*) Show that, if P is similar to Q, then Q is similar to P.
    c) (*Transitive.*) Show that, if P is similar to Q and Q is similar to R, then P is similar to R.

17. Show that, for every square matrix A all of whose eigenvalues are real, the product of its eigenvalues is $\det(A)$.

18. Show that similar square matrices have the same eigenvalues with the same algebraic multiplicities.

19. Let A and C be $n \times n$ matrices, and let C be invertible. Show that, if $\mathbf{v}$ is an eigenvector of A with corresponding eigenvalue λ, then $C^{-1}\mathbf{v}$ is an eigenvector of $C^{-1}AC$ with corresponding eigenvalue λ. Then show that all eigenvectors of $C^{-1}AC$ are of the form $C^{-1}\mathbf{v}$, where $\mathbf{v}$ is an eigenvector of A.

20. Explain how we can deduce from Exercise 19 that, if A and B are similar square matrices, each eigenvalue of A has the same geometric multiplicity for A that it has for B. (See Exercise 18 for the corresponding statement on algebraic multiplicities.)

21. Show that, if $\lambda_1, \lambda_2, \ldots, \lambda_m$ are distinct real eigenvalues of an $n \times n$ matrix A and if B_i is a basis for the eigenspace E_{λ_i}, then the union of the bases B_i is an independent set of vectors in $\mathbb{R}^n$. [HINT: Make use of Theorem 4.3.]

22. Let $T: V \to V$ be a linear transformation of a vector space V into itself. Show that, if $\mathbf{v}_1, \mathbf{v}_2, \ldots, \mathbf{v}_k$ are eigenvectors of T corresponding to distinct nonzero eigenvalues $\lambda_1, \lambda_2, \ldots, \lambda_k$, then $T(\mathbf{v}_1)$, $T(\mathbf{v}_2), \ldots, T(\mathbf{v}_k)$ are independent. [HINT: See Exercise 39 of Section 4.1.]

23. (*Calculus*) Show that the functions $e^{\lambda_1 x}, e^{\lambda_2 x}, \ldots, e^{\lambda_k x}$, where the λ_i are distinct, are independent in the vector space W of all functions mapping $\mathbb{R}$ into $\mathbb{R}$ and having derivatives of all orders. [HINT: See Exercise 39 of Section 4.1.]

24. (*Calculus*) Using Exercise 23, show that the infinite set $\{e^{kx} \mid k \in \mathbb{R}\}$ is an independent set in the vector space W described in Exercise 23. [HINT: How many vectors are involved in any dependence relation?]

25. Using MATCOMP, compute C^{-1} to check that $C^{-1}AC = D$ in Example 2.

In Exercises 26–35, use MATCOMP or similar software to determine whether the given matrix is diagonalizable.

26. $\begin{bmatrix} 18 & 25 & -25 \\ 1 & 6 & -1 \\ 18 & 34 & -25 \end{bmatrix}$ 27. $\begin{bmatrix} 8.3 & 8.0 & -6.0 \\ -2.0 & 0.3 & 3.0 \\ 0.0 & 0.0 & 4.3 \end{bmatrix}$

28. $\begin{bmatrix} 13.7 & 34.8 & -11.6 \\ -5.8 & -15.3 & 5.8 \\ 0.0 & 0.0 & 2.1 \end{bmatrix}$

29. $\begin{bmatrix} 24.55 & 46.60 & 46.60 \\ -4.66 & -8.07 & -9.32 \\ -9.32 & -18.64 & -17.39 \end{bmatrix}$

30. $\begin{bmatrix} 7 & -20 & -5 & 5 \\ 5 & -13 & -5 & 0 \\ -5 & 10 & 7 & 5 \\ 5 & -10 & -5 & -3 \end{bmatrix}$

31. $\begin{bmatrix} -22.7 & -26.9 & -6.3 & -46.5 \\ -59.7 & -40.9 & 20.9 & -99.5 \\ 15.9 & 9.6 & -8.4 & 26.5 \\ 43.8 & 36.5 & -7.3 & 78.2 \end{bmatrix}$

32. $\begin{bmatrix} 66.2 & 58.0 & -11.6 & 116.0 \\ 120.6 & 89.6 & -42.6 & 201.0 \\ -21.0 & -15.0 & 7.6 & -35.0 \\ -99.6 & -79.0 & 28.6 & -169.4 \end{bmatrix}$

33. $\begin{bmatrix} -253 & -232 & -96 & 1088 & 280 \\ 213 & 204 & 93 & -879 & -225 \\ -90 & -90 & -47 & 360 & 90 \\ -38 & -36 & -18 & 162 & 40 \\ 62 & 64 & 42 & -251 & -57 \end{bmatrix}$

34. $\begin{bmatrix} 154 & -24 & -36 & -1608 & -336 \\ -126 & 16 & 18 & 1314 & 270 \\ 54 & 0 & 4 & -540 & -108 \\ 24 & 0 & 0 & -236 & -48 \\ -42 & -12 & -18 & 366 & 70 \end{bmatrix}$

35. $\begin{bmatrix} -2513 & 596 & -414 & -2583 & 1937 \\ 127 & -32 & 33 & 132 & -81 \\ -421 & 94 & -83 & -434 & 306 \\ 2610 & -615 & 443 & 2684 & -1994 \\ 90 & -19 & 29 & 94 & -50 \end{bmatrix}$

TWO APPLICATIONS

In Section 4.1, we were motivated to introduce eigenvalues by our desire to compute $A^k\mathbf{x}$. Recall that we regard $\mathbf{x}$ as an initial information vector of some process, and A is a matrix that transforms, by left multiplication, an information vector at any stage of the process into the information vector at the next stage. In our first application, we examine this computation of $A^k\mathbf{x}$ and the significance of the eigenvalues of A. As illustration, we determine the behavior of the terms F_n of the Fibonacci sequence for large values of n. Our second application shows how diagonalization of matrices can be used to solve some linear systems of differential equations.

Application: Computing $A^k\mathbf{x}$

Let A be a diagonalizable $n \times n$ matrix, and let $\lambda_1, \lambda_2, \ldots, \lambda_n$ be the n not necessarily distinct eigenvalues of A. That is, each eigenvalue of A is repeated in this list in accord with its algebraic multiplicity. Let $B = (\mathbf{v}_1, \mathbf{v}_2, \ldots, \mathbf{v}_n)$ be an ordered basis for $\mathbb{R}^n$, where $\mathbf{v}_j$ is an eigenvector for λ_j. We have seen that, if C is the matrix having $\mathbf{v}_j$ as jth column vector, then

$$C^{-1}AC = D = \begin{bmatrix} \lambda_1 & & & \\ & \lambda_2 & & \mathbf{0} \\ & & \ddots & \\ \mathbf{0} & & & \lambda_n \end{bmatrix}.$$

For any vector $\mathbf{x}$ in $\mathbb{R}^n$, let $\mathbf{d}$ be its coordinate vector relative to the basis B. Thus,

$$\mathbf{x} = d_1\mathbf{v}_1 + d_2\mathbf{v}_2 + \cdots + d_n\mathbf{v}_n.$$

Then

$$A^k\mathbf{x} = d_1\lambda_1{}^k\mathbf{v}_1 + d_2\lambda_2{}^k\mathbf{v}_2 + \cdots + d_n\lambda_n{}^k\mathbf{v}_n. \tag{1}$$

Equation (1) expresses $A^k\mathbf{x}$ as a linear combination of the eigenvectors $\mathbf{v}_j$.

Let us regard $\mathbf{x}$ as an initial *information vector* in a process in which the information vector at the next stage of the process is found by multiplying the present information vector on the left by a matrix A. Illustrations of this situation are provided by Markov chains in Section 1.6 and the generation of the Fibonacci sequence in Section 4.1. We are interested here in the long-term outcome of the process. That is, we wish to study $A^k\mathbf{x}$ for large values of k.

Let us number our eigenvalues and eigenvectors in Eq. (1) so that $|\lambda_i| \geq |\lambda_j|$ if $i < j$; that is, the eigenvalues are arranged in order of decreasing magnitude. Suppose that $|\lambda_1| > |\lambda_2|$ so that λ_1 is the unique eigenvalue of maximum magnitude. Equation (1) may be written

$$A^k\mathbf{x} = \lambda_1{}^k(d_1\mathbf{v}_1 + d_2(\lambda_2/\lambda_1)^k\mathbf{v}_2 + \cdots + d_n(\lambda_n/\lambda_1)^k\mathbf{v}_n).$$

Thus, if k is large and $d_1 \neq 0$, the vector $A^k\mathbf{x}$ is approximately equal to $d_1\lambda_1{}^k\mathbf{v}_1$ in the sense that $\|A^k\mathbf{x} - d_1\lambda_1{}^k\mathbf{v}_1\|$ is small compared with $\|A^k\mathbf{x}\|$.

EXAMPLE 1 Show that a diagonalizable transition matrix T for a Markov chain has no eigenvalues of magnitude > 1.

Solution Example 3 of Section 4.1 shows that 1 is an eigenvalue for every transition matrix of a Markov chain. For every choice of population distribution vector $\mathbf{p}$, the vector $T^k \mathbf{p}$ is again a vector with nonnegative entries having sum 1. The preceding discussion shows that all eigenvalues of T must have magnitude ≤ 1; otherwise, entries in some $T^k \mathbf{p}$ would have very large magnitude as k increases. ▪

EXAMPLE 2 Find the order of magnitude of the term F_k of the Fibonacci sequence $F_0, F_1, F_2, F_3, \ldots$, that is, the sequence

$$0, 1, 1, 2, 3, 5, 8, 13, \ldots$$

for large values of k.

Solution We saw in Section 4.1 that, if we let

$$\mathbf{x}_k = \begin{bmatrix} F_k \\ F_{k-1} \end{bmatrix},$$

then

$$\mathbf{x}_k = \begin{bmatrix} 1 & 1 \\ 1 & 0 \end{bmatrix} \mathbf{x}_{k-1}.$$

We compute relation (1) for

$$A = \begin{bmatrix} 1 & 1 \\ 1 & 0 \end{bmatrix} \quad \text{and} \quad \mathbf{x} = \mathbf{x}_1 = \begin{bmatrix} 1 \\ 0 \end{bmatrix}.$$

The characteristic equation of A is

$$(1 - \lambda)(-\lambda) - 1 = \lambda^2 - \lambda - 1 = 0.$$

Using the quadratic formula, we find the eigenvalues

$$\lambda_1 = \frac{1 + \sqrt{5}}{2} \quad \text{and} \quad \lambda_2 = \frac{1 - \sqrt{5}}{2}.$$

Reducing $A - \lambda_1 I$, we obtain

$$A - \lambda_1 I \sim \begin{bmatrix} \dfrac{1 - \sqrt{5}}{2} & 1 \\ 0 & 0 \end{bmatrix} \quad \text{so} \quad \mathbf{v}_1 = \begin{bmatrix} 2 \\ \sqrt{5} - 1 \end{bmatrix}$$

is an eigenvector of λ_1. In an analogous fashion, we find that

$$\mathbf{v}_2 = \begin{bmatrix} -2 \\ \sqrt{5} + 1 \end{bmatrix}$$

is an eigenvalue for λ_2. Thus we take

$$C = \begin{bmatrix} 2 & -2 \\ \sqrt{5} - 1 & \sqrt{5} + 1 \end{bmatrix}.$$

To find the coordinate vector $\mathbf{d}$ of $\mathbf{x}_1$ relative to the basis $(\mathbf{v}_1, \mathbf{v}_2)$, we observe that $\mathbf{x}_1 = C\mathbf{d}$. We find that

$$C^{-1} = (1/(4\sqrt{5}))\begin{bmatrix} \sqrt{5} + 1 & 2 \\ 1 - \sqrt{5} & 2 \end{bmatrix};$$

thus,

$$\begin{bmatrix} d_1 \\ d_2 \end{bmatrix} = C^{-1}\mathbf{x}_1 = C^{-1}\begin{bmatrix} 1 \\ 0 \end{bmatrix} = (1/(4\sqrt{5}))\begin{bmatrix} \sqrt{5} + 1 \\ 1 - \sqrt{5} \end{bmatrix}.$$

Equation 1 takes the form

$$A^k\mathbf{x}_1 = \begin{bmatrix} F_{k+1} \\ F_k \end{bmatrix} = \left(\frac{\sqrt{5} + 1}{4\sqrt{5}}\right)\left(\frac{1 + \sqrt{5}}{2}\right)^k\begin{bmatrix} 2 \\ \sqrt{5} - 1 \end{bmatrix}$$
$$- \left(\frac{\sqrt{5} - 1}{4\sqrt{5}}\right)\left(\frac{1 - \sqrt{5}}{2}\right)^k\begin{bmatrix} -2 \\ \sqrt{5} + 1 \end{bmatrix}. \quad (2)$$

For large k, the kth power of the eigenvalue $\lambda_1 = (1 + \sqrt{5})/2$ dominates, so $A^k\mathbf{x}_1$ is approximately equal to the shaded portion of Eq. (2). Computing the second component of $A^k\mathbf{x}_1$, we find that

$$F_k = (1/\sqrt{5})\left(\left(\frac{1 + \sqrt{5}}{2}\right)^k - \left(\frac{1 - \sqrt{5}}{2}\right)^k\right). \quad (3)$$

Thus,

$$F_k \approx (1/\sqrt{5})\left(\frac{1 + \sqrt{5}}{2}\right)^k \qquad \text{for large } k. \quad (4)$$

Indeed, since $|\lambda_2| = |(1 - \sqrt{5})/2| < 1$, we see that the contribution from this second eigenvalue to the right-hand side of Eq. (3) approaches zero as k increases. Since $|\lambda_2^k/\sqrt{5}| < \frac{1}{2}$ for $k = 1$ and hence for all k, we see that F_k can be characterized as the closest integer to $(1/\sqrt{5})((1 + \sqrt{5})/2)^k$ for all k. Approximation (4) verifies that F_k increases exponentially with k, as expected for a population of rabbits. ■

   Example 2 is a typical analysis of a process in which an information vector after the kth stage is equal to $A^k\mathbf{x}_1$ for an initial information vector $\mathbf{x}_1$ and a diagonalizable matrix A. An important consideration is whether any of the eigenvalues of A have magnitude greater than 1. When this is the case, the components of the information vector may grow exponentially in magnitude, as illustrated by the Fibonacci sequence in Example 2, where $|\lambda_1| > 1$. On the other hand, if all the eigenvalues have magnitude less than 1, the components of the information vector must approach zero as k increases.

Stability

The process just described is called **unstable** if A has an eigenvalue of magnitude greater than 1, **stable** if all eigenvalues have magnitude less than 1, and **neutrally stable** if the maximum magnitude of the eigenvalues is 1.

Thus a Markov chain is a neutrally stable process, whereas generation of the Fibonacci sequence is unstable. The eigenvectors are called the **normal modes** of the process.

In the type of process just described, we study information at evenly spaced time intervals. If we study such a process as the number of time intervals increases and their duration approaches zero, we find ourselves in calculus. Eigenvalues and eigenvectors play an important role in applications of calculus, especially in studying any sort of vibration. In these applications of calculus, components of vectors are functions of time. Our second application illustrates this.

Application: Systems of Linear Differential Equations

In calculus, we see the importance of the differential equation

$$\frac{dx}{dt} = kx$$

in simple rate of growth problems involving the time derivative of the amount x present of a single quantity. We may also write this equation as $x' = kx$, where we understand that x is a function of the time variable t. In more complex growth situations, n quantities may be present in amounts $x_1, x_2, \ldots, x_n$. The rate of change of x_i may depend not only on the amount of x_i present, but also on the amounts of the other $n - 1$ quantities present at time t. We consider a situation in which the rate of growth of each x_i depends linearly on the amounts present of the n quantities. This leads to a system of linear differential equations

$$
\begin{aligned}
x_1' &= a_{11}x_1 + a_{12}x_2 + \cdots + a_{1n}x_n \\
x_2' &= a_{21}x_1 + a_{22}x_2 + \cdots + a_{2n}x_n \\
&\ \ \vdots \\
x_n' &= a_{n1}x_1 + a_{n2}x_2 + \cdots + a_{nn}x_n,
\end{aligned}
\tag{5}
$$

where each x_i is a differentiable function of the real variable t and each a_{ij} is a scalar. The simplest such system is the single differential equation

$$x' = ax, \tag{6}$$

which has the general solution

$$x(t) = ke^{at}, \tag{7}$$

where k is a scalar. (See Example 8 on page 251.) Direct computation verifies that function (7) is the general solution of Eq. (6).

Turning to the solution of system (5), we write it in matrix form as

$$\mathbf{x}' = A\mathbf{x}, \tag{8}$$

where

$$
\mathbf{x} = \begin{bmatrix} x_1(t) \\ x_2(t) \\ \vdots \\ x_n(t) \end{bmatrix}, \quad
\mathbf{x}' = \begin{bmatrix} x_1'(t) \\ x_2'(t) \\ \vdots \\ x_n'(t) \end{bmatrix}, \quad \text{and} \quad A = [a_{ij}].
$$

If the matrix A is a diagonal matrix so that $a_{ij} = 0$ for $i \neq j$, then system (1) reduces to a system of n equations, each like Eq. (6), namely:

$$
\begin{aligned}
x_1' &= a_{11}x_1 \\
x_2' &= a_{22}x_2 \\
&\;\;\vdots \\
x_n' &= a_{nn}x_n.
\end{aligned}
\tag{9}
$$

The general solution is given by

$$
\mathbf{x} = \begin{bmatrix} x_1 \\ x_2 \\ \vdots \\ x_n \end{bmatrix} = \begin{bmatrix} k_1 e^{a_{11}t} \\ k_2 e^{a_{22}t} \\ \vdots \\ k_n e^{a_{nn}t} \end{bmatrix}.
$$

In the general case, we try to diagonalize A and reduce system (8) to a system like (9). If A is diagonalizable, we have

$$
D = C^{-1}AC = \begin{bmatrix} \lambda_1 & & & \mathbf{0} \\ & \lambda_2 & & \\ & & \ddots & \\ \mathbf{0} & & & \lambda_n \end{bmatrix}
$$

for some invertible $n \times n$ matrix C. If we substitute $A = CDC^{-1}$, Eq. (8) takes the form $C^{-1}\mathbf{x}' = D(C^{-1}\mathbf{x})$, or

$$
\mathbf{y}' = D\mathbf{y},
\tag{10}
$$

where

$$
\mathbf{x} = C\mathbf{y}.
\tag{11}
$$

(If we let $\mathbf{x} = C\mathbf{y}$, we can confirm that $\mathbf{x}' = C\mathbf{y}'$.) The general solution of system (10) is

$$
\mathbf{y} = \begin{bmatrix} y_1 \\ y_2 \\ \vdots \\ y_n \end{bmatrix} = \begin{bmatrix} k_1 e^{\lambda_1 t} \\ k_2 e^{\lambda_2 t} \\ \vdots \\ k_n e^{\lambda_n t} \end{bmatrix},
$$

where the λ_j are eigenvalues of the diagonalizable matrix A. The general solution $\mathbf{x}$ of system (8) is then obtained from Eq. (11), using the corresponding eigenvectors $\mathbf{v}_j$ of A as columns of the matrix C.

We emphasize that we have described the general solution of system (8) only in the case where the matrix A is diagonalizable. The algebraic multiplicity of each eigenvalue of A must equal its geometric multiplicity. The following example illustrates that the eigenvalues need not be distinct, as long as their algebraic and geometric multiplicities are the same.

EXAMPLE 3   Solve the linear differential system

$$x_1' = x_1 - x_2 - x_3$$
$$x_2' = -x_1 + x_2 - x_3$$
$$x_3' = -x_1 - x_2 + x_3.$$

Solution   The first step is to diagonalize the matrix

$$A = \begin{bmatrix} 1 & -1 & -1 \\ -1 & 1 & -1 \\ -1 & -1 & 1 \end{bmatrix}.$$

Expanding the determinant $|A - \lambda I|$ across the first row, we obtain

$$|A - \lambda I| = \begin{vmatrix} 1 - \lambda & -1 & -1 \\ -1 & 1 - \lambda & -1 \\ -1 & -1 & 1 - \lambda \end{vmatrix}$$

$$= (1 - \lambda)\begin{vmatrix} 1 - \lambda & -1 \\ -1 & 1 - \lambda \end{vmatrix} + \begin{vmatrix} -1 & -1 \\ -1 & 1 - \lambda \end{vmatrix} - \begin{vmatrix} -1 & 1 - \lambda \\ -1 & -1 \end{vmatrix}$$

$$= (1 - \lambda)(\lambda^2 - 2\lambda) + 2(\lambda - 2)$$

$$= (\lambda - 2)((1 - \lambda)\lambda + 2)$$

$$= -(\lambda + 1)(\lambda - 2)^2.$$

This yields eigenvalues $\lambda_1 = -1$ and $\lambda_2 = \lambda_3 = 2$.

Next we compute eigenvectors. For $\lambda_1 = -1$, we have

$$A + I = \begin{bmatrix} 2 & -1 & -1 \\ -1 & 2 & -1 \\ -1 & -1 & 2 \end{bmatrix} \sim \begin{bmatrix} 1 & 1 & -2 \\ 0 & 3 & -3 \\ 0 & 0 & 0 \end{bmatrix},$$

which gives the eigenvector

$$\mathbf{v}_1 = \begin{bmatrix} 1 \\ 1 \\ 1 \end{bmatrix}.$$

For $\lambda_2 = \lambda_3 = 2$, we have

$$A - 2I = \begin{bmatrix} -1 & -1 & -1 \\ -1 & -1 & -1 \\ -1 & -1 & -1 \end{bmatrix} \sim \begin{bmatrix} 1 & 1 & 1 \\ 0 & 0 & 0 \\ 0 & 0 & 0 \end{bmatrix},$$

which gives the independent eigenvectors

$$\mathbf{v}_2 = \begin{bmatrix} -1 \\ 1 \\ 0 \end{bmatrix} \quad \text{and} \quad \mathbf{v}_3 = \begin{bmatrix} -1 \\ 0 \\ 1 \end{bmatrix}.$$

A diagonalizing matrix is then

$$C = \begin{bmatrix} 1 & -1 & -1 \\ 1 & 1 & 0 \\ 1 & 0 & 1 \end{bmatrix},$$

so that

$$D = C^{-1}AC = \begin{bmatrix} -1 & 0 & 0 \\ 0 & 2 & 0 \\ 0 & 0 & 2 \end{bmatrix}.$$

Putting $\mathbf{x} = C\mathbf{y}$, we see that the system becomes

$$\begin{aligned} y_1' &= -y_1 \\ y_2' &= 2y_2 \\ y_3' &= 2y_3, \end{aligned}$$

whose solution is given by

$$\begin{bmatrix} y_1 \\ y_2 \\ y_3 \end{bmatrix} = \begin{bmatrix} k_1 e^{-t} \\ k_2 e^{2t} \\ k_3 e^{2t} \end{bmatrix}.$$

Therefore, the solution to the original system is

$$\begin{bmatrix} x_1 \\ x_2 \\ x_3 \end{bmatrix} = \begin{bmatrix} 1 & -1 & -1 \\ 1 & 1 & 0 \\ 1 & 0 & 1 \end{bmatrix} \begin{bmatrix} y_1 \\ y_2 \\ y_3 \end{bmatrix} = \begin{bmatrix} k_1 e^{-t} - k_2 e^{2t} - k_3 e^{2t} \\ k_1 e^{-t} + k_2 e^{2t} \\ k_1 e^{-t} + k_3 e^{2t} \end{bmatrix}. \qquad ■$$

In Section 8.4, we show that every square matrix A is similar to a matrix J in *Jordan canonical form*, so that $J = C^{-1}AC$. The Jordan canonical form J and the invertible matrix C may have complex entries. In the Jordan canonical form J, all nondiagonal entries are zero except for some entries 1 immediately above the diagonal entries. If the Jordan form $J = C^{-1}AC$ is known, the substitution $\mathbf{x} = C\mathbf{y}$ again reduces a system $\mathbf{x}' = A\mathbf{x}$ of linear differential equations to one that can be solved easily for $\mathbf{y}$. The solution of the original system is computed as $\mathbf{x} = C\mathbf{y}$, just as was the solution in Example 3. This is an extremely important technique in the study of differential equations.

SUMMARY

1. Let A be diagonalizable by a matrix C, let $\mathbf{x}$ be any column vector, and let $\mathbf{d} = C^{-1}\mathbf{x}$. Then

$$A^k\mathbf{x} = d_1\lambda_1{}^k\mathbf{v}_1 + d_2\lambda_2{}^k\mathbf{v}_2 + \cdots + d_n\lambda_n{}^k\mathbf{v}_n,$$

   where $\mathbf{v}_j$ is the jth column vector of C.

2. Let multiplication of a column information vector by A give the information vector for the next stage of a process, as described on page 266. The process is stable, neutrally stable, or unstable, according as the maximum magnitude of the eigenvalues of A is less than 1, equal to 1, or greater than 1, respectively.

3. The system $\mathbf{x}' = A\mathbf{x}$ of linear differential equations can be solved, if A is diagonalizable, using the following three steps.

Step 1 Find a matrix C so that $D = C^{-1}AC$ is a diagonal matrix.

Step 2 Solve the simpler diagonal system $\mathbf{y}' = D\mathbf{y}$.

Step 3 The solution of the original system is $\mathbf{x} = C\mathbf{y}$.

EXERCISES

1. Let the sequence $a_0, a_1, a_2, \ldots$ be given by $a_0 = 0$, $a_1 = 1$, and $a_k = (a_{k-1} + a_{k-2})/2$ for $k \geq 2$.
   a) Find the matrix A that can be used to generate this sequence as we used a matrix to generate the Fibonacci sequence in Section 4.1.
   b) Classify this generation process as stable, neutrally stable, or unstable.
   c) Compute expression (1) for $\mathbf{x} = \begin{bmatrix} a_1 \\ a_0 \end{bmatrix}$ for this process. Check computations with the first few terms of the sequence.
   d) Use the answer to part (c) to estimate a_k for large k.

2. Repeat Exercise 1 if $a_k = a_{k-1} - \left(\frac{3}{16}\right)a_{k-2}$ for $k \geq 2$.

3. Repeat Exercise 1 but change the initial data to $a_0 = 1$, $a_1 = 0$.

4. Repeat Exercise 1 if $a_k = \left(\frac{1}{2}\right)a_{k-1} + \left(\frac{3}{16}\right)a_{k-2}$ for $k \geq 2$.

5. Repeat Exercise 1 if $a_k = a_{k-1} + \left(\frac{3}{4}\right)a_{k-2}$ for $k \geq 2$.

In Exercises 6–13, solve the given system of linear differential equations as outlined in the summary.

6. $\begin{aligned} x_1' &= 3x_1 - 5x_2 \\ x_2' &= \quad\quad 2x_2 \end{aligned}$

7. $\begin{aligned} x_1' &= x_1 + 4x_2 \\ x_2' &= 3x_1 \end{aligned}$

8. $\begin{aligned} x_1' &= x_1 + 2x_2 \\ x_2' &= 2x_1 + x_2 \end{aligned}$

9. $\begin{aligned} x_1' &= 2x_1 + 2x_2 \\ x_2' &= x_1 + 3x_2 \end{aligned}$

10. $\begin{aligned} x_1' &= 6x_1 + 3x_2 - 3x_3 \\ x_2' &= -2x_1 - x_2 + 2x_3 \\ x_3' &= 16x_1 + 8x_2 - 7x_3 \end{aligned}$

11. $\begin{aligned} x_1' &= -3x_1 + 10x_2 - 6x_3 \\ x_2' &= \quad\quad 7x_2 - 6x_3 \\ x_3' &= \quad\quad\quad\quad x_3 \end{aligned}$

12. $\begin{aligned} x_1' &= -3x_1 + 5x_2 - 20x_3 \\ x_2' &= 2x_1 \quad\quad + 8x_3 \\ x_3' &= 2x_1 + x_2 + 7x_3 \end{aligned}$

13. $\begin{aligned} x_1' &= -2x_1 \quad\quad - x_3 \\ x_2' &= \quad\quad 2x_2 \\ x_3' &= 3x_1 \quad\quad + 2x_3 \end{aligned}$

5 ORTHOGONALITY

We are accustomed to working in $\mathbb{R}^2$ with coordinates of vectors relative to the standard ordered basis $(\mathbf{e}_1, \mathbf{e}_2) = (\mathbf{i}, \mathbf{j})$. These basis vectors are orthogonal (perpendicular) and have length 1. The vectors in the standard ordered basis $(\mathbf{e}_1, \mathbf{e}_2, \ldots, \mathbf{e}_n)$ of $\mathbb{R}^n$ have these same two properties. It is precisely these two properties of the standard bases that make computations using coordinates relative to them quite easy. For example, computations of angles and of length are easy when we use coordinates relative to standard bases. Throughout linear algebra, computations using coordinates relative to bases consisting of orthogonal unit vectors are generally the easiest to perform, and they generate less error in work with a computer.

This chapter is devoted to orthogonality. Orthogonal projection, which is the main tool for finding a basis of orthogonal vectors, is developed in Section 5.1. Section 5.2 shows that every finite-dimensional inner-product space has a basis of orthogonal unit vectors, and shows how to construct such a basis from any given basis. Section 5.3 deals with orthogonal matrices and orthogonal linear transformations. In Section 5.4 we show that orthogonal projection can be achieved by matrix multiplication. To conclude the chapter, we give applications to the method of least squares and over-determined linear systems in Section 5.5.

5.1 PROJECTIONS

The Projection of b on sp(a)

For convenience, we develop the ideas in this section for the vector space $\mathbb{R}^n$, using the dot product. The work is valid in general finite-dimensional spaces, using the more cumbersome notation $\langle \mathbf{u}, \mathbf{v} \rangle$ for the inner product of $\mathbf{u}$ and $\mathbf{v}$. An illustration in function spaces, using an inner product defined by an integral, appears at the end of this section.

A practical concern in vector applications involves determining what portion of a vector **b** can be considered to act in the direction given by another vector **a**. A force vector acting in a certain direction may be moving a body along a line having a different direction. For example, suppose that you are trying to roll your stalled car off the road by pushing on the door jamb at the side, so you can reach in and control the steering wheel when necessary. You are not applying the force in precisely the direction in which the car moves, as you would be if you could push from directly behind the car. Such considerations lead the physicist to consider the projection **p** of the force vector **F** on a direction vector **a**, as shown in Figure 5.1. In Figure 5.1, the vector **p** is found by dropping a perpendicular from the tip of **F** to the vector **a**.

Figure 5.2 shows the situation where the projection of **F** on **a** has a direction opposite to the direction of **a**; in terms of our car example, you would be applying the force to move the car backward rather than forward. Figure 5.2 suggests that it is preferable to speak of *projection on the subspace* sp(**a**) (which in this example is a line) than to speak of *projection on* **a**.

We derive geometrically a formula for the projection **p** of the force vector **F** on sp(**a**), based on Figures 5.1 and 5.2. We see that **p** is a multiple of **a**. Now $(1/\|\mathbf{a}\|)\mathbf{a}$ is a unit vector having the same direction as **a**, so **p** is a scalar multiple of this unit vector. We need only find the appropriate scalar. Referring to Figure 5.1, we see that the appropriate scalar is $\|\mathbf{F}\| \cos \theta$, since it is the length of the leg of the right triangle labeled **p**. This same formula also gives the appropriate negative scalar for the case shown in Figure 5.2, since $\cos \theta$ is negative for this angle θ lying between 90° and 180°. Thus we obtain

$$\mathbf{p} = \left(\frac{\|\mathbf{F}\| \cos \theta}{\|\mathbf{a}\|}\right)\mathbf{a} = \frac{\|\mathbf{F}\|\|\mathbf{a}\| \cos \theta}{\|\mathbf{a}\|\|\mathbf{a}\|} = \left(\frac{\mathbf{F} \cdot \mathbf{a}}{\mathbf{a} \cdot \mathbf{a}}\right)\mathbf{a}.$$

Of course, we assume that $\mathbf{a} \neq \mathbf{0}$. We replace our force vector **F** by a general vector **b** and box this formula.

Projection p of b on sp(a) in $\mathbb{R}^n$

$$\mathbf{p} = \left(\frac{\mathbf{a} \cdot \mathbf{b}}{\mathbf{a} \cdot \mathbf{a}}\right)\mathbf{a}$$

(1)

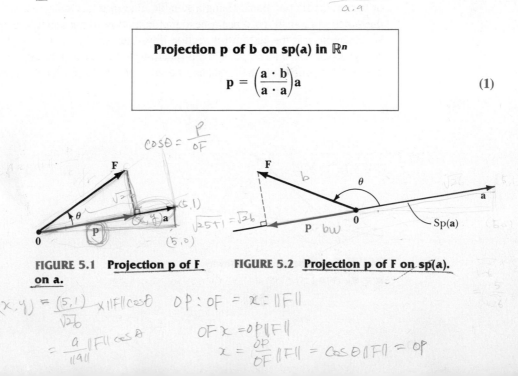

FIGURE 5.1 Projection p of F on a.

FIGURE 5.2 Projection p of F on sp(a).

EXAMPLE 1   Find the projection **p** of the vector $(1, 2, 3)$ on $\mathrm{sp}((2, 4, 3))$ in $\mathbb{R}^3$.

Solution   We let $\mathbf{a} = (2, 4, 3)$ and $\mathbf{b} = (1, 2, 3)$ in formula (1), obtaining

$$\mathbf{p} = \left(\frac{\mathbf{a} \cdot \mathbf{b}}{\mathbf{a} \cdot \mathbf{a}}\right)\mathbf{a} = \left(\frac{2 + 8 + 9}{4 + 16 + 9}\right)\mathbf{a} = \left(\frac{19}{29}\right)(2, 4, 3). \qquad ■$$

The Concept of Projection

We now explain what is meant by the *projection* of a vector **b** in $\mathbb{R}^n$ on a general subspace W of $\mathbb{R}^n$. We will show that there are unique vectors $\mathbf{b}_W$ and $\mathbf{b}_{W^\perp}$ such that

1. $\mathbf{b}_W$ is in the subspace W;
2. $\mathbf{b}_{W^\perp}$ is orthogonal to every vector in W; and
3. $\mathbf{b} = \mathbf{b}_W + \mathbf{b}_{W^\perp}$.

The symbol $W^\perp$ denotes the set of all vectors in $\mathbb{R}^n$ that are perpendicular to every vector in W. Properties of $W^\perp$ will appear shortly in Theorem 5.1. Figure 5.3 gives a symbolic illustration of this decomposition of **b** into a sum of a vector in W and a vector orthogonal to W. Once we have demonstrated the existence of this decomposition, we will define the *projection of* **b** *on* W to be the vector $\mathbf{b}_W$.

The projection $\mathbf{b}_W$ of **b** on W is the vector **w** in W that is closest to **b**. That is, $\mathbf{w} = \mathbf{b}_W$ *minimizes* the distance $\|\mathbf{b} - \mathbf{w}\|$ from **b** to W for all **w** in W. This seems reasonable, since we have $\mathbf{b} = \mathbf{b}_W + \mathbf{b}_{W^\perp}$ and since $\mathbf{b}_{W^\perp}$ is orthogonal to every vector in W. Of course, this minimum distance from **b** to W is just $\|\mathbf{b}_{W^\perp}\|$, as indicated by Figure 5.3.

The fact that projection solves this minimization problem is useful in many contexts. For an example involving function spaces, suppose that f is a complicated function and that W consists of functions that are easily handled, such as polynomials or trigonometric functions. Using a suitable inner product and projection, we find that the function f_W in W becomes a best approximation to the function f by functions in W. Example 6 at the end of this section illustrates this by approximating the function $f(x) = x$ over the interval $[0, 1]$ by a function in the space W of all constant functions on $[0, 1]$. Visualizing their graphs, we are not surprised that the constant function

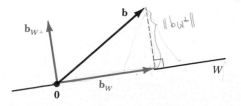

FIGURE 5.3   The decomposition
$\mathbf{b} = \mathbf{b}_W + \mathbf{b}_W{}^\perp.$

$p(x) = \frac{1}{2}$ turns out to be the best approximation to x in W, using the inner product we defined on function spaces in Section 2.10. In Section 5.5, we will use this minimization feature again to find a best approximate solution of an inconsistent overdetermined linear system.

For another physical illustration, suppose that you are pushing a box across a floor by pushing forward and downward on the top edge. (See Fig. 5.4.) We take as origin the point where the force is applied to the box and as W the plane through that origin parallel to the floor. If $\mathbf{F}$ is the force vector applied at the origin, then $\mathbf{F}_W$ is the portion of the force vector that actually moves the body along the floor, and $\mathbf{F}_{W^\perp}$ (which is directed straight down) is the portion of the force vector that attempts to push the box into the floor and thereby increases the friction between the box and the floor.

Orthogonal Complement of a Subspace

As a preliminary to proving the existence and uniqueness of the vectors $\mathbf{b}_W$ and $\mathbf{b}_{W^\perp}$ just described, we consider all vectors in $\mathbb{R}^n$ that are orthogonal to every vector in a subspace W.

DEFINITION 5.1 Orthogonal Complement

Let W be a subspace of $\mathbb{R}^n$. The set of all vectors in $\mathbb{R}^n$ that are orthogonal to every vector in W is the **orthogonal complement** of W, and is denoted by $W^\perp$.

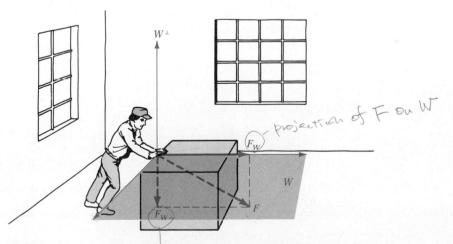

FIGURE 5.4 **The decomposition of a force vector.**

It is not difficult to find the orthogonal complement of W if we know a generating set for W. Let $\{\mathbf{v}_1, \mathbf{v}_2, \ldots, \mathbf{v}_k\}$ be a generating set for W. Let A be the $k \times n$ matrix having $\mathbf{v}_i$ as its ith row vector. That is,

$$A = \begin{bmatrix} \rule{1cm}{0.4pt} \mathbf{v}_1 \rule{1cm}{0.4pt} \\ \rule{1cm}{0.4pt} \mathbf{v}_2 \rule{1cm}{0.4pt} \\ \vdots \\ \rule{1cm}{0.4pt} \mathbf{v}_k \rule{1cm}{0.4pt} \end{bmatrix}.$$

Thus, W is the row space of A. Now the nullspace of A consists of all vectors $\mathbf{x}$ in $\mathbb{R}^n$ that are solutions of the homogeneous system $A\mathbf{x} = \mathbf{0}$. But $A\mathbf{x} = \mathbf{0}$ if and only if $\mathbf{v}_i \cdot \mathbf{x} = 0$ for $i = 1, 2, \ldots, k$. Therefore, the nullspace of A is the set of all vectors $\mathbf{x}$ in $\mathbb{R}^n$ that are orthogonal to each of the rows of A, and hence to the row space of A. In other words, the orthogonal complement of the row space of A is the nullspace of A. Thus we have found $W^\perp$. We summarize this procedure in a box. Notice that this procedure marks one of the rare occasions in this text when vectors are placed as rows of a matrix.

need exp. why the null space = orthogonal complement.

Finding the Orthogonal Complement of a Subspace W of $\mathbb{R}^n$

1. Find a matrix A having as *row* vectors a generating set for W.
2. Find the nullspace of A—that is, the solution space of $A\mathbf{x} = \mathbf{0}$. This nullspace is $W^\perp$.

EXAMPLE 2 Find a basis for the orthogonal complement in $\mathbb{R}^4$ of the subspace

$$W = \text{sp}((1, 2, 2, 1), (3, 4, 2, 3), (0, 1, 3, 2)).$$

Solution We find the nullspace of the matrix

$$A = \begin{bmatrix} 1 & 2 & 2 & 1 \\ 3 & 4 & 2 & 3 \\ 0 & 1 & 3 & 2 \end{bmatrix}.$$

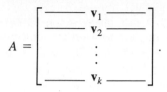

Reducing A, we have

$$\begin{bmatrix} 1 & 2 & 2 & 1 \\ 3 & 4 & 2 & 3 \\ 0 & 1 & 3 & 2 \end{bmatrix} \sim \begin{bmatrix} 1 & 2 & 2 & 1 \\ 0 & -2 & -4 & 0 \\ 0 & 1 & 3 & 2 \end{bmatrix} \sim \begin{bmatrix} 1 & 0 & -2 & 1 \\ 0 & 1 & 2 & 0 \\ 0 & 0 & 1 & 2 \end{bmatrix}.$$

Therefore, the nullspace of A, which is the orthogonal complement of W, is the set of vectors of the form

$$(-5r, 4r, -2r, r) \quad \text{for any scalar } r.$$

Thus, $\{(-5, 4, -2, 1)\}$ is a basis for $W^\perp$. ■

1 2 2 1
3 4 2 3
0 1 3 2
-5 4 -2 1

1 2 2 1
0 1 2 0
0 0 1 2
0 0 0 46

means $W^\perp$ is

We now show that the orthogonal complement of a subspace W has some very nice properties. In particular, we exhibit the decomposition $\mathbf{b} = \mathbf{b}_W + \mathbf{b}_{W^\perp}$ described earlier and show that it is unique.

THEOREM 5.1 Properties of $W^\perp$

The orthogonal complement $W^\perp$ of a subspace W of $\mathbb{R}^n$ has the following properties:

(i) $W^\perp$ is a subspace of $\mathbb{R}^n$.

(ii) $\dim(W^\perp) = n - \dim(W)$.

(iii) $(W^\perp)^\perp = W$; that is, the orthogonal complement of $W^\perp$ is W.

(iv) Each vector $\mathbf{b}$ in $\mathbb{R}^n$ can be expressed uniquely in the form $\mathbf{b} = \mathbf{b}_W + \mathbf{b}_{W^\perp}$ for $\mathbf{b}_W$ in W and $\mathbf{b}_{W^\perp}$ in $W^\perp$.

PROOF Let $\dim(W) = k$, and let $\{\mathbf{v}_1, \mathbf{v}_2, \ldots, \mathbf{v}_k\}$ be a basis for W. Let A be the $k \times n$ matrix having $\mathbf{v}_i$ as its ith row vector for $i = 1, \ldots, k$.

For property (i), we have seen that $W^\perp$ is the nullspace of the matrix A, so it is a subspace of $\mathbb{R}^n$.

For property (ii), consider the rank equation of A:

$$\text{rank}(A) + \text{nullity}(A) = n.$$

Since $\dim(W) = \text{rank}(A)$ and since $W^\perp$ is the nullspace of A, we see that $\dim(W^\perp) = n - \dim(W)$.

For property (iii), applying property (ii) with the subspace $W^\perp$, we find that

$$\dim(W^\perp)^\perp = n - \dim(W^\perp) = n - (n - k) = k.$$

However, every vector in W is orthogonal to every vector in $W^\perp$, so W is a subspace of $(W^\perp)^\perp$. Since both W and $(W^\perp)^\perp$ have the same dimension k, we conclude that W must be equal to $(W^\perp)^\perp$. (See Exercise 22 in Section 2.6.)

For property (iv), let $\{\mathbf{v}_{k+1}, \mathbf{v}_{k+2}, \ldots, \mathbf{v}_n\}$ be a basis for $W^\perp$. We claim that the set

$$\{\mathbf{v}_1, \mathbf{v}_2, \ldots, \mathbf{v}_n\} \tag{2}$$

is a basis for $\mathbb{R}^n$. Consider a relation

$$r_1\mathbf{v}_1 + r_2\mathbf{v}_2 + \cdots + r_k\mathbf{v}_k + s_{k+1}\mathbf{v}_{k+1} + s_{k+2}\mathbf{v}_{k+2} + \cdots + s_n\mathbf{v}_n = \mathbf{0}.$$

Rewrite the relation as

$$r_1\mathbf{v}_1 + r_2\mathbf{v}_2 + \cdots + r_k\mathbf{v}_k = -s_{k+1}\mathbf{v}_{k+1} - s_{k+2}\mathbf{v}_{k+2} - \cdots - s_n\mathbf{v}_n. \tag{3}$$

The sum on the left-hand side is in W, and the sum on the right-hand side is in $W^\perp$. Since these sums are equal, they represent a vector that is in both W and $W^\perp$ and must be orthogonal to itself. The only such vector is the zero vector, so both sides of Eq. (3) must equal $\mathbf{0}$. Since the $\mathbf{v}_i$ are independent for $1 \le i \le k$ and the $\mathbf{v}_j$ are independent for $k + 1 \leq j \leq n$, we see that all the scalars r_i and s_j are zero. This shows that set (2) is independent; since it contains n vectors, it must be a basis for $\mathbb{R}^n$. Therefore, for every vector $\mathbf{b}$ in $\mathbb{R}^n$ we can express $\mathbf{b}$ in the form

$$\mathbf{b} = \underbrace{r_1\mathbf{v}_1 + r_2\mathbf{v}_2 + \cdots + r_k\mathbf{v}_k}_{\mathbf{b}_W} + \underbrace{s_{k+1}\mathbf{v}_{k+1} + s_{k+2}\mathbf{v}_{k+2} + \cdots + s_n\mathbf{v}_n}_{\mathbf{b}_{W^\perp}}.$$

This shows that $\mathbf{b}$ can indeed be expressed as a sum of a vector in W and a vector in $W^\perp$.

It remains for us to show the uniqueness of such a sum. Suppose that we have two such sums

$$\mathbf{b} = \mathbf{w} + \mathbf{v} = \mathbf{w}' + \mathbf{v}'$$

for $\mathbf{w}, \mathbf{w}'$ in W and for $\mathbf{v}, \mathbf{v}'$ in $W^\perp$. Then

$$\mathbf{w} - \mathbf{w}' = \mathbf{v}' - \mathbf{v}.$$

As we argued previously, the two sides of this equation represent a vector that is in both W and $W^\perp$ *and must therefore be the zero vector.* Thus, $\mathbf{w} = \mathbf{w}'$ and $\mathbf{v} = \mathbf{v}'$, so the expression for $\mathbf{b}$ as a sum of vectors in W and in $W^\perp$ *is indeed unique.* ▲

Projection of a Vector on a Subspace

Now that the groundwork is firmly established, we can define the projection of a vector in $\mathbb{R}^n$ on a subspace and then illustrate with some examples.

DEFINITION 5.2 Projection of b on W

Let $\mathbf{b}$ be a vector in $\mathbb{R}^n$, and let W be a subspace of $\mathbb{R}^n$. Let

$$\mathbf{b} = \mathbf{b}_W + \mathbf{b}_{W^\perp},$$

as described in Theorem 5.1. Then $\mathbf{b}_W$ is the **projection of b on** W.

Theorem 5.1 shows that the projection of $\mathbf{b}$ on W is unique. It also shows one way in which it can be computed.

Steps to Find the Projection of b on W

1. Select a basis $\{\mathbf{v}_1, \mathbf{v}_2, \ldots, \mathbf{v}_k\}$ for the subspace W. (Often this is given.)
2. Find a basis $\{\mathbf{v}_{k+1}, \mathbf{v}_{k+2}, \ldots, \mathbf{v}_n\}$ for $W^\perp$, as in Example 2.
3. Find the coordinate vector $\mathbf{r} = (r_1, r_2, \ldots, r_n)$ of $\mathbf{b}$ relative to the *basis* $(\mathbf{v}_1, \mathbf{v}_2, \ldots, \mathbf{v}_n)$ so that $\mathbf{b} = r_1\mathbf{v}_1 + r_2\mathbf{v}_2 + \cdots + r_n\mathbf{v}_n$. (See the box on page 155.)
4. Then $\mathbf{b}_W = r_1\mathbf{v}_1 + r_2\mathbf{v}_2 + \cdots + r_k\mathbf{v}_k$.

EXAMPLE 3 Find the projection of $\mathbf{b} = (2, 1, 5)$ on the subspace $W = \text{sp}((1, 2, 1), (2, 1, -1))$.

Solution We follow the boxed procedure.

Step 1: Since $\mathbf{v}_1 = (1, 2, 1)$ and $\mathbf{v}_2 = (2, 1, -1)$ are independent, they form a basis for W.

Step 2: A basis for $W^\perp$ can be found by obtaining the nullspace of the matrix

$$A = \begin{bmatrix} 1 & 2 & 1 \\ 2 & 1 & -1 \end{bmatrix}.$$

An echelon form of A is

$$\begin{bmatrix} 1 & 2 & 1 \\ 0 & -3 & -3 \end{bmatrix},$$

and the nullspace of A is the set of vectors $(r, -r, r)$, where r is any scalar. Let us take $\mathbf{v}_3 = (1, -1, 1)$ to form the basis $\{\mathbf{v}_3\}$ of $W^\perp$. (Alternatively, we could have computed $\mathbf{v}_1 \times \mathbf{v}_2$ to find a suitable $\mathbf{v}_3$.)

Step 3: To find the coordinate vector $\mathbf{r}$ of $\mathbf{b}$ relative to the ordered basis $(\mathbf{v}_1, \mathbf{v}_2, \mathbf{v}_3)$, we proceed as described in Section 2.6, and perform the reduction

$$\begin{bmatrix} 1 & 2 & 1 & 2 \\ 2 & 1 & -1 & 1 \\ 1 & -1 & 1 & 5 \end{bmatrix} \sim \begin{bmatrix} 1 & 2 & 1 & 2 \\ 0 & -3 & -3 & -3 \\ 0 & -3 & 0 & 3 \end{bmatrix} \sim \begin{bmatrix} 1 & 0 & 1 & 4 \\ 0 & 1 & 0 & -1 \\ 0 & 0 & -3 & -6 \end{bmatrix}$$

$$\underbrace{\mathbf{v}_1 \quad \mathbf{v}_2}_{b_W} \quad \underbrace{\mathbf{v}_3}_{b_{W^\perp}} \quad \mathbf{b}$$

$$\sim \begin{bmatrix} 1 & 0 & 0 & 2 \\ 0 & 1 & 0 & -1 \\ 0 & 0 & 1 & 2 \end{bmatrix} \Big\} b_W$$

Thus, $\mathbf{r} = (2, -1, 2)$.

Step 4: The projection of $\mathbf{b}$ on W is

$$\mathbf{b}_W = 2\mathbf{v}_1 - \mathbf{v}_2 = 2(1, 2, 1) - (2, 1, -1) = (0, 3, 3).$$

As a check, notice that $2\mathbf{v}_3 = (2, -2, 2)$ is the projection of $\mathbf{b}$ on $W^\perp$, and $\mathbf{b} = \mathbf{b}_W + \mathbf{b}_{W^\perp} = (0, 3, 3) + (2, -2, 2) = (2, 1, 5)$. ■

We have seen that if W is a 1-dimensional subspace of $\mathbb{R}^n$, then we can readily find the projection $\mathbf{b}_W$ of $\mathbf{b} \in \mathbb{R}^n$ on W. We simply use Eq. (1). On the other hand, if $W^\perp$ is 1-dimensional, we can use Eq. (1) to find the projection of $\mathbf{b}$ on $W^\perp$, and then find $\mathbf{b}_W$ using the relation $\mathbf{b}_W = \mathbf{b} - \mathbf{b}_{W^\perp}$. Our next example illustrates this.

EXAMPLE 4 Find the projection of the vector $(3, -1, 2)$ on the plane $x + y + z = 0$ through the origin in $\mathbb{R}^3$.

Solution Let W be the subspace of $\mathbb{R}^3$ given by the plane $x + y + z = 0$. Then $W^\perp$ is 1-dimensional, and a generating vector for $W^\perp$ is $\mathbf{a} = (1, 1, 1)$, obtained by taking the coefficients of x, y, and z in this equation. Let $\mathbf{b} = (3, -1, 2)$. By Eq. (1), we have

$$\mathbf{b}_{W^\perp} = \frac{\mathbf{b} \cdot \mathbf{a}}{\mathbf{a} \cdot \mathbf{a}} \mathbf{a} = \frac{3 - 1 + 2}{1 + 1 + 1}(1, 1, 1) = \frac{4}{3}(1, 1, 1).$$

Thus $\mathbf{b}_W = \mathbf{b} - \mathbf{b}_{W^\perp} = (3, -1, 2) - \dfrac{4}{3}(1, 1, 1) = \left(\dfrac{5}{3}, -\dfrac{7}{3}, \dfrac{2}{3}\right).$ ■

Our next example shows that the procedure described in the box preceding Example 3 yields formula (1) for the projection of a vector $\mathbf{b}$ on sp($\mathbf{a}$).

EXAMPLE 5   Let $\mathbf{a} \neq \mathbf{0}$ and $\mathbf{b}$ be vectors in $\mathbb{R}^n$. Find the projection of $\mathbf{b}$ on sp($\mathbf{a}$), using the same boxed procedure we have been applying.

Solution   We project $\mathbf{b}$ on the subspace $W = $ sp($\mathbf{a}$) of $\mathbb{R}^n$. Let $\{\mathbf{v}_2, \mathbf{v}_3, \ldots, \mathbf{v}_n\}$ be a basis for $W^\perp$, and let $\mathbf{r} = (r_1, r_2, \ldots, r_n)$ be the coordinate vector of $\mathbf{b}$ relative to the ordered basis $(\mathbf{a}, \mathbf{v}_2, \ldots, \mathbf{v}_n)$. Then

$$\mathbf{b} = r_1\mathbf{a} + \underbrace{r_2\mathbf{v}_2 + \cdots + r_n\mathbf{v}_n}.$$

Since $\mathbf{v}_i \cdot \mathbf{a} = 0$ for $i = 2, \ldots, n$, we see that $\mathbf{b} \cdot \mathbf{a} = r_1\mathbf{a} \cdot \mathbf{a} = r_1(\mathbf{a} \cdot \mathbf{a})$. Since $\mathbf{a} \neq \mathbf{0}$, we can write

$$r_1 = \frac{\mathbf{b} \cdot \mathbf{a}}{\mathbf{a} \cdot \mathbf{a}}.$$

$b \cdot a = r_1 (a \cdot a)$

$r_1 = \dfrac{b \cdot a}{a \cdot a}$

The projection of $\mathbf{b}$ on sp($\mathbf{a}$) is then $r_1\mathbf{a} = \left(\dfrac{\mathbf{b} \cdot \mathbf{a}}{\mathbf{a} \cdot \mathbf{a}}\right)\mathbf{a}.$ ■

Everything we have done in this section is equally valid for any finite-dimensional inner-product space. For example, formula (1) for projecting one vector on a one-dimensional subspace takes the form

$$\mathbf{p} = \left(\frac{\langle \mathbf{a}, \mathbf{b} \rangle}{\langle \mathbf{a}, \mathbf{a} \rangle}\right)\mathbf{a}. \tag{4}$$

We conclude with an example in the space P_2 of polynomial functions, using projection to find the best constant function approximation of a nonconstant function.

EXAMPLE 6   (*Calculus*) Let the inner product of two polynomials $p(x)$ and $q(x)$ in the space P_2 of polynomial functions of degree ≤ 2 be defined by

$$\langle p(x), q(x) \rangle = \int_0^1 p(x)q(x)\ dx.$$

(See Example 5 of Section 2.10.) Find the projection of $f(x) = x$ on sp(1), using formula (4). Then find the projection of x on sp(1)$^\perp$.

Solution   By formula (4), the projection of x on sp(1) is

$b_W =$ $\left(\dfrac{\langle x, 1 \rangle}{\langle 1, 1 \rangle}\right)1 = \dfrac{\displaystyle\int_0^1 (x)(1)\ dx}{\displaystyle\int_0^1 (1)(1)\ dx} = \left(\dfrac{1/2}{1}\right)1 = \dfrac{1}{2}.$

$\dfrac{x^2}{2} \cdot x \Big|_0^1 = \dfrac{2}{2} \cdot \dfrac{1}{x^2} = \dfrac{x}{2}\Big|_0^1$

The projection of x on sp(1)$^\perp$ is obtained by subtracting the projection on sp(1) from x. We obtain $x - \dfrac{1}{2}$. As a check, we should then have $\langle \dfrac{1}{2}, x - \dfrac{1}{2} \rangle = 0$, and a computation of $\int_0^1 \left(\dfrac{1}{2}\right)\left(x - \dfrac{1}{2}\right)dx$ shows that this is indeed so. ■

SUMMARY

1. The projection of **b** in $\mathbb{R}^n$ on sp(**a**) for a nonzero vector **a** in $\mathbb{R}^n$ is given by $((\mathbf{a} \cdot \mathbf{b})/(\mathbf{a} \cdot \mathbf{a}))\mathbf{a}$.

2. The orthogonal complement $W^{\perp}$ of a subspace W of $\mathbb{R}^n$ is the set of all vectors in $\mathbb{R}^n$ that are orthogonal to every vector in W. Further, $W^{\perp}$ is a subspace of $\mathbb{R}^n$ of dimension $n - \dim(W)$, and $(W^{\perp})^{\perp} = W$.

3. The row space and the nullspace of an $m \times n$ matrix A are orthogonal complements of each other. In particular, $W^{\perp}$ can be computed as the nullspace of a matrix A having as its rows the vectors in a generating set for W.

4. Let W be a subspace of $\mathbb{R}^n$. Each vector **b** in $\mathbb{R}^n$ can be expressed uniquely in the form $\mathbf{b} = \mathbf{b}_W + \mathbf{b}_{W^{\perp}}$ for $\mathbf{b}_W$ in W and $\mathbf{b}_{W^{\perp}}$ in $W^{\perp}$.

5. The vectors $\mathbf{b}_W$ and $\mathbf{b}_{W^{\perp}}$ are the projections of **b** on W and on $W^{\perp}$, respectively. They can be computed following the boxed procedure on page 280.

EXERCISES

In Exercises 1–6, find the indicated projections.

1. The projection of $(2, 1)$ on sp$((3, 4))$ in $\mathbb{R}^2$

2. The projection of $(3, 4)$ on sp$((2, 1))$ in $\mathbb{R}^2$

3. The projection of $(1, 2, 1)$ on each of the unit coordinate vectors in $\mathbb{R}^3$

4. The projection of $(1, 2, 1)$ on the line with parametric equations $x = 3t$, $y = t$, $z = 2t$ in $\mathbb{R}^3$

5. The projection of $(-1, 2, 0, 1)$ on sp$((2, -3, 1, 2))$ in $\mathbb{R}^4$

6. The projection of $(2, -1, 3, -5)$ on the line in $\mathbb{R}^4$ with parametric equations $x_1 = t$, $x_2 = 0$, $x_3 = -t$, $x_4 = 2t$

In Exercises 7–12, find the orthogonal complement of the given subspace.

7. The subspace sp$((1, 2, -1))$ in $\mathbb{R}^3$

8. The line with parametric equations $x_1 = 2t$, $x_2 = -t$, $x_3 = 0$, $x_4 = -3t$ in $\mathbb{R}^4$

9. The subspace sp$((1, 3, 0), (2, 1, 4))$ in $\mathbb{R}^3$

10. The plane $2x + y + 3z = 0$ in $\mathbb{R}^3$

11. The subspace sp$((2, 1, 3, 4), (1, 0, -2, 1))$ in $\mathbb{R}^4$

12. The subspace (hyperplane) $ax_1 + bx_2 + cx_3 + dx_4 = 0$ in $\mathbb{R}^4$ [HINT: Recall how to find a normal vector to a hyperplane.]

13. Find a nonzero vector in $\mathbb{R}^3$ perpendicular to $(1, 1, 2)$ and $(2, 3, 1)$ by
   a) the methods of this section,
   b) computing a determinant.

14. Find a nonzero vector in $\mathbb{R}^4$ perpendicular to $(1, 0, -1, 1)$, $(0, 0, -1, 1)$, and $(2, -1, 2, 0)$ by
   a) the methods of this section,
   b) computing a determinant.

In Exercises 15–22, find the indicated projections.

15. The projection of $(1, 2, 1)$ on the subspace sp$((3, 1, 2), (1, 0, 1))$ of $\mathbb{R}^3$

16. The projection of $(1, 2, 1)$ on the plane $x + y + z = 0$ in $\mathbb{R}^3$

17. The projection of $(1, 0, 0)$ on the subspace sp$((2, 1, 1), (1, 0, 2))$ of $\mathbb{R}^3$

18. The projection of $(-1, 0, 1)$ on the plane $x + y = 0$ in $\mathbb{R}^3$

19. The projection of $(0, 0, 1)$ on the plane $2x - y - z = 0$ in $\mathbb{R}^3$

20. The projection in $\mathbb{R}^4$ of $(-2, 1, 3, -5)$ on
   a) the subspace $\mathrm{sp}(\mathbf{e}_3)$
   b) the subspace $\mathrm{sp}(\mathbf{e}_1, \mathbf{e}_4)$
   c) the subspace $\mathrm{sp}(\mathbf{e}_1, \mathbf{e}_3, \mathbf{e}_4)$
   d) $\mathbb{R}^4$

21. The projection of $(1, 0, -1, 1)$ on the subspace $\mathrm{sp}((1, 0, 0, 0), (0, 1, 1, 0), (0, 0, 1, 1))$ in $\mathbb{R}^4$

22. The projection of $(0, 1, -1, 0)$ on the subspace (hyperplane) $x_1 - x_2 + x_3 + x_4 = 0$ in $\mathbb{R}^4$ [HINT: See Example 4.]

23. (Calculus) Referring to Example 6, find the projection of $f(x) = 1$ on $\mathrm{sp}(x)$ in P_2.

24. (Calculus) Referring to Example 6, find the projection of $f(x) = x$ on $\mathrm{sp}(1 + x)$.

25. Assume that $\mathbf{a}$, $\mathbf{b}$, and $\mathbf{c}$ are vectors in $\mathbb{R}^n$ and that W is a subspace of $\mathbb{R}^n$. Mark each of the following True or False.
   ___ a) The projection of $\mathbf{b}$ on $\mathrm{sp}(\mathbf{a})$ is a scalar multiple of $\mathbf{b}$.
   ___ b) The projection of $\mathbf{b}$ on $\mathrm{sp}(\mathbf{a})$ is a scalar multiple of $\mathbf{a}$.
   ___ c) The set of all vectors in $\mathbb{R}^n$ orthogonal to every vector in W is a subspace of $\mathbb{R}^n$.
   ___ d) If S is any set of vectors in an inner-product space V, then the set of all vectors in V that are orthogonal to every vector in S is a subspace of V.
   ___ e) If the projection of $\mathbf{b}$ on W is $\mathbf{b}$ itself, then $\mathbf{b}$ is orthogonal to every vector in W.
   ___ f) If the projection of $\mathbf{b}$ on W is $\mathbf{b}$ itself, then $\mathbf{b}$ is in W.
   ___ g) The vector $\mathbf{b}$ is orthogonal to every vector in W if and only if $\mathbf{b}_W = \mathbf{0}$.
   ___ h) The intersection of W and $W^{\perp}$ is empty.
   ___ i) If $\mathbf{b}$ and $\mathbf{c}$ have the same projection on W, then $\mathbf{b} = \mathbf{c}$.
   ___ j) If $\mathbf{b}$ and $\mathbf{c}$ have the same projection on every subspace of $\mathbb{R}^n$, then $\mathbf{b} = \mathbf{c}$.

26. Let $\mathbf{a}$ and $\mathbf{b}$ be nonzero vectors in $\mathbb{R}^n$, and let θ be the angle between $\mathbf{a}$ and $\mathbf{b}$. The scalar $\|\mathbf{b}\| \cos \theta$ is called the **scalar component** of $\mathbf{b}$ along $\mathbf{a}$. Interpret this scalar graphically (see Figures 5.1 and 5.2), and give a formula for it in terms of the dot product.

27. Let W be a subspace of $\mathbb{R}^n$ and let $\mathbf{b}$ be a vector in $\mathbb{R}^n$. Show that there is one and only one vector $\mathbf{p}$ in W such that $\mathbf{b} - \mathbf{p}$ is perpendicular to every vector in W. [HINT: Suppose that $\mathbf{p}_1$ and $\mathbf{p}_2$ are two such vectors, and show that $\mathbf{p}_1 - \mathbf{p}_2$ is in $W^{\perp}$.]

28. Let A be an $m \times n$ matrix.
   a) Show that the set W of row vectors $\mathbf{x}$ in $\mathbb{R}^m$ such that $\mathbf{x}A = \mathbf{0}$ is a subspace of $\mathbb{R}^m$.
   b) Show that the subspace W in part (a) and the column space of A are orthogonal complements.

29. Subspaces U and W of $\mathbb{R}^n$ are **orthogonal** if $\mathbf{u} \cdot \mathbf{w} = 0$ for all $\mathbf{u}$ in U and all $\mathbf{w}$ in W. Let U and W be orthogonal subspaces of $\mathbb{R}^n$, and let $\dim(U) = n - \dim(W)$. Show that each subspace is the orthogonal complement of the other.

30. Let S and T be nonempty subsets of an inner-product space V with the property that every vector in S is orthogonal to every vector in T. Show that the span of S and the span of T are orthogonal subspaces of V.

31. Let W be a subspace of $\mathbb{R}^n$ with orthogonal complement $W^{\perp}$. Writing $\mathbf{a} = \mathbf{a}_W + \mathbf{a}_{W^{\perp}}$, as in Theorem 5.1, prove that

$$\|\mathbf{a}\| = \sqrt{\|\mathbf{a}_W\|^2 + \|\mathbf{a}_{W^{\perp}}\|^2}.$$

[HINT: Use the formula $\|\mathbf{a}\|^2 = \mathbf{a} \cdot \mathbf{a}$.]

32. (*Distance from a point to a subspace*). Let W be a subspace of $\mathbb{R}^n$. Figure 5.5 suggests that the distance from a point $\mathbf{a}$ in $\mathbb{R}^n$ to the subspace W is equal to the magnitude of the projection of the vector $\mathbf{a}$ on the orthogonal complement of W. Find the distance from the point $(1, 2, 3)$ in $\mathbb{R}^3$ to the subspace (plane) $\mathrm{sp}((2, 2, 1), (1, 2, 1))$.

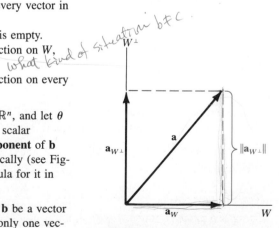

FIGURE 5.5 The distance from a to W is $\|\mathbf{a}_W{}^{\perp}\|$.

33. Find the distance from the point (2, 1, 3, 1) in $\mathbb{R}^4$ to the plane $sp((1, 0, 1, 0), (1, -1, 1, 1))$. [HINT: See Exercise 32.]

*In Exercises 34–39, use the idea in Exercise 32 to find the distance from the given point **a** to the given one-dimensional subspace (line). [NOTE: To calculate $\|\mathbf{a}_{W^\perp}\|$, first calculate $\|\mathbf{a}_W\|$ and then use Exercise 31.]*

34. $\mathbf{a} = (1, 2, 1)$,
    $W = sp((2, 1, 0))$ in $\mathbb{R}^3$

35. $\mathbf{a} = (2, -1, 3)$,
    $W = sp((1, 2, 4))$ in $\mathbb{R}^3$

36. $\mathbf{a} = (1, 2, -1, 0)$,
    $W = sp((3, 1, 4, -1))$ in $\mathbb{R}^4$

37. $\mathbf{a} = (2, 1, 1, 2)$,
    $W = sp((1, 2, 1, 3))$ in $\mathbb{R}^4$

38. $\mathbf{a} = (1, 2, 3, 4, 5)$,
    $W = sp((1, 1, 1, 1, 1))$ in $\mathbb{R}^5$

39. $\mathbf{a} = (1, 0, 1, 0, 1, 0, 1)$,
    $W = sp((1, 2, 3, 4, 3, 2, 1))$ in $\mathbb{R}^7$

5.2 THE GRAM–SCHMIDT PROCESS

In the preceding section, we saw how to project a vector **b** on a subspace W of $\mathbb{R}^n$. The calculations can be somewhat tedious. We open this section by observing that, if we know a basis for W consisting of mutually perpendicular vectors, the computational burden can be eased. We then present the Gram–Schmidt algorithm, showing how such a nice basis for W can be found.

Orthogonal and Orthonormal Bases

A set $\{\mathbf{v}_1, \mathbf{v}_2, \ldots, \mathbf{v}_k\}$ of nonzero vectors in $\mathbb{R}^n$ is **orthogonal** if the vectors $\mathbf{v}_j$ are mutually perpendicular—that is, if $\mathbf{v}_i \cdot \mathbf{v}_j = 0$ for $i \neq j$. Our next theorem shows that an orthogonal generating set for a subspace of $\mathbb{R}^n$ is sure to be a basis.

THEOREM 5.2 Orthogonal Bases

Let $\{\mathbf{v}_1, \mathbf{v}_2, \ldots, \mathbf{v}_k\}$ be an orthogonal set of nonzero vectors in $\mathbb{R}^n$. Then this set is independent and consequently is a basis for $sp(\mathbf{v}_1, \mathbf{v}_2, \ldots, \mathbf{v}_k)$.

PROOF To show that the orthogonal set $\{\mathbf{v}_1, \mathbf{v}_2, \ldots, \mathbf{v}_k\}$ is independent, let us suppose that

$$\mathbf{v}_j = s_1\mathbf{v}_1 + s_2\mathbf{v}_2 + \cdots + s_{j-1}\mathbf{v}_{j-1}.$$

Taking the dot product of both sides of this equation with $\mathbf{v}_j$ yields $\mathbf{v}_j \cdot \mathbf{v}_j = 0$, which contradicts the hypothesis that $\mathbf{v}_j \neq \mathbf{0}$. Thus, no $\mathbf{v}_j$ is a linear combination of its predecessors, so $\{\mathbf{v}_1, \mathbf{v}_2, \ldots, \mathbf{v}_k\}$ is independent and thus is a basis for $sp(\mathbf{v}_1, \mathbf{v}_2, \ldots, \mathbf{v}_k)$. ▲

EXAMPLE 1 Find an orthogonal basis for the plane $2x - y + z = 0$ in $\mathbb{R}^3$.

Solution The given plane contains the origin and hence is a subspace of $\mathbb{R}^3$. We need only find two perpendicular vectors $\mathbf{v}_1$ and $\mathbf{v}_2$ in this plane. Letting $y = 0$ and $z = 2$ we find that $x = -1$ in the given equation, so $\mathbf{v}_1 = (-1, 0, 2)$ lies in the plane. Since the vector $(2, -1, 1)$ of coefficients is perpendicular to the plane, we compute a cross product, and let

$$\mathbf{v}_2 = (-1, 0, 2) \times (2, -1, 1) = (2, 5, 1).$$

This vector is perpendicular to the coefficient vector $(2, -1, 1)$, so it lies in the plane; and of course, it is also perpendicular to the vector $(-1, 0, 2)$. Thus, $\{\mathbf{v}_1, \mathbf{v}_2\}$ is an orthogonal basis for the plane. ■

Now we show how easy it is to project a vector $\mathbf{b}$ on a subspace W in $\mathbb{R}^n$ if we know an orthogonal basis $\{\mathbf{v}_1, \mathbf{v}_2, \ldots, \mathbf{v}_k\}$ for W. Recall from Section 5.1 that

$$\mathbf{b} = \mathbf{b}_W + \mathbf{b}_{W^\perp}, \tag{1}$$

where $\mathbf{b}_W$ is the projection of $\mathbf{b}$ on W, and $\mathbf{b}_{W^\perp}$ is the projection of $\mathbf{b}$ on $W^\perp$. Since $\mathbf{b}_W$ lies in W, we have

$$\mathbf{b}_W = r_1\mathbf{v}_1 + r_2\mathbf{v}_2 + \cdots + r_k\mathbf{v}_k \tag{2}$$

for some choice of scalars r_i. Computing the dot product of $\mathbf{b}$ with $\mathbf{v}_i$ and using Eqs. (1) and (2), we have

$$\mathbf{b} \cdot \mathbf{v}_i = (\mathbf{b}_W \cdot \mathbf{v}_i) + (\mathbf{b}_{W^\perp} \cdot \mathbf{v}_i)$$
$$= (r_1\mathbf{v}_1 \cdot \mathbf{v}_i + r_2\mathbf{v}_2 \cdot \mathbf{v}_i + \cdots + r_k\mathbf{v}_k \cdot \mathbf{v}_i) + 0 \quad \mathbf{v}_i \text{ is in } W$$
$$= r_i\mathbf{v}_i \cdot \mathbf{v}_i. \qquad\qquad \mathbf{v}_i \cdot \mathbf{v}_j = 0 \text{ for } i \neq j$$

Therefore, $r_i = (\mathbf{b} \cdot \mathbf{v}_i)/(\mathbf{v}_i \cdot \mathbf{v}_i)$, so

$$r_i\mathbf{v}_i = \left(\frac{\mathbf{b} \cdot \mathbf{v}_i}{\mathbf{v}_i \cdot \mathbf{v}_i}\right)\mathbf{v}_i,$$

which is just the projection of $\mathbf{b}$ on $\mathbf{v}_i$. In other words, to project $\mathbf{b}$ on W, we need only project $\mathbf{b}$ on each of the orthogonal basis vectors, and then add! We summarize this in a theorem.

THEOREM 5.3   Projection Using an Orthogonal Basis

Let $\{\mathbf{v}_1, \mathbf{v}_2, \ldots, \mathbf{v}_k\}$ be an orthogonal basis for a subspace W of $\mathbb{R}^n$, and let $\mathbf{b}$ be any vector in $\mathbb{R}^n$. The projection of $\mathbf{b}$ on W is

$$\mathbf{b}_W = \left(\frac{\mathbf{b} \cdot \mathbf{v}_1}{\mathbf{v}_1 \cdot \mathbf{v}_1}\right)\mathbf{v}_1 + \left(\frac{\mathbf{b} \cdot \mathbf{v}_2}{\mathbf{v}_2 \cdot \mathbf{v}_2}\right)\mathbf{v}_2 + \cdots + \left(\frac{\mathbf{b} \cdot \mathbf{v}_k}{\mathbf{v}_k \cdot \mathbf{v}_k}\right)\mathbf{v}_k. \tag{3}$$

EXAMPLE 2   Find the projection of $\mathbf{b} = (3, -2, 2)$ on the plane $2x - y + z = 0$ in $\mathbb{R}^3$.

Solution   In Example 1, we found an orthogonal basis for the given plane, consisting of the vectors $\mathbf{v}_1 = (-1, 0, 2)$ and $\mathbf{v}_2 = (2, 5, 1)$. Thus, the plane may be expressed as $W = \text{sp}(\mathbf{v}_1, \mathbf{v}_2)$. Using Eq. (3), we have

$$\mathbf{b}_W = \left(\frac{\mathbf{b} \cdot \mathbf{v}_1}{\mathbf{v}_1 \cdot \mathbf{v}_1}\right)\mathbf{v}_1 + \left(\frac{\mathbf{b} \cdot \mathbf{v}_2}{\mathbf{v}_2 \cdot \mathbf{v}_2}\right)\mathbf{v}_2$$

$$= \tfrac{1}{5}(-1, 0, 2) + \left(-\tfrac{2}{30}\right)(2, 5, 1) = \tfrac{3}{15}(-1, 0, 2) - \tfrac{1}{15}(2, 5, 1)$$

$$= \left(-\tfrac{1}{3}, -\tfrac{1}{3}, \tfrac{1}{3}\right). \qquad ■$$

It is sometimes desirable to *normalize* the vectors in an orthogonal basis, converting each basis vector to one parallel to it but of unit length. The result remains a basis for the same subspace. Notice that the standard basis in $\mathbb{R}^n$ consists of such *perpendicular unit vectors*. Such bases are extremely useful and merit a formal definition.

DEFINITION 5.3   Orthonormal Basis

Let W be a subspace of $\mathbb{R}^n$. A basis $\{\mathbf{q}_1, \mathbf{q}_2, \ldots, \mathbf{q}_k\}$ for W is **orthonormal** if

(i) $\mathbf{q}_i \cdot \mathbf{q}_j = 0$ for $i \neq j$, and   **Mutually perpendicular**

(ii) $\mathbf{q}_i \cdot \mathbf{q}_i = 1$.   **Length 1**

The standard basis for $\mathbb{R}^n$ is just one of many orthonormal bases for $\mathbb{R}^n$ if $n > 1$. For example, any two perpendicular vectors $\mathbf{v}_1$ and $\mathbf{v}_2$ on the unit circle (illustrated in Fig. 5.6) form an orthonormal basis for $\mathbb{R}^2$.

For the projection of a vector on a subspace that has a known orthonormal basis, Eq. (3) in Theorem 5.3 assumes a simpler form:

Projection of b on W with Orthonormal Basis $\{\mathbf{q}_1, \mathbf{q}_2, \ldots, \mathbf{q}_k\}$

$$\mathbf{b}_W = (\mathbf{b} \cdot \mathbf{q}_1)\mathbf{q}_1 + (\mathbf{b} \cdot \mathbf{q}_2)\mathbf{q}_2 + \cdots + (\mathbf{b} \cdot \mathbf{q}_k)\mathbf{q}_k \qquad (4)$$

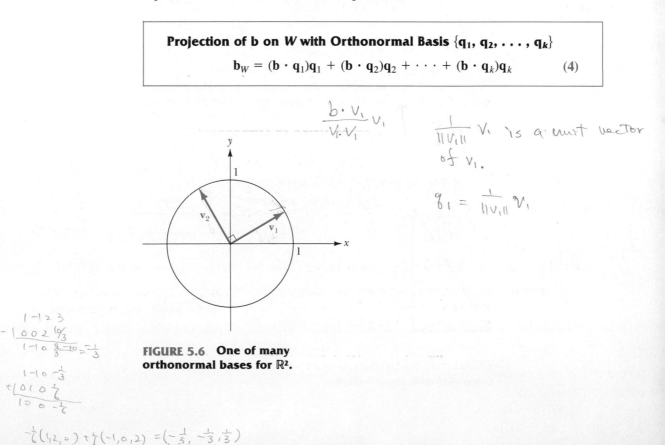

$$\frac{b \cdot v_1}{v_1 \cdot v_1} v_1$$

$\dfrac{1}{\|v_1\|} v_1$ is a unit vector of v_1.

$$q_1 = \frac{1}{\|v_1\|} v_1$$

FIGURE 5.6 One of many orthonormal bases for $\mathbb{R}^2$.

EXAMPLE 3 Find an orthonormal basis for $W = \text{sp}(\mathbf{v}_1, \mathbf{v}_2, \mathbf{v}_3)$ in $\mathbb{R}^4$ if $\mathbf{v}_1 = (1, 1, 1, 1)$, $\mathbf{v}_2 = (-1, 1, -1, 1)$, and $\mathbf{v}_3 = (1, -1, -1, 1)$. Then find the projection of $\mathbf{b} = (1, 2, 3, 4)$ on W.

Solution We see that $\mathbf{v}_1 \cdot \mathbf{v}_2 = \mathbf{v}_2 \cdot \mathbf{v}_3 = \mathbf{v}_1 \cdot \mathbf{v}_3 = 0$. Since $\|\mathbf{v}_1\| = \|\mathbf{v}_2\| = \|\mathbf{v}_3\| = 2$, we let $\mathbf{q}_i = \frac{1}{2}\mathbf{v}_i$ for $i = 1, 2, 3$, to obtain an orthonormal basis $\{\mathbf{q}_1, \mathbf{q}_2, \mathbf{q}_3\}$ for W.

To find the projection of $\mathbf{b} = (1, 2, 3, 4)$ on W, we use Eq. (4) and we obtain

$$\mathbf{b}_W = (\mathbf{b} \cdot \mathbf{q}_1)\mathbf{q}_1 + (\mathbf{b} \cdot \mathbf{q}_2)\mathbf{q}_2 + (\mathbf{b} \cdot \mathbf{q}_3)\mathbf{q}_3$$

$$= 5\mathbf{q}_1 + \mathbf{q}_2 + 0\mathbf{q}_3 = (2, 3, 2, 3). \qquad ■$$

The Gram–Schmidt Process

We now describe a computational technique for creating an orthonormal basis from a given basis of a subspace W of $\mathbb{R}^n$. The theorem that follows asserts the existence of such a basis; its proof is constructive. That is, the proof shows how an orthonormal basis can be constructed.

THEOREM 5.4 Orthonormal Basis (Gram–Schmidt) Theorem

Let W be a subspace of $\mathbb{R}^n$, let $\{\mathbf{a}_1, \mathbf{a}_2, \ldots, \mathbf{a}_k\}$ be any basis for W, and let

$$W_j = \text{sp}(\mathbf{a}_1, \mathbf{a}_2, \ldots, \mathbf{a}_j) \qquad \text{for} \qquad j = 1, 2, \ldots, k.$$

Then there is an orthonormal basis $\{\mathbf{q}_1, \mathbf{q}_2, \ldots, \mathbf{q}_k\}$ for W such that $W_j = \text{sp}(\mathbf{q}_1, \mathbf{q}_2, \ldots, \mathbf{q}_j)$.

PROOF Let $\mathbf{v}_1 = \mathbf{a}_1$. For $j = 2, \ldots, k$, let $\mathbf{p}_j$ be the projection of $\mathbf{a}_j$ on W_{j-1}, and let $\mathbf{v}_j = \mathbf{a}_j - \mathbf{p}_j$. That is, $\mathbf{v}_j$ is obtained by subtracting from $\mathbf{a}_j$ its projection on the subspace generated by its predecessors. Figure 5.7 gives a symbolic illustration. The decomposition

$$\mathbf{a}_j = \mathbf{p}_j + (\mathbf{a}_j - \mathbf{p}_j)$$

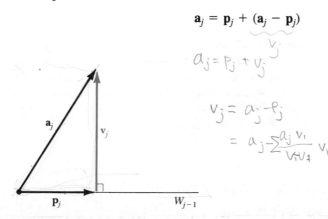

FIGURE 5.7 The vector $\mathbf{v}_j$ in the Gram–Schmidt construction.

is the unique expression for $\mathbf{a}_j$ as the sum of the vector $\mathbf{p}_j$ in W_{j-1} and the vector $\mathbf{a}_j - \mathbf{p}_j$ in $(W_{j-1})^\perp$, described in Theorem 5.1. Since $\mathbf{a}_j$ is in W_j and since $\mathbf{p}_j$ is in W_{j-1}, which is itself contained in W_j, we see that $\mathbf{v}_j = \mathbf{a}_j - \mathbf{p}_j$ lies in the subspace W_j. Now $\mathbf{v}_j$ is perpendicular to every vector in W_{j-1}. Consequently, $\mathbf{v}_j$ is perpendicular to $\mathbf{v}_1, \mathbf{v}_2, \ldots, \mathbf{v}_{j-1}$. We conclude that each vector in the set

$$\{\mathbf{v}_1, \mathbf{v}_2, \ldots, \mathbf{v}_j\} \tag{5}$$

is perpendicular to each of its predecessors. Thus the set (5) consists of j mutually perpendicular nonzero vectors in the j-dimensional subspace W_j, and so the set constitutes an orthogonal basis for W_j. It follows that, if we set $\mathbf{q}_i = (1/\|\mathbf{v}_i\|)\mathbf{v}_i$ for $i = 1, 2, \ldots, j$, then $W_j = \text{sp}(\mathbf{q}_1, \mathbf{q}_2, \ldots, \mathbf{q}_j)$. Taking $j = k$, we see that

$$\{\mathbf{q}_1, \mathbf{q}_2, \ldots, \mathbf{q}_k\}$$

is an orthonormal basis for W. ▲

The proof of Theorem 5.4 was computational, providing us with a technique for constructing an orthonormal basis for a subspace W of $\mathbb{R}^n$. The technique is known as the *Gram–Schmidt Process*, and we have boxed it for easy reference.

Gram–Schmidt Process

To find an orthonormal basis for a subspace W of $\mathbb{R}^n$:

1. Find a basis $\{\mathbf{a}_1, \mathbf{a}_2, \ldots, \mathbf{a}_k\}$ for W.
2. Let $\mathbf{v}_1 = \mathbf{a}_1$. For $j = 2, \ldots, k$, let $\mathbf{v}_j$ be obtained by subtracting from $\mathbf{a}_j$ its projection on the subspace generated by its predecessors.
3. The $\mathbf{v}_j$ so obtained form an orthogonal basis for W, and they may be normalized to yield an orthonormal basis.

When actually executing the Gram–Schmidt process, we project vectors on subspaces, as described in step 2 of the box. We know that it is best to work with an orthogonal or orthonormal basis for a subspace when projecting on it; and since the subspace $W_{j-1} = \text{sp}(\mathbf{a}_1, \mathbf{a}_2, \ldots, \mathbf{a}_{j-1})$ is also the subspace generated by the orthogonal set $\{\mathbf{v}_1, \mathbf{v}_2, \ldots, \mathbf{v}_{j-1}\}$, it is surely best to work with the latter basis for W_{j-1} when computing the desired projection of $\mathbf{a}_j$ on W_{j-1}. Step 2 in the box and Eq. (3) show that the specific formula for $\mathbf{v}_j$ is as follows:

General Gram–Schmidt Formula

$$\mathbf{v}_j = \mathbf{a}_j - \left(\left(\frac{\mathbf{a}_j \cdot \mathbf{v}_1}{\mathbf{v}_1 \cdot \mathbf{v}_1}\right)\mathbf{v}_1 + \left(\frac{\mathbf{a}_j \cdot \mathbf{v}_2}{\mathbf{v}_2 \cdot \mathbf{v}_2}\right)\mathbf{v}_2 + \cdots + \left(\frac{\mathbf{a}_j \cdot \mathbf{v}_{j-1}}{\mathbf{v}_{j-1} \cdot \mathbf{v}_{j-1}}\right)\mathbf{v}_{j-1}\right). \tag{6}$$

One may normalize the $\mathbf{v}_j$, forming the vector $\mathbf{q}_j = (1/\|\mathbf{v}_j\|)\mathbf{v}_j$, to obtain a vector of length 1 at each step of the construction. In that case, Eq. (6) can be replaced by the following simple form:

Normalized Gram–Schmidt Formula

$$\mathbf{v}_j = \mathbf{a}_j - ((\mathbf{a}_j \cdot \mathbf{q}_1)\mathbf{q}_1 + (\mathbf{a}_j \cdot \mathbf{q}_2)\mathbf{q}_2 + \cdots + (\mathbf{a}_j \cdot \mathbf{q}_{j-1})\mathbf{q}_{j-1}) \qquad (7)$$

The arithmetic using Eq. (6) and that using Eq. (7) are similar, but Eq. (6) postpones the introduction of the radicals from normalizing until the entire orthogonal basis is obtained. We shall use Eq. (6) in our work. However, a computer will generate less error if it normalizes as it goes along. This is indicated in the next section.

EXAMPLE 4   Find an orthonormal basis for the subspace $W = \text{sp}((1, 0, 1), (1, 1, 1))$ of $\mathbb{R}^3$.

Solution   We use the Gram–Schmidt process with formula (6), finding first an orthogonal basis for W. We take $\mathbf{v}_1 = (1, 0, 1)$. From formula (6) with $\mathbf{v}_1 = (1, 0, 1)$ and $\mathbf{a}_2 = (1, 1, 1)$, we have

$$\mathbf{v}_2 = \mathbf{a}_2 - \left(\frac{\mathbf{a}_2 \cdot \mathbf{v}_1}{\mathbf{v}_1 \cdot \mathbf{v}_1}\right)\mathbf{v}_1 = (1, 1, 1) - (2/2)(1, 0, 1) = (0, 1, 0).$$

An orthogonal basis for W is $\{(1, 0, 1), (0, 1, 0)\}$, and an orthonormal basis is $\{(1/\sqrt{2}, 0, 1/\sqrt{2}), (0, 1, 0)\}$.   ■

Referring to the proof of Theorem 5.4, we see that the sets $\{\mathbf{v}_1, \mathbf{v}_2, \ldots, \mathbf{v}_j\}$ and $\{\mathbf{q}_1, \mathbf{q}_2, \ldots, \mathbf{q}_j\}$ are both bases for the subspace $W_j = \text{sp}(\mathbf{a}_1, \mathbf{a}_2, \ldots, \mathbf{a}_j)$. Consequently, the vector $\mathbf{a}_j$ can be expressed as a linear combination

$$\mathbf{a}_j = r_{1j}\mathbf{q}_1 + r_{2j}\mathbf{q}_2 + \cdots + r_{jj}\mathbf{q}_j. \qquad (8)$$

THE GRAM–SCHMIDT PROCESS is named for the Danish mathematician Jorgen P. Gram (1850–1916) and the German Erhard Schmidt (1876–1959). It was first published by Gram in 1883 in a paper entitled "Series Development Using the Method of Least Squares." It was published again with a careful proof by Schmidt in 1907 in a work on integral equations. In fact, Schmidt even referred to Gram's result. For Schmidt, as for Gram, the vectors were continuous functions defined on an interval $[a, b]$ with the inner product of two such functions ϕ, ψ being given as $\int_a^b \phi(x)\psi(x)\, dx$. Schmidt was more explicit than Gram, however, writing out the process in great detail and proving that the set of functions ψ_i derived from his original set ϕ_i was in fact an orthonormal set.

Schmidt, who was at the University of Berlin from 1917 until his death, is best known for his definitive work on Hilbert spaces—spaces of square summable sequences of complex numbers. In fact, he applied the Gram–Schmidt process to sets of vectors in these spaces to help develop necessary and sufficient conditions for such sets to be linearly independent.

Eqs. (8) for $j = 1, 2, \ldots, k$ can be written in matrix form as

$$\begin{bmatrix} | & | & & | \\ \mathbf{a}_1 & \mathbf{a}_2 & \cdots & \mathbf{a}_k \\ | & | & & | \end{bmatrix} = \begin{bmatrix} | & | & & | \\ \mathbf{q}_1 & \mathbf{q}_2 & \cdots & \mathbf{q}_k \\ | & | & & | \end{bmatrix} \begin{bmatrix} r_{11} & r_{12} & \cdots & r_{1k} \\ & r_{22} & \cdots & r_{2k} \\ & & \ddots & \vdots \\ \mathbf{0} & & & r_{kk} \end{bmatrix},$$

$$\underbrace{}_{A} \qquad \underbrace{}_{Q} \qquad \underbrace{}_{R}$$

so $A = QR$ for the indicated matrices. Since each $\mathbf{a}_j$ is in W_j but not in W_{j-1}, we see that no r_{jj} is zero, so R is an invertible $k \times k$ matrix. This factorization $A = QR$ is important in numerical linear algebra, and we state it as a corollary. We will find use for it in Sections 5.5 and 7.4.

COROLLARY 1   *QR* Factorization

Let A be an $n \times k$ matrix with independent column vectors in $\mathbb{R}^n$. There exists an $n \times k$ matrix Q with orthonormal column vectors and an upper-triangular invertible $k \times k$ matrix R such that $A = QR$.

EXAMPLE 5   Let

$$A = \begin{bmatrix} 1 & 1 \\ 0 & 1 \\ 1 & 1 \end{bmatrix}.$$

Factor A in the form $A = QR$ described in Corollary 1 of Theorem 5.4, using the computations in Example 4.

Solution   From Example 4, we see that we can take

$$Q = \begin{bmatrix} 1/\sqrt{2} & 0 \\ 0 & 1 \\ 1/\sqrt{2} & 0 \end{bmatrix}$$

and solve $QR = A$ for the matrix R. That is, we solve the matrix equation

$$\begin{bmatrix} 1/\sqrt{2} & 0 \\ 0 & 1 \\ 1/\sqrt{2} & 0 \end{bmatrix} \begin{bmatrix} r_{11} & r_{12} \\ 0 & r_{22} \end{bmatrix} = \begin{bmatrix} 1 & 1 \\ 0 & 1 \\ 1 & 1 \end{bmatrix}$$

for the entries $r_{11}, r_{12},$ and r_{22}. This corresponds to two linear systems of three equations each, but by inspection we see that $r_{11} = \sqrt{2}$, $r_{12} = \sqrt{2}$, and $r_{22} = 1$. Thus,

$$A = \begin{bmatrix} 1 & 1 \\ 0 & 1 \\ 1 & 1 \end{bmatrix} = \begin{bmatrix} 1/\sqrt{2} & 0 \\ 0 & 1 \\ 1/\sqrt{2} & 0 \end{bmatrix} \begin{bmatrix} \sqrt{2} & \sqrt{2} \\ 0 & 1 \end{bmatrix} = QR. \qquad ■$$

We give another illustration of the Gram–Schmidt process, this time requiring two applications of formula (6).

EXAMPLE 6 Using the Gram–Schmidt process, transform the basis $\{(1, 2, 1), (2, 1, 1), (1, 1, 2)\}$ of $\mathbb{R}^3$ into an orthonormal one.

Solution First we find an orthogonal basis, using formula (6). We take $\mathbf{v}_1 = (1, 2, 1)$, and we compute $\mathbf{v}_2$ by subtracting from $\mathbf{a}_2 = (2, 1, 1)$ its projection on $\mathbf{v}_1$:

$$\mathbf{v}_2 = \mathbf{a}_2 - \left(\frac{\mathbf{v}_1 \cdot \mathbf{a}_2}{\mathbf{v}_1 \cdot \mathbf{v}_1}\right)\mathbf{v}_1 = (2, 1, 1) - \tfrac{5}{6}(1, 2, 1)$$

$$= \left(\tfrac{7}{6}, -\tfrac{4}{6}, \tfrac{1}{6}\right).$$

To ease computations, we replace $\mathbf{v}_2$ by $6\mathbf{v}_2$, obtaining $\mathbf{v}_2 = (7, -4, 1)$. Finally we subtract from $\mathbf{a}_3 = (1, 1, 2)$ its projection on the subspace $\mathrm{sp}(\mathbf{v}_1, \mathbf{v}_2)$, obtaining

$$\mathbf{v}_3 = \mathbf{a}_3 - \left(\left(\frac{\mathbf{a}_3 \cdot \mathbf{v}_1}{\mathbf{v}_1 \cdot \mathbf{v}_1}\right)\mathbf{v}_1 + \left(\frac{\mathbf{a}_3 \cdot \mathbf{v}_2}{\mathbf{v}_2 \cdot \mathbf{v}_2}\right)\mathbf{v}_2\right)$$

$$= (1, 1, 2) - \left(\tfrac{5}{6}(1, 2, 1) + \tfrac{5}{66}(7, -4, 1)\right)$$

$$= (1, 1, 2) - \tfrac{1}{66}(90, 90, 60)$$

$$= \left(-\tfrac{24}{66}, -\tfrac{24}{66}, \tfrac{72}{66}\right) = \tfrac{4}{11}(-1, -1, 3).$$

Replacing $\mathbf{v}_3$ by $\tfrac{11}{4}\mathbf{v}_3$, we see that the basis $\{(1, 2, 1), (7, -4, 1), (-1, -1, 3)\}$ is orthogonal. An orthonormal basis is

$$\{(1/\sqrt{6})(1, 2, 1), (1/\sqrt{66})(7, -4, 1), (1/\sqrt{11})(-1, -1, 3)\}. \qquad ■$$

As you can see, the arithmetic involved in the Gram–Schmidt process can be a bit tedious with pencil and paper, but it is very easy to implement the construction on a computer.

We know that any independent set of vectors in $\mathbb{R}^n$ can be extended to a basis for $\mathbb{R}^n$. Using Theorem 5.4, we can prove a similar result for orthogonal sets.

COROLLARY 2 Expansion of an Orthogonal Set to an Orthogonal Basis

Every orthogonal set of vectors in a subspace W of $\mathbb{R}^n$ can be expanded if necessary to an orthogonal basis for W.

PROOF An orthogonal set $\{\mathbf{v}_1, \mathbf{v}_2, \ldots, \mathbf{v}_r\}$ of vectors in W is an independent set by Theorem 5.2, and can be expanded to a basis $\{\mathbf{v}_1, \ldots, \mathbf{v}_r, \mathbf{a}_1, \ldots, \mathbf{a}_s\}$ of W by Theorem 2.12. We apply the Gram–Schmidt process to this basis for W. Since the $\mathbf{v}_j$ are already mutually perpendicular, none of them will be changed by the Gram–Schmidt process, which thus yields an orthogonal basis containing the given vectors $\mathbf{v}_j$ for $j = 1, \ldots, r$. ▲

EXAMPLE 7 Expand $\{(1, 1, 0), (1, -1, 1)\}$ to an orthogonal basis for $\mathbb{R}^3$, and then transform this to an orthonormal basis for $\mathbb{R}^3$.

Solution First we expand the given set to a basis $\{\mathbf{a}_1, \mathbf{a}_2, \mathbf{a}_3\}$ for $\mathbb{R}^3$. We take $\mathbf{a}_1 = (1, 1, 0)$,

$a_2 = (1, -1, 1)$, and $a_3 = (1, 0, 0)$, which we can see form a basis for $\mathbb{R}^3$. (See Theorem 2.13.)

Now we use the Gram–Schmidt process with formula (6). Since a_1 and a_2 are perpendicular, we let $v_1 = a_1 = (1, 1, 0)$ and $v_2 = a_2 = (1, -1, 1)$. From formula (6), we have

$$[1, 1, 0]\begin{bmatrix} 1 \\ -1 \\ -1 \end{bmatrix} = 1 - 1 = 0$$

$$v_3 = a_3 - \left(\left(\frac{a_3 \cdot v_1}{v_1 \cdot v_1}\right)v_1 + \left(\frac{a_3 \cdot v_2}{v_2 \cdot v_2}\right)v_2\right)$$

$$= (1, 0, 0) - \left(\tfrac{1}{2}(1, 1, 0) + \tfrac{1}{3}(1, -1, 1)\right)$$

$$= (1, 0, 0) - \left(\tfrac{5}{6}, \tfrac{1}{6}, \tfrac{1}{3}\right) = \left(\tfrac{1}{6}, -\tfrac{1}{6}, -\tfrac{1}{3}\right).$$

Multiplying this vector by -6, we replace v_3 by $(-1, 1, 2)$. Thus we have expanded the given set to an orthogonal basis

$$[1, 1, 0]\begin{bmatrix} -1 \\ 1 \\ 2 \end{bmatrix} = -1 + 1 = 0$$

$$\{(1, 1, 0), (1, -1, 1), (-1, 1, 2)\}$$

of $\mathbb{R}^3$. Normalizing these vectors to unit length, we obtain

$$[1, -1, 1]\begin{bmatrix} -1 \\ 1 \\ 2 \end{bmatrix} = -1 - 1 + 2 = 0$$

$$\{(1/\sqrt{2}, 1/\sqrt{2}, 0), (1/\sqrt{3}, -1/\sqrt{3}, 1/\sqrt{3}), (-1/\sqrt{6}, 1/\sqrt{6}, 2/\sqrt{6})\}$$

as an orthonormal basis. ■

The results in this section easily extend to any inner-product space. We have the notions of an orthogonal set, an orthogonal basis, and an orthonormal basis, with essentially the same definitions given earlier. The Gram–Schmidt process is still valid. In the exercises, we ask you to use the Gram–Schmidt process with the polynomial space P_2, transforming the basis $\{1, x, x^2\}$ into an orthonormal basis, with the inner product given in Example 6 of Section 5.1.

SUMMARY

1. A basis for a subspace W is orthogonal if the basis vectors are mutually perpendicular, and it is orthonormal if the vectors also have length 1.

2. Any orthogonal set of vectors in $\mathbb{R}^n$ is a basis for the subspace it generates.

3. Let W be a subspace of $\mathbb{R}^n$ with an orthogonal basis. The projection of a vector b in $\mathbb{R}^n$ on W is equal to the sum of the projections of b on each basis vector.

4. Every nonzero subspace W of $\mathbb{R}^n$ has an orthonormal basis. Any basis can be transformed into an orthogonal basis by means of the Gram–Schmidt process, in which each vector a_j of the given basis is replaced by the vector v_j obtained by subtracting from a_j its projection on the subspace generated by its predecessors.

5. Any orthogonal set of vectors in a subspace W of $\mathbb{R}^n$ can be expanded, if necessary, to an orthogonal basis for W.

6. Let A be an $n \times k$ matrix of rank k. Then A can be factored as QR, where Q is an $n \times k$ matrix with orthonormal column vectors and R is a $k \times k$ upper-triangular invertible matrix.

EXERCISES

In Exercises 1–4, verify that the generating set of the given subspace W is orthogonal, and find the projection of the given vector **b** on W.

1. $W = \text{sp}((2, 3, 1), (-1, 1, -1))$; **b** = (2, 1, 4)

2. $W = \text{sp}((-1, 0, 1), (1, 1, 1))$; **b** = (1, 2, 3)

3. $W = \text{sp}((1, -1, -1, 1), (1, 1, 1, 1), (-1, 0, 0, 1))$; **b** = (2, 1, 3, 1)

4. $W = \text{sp}((1, -1, 1, 1), (-1, 1, 1, 1), (1, 1, -1, 1))$; **b** = (1, 4, 1, 2)

5. Find an orthonormal basis for the plane $2x + 3y + z = 0$.

6. Find an orthonormal basis for the subspace
$W = \{(x_1, x_2, x_3, x_4) \mid x_1 = x_2 + 2x_3,$
$\qquad\qquad x_4 = -x_2 + x_3\}$
of $\mathbb{R}^4$.

7. Find an orthonormal basis for the subspace $\text{sp}((0, 1, 0), (1, 1, 1))$ of $\mathbb{R}^3$.

8. Find an orthonormal basis for the subspace $\text{sp}((1, 1, 0), (-1, 2, 1))$ of $\mathbb{R}^3$.

9. Transform the basis $\{(1, 0, 1), (0, 1, 2), (2, 1, 0)\}$ for $\mathbb{R}^3$ into an orthonormal basis, using the Gram–Schmidt process.

10. Repeat Exercise 9, using the basis $\{(1, 1, 1), (1, 0, 1), (0, 1, 1)\}$ for $\mathbb{R}^3$.

11. Find an orthonormal basis for the subspace of $\mathbb{R}^4$ spanned by (1, 0, 1, 0), (1, 1, 1, 0), and (1, -1, 0, 1).

12. Find an orthonormal basis for the subspace of $\mathbb{R}^5$ spanned by (1, -1, 1, 0, 0), (-1, 0, 0, 0, 1), (0, 0, 1, 0, 1), and (1, 0, 0, 1, 1).

13. Find the projection of (5, -3, 4) on the subspace in Exercise 7, using the orthonormal basis found there.

14. Repeat Exercise 13, but use the subspace in Exercise 8.

15. Find the projection of (2, 0, -1, 1) on the subspace in Exercise 11, using the orthonormal basis found there.

16. Find the projection of (-1, 0, 0, 1, -1) on the subspace in Exercise 12, using the orthonormal basis found there.

17. Find an orthonormal basis for $\mathbb{R}^4$ that contains an orthonormal basis for the subspace $\text{sp}((1, 0, 1, 0), (0, 1, 1, 0))$.

18. Find an orthogonal basis for the orthogonal complement of $\text{sp}((1, -1, 3))$ in $\mathbb{R}^3$.

19. Find an orthogonal basis for the nullspace of the matrix
$$\begin{bmatrix} 1 & 2 & 1 & 1 \\ 0 & 1 & -1 & 2 \\ 2 & 5 & 1 & 4 \\ 1 & 1 & 2 & -1 \end{bmatrix}.$$

20. Find an orthonormal basis for $\mathbb{R}^3$ that contains the vector $(1/\sqrt{3})(1, 1, 1)$.

21. Find an orthonormal basis for $\text{sp}((2, 1, 1), (1, -1, 2))$ that contains $(1/\sqrt{6})(2, 1, 1)$.

22. Find an orthogonal basis for $\text{sp}((1, 2, 1, 2), (2, 1, 2, 0))$ that contains (1, 2, 1, 2).

23. Find an orthogonal basis for $\text{sp}((2, 1, -1, 1), (1, 1, 3, 0), (1, 1, 1, 1))$ that contains (2, 1, -1, 1) and (1, 1, 3, 0).

24. Let B be the ordered orthonormal basis $\left(\left(\frac{1}{3}, \frac{2}{3}, -\frac{2}{3}\right), \left(\frac{2}{3}, \frac{1}{3}, \frac{2}{3}\right), \left(-\frac{2}{3}, \frac{2}{3}, \frac{1}{3}\right)\right)$ for $\mathbb{R}^3$.
   a) Find the coordinate vectors (c_1, c_2, c_3) for (1, 2, -4) and (d_1, d_2, d_3) for (5, -3, 2), relative to the ordered basis B.
   b) Compute $(1, 2, -4) \cdot (5 - 3, 2)$, and then compute $(c_1, c_2, c_3) \cdot (d_1, d_2, d_3)$. What do you notice?

25. Mark each of the following True or False.
   ___ a) All vectors in an orthogonal basis have length 1.
   ___ b) All vectors in an orthonormal basis have length 1.
   ___ c) Every nontrivial finite-dimensional inner-product space has an orthonormal basis.
   ___ d) Every vector in a finite-dimensional inner-product space is in some orthonormal basis.
   ___ e) Every nonzero vector in a finite-dimensional inner-product space is in some orthonormal basis.
   ___ f) Every unit vector in a finite-dimensional inner-product space is in some orthonormal basis.

___ g) Every $n \times k$ matrix A has a factorization $A = QR$, where the column vectors of Q form an orthonormal set and R is an invertible $k \times k$ matrix.

___ h) Every $n \times k$ matrix A of rank k has a factorization $A = QR$, where the column vectors of Q form an orthonormal set and R is an invertible $k \times k$ matrix.

___ i) It is advantageous to work with an orthogonal basis for W when projecting a vector $\mathbf{b}$ in $\mathbb{R}^n$ on a subspace W of $\mathbb{R}^n$.

___ j) It is even more advantageous to work with an orthonormal basis for W when performing the projection in part (i).

In Exercises 26–28, use the text answers for the indicated earlier exercise to find a QR-factorization of the matrix having as column vectors the transposes of the row vectors given in that exercise.

26. Exercise 7   **27.** Exercise 9   **28.** Exercise 11

29. Let V be an inner-product space of dimension n, and let B be an ordered orthonormal basis for V. Show that, for any vectors $\mathbf{a}$ and $\mathbf{b}$ in V, the inner product $\langle \mathbf{a}, \mathbf{b} \rangle$ is equal to the dot product of the coordinate vectors of $\mathbf{a}$ and $\mathbf{b}$ relative to B. (See Exercise 24 for an illustration.)

(Calculus) In Exercises 30–32, let V be the vector space of continuous functions mapping $\mathbb{R}$ into $\mathbb{R}$.

30. Find an orthonormal basis for $\text{sp}(\sin x, \cos x)$

if the inner product on V is defined by $\langle f, g \rangle = \int_0^\pi f(x)g(x)\,dx$.

31. Find an orthonormal basis for the subspace $P_2 = \text{sp}(1, x, x^2)$ of V if the inner product on V is defined by $\langle f, g \rangle = \int_{-1}^1 f(x)g(x)\,dx$.

32. Find an orthonormal basis for the subspace $\text{sp}(1, e^x)$ of V if the inner product on V is defined by $\langle f, g \rangle = \int_0^1 f(x)g(x)\,dx$.

33. Let A be an $n \times k$ matrix. Show that the column vectors of A are orthonormal if and only if $A^T A = I$.

34. Let A be an $n \times n$ matrix. Show that A has orthonormal column vectors if and only if A is invertible with inverse $A^{-1} = A^T$.

35. Let A be an $n \times n$ matrix. Show that the column vectors of A are orthonormal if and only if the row vectors of A are orthonormal. [HINT: Use Exercise 34 and the fact that A commutes with its inverse.]

▌ *The supplied program QRFACTOR allows the user to enter k independent vectors in $\mathbb{R}^n$ for n and k at most 10. The program can then be used to find an orthonormal set of vectors spanning the same subspace. It will also exhibit a QR-factorization of the $n \times k$ matrix A having the entered vectors as column vectors. Use this program for the remaining exercises.*

36. Check the answers you gave for Exercises 7–12.

37. Check the answers you gave for Exercises 26–28.

5.3   ORTHOGONAL MATRICES

Let A be the $n \times n$ matrix with column vectors $\mathbf{a}_1, \mathbf{a}_2, \ldots, \mathbf{a}_n$. Recall that these vectors form an orthonormal basis for $\mathbb{R}^n$ if and only if

$$\mathbf{a}_i \cdot \mathbf{a}_j = \begin{cases} 0 & \text{if} & i \neq j, & \mathbf{a}_i \perp \mathbf{a}_j \\ 1 & \text{if} & i = j. & \|\mathbf{a}_j\| = 1 \end{cases}$$

Since

$$A^T A = \begin{bmatrix} \rule{1.5em}{0.4pt}\ \mathbf{a}_1\ \rule{1.5em}{0.4pt} \\ \rule{1.5em}{0.4pt}\ \mathbf{a}_2\ \rule{1.5em}{0.4pt} \\ \vdots \\ \rule{1.5em}{0.4pt}\ \mathbf{a}_n\ \rule{1.5em}{0.4pt} \end{bmatrix} \begin{bmatrix} \big| & \big| & & \big| \\ \mathbf{a}_1 & \mathbf{a}_2 & \cdots & \mathbf{a}_n \\ \big| & \big| & & \big| \end{bmatrix}$$

has $\mathbf{a}_i \cdot \mathbf{a}_j$ in the ith row and jth column, we see that the columns of A form an orthonormal basis of $\mathbb{R}^n$ if and only if

$$A^T A = I. \tag{1}$$

In computations with matrices using a computer, it is desirable to use matrices satisfying Eq. (1) as much as possible, as we discuss later in this section.

DEFINITION 5.4 Orthogonal Matrix

An $n \times n$ matrix A is **orthogonal** if $A^T A = I$.

The term *orthogonal* applied to a matrix is just a bit misleading. For example $\begin{bmatrix} 1 & -2 \\ 2 & 1 \end{bmatrix}$ is not an orthogonal matrix. For an orthogonal matrix, not only must the columns be mutually orthogonal, they must also be unit vectors; that is, they must have length 1. This is not indicated by the name. It is unfortunate that the conventional name is *orthogonal matrix* rather than *orthonormal matrix*, but it is very difficult to change established terminology.

From Definition 5.4 we see that an $n \times n$ matrix A is orthogonal if and only if A is invertible and $A^{-1} = A^T$. Since every invertible matrix commutes with its inverse, it follows that $AA^T = I$ too; that is, $(A^T)^T A^T = I$. This means that the column vectors of A^T, which are the row vectors of A, also form an orthonormal basis for $\mathbb{R}^n$. Conversely, if the row vectors of an $n \times n$ matrix A form an orthonormal basis for $\mathbb{R}^n$, then so do the column vectors. We summarize these remarks in a theorem.

THEOREM 5.5 Characterizing Properties of an Orthogonal Matrix

Let A be an $n \times n$ matrix. The following conditions are equivalent:

1. The rows of A form an orthonormal basis for $\mathbb{R}^n$.
2. The columns of A form an orthonormal basis for $\mathbb{R}^n$.
3. The matrix A is orthogonal—that is, invertible with $A^{-1} = A^T$.

EXAMPLE 1 Verify that the matrix

$$A = \frac{1}{7}\begin{bmatrix} 2 & 3 & 6 \\ 3 & -6 & 2 \\ 6 & 2 & -3 \end{bmatrix}$$

is an orthogonal matrix, and find A^{-1}.

Solution We have

$$A^T A = \frac{1}{49}\begin{bmatrix} 2 & 3 & 6 \\ 3 & -6 & 2 \\ 6 & 2 & -3 \end{bmatrix}\begin{bmatrix} 2 & 3 & 6 \\ 3 & -6 & 2 \\ 6 & 2 & -3 \end{bmatrix} = \begin{bmatrix} 1 & 0 & 0 \\ 0 & 1 & 0 \\ 0 & 0 & 1 \end{bmatrix}.$$

In this example, A is symmetric, so

$$A^{-1} = A^T = A = \frac{1}{7}\begin{bmatrix} 2 & 3 & 6 \\ 3 & -6 & 2 \\ 6 & 2 & -3 \end{bmatrix}.$$

∎

We now present the properties of orthogonal matrices that make them especially desirable for use in matrix computations.

THEOREM 5.6 Properties of *Ax* for an Orthogonal Matrix *A*

Let A be an orthogonal $n \times n$ matrix and let $\mathbf{x}$ and $\mathbf{y}$ be any column vectors in $\mathbb{R}^n$.

(i) $(A\mathbf{x}) \cdot (A\mathbf{y}) = \mathbf{x} \cdot \mathbf{y}$.   Preservation of dot product

(ii) $\|A\mathbf{x}\| = \|\mathbf{x}\|$.   Preservation of length

(iii) The angle between $\mathbf{x}$ and $\mathbf{y}$ equals the angle between $A\mathbf{x}$ and $A\mathbf{y}$.   Preservation of angle

PROOF For property (i), we need only recall that the dot product $\mathbf{x} \cdot \mathbf{y}$ of two *column* vectors can be found by using the matrix multiplication $(\mathbf{x}^T)\mathbf{y}$. Since A is orthogonal, we know that $A^TA = I$, so

$$[(A\mathbf{x}) \cdot (A\mathbf{y})] = (A\mathbf{x})^T A\mathbf{y} = \mathbf{x}^T A^T A\mathbf{y} = \mathbf{x}^T I \mathbf{y} = \mathbf{x}^T \mathbf{y} = [\mathbf{x} \cdot \mathbf{y}].$$

For property (ii), the length of a vector can be defined in terms of the dot product— namely, $\|\mathbf{x}\| = \sqrt{\mathbf{x} \cdot \mathbf{x}}$. Since multiplication by A preserves the dot product, it must preserve the length.

$$\|A\mathbf{x}\|^2 = (A\mathbf{x})^T \cdot (A\mathbf{x}) = \mathbf{x}^T A^T A \mathbf{x} = \mathbf{x}^T \cdot \mathbf{x} = [\mathbf{x} \cdot \mathbf{x}] = \|\mathbf{x}\|^2.$$

PROPERTIES OF ORTHOGONAL MATRICES for square systems of coefficients appear in various works of the early nineteenth century. For example, in 1833 Carl Gustav Jacob Jacobi (1804–1851) sought to find a linear substitution

$$y_1 = \sum \alpha_{1i}x_i, \, y_2 = \sum \alpha_{2i}x_i, \ldots, y_n = \sum \alpha_{ni}x_i$$

such that $\Sigma y_i^2 = \Sigma x_i^2$. He found that the coefficients of the substitution must satisfy the orthogonality property

$$\sum_i \alpha_{ij}\alpha_{ik} = \begin{cases} 0, & j \neq k, \\ 1, & j = k. \end{cases}$$

One can even trace orthogonal systems of coefficients back to seventeenth- and eighteenth-century works in analytic geometry, when rotations of the plane or of 3-space are given in order to transform the equations of curves or surfaces. Expressed as matrices, these rotations would give orthogonal ones.

The formal definition of an orthogonal matrix, however, and a comprehensive discussion appeared in an 1878 paper of Georg Ferdinand Frobenius (1849–1917) entitled "On Linear Substitutions and Bilinear Forms." In particular, Frobenius dealt with the eigenvalues of such a matrix.

Frobenius, who was a full professor in Zurich and later in Berlin, made his major mathematical contribution in the area of group theory. He was instrumental in developing the concept of an abstract group, as well as in investigating the theory of finite matrix groups and group characters.

For property (iii), the angle between nonzero vectors $\mathbf{x}$ and $\mathbf{y}$ can be defined in terms of the dot product—namely, as

$$\cos^{-1}\left(\frac{\mathbf{x} \cdot \mathbf{y}}{\sqrt{\mathbf{x} \cdot \mathbf{x}}\sqrt{\mathbf{y} \cdot \mathbf{y}}}\right).$$

Since multiplication by A preserves the dot product, it must preserve angles. ▲

EXAMPLE 2 Let $\mathbf{v}$ be a vector in $\mathbb{R}^3$ with coordinate vector $(2, 3, 5)$ relative to some ordered orthonormal basis $(\mathbf{a}_1, \mathbf{a}_2, \mathbf{a}_3)$ of $\mathbb{R}^3$. Find $\|\mathbf{v}\|$.

Solution We have $\mathbf{v} = 2\mathbf{a}_1 + 3\mathbf{a}_2 + 5\mathbf{a}_3$, which can be expressed as

$$\mathbf{v} = A\mathbf{x} \quad \text{where} \quad A = \begin{bmatrix} | & | & | \\ \mathbf{a}_1 & \mathbf{a}_2 & \mathbf{a}_3 \\ | & | & | \end{bmatrix} \quad \text{and} \quad \mathbf{x} = \begin{bmatrix} 2 \\ 3 \\ 5 \end{bmatrix}.$$

Using property (ii) of Theorem 5.6, we obtain

$$\|\mathbf{v}\| = \|A\mathbf{x}\| = \|\mathbf{x}\| = \sqrt{4 + 9 + 25} = \sqrt{38}. \qquad ▪$$

Property (ii) of Theorem 5.6 is the reason that it is desirable to use orthogonal matrices in matrix computations on a computer. Suppose, for example, that we have occasion to perform a multiplication $A\mathbf{x}$ for a square matrix A and a column vector $\mathbf{x}$ whose components are quantities we have to measure. Our measurements are apt to have some error, so rather than using the true vector $\mathbf{x}$ for these measured quantities, we are likely to work with $\mathbf{x} + \mathbf{e}$, where $\mathbf{e}$ is a nonzero error vector. Upon multiplication by A, we then obtain

$$A(\mathbf{x} + \mathbf{e}) = A\mathbf{x} + A\mathbf{e}.$$

The new error vector is $A\mathbf{e}$. If the matrix A is orthogonal, we know that $\|A\mathbf{e}\| = \|\mathbf{e}\|$, so the magnitude of the error vector remains the same under multiplication by A. We express this important fact as follows:

> Multiplication by orthogonal matrices is a *stable* operation.

If A is not orthogonal, $\|A\mathbf{e}\|$ can be a great deal larger than $\|\mathbf{e}\|$. Repeated multiplication by nonorthogonal matrices can cause the original error vector to blow up to such an extent that the final answer is meaningless. To take advantage of this stability, scientists try to *orthogonalize* computational algorithms with matrices, in order to produce reliable results.

Orthogonal Diagonalization of Symmetric Matrices

Recall that a symmetric matrix is an $n \times n$ square matrix A in which the kth row vector is equal to the kth column vector for each $k = 1, 2, \ldots, n$. Equivalently, A is symmetric if and only if it is equal to its transpose A^T. The problem of diagonalizing a symmetric matrix arises in many applications. (See Section 7.1, for example.) As we stated in Theorem 4.5, this diagonalization can always be achieved. That is, there is an invertible matrix C such that $C^{-1}AC = D$ is a diagonal matrix. In fact, C can be chosen to be an *orthogonal matrix*, as we will show. This means that diagonalization of symmetric matrices is a computationally stable process. We begin by proving the perpendicularity of eigenvectors of a symmetric matrix that correspond to distinct eigenvalues.

THEOREM 5.7 Orthogonality of Eigenspaces of a Symmetric Matrix

Eigenvectors of a symmetric matrix that correspond to different eigenvalues are orthogonal. That is, the eigenspaces of a symmetric matrix are orthogonal.

PROOF Let A be an $n \times n$ symmetric matrix, and let $\mathbf{v}_1$ and $\mathbf{v}_2$ be eigenvectors corresponding to distinct eigenvalues λ_1 and λ_2, respectively. Writing vectors as column vectors, we have

$$A\mathbf{v}_1 = \lambda_1\mathbf{v}_1 \quad \text{and} \quad A\mathbf{v}_2 = \lambda_2\mathbf{v}_2.$$

To show that $\mathbf{v}_1$ and $\mathbf{v}_2$ are orthogonal, we compute

$$\lambda_1(\mathbf{v}_1 \cdot \mathbf{v}_2) = (\lambda_1\mathbf{v}_1) \cdot \mathbf{v}_2 = (A\mathbf{v}_1) \cdot \mathbf{v}_2.$$

The final dot product can be written in matrix form as

$$(A\mathbf{v}_1)^T\mathbf{v}_2 = (\mathbf{v}_1^T A^T)\mathbf{v}_2.$$

Therefore,

$$[\lambda_1(\mathbf{v}_1 \cdot \mathbf{v}_2)] = \mathbf{v}_1^T A^T \mathbf{v}_2. \tag{2}$$

On the other hand,

$$[\lambda_2(\mathbf{v}_1 \cdot \mathbf{v}_2)] = [\mathbf{v}_1 \cdot (\lambda_2\mathbf{v}_2)] = \mathbf{v}_1^T A\mathbf{v}_2. \tag{3}$$

Since $A = A^T$, Eqs. (2) and (3) show that

$$\lambda_1(\mathbf{v}_1 \cdot \mathbf{v}_2) = \lambda_2(\mathbf{v}_1 \cdot \mathbf{v}_2) \quad \text{or} \quad (\lambda_1 - \lambda_2)(\mathbf{v}_1 \cdot \mathbf{v}_2) = 0.$$

Since $\lambda_1 - \lambda_2 \neq 0$, we conclude that $\mathbf{v}_1 \cdot \mathbf{v}_2 = 0$, which shows that $\mathbf{v}_1$ is orthogonal to $\mathbf{v}_2$. ▲

The results stated for symmetric matrices in Section 4.2 tell us that an $n \times n$ real symmetric matrix A has only real numbers as the roots of its characteristic polynomial, and that the algebraic multiplicity of each eigenvalue is equal to its geometric multiplicity; therefore, we can find a basis for $\mathbb{R}^n$ consisting of eigenvectors of A. Using the Gram–Schmidt process, we can modify the vectors of the basis in each *eigenspace* to be an orthonormal set. Theorem 5.7 then tells us that the basis vectors from different eigenspaces are also perpendicular, so we obtain a basis of mutually perpendicular real eigenvectors of unit length. We can take as the diagonalizing matrix C, such that $C^{-1}AC = D$, an orthogonal matrix whose column vectors consist of the vectors in this orthonormal basis for $\mathbb{R}^n$. We summarize our discussion in a theorem.

THEOREM 5.8 Fundamental Theorem of Real Symmetric Matrices

Every real symmetric matrix A is diagonalizable. The diagonalization $C^{-1}AC = D$ can be achieved by using a real orthogonal matrix C.

The converse of Theorem 5.8 is also true. If $D = C^{-1}AC$ is a diagonal matrix and C is an orthogonal matrix, then A is symmetric. (See Exercise 24.) The equation $D = C^{-1}AC$ is said to be an **orthogonal diagonalization** of A.

EXAMPLE 3   Find an orthogonal diagonalization of the matrix

$$A = \begin{bmatrix} 1 & 2 \\ 2 & 4 \end{bmatrix}.$$

Solution   The eigenvalues of A are the roots of the characteristic equation

$$\begin{vmatrix} 1 - \lambda & 2 \\ 2 & 4 - \lambda \end{vmatrix} = \lambda^2 - 5\lambda = 0.$$

They are $\lambda_1 = 0$ and $\lambda_2 = 5$. We proceed to find the corresponding eigenvectors.
For $\lambda_1 = 0$, we have

$$A - \lambda_1 I = A = \begin{bmatrix} 1 & 2 \\ 2 & 4 \end{bmatrix} \sim \begin{bmatrix} 1 & 2 \\ 0 & 0 \end{bmatrix},$$

which yields the eigenspace

$$E_0 = \text{sp}\left(\begin{bmatrix} -2 \\ 1 \end{bmatrix}\right).$$

For $\lambda_2 = 5$, we have

$$A - \lambda_2 I = A - 5I = \begin{bmatrix} -4 & 2 \\ 2 & -1 \end{bmatrix} \sim \begin{bmatrix} 2 & -1 \\ 0 & 0 \end{bmatrix},$$

which yields the eigenspace

$$E_5 = \text{sp}\left(\begin{bmatrix} 1 \\ 2 \end{bmatrix}\right).$$

Thus,

$$\begin{bmatrix} 0 & 0 \\ 0 & 5 \end{bmatrix} = \begin{bmatrix} -2 & 1 \\ 1 & 2 \end{bmatrix}^{-1} \begin{bmatrix} 1 & 2 \\ 2 & 4 \end{bmatrix} \begin{bmatrix} -2 & 1 \\ 1 & 2 \end{bmatrix}$$

is a diagonalization of A, and

$$\begin{bmatrix} 0 & 0 \\ 0 & 5 \end{bmatrix} = \begin{bmatrix} -2/\sqrt{5} & 1/\sqrt{5} \\ 1/\sqrt{5} & 2/\sqrt{5} \end{bmatrix}^{-1} \begin{bmatrix} 1 & 2 \\ 2 & 4 \end{bmatrix} \begin{bmatrix} -2/\sqrt{5} & 1/\sqrt{5} \\ 1/\sqrt{5} & 2/\sqrt{5} \end{bmatrix}$$

is an orthogonal diagonalization of A. ■

EXAMPLE 4 Find an orthogonal diagonalization of the matrix

$$A = \begin{bmatrix} 1 & -1 & -1 \\ -1 & 1 & -1 \\ -1 & -1 & 1 \end{bmatrix},$$

that is, find an orthogonal matrix C such that $C^{-1}AC$ is a diagonal matrix D.

Solution Example 3 of Section 4.3 shows that the eigenvalues and associated eigenspaces of A are

$V_1 \cdot V_2 = \begin{bmatrix} 1 & 1 & 1 \end{bmatrix}\begin{bmatrix} -1 \\ 1 \\ 0 \end{bmatrix} = 0$

$V_1 \cdot V_3 = \begin{bmatrix} 1 & 1 & 1 \end{bmatrix}\begin{bmatrix} -1 \\ 0 \\ 1 \end{bmatrix} = 0,$

$$\lambda_1 = -1, \; E_1 = \text{sp}\left(\begin{bmatrix} 1 \\ 1 \\ 1 \end{bmatrix}\right),$$
$$\underset{\mathbf{v}_1}{}$$

$$\lambda_2 = \lambda_3 = 2, \; E_2 = \text{sp}\left(\begin{bmatrix} -1 \\ 1 \\ 0 \end{bmatrix}, \begin{bmatrix} -1 \\ 0 \\ 1 \end{bmatrix}\right).$$
$$\quad\quad\quad\quad\quad \underset{\mathbf{v}_2}{} \quad \underset{\mathbf{v}_3}{}$$

Notice that vectors $\mathbf{v}_2$ and $\mathbf{v}_3$ in E_2 are orthogonal to the vector $\mathbf{v}_1$ in E_1, as must be the case for this symmetric matrix A. The vectors $\mathbf{v}_2$ and $\mathbf{v}_3$ in E_2 are not orthogonal, but we can use the Gram–Schmidt process to find an orthogonal basis for E_2. We replace $\mathbf{v}_3$ by

$$\mathbf{v}_3 - \frac{\mathbf{v}_3 \cdot \mathbf{v}_2}{\mathbf{v}_2 \cdot \mathbf{v}_2}\, \mathbf{v}_2 = \begin{bmatrix} -1 \\ 0 \\ 1 \end{bmatrix} - \frac{1}{2}\begin{bmatrix} -1 \\ 1 \\ 0 \end{bmatrix} = \begin{bmatrix} -1/2 \\ -1/2 \\ 1 \end{bmatrix}, \text{ or by } \begin{bmatrix} -1 \\ -1 \\ 2 \end{bmatrix}.$$

Thus $\{(1, 1, 1), (-1, 1, 0), (-1, -1, 2)\}$ is an orthgonal basis for $\mathbb{R}^3$ of eigenvectors of A. An orthgonal diagonalizing matrix C is obtained by normalizing these vectors and taking the vectors in the resulting orthonormal basis as column vectors in C. We obtain

$$C = \begin{bmatrix} 1/\sqrt{3} & -1/\sqrt{2} & -1/\sqrt{6} \\ 1/\sqrt{3} & 1/\sqrt{2} & -1/\sqrt{6} \\ 1/\sqrt{3} & 0 & 2/\sqrt{6} \end{bmatrix}. \qquad ■$$

Orthogonal Linear Transformations

Every theorem about matrices has an interpretation for linear transformations. Let A be an orthogonal $n \times n$ matrix, and let $T: \mathbb{R}^n \to \mathbb{R}^n$ be defined by $T(\mathbf{x}) = A\mathbf{x}$. Theorem 5.6 immediately establishes the following properties of T:

(i) $T(\mathbf{x}) \cdot T(\mathbf{y}) = \mathbf{x} \cdot \mathbf{y}$; **Preservation of dot product**

(ii) $\|T(\mathbf{x})\| = \|\mathbf{x}\|$; **Preservation of length**

(iii) The angle between $T(\mathbf{x})$ and $T(\mathbf{y})$ equals the angle between $\mathbf{x}$ and $\mathbf{y}$. **Preservation of angle**

The first property is commonly used to define an orthogonal linear transformation $T: V \rightarrow V$ for a general inner-product space V that is not necessarily finite-dimensional and where $\langle \mathbf{x}, \mathbf{y} \rangle$ is used in place of $\mathbf{x} \cdot \mathbf{y}$. (See Section 2.9.)

DEFINITION 5.5 Orthogonal Linear Transformation

Let V be an inner-product space. A linear transformation $T: V \rightarrow V$ mapping V into itself is **orthogonal** if it satisfies $\langle T(\mathbf{v}), T(\mathbf{w}) \rangle = \langle \mathbf{v}, \mathbf{w} \rangle$ for all vectors $\mathbf{v}$ and $\mathbf{w}$ in V.

For example, the linear transformation that reflects the plane $\mathbb{R}^2$ through a line containing the origin clearly preserves both the angle θ between vectors $\mathbf{u}$ and $\mathbf{v}$ and the magnitudes of the vectors. Since the dot product in $\mathbb{R}^2$ satisfies

$$\mathbf{u} \cdot \mathbf{v} = \|\mathbf{u}\|\|\mathbf{v}\| \cos \theta,$$

it follows that dot products are also preserved. Therefore, this reflection of the plane is an orthogonal linear transformation.

We just showed that every orthogonal matrix gives rise to an orthogonal linear transformation of $\mathbb{R}^n$ into itself. The converse is also true.

THEOREM 5.9 Orthogonal Transformations vis a vis Matrices

A linear transformation T of $\mathbb{R}^n$ into itself is orthogonal if and only if its standard matrix representation A is an orthogonal matrix.

PROOF It remains for us to show that A is an orthogonal matrix if T preserves the dot product. The columns of A are the vectors $T(\mathbf{e}_1), T(\mathbf{e}_2), \ldots, T(\mathbf{e}_n)$, where $\mathbf{e}_j$ is the jth unit coordinate vector of $\mathbb{R}^n$. We have

$$T(\mathbf{e}_i) \cdot T(\mathbf{e}_j) = \mathbf{e}_i \cdot \mathbf{e}_j = \begin{cases} 0 & \text{if} & i \neq j, \\ 1 & \text{if} & i = j, \end{cases}$$

showing that the columns of A form an orthonormal basis of $\mathbb{R}^n$. Thus, A is an orthogonal matrix. ▲

EXAMPLE 5 Show that the linear transformation $T: \mathbb{R}^3 \rightarrow \mathbb{R}^3$ defined by $T(x_1, x_2, x_3) = (x_1/\sqrt{2} + x_3/\sqrt{2}, x_2, -x_1/\sqrt{2} + x_3/\sqrt{2})$ is orthogonal.

Solution The orthogonality of the transformation follows from the fact that the standard matrix representation

$$A = \begin{bmatrix} 1/\sqrt{2} & 0 & 1/\sqrt{2} \\ 0 & 1 & 0 \\ -1/\sqrt{2} & 0 & 1/\sqrt{2} \end{bmatrix}$$

is an orthogonal matrix. ■

In Exercise 40, we ask you to show that a linear transformation $T: \mathbb{R}^n \to \mathbb{R}^n$ is orthogonal if and only if T maps unit vectors into unit vectors. Sometimes, this is an easy condition to verify, as our next example illustrates.

EXAMPLE 6   Show that the linear transformation that rotates the plane counterclockwise through any angle is an orthogonal linear transformation.

Solution   A rotation of the plane preserves the lengths of all vectors—in particular, unit vectors. Example 2 of Section 2.9 shows that a rotation is a linear transformation, and Exercise 40 then shows that the transformation is orthogonal. ■

You might conjecture that a linear transformation $T: \mathbb{R}^n \to \mathbb{R}^n$ is orthogonal if and only if it preserves the angle between vectors. Exercise 38 asks you to give a counterexample, showing that this is *not* the case.

SUMMARY

1. A square $n \times n$ matrix A is orthogonal if it satisfies any one (and hence all) of these three equivalent conditions:
    (i) The rows of A form an orthonormal basis for $\mathbb{R}^n$.
    (ii) The columns of A form an orthonormal basis for $\mathbb{R}^n$.
    (iii) The matrix A is invertible, and $A^{-1} = A^T$.

2. Multiplication of column vectors in $\mathbb{R}^n$ on the left by an $n \times n$ orthogonal matrix preserves length, dot product, and the angle between vectors. Such multiplication is computationally stable.

3. A linear transformation of $\mathbb{R}^n$ into itself is orthogonal if and only if it preserves the dot product, or (equivalently) if and only if its standard matrix representation is orthogonal, or (equivalently) if and only if it maps unit vectors into unit vectors.

4. The eigenspaces of a symmetric matrix A are mutually orthogonal, and A has n mutually perpendicular eigenvectors.

5. A symmetric matrix A is diagonalizable by an orthogonal matrix C. That is, there exists an orthogonal matrix C such that $D = C^{-1}AC$ is a diagonal matrix.

EXERCISES

In Exercises 1–4, verify that the given matrix is orthogonal, and find its inverse.

1. $(1/\sqrt{2}) \begin{bmatrix} 1 & 1 \\ -1 & 1 \end{bmatrix}$

2. $\begin{bmatrix} \frac{3}{5} & 0 & \frac{4}{5} \\ -\frac{4}{5} & 0 & \frac{3}{5} \\ 0 & 1 & 0 \end{bmatrix}$

3. $\frac{1}{7} \begin{bmatrix} 2 & -3 & 6 \\ 3 & 6 & 2 \\ -6 & 2 & 3 \end{bmatrix}$

4. $\frac{1}{2} \begin{bmatrix} 1 & -1 & 1 & 1 \\ -1 & 1 & 1 & 1 \\ 1 & 1 & -1 & 1 \\ 1 & 1 & 1 & -1 \end{bmatrix}$

If A and D are square and AD is orthogonal then $(AD)^{-1} = (AD)^T$, $D^{-1}A^{-1} = D^TA^T$ *so that* $A^{-1} =$

$DD^T A^T = D^2 A^T$. *In Exercises 5–8, find the inverse of each matrix A by first finding a diagonal matrix D so that AD has column vectors of length 1, and then applying the formula* $A^{-1} = D^2 A^T$.

5. $\begin{bmatrix} 1 & 3 \\ -1 & 3 \end{bmatrix}$

6. $\begin{bmatrix} 3 & 0 & 8 \\ -4 & 0 & 6 \\ 0 & 1 & 0 \end{bmatrix}$

7. $\begin{bmatrix} 4 & -3 & 6 \\ 6 & 6 & 2 \\ -12 & 2 & 3 \end{bmatrix}$

8. $\begin{bmatrix} 2 & -1 & 3 & 1 \\ -2 & 1 & 3 & 1 \\ 2 & 1 & -3 & 1 \\ 2 & 1 & 3 & -1 \end{bmatrix}$

9. Supply a third column vector so that the matrix
$\begin{bmatrix} 1/\sqrt{3} & 1/\sqrt{2} & \\ 1/\sqrt{3} & 0 & \\ 1/\sqrt{3} & -1/\sqrt{2} & \end{bmatrix}$ is orthogonal.

10. Repeat Exercise 9 for the matrix
$\begin{bmatrix} \frac{2}{7} & 3/\sqrt{13} & \\ \frac{3}{7} & -2/\sqrt{13} & \\ \frac{6}{7} & 0 & \end{bmatrix}$.

11. Let $(\mathbf{a}_1, \mathbf{a}_2, \mathbf{a}_3)$ be an ordered orthonormal basis for $\mathbb{R}^3$, and let $\mathbf{b}$ be a unit vector with coordinate vector $\left(\frac{1}{2}, \frac{1}{3}, c\right)$ relative to this basis. Find all possible values for c.

12. Let $(\mathbf{a}_1, \mathbf{a}_2, \mathbf{a}_3, \mathbf{a}_4)$ be an ordered orthonormal basis for $\mathbb{R}^4$, and let $(2, 1, 4, -3)$ be the coordinate vector of a vector $\mathbf{b}$ in $\mathbb{R}^4$ relative to this basis. Find $\|\mathbf{b}\|$.

In Exercises 13–18, find a matrix C such that D = $C^{-1}AC$ is an orthogonal diagonalization of the given symmetric matrix A.

13. $\begin{bmatrix} 1 & 2 \\ 2 & 1 \end{bmatrix}$

14. $\begin{bmatrix} 3 & 2 \\ 2 & 0 \end{bmatrix}$

15. $\begin{bmatrix} 2 & 1 & 0 \\ 1 & 2 & 1 \\ 0 & 1 & 2 \end{bmatrix}$

16. $\begin{bmatrix} 0 & 1 & 1 \\ 1 & 2 & 1 \\ 1 & 1 & 0 \end{bmatrix}$

17. $\begin{bmatrix} 0 & 1 & 1 & 0 \\ 1 & -2 & 2 & 1 \\ 1 & 2 & -2 & 1 \\ 0 & 1 & 1 & 0 \end{bmatrix}$

18. $\begin{bmatrix} 1 & -1 & 0 & 0 \\ -1 & 1 & 0 & 0 \\ 0 & 0 & 1 & 3 \\ 0 & 0 & 3 & 1 \end{bmatrix}$

19. Mark each of the following True or False.

____ a) A square matrix is orthogonal if its column vectors are orthogonal.

____ b) Every orthogonal matrix has nullspace $\{\mathbf{0}\}$.

____ c) If A^T is orthogonal, then A is orthogonal.

____ d) If A is an $n \times n$ symmetric orthogonal matrix, then $A^2 = I$.

____ e) If A is an $n \times n$ symmetric matrix such that $A^2 = I$, then A is orthogonal.

____ f) If A and B are orthogonal $n \times n$ matrices, then AB is orthogonal.

____ g) Every orthogonal linear transformation carries every unit vector into a unit vector.

____ h) Every linear transformation that carries each unit vector into a unit vector is orthogonal.

____ i) Every map of the plane into itself that is an *isometry* (preserves distance between points) is given by an orthogonal linear transformation.

____ j) Every map of the plane into itself that is an *isometry* and that leaves the origin fixed is given by an orthogonal linear transformation.

20. Let A be an orthogonal $n \times n$ matrix. Show that $\|A\mathbf{x}\| = \|A^{-1}\mathbf{x}\|$ for any vector $\mathbf{x}$ in $\mathbb{R}^n$.

21. Let A be an orthogonal matrix. Show that A^2 is an orthogonal matrix, too.

22. Show that, if A is an orthogonal matrix, then $\det(A) = \pm 1$.

23. Find a 2×2 matrix with determinant 1 that is not an orthogonal matrix.

24. Let $D = C^{-1}AC$ be a diagonal matrix, where C is an orthogonal matrix. Show that A is symmetric.

25. Let A be an $n \times n$ matrix such that $A\mathbf{x} \cdot A\mathbf{y} = \mathbf{x} \cdot \mathbf{y}$ for all vectors $\mathbf{x}$ and $\mathbf{y}$ in $\mathbb{R}^n$. Show that A is an orthogonal matrix.

26. Let A be an $n \times n$ matrix such that $\|A\mathbf{x}\| = \|\mathbf{x}\|$ for all vectors $\mathbf{x}$ in $\mathbb{R}^n$. Show that A is an orthogonal matrix. [HINT: Show that $\mathbf{x} \cdot \mathbf{y} = \frac{1}{4}(\|x + y\|^2 - \|x - y\|^2)$, and then use Exercise 25.]

27. Show that the real eigenvalues of an orthogonal matrix must be equal to 1 or -1. [HINT: Think in terms of linear transformations.]

28. Describe all real diagonal orthogonal matrices.

29. a) Show that a row-interchange elementary matrix is orthogonal.

    b) Let A be a matrix obtained by permuting (that is, changing the order of) the rows of the $n \times n$ identity matrix. Show that A is an orthogonal matrix.

30. Let $\{\mathbf{a}_1, \mathbf{a}_2, \ldots, \mathbf{a}_n\}$ be an orthonormal basis of column vectors for $\mathbb{R}^n$, and let C be an orthogonal $n \times n$ matrix. Show that

$$\{C\mathbf{a}_1, C\mathbf{a}_2, \ldots, C\mathbf{a}_n\}$$

    is also an orthonormal basis for $\mathbb{R}^n$.

31. Let A and C be orthogonal $n \times n$ matrices. Show that $C^{-1}AC$ is orthogonal.

In Exercises 32–37, determine whether the given linear transformation is orthogonal.

32. $T: \mathbb{R}^2 \to \mathbb{R}^2$ defined by $T(x, y) = (y, x)$

33. $T: \mathbb{R}^3 \to \mathbb{R}^3$ defined by $T(x, y, z) = (x, y, 0)$

34. $T: \mathbb{R}^2 \to \mathbb{R}^2$ defined by $T(x, y) = (2x, y)$

35. $T: \mathbb{R}^2 \to \mathbb{R}^2$ defined by $T(x, y) = (x, -y)$

36. $T: \mathbb{R}^2 \to \mathbb{R}^2$ defined by $T(x, y) =$

$(x/2 - \sqrt{3}y/2, -\sqrt{3}x/2 + y/2)$

37. $T: \mathbb{R}^3 \to \mathbb{R}^3$ defined by $T(x, y, z) = (x/3 + 2y/3 + 2z/3, -2x/3 - y/3 + 2z/3, -2x/3 + 2y/3 - z/3)$

38. Find a linear transformation $T: \mathbb{R}^n \to \mathbb{R}^n$ that preserves the angle between vectors but is not an orthogonal transformation.

39. Show that every 2×2 orthogonal matrix is of one of two forms: either

$$\begin{bmatrix} \cos\theta & \sin\theta \\ \sin\theta & -\cos\theta \end{bmatrix} \quad \text{or} \quad \begin{bmatrix} \cos\theta & -\sin\theta \\ \sin\theta & \cos\theta \end{bmatrix},$$

    for some angle θ.

40. Let $T: \mathbb{R}^n \to \mathbb{R}^n$ be a linear transformation. Show that T is orthogonal if and only if T maps unit vectors to unit vectors. [HINT: Use Exercise 26.]

41. (*Real Householder matrix*) Let $\mathbf{v}$ be a nonzero column vector in $\mathbb{R}^n$. Show that $C_\mathbf{v} =$ $I - \dfrac{2}{\mathbf{v} \cdot \mathbf{v}}(\mathbf{v}\mathbf{v}^T)$ is an orthogonal matrix. (These Householder matrices can be used to perform certain important stable reductions of matrices.)

5.4   THE PROJECTION MATRIX

Let W be a subspace of $\mathbb{R}^n$. Projection of vectors in $\mathbb{R}^n$ on W gives a mapping T of $\mathbb{R}^n$ into itself. Figure 5.8 illustrates the projection $T(\mathbf{a} + \mathbf{b})$ of $\mathbf{a} + \mathbf{b}$ on W, and Figure 5.9 illustrates the projection $T(r\mathbf{a})$ of $r\mathbf{a}$ on W. These figures suggest that

$$T(\mathbf{a} + \mathbf{b}) = T(\mathbf{a}) + T(\mathbf{b})$$

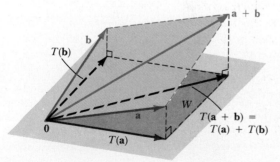

FIGURE 5.8   Projection of a + b on W.

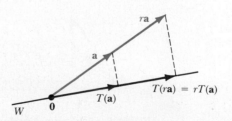

FIGURE 5.9   Projection of ra on W.

and

$$T(r\mathbf{a}) = rT(\mathbf{a}).$$

Thus we expect that T is a linear transformation. If this is the case, there must be a matrix P such that $T(\mathbf{x}) = P\mathbf{x}$, namely, the standard matrix representation of T. Recall that in Section 5.1 we showed how to find the projection of a vector $\mathbf{b}$ on W by finding a basis for W, then finding a basis for its orthogonal complement $W^\perp$, then finding coordinates of $\mathbf{b}$ with respect to the resulting basis of $\mathbb{R}^n$, and so on. It is a somewhat involved process. It would be nice to have a matrix P such that projection of $\mathbf{b}$ on W could be found by computing $P\mathbf{b}$.

In this section, we start by showing that there is indeed a matrix P such that the projection of $\mathbf{b}$ on W is equal to $P\mathbf{b}$. Of course, this then shows that projection is a linear transformation, since we know that multiplication of column vectors in $\mathbb{R}^n$ on the left by any $n \times n$ matrix A gives a linear transformation of $\mathbb{R}^n$ into itself. We derive a formula for this *projection matrix* P corresponding to a subspace W. The formula involves choosing a basis for W, but the final matrix P obtained is independent of the basis chosen.

We have seen that projection on a subspace W is less difficult to compute when an orthonormal basis for W is used. The formula for the matrix P also becomes quite simple when an orthonormal basis is used, and consequently projection becomes an easy computation. Section 5.5 gives an application of these ideas to data analysis.

The Formula for the Projection Matrix

Let $W = \text{sp}(\mathbf{a}_1, \mathbf{a}_2, \ldots, \mathbf{a}_k)$ be a subspace of $\mathbb{R}^n$, where the vectors $\mathbf{a}_1, \mathbf{a}_2, \ldots, \mathbf{a}_k$ are independent. Let $\mathbf{b}$ be a vector in $\mathbb{R}^n$, and let $\mathbf{p} = \mathbf{b}_W$ be its projection on W. From the unique decomposition

$$\mathbf{b} = \mathbf{p} + (\mathbf{b} - \mathbf{p})$$
$$\quad \mathbf{b}_W \qquad \mathbf{b}_{W^\perp}$$

explained in Section 5.1, we see that the projection vector $\mathbf{p}$ is the unique vector satisfying the following two properties. (See Fig. 5.10 for an illustration when $\dim(W) = 2$.)

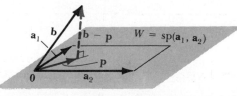

FIGURE 5.10 The projection of b on sp($\mathbf{a}_1$, $\mathbf{a}_2$).

Properties of the Projection p of Vector b on the Subspace W

1. The vector **p** must lie in the subspace W.
2. The vector **b** − **p** must be perpendicular to every vector in W.

If we write our vectors in $\mathbb{R}^n$ as column vectors, the subspace W is the column space of the $n \times k$ matrix A, whose columns are $\mathbf{a}_1, \mathbf{a}_2, \ldots, \mathbf{a}_k$. All vectors in $W = \text{sp}(\mathbf{a}_1, \mathbf{a}_2, \ldots, \mathbf{a}_k)$ have the form $A\mathbf{x}$, where

$$\mathbf{x} = \begin{bmatrix} x_1 \\ x_2 \\ \vdots \\ x_k \end{bmatrix} \qquad \text{for any scalars } x_1, x_2, \ldots, x_k.$$

Since **p** lies in the space W, we see that $\mathbf{p} = A\mathbf{r}$, where

$$\mathbf{r} = \begin{bmatrix} r_1 \\ r_2 \\ \vdots \\ r_k \end{bmatrix} \qquad \text{for some scalars } r_1, r_2, \ldots, r_k.$$

Since $\mathbf{b} - A\mathbf{r}$ must be perpendicular to each vector in W, the dot product of $\mathbf{b} - A\mathbf{r}$ and $A\mathbf{x}$ must be zero for all vectors $\mathbf{x}$. This dot-product condition can be written as the matrix equation

$$(A\mathbf{x})^T(\mathbf{b} - A\mathbf{r}) = \mathbf{x}^T(A^T\mathbf{b} - A^TA\mathbf{r}) = O.$$

In other words, the dot product of the vectors $\mathbf{x}$ and $A^T\mathbf{b} - A^TA\mathbf{r}$ must be zero *for all vectors* $\mathbf{x}$. This can only happen if the vector $A^T\mathbf{b} - A^TA\mathbf{r}$ is itself the zero vector. (See Exercise 41 of Section 1.1.) Therefore, we have

$$A^T\mathbf{b} - A^TA\mathbf{r} = \mathbf{0}. \tag{1}$$

Now the $k \times k$ matrix A^TA appearing in Eq. (1) is invertible, because it has the same rank as A (see Theorem 2.16); and A has rank k, because its columns are independent. Solving Eq. (1) for **r**, we obtain

$$\mathbf{r} = (A^TA)^{-1}A^T\mathbf{b}. \tag{2}$$

Denoting the projection $\mathbf{p} = A\mathbf{r}$ of **b** on W by $\mathbf{p} = \mathbf{b}_W$ and writing **b** as a column vector, we obtain formula (3) in the following box.

Projection b_W of b on the Subspace W

Let $W = \text{sp}(\mathbf{a}_1, \mathbf{a}_2, \ldots, \mathbf{a}_k)$ be a k–dimensional subspace of $\mathbb{R}^n$, and let A have as columns the vectors $\mathbf{a}_1, \mathbf{a}_2, \ldots, \mathbf{a}_k$. The projection of **b** in $\mathbb{R}^n$ on W is given by

$$\mathbf{b}_W = A(A^TA)^{-1}A^T\mathbf{b}. \tag{3}$$

We leave as Exercise 12 the demonstration that the formula in Section 5.1 for projecting **b** on sp(**a**) can be written in the form of Eq. (3), using matrices of appropriate size. Our first illustration reworks Example 1 of Section 5.1, using formula (3).

EXAMPLE 1 Using Eq. (3), find the projection of the vector **b** on the subspace $W = \text{sp}(\mathbf{a})$, where

$$\mathbf{b} = \begin{bmatrix} 1 \\ 2 \\ 3 \end{bmatrix} \quad \text{and} \quad \mathbf{a} = \begin{bmatrix} 2 \\ 4 \\ 3 \end{bmatrix}.$$

Solution Let A be the matrix whose single column is **a**. Then

$$A^T A = [2 \quad 4 \quad 3] \begin{bmatrix} 2 \\ 4 \\ 3 \end{bmatrix} = [29].$$

Putting this into Eq. (3), we have

$$\mathbf{b}_W = A(A^T A)^{-1} A^T \mathbf{b} = \begin{bmatrix} 2 \\ 4 \\ 3 \end{bmatrix} \left(\tfrac{1}{29}\right) [2 \quad 4 \quad 3] \begin{bmatrix} 1 \\ 2 \\ 3 \end{bmatrix} = \left(\tfrac{1}{29}\right) \begin{bmatrix} 4 & 8 & 6 \\ 8 & 16 & 12 \\ 6 & 12 & 9 \end{bmatrix} \begin{bmatrix} 1 \\ 2 \\ 3 \end{bmatrix} = \left(\tfrac{19}{29}\right) \begin{bmatrix} 2 \\ 4 \\ 3 \end{bmatrix}.$$

■

We refer to the matrix $A(A^T A)^{-1} A^T$ in Eq. (3) as the **projection matrix for the subspace** W. It takes any vector **b** in $\mathbb{R}^n$ and, by left multiplication, projects it onto the vector $\mathbf{b}_W$, which lies in W. We will show shortly that this matrix is uniquely determined by W, which allows us to use the definite article *the* when talking about it. Before demonstrating this uniqueness, we box the formula and give two more examples.

The Projection Matrix *P* for the Subspace *W*

Let $W = \text{sp}(\mathbf{a}_1, \mathbf{a}_2, \ldots, \mathbf{a}_k)$ be a k-dimensional subspace of $\mathbb{R}^n$, and let A have as columns the vectors $\mathbf{a}_1, \mathbf{a}_2, \ldots, \mathbf{a}_k$. The projection matrix for the subspace W is given by

$$P = A(A^T A)^{-1} A^T.$$

EXAMPLE 2 Find the projection matrix for the x_2, x_3-plane in $\mathbb{R}^3$.

Solution The x_2, x_3-plane is the subspace $W = \text{sp}(\mathbf{e}_2, \mathbf{e}_3)$, where $\mathbf{e}_2$ and $\mathbf{e}_3$ are the column vectors of the matrix

$$A = \begin{bmatrix} 0 & 0 \\ 1 & 0 \\ 0 & 1 \end{bmatrix}.$$

We find that $A^TA = I$, the 2×2 identity matrix, and that the projection matrix is

$$P = A(A^TA)^{-1}A^T = \begin{bmatrix} 0 & 0 \\ 1 & 0 \\ 0 & 1 \end{bmatrix} \begin{bmatrix} 0 & 1 & 0 \\ 0 & 0 & 1 \end{bmatrix} = \begin{bmatrix} 0 & 0 & 0 \\ 0 & 1 & 0 \\ 0 & 0 & 1 \end{bmatrix}.$$

Notice that P projects each vector

$$\begin{bmatrix} b_1 \\ b_2 \\ b_3 \end{bmatrix} \quad \text{in } \mathbb{R}^3 \text{ onto} \quad P\begin{bmatrix} b_1 \\ b_2 \\ b_3 \end{bmatrix} = \begin{bmatrix} 0 \\ b_2 \\ b_3 \end{bmatrix}$$

in the x_2,x_3-plane, as expected. ■

EXAMPLE 3   Find the matrix that projects vectors in $\mathbb{R}^3$ on the plane $2x - y - 3z = 0$. Also, find the projection of each vector **b** in $\mathbb{R}^3$ on this plane.

Solution   We observe that the given plane contains the zero vector and can therefore be written as the subspace $W = \text{sp}(\mathbf{a}_1, \mathbf{a}_2)$, where $\mathbf{a}_1$ and $\mathbf{a}_2$ are any two nonzero and nonparallel vectors in the plane. We choose

$$\mathbf{a}_1 = \begin{bmatrix} 0 \\ 3 \\ -1 \end{bmatrix} \quad \text{and} \quad \mathbf{a}_2 = \begin{bmatrix} 1 \\ 2 \\ 0 \end{bmatrix},$$

so that

$$A = \begin{bmatrix} 0 & 1 \\ 3 & 2 \\ -1 & 0 \end{bmatrix}.$$

Then

$$(A^TA)^{-1} = \begin{bmatrix} 10 & 6 \\ 6 & 5 \end{bmatrix}^{-1} = \tfrac{1}{14}\begin{bmatrix} 5 & -6 \\ -6 & 10 \end{bmatrix}$$

and the desired matrix is

$$P = \tfrac{1}{14}\begin{bmatrix} 0 & 1 \\ 3 & 2 \\ -1 & 0 \end{bmatrix}\begin{bmatrix} 5 & -6 \\ -6 & 10 \end{bmatrix}\begin{bmatrix} 0 & 3 & -1 \\ 1 & 2 & 0 \end{bmatrix}$$

$$= \tfrac{1}{14}\begin{bmatrix} -6 & 10 \\ 3 & 2 \\ -5 & 6 \end{bmatrix}\begin{bmatrix} 0 & 3 & -1 \\ 1 & 2 & 0 \end{bmatrix} = \tfrac{1}{14}\begin{bmatrix} 10 & 2 & 6 \\ 2 & 13 & -3 \\ 6 & -3 & 5 \end{bmatrix}.$$

Each vector **b** in $\mathbb{R}^3$ projects onto the vector

$$\mathbf{b}_W = P\mathbf{b} = \tfrac{1}{14}\begin{bmatrix} 10 & 2 & 6 \\ 2 & 13 & -3 \\ 6 & -3 & 5 \end{bmatrix}\begin{bmatrix} b_1 \\ b_2 \\ b_3 \end{bmatrix} = \tfrac{1}{14}\begin{bmatrix} 10b_1 + 2b_2 + 6b_3 \\ 2b_1 + 13b_2 - 3b_3 \\ 6b_1 - 3b_2 + 5b_3 \end{bmatrix}. \qquad ■$$

Uniqueness and Properties of the Projection Matrix

It might appear from the formula $P = A(A^TA)^{-1}A^T$ that a projection matrix P for the subspace W of $\mathbb{R}^n$ depends on the particular choice of basis for W that is used for the column vectors of A. However, this is not so. Suppose that C is another $n \times n$ matrix whose column vectors form a basis for W. Then $\bar{P} = C(C^TC)^{-1}C^T$ has the property that $\bar{P}\mathbf{x}$ is the unique projection of $\mathbf{x}$ on W for all $\mathbf{x}$ in $\mathbb{R}^n$. That is, $\bar{P}\mathbf{x} = P\mathbf{x}$ for all $\mathbf{x}$ in $\mathbb{R}^n$. It follows that $\bar{P} = P$. (See Exercise 41, Section 1.1). In view of this uniqueness, we may refer to the matrix $P = A(A^TA)^{-1}A^T$ as *the* projection matrix for the subspace W. We summarize our work in a theorem.

THEOREM 5.10 Projection Matrix

Let W be a subspace of $\mathbb{R}^n$. There is a unique $n \times n$ matrix P such that, for each column vector $\mathbf{b}$ in $\mathbb{R}^n$, the vector $P\mathbf{b}$ is the projection of $\mathbf{b}$ on W. This projection matrix P can be found by selecting any basis $\{\mathbf{a}_1, \mathbf{a}_2, \ldots, \mathbf{a}_k\}$ for W and computing $P = A(A^TA)^{-1}A^T$, where A is the $n \times k$ matrix having column vectors $\mathbf{a}_1, \mathbf{a}_2, \ldots, \mathbf{a}_k$.

Exercise 16 indicates that the projection matrix P given in Theorem 5.10 satisfies two properties:

Properties of a Projection Matrix P

1. $P^2 = P$. P is **idempotent**.
2. $P^T = P$. P is symmetric.

We can use property 2 as a partial check for errors in long computations that lead to P, as in Example 3. These two properties completely characterize the projection matrices, as we now show.

THEOREM 5.11 Characterization of Projection Matrices

The projection matrix P for a subspace W of $\mathbb{R}^n$ is both idempotent and symmetric. Conversely, every $n \times n$ matrix that is both idempotent and symmetric is a projection matrix: specifically, it is the projection matrix for its column space.

PROOF Exercise 16 indicates that a projection matrix is both idempotent and symmetric.

To establish the converse, let P be an $n \times n$ matrix that is both symmetric and idempotent. We show that P is the projection matrix for its own column space W. Let

b be any vector in $\mathbb{R}^n$. By Theorem 5.10, we need show only that $P\mathbf{b}$ satisfies the characterizing properties of the projection of **b** on W given in the box on page 307. Now $P\mathbf{b}$ surely lies in the column space W of P, since W consists of all vectors $P\mathbf{x}$ for any vector **x** in $\mathbb{R}^n$. The second requirement is that $\mathbf{b} - P\mathbf{b}$ must be perpendicular to each vector $P\mathbf{x}$ in the column space of P. Writing the dot product of $\mathbf{b} - P\mathbf{b}$ and $P\mathbf{x}$ in matrix form, and using the hypotheses $P^2 = P = P^T$, we have

$$(\mathbf{b} - P\mathbf{b})^T P\mathbf{x} = ((I - P)\mathbf{b})^T P\mathbf{x} = \mathbf{b}^T (I - P)^T P\mathbf{x}$$
$$= \mathbf{b}^T (I - P) P\mathbf{x} = \mathbf{b}^T (P - P^2)\mathbf{x}$$
$$= \mathbf{b}^T (P - P)\mathbf{x} = \mathbf{b}^T O\mathbf{x} = O.$$

Since their dot product is zero, we see that $\mathbf{b} - P\mathbf{b}$ and $P\mathbf{x}$ are indeed perpendicular, and our proof is complete. ▲

The Orthonormal Case

In Example 2, we saw that the projection matrix for the x_2,x_3-plane in $\mathbb{R}^3$ has the very simple description

$$P = \begin{bmatrix} 0 & 0 & 0 \\ 0 & 1 & 0 \\ 0 & 0 & 1 \end{bmatrix}.$$

The usually complicated formula $P = A(A^TA)^{-1}A^T$ simplified when we used the standard unit coordinate vectors in our basis for W. Namely, we had

$$A = \begin{bmatrix} 0 & 0 \\ 1 & 0 \\ 0 & 1 \end{bmatrix} \quad \text{so} \quad A^TA = \begin{bmatrix} 0 & 1 & 0 \\ 0 & 0 & 1 \end{bmatrix} \begin{bmatrix} 0 & 0 \\ 1 & 0 \\ 0 & 1 \end{bmatrix} = \begin{bmatrix} 1 & 0 \\ 0 & 1 \end{bmatrix} = I.$$

Thus, $(A^TA)^{-1} = I$, too, which simplifies what is normally the worst part of the computation in our formula for P. This simplification can be made in computing P for any subspace W of $\mathbb{R}^n$, provided that we know an *orthonormal basis* $\{\mathbf{a}_1, \mathbf{a}_2, \ldots, \mathbf{a}_k\}$ for W. If A is the $n \times k$ matrix whose columns are $\mathbf{a}_1, \mathbf{a}_2, \ldots, \mathbf{a}_k$, then we know that $A^TA = I$. The formula for the projection matrix becomes

$$P = A(A^TA)^{-1}A^T = AIA^T = AA^T.$$

We box this result for easy reference.

Projection Matrix: Orthonormal Case

Let $\{\mathbf{a}_1, \mathbf{a}_2, \ldots, \mathbf{a}_k\}$ be an orthonormal basis for a subspace W of $\mathbb{R}^n$. The projection matrix for W is

$$P = AA^T, \tag{4}$$

where A is the $n \times k$ matrix having column vectors $\mathbf{a}_1, \mathbf{a}_2, \ldots, \mathbf{a}_k$.

EXAMPLE 4 Find the projection matrix for the subspace $W = \text{sp}(\mathbf{a}_1, \mathbf{a}_2)$ of $\mathbb{R}^3$ if

$$\mathbf{a}_1 = \begin{bmatrix} 1/\sqrt{3} \\ -1/\sqrt{3} \\ 1/\sqrt{3} \end{bmatrix} \quad \text{and} \quad \mathbf{a}_2 = \begin{bmatrix} 1/\sqrt{2} \\ 1/\sqrt{2} \\ 0 \end{bmatrix}.$$

Find the projection of each vector $\mathbf{b}$ in $\mathbb{R}^3$ on W.

Solution Let

$$A = \begin{bmatrix} 1/\sqrt{3} & 1/\sqrt{2} \\ -1/\sqrt{3} & 1/\sqrt{2} \\ 1/\sqrt{3} & 0 \end{bmatrix}.$$

Then

$$A^T A = \begin{bmatrix} 1 & 0 \\ 0 & 1 \end{bmatrix},$$

so

$$P = AA^T = \begin{bmatrix} \frac{5}{6} & \frac{1}{6} & \frac{1}{3} \\ \frac{1}{6} & \frac{5}{6} & -\frac{1}{3} \\ \frac{1}{3} & -\frac{1}{3} & \frac{1}{3} \end{bmatrix}.$$

Each column vector $\mathbf{b}$ in $\mathbb{R}^3$ projects onto

$$\mathbf{b}_W = P\mathbf{b} = \begin{bmatrix} \frac{5}{6} & \frac{1}{6} & \frac{1}{3} \\ \frac{1}{6} & \frac{5}{6} & -\frac{1}{3} \\ \frac{1}{3} & -\frac{1}{3} & \frac{1}{3} \end{bmatrix} \begin{bmatrix} b_1 \\ b_2 \\ b_3 \end{bmatrix} = \frac{1}{6} \begin{bmatrix} 5b_1 + b_2 + 2b_3 \\ b_1 + 5b_2 - 2b_3 \\ 2b_1 - 2b_2 + 2b_3 \end{bmatrix}.$$

Alternatively, we can compute $\mathbf{b}_W$ directly, using boxed Eq. (4) in Section 5.2. We obtain

$$\mathbf{b}_W = (\mathbf{b} \cdot \mathbf{a}_1)\mathbf{a}_1 + (\mathbf{b} \cdot \mathbf{a}_2)\mathbf{a}_2 = \frac{1}{3} \begin{bmatrix} b_1 - b_2 + b_3 \\ -b_1 + b_2 - b_3 \\ b_1 - b_2 + b_3 \end{bmatrix} + \frac{1}{2} \begin{bmatrix} b_1 + b_2 \\ b_1 + b_2 \\ 0 \end{bmatrix}$$

$$= \frac{1}{6} \begin{bmatrix} 5b_1 + b_2 + 2b_3 \\ b_1 + 5b_2 - 2b_3 \\ 2b_1 - 2b_2 + 2b_3 \end{bmatrix}. \quad ■$$

SUMMARY

Let $\{\mathbf{a}_1, \mathbf{a}_2, \ldots, \mathbf{a}_k\}$ be a basis for a subspace W of $\mathbb{R}^n$, and let A be the $n \times k$ matrix having $\mathbf{a}_j$ as jth column vector, so that W is the column space of A.

1. The projection of a column vector $\mathbf{b}$ in $\mathbb{R}^n$ on W is $\mathbf{b}_W = A(A^TA)^{-1}A^T\mathbf{b}$.

2. The matrix $P = A(A^TA)^{-1}A^T$ is the projection matrix for the subspace W. It is the unique matrix P such that, for every vector $\mathbf{b}$ in $\mathbb{R}^n$, the vector $P\mathbf{b}$ lies in W and the vector $\mathbf{b} - P\mathbf{b}$ is perpendicular to every vector in W.

3. If the basis $\{\mathbf{a}_1, \mathbf{a}_2, \ldots, \mathbf{a}_k\}$ of the subspace W is orthonormal, the projection matrix for W is $P = AA^T$.

4. The projection matrix P of a subspace W is idempotent and symmetric. Every symmetric idempotent matrix is the projection matrix for its column space.

EXERCISES

In Exercises 1–8, find the projection matrix for the given subspace, and find the projection of the indicated vector on the subspace.

1. $(1, 2, 1)$ on $\text{sp}(2, 1, -1)$ in $\mathbb{R}^3$

2. $(1, 3, 4)$ on $\text{sp}(1, -1, 2)$ in $\mathbb{R}^3$

3. $(2, -1, 3)$ on $\text{sp}((2, 1, 1), (-1, 2, 1))$ in $\mathbb{R}^3$

4. $(1, 2, 1)$ on $\text{sp}((3, 0, 1), (1, 1, 1))$ in $\mathbb{R}^3$

5. $(1, 3, 1)$ on the plane $x + y - 2z = 0$ in $\mathbb{R}^3$

6. $(4, 2, -1)$ on the plane $3x + 2y + z = 0$ in $\mathbb{R}^3$

7. $(1, 2, 1, 3)$ on $\text{sp}((1, 2, 1, 1), (-1, 1, 0, -1))$ in $\mathbb{R}^4$

8. $(1, 1, 2, 1)$ on $\text{sp}((1, 1, 1, 1), (1, -1, 1, -1), (-1, 1, 1, -1))$ in $\mathbb{R}^4$

9. Find the projection matrix for the x_1, x_2-plane in $\mathbb{R}^3$.

10. Find the projection matrix for the x_1, x_3-coordinate subspace of $\mathbb{R}^4$.

11. Find the projection matrix for the x_1, x_2, x_4-coordinate subspace of $\mathbb{R}^4$.

12. Show that boxed Eq. (3) of this section reduces to Eq. (1) of Section 5.1 for projecting $\mathbf{b}$ on $\text{sp}(\mathbf{a})$.

13. Give a geometric argument indicating that every projection matrix is idempotent.

14. Let $\mathbf{a}$ be a unit column vector in $\mathbb{R}^n$. Show that $\mathbf{a}\mathbf{a}^T$ is the projection matrix for the subspace $\text{sp}(\mathbf{a})$.

15. Mark each of the following True or False.

___ a) A subspace W of dimension k in $\mathbb{R}^n$ has associated with it a $k \times k$ projection matrix.

___ b) Every subspace W of $\mathbb{R}^n$ has associated with it an $n \times n$ projection matrix.

___ c) Projection of $\mathbb{R}^n$ on a subspace W is a linear transformation of $\mathbb{R}^n$ into itself.

___ d) Two different subspaces of $\mathbb{R}^n$ may have the same projection matrix.

___ e) Two different matrices may be projection matrices for the same subspace of $\mathbb{R}^n$.

___ f) Every projection matrix is symmetric.

___ g) Every symmetric matrix is a projection matrix.

___ h) An $n \times n$ symmetric matrix A is a projection matrix if and only if $A^2 = I$.

___ i) Every symmetric idempotent matrix is the projection matrix for its column space.

___ j) Every symmetric idempotent matrix is the projection matrix for its row space.

16. Show that the projection matrix $P = A(A^TA)^{-1}A^T$ given in Theorem 5.10 satisfies the following two conditions:

a) $P^2 = P$,   b) $P^T = P$.

17. What is the projection matrix for the subspace $\mathbb{R}^n$ of $\mathbb{R}^n$?

18. Let U be a subspace of W, which is a subspace of $\mathbb{R}^n$. Let P be the projection matrix for W, and let R be the projection matrix for U. Find PR and RP. [HINT: Argue geometrically.]

19. Let P be the projection matrix for a k-dimensional subspace of $\mathbb{R}^n$.

a) Find all eigenvalues of P.

b) Find the algebraic multiplicity and the geometric multiplicity of each eigenvalue found in part (a).

c) Explain how we can deduce that P is diagonalizable, without using the fact that P is a symmetric matrix.

20. Show that every symmetric matrix whose only eigenvalues are 0 and 1 is a projection matrix.

21. Find all invertible projection matrices.

In Exercises 22–28, find the projection matrix for the subspace W having the given orthonormal basis. The vectors are given in row notation to save space in printing.

22. $W = \text{sp}(\mathbf{a}_1, \mathbf{a}_2)$ in $\mathbb{R}^3$, where
$\mathbf{a}_1 = (1/\sqrt{2}, 0, -1/\sqrt{2})$ and
$\mathbf{a}_2 = (1/\sqrt{3}, -1/\sqrt{3}, 1/\sqrt{3})$

23. $W = \text{sp}(\mathbf{a}_1, \mathbf{a}_2)$ in $\mathbb{R}^3$, where $\mathbf{a}_1 = \left(\frac{3}{5}, \frac{4}{5}, 0\right)$ and $\mathbf{a}_2 = (0, 0, 1)$

24. $W = \text{sp}(\mathbf{a}_1, \mathbf{a}_2)$ in $\mathbb{R}^4$, where
$\mathbf{a}_1 = \left(3/(5\sqrt{2}), 4/(5\sqrt{2}), \frac{1}{2}, -\frac{1}{2}\right)$ and
$\mathbf{a}_2 = \left(4/(5\sqrt{2}), -3/(5\sqrt{2}), \frac{1}{2}, \frac{1}{2}\right)$

25. $W = \text{sp}(\mathbf{a}_1, \mathbf{a}_2)$ in $\mathbb{R}^4$, where $\mathbf{a}_1 = \left(\frac{2}{7}, 0, \frac{3}{7}, -\frac{6}{7}\right)$ and $\mathbf{a}_2 = \left(-\frac{3}{7}, \frac{6}{7}, \frac{2}{7}, 0\right)$

26. $W = \text{sp}(\mathbf{a}_1, \mathbf{a}_2)$ in $\mathbb{R}^4$, where
$\mathbf{a}_1 = \left(0, -\frac{2}{3}, \frac{2}{3}, \frac{1}{3}\right)$ and $\mathbf{a}_2 = \left(\frac{2}{3}, 0, -\frac{1}{3}, \frac{2}{3}\right)$

27. $W = \text{sp}(\mathbf{a}_1, \mathbf{a}_2, \mathbf{a}_3)$ in $\mathbb{R}^4$, where
$\mathbf{a}_1 = (1/\sqrt{3}, 0, 1/\sqrt{3}, 1/\sqrt{3})$,
$\mathbf{a}_2 = (1/\sqrt{3}, 1/\sqrt{3}, -1/\sqrt{3}, 0)$, and
$\mathbf{a}_3 = (1/\sqrt{3}, -1/\sqrt{3}, 0, -1/\sqrt{3})$

28. $W = \text{sp}(\mathbf{a}_1, \mathbf{a}_2, \mathbf{a}_3)$ in $\mathbb{R}^4$, where
$\mathbf{a}_1 = \left(\frac{1}{2}, \frac{1}{2}, \frac{1}{2}, \frac{1}{2}\right)$, $\mathbf{a}_2 = \left(-\frac{1}{2}, \frac{1}{2}, -\frac{1}{2}, \frac{1}{2}\right)$, and
$\mathbf{a}_3 = \left(\frac{1}{2}, \frac{1}{2}, -\frac{1}{2}, -\frac{1}{2}\right)$

*In Exercises 29–32, find the projection of **b** on W.*

29. The subspace W in Exercise 22;
$\mathbf{b} = (6, -12, -6)$

30. The subspace W in Exercise 23;
$\mathbf{b} = (20, -15, 5)$

31. The subspace W in Exercise 26;
$\mathbf{b} = (9, 0, -9, 18)$

32. The subspace W in Exercise 28;
$\mathbf{b} = (4, -12, -4, 0)$

33. Let W be a subspace of $\mathbb{R}^n$, and let P be the projection matrix for W. **Reflection of $\mathbb{R}^n$ through W** is the mapping of $\mathbb{R}^n$ into itself that carries each vector $\mathbf{b}$ in $\mathbb{R}^n$ into its reflection $\mathbf{b}_r$, according to the following geometric description:

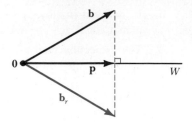

FIGURE 5.11 Reflection of $\mathbb{R}^n$ through W.

Let $\mathbf{p}$ be the projection of $\mathbf{b}$ on W. Starting at the point $\mathbf{b}$, travel in a straight line to the point $\mathbf{p}$, and then continue in the same direction an equal distance to arrive at $\mathbf{b}_r$. (See Fig. 5.11.)

Show that $\mathbf{b}_r = (2P - I)\mathbf{b}$. (Notice that, since reflection can be accomplished by matrix multiplication, this reflection must be a linear transformation of $\mathbb{R}^n$ into itself.)

The formula $A(A^T A)^{-1} A^T$ for a projection matrix can be tedious to compute using pencil and paper, but the software MATCOMP can do it easily. In Exercises 34–38, use MATCOMP, or similar software, to find the indicated vector projections.

34. The projections in $\mathbb{R}^6$ of $(-1, 2, 3, 1, 6, 2)$ and $(2, 0, 3, -1, 4, 5)$ on $\text{sp}((1, -2, 3, 1, 4, 0))$

35. The projections in $\mathbb{R}^3$ of $(1, -1, 4)$, $(3, 3, -1)$, and $(-2, 4, 7)$ on $\text{sp}((1, 3, -4), (2, 0, 3))$

36. The projections in $\mathbb{R}^4$ of $(-1, 3, 2, 0)$ and $(4, -1, 1, 5)$ on $\text{sp}((0, 1, 2, 1), (-1, 2, 1, 4))$

37. The projections in $\mathbb{R}^4$ of $(2, 1, 0, 3)$, $(1, 1, -1, 2)$, and $(4, 3, 1, 3)$ on $\text{sp}((1, 0, -1, 0), (1, 2, -1, 4), (2, 1, 3, -1))$

38. The projections in $\mathbb{R}^5$ of $(2, 1, -3, 2, 4)$ and $(1, -4, 0, 1, 5)$ on $\text{sp}((3, 1, 4, 0, 1), (2, 1, 3, -5, 1))$

39. Work with Topic 3 of the available program VECTGRPH until you are able to get a score of at least 80% most of the time.

5.5 THE METHOD OF LEAST SQUARES

The Nature of the Problem

In this section we apply our work on projections to problems of data analysis. Suppose that data measurements of the form (a_i, b_i) are obtained from observation or experimentation and are plotted as data points in the x,y-plane. It is desirable to find a mathematical relationship $y = f(x)$ that represents the data reasonably well, so that we can make predictions of data values that were not measured. Geometrically, this means that we would like the graph of $y = f(x)$ in the plane to pass very close to our data points. Depending on the nature of the experiment and the configuration of the plotted data points, we might decide on an appropriate type of function $y = f(x)$ such as a linear function, a quadratic function, or an exponential function. We illustrate with three types of problem.

PROBLEM 1   According to Hooke's law, the distance that a spring stretches is proportional to the force applied. Suppose that we attach four different weights a_1, a_2, a_3, and a_4 in turn to the bottom of a spring suspended vertically. We measure the four lengths b_1, b_2, b_3, and b_4 of the stretched spring, and the data in Table 5.1 are obtained. Because of Hooke's law, we expect the data points (a_i, b_i) to be close to some line with equation

$$y = f(x) = r_0 + r_1 x,$$

where r_0 is the length of the spring and r_1 is the *spring constant*. That is, if our measurements were exact and the spring ideal, we would have $b_i = r_0 + r_1 a_i$ for specific values r_0 and r_1.   ■

TABLE 5.1

a_i = Weight in ounces	2.0	4.0	5.0	6.0
b_i = Length in inches	6.5	8.5	11.0	12.5

In Problem 1, we have only the two unknowns r_0 and r_1; and in theory, just two measurements should suffice to find them. In practice, however, we expect to have some error in physical measurements. It is standard procedure to make more measurements than are theoretically necessary in the hope that the errors will roughly cancel each other out, in accordance with the laws of probability. Substitution of each data point (a_i, b_i) from Problem 1 into the equation $y = r_0 + r_1 x$ gives a single linear equation in the two unknowns r_0 and r_1. The four data points of Problem 1 thus give rise to a linear system of four equations in only two unknowns. Such a linear system with more equations than unknowns is called **overdetermined**, and one expects to find that the system is *inconsistent*, having no actual solution. It will be our task to find values for the unknowns r_0 and r_1 that will come as close as possible, in some sense, to satisfying all four of the equations.

We have used the illustration presented in Problem 1 to introduce our goal in this section, and we will solve the problem in a moment. We first present two more hypothetical problems, which we will also solve later in the section.

PROBLEM 2   At a recent boat show, the observations listed in Table 5.2 were made relating the prices b_i of sailboats and their weights a_i. Plotting the data points (a_i, b_i), as shown in Figure 5.12, we might expect a quadratic function of the form

$$y = f(x) = r_0 + r_1 x + r_2 x^2$$

to fit the data fairly well.   ■

TABLE 5.2

a_i = Weight in tons	2	4	5	8
b_i = Price in units of \$10,000	1	3	5	12

PROBLEM 3   A population of rabbits on a large island was estimated each year from 1981 to 1984, giving the data in Table 5.3. Knowing that population growth is exponential in the absence of disease, predators, famine, and so on, we expect an exponential function

$$y = f(x) = re^{sx}$$

to provide the best representation of these data. Notice that, by using logarithms, we can convert this exponential function into a form linear in x:

$$\ln(y) = \ln(r) + sx.$$ ■

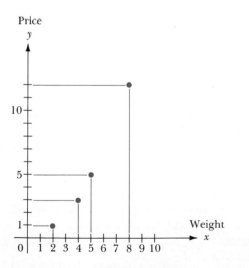

FIGURE 5.12   Problem 2 data.

TABLE 5.3

$\mathbf{a}_i$ = (Year observed) − 1980	1	2	3	4
$\mathbf{b}_i$ = Number of rabbits in units of 1000	3	4.5	8	17

The Method of Least Squares

Consider now the problem of finding a linear function $f(x) = r_0 + r_1 x$ that best fits data points (a_i, b_i) for $i = 1, 2, \ldots, m$, where $m > 2$. Geometrically, this amounts to finding the line in the plane that comes closest, in some sense, to passing through the m data points. If there were no error in our measurements and our data were truly linear, then for some r_0 and r_1 we would have

$$b_i = r_0 + r_1 a_i \qquad \text{for} \qquad i = 1, 2, \ldots, m.$$

These m linear equations in the two unknowns r_0 and r_1 form an overdetermined system of equations that probably has no solution. Our data points actually satisfy a system of linear approximations, which can be expressed in matrix form as

$$\begin{bmatrix} b_1 \\ b_2 \\ . \\ . \\ . \\ b_m \end{bmatrix} \approx \begin{bmatrix} 1 & a_1 \\ 1 & a_2 \\ . & . \\ . & . \\ . & . \\ 1 & a_m \end{bmatrix} \begin{bmatrix} r_0 \\ r_1 \end{bmatrix} \qquad (1)$$

$$\begin{matrix} \mathbf{b} & A & \mathbf{r} \end{matrix}$$

or simply as $\mathbf{b} \approx A\mathbf{r}$. We try to find an optimal solution vector $\bar{\mathbf{r}}$ for the system (1) of approximations. For each vector $\mathbf{r}$, the **error vector** $A\mathbf{r} - \mathbf{b}$ measures how far our system (1) is from being a system of equations with solution vector $\mathbf{r}$. The abso-

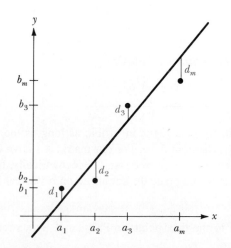

FIGURE 5.13 The distances d_i.

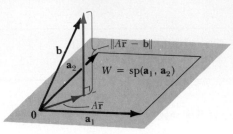

FIGURE 5.14   The length $\|A\bar{\mathbf{r}} - \mathbf{b}\|$.

lute values of the components of the vector $A\mathbf{r} - \mathbf{b}$ represent the vertical distances $d_i = |r_0 + r_1 a_i - b_i|$, shown in Figure 5.13.

We want to minimize, in some sense, our error vector $A\mathbf{r} - \mathbf{b}$. A number of different methods for minimization are very useful. For example, one might want to minimize the maximum of the distances d_i. We study just one sense of minimization, the one that probably seems most natural at this point. We will minimize the *length* $\|A\mathbf{r} - \mathbf{b}\|$ of our error vector. Minimizing $\|A\mathbf{r} - \mathbf{b}\|$ is equivalent to minimizing $\|A\mathbf{r} - \mathbf{b}\|^2$, which means minimizing the sum

$$d_1{}^2 + d_2{}^2 + \cdots + d_m{}^2 \tag{2}$$

of the squares of the distances in Figure 5.13. Hence the name **method of least squares** given to this procedure.

If $\mathbf{a}_1$ and $\mathbf{a}_2$ denote the columns of A in system (1), the vector $A\mathbf{r} = r_0\mathbf{a}_1 + r_1\mathbf{a}_2$ lies in the column space $W = \mathrm{sp}(\mathbf{a}_1, \mathbf{a}_2)$ of A. From Figure 5.14, we see geometrically that, of all the vectors $A\mathbf{r}$ in W, the one that minimizes $\|A\mathbf{r} - \mathbf{b}\|$ is the *projection* $\mathbf{b}_W = A\bar{\mathbf{r}}$ of $\mathbf{b}$ on W. Equation (3) of Section 5.4 shows that then $A\bar{\mathbf{r}} = A(A^TA)^{-1}A^T\mathbf{b}$ and provides us with this boxed formula for the solution vector $\bar{\mathbf{r}}$, which is optimal in this least-squares sense:

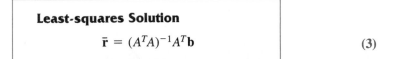

Least-squares Solution

$$\bar{\mathbf{r}} = (A^TA)^{-1}A^T\mathbf{b} \tag{3}$$

The 2×2 matrix A^TA in formula (3) is invertible as long as the columns of A are independent, as shown by Theorem 2.16. For our matrix A shown in system (1), this just means that not all the values a_i are the same. Geoemetrically, this corresponds to saying that our data points in the plane do not all lie on a vertical line.

EXAMPLE 1   Find the least-squares fit to the data points in Problem 1 by a straight line—that is, by a linear function $y = r_0 + r_1x$.

Solution   We form the system $\mathbf{b} \approx A\mathbf{r}$ in system (1) for the data in Table 5.1:

$$\underbrace{\begin{bmatrix} 6.5 \\ 8.5 \\ 11.0 \\ 12.5 \end{bmatrix}}_{\mathbf{b}} \approx \underbrace{\begin{bmatrix} 1 & 2 \\ 1 & 4 \\ 1 & 5 \\ 1 & 6 \end{bmatrix}}_{A} \underbrace{\begin{bmatrix} r_0 \\ r_1 \end{bmatrix}}_{\mathbf{r}}.$$

We have

$$A^T A = \begin{bmatrix} 1 & 1 & 1 & 1 \\ 2 & 4 & 5 & 6 \end{bmatrix} \begin{bmatrix} 1 & 2 \\ 1 & 4 \\ 1 & 5 \\ 1 & 6 \end{bmatrix} = \begin{bmatrix} 4 & 17 \\ 17 & 81 \end{bmatrix}$$

and

$$(A^T A)^{-1} = \tfrac{1}{35} \begin{bmatrix} 81 & -17 \\ -17 & 4 \end{bmatrix}$$

From Eq. (3), we obtain

$$\bar{\mathbf{r}} = (A^T A)^{-1} A^T \mathbf{b} = \tfrac{1}{35} \begin{bmatrix} 81 & -17 \\ -17 & 4 \end{bmatrix} \begin{bmatrix} 1 & 1 & 1 & 1 \\ 2 & 4 & 5 & 6 \end{bmatrix} \begin{bmatrix} 6.5 \\ 8.5 \\ 11.0 \\ 12.5 \end{bmatrix}$$

$$= \tfrac{1}{35} \begin{bmatrix} 47 & 13 & -4 & -21 \\ -9 & -1 & 3 & 7 \end{bmatrix} \begin{bmatrix} 6.5 \\ 8.5 \\ 11.0 \\ 12.5 \end{bmatrix} = \tfrac{1}{35} \begin{bmatrix} 109.5 \\ 53.5 \end{bmatrix} \approx \begin{bmatrix} 3.1 \\ 1.5 \end{bmatrix}.$$

Therefore, the equation of the line that best fits the data in the least-squares sense is $y = 1.5x + 3.1$. This line and the data points are shown in Figure 5.15.   ▪

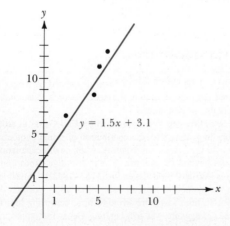

FIGURE 5.15   The least-squares fit of data points.

EXAMPLE 2   Use the method of least squares to fit the data in Problem 3 by an exponential function $y = f(x) = re^{sx}$.

Solution   We use logarithms and convert the exponential equation to an equation that is linear in x:

$$\ln(y) = \ln(r) + sx.$$

TABLE 5.4

$x = a_i$	1	2	3	4
$y = b_i$	3	4.5	8	17
$z = \ln(b_i)$	1.10	1.50	2.08	2.83

From Table 5.3, we obtain the data in Table 5.4 to use for our logarithmic equation. We obtain

$$A^T A = \begin{bmatrix} 1 & 1 & 1 & 1 \\ 1 & 2 & 3 & 4 \end{bmatrix} \begin{bmatrix} 1 & 1 \\ 1 & 2 \\ 1 & 3 \\ 1 & 4 \end{bmatrix} = \begin{bmatrix} 4 & 10 \\ 10 & 30 \end{bmatrix}$$

$$(A^T A)^{-1} = \begin{bmatrix} \frac{3}{2} & -\frac{1}{2} \\ -\frac{1}{2} & \frac{1}{5} \end{bmatrix}$$

$$(A^T A)^{-1} A^T = \begin{bmatrix} \frac{3}{2} & -\frac{1}{2} \\ -\frac{1}{2} & \frac{1}{5} \end{bmatrix} \begin{bmatrix} 1 & 1 & 1 & 1 \\ 1 & 2 & 3 & 4 \end{bmatrix} = \begin{bmatrix} 1 & \frac{1}{2} & 0 & -\frac{1}{2} \\ -\frac{3}{10} & -\frac{1}{10} & \frac{1}{10} & \frac{3}{10} \end{bmatrix}.$$

Multiplying this last matrix on the right by

$$\begin{bmatrix} 1.10 \\ 1.50 \\ 2.08 \\ 2.83 \end{bmatrix},$$

A TECHNIQUE VERY CLOSE TO THAT OF LEAST SQUARES was developed by Roger Cotes (1682–1716), the gifted mathematician who edited the second edition of Isaac Newton's *Principia*, in a work dealing with errors in astronomical observations, written around 1715.

The complete principle, however, was first formulated by Carl Gauss at around the age of 16 while he was adjusting approximations in dealing with the distribution of prime numbers. Gauss later stated that he used the method frequently over the years—for example, when he did calculations concerning the orbits of asteroids. Gauss published the method in 1809 and gave a definitive exposition 14 years later.

On the other hand, it was Adrien-Marie Legendre (1752–1833), founder of the theory of elliptic functions, who first published the method of least squares, in an 1806 work on determining the orbits of comets. After Gauss's 1809 publication, Legendre wrote to him, censuring him for claiming the method as his own. Even as late as 1827, Legendre was still berating Gauss for "appropriating the discoveries of others." In fact, the problem lay in Gauss's failure to publish many of his discoveries promptly; he mentioned them only after they were published by others.

TABLE 5.5

a_i	b_i	$f(a_i)$
1	3	2.7
2	4.5	4.9
3	8	8.7
4	17	15.5

we obtain, from Eq. (3),

$$\bar{\mathbf{r}} = \begin{bmatrix} \ln r \\ s \end{bmatrix} \approx \begin{bmatrix} .435 \\ .577 \end{bmatrix}.$$

Thus $r = e^{.435} \approx 1.54$, and we obtain $y = f(x) = 1.54e^{.577x}$ as a fitting exponential function.

The graph of the function and of the data points in Table 5.5 is shown in Figure 5.16. On the basis of the function $f(x)$ obtained, we can project the population of rabbits on the island in 1990 to be about

$$f(10) \cdot 1000 \approx 494,000 \text{ rabbits,}$$

unless predators, disease, or lack of food interferes. ■

In Example 2, we used the least-squares method with the *logarithm* of our original y-coordinate data. Thus, it is the equation $\ln y = r + sx$ that is the least-squares fit of the logarithmic data. Using the least-squares method to fit the logarithm of the y-coordinate data produces an exponential function that approximates the smaller y-values in the data better than it does the larger y-values, as illustrated by Table 5.5.

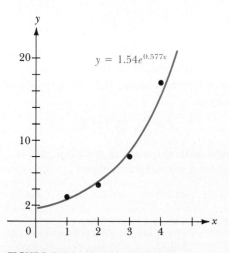

FIGURE 5.16 Data points and the exponential fit.

It can be shown that the fit $y = f(x) = re^{sx}$ amounts roughly to minimizing the *percent of error* in taking $f(a_i)$ for b_i. The least-squares fit of logarithmic data does not yield the least-squares fit of the original data. The supplied program YOUFIT can be used to illustrate this. (See Exercises 30 and 31.)

Least-squares Solutions for Larger Systems

It should be clear that data points of more than two components may lead to over-determined linear systems in more than two unknowns. Suppose that an experiment is repeated m times, and data values

$$b_i, a_{i1}, \ldots, a_{in}$$

are obtained from measurements on the ith experiment. For example, the data values a_{ij} might be ones that can be controlled and the value b_i measured. Of course, there may be errors in the controlled as well as in the measured values. Suppose that we have reason to believe that the data obtained for each experiment should *theoretically* satisfy the same linear relation

$$y = r_0 + r_1 x_1 + r_2 x_2 + \cdots + r_n x_n, \tag{4}$$

with $y = b_i$ and $x_j = a_{ij}$. We then obtain a system of m linear approximations in $n + 1$ unknowns $r_0, r_1, r_2, \ldots, r_n$:

$$\begin{bmatrix} b_1 \\ b_2 \\ \vdots \\ b_m \end{bmatrix} \approx \begin{bmatrix} 1 & a_{11} & a_{12} & \cdots & a_{1n} \\ 1 & a_{21} & a_{22} & \cdots & a_{2n} \\ \vdots & \vdots & \vdots & & \vdots \\ 1 & a_{m1} & a_{m2} & \cdots & a_{mn} \end{bmatrix} \begin{bmatrix} r_0 \\ r_1 \\ r_2 \\ \vdots \\ r_n \end{bmatrix}. \tag{5}$$
$$\quad\; \mathbf{b} \qquad\qquad\qquad A \qquad\qquad\quad \mathbf{r}$$

If $m > n + 1$, then system (5) corresponds to an overdetermined linear system $\mathbf{b} = A\mathbf{r}$, which probably has no exact solution. If the rank of A is $n + 1$, then a repetition of our geometric argument above indicates that the least-squares solution $\bar{\mathbf{r}}$ for the system $\mathbf{b} \approx A\mathbf{r}$ is given by

$$\bar{\mathbf{r}} = (A^T A)^{-1} A^T \mathbf{b}. \tag{6}$$

A linear system of the form (5) arises if m data points (a_i, b_i) are found and a least-squares fit is sought for a polynomial function

$$y = r_0 + r_1 x + \cdots + r_{n-1} x^{n-1} + r_n x^n.$$

The data point (a_i, b_i) leads to the approximation

$$b_i \approx r_0 + r_1 a_i + \cdots + r_{n-1} a_i^{n-1} + r_n a_i^n.$$

The m data points thus give rise to a linear system of the form (5), where the matrix

A is given by

$$A = \begin{bmatrix} 1 & a_1 & a_1{}^2 & \cdots & a_1{}^n \\ 1 & a_2 & a_2{}^2 & \cdots & a_2{}^n \\ \vdots & \vdots & \vdots & & \vdots \\ 1 & a_m & a_m{}^2 & \cdots & a_m{}^n \end{bmatrix}. \tag{7}$$

EXAMPLE 3 Find the least-squares fit to the data in Problem 2 by a parabola—that is, by a quadratic function

$$y = r_0 + r_1 x + r_2 x^2.$$

Solution We write the data in the form $\mathbf{b} \approx A\mathbf{r}$, where A has the form (7):

$$\begin{bmatrix} 1 \\ 3 \\ 5 \\ 12 \end{bmatrix} \approx \begin{bmatrix} 1 & 2 & 4 \\ 1 & 4 & 16 \\ 1 & 5 & 25 \\ 1 & 8 & 64 \end{bmatrix} \begin{bmatrix} r_0 \\ r_1 \\ r_2 \end{bmatrix}.$$

Then

$$A^T A = \begin{bmatrix} 1 & 1 & 1 & 1 \\ 2 & 4 & 5 & 8 \\ 4 & 16 & 25 & 64 \end{bmatrix} \begin{bmatrix} 1 & 2 & 4 \\ 1 & 4 & 16 \\ 1 & 5 & 25 \\ 1 & 8 & 64 \end{bmatrix} = \begin{bmatrix} 4 & 19 & 109 \\ 19 & 109 & 709 \\ 109 & 709 & 4993 \end{bmatrix}$$

and

$$(A^T A)^{-1} = \frac{1}{5400} \begin{bmatrix} 41556 & -17586 & 1590 \\ -17586 & 8091 & -765 \\ 1590 & -765 & 75 \end{bmatrix}.$$

We compute

$$(A^T A)^{-1} A^T = \frac{1}{5400} \begin{bmatrix} 41556 & -17586 & 1590 \\ -17586 & 8091 & -765 \\ 1590 & -765 & 75 \end{bmatrix} \begin{bmatrix} 1 & 1 & 1 & 1 \\ 2 & 4 & 5 & 8 \\ 4 & 16 & 25 & 64 \end{bmatrix}$$

$$= \frac{1}{5400} \begin{bmatrix} 12744 & -3348 & -6624 & 2628 \\ -4464 & 2538 & 3744 & -1818 \\ 360 & -270 & -360 & 270 \end{bmatrix}$$

$$= \begin{bmatrix} 2.36 & -.62 & -1.22667 & .4866667 \\ -.826667 & .47 & .6933333 & -.3366667 \\ .06666667 & -.05 & -.06666667 & .05 \end{bmatrix}.$$

Then $\bar{\mathbf{r}}$ is obtained by multiplying this last matrix on the right by

$$\begin{bmatrix} 1 \\ 3 \\ 5 \\ 12 \end{bmatrix} \qquad \text{resulting in} \qquad \bar{\mathbf{r}} \approx \begin{bmatrix} .207 \\ .01 \\ .183 \end{bmatrix}.$$

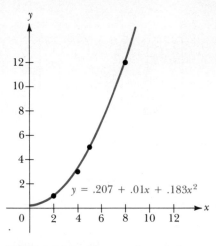

FIGURE 5.17 The graph and data points for Example 3.

Thus, the quadratic function that best approximates the data in the least-squares sense is

$$y \approx f(x) = .207 + .01x + .183x^2.$$

In Figure 5.17, we show the graph of this quadratic function and plot the data points. Data are given in Table 5.6. On the basis of the least-squares fit, we might project that the price of a 10-ton sailing yacht would be about 18.6 times $10,000, or $186,000, and that the price of a 20-ton yacht would be about $f(20)$ times $10,000, or about $736,000. However, one should be wary of using a function $f(x)$ that seems to fit data well for measured values of x to project data for values of x far from those measured values. Our quadratic function seems to fit our data quite well, but we should not expect the cost we might project for a 100-ton ship to be at all accurate. ■

TABLE 5.6

a_i	b_i	$f(a_i)$
2	1	.959
4	3	3.17
5	5	4.83
8	12	12.0

EXAMPLE 4   Show that the least-squares linear fit to data points (a_i, b_i) for $i = 1, 2, \ldots, m$ by a constant function $f(x) = r_0$ is achieved when r_0 is the average y-value

$$(b_1 + b_2 + \cdots + b_m)/m$$

of the data values b_i.

Solution Such a constant function $y = f(x) = r_0$ has a horizontal line as its graph. In this situation, we have $n = 1$ as the number of unknowns and system (5) becomes

$$\underbrace{\begin{bmatrix} b_1 \\ b_2 \\ \vdots \\ \vdots \\ b_m \end{bmatrix}}_{\mathbf{b}} \approx \underbrace{\begin{bmatrix} 1 \\ 1 \\ \vdots \\ \vdots \\ 1 \end{bmatrix}}_{A} \underbrace{[r_0]}_{\mathbf{r}}.$$

Thus A is the $m \times 1$ matrix with all entries 1. We easily find that $A^T A = [m]$, so $(A^T A)^{-1} = [1/m]$. We also find that $A^T \mathbf{b} = b_1 + b_2 + \cdots + b_m$. Thus, the least-squares solution (6) is

$$\bar{\mathbf{r}} = (b_1 + b_2 + \cdots + b_m)/m,$$

as asserted. ■

Overdetermined Systems of Linear Equations

We have presented examples in which least-squares solutions were found to systems of linear approximations arising from applications. We used $\mathbf{r}$ as the column vector of unknowns in those examples, since we were using x as the independent variable in the formula for a fitting function $f(x)$. We now discuss the mathematics of overdetermined systems, and we use $\mathbf{x}$ as the column vector of unknowns again.

Consider a general overdetermined system of linear equations

$$A\mathbf{x} = \mathbf{b},$$

where A is an $m \times n$ matrix of rank n and $m > n$. We expect such a system to be inconsistent—that is, to have no solution. As we have seen, the system of linear approximations $A\mathbf{x} \approx \mathbf{b}$ has least-squares solution

$$\bar{\mathbf{x}} = (A^T A)^{-1} A^T \mathbf{b}. \tag{8}$$

The vector $\bar{\mathbf{x}}$ is thus the **least-squares solution of the overdetermined system** $A\mathbf{x} = \mathbf{b}$. Equation (8) can be written in the equivalent form

$$A^T A \bar{\mathbf{x}} = A^T \mathbf{b}, \tag{9}$$

which exhibits $\bar{\mathbf{x}}$ as the solution of the consistent linear system $A^T A\mathbf{x} = A^T \mathbf{b}$.

EXAMPLE 5 Find the least-squares approximate solution of the overdetermined linear system

$$\begin{bmatrix} 1 & 0 & 1 \\ 2 & 1 & 1 \\ 1 & 3 & 0 \\ 0 & 2 & 1 \\ -1 & 2 & -1 \end{bmatrix} \begin{bmatrix} x_1 \\ x_2 \\ x_3 \end{bmatrix} = \begin{bmatrix} 1 \\ 0 \\ 1 \\ 2 \\ 0 \end{bmatrix},$$

by transforming it into the consistent form (9).

Solution The consistent system (9) is

$$
\begin{bmatrix} 1 & 2 & 1 & 0 & -1 \\ 0 & 1 & 3 & 2 & 2 \\ 1 & 1 & 0 & 1 & -1 \end{bmatrix}
\begin{bmatrix} 1 & 0 & 1 \\ 2 & 1 & 1 \\ 1 & 3 & 0 \\ 0 & 2 & 1 \\ -1 & 2 & -1 \end{bmatrix}
\begin{bmatrix} x_1 \\ x_2 \\ x_3 \end{bmatrix}
=
\begin{bmatrix} 1 & 2 & 1 & 0 & -1 \\ 0 & 1 & 3 & 2 & 2 \\ 1 & 1 & 0 & 1 & -1 \end{bmatrix}
\begin{bmatrix} 1 \\ 0 \\ 1 \\ 2 \\ 0 \end{bmatrix}
$$

or

$$
\begin{bmatrix} 7 & 3 & 4 \\ 3 & 18 & 1 \\ 4 & 1 & 4 \end{bmatrix}
\begin{bmatrix} x_1 \\ x_2 \\ x_3 \end{bmatrix}
=
\begin{bmatrix} 2 \\ 7 \\ 3 \end{bmatrix},
$$

whose solution is found to be

$$
\begin{bmatrix} x_1 \\ x_2 \\ x_3 \end{bmatrix}
\approx
\begin{bmatrix} -.614 \\ .421 \\ 1.259 \end{bmatrix},
$$

accurate to three decimal places. This is the least-squares approximate solution $\bar{\mathbf{x}}$ to the given overdetermined system. ■

The Orthogonal Case

To obtain the least-squares linear fit $y = r_0 + r_1 x$ of k data points (a_i, b_i), we form the matrix

$$
A = \begin{bmatrix} 1 & a_1 \\ 1 & a_2 \\ & \vdots \\ 1 & a_k \end{bmatrix} \tag{10}
$$

and compute the least-squares solution vector $\bar{\mathbf{r}} = (A^T A)^{-1} A^T \mathbf{b}$. If the set of column vectors in A were orthonormal, then $A^T A$ would be the 2×2 identity matrix I. Of course, this is never the case, since the first column vector of A has length $\sqrt{k}$. However, it may happen that the column vectors in A are mutually perpendicular; we can see then that

$$
A^T A = \begin{bmatrix} k & 0 \\ 0 & \mathbf{a} \cdot \mathbf{a} \end{bmatrix}, \quad \text{so} \quad (A^T A)^{-1} = \begin{bmatrix} 1/k & 0 \\ 0 & 1/(\mathbf{a} \cdot \mathbf{a}) \end{bmatrix}.
$$

The matrix A has this property if the x-values $a_1, a_2, \ldots, a_k$ for the data are symmetrically positioned about zero. We illustrate with an example.

EXAMPLE 6 Find the least-squares linear fit of the data points $(-3, 8)$, $(-1, 5)$, $(1, 3)$, and $(3, 0)$.

Solution The matrix A is given by

$$
A = \begin{bmatrix} 1 & -3 \\ 1 & -1 \\ 1 & 1 \\ 1 & 3 \end{bmatrix}.
$$

We can see why the symmetry of the x-values about zero causes the column vectors of this matrix to be orthogonal. We find that

$$A^TA = \begin{bmatrix} 1 & 1 & 1 & 1 \\ -3 & -1 & 1 & 3 \end{bmatrix}\begin{bmatrix} 1 & -3 \\ 1 & -1 \\ 1 & 1 \\ 1 & 3 \end{bmatrix} = \begin{bmatrix} 4 & 0 \\ 0 & 20 \end{bmatrix}.$$

Then

$$\bar{\mathbf{r}} = \begin{bmatrix} r_0 \\ r_1 \end{bmatrix} = (A^TA)^{-1}A^T\mathbf{b} = \begin{bmatrix} \frac{1}{4} & 0 \\ 0 & \frac{1}{20} \end{bmatrix}\begin{bmatrix} 1 & 1 & 1 & 1 \\ -3 & -1 & 1 & 3 \end{bmatrix}\begin{bmatrix} 8 \\ 5 \\ 3 \\ 0 \end{bmatrix}$$

$$= \begin{bmatrix} \frac{1}{4} & 0 \\ 0 & \frac{1}{20} \end{bmatrix}\begin{bmatrix} 16 \\ -26 \end{bmatrix} = \begin{bmatrix} 4 \\ -1.3 \end{bmatrix}.$$

Thus, the least-squares linear fit is given by $y = 4 - 1.3x$. ▪

Suppose now that the x-values in the data are not symmetrically positioned about zero, but that there is a number $x = c$ about which they are symmetrically located. If we make the variable transformation $t = x - c$, then the t-values for the data are symmetrically positioned about $t = 0$. We can find the t,y-equation for the least-squares linear fit, and then replace t by $x - c$ to obtain the x,y-equation. (Exercises 14 and 15 refine this idea still further.)

EXAMPLE 7   The number of sales of a particular item by a manufacturer in each of the first four months of 1985 is given in Table 5.7. Find the least-squares fit of these data, and use it to project sales in the fifth month.

TABLE 5.7

$\mathbf{a}_i$ = Month	1	2	3	4
$\mathbf{b}_i$ = Thousands sold	2.5	3	3.8	4.5

Solution   Our x-values a_i are symmetrically located about $c = 2.5$. If we take $t = x - 2.5$, the data points (t, y) become

$$(-1.5, 2.5), \quad (-.5, 3), \quad (.5, 3.8), \quad (1.5, 4.5).$$

We find that

$$A^TA = \begin{bmatrix} 1 & 1 & 1 & 1 \\ -1.5 & -.5 & .5 & 1.5 \end{bmatrix}\begin{bmatrix} 1 & -1.5 \\ 1 & -.5 \\ 1 & .5 \\ 1 & 1.5 \end{bmatrix} = \begin{bmatrix} 4 & 0 \\ 0 & 5 \end{bmatrix}.$$

Thus,

$$(A^TA)^{-1}A^T\mathbf{b} = \begin{bmatrix} \frac{1}{4} & 0 \\ 0 & \frac{1}{5} \end{bmatrix} \begin{bmatrix} 1 & 1 & 1 & 1 \\ -1.5 & -.5 & .5 & 1.5 \end{bmatrix} \begin{bmatrix} 2.5 \\ 3 \\ 3.8 \\ 4.5 \end{bmatrix}$$

$$= \begin{bmatrix} \frac{1}{4} & 0 \\ 0 & \frac{1}{5} \end{bmatrix} \begin{bmatrix} 13.8 \\ 3.4 \end{bmatrix} = \begin{bmatrix} 3.45 \\ .68 \end{bmatrix}.$$

Thus, the t,y-equation is $y = 3.45 + .68t$. Replacing t by $x - 2.5$, we obtain

$$y = 3.45 + .68(x - 2.5) = 1.75 + .68x.$$

Setting $x = 5$, we estimate sales in the fifth month to be 5.15 units, or 5150 items.

■

Using the QR-Factorization

The QR-factorization discussed at the end of Section 5.2 has application to the method of least squares. Consider an overdetermined system $A\mathbf{x} = \mathbf{b}$, where the matrix A has independent column vectors. We factor A as QR and remember that $Q^TQ = I$, since Q has orthonormal column vecctors. Then we obtain

$$(A^TA)^{-1}A^T\mathbf{b} = ((QR)^TQR)^{-1}(QR)^T\mathbf{b} = (R^TQ^TQR)^{-1}R^TQ^T\mathbf{b}$$
$$= (R^TR)^{-1}R^TQ^T\mathbf{b} = R^{-1}(R^T)^{-1}R^TQ^T\mathbf{b} = R^{-1}Q^T\mathbf{b}. \qquad (11)$$

To find the least-squares solution vector $R^{-1}Q^T\mathbf{b}$ once Q and R have been found, we first multiply $\mathbf{b}$ by Q^T, which is a stable computation because Q is orthogonal. We then solve the upper-triangular system $R\mathbf{x} = Q^T\mathbf{b}$ by back substitution. Since R is already upper triangular, no matrix reduction is needed! The available program QRFACTOR has an option for this least-squares computation. The program demonstrates how quickly the least-squares solution vector can be found once a QR factorization of A is known.

Suppose that we believe that the output y of a process should be a linear function of time t. We measure the y-values at specified times during the process. We can then find the least-squares linear fit $y = r_1 + s_1t$ from the measured data by finding the QR-factorization of an appropriate matrix A and subsequently finding the solution vector (11) in the way we just described. If we now repeat the process and make measurements again at the same times, the matrix A will not change, so we can find another least-squares linear fit using the solution vector (11) with the same matrices Q and R. We can repeat this many times, obtaining linear fits $r_1 + s_1t$, $r_2 + s_2t$, ..., $r_m + s_mt$. For our final estimate of the proper linear fit for this process, we might take $r + st$, where r is the average of the m values r_i and s is the average of the m values s_i. We expect that errors made in measurements will roughly cancel each other out over numerous repetitions of the process, so that our estimated linear fit $r + st$ will be fairly accurate.

SUMMARY

1. A set of k data points (a_i, b_i) whose first coordinates are not all equal can be fitted with a polynomial function or with an exponential function, using the method of least squares illustrated in Examples 1 through 3.

2. Suppose that the x-values a_i in a set of k data points (a_i, b_i) are symmetrically positioned about zero. Let A be the $k \times 2$ matrix with first column vector having all entries 1 and second column vector $\mathbf{a}$. The columns of A are orthogonal, and

$$A^T A = \begin{bmatrix} k & 0 \\ 0 & \mathbf{a} \cdot \mathbf{a} \end{bmatrix}.$$

Computation of the least-squares linear fit for the data points is then simplified. If the x-values of the data points are symmetrically positioned about $x = c$, then the substitution $t = x - c$ gives data points with t-values symmetrically positioned about zero, and the above simplification applies. See Example 7.

Let $A\mathbf{x} = \mathbf{b}$ be a linear system of m equations in n unknowns, where $m > n$ (an overdetermined system) and where the rank of A is n.

3. The least-squares solution of the corresponding system $A\mathbf{x} \approx \mathbf{b}$ of linear approximations is the vector $\mathbf{x} = \bar{\mathbf{x}}$, which minimizes the magnitude of the error vectors $A\mathbf{x} - \mathbf{b}$ for $\mathbf{x} \in \mathbb{R}^n$.

4. The least-squares solution $\bar{\mathbf{x}}$ of $A\mathbf{x} \approx \mathbf{b}$ and of $A\mathbf{x} = \mathbf{b}$ is given by the formula

$$\bar{\mathbf{x}} = (A^T A)^{-1} A^T \mathbf{b}.$$

Geometrically, $A\bar{\mathbf{x}}$ is the projection of $\mathbf{b}$ on the column space of A.

5. An alternative to using the formula for $\bar{\mathbf{x}}$ in summary item 4 is to convert the overdetermined system $A\mathbf{x} = \mathbf{b}$ to the consistent system $(A^T A)\mathbf{x} = A^T \mathbf{b}$ by multiplying both sides of $A\mathbf{x} = \mathbf{b}$ by A^T, and then to find its unique solution, which is the least-squares solution $\bar{\mathbf{x}}$ of $A\mathbf{x} = \mathbf{b}$.

EXERCISES

1. Let the length b_i of a spring with an attached weight a_i be determined by measurements, as shown in Table 5.8.
   a) Find the least-squares linear fit, in accordance with Hooke's law.
   b) Use the answer to part (a) to estimate the length of the spring if a weight of 5 ounces is attached.

TABLE 5.8

$\mathbf{a}_i$ = Weight in ounces	1	2	4	6
$\mathbf{b}_i$ = Length in inches	3	4.1	5.9	8.2

2. A company had profits (in units of \$10,000) of 0.5 in 1980, 1 in 1982, and 2 in 1985. Let time t

be measured in years, with $t = 0$ in 1980.

a) Find the least-squares linear fit of the data.

b) Using the answer to part (a), estimate the profit in 1986.

3. Repeat Exercise 2, but find an exponential fit of the data, working with logarithms as explained in Example 2.

4. A publishing company specializing in college texts starts with a field sales force of 10 people, and it has profits of $100,000. On increasing this sales force to 20, it has profits of $300,000; and increasing its sales force to 30 produces profits of $400,000.

a) Find the least-squares linear fit for these data. [HINT: Express the numbers of salespeople in multiples of 10 and the profit in multiples of $100,000.]

b) Use the answer to part (a) to estimate the profit if the sales force is reduced to 25.

c) Does the profit obtained using the answer to part (a) for a sales force of 0 people seem in any way plausible?

In Exercises 5–7, find the least-squares fit to the given data by a linear function $f(x) = r_0 + r_1 x$, having a straight line as graph in the plane. Graph the line and the data points.

5. (0, 1), (2, 6), (3, 11), (4, 12)

6. (1, 1), (2, 4), (3, 6), (4, 9)

7. (0, 0), (1, 1), (2, 3), (3, 8)

8. Find the least-squares fit to the data in Exercise 6 by a parabola (a quadratic polynomial function).

9. Repeat Exercise 8, but use the data in Exercise 7 instead.

In Exercises 10–13, use the technique illustrated in Examples 6 and 7 to solve the least-squares problem.

10. Find the least-squares linear fit to the data points $(-4, -2)$, $(-2, 0)$, $(0, 1)$, $(2, 4)$, $(4, 5)$.

11. Find the least-squares linear fit to the data points $(0, 1)$, $(1, 4)$, $(2, 6)$, $(3, 8)$, $(4, 9)$.

12. The gallons of maple syrup made from the sugar bush of a Vermont farmer over the past five years were:

80 gallons five years ago,
70 gallons four years ago,

75 gallons three years ago,
65 gallons two years ago,
60 gallons last year.

Use a least-squares linear fit of these data to project the number of gallons that will be produced this year. (Does this problem make practical sense? Why?)

13. The minutes required for a rat to find its way out of a maze on each repeated attempt were 8, 8, 6, 5, and 6 on its first five tries. Use a least-squares linear fit of these data to project the time the rat will take on its sixth try.

14. Let (a_1, b_1), (a_2, b_2), . . . , (a_m, b_m) be data points. If $\sum_{i=1}^{m} a_i = 0$, show that the line that best fits the data in the least-squares sense is given by $r_0 + r_1 x$, where

$$r_0 = \left(\sum_{i=1}^{m} b_i \right) \bigg/ m$$

and

$$r_1 = \left(\sum_{i=1}^{m} a_i b_i \right) \bigg/ \left(\sum_{i=1}^{m} a_i^2 \right).$$

15. Repeat Exercise 14, but do not assume that $\sum_{i=1}^{m} a_i = 0$. Show that the least-squares linear fit of the data is given by $y = r_0 + r_1(x - c)$, where $c = (\sum_{i=1}^{m} a_i)/m$ and r_0 and r_1 have the values given in Exercise 14.

In Exercises 16–20, find the least-squares solution of the given overdetermined system $A\mathbf{x} = \mathbf{b}$ by converting it to a consistent system and then solving, as illustrated in Example 5.

16. $\begin{bmatrix} 1 & 2 \\ 1 & 1 \\ 2 & 3 \end{bmatrix} \begin{bmatrix} x_1 \\ x_2 \end{bmatrix} = \begin{bmatrix} 3 \\ 1 \\ 3 \end{bmatrix}$

17. $\begin{bmatrix} 3 & 1 \\ 1 & 2 \\ 2 & -1 \end{bmatrix} \begin{bmatrix} x_1 \\ x_2 \end{bmatrix} = \begin{bmatrix} 1 \\ 0 \\ -2 \end{bmatrix}$

18. $\begin{bmatrix} 1 & 1 & 1 \\ -1 & 0 & 1 \\ 1 & -1 & 0 \\ 0 & 1 & -1 \end{bmatrix} \begin{bmatrix} x_1 \\ x_2 \\ x_3 \end{bmatrix} = \begin{bmatrix} 0 \\ 1 \\ -1 \\ -2 \end{bmatrix}$

19. $\begin{bmatrix} 1 & 1 & 3 \\ -1 & 0 & 5 \\ 0 & 1 & -2 \\ 1 & -1 & 1 \\ 1 & 0 & 1 \end{bmatrix} \begin{bmatrix} x_1 \\ x_2 \\ x_3 \end{bmatrix} = \begin{bmatrix} 1 \\ -1 \\ 3 \\ -2 \\ 0 \end{bmatrix}$

20. $\begin{bmatrix} 2 & -1 & 1 \\ 1 & 0 & 1 \\ -1 & 0 & 1 \\ 0 & -5 & -1 \\ 1 & 2 & -2 \end{bmatrix} \begin{bmatrix} x_1 \\ x_2 \\ x_3 \end{bmatrix} = \begin{bmatrix} 2 \\ 1 \\ 2 \\ 5 \\ 0 \end{bmatrix}$

21. Mark each of the following True or False.

____ a) There is a unique polynomial function of degree k with graph passing through any given k points in $\mathbb{R}^2$.

____ b) There is a unique polynomial function of degree k with graph passing through any k points in $\mathbb{R}^2$ having distinct first coordinates.

____ c) There is a unique polynomial function of degree at most $k - 1$ with graph passing through any k points in $\mathbb{R}^2$ having distinct first coordinates.

____ d) The least-squares solution of $A\mathbf{x} = \mathbf{b}$ is unique.

____ e) The least-squares solution of $A\mathbf{x} = \mathbf{b}$ can be an actual solution only if A is a square matrix.

____ f) The least-squares solution vector of $A\mathbf{x} = \mathbf{b}$ is the projection of $\mathbf{b}$ on the column space of A.

____ g) The least-squares solution vector of $A\mathbf{x} = \mathbf{b}$ is the vector $\bar{\mathbf{x}}$ such that $A\bar{\mathbf{x}}$ is the projection of $\mathbf{b}$ on the column space of A.

____ h) The least-squares solution vector of $A\mathbf{x} = \mathbf{b}$ is the vector $\mathbf{x} = \bar{\mathbf{x}}$ that minimizes the magnitude of $A\mathbf{x} - \mathbf{b}$.

____ i) A linear system has a least-squares solution only if the number of equations is greater than the number of unknowns.

____ j) Every linear system has a least-squares solution.

■ *The formula* $\mathbf{x} = (A^T A)^{-1} A^T \mathbf{b}$ *for the least-squares solution of an overdetermined system* $A\mathbf{x} = \mathbf{b}$ *can easily be evaluated using MATCOMP. In Exercises 22–27, find the least-squares solution requested in the exercise indicated, using MATCOMP.*

22. Exercise 1
23. Exercise 2
24. Exercise 4
25. Exercise 8
26. Exercise 9
27. Exercise 20

■ *The supplied program YOUFIT can be used to illustrate graphically the fitting of data points by linear, quadratic, or exponential functions. In Exercises 28–32, use YOUFIT to try visually to fit the given data with the indicated type of graph. When this is done, enter the zero data suggested to see the computer's fit. Run twice more with the same data points but without trying to fit the data visually, and determine whether the data is best fitted by a linear, quadratic, or (logarithmically fitted) exponential function, by comparing the least-squares sums for the three cases.*

28. Fit $(1, 2)$, $(4, 6)$, $(7, 10)$, $(10, 14)$, $(14, 19)$ by a linear function.

29. Fit $(2, 2)$, $(6, 10)$, $(10, 12)$, $(16, 2)$ by a quadratic function.

30. Fit $(1, 1)$, $(10, 8)$, $(14, 12)$, $(16, 20)$ by an exponential function. Try to achieve a lower squares sum than the computer obtains with its least-squares fit that uses logarithms of y-values.

31. Repeat Exercise 30 with data $(1, 9)$, $(5, 1)$, $(6, .5)$, $(9, .01)$.

32. Fit $(2, 9)$, $(4, 6)$, $(7, 1)$, $(8, .1)$ by a linear function.

■ *The supplied program QRFACTOR has an option to use the QR-factorization to give the least-squares solution of a linear system* $A\mathbf{x} = \mathbf{b}$. *Use this program for the remaining exercises.*

33. Find the least-squares linear fit for the data points $(-3, 10)$, $(-2, 8)$, $(-1, 7)$, $(0, 6)$, $(1, 4)$, $(2, 5)$, $(3, 6)$.

34. Find the least-squares quadratic fit for the data points in Exercise 33.

35. Find the least-squares cubic fit for the data points in Exercise 33.

36. Find the least-squares quartic fit for the data points in Exercise 33.

37. Find the quadratic polynomial function whose graph passes through the points $(1, 4)$, $(2, 15)$, $(3, 32)$.

38. Find the cubic polynomial function whose graph passes through the points $(-1, 12)$, $(0, -5)$, $(2, 15)$, $(3, 52)$.

LINEAR TRANSFORMATIONS AND SIMILARITY

Most of the linear algebra in the preceding chapters has been developed in terms of matrices rather than in terms of linear transformations. As a rectangular array of numbers, a matrix seems to be a more concrete and easily grasped concept than a linear transformation. However, we know that an $m \times n$ matrix A does correspond naturally to a linear transformation $T: \mathbb{R}^n \to \mathbb{R}^m$ defined by $T(\mathbf{x}) = A\mathbf{x}$. Consequently, every idea and computation we have discussed involving matrices has a parallel in terms of a linear transformation. We mentioned some of these parallels as we worked with matrices. For example, we defined the notions of an *eigenvalue* and of an *eigenvector* of a linear transformation in Section 4.1, after we introduced these concepts for matrices.

When using matrices, we are dealing only with finite-dimensional vector spaces and working with coordinates. Using linear transformations frees us from these restrictions. For example, after defining the notions of *eigenvalue* and *eigenvector* for linear transformations, we can view the function e^{kx} as an eigenvector with corresponding eigenvalue k for the differentiation transformation. Here the vector space is the infinite-dimensional space of functions having derivatives of all orders, and no coordinates appear. (See Example 8 in Section 4.1.)

In Section 6.1, we recall the definition of a general linear transformation and then review more of our matrix concepts in terms of linear transformations. The conventional terminology for an idea in the language of transformations is often quite different from the corresponding matrix terminology. For example, the parallel of the *nullspace* of a matrix is often called the *kernel* of the associated transformation, and the parallel of the *column space* of the matrix is the *range* of the transformation. It is important to realize that we are talking about the same ideas in different settings, and that the setting of linear transformations is more general than the matrix setting.

We have associated with every linear transformation $T: \mathbb{R}^n \to \mathbb{R}^m$ its standard matrix representation, the $m \times n$ matrix

$$A = \begin{bmatrix} | & | & & | \\ T(\mathbf{e}_1) & T(\mathbf{e}_2) & \cdots & T(\mathbf{e}_n) \\ | & | & & | \end{bmatrix}.$$

For any $\mathbf{x}$ in $\mathbb{R}^n$, we have

$$T(\mathbf{x}) = A\mathbf{x}. \tag{1}$$

Clearly the standard ordered basis for $\mathbb{R}^n$ plays an important role in determining the matrix A. But the standard ordered basis is just one of an infinite number of possible ordered bases for $\mathbb{R}^n$. In Sections 4.2 and 4.3, we saw that it can be very advantageous to work with a basis of eigenvectors, if such a basis exists. Section 6.2 explains how to represent a linear transformation $T: V \rightarrow V'$ of finite-dimensional vector spaces by a matrix R relative to any ordered base B and B' respectively; the matrix R represents T in the sense that

$$T(\mathbf{x})_{B'} = R\mathbf{x}_B, \tag{2}$$

where the subscripts denote coordinate vectors relative to ordered bases. Clearly Eq. (2) is conceptually the same as Eq. (1); only the bases have changed.

Section 6.3 discusses the relationship between matrix representations A and R of the same linear transformation $T: V \rightarrow V$ relative to possibly different ordered bases. We will see that A and R are *similar matrices*, a concept introduced in Section 4.2 when we wished to work with a basis of eigenvectors rather than with the standard basis. Conversely, we will see that any two similar matrices do represent the same transformation but relative to possibly different ordered bases. This is a very significant result in studying the structure of matrices. For example, it follows at once that any two similar matrices must have the same eigenvalues.

In summary, Chapter 6 develops further the relationship between linear transformations and matrices.

6.1 PROPERTIES OF LINEAR TRANSFORMATIONS

Review of Function Terminology

Linear transformations are functions, and consequently we begin by reviewing a bit of function terminology. A function $f: X \rightarrow Y$ mapping a set X into a set Y assigns to each element x of the set X a *single* element y of the set Y. The notation $y = f(x)$ indicates that y is the element assigned to x by f. The set X is the **domain** of f, and the set Y is the **codomain** of f. Let S be a subset of X, and let U be a subset of Y. As illustrated in Figure 6.1,

$$f[S] = \{f(x) \mid x \text{ in } S\} \qquad \underline{\text{Image of } S \text{ under } f}$$

and

$$f^{-1}[U] = \{x \text{ in } X \mid f(x) \text{ is in } U\}. \qquad \underline{\text{Inverse image of } U \text{ under } f}$$

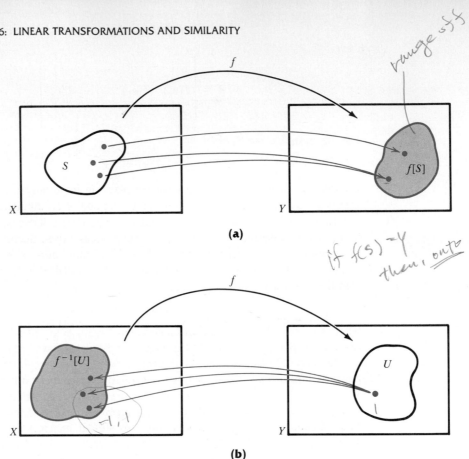

range of f

if f(s) =Y then, onto

FIGURE 6.1 (a) The image of *S* under *f*; (b) the inverse image of *U* under *f*.

For example, if $f: \mathbb{R} \to \mathbb{R}$ is the squaring function, so that $f(x) = x^2$, then

$$f[\{1, 2, 3\}] = \{1, 4, 9\},$$

elements assigned to x by f.

while

$$f^{-1}[\{1, 4, 9\}] = \{1, -1, 2, -2, 3, -3\}.$$

If $f: X \to Y$, then the image $f[X]$ of X is the **range** of f. If $f[X] = Y$, then f is **onto** Y.

A function $f: X \to Y$ is **one-to-one** if $f(x_1) \neq f(x_2)$ unless $x_1 = x_2$ in X. When f is one-to-one, each y_1 in the range $f[X]$ is equal to $f(x_1)$ for a *unique* x_1 in X. The **inverse function** $f^{-1}: f[X] \to X$ is then defined by $f^{-1}(y_1) = x_1$. (See Fig. 6.2.)

Review of Basic Concepts Presented in Section 2.9

Let V and V' be vector spaces. A **linear transformation** is a function $T: V \to V'$ satisfying two linearity conditions,

$$T(\mathbf{v}_1 + \mathbf{v}_2) = T(\mathbf{v}_1) + T(\mathbf{v}_2) \qquad \textbf{Preservation of addition} \qquad (3)$$

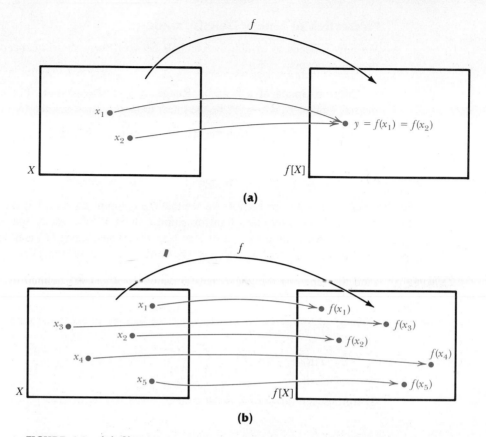

FIGURE 6.2 **(a)** *f* is not one-to-one, so f^{-1} is not well-defined; **(b)** *f* is one-to-one, so f^{-1} is well-defined.

$$T(r\mathbf{v}_1) = rT(\mathbf{v}_1), \qquad \text{Preservation of scalar multiplication} \qquad (4)$$

for all vectors $\mathbf{v}_1$ and $\mathbf{v}_2$ in V and for all scalars r in $\mathbb{R}$. The two defining conditions are equivalent to the single condition.

$$T(r_1\mathbf{v}_1 + r_2\mathbf{v}_2) = r_1T(\mathbf{v}_1) + r_2T(\mathbf{v}_2) \qquad \text{Preservation of linear combinations} \qquad (5)$$

for all vectors $\mathbf{v}_1$ and $\mathbf{v}_2$ in V and for all scalars r_1, r_2 in $\mathbb{R}$. (See Exercise 26 in Section 2.9.) As special cases of this equation, we obtain

$$T(\mathbf{0}) = \mathbf{0}' \qquad \text{Preservation of zero}$$

$$T(\mathbf{v}_1 - \mathbf{v}_2) = T(\mathbf{v}_1) - T(\mathbf{v}_2). \qquad \text{Preservation of subtraction}$$

In general, we can expect any algebraic property of the vector space V to be *preserved* by the linear transformation T, as long as that property ultimately involves only vector addition and scalar multiplication. For example, the subtraction $\mathbf{v}_1 - \mathbf{v}_2$ can be expressed as $\mathbf{v}_1 + (-1)\mathbf{v}_2$. Hence, $T(\mathbf{v}_1 - \mathbf{v}_2) = T(\mathbf{v}_1) - T(\mathbf{v}_2)$.

Properties of Linear Transformations

We now proceed to exhibit the parallels between some matrix concepts and transformation concepts.

1. Column Space of a Matrix: Range of Transformation. Let A be an $m \times n$ matrix, and let $T: \mathbb{R}^n \to \mathbb{R}^m$ be the linear transformation defined by $T(\mathbf{x}) = A\mathbf{x}$. Then

$$\text{Column space of } A = \{A\mathbf{x} \mid \mathbf{x} \in \mathbb{R}^n\}$$

and

$$\text{Range of } T = \{T(\mathbf{x}) \mid \mathbf{x} \in \mathbb{R}^n\}.$$

Since $T(\mathbf{x}) = A\mathbf{x}$ for each $\mathbf{x}$, we see that the column space of A is the same set as the range of T. For any linear transformation $T: V \to V'$ of a vector space V into a vector space V', we write the range of T as range(T). Thus, range (T) and the column space of a matrix are parallel concepts. The **rank** of T is dim(range(T)).

EXAMPLE 1   Find the range of the linear transformation $T: \mathbb{R}^3 \to \mathbb{R}^2$ defined by $T(x_1, x_2, x_3) = (x_1 - 2x_2, x_2 + 3x_3)$.

Solution   Writing vectors as column vectors, we see that the linear transformation has the form

$$T(\mathbf{x}) = A\mathbf{x} = \begin{bmatrix} 1 & -2 & 0 \\ 0 & 1 & 3 \end{bmatrix} \begin{bmatrix} x_1 \\ x_2 \\ x_3 \end{bmatrix} = x_1 \begin{bmatrix} 1 \\ 0 \end{bmatrix} + x_2 \begin{bmatrix} -2 \\ 1 \end{bmatrix} + x_3 \begin{bmatrix} 0 \\ 3 \end{bmatrix}.$$

Therefore, the range of T is the column space of the matrix

$$A = \begin{bmatrix} 1 & -2 & 0 \\ 0 & 1 & 3 \end{bmatrix}.$$

Since the first two columns of A are linearly independent, we conclude that range(T) = $\mathbb{R}^2$. ▪

Let us pursue some consequences of the observation that the *column space* of a matrix is the *range* of the associated linear transformation. One consequence is that we can use matrices to find the range of a transformation of finite-dimensional vector spaces, as illustrated in Example 1. For an illustration of a theoretical nature, recall that the column space of A is a subspace of $\mathbb{R}^m$, because it is the subspace of $\mathbb{R}^m$ generated by the column vectors in A. Then the range of the associated linear transformation is a *subspace* of $\mathbb{R}^m$. We now extend this result to linear transformations, free from finite-dimensional restrictions.

THEOREM 6.1   Preservation of Subspaces

Let V and V' be vector spaces, and let $T: V \to V'$ be a linear transformation.

(i) If W is a subspace of V, then $T[W]$ is a subspace of V'.

(ii) If S' is a subspace of V', then $T^{-1}[S']$ is a subspace of V.

PROOF (i) Since $T(0) = 0'$, we need only show that $T[W]$ is closed under vector addition and under multiplication by scalars. Let $T(\mathbf{v}_1)$ and $T(\mathbf{v}_2)$ be any vectors in $T[W]$, where $\mathbf{v}_1$ and $\mathbf{v}_2$ are vectors in W. Then

$$T(\mathbf{v}_1) + T(\mathbf{v}_2) = T(\mathbf{v}_1 + \mathbf{v}_2),$$

by preservation of addition. Now $\mathbf{v}_1 + \mathbf{v}_2$ is in W since W is itself closed under addition, so $T(\mathbf{v}_1 + \mathbf{v}_2)$ is in $T[W]$. This shows that $T[W]$ is closed under vector addition. If r is any scalar, then $r\mathbf{v}_1$ is in W, and

$$rT(\mathbf{v}_1) = T(r\mathbf{v}_1).$$

This shows that $rT(\mathbf{v}_1)$ is in $T[W]$, so $T[W]$ is closed under multiplication by scalars. Thus $T[W]$ is a subspace of V'.

(ii) Notice that $0 \in T^{-1}[S']$. Let $\mathbf{x}$ and $\mathbf{y}$ be any vectors in $T^{-1}[S']$, so that $T(\mathbf{x})$ and $T(\mathbf{y})$ are in S'. Then

$$T(\mathbf{x} + \mathbf{y}) = T(\mathbf{x}) + T(\mathbf{y})$$

is also in the subspace S', so $\mathbf{x} + \mathbf{y}$ is in $T^{-1}[S']$. For any scalar r, we know that

$$rT(\mathbf{x}) = T(r\mathbf{x}),$$

and $rT(\mathbf{x})$ is in S'. Thus, $r\mathbf{x}$ is also in $T^{-1}[S']$. This shows that $T^{-1}[S']$ is closed under addition and under scalar multiplication, so $T^{-1}[S']$ is a subspace of V. ▲

For example, the linear transformation $T: \mathbb{R}^2 \to \mathbb{R}^2$ defined by $T(x, y) = (x, 0)$ projects the x,y-plane $\mathbb{R}^2$ onto the x-axis. If W_1 is the line given by $y = -x$, then $T[W_1]$ is the x-axis and $T^{-1}[x\text{-axis}]$ is the entire plane. If W_2 is the y-axis, then $T[W_2] = \{0\}$ and $T^{-1}[\{0\}] = W_2$.

Theorem 6.1 has many uses. Here is one illustration.

EXAMPLE 2 Let F be the vector space of all functions $f: \mathbb{R} \to \mathbb{R}$, and let c be a scalar in $\mathbb{R}$. Determine whether the mapping $T: F \to \mathbb{R}$ defined by $T(f) = |f(c)|$ is a linear transformation.

Solution Since $T[F]$ contains positive numbers but no negative numbers, it is not a subspace of the one-dimensional space $\mathbb{R}$. Therefore, T is not a linear transformation, as shown by Theorem 6.1. ■

Let W be a subspace of $\mathbb{R}^n$, and let A be an $m \times n$ matrix. Theorem 6.1 shows that $\{A\mathbf{x} \mid \mathbf{x} \in W\}$ is a subspace of the column space of A, because this set corresponds precisely to $T[W]$ in the case $T(\mathbf{x}) = A\mathbf{x}$. We never proved such a theorem about matrices, but it follows at once from Theorem 6.1. In Exercise 30, we ask you to state for matrices the analogue of the statement that $T^{-1}[S']$ is a subspace of V in Theorem 6.1.

EXAMPLE 3 (*Calculus*) Let T be the differentiation transformation mapping the space P of all polynomial functions into itself. Describe the subspace $T^{-1}[P_2]$, where P_2 is the subspace of all polynomial functions of degree at most 2.

Solution By calculus, we know that the derivative $T(p(x)) = p'(x)$ of a polynomial $p(x)$ has degree at most 2 if and only if $p(x)$ has degree at most 3, so we see that $T^{-1}[P_2] = \{p(x) \in P \mid p'(x) \in P_2\} = P_3$, which is again a subspace of P. ■

2. Nullspace of a Matrix: Kernel of a Linear Transformation. Let A be an $m \times n$ matrix and let $T: \mathbb{R}^n \to \mathbb{R}^m$ be the linear transformation defined by $T(\mathbf{x}) = A(\mathbf{x})$. We know that

$$\text{Nullspace of } A = \{\mathbf{x} \in \mathbb{R}^n \mid A\mathbf{x} = \mathbf{0}\}$$
$$= \{\mathbf{x} \in \mathbb{R}^n \mid T(\mathbf{x}) = \mathbf{0}\}.$$

In the general case, if $T: V \to V'$ is a linear transformation, then $\{\mathbf{v} \in V \mid T(\mathbf{v}) = \mathbf{0}\}$ is referred to both as the **nullspace of** T and as the **kernel of** T, and is written $\underline{\ker(T)}$. The fact that $\ker(T)$ is a subspace of V follows from Theorem 6.1, since $\ker(T) = T^{-1}[\{\mathbf{0}\}]$. The **nullity** of T is $\dim(\ker(T))$. We have

$$\text{rank}(T) + \text{nullity}(T) = \dim(V). \quad \textbf{rank equation}$$

EXAMPLE 4 Find the kernel of the linear transformation of Example 1.

Solution The linear transformation in Example 1 has the form $T(\mathbf{x}) = A\mathbf{x}$, where

$$A = \begin{bmatrix} 1 & -2 & 0 \\ 0 & 1 & 3 \end{bmatrix} \quad \text{and} \quad \mathbf{x} = \begin{bmatrix} x_1 \\ x_2 \\ x_3 \end{bmatrix}.$$

Since A has an echelon form, we see that its nullspace is the set of all vectors of the form

$$\mathbf{x} = \begin{bmatrix} -6s \\ -3s \\ s \end{bmatrix}$$

for any scalar s. This nullspace is the kernel of T. ■

EXAMPLE 5 (*Calculus*) Let F be the vector space of all functions mapping $\mathbb{R}$ into $\mathbb{R}$, and let W be the subspace of F consisting of all differentiable functions. Find the kernel of the linear transformation $T: W \to F$ defined by $T(f) = f'$, the derivative of f.

Solution We see that $\ker(T) = \{f \in W \mid f' = 0\}$, which is the subspace of W consisting of all constant functions. ■

We now present a theoretical consequence of this correspondence between the nullspace of a matrix and the kernel of the associated linear transformation. Recall from Chapter 1 that, for a system of linear equations, the following two conditions are equivalent:

(i) A consistent system $A\mathbf{x} = \mathbf{b}$ has a unique solution.

(ii) The homogeneous system $A\mathbf{x} = \mathbf{0}$ has only the trivial solution.

In terms of the linear transformation $T(\mathbf{x}) = A\mathbf{x}$, condition (i) tells us that there is *at most one* $\mathbf{x}$ in $\mathbb{R}^n$ such that $T(\mathbf{x}) = \mathbf{b}$; that is, T is *one-to-one*. This must be equivalent to the transformation analogue of condition (ii)—namely, that the kernel of T is the zero subspace $\{\mathbf{0}\}$. We prove this result in the context of general vector spaces.

THEOREM 6.2 Role of the Kernel

Let V and V' be vector spaces, and let $T: V \rightarrow V'$ be a linear transformation. Then T is one-to-one if and only if $\ker(T) = \{\mathbf{0}\}$, the zero subspace of V. If T is one-to-one and has range V', then the inverse map $T^{-1}: V' \rightarrow V$ is also a linear transformation.

P180

PROOF Suppose that $\ker(T) = \{\mathbf{0}\}$, and let $T(\mathbf{v}_1) = T(\mathbf{v}_2)$ for $\mathbf{v}_1$ and $\mathbf{v}_2$ in V. Then $T(\mathbf{v}_1 - \mathbf{v}_2) = T(\mathbf{v}_1) - T(\mathbf{v}_2) = \mathbf{0}$, so $\mathbf{v}_1 - \mathbf{v}_2$ is in $\ker(T)$, and hence $\mathbf{v}_1 - \mathbf{v}_2 = \mathbf{0}$. We conclude that $\mathbf{v}_1 = \mathbf{v}_2$, showing that T is one-to-one.

Conversely, suppose that T is one-to-one, and that $\mathbf{v}$ lies in $\ker(T)$ so that $T(\mathbf{v}) = \mathbf{0}$. Now $T(\mathbf{0}) = \mathbf{0}$, so $T(\mathbf{v}) = T(\mathbf{0})$. Since T is one-to-one, we must have $\mathbf{v} = \mathbf{0}$. This shows that $\ker(T) = \{\mathbf{0}\}$.

The final statement of Theorem 6.2 was established in Theorem 2.19. ▲

3. Invertible Matrix: Isomorphism. Recall from Section 2.9 that a linear transformation $T: V \rightarrow V'$ is an *isomorphism* if it is one-to-one and has as its range all of V'. In the case of a linear transformation $T: \mathbb{R}^n \rightarrow \mathbb{R}^m$ given by $T(\mathbf{x}) = A\mathbf{x}$, an isomorphism is impossible unless $n = m$ and A is invertible. Using the rank equation $\mathrm{rank}(A) + \mathrm{nullity}(A) = n$, we see that T is one-to-one (nullity $(A) = 0$) if and only if T has as its range all of $\mathbb{R}^n$ (colrank$(A) = n$). Thus, when $n = m$, it is only necessary to verify one of these two requirements (T is one-to-one, range$(T) = \mathbb{R}^n$) to show that T is an isomorphism. An analogous result is true for a linear transformation $T: V \rightarrow V'$ if V is finite-dimensional and if $\dim(V') = \dim(V)$. (See Exercise 36.)

EXAMPLE 6   Show that the linear transformation $T: \mathbb{R}^2 \rightarrow \mathbb{R}^2$ defined by

$$T(x, y) = (x - 2y, x + y)$$

$T(x,y) = \begin{bmatrix} 1 & -2 \\ 1 & 1 \end{bmatrix}\begin{bmatrix} x \\ y \end{bmatrix}$

is an isomorphism, and find a formula for the inverse transformation T^{-1}.

Solution   Writing vectors in column form, we have

$$T\left(\begin{bmatrix} x \\ y \end{bmatrix}\right) = \begin{bmatrix} 1 & -2 \\ 1 & 1 \end{bmatrix}\begin{bmatrix} x \\ y \end{bmatrix}.$$

Since

$$A = \begin{bmatrix} 1 & -2 \\ 1 & 1 \end{bmatrix}$$

has a nonzero determinant, it is invertible, with nullspace $\{\mathbf{0}\}$. Thus, $\ker(T) = \{\mathbf{0}\}$ so T is one-to-one and consequently is an isomorphism. The inverse of T is given by

$$T^{-1}\left(\begin{bmatrix} u \\ v \end{bmatrix}\right) = A^{-1}\begin{bmatrix} u \\ v \end{bmatrix} = \frac{1}{3}\begin{bmatrix} 1 & 2 \\ -1 & 1 \end{bmatrix}\begin{bmatrix} u \\ v \end{bmatrix} = \frac{1}{3}\begin{bmatrix} u + 2v \\ -u + v \end{bmatrix}.$$

Switching back to row notation, we have

$$T^{-1}(u, v) = \left(\frac{u}{3} + \frac{2v}{3}, -\frac{u}{3} + \frac{v}{3}\right).$$ ■

4. Eigenvalues and Eigenvectors. In this case, the terminology for transformations is the same as for matrices. In Section 4.1, we defined the notion of an eigenvector **v** and corresponding eigenvalue λ for both an $n \times n$ matrix A and a linear transformation $T: V \to V$, as follows:

Matrix	Transformation
$A\mathbf{v} = \lambda\mathbf{v}, \quad \mathbf{v} \neq \mathbf{0}$	$T(\mathbf{v}) = \lambda\mathbf{v}, \quad \mathbf{v} \neq \mathbf{0}$

All the notions relating to eigenvalues and eigenvectors for matrices can be carried over to any linear transformation of an n-dimensional vector space into itself, since we showed in Section 2.9 that any such vector space is isomorphic to $\mathbb{R}^n$. For example, we have the notions of *algebraic multiplicity*, of *eigenspace*, and of *geometric multiplicity* for matrices. Consequently, we use these same terms for any linear transformation of a finite-dimensional vector space into itself. In Exercise 31, we ask you to show that the notion of *eigenspace* is valid in the infinite-dimensional case.

The following box presents one strategy for answering questions about a linear transformation of an n-dimensional vector space V into itself by working in $\mathbb{R}^n$, using matrices.

Coordinatization Strategy for a Finite-dimensional Vector Space V

Step 1 Choose a convenient ordered basis, and *coordinatize V*, making it look just like $\mathbb{R}^n$.

Step 2 If $T: V \to V$ is a linear transformation, regard T as acting on the coordinates of each vector in V, so that T induces in a natural way a linear transformation $\bar{T}: \mathbb{R}^n \to \mathbb{R}^n$, as illustrated in Example 7, which follows. Find the standard matrix representation of $\bar{T}$.

Step 3 Perform computations, using matrices, to answer the questions about this transformation $\bar{T}: \mathbb{R}^n \to \mathbb{R}^n$.

Step 4 Interpret the results in the context of the original vector space.

Here is an illustration of this strategy.

EXAMPLE 7 For the space P_2 of polynomials of degree at most 2, let $T: P_2 \to P_2$ be defined by $T(p(x)) = p(-x) + p(x + 1)$. We can see that T is a linear transformation. Assuming this, find the eigenvalues and eigenvectors of T.

Solution *Step 1.* We choose the ordered basis $B = (x^2, x, 1)$ to coordinatize P_2, so that the polynomial $ax^2 + bx + c$ has coordinate vector (a, b, c).

Step 2. Now

$$T(x^2) = (-x)^2 + (x + 1)^2 = 2x^2 + 2x + 1,$$

$$T(x) = -x + (x + 1) = 1,$$

$$T(1) = 1 + 1 = 2.$$

Of course the coordinate vectors of x^2, x, and 1 relative to B are given by $(1, 0, 0)$, $(0, 1, 0)$, and $(0, 0, 1)$, respectively, and the equations listed in this step show that the induced transformation $\bar{T}: \mathbb{R}^n \rightarrow \mathbb{R}^n$ is determined by

$$\bar{T}(1, 0, 0) = (2, 2, 1), \qquad \bar{T}(0, 1, 0) = (0, 0, 1), \qquad \bar{T}(0, 0, 1) = (0, 0, 2).$$

The standard matrix representation of $\bar{T}$ is thus

$$A = \begin{bmatrix} 2 & 0 & 0 \\ 2 & 0 & 0 \\ 1 & 1 & 2 \end{bmatrix}.$$

$$\begin{vmatrix} 2-\lambda & 0 & 0 \\ 2 & -\lambda & 0 \\ 1 & 1 & 2-\lambda \end{vmatrix} = (2-\lambda)(-\lambda)(2-\lambda)$$

$$\lambda_1 = 0, \quad \lambda_2 = \lambda_3 = 2$$

Step 3. Since A is triangular, we see that 2 is an eigenvalue of A of algebraic multiplicity 2, and that 0 is an eigenvalue of algebraic multiplicity 1. The eigenvectors corresponding to the eigenvalue 2 are the nontrivial solutions of the homogeneous system with augmented matrix

$\lambda = 2$.

$$\left[\begin{array}{ccc|c} 0 & 0 & 0 & 0 \\ 2 & -2 & 0 & 0 \\ 1 & 1 & 0 & 0 \end{array}\right] \sim \left[\begin{array}{ccc|c} 1 & 1 & 0 & 0 \\ 0 & -4 & 0 & 0 \\ 0 & 0 & 0 & 0 \end{array}\right],$$

which has as solutions

$$c\begin{bmatrix} 0 \\ 0 \\ 1 \end{bmatrix} \qquad \text{for any scalar } c.$$

The eigenvectors corresponding to the eigenvalue 0 are the nontrivial solutions of the system with augmented matrix

$\lambda = 0$

$$\left[\begin{array}{ccc|c} 2 & 0 & 0 & 0 \\ 2 & 0 & 0 & 0 \\ 1 & 1 & 2 & 0 \end{array}\right] \sim \left[\begin{array}{ccc|c} 1 & 1 & 2 & 0 \\ 0 & -2 & -4 & 0 \\ 0 & 0 & 0 & 0 \end{array}\right] \sim \left[\begin{array}{ccc|c} 1 & 0 & 0 & 0 \\ 0 & 1 & 2 & 0 \\ 0 & 0 & 0 & 0 \end{array}\right],$$

which has as solutions

$$d\begin{bmatrix} 0 \\ 2 \\ -1 \end{bmatrix} \qquad \text{for any scalar } d.$$

Thus both eigenvalues have geometric multiplicity 1.

Step 4. Interpreting these results for the space P_2, we see that the eigenspace of T corresponding to the eigenvalue 2 consists of the constant polynomials c, while the eigenspace corresponding to the eigenvalue 0 consists of polynomials of the form $(2d)x - d$. ■

$\lambda = 2$

$$c\begin{bmatrix} 0 & 0 & 1 \end{bmatrix}\begin{bmatrix} x_2 \\ x \\ 1 \end{bmatrix} = c$$

$\lambda = 0$

$$d\begin{bmatrix} 0 & 2 & -1 \end{bmatrix}\begin{bmatrix} x_2 \\ x \\ 1 \end{bmatrix} = 2dx - d$$

Constant polynomials

SUMMARY

Let V and V' be vector spaces, and let $T: V \to V'$ be a linear transformation.

1. If W is a subspace of V, then $T[W]$ is a subspace of V'. If S' is a subspace of V', then $T^{-1}[S']$ is a subspace of V.

2. The range of T is the subspace $\{T(\mathbf{v}) \mid \mathbf{v} \text{ in } V\}$ of V', and the kernel $\ker(T)$ is the subspace $\{\mathbf{v} \in V \mid T(\mathbf{v}) = \mathbf{0}\}$ of V.

3. T is one-to-one if and only if $\ker(T) = \{\mathbf{0}\}$. If T is one-to-one and has as its range all of V', then $T^{-1}: V' \to V$ is well-defined and is a linear transformation. In this case, both T and T^{-1} are isomorphisms.

EXERCISES

In Exercises 1–7, describe the range of the given linear transformation.

1. $T: \mathbb{R}^2 \to \mathbb{R}^2$ defined by $T(x, y) = (x - y, x + y)$

2. $T: \mathbb{R}^2 \to \mathbb{R}^2$ defined by $T(x, y) = (x + y, 2x + y)$

3. $T: \mathbb{R}^3 \to \mathbb{R}^3$ defined by $T(x_1, x_2, x_3) = (2x_1 + x_3, 3x_1 + x_2, x_1 + x_2 - x_3)$

4. $T: \mathbb{R}^3 \to \mathbb{R}^3$ defined by $T(x_1, x_2, x_3) = (x_1 - x_2, x_1 + x_2, x_1 + x_2 + x_3)$

5. $T: \mathbb{R}^3 \to \mathbb{R}^2$ defined by $T(x_1, x_2, x_3) = (x_1 + x_2, x_2 + x_3)$

6. $T: \mathbb{R}^4 \to \mathbb{R}^4$ defined by $T(x_1, x_2, x_3, x_4) = (x_4, x_3, x_2, x_1)$

7. $T: F \to \mathbb{R}$ defined by $T(f) = f(3)$, where F is the space of all functions mapping $\mathbb{R}$ into $\mathbb{R}$

In Exercises 8–14, find the kernel of the given linear transformation. Use Theorem 6.2 to determine whether the linear transformation is one-to-one. For those that are one-to-one, find a formula for the inverse transformation, as in Example 6.

8. Exercise 1
9. Exercise 2
10. Exercise 3
11. Exercise 4
12. Exercise 5
13. Exercise 6
14. Exercise 7

In Exercises 15–19, determine whether the given linear transformation is an isomorphism.

15. $T: \mathbb{R}^2 \to \mathbb{R}^2$ defined by $T(x, y) = (x + y, 2y)$

16. $T: \mathbb{R}^2 \to \mathbb{R}^2$ defined by $T(x, y) = (x + y, x + y)$

17. $T: \mathbb{R}^3 \to \mathbb{R}^3$ defined by $T(x_1, x_2, x_3) = (x_3, x_1 + x_2 + x_3, x_1 + x_2)$

18. $T: \mathbb{R}^3 \to \mathbb{R}^3$ defined by $T(x_1, x_2, x_3) = (x_1 + x_2, x_2 + x_3, x_1 + x_3)$

19. $T: \mathbb{R}^4 \to \mathbb{R}^4$ defined by $T(x_1, x_2, x_3, x_4) = (x_1, x_1 + x_2 + x_3, x_3, x_1 + x_3 + x_4)$

In Exercises 20–22, let F be the vector space of all functions $f: \mathbb{R} \to \mathbb{R}$. Find the kernel of the given linear transformation.

20. $T: F \to F$ defined by $T(f) = f(-4)$

21. $T: F \to F$ defined by $T(f) = f + f$

22. $T: F \to F$ defined by $T(f) = -f$

In Exercises 23–28, follow the four-step strategy given in the box preceding Example 7 to solve the problem.

23. Find all eigenvalues and eigenvectors of the linear transformation $T: P_1 \to P_1$, where $T(p(x)) = p(x + 1) + p(2)$.

24. Find all eigenvalues and eigenvectors of the linear transformation $T: P_1 \to P_1$, where $T(p(x)) = p(x + 2) + p(1) + 2p(0)x$.

25. Find all eigenvalues and eigenvectors of the linear transformation mapping $sp(\sin x, \cos x)$ into itself, where $T(a \sin x + b \cos x) = (3a - 2b)\sin x + (2a - b)\cos x$.

26. Find all eigenvalues and eigenvectors of the linear transformation mapping $T: P_2 \to P_2$, where $T(p(x)) = p(x + 1) + p(-x + 1)$.

27. Find all scalars c and polynomials $p(x)$ in P_2 such that $p(1 - x) = cp(x)$.

28. Find all scalars c such that the equation $p(2x - 1) = cp(x)$ has some nonzero polynomial solutions in P_3.

29. Let $T: \mathbb{R}^n \to \mathbb{R}^m$ be a linear transformation with standard matrix representation A. Mark each of the following True of False.

____ a) The range of T is the row space of A.

____ b) The dimension of the range of T is equal to the row rank of A.

____ c) The range of T is the column space of A.

____ d) The transformation T is one-to-one if and only if the nullspace of A is $\{\mathbf{0}\}$.

____ e) The matrix A is invertible if T is one-to-one.

____ f) The matrix A is invertible if $m = n$ and T is one-to-one.

____ g) If A is invertible, then T is one-to-one.

____ h) $\dim(\ker(T)) + \dim(\text{range}(T)) = m$.

____ i) If $m > n$, then $\ker(T) \neq \mathbb{R}^n$.

____ j) If $m < n$, then $\ker(T) \neq \{\mathbf{0}\}$.

30. State for matrices the analogue of the statement that $T^{-1}[S']$ is a subspace of V in Theorem 6.1.

31. Show that, if λ is an eigenvalue of a linear transformation $T: V \to V'$, then $E_\lambda = \{\mathbf{v} \in V \mid T(\mathbf{v}) = \lambda\mathbf{v}\}$ is a subspace of V (the eigenspace of T corresponding to the eigenvalue λ).

Exercises 36–38 of Section 2.9 show that the set of all linear transformations of $\mathbb{R}^n$ into $\mathbb{R}^m$ has a natural vector-space structure. In particular, it makes sense to add two such linear transformations and to multiply a linear transformation by a scalar. Use this fact in Exercises 32–35.

32. Let T_1 and T_2 be linear transformations of $\mathbb{R}^n$ into $\mathbb{R}^m$ having standard matrix representations A_1 and A_2, respectively. What is the standard matrix representation of $T_1 + T_2$? Illustrate your claim with two different transformations of your choice mapping $\mathbb{R}^3$ into $\mathbb{R}^2$.

33. Let $T: \mathbb{R}^n \to \mathbb{R}^m$ be a linear transformation with standard matrix representation A, and let r be any scalar. What is the standard matrix representation of rT? Illustrate your claim with a linear transformation of your choice mapping $\mathbb{R}^2$ into $\mathbb{R}^3$.

34. Show that, for any five linear transformations T_1, T_2, T_3, T_4, T_5 mapping $\mathbb{R}^2$ into $\mathbb{R}^2$, there exist scalars c_1, c_2, c_3, c_4, c_5 (not all of which are zero) such that $T = c_1T_1 + c_2T_2 + c_3T_3 + c_4T_4 + c_5T_5$ has the property that $T(\mathbf{x}) = \mathbf{0}$ for all $\mathbf{x}$ in $\mathbb{R}^2$.

35. Let $T: \mathbb{R}^n \to \mathbb{R}^n$ be a linear transformation. Show that, if $T(T(\mathbf{x})) = T(\mathbf{x}) + T(\mathbf{x}) + 3\mathbf{x}$ for all $\mathbf{x}$ in $\mathbb{R}^n$, then T is a one-to-one mapping of $\mathbb{R}^n$ into $\mathbb{R}^n$.

36. Let V and V' be vector spaces having the same finite dimension, and let $T: V \to V'$ be a linear transformation. Show that T is one-to-one if and only if $\text{range}(T) = V'$. [HINT: Use Exercise 22 in Section 2.6.]

37. Give an example of a vector space V and a linear transformation $T: V \to V$ such that T is one-to-one but $\text{range}(T) \neq V$. [HINT: By Exercise 36, what must be true of the dimension of V?]

38. Repeat Exercise 37, but this time make $\text{range}(T) = V$ for a transformation T that is not one-to-one.

39. Work with Topic 4 of the supplied program VECTGRPH until you can consistently achieve a score of at least 80%.

40. Work with Topic 5 of the supplied program VECTGRPH until you can regularly attain a score of at least 80%.

41. Work with Topic 8 of the available program VECTGRPH until you can always get a grade of at least 80%.

6.2 MATRIX REPRESENTATIONS

Review of Standard Matrix Representations Presented in Section 2.9

We quickly review a few notions basic to this section. Let

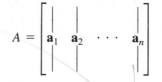

$$A = \begin{bmatrix} | & | & & | \\ \mathbf{a}_1 & \mathbf{a}_2 & \cdots & \mathbf{a}_n \\ | & | & & | \end{bmatrix}$$

be an $m \times n$ matrix with column vectors $\mathbf{a}_1, \mathbf{a}_2, \ldots, \mathbf{a}_n$, and let $\mathbf{x}$ be a column vector in $\mathbb{R}^n$. The vector $A\mathbf{x}$ is a linear combination of the column vectors of A, namely,

$$A\mathbf{x} = x_1\mathbf{a}_1 + x_2\mathbf{a}_2 + \cdots + x_n\mathbf{a}_n.$$

Now let $T: \mathbb{R}^n \to \mathbb{R}^m$ be a linear transformation, and let A be the matrix with column vectors $T(\mathbf{e}_1), T(\mathbf{e}_2), \ldots, T(\mathbf{e}_n)$. Then $T(\mathbf{x}) = A\mathbf{x}$ for each $\mathbf{x}$ in $\mathbb{R}^n$, because

$$A\mathbf{x} = \begin{bmatrix} | & | & & | \\ T(\mathbf{e}_1) & T(\mathbf{e}_2) & \cdots & T(\mathbf{e}_n) \\ | & | & & | \end{bmatrix} \begin{bmatrix} x_1 \\ x_2 \\ \vdots \\ x_n \end{bmatrix}$$

$$= x_1T(\mathbf{e}_1) + x_2T(\mathbf{e}_2) + \cdots + x_nT(\mathbf{e}_n) = T(\mathbf{x}).$$

This matrix A is the *standard matrix representation* of T; as we observed in Section 6.1, it is useful in studying linear transformations of finite-dimensional vector spaces.

Review of Change of Coordinates Presented in Section 2.7

Let $B = (\mathbf{b}_1, \mathbf{b}_2, \ldots, \mathbf{b}_n)$ be an ordered basis for $\mathbb{R}^n$, and let $\mathbf{x}$ be a column vector in $\mathbb{R}^n$. The *coordinate vector of* $\mathbf{x}$ *relative to the basis B* is the unique column vector $\mathbf{c}$ with components $c_1, c_2, \ldots, c_n$ such that

$$\mathbf{x} = c_1\mathbf{b}_1 + c_2\mathbf{b}_2 + \cdots + c_n\mathbf{b}_n. \tag{1}$$

We denote this coordinate vector of $\mathbf{x}$ relative to B by $\mathbf{x}_B$. Examples 1 and 2 of Section 2.7 illustrate this notion of *coordinate vector*.

Now form the *basis matrix* M_B of B, symbolically,

$$M_B = \begin{bmatrix} | & | & & | \\ \mathbf{b}_1 & \mathbf{b}_2 & \cdots & \mathbf{b}_n \\ | & | & & | \end{bmatrix}. \tag{2}$$

Equation (1) shows that

$$M_B\mathbf{x}_B = \mathbf{x}. \tag{3}$$

If B' is another basis for $\mathbb{R}^n$, then, as in Eq. (3), we also have $M_{B'}\mathbf{x}_{B'} = \mathbf{x}$. Consequently, we obtain the relation

$$M_{B'}\mathbf{x}_{B'} = M_B\mathbf{x}_B \qquad (4)$$

which expresses **x** in terms of the basis B' as well as in terms of the basis B. Eqs. (3) and (4) are basic to this section. If you understand the background for them that we have just reviewed, everything falls into place. We now write Eq. (4) in a form that tells us how to change from coordinates relative to a basis B to coordinates relative to a basis B', namely,

$$\mathbf{x}_{B'} = (M_{B'}^{-1}M_B)\mathbf{x}_B. \qquad (5)$$

$$B' \text{ coordinates} \qquad C_{B,B'} \qquad B \text{ coordinates}$$

The matrix $C_{B,B'} = M_{B'}^{-1}M_B$ in Eq. (5) is the *change-of-coordinates matrix from B to B'*. Notice how easily the formula for this change-of-coordinates matrix is obtained from Eq. (4). We suggest that you base your work in this section on Eq. (4).

Let us now recall how to compute a change-of-coordinates matrix $C_{B,B'} = M_{B'}^{-1}M_B$. Notice that $M_{B'}C_{B,B'} = M_B$, so $C_{B,B'}$ is the matrix solution of the equation $M_{B'}X = M_B$. To find the solution $X = A^{-1}$ of the matrix equation $AX = I$, where A is an invertible matrix, we form the augmented matrix $[A \mid I]$ and reduce it to the matrix $[I \mid A^{-1}]$. In exactly the same way, we find the solution $X = C_{B,B'}$ of $M_{B'}X = M_B$ by forming the augmented matrix $[M_{B'} \mid M_B]$ and reducing it to $[I \mid C_{B,B'}]$. (See Exercise 27 in Section 1.4.) Notice the reversal of order of the subscripts B and B' in these last two augmented matrices. (See Examples 4 and 5 in Section 2.7 for illustrations.)

This completes our review of earlier topics. $M_B^{-1}M_B = C_{B,B'}$

General Matrix Representations $M_B = M_{B'}C_{B,B'}$

Let $T: \mathbb{R}^n \to \mathbb{R}^m$ be a linear transformation. Thus far, we have only worked with the standard matrix representation A of T. We proceed to show that T has many other matrix representations. In fact, we will show that it has a unique matrix representation $R_{B,B'}$ such that

$$T(\mathbf{x})_{B'} = R_{B,B'}\mathbf{x}_B \qquad (6)$$

for each choice of an ordered basis B for $\mathbb{R}^n$ and of an ordered basis B' for $\mathbb{R}^m$. That is, the coordinate vector of $T(\mathbf{x})$ in $\mathbb{R}^m$ relative to B' is the coordinate vector of **x** in $\mathbb{R}^n$ relative to B, multiplied on the left by the $m \times n$ matrix $R_{B,B'}$. The fact that this matrix is unique follows from Exercise 41 of Section 1.1.

DEFINITION 6.1 Matrix Representation of T Relative to B,B'

Let $T: \mathbb{R}^n \to \mathbb{R}^m$ be a linear transformation, and let B and B' be ordered bases for $\mathbb{R}^n$ and $\mathbb{R}^m$, respectively. The unique matrix $R_{B,B'}$ such that $T(\mathbf{x})_{B'} = R_{B,B'}\mathbf{x}_B$ for each **x** in $\mathbb{R}^n$ is the **matrix representation of T relative to B,B'**.

From this more general viewpoint, we see that our standard matrix representation A is the matrix $R_{E,E'}$, where E and E' are the standard ordered bases of $\mathbb{R}^n$ and $\mathbb{R}^m$, respectively. In this case, Eq. (6) has the familiar form $T(\mathbf{x}) = A\mathbf{x}$.

Using Eqs. (3) and (4), we can easily see how to find the matrix $R_{B,B'}$ satisfying Eq. (6). We wish to compute $T(\mathbf{x})$ *starting with* the coordinate vector $\mathbf{x}_B$ of $\mathbf{x}$, *ending with* the coordinate vector $T(\mathbf{x})_{B'}$ of $T(\mathbf{x})$, and performing exclusively matrix multiplications. Here is a diagram of the steps in the computation:

$Ax = \boxed{AM_Bx_B}$

$$M_B\mathbf{x}_B = \mathbf{x} \quad A\mathbf{x} = T(\mathbf{x}) \quad M_{B'}^{-1}T(\mathbf{x}) = T(\mathbf{x})_{B'} \quad = T(x)$$

$$\mathbf{x}_B \xrightarrow{ M_B } \mathbf{x} \xrightarrow{ A } T(\mathbf{x}) \xrightarrow{ M_{B'}^{-1} } T(\mathbf{x})_{B'}. \tag{7}$$

Under each arrow, we have written the matrix by which we should multiply the column vector at the tail of the arrow to produce the column vector at the tip of the arrow. Thus, the desired matrix $R_{B,B'}$ is given by

$$R_{B,B'} = M_{B'}^{-1}AM_B. \tag{8}$$

The *order* of the matrices on the right-hand side in Eq. (8) is the *opposite* of the order of the matrices under the arrows in diagram (7). This occurs because we multiply our column vectors on the *left* successively by matrices that consequently accumulate from right to left.

Writing Eq. (8) as $M_{B'}R_{B,B'} = AM_B$, we see that $R_{B,B'}$ can be computed by forming the augmented matrix $[M_{B'} \mid AM_B]$ and reducing it to $[I \mid R_{B,B'}]$. (See Exercise 27 in Section 1.4.) Since A is the standard matrix representation of T, the column vectors of the matrix AM_B are $T(\mathbf{b}_1), T(\mathbf{b}_2), \ldots, T(\mathbf{b}_n)$. Written out, the augmented matrix $[M_{B'} \mid AM_B]$ becomes

$$\left[\begin{array}{cccc|cccc} \big| & \big| & & \big| & \big| & \big| & & \big| \\ \mathbf{b}_1' & \mathbf{b}_2' & \cdots & \mathbf{b}_m' & T(\mathbf{b}_1) & T(\mathbf{b}_2) & \cdots & T(\mathbf{b}_n) \\ \big| & \big| & & \big| & \big| & \big| & & \big| \end{array}\right]. \tag{9}$$

$$\underbrace{}_{M_{B'}} \qquad \underbrace{}_{AM_B}$$

Thus we can form the right-hand side of the partitioned matrix by using a formula for T (which is often given), rather than by actually finding A and performing matrix multiplication.

By using diagrams such as (7), we can find appropriate matrices to handle not only this mapping situation, but all others that arise. We summarize the procedure for finding $R_{B,B'}$ in a box.

Finding the Matrix Representation of $T: \mathbb{R}^n \to \mathbb{R}^m$ Relative to Ordered Bases B, B'

Step 1 Form the partitioned matrix (9).

Step 2 Use a Gauss–Jordan reduction to obtain the partitioned matrix $[I \mid R_{B,B'}]$, where I is the $m \times m$ identity matrix and where $R_{B,B'}$ is the desired matrix representation.

EXAMPLE 1 Let $T: \mathbb{R}^3 \to \mathbb{R}^3$ be the linear transformation defined by

$$T(x_1, x_2, x_3) = (2x_1 - x_2 + 2x_3, \; x_1 + x_2, \; 3x_1 - x_2 - x_3).$$

Find its matrix representation $R_{B,B'}$ relative to ordered bases B, B', where

$$B = ((1, 0, 0), (1, 1, 1), (1, -1, 1))$$

and

$$B' = ((1, 1, 2), (1, 2, 1), (2, 1, 1)).$$

Solution Writing vectors in column form, we obtain

$$T\left(\begin{bmatrix}1\\0\\0\end{bmatrix}\right) = \begin{bmatrix}2\\1\\3\end{bmatrix}, \qquad T\left(\begin{bmatrix}1\\1\\1\end{bmatrix}\right) = \begin{bmatrix}3\\2\\1\end{bmatrix}, \qquad T\left(\begin{bmatrix}1\\-1\\1\end{bmatrix}\right) = \begin{bmatrix}5\\0\\3\end{bmatrix}.$$

Therefore,

$$AM_B = \begin{bmatrix}2 & 3 & 5\\1 & 2 & 0\\3 & 1 & 3\end{bmatrix}, \tag{10}$$

and the reduction of the augmented matrix (9) becomes

$$\begin{bmatrix}1 & 1 & 2 & | & 2 & 3 & 5\\1 & 2 & 1 & | & 1 & 2 & 0\\2 & 1 & 1 & | & 3 & 1 & 3\end{bmatrix} \sim \begin{bmatrix}1 & 1 & 2 & | & 2 & 3 & 5\\0 & 1 & -1 & | & -1 & -1 & -5\\0 & -1 & -3 & | & -1 & -5 & -7\end{bmatrix}$$

$$\underbrace{\phantom{\begin{bmatrix}1\\1\\2\end{bmatrix}}}_{M_{B'}} \quad \underbrace{\phantom{\begin{bmatrix}1\\1\\2\end{bmatrix}}}_{AM_B}$$

$$\sim \begin{bmatrix}1 & 0 & 3 & | & 3 & 4 & 10\\0 & 1 & -1 & | & -1 & -1 & -5\\0 & 0 & -4 & | & -2 & -6 & -12\end{bmatrix} \sim \begin{bmatrix}1 & 0 & 0 & | & \frac{3}{2} & -\frac{1}{2} & 1\\0 & 1 & 0 & | & -\frac{1}{2} & \frac{1}{2} & -2\\0 & 0 & 1 & | & \frac{1}{2} & \frac{3}{2} & 3\end{bmatrix}.$$

$$\underbrace{}_{I} \qquad \underbrace{\phantom{R_{B,B'}}}_{R_{B,B'}}$$

The right-hand part of the final partitioned matrix is the desired result. ▪

EXAMPLE 2 Let $T: \mathbb{R}^3 \to \mathbb{R}^3$ be the linear transformation in Example 1, let B, B' be the ordered bases given there, and let $\mathbf{v} = (2, 1, 5)$. Find the coordinate vector of $T(2, 1, 5)$ relative to the basis B' in two ways: (a) by computing $T(2, 1, 5)$ and then finding its coordinates relative to B'; and (b) by finding the coordinate vector $\mathbf{v}_B$ and using the matrix representation $R_{B,B'}$ of T found in Example 1.

Solution (a) Evaluating $T(2, 1, 5)$, using the formula

$$T(x_1, x_2, x_3) = (2x_1 - x_2 + 2x_3, \; x_1 + x_2, \; 3x_1 - x_2 - x_3),$$

we compute $T(2, 1, 5) = (13, 3, 0)$. We find $T(2, 1, 5)_{B'}$ as follows:

$$\begin{bmatrix}1 & 1 & 2 & | & 13\\1 & 2 & 1 & | & 3\\2 & 1 & 1 & | & 0\end{bmatrix} \sim \begin{bmatrix}1 & 1 & 2 & | & 13\\0 & 1 & -1 & | & -10\\0 & -1 & -3 & | & -26\end{bmatrix}$$

$$\underset{\mathbf{b}_1' \quad \mathbf{b}_2' \quad \mathbf{b}_3'}{} \quad \underset{T(\mathbf{v})}{}$$

$$\sim \begin{bmatrix} 1 & 0 & 3 & | & 23 \\ 0 & 1 & 1 & | & -10 \\ 0 & 0 & -4 & | & -36 \end{bmatrix} \sim \begin{bmatrix} 1 & 0 & 0 & | & -4 \\ 0 & 1 & 0 & | & -1 \\ 0 & 0 & 1 & | & 9 \end{bmatrix}.$$

$$\underset{T(\mathbf{v})_{B'}}{}$$

(b) We first find $\mathbf{v}_B$:

$$\begin{bmatrix} 1 & 1 & 1 & | & 2 \\ 0 & 1 & -1 & | & 1 \\ 0 & 1 & 1 & | & 5 \end{bmatrix} \sim \begin{bmatrix} 1 & 0 & 2 & | & 1 \\ 0 & 1 & -1 & | & 1 \\ 0 & 0 & 2 & | & 4 \end{bmatrix} \sim \begin{bmatrix} 1 & 0 & 0 & | & -3 \\ 0 & 1 & 0 & | & 3 \\ 0 & 0 & 1 & | & 2 \end{bmatrix}.$$

$$\underset{\mathbf{b}_1}{} \quad \underset{\mathbf{b}_2}{} \quad \underset{\mathbf{b}_3}{} \quad \underset{\mathbf{v}}{} \qquad\qquad\qquad\qquad\qquad \underset{\mathbf{v}_B}{}$$

Then, using the matrix $R_{B,B'}$ found in Example 1, we have

$$T(\mathbf{v})_{B'} = R_{B,B'}\mathbf{v}_B = \begin{bmatrix} \frac{3}{2} & -\frac{1}{2} & 1 \\ -\frac{1}{2} & \frac{1}{2} & -2 \\ \frac{1}{2} & \frac{3}{2} & 3 \end{bmatrix} \begin{bmatrix} -3 \\ 3 \\ 2 \end{bmatrix} = \begin{bmatrix} -4 \\ -1 \\ 9 \end{bmatrix}. \qquad ■$$

THEOREM 6.3 General Matrix Representation of T

Let $T: \mathbb{R}^n \to \mathbb{R}^m$ be a linear transformation, and let B and B' be ordered bases of $\mathbb{R}^n$ and $\mathbb{R}^m$, respectively. Then the matrix representation $R_{B,B'}$ of T has as column vectors $T(\mathbf{b}_1)_{B'}, T(\mathbf{b}_2)_{B'}, \ldots, T(\mathbf{b}_n)_{B'}$; that is,

$$R_{B,B'} = \begin{bmatrix} | & | & & | \\ T(\mathbf{b}_1)_{B'} & T(\mathbf{b}_2)_{B'} & \cdots & T(\mathbf{b}_n)_{B'} \\ | & | & & | \end{bmatrix}. \qquad (11)$$

PROOF We know that $R_{B,B'}$ can be obtained by reducing the matrix

$$\begin{bmatrix} | & | & & | & | & | & & | \\ \mathbf{b}_1' & \mathbf{b}_2' & \cdots & \mathbf{b}_m' & T(\mathbf{b}_1) & T(\mathbf{b}_2) & \cdots & T(\mathbf{b}_n) \\ | & | & & | & | & | & & | \end{bmatrix}.$$

to the form $[I \mid R_{B,B'}]$. But this is precisely the augmented matrix reduction we would perform to find the coordinates of all the column vectors on the right-hand side of the partition relative to the ordered basis of column vectors to the left of the partition. Thus the reduction produces on the right-hand side the coordinate vectors of each $T(\mathbf{b}_j)$ relative to the basis B', that is, the vectors $T(\mathbf{b}_j)_{B'}$. ▲

EXAMPLE 3 Rotation of the plane counterclockwise through 90° is a linear transformation T, where $T(x, y) = (-y, x)$. Use the observation in the preceding example to find the matrix representation $R_{B,B}$ of this transformation relative to B,B, where B is the ordered basis $((1, 1), (-1, 0))$.

Solution We can find the coordinates of $T(1, 1)$ and of $T(-1, 0)$ relative to B by inspection (see Fig. 6.3):

$$A = \begin{bmatrix} 0 & -1 \\ 1 & 0 \end{bmatrix} \quad Ax = \begin{bmatrix} 0 & -1 \\ 1 & 0 \end{bmatrix}\begin{bmatrix} x \\ y \end{bmatrix} \qquad T\left(\begin{bmatrix} 1 \\ 1 \end{bmatrix}\right) = \begin{bmatrix} -1 \\ 1 \end{bmatrix} = 1\begin{bmatrix} 1 \\ 1 \end{bmatrix} + 2\begin{bmatrix} -1 \\ 0 \end{bmatrix} \quad T\begin{bmatrix} 1 \\ 1 \end{bmatrix}_B$$

and

$$T\left(\begin{bmatrix} -1 \\ 0 \end{bmatrix}\right) = \begin{bmatrix} 0 \\ -1 \end{bmatrix} = -1\begin{bmatrix} 1 \\ 1 \end{bmatrix} + (-1)\begin{bmatrix} -1 \\ 0 \end{bmatrix}. \quad T\begin{bmatrix} -1 \\ 0 \end{bmatrix}_B$$

Thus, $R_{B,B}$ is the matrix having as column vectors the coordinate vectors of $T(1, 1)$ and $T(-1, 0)$ relative to B, namely,

$$\begin{array}{cc} 1 & -1 \\ 1 & 0 \end{array}\begin{array}{cc} -1 & 0 \\ 1 & -1 \end{array} \sim \begin{array}{cc} 1 & 0 \\ 0 & 1 \end{array}\begin{array}{cc} 1 & -1 \\ 2 & -1 \end{array} \qquad R_{B,B} = \begin{bmatrix} 1 & -1 \\ 2 & -1 \end{bmatrix}. \; = \begin{bmatrix} T(b_1) & T(b_2) \end{bmatrix} \qquad ■$$

$$C_{B,E} \quad R_{B,E} \qquad C_{B,E}^{-1} \; C_{B,E} \qquad C_{B,E} \; R_{B,E}$$
$$= I \qquad = C_{E,B} R_{B,E} = R_{B,B}$$

EXAMPLE 4 Find a formula for the linear transformation T that reflects vectors in the plane $\mathbb{R}^2$ through the line $x + 2y = 0$.

Solution In order to find a formula for T, we find its standard matrix representation $A = R_{E,E}$. From Figure 6.4, we see that $T(1, 2) = (-1, -2)$ and $T(2, -1) = (2, -1)$. Letting $B = ((1, 2), (2, -1))$, we see that

$$R_{B,E} = \begin{bmatrix} -1 & 2 \\ -2 & -1 \end{bmatrix}.$$

$$\begin{array}{cc} 1 & 2 \\ 2 & -1 \end{array}\begin{array}{cc} 1 & 0 \\ 0 & 1 \end{array}$$

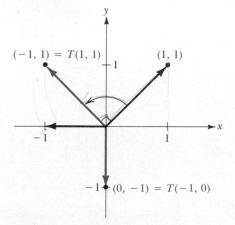

FIGURE 6.3 Rotation counter-clockwise through 90°.

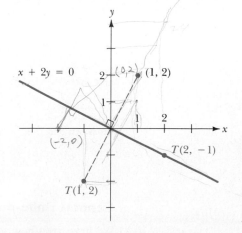

FIGURE 6.4 Reflection through the line $x + 2y = 0$.

The diagram

$$\mathbf{x} \xrightarrow[C_{E,B}]{} \mathbf{x}_B \xrightarrow[R_{B,E}]{} T(\mathbf{x})_E$$

now shows how to find the standard matrix representation of T, namely, $R_{E,E} = R_{B,E}C_{E,B}$. We compute

$$\left[R \right]_B^E \left[C \right]_E^B \qquad C_{E,B} = C_{B,E}^{-1} = \begin{bmatrix} 1 & 2 \\ 2 & -1 \end{bmatrix}^{-1} = -\frac{1}{5}\begin{bmatrix} -1 & -2 \\ -2 & 1 \end{bmatrix}.$$

$$E \to B \qquad B \to E$$

Thus,

$$A = R_{E,E} = -\frac{1}{5}\begin{bmatrix} -1 & 2 \\ -2 & -1 \end{bmatrix}\begin{bmatrix} -1 & -2 \\ -2 & 1 \end{bmatrix} = \frac{1}{5}\begin{bmatrix} 3 & -4 \\ -4 & -3 \end{bmatrix}.$$

A formula for T is therefore $T(\mathbf{x}) = A\mathbf{x}$, which in row notation is $T(x, y) = \frac{1}{5}(3x - 4y, -4x - 3y)$. ■

$$T(1, 2) = \\ = \frac{1}{5}(3 - 8, -4 - 6) = (-1, -2)$$

Let $T: \mathbb{R}^n \to \mathbb{R}^m$ and $T': \mathbb{R}^m \to \mathbb{R}^s$ be linear transformations. Section 1.5 showed that composition of linear transformations corresponds to multiplication of their standard matrix representations; that is, for standard bases we have

$$\text{Matrix for } (T' \circ T) = (\text{Matrix for } T')(\text{Matrix for } T). \tag{12}$$

The analogous property holds for any choice of bases. Consider the linear transformations and bases shown by the diagram

$$\begin{array}{ccc} & T & & T' & \\ \mathbb{R}^n & \longrightarrow & \mathbb{R}^m & \longrightarrow & \mathbb{R}^s. \\ B & & B' & & B'' \end{array} \tag{13}$$

The element and matrix counterpart diagram

$$R_{B,B''}$$
$$\mathbf{x}_B \xrightarrow[R_{B,B'}]{} T(\mathbf{x})_{B'} \xrightarrow[R_{B',B''}]{} T'(T(\mathbf{x}))_{B''} \tag{14}$$

and associativity of matrix multiplication show that

$$(T' \circ T)(\mathbf{x})_{B''} = T'(T(\mathbf{x}))_{B''} = R_{B',B''}(R_{B,B'}\mathbf{x}_B) = (R_{B',B''}R_{B,B'})\mathbf{x}_B.$$

Thus we see that

$$R_{B,B''} = R_{B',B''}R_{B,B'}. \tag{15}$$

General Finite-dimensional Vector Spaces

We have seen how to find the matrix representation $R_{B,B'}$ of a linear transformation of $\mathbb{R}^n$ into $\mathbb{R}^m$ relative to ordered bases B for $\mathbb{R}^n$ and B' for $\mathbb{R}^m$. This matrix representation fulfills our desire to be able to compute T by matrix multiplication using coordinates relative to any bases. That is,

$$T(\mathbf{x})_{B'} = R_{B,B'}\mathbf{x}_B \qquad \text{for all } \mathbf{x} \in \mathbb{R}^n. \tag{16}$$

Now let $T: V \to V'$ be a linear transformation of V into V', where V and V' are any finite-dimensional vector spaces with ordered bases B and B', respectively. Using the coordinates relative to these bases, we wish to find a matrix $R_{B,B'}$ such that

$$T(\mathbf{v})_{B'} = R_{B,B'}\mathbf{v}_B \qquad \text{for all } \mathbf{v} \in V, \tag{17}$$

in analogy with Eq. (16). The matrix $R_{B,B'}$ with jth column vector $T(\mathbf{b}_j)_{B'}$ given in Eq. (11) fulfills this requirement, and is called the *matrix representation of T relative to B,B'*. To check that $R_{B,B'}$ satisfies Eq. (17), we need only check that the equation holds when $\mathbf{v}$ is any of the vectors $\mathbf{b}_j$ in the basis B. But $(\mathbf{b}_j)_B$ is just the vector $\mathbf{e}_j$ in the standard ordered basis of $\mathbb{R}^n$. Therefore, if $\mathbf{v} = \mathbf{b}_j$, the right-hand side of Eq. (17) is $R_{B,B'}\mathbf{e}_j$, the jth column of $R_{B,B'}$. This agrees with the left-hand side of Eq. (17). Exercise 41 of Section 1.1 shows that $R_{B,B'}$ is the *unique* matrix satisfying Eq. (17).

DEFINITION 6.2 General Matrix Representations

Let $T: V \to V'$ be a linear transformation, where V and V' are finite-dimensional vector spaces with ordered bases B and B', respectively. The unique matrix $R_{B,B'}$ satisfying Eq. (17) is the **matrix representation of T relative to B,B'**. It is given by Eq. (11).

For general finite-dimensional vector spaces, composition of linear transformations again corresponds to multiplication of matrix representations relative to any chosen bases. That is, the analogous Eq. (15) is still valid. The argument, too, is essentially the same; in diagram (14), each vector $\mathbf{x}$ would simply become subscripted to denote coordinates with respect to the appropriate basis. (See Exercise 22.)

Our next example illustrates Definition 6.2. The example may amuse readers who have taken calculus; we differentiate a polynomial function by matrix multiplication.

EXAMPLE 5 (*Calculus*) Let $B = (x^4, x^3, x^2, x, 1)$, which is an ordered basis for P_4, and let $T: P_4 \to P_4$ be the differentiation transformation. Find the matrix representation $R_{B,B}$ of T relative to B,B, and illustrate how it can be used to differentiate the polynomial function

$$3x^4 - 5x^3 + 7x^2 - 8x + 2.$$

Then use calculus to show that $(R_{B,B})^5 = O$.

Solution Since $T(x^k) = kx^{k-1}$, we see that the matrix representation $R_{B,B}$ in formula (11) is

$$R_{B,B} = \begin{bmatrix} 0 & 0 & 0 & 0 & 0 \\ 4 & 0 & 0 & 0 & 0 \\ 0 & 3 & 0 & 0 & 0 \\ 0 & 0 & 2 & 0 & 0 \\ 0 & 0 & 0 & 1 & 0 \end{bmatrix}.$$

Now the coordinate vector of $p(x) = 3x^4 - 5x^3 + 7x^2 - 8x + 2$ relative to B is

$$p(x)_B = \begin{bmatrix} 3 \\ -5 \\ 7 \\ -8 \\ 2 \end{bmatrix}.$$

Applying Eq. (17) with $B' = B$, we find that

$$T(p(x))_B = R_{B,B}\, p(x)_B = \begin{bmatrix} 0 & 0 & 0 & 0 & 0 \\ 4 & 0 & 0 & 0 & 0 \\ 0 & 3 & 0 & 0 & 0 \\ 0 & 0 & 2 & 0 & 0 \\ 0 & 0 & 0 & 1 & 0 \end{bmatrix} \begin{bmatrix} 3 \\ -5 \\ 7 \\ -8 \\ 2 \end{bmatrix} = \begin{bmatrix} 0 \\ 12 \\ -15 \\ 14 \\ -8 \end{bmatrix}.$$

Thus, $p'(x) = T(p(x)) = 12x^3 - 15x^2 + 14x - 8$.

The discussion surrounding Eq. (12) shows that the linear transformation of P_4 that computes the fifth derivative has matrix representation $(R_{B,B})^5$. Since the fifth derivative of any polynomial in P_4 is zero, we see that $(R_{B,B})^5 = O$. ■

SUMMARY

Let T be a linear transformation of $\mathbb{R}^n$ into $\mathbb{R}^m$, and let $B = (\mathbf{b}_1, \mathbf{b}_2, \ldots, \mathbf{b}_n)$ and $B' = (\mathbf{b}'_1, \mathbf{b}'_2, \ldots, \mathbf{b}'_m)$ be ordered bases for $\mathbb{R}^n$ and $\mathbb{R}^m$, respectively.

1. The standard matrix representation of T is the $m \times n$ matrix A whose jth column vector is $T(\mathbf{e}_j)$, where $\mathbf{e}_j$ is the jth vector in the standard basis E for $\mathbb{R}^n$.

2. Let M_B be the basis matrix for B, having the vectors in B as its column vectors. Then $M_B\mathbf{x}_B = \mathbf{x}$ for all $\mathbf{x}$ in $\mathbb{R}^n$.

3. The matrix representation $R_{B,B'}$ of T relative to bases B,B' may be found by reducing the augmented matrix $[M_{B'} \mid AM_B]$ to the form $[I \mid R_{B,B'}]$. This matrix representation has the property that $R_{B,B'}\mathbf{x}_B = T(\mathbf{x})_{B'}$ for all $\mathbf{x}$ in $\mathbb{R}^n$.

4. The matrix AM_B appearing in the preceding summary item has as its column vectors the images $T(\mathbf{b}_1), T(\mathbf{b}_2), \ldots, T(\mathbf{b}_n)$ of the basis vectors in B, which can be computed directly if a formula for T is given.

5. Let T be a linear transformation of V into V', where V and V' are finite-dimensional vector spaces having ordered bases B and B', respectively. The matrix representation of T relative to B,B' is the matrix $R_{B,B'}$ having as its column vectors the coordinate vectors $T(\mathbf{b}_1)_{B'}, T(\mathbf{b}_2)_{B'}, \ldots, T(\mathbf{b}_n)_{B'}$ of the images of the vectors in the basis B relative to the basis B'.

EXERCISES

In Exercises 1–6, find (a) the standard matrix representation of the given linear transformation T, and (b) the matrix representation of T relative to B,B'.

1. $T: \mathbb{R}^2 \to \mathbb{R}^2$ defined by $T(x, y) = (2x - 3y, x + y)$; $B = ((1, 1), (1, 0))$, $B' = ((0, 1), (1, 1))$

2. $T: \mathbb{R}^2 \to \mathbb{R}^3$ defined by $T(x, y) = (2x + y, x + 2y, x - 3y)$; $B = ((1, 0), (0, 1))$, $B' = ((1, 1, 1), (1, 1, 0), (1, 0, 0))$

3. $T: \mathbb{R}^3 \to \mathbb{R}^3$ defined by $T(x_1, x_2, x_3) = (x_1 + x_2 + x_3, x_1 - x_2 - x_3, -x_1 - x_2 + x_3)$; $B = ((2, 3, 1), (1, 2, 0), (2, 0, 3))$, $B' = ((1, 0, 0), (0, 1, 0), (0, 0, 1))$

4. $T: \mathbb{R}^3 \to \mathbb{R}^3$ defined by $T(x_1, x_2, x_3) = (x_1 + 2x_2 + x_3, x_1, x_2 + x_3)$; $B = ((1, 0, 1), (1, 1, 0), (0, 1, 1))$, $B' = ((0, 1, 1), (1, 1, 0), (1, 0, 1))$

5. $T: \mathbb{R}^3 \to \mathbb{R}^4$ defined by $T(x_1, x_2, x_3) = (x_1 + x_2, x_2 + x_3, x_1 + x_3, x_1 + x_2 + x_3)$; $B = ((1, 1, 1), (1, 1, 0), (1, 0, 0))$, $B' = ((1, 1, 1, 1), (1, 1, 1, 0), (1, 1, 0, 0), (1, 0, 0, 0))$

6. $T: \mathbb{R}^3 \to \mathbb{R}^2$ defined by $T(x_1, x_2, x_3) = (x_1 + 2x_2 + 3x_3, x_1)$; $B = ((1, 1, 1), (1, 1, 0), (1, 0, 0))$, $B' = ((1, 1), (1, 2))$

7. Show that the rank and nullity of a linear transformation $T: \mathbb{R}^n \to \mathbb{R}^m$ are equal to the rank and nullity of each matrix representation $R_{B,B'}$, where B and B' are ordered bases for $\mathbb{R}^n$ and $\mathbb{R}^m$, respectively.

In Exercises 8–13, find the rank and nullity of the linear transformation in the indicated earlier exercise.

8. Exercise 1
9. Exercise 2
10. Exercise 3
11. Exercise 4
12. Exercise 5
13. Exercise 6

Exercises 14–19 deal with reflections of $\mathbb{R}^2$ and of $\mathbb{R}^3$. The reflection of $\mathbb{R}^3$ through a plane passing through the origin is the obvious analogue of the reflection of $\mathbb{R}^2$ through a line passing through the origin.

14. Find the matrix representation of the reflection of $\mathbb{R}^2$ through the line $y = x$ relative to B,B, where $B = ((1, 1), (-1, 1))$.

15. Repeat Exercise 14 for the same basis B but for reflection through the x-axis.

16. Find the matrix representation of the reflection of $\mathbb{R}^2$ through the y-axis relative to B,B, where $B = ((1, 1), (1, 0))$.

17. Find the standard matrix representation of the reflection of $\mathbb{R}^2$ through the line $y = mx$. [HINT: Use a method like that in Example 4.]

18. Find the matrix representation of the reflection of $\mathbb{R}^3$ through the plane $x_2 = 0$ relative to B,B, where $B = ((1, 1, 1), (1, 1, 0), (1, 0, 0))$.

19. Find the matrix representation of the reflection of $\mathbb{R}^3$ through the plane $x_3 = 0$ relative to B,B, where $B = ((1, -1, 0), (-1, 1, 2), (1, 1, 1))$.

20. Consider the ordered basis $B = ((1, 1), (1, -1))$ of $\mathbb{R}^2$. Let T be the linear transformation corresponding to a rotation of the plane counterclockwise through an angle of $45°$. Find the standard matrix representation of T, and find the matrix representation $R_{B,B}$ of T relative to B,B.

21. Mark each of the following True or False.

___ a) For an ordered basis B of $\mathbb{R}^n$, the basis matrix M_B having vectors from B as its columns is the change-of-coordinates matrix from B to E, where E is the standard ordered basis.

___ b) For ordered bases B and B' of $\mathbb{R}^n$, the change-of-coordinates matrix from B to B' is $M_{B'}^{-1}M_B$.

___ c) The change-of-coordinates matrix described in part (b) is $M_B^{-1}M_{B'}$.

___ d) The matrix representation of a linear transformation $T: \mathbb{R}^n \to \mathbb{R}^m$ relative to ordered bases B,B' is an $m \times n$ matrix.

___ e) If $T: \mathbb{R}^n \to \mathbb{R}^n$ is the identity map $T(\mathbf{x}) = \mathbf{x}$, then the matrix representation $R_{B,B}$ of T relative to B,B is the identity $n \times n$ matrix, for any choice of ordered basis B of $\mathbb{R}^n$.

___ f) If $T: \mathbb{R}^n \to \mathbb{R}^n$ is the identity map, the matrix representation $R_{B,B'}$ of T relative to

B,B' is the identity $n \times n$ matrix, for any choice of ordered bases B,B' of $\mathbb{R}^n$.

____ g) If $T: \mathbb{R}^n \to \mathbb{R}^n$ is the identity map, the matrix representation $R_{B,B'}$ of T relative to B,B' is the change-of-coordinates matrix from B to B' for any choice of ordered bases B,B' of $\mathbb{R}^n$.

____ h) The collection of all linear transformations mapping $\mathbb{R}^n$ into $\mathbb{R}^m$ can be coordinatized in a unique way.

____ i) The collection of all linear transformations mapping $\mathbb{R}^n$ into $\mathbb{R}^m$ can be coordinatized in a unique way relative to bases B of $\mathbb{R}^n$ and B' of $\mathbb{R}^m$.

____ j) The collection of all linear transformations mapping $\mathbb{R}^n$ into $\mathbb{R}^m$ can be coordinatized in a unique way relative to ordered bases B of $\mathbb{R}^n$ and B' of $\mathbb{R}^m$.

22. (*Calculus*) This exercise illustrates Eq. (15). Let P_4 be the vector space of polynomials of degree at most 4, and let $T: P_4 \to P_4$ be the differentiation transformation. Find the *second* derivative of the polynomial $6x^4 + 7x^3 - 5x^2 + 2x - 1$ by using the matrix $R_{B,B}$ found in Example 5.

23. (*Calculus*) This exercise illustrates Eq. (15). Let $V = \text{sp}(e^{2x}, xe^{2x})$.

a) Find the matrix representation of the differentiation transformation T mapping V into V relative to the basis B,B, where $B = (e^{2x}, xe^{2x})$.

b) Use your answer to part (a) to compute the *fourth* derivative of xe^{2x}.

24. If the matrix representation $R_{B,B}$ of $T: \mathbb{R}^n \to \mathbb{R}^n$ relative to B,B is a diagonal matrix, describe the effect of T on the basis vectors in B.

25. Consider the ordered basis $B = ((1, 1), (-1, 1))$ of $\mathbb{R}^2$. Let (u, v) denote the coordinate vector of (x, y) in $\mathbb{R}^2$ relative to B. Find the u,v-equation describing the curve in the plane having x,y-equation $xy = 1$.

26. Repeat Exercise 25, but first normalize B to obtain an orthonormal basis. Why is it geometrically desirable to normalize B?

27. Let V and V' be vector spaces having ordered bases $B = (\mathbf{b}_1, \mathbf{b}_2, \mathbf{b}_3)$ and $B' = (\mathbf{b}_1', \mathbf{b}_2', \mathbf{b}_3', \mathbf{b}_4')$, respectively. Let $T: V \to V'$ be a linear transformation such that

$$T(\mathbf{b}_1) = 3\mathbf{b}_1' + \mathbf{b}_2' + 4\mathbf{b}_3' - \mathbf{b}_4'$$
$$T(\mathbf{b}_2) = \mathbf{b}_1' + 2\mathbf{b}_2' - \mathbf{b}_3' + 2\mathbf{b}_4'$$
$$T(\mathbf{b}_3) = -2\mathbf{b}_1' - \mathbf{b}_2' + 2\mathbf{b}_3'.$$

Find the matrix representation $R_{B,B'}$ of T relative to B,B'.

28. Let V be a vector space having ordered basis $B = (\mathbf{b}_1, \mathbf{b}_2, \mathbf{b}_3)$, and let $T: V \to V$ be a linear transformation such that

$$T(\mathbf{b}_1) = 3\mathbf{b}_1 + \mathbf{b}_2 + \mathbf{b}_3$$
$$T(\mathbf{b}_2) = \mathbf{b}_1 + 2\mathbf{b}_2 - \mathbf{b}_3$$
$$T(\mathbf{b}_3) = \mathbf{b}_1 - \mathbf{b}_2 + 2\mathbf{b}_3.$$

Find the matrix representation $R_{B,B}$ of T relative to B,B. Then show that there exists an ordered basis B' of V such that the matrix representation $R_{B',B'}$ of T relative to B',B' is diagonal.

29. Let B and B' each be an ordered basis for $\mathbb{R}^n$, and let $T: \mathbb{R}^n \to \mathbb{R}^n$ be a linear transformation. Let $R_{B,B}$ and $R_{B',B'}$ be the matrix representations of T relative to B,B and B',B', respectively. Express $R_{B',B'}$ as a product involving $R_{B,B}$, the basis matrices of B and B', or inverses of these basis matrices.

Exercises 30 and 31 deal with the vector space P_2 of polynomials of degree at most 2 and with its ordered bases $B = (x^2, x, 1)$ and $B' = (1, x, x^2)$.

30. Let $T: P_2 \to P_2$ be the linear transformation defined by $T(p(x)) = p(x + 1)$ for any $p(x)$ in P_2.

a) Find $R_{B,B}$, b) Find $R_{B',B'}$,
c) Find $R_{B,B'}$, d) Find $R_{B',B}$.

31. Repeat Exercise 30, given that $T: P_2 \to P_2$ is the linear transformation defined by $T(p(x)) = p(x - 2) + p(x + 2) + p(0)$.

In Exercises 32–34, let V and V' be vector spaces with ordered bases $B = (\mathbf{b}_1, \mathbf{b}_2, \mathbf{b}_3)$ and $B' = (\mathbf{b}_1', \mathbf{b}_2', \mathbf{b}_3', \mathbf{b}_4')$, respectively, and let $T: V \to V'$ be the linear transformation having the given matrix $R_{B,B'}$ as matrix representation relative to B,B'. Find $T(\mathbf{v})$ for the given vector $\mathbf{v}$.

32. $R_{B,B'} = \begin{bmatrix} 4 & 1 & -1 \\ 2 & 2 & 0 \\ 0 & 6 & 1 \\ 2 & 1 & 3 \end{bmatrix}$, $\mathbf{v} = \mathbf{b}_1 + \mathbf{b}_2 + \mathbf{b}_3$

33. $R_{B,B'}$ as in Exercise 32, $\mathbf{v} = 3\mathbf{b}_3 - \mathbf{b}_1$

34. $R_{B,B'} = \begin{bmatrix} 0 & 4 & -1 \\ 1 & 1 & 2 \\ 2 & 0 & 1 \\ 0 & 1 & 1 \end{bmatrix}$, $\mathbf{v} = 6\mathbf{b}_1 - 4\mathbf{b}_2 + \mathbf{b}_3$

Exercises 35–42 involve calculus. We let D denote differentiating once, D^2 denote differentiating twice, and so on.

35. Let $T: P_3 \rightarrow P_3$ be the linear transformation defined by $T(p(x)) = D^2(p(x)) - 4D(p(x)) + p(x)$. Find the matrix representation $R_{B,B}$ of T, where $B = (x, 1 + x, x + x^2, x^3)$.

36. Let $W = \text{sp}(e^{2x}, e^{4x}, e^{8x})$ be the subspace of the vector space of all real-valued functions with domain $\mathbb{R}$, and let $B = (e^{2x}, e^{4x}, e^{8x})$. Find the matrix representation $R_{B,B}$ of the linear transformation $T: W \rightarrow W$ defined by $T(f) = D^2(f) + 2D(f) + f$.

37. For W and B in Exercise 36, find the matrix representation $R_{B,B}$ of the linear transformation $T: W \rightarrow W$ defined by $T(f) = \int_{-\infty}^{x} f(t)\, dt$.

38. For W and B in Exercise 36, find $T(ae^{2x} + be^{4x} + ce^{8x})$ for the linear transformation T whose matrix representation relative to B,B is

$$R_{B,B} = \begin{bmatrix} 1 & 0 & 1 \\ 0 & 1 & 0 \\ 1 & 0 & 1 \end{bmatrix}.$$

39. Repeat Exercise 38, given that

$$R_{B,B} = \begin{bmatrix} 6 & 0 & 0 \\ 0 & 5 & 0 \\ 0 & 0 & -3 \end{bmatrix}.$$

40. Let $W = \text{sp}(\sin 2x, \cos 2x)$ be the subspace of the vector space of all real-valued functions with domain $\mathbb{R}$, and let $B = (\sin 2x, \cos 2x)$. Find the matrix representation $R_{B,B}$ for the linear transformation $T: W \rightarrow W$ defined by $T(f) = D^2(f) + 2D(f) + f$.

41. For W and B in Exercise 40, find $T(a \sin 2x + b \cos 2x)$ for the linear transformation $T: W \rightarrow W$ whose matrix representation is

$$R_{B,B} = \begin{bmatrix} 1 & 1 \\ 1 & -1 \end{bmatrix}.$$

42. For W and B in Exercise 40, find $T(a \sin 2x + b \cos 2x)$ for the linear transformation $T: W \rightarrow W$ whose matrix representation is

$$R_{B,B} = \begin{bmatrix} 0 & -2 \\ 2 & 0 \end{bmatrix}.$$

43. Work with Option 6 of the available program VECTGRPH until you can get a grade of 80% or better reliably.

6.3 CHANGE OF BASIS AND SIMILARITY

Let V be a finite-dimensional vector space with ordered basis B, and let $T: V \rightarrow V$ be a linear transformation. We are interested in the $n \times n$ matrix representation $R_{B,B}$ of T relative to B,B. We refer to this matrix simply as the matrix representation of T relative to B, and we denote it by R_B. Thus,

$$T(\mathbf{v})_B = R_B \mathbf{v}_B \qquad \text{for all } \mathbf{v} \in V.$$

We regard the matrix R_B as *coordinatizing the transformation* T relative to B. If we select a different ordered basis B' of V, we obtain a different matrix representation $R_{B'}$ of T. In this section, we examine the relationship between R_B and $R_{B'}$—that is, between matrix representations of the same linear transformation but relative to different ordered bases. We will see that two such matrix representations are *similar matrices*, and conversely, that two similar matrices can be regarded as representing the same linear transformation but relative to different bases.

Similarity of Representations Relative to Different Bases

For simplicity of notation, we work in $\mathbb{R}^n$. This involves no loss of generality, since we know that any vector space of dimension n is isomorphic to $\mathbb{R}^n$. Let $T: \mathbb{R}^n \rightarrow \mathbb{R}^n$ be a linear transformation of $\mathbb{R}^n$ into itself, and let B and B' be ordered bases of $\mathbb{R}^n$.

We wish to find the relationship between the matrix representation R_B of T relative to B and the matrix representation $R_{B'}$ of T relative to B'. A diagram for computing $T(\mathbf{x})_{B'}$ from $\mathbf{x}_{B'}$ similar to diagram (7) in Section 6.2 solves this problem:

$$\mathbf{x}_{B'} \xrightarrow{\quad} \mathbf{x} \xrightarrow{\quad} \mathbf{x}_B \xrightarrow{\quad} T(\mathbf{x})_B \xrightarrow{\quad} T(\mathbf{x}) \xrightarrow{\quad} T(\mathbf{x})_{B'}. \qquad (1)$$
$$\quad M_{B'} \qquad M_B^{-1} \qquad R_B \qquad M_B \qquad M_{B'}^{-1}$$

Again, we write under each arrow the matrix that performs that operation via left multiplication. Recall that the matrix $M_B^{-1}M_{B'}$ is the change-of-coordinates matrix $C_{B',B}$ from B' to B, and likewise $M_{B'}^{-1}M_B = C_{B,B'}$. From diagram (1), remembering to reverse the order, we find that

$$R_{B'} = (M_{B'}^{-1}M_B)R_B(M_B^{-1}M_{B'})$$
$$= C_{B,B'}R_BC_{B',B}.$$

Reading the matrix product in the equation

$$R_{B'} = C_{B,B'}R_BC_{B'B} \qquad (2)$$

from right to left, we see that, in order to compute $T(\mathbf{x})_{B'}$ from $x_{B'}$ if we know R_B, we first change to B coordinates; then we compute the transformation relative to B coordinates; and then we change back to B' coordinates. This procedure is easy to remember.

The matrices $C_{B,B'} = M_B^{-1}M_{B'}$ and $C_{B',B} = M_{B'}^{-1}M_B$ are inverses of each other. If we drop subscripts for a moment and let $C = C_{B',B}$, then Eq. (2) becomes

$$R_{B'} = C^{-1}R_BC. \qquad (3)$$

Recall that two $n \times n$ matrices A and R are *similar* if there exists an invertible $n \times n$ matrix C such that $R = C^{-1}AC$. (See Definition 4.4.) We have shown that matrix representations of the same transformation relative to different bases are similar. We state this as a theorem in the context of a general finite-dimensional vector space.

THEOREM 6.4 Similarity of Matrix Representations of *T*

Let T be a linear transformation of a finite-dimensional vector space V into itself, and let B and B' be ordered bases of V. Let R_B and $R_{B'}$ be the matrix representations of T relative to B and B', respectively. Then

$$R_{B'} = C^{-1}R_BC,$$

where $C = C_{B',B}$ is the change-of-coordinates matrix from B' to B. Consequently, $R_{B'}$ and R_B are similar matrices.

EXAMPLE 1   Consider the linear transformation $T: \mathbb{R}^3 \to \mathbb{R}^3$ defined by

$$T(x_1, x_2, x_3) = (x_1 + x_2 + x_3, x_1 + x_2, x_3).$$

Find the standard matrix representation A of T, and find the matrix representation R_B of T relative to B, where

$$B = ((1, 1, 0), (1, 0, 1), (0, 1, 1)).$$

In addition, find an invertible matrix C such that $R_B = C^{-1}AC$.

Solution   Here the standard ordered basis E plays the role that the basis B played in Theorem 6.4, and the basis B here plays the role played by B' in the theorem. Of course the standard matrix representation of T is

$$A = \begin{bmatrix} 1 & 1 & 1 \\ 1 & 1 & 0 \\ 0 & 0 & 1 \end{bmatrix}.$$

We compute R_B as follows:

$$\begin{bmatrix} 1 & 1 & 0 & 2 & 2 & 2 \\ 1 & 0 & 1 & 2 & 1 & 1 \\ 0 & 1 & 1 & 0 & 1 & 1 \end{bmatrix} \sim \begin{bmatrix} 1 & 1 & 0 & 2 & 2 & 2 \\ 0 & -1 & 1 & 0 & -1 & -1 \\ 0 & 1 & 1 & 0 & 1 & 1 \end{bmatrix}$$
$$\quad\; \mathbf{b_1} \;\; \mathbf{b_2} \;\; \mathbf{b_3} \quad T(\mathbf{b_1}) \;\; T(\mathbf{b_2}) \;\; T(\mathbf{b_3})$$

$$\sim \begin{bmatrix} 1 & 0 & 1 & 2 & 1 & 1 \\ 0 & 1 & -1 & 0 & 1 & 1 \\ 0 & 0 & 2 & 0 & 0 & 0 \end{bmatrix} \sim \begin{bmatrix} 1 & 0 & 0 & 2 & 1 & 1 \\ 0 & 1 & 0 & 0 & 1 & 1 \\ 0 & 0 & 1 & 0 & 0 & 0 \end{bmatrix}.$$
$$\qquad\qquad\qquad\qquad\qquad\qquad I \qquad\qquad R_B$$

Thus,

$$R_B = \begin{bmatrix} 2 & 1 & 1 \\ 0 & 1 & 1 \\ 0 & 0 & 0 \end{bmatrix}.$$

An invertible matrix C such that $R_B = C^{-1}AC$ is $C = C_{B,E}$. We find $C = C_{B,E}$ by reducing the augmented matrix $[M_E \mid M_B] = [I \mid M_B]$. Since this matrix is already reduced, we see that $C = M_B$ in this case. The matrices A and R_B are similar matrices, both representing the given linear transformation. As a check, we could compute that $AC = CR_B$.   ■

We give an example illustrating Theorem 6.4 in the case of a finite-dimensional vector space other than $\mathbb{R}^n$.

EXAMPLE 2   For the space P_2 of polynomials of degree at most 2, let $T: P_2 \to P_2$ be defined by $T(p(x)) = p(x - 1)$. Consider the ordered bases $B = (x^2, x, 1)$ and $B' = (x, x + 1, x^2 - 1)$. Find the matrix representations R_B and $R_{B'}$ of T and a matrix C such that $R_{B'} = C^{-1}R_BC$.

Solution   Since

$$T(x^2) = (x - 1)^2 = x^2 - 2x + 1,$$

$$T(x) = x - 1,$$

$$T(1) = 1,$$

the matrix representation of T relative to $B = (x^2, x, 1)$ is

$$R_B = \begin{bmatrix} 1 & 0 & 0 \\ -2 & 1 & 0 \\ 1 & -1 & 1 \end{bmatrix}.$$

Next we compute the change-of-coordinates matrices $C_{B,B'}$ and $C_{B',B}$. We see that

$$C = C_{B',B} = \begin{bmatrix} 0 & 0 & 1 \\ 1 & 1 & 0 \\ 0 & 1 & -1 \end{bmatrix}.$$

Moreover

$$x^2 = -1(x) + 1(x + 1) + 1(x^2 - 1),$$

$$x = \quad 1(x) + 0(x + 1) + 0(x^2 - 1),$$

$$1 = -1(x) + 1(x + 1) + 0(x^2 - 1),$$

so

$$C_{B,B'} = \begin{bmatrix} -1 & 1 & -1 \\ 1 & 0 & 1 \\ 1 & 0 & 0 \end{bmatrix}.$$

Notice that $C_{B,B'}$ can be computed as $(C_{B',B})^{-1}$. We now have

$$R_{B'} = \underbrace{\begin{bmatrix} -1 & 1 & -1 \\ 1 & 0 & 1 \\ 1 & 0 & 0 \end{bmatrix}}_{C_{B,B'}} \underbrace{\begin{bmatrix} 1 & 0 & 0 \\ -2 & 1 & 0 \\ 1 & -1 & 1 \end{bmatrix}}_{R_B} \underbrace{\begin{bmatrix} 0 & 0 & 1 \\ 1 & 1 & 0 \\ 0 & 1 & -1 \end{bmatrix}}_{C_{B',B}}$$

$$= \begin{bmatrix} -1 & 1 & -1 \\ 1 & 0 & 1 \\ 1 & 0 & 0 \end{bmatrix} \begin{bmatrix} 0 & 0 & 1 \\ 1 & 1 & -2 \\ -1 & 0 & 0 \end{bmatrix} = \begin{bmatrix} 2 & 1 & -3 \\ -1 & 0 & 1 \\ 0 & 0 & 1 \end{bmatrix}.$$

Alternatively, $R_{B'}$ can be computed directly as

$$R_{B'} = \begin{bmatrix} \mid & \mid & \mid \\ T(\mathbf{b}_1')_{B'} & T(\mathbf{b}_2')_{B'} & T(\mathbf{b}_3')_{B'} \\ \mid & \mid & \mid \end{bmatrix}. \qquad ■$$

The Similarity Relationship

We have seen that matrix representations of the same transformation relative to different bases are similar. Conversely, any two similar matrices can be viewed as representations of the same transformation relative to different bases. To see this, let A be an $n \times n$ matrix, and let C be any invertible $n \times n$ matrix. Let $T: \mathbb{R}^n \to \mathbb{R}^n$ be defined by $T(\mathbf{x}) = A\mathbf{x}$ so that A is the standard matrix representation of T. Since C is invertible, its column vectors are independent and form a basis for $\mathbb{R}^n$. Let B be the ordered basis having as *j*th vector the *j*th column vector of *C*. Then *C* is precisely the change-of-coordinates matrix from *B* to the standard ordered basis *E*. That is, $C = C_{B,E}$. Consequently, $C^{-1}AC = C_{E,B}AC_{B,E}$ is the matrix representation of *T* relative to *B*.

Significance of the Similarity Relationship for Matrices

Two $n \times n$ matrices are similar if and only if they are matrix representations of the same linear transformation *T* relative to suitable ordered bases.

Let *V* be a finite-dimensional vector space, and let $T: V \rightarrow V$ be a linear transformation. Suppose that *T* has a property that can be characterized in terms of the mapping, without reference to coordinates relative to any basis. For example, *T* has λ as an eigenvalue if and only if

$$T(\mathbf{v}) = \lambda\mathbf{v} \qquad \text{for some nonzero vector } \mathbf{v} \text{ in } V. \qquad (4)$$

This statement makes no reference to coordinates relative to a basis. It follows immediately that, if λ is an eigenvalue of *T*, then λ is an eigenvalue of *every* matrix representation of *T*.

Now any two similar matrices can be viewed as matrix representations of the same linear transformation *T*, relative to suitable bases. Consequently, similar matrices must share any properties that can be described for the transformation in a coordinate-free fashion. In particular we obtain, with no additional work, this nice result:

Similar matrices have the same eigenvalues.

We take a moment to expand on the ideas we have just introduced. Let *V* be a vector space of finite dimension *n*. The study of linear transformations $T: V \rightarrow V$ is essentially the same as the study of products $A\mathbf{x}$ of vectors $\mathbf{x}$ in $\mathbb{R}^n$ by $n \times n$ matrices *A*. We should understand this relationship thoroughly and be able to bounce back and forth at will from $T: V \rightarrow V$ to $A\mathbf{x}$ for $\mathbf{x}$ in $\mathbb{R}^n$. Sometimes a theorem that is not immediately obvious from one point of view is quite easy from the other. For example, from the matrix point of view, it is not immediately apparent that similar matrices have the same eigenvalues; we ask for a matrix proof in Exercise 27. However, the truth of this statement becomes obvious from the linear transformation point of view. On the other hand, it is not obvious from Eq. (4) concerning an eigenvalue of a linear transformation $T: \mathbb{R}^n \rightarrow \mathbb{R}^n$ that a linear transformation of *V* can have at most *n* eigenvalues. However, this is easy to establish from the matrix point of view: $A\mathbf{x} = \lambda\mathbf{x}$ has a nontrivial solution if and only if the coefficient matrix $A - \lambda I$ of the system $(A - \lambda I)\mathbf{x} = \mathbf{0}$ has determinant zero, and $\det(A - \lambda I) = 0$ is a polynomial equation of degree *n*.

When making arguments like the preceding one, we must make certain that no concepts of coordinates are involved. For example, a common error is to say, "Similar matrices have the same eigenvectors, since they represent the same linear transformation T." That statement is *false*. For example, every diagonal matrix has the standard basis of $\mathbb{R}^n$ as eigenvectors, but this is not true of every diagonalizable matrix. An eigenvector of T has different coordinates relative to an ordered basis B than it has relative to a different ordered basis B'. Thus the matrix representations of T relative to B and to B' have different eigenvectors. However, assertions about $T: V \to V$ of the form

There is a basis of eigenvectors of T,

The nullspace of T has dimension 4,

The eigenspace E_5 of T has dimension 2,

are *coordinate-free* assertions, so any of these properties that are true when T is replaced by a matrix A must also hold for any matrix similar to A.

We now state a theorem relating eigenvalues and eigenvectors of similar matrices. Exercises 24 and 25 ask for proofs of the second and third statements in the theorem. We already proved the first statement.

THEOREM 6.5 Eigenvalues and Eigenvectors of Similar Matrices

Let A and R be similar $n \times n$ matrices, so that $R = C^{-1}AC$ for some invertible $n \times n$ matrix C. Let the eigenvalues of A be the (not necessarily distinct) numbers $\lambda_1, \lambda_2, \ldots, \lambda_n$.

1. The eigenvalues of R are also $\lambda_1, \lambda_2, \ldots, \lambda_n$.
2. The algebraic and geometric multiplicity of each λ_i as an eigenvalue of A remains the same as when it is viewed as an eigenvalue of R.
3. If $\mathbf{v}_i$ in $\mathbb{R}^n$ is an eigenvector of the matrix A corresponding to λ_i, then $C^{-1}\mathbf{v}_i$ is an eigenvector of the matrix R corresponding to λ_i.

We give one more illustration of the usefulness of this interplay between linear transformations and matrices, and of the significance of similarity. Let V be a finite-dimensional vector space, and let $T: V \to V$ be a linear transformation. It was easy to define the notions of *eigenvalue* and *eigenvector* in terms of T. We can now define the **geometric multiplicity** of an eigenvalue λ to be the dimension of the eigenspace $E_\lambda = \{\mathbf{v} \in V \,|\, T(\mathbf{v}) = \lambda\mathbf{v}\}$. However, at this time there is no obvious way to characterize the algebraic multiplicity of λ exclusively in terms of the mapping T, without coordinatization. Consequently, we define the **algebraic multiplicity** of λ as its algebraic multiplicity as an eigenvalue of a matrix represenation of λ. This makes sense because this algebraic multiplicity of λ is the same for *all* matrix representations of T. Property (ii) of Theorem 6.5 assures us that this is the case.

Diagonalization

Let $T: V \rightarrow V$ be a linear transformation of an n-dimensional vector space into itself. Suppose that there exists an ordered basis $B = (\mathbf{b}_1, \mathbf{b}_2, \ldots, \mathbf{b}_n)$ of V composed of eigenvectors of T. Let the eigenvalue corresponding to $\mathbf{b}_i$ be λ_i. Then the matrix representation of T relative to B has the simple diagonal form

$$D = \begin{bmatrix} \lambda_1 & & & \\ & \lambda_2 & & \mathbf{0} \\ & & \cdot & \\ \mathbf{0} & & \cdot & \\ & & & \lambda_n \end{bmatrix}.$$

We give a definition of diagonalization for linear transformations that clearly parallels one for matrices.

DEFINITION 6.3 Diagonalizable Transformation

A linear transformation T of a finite-dimensional vector space V into itself is **diagonalizable** if V has an ordered basis consisting of eigenvectors of T.

For example, reflection of the plane through the line $x + 2y = 0$ is a diagonalizable transformation. We made use of this fact in Example 4 of Section 6.2.

If V has a known ordered basis B of eigenvectors, it becomes easy to compute the k-fold composition $T^k(\mathbf{v})$ for any positive integer k and for any vector $\mathbf{v}$ in V. We need only find the coordinate vector $\mathbf{d}$ in $\mathbb{R}^n$ of $\mathbf{v}$ relative to B, so that

$$\mathbf{v} = d_1\mathbf{b}_1 + d_2\mathbf{b}_2 + \cdots + d_n\mathbf{b}_n.$$

Then

$$T^k(\mathbf{v}) = d_1\lambda_1{}^k\mathbf{b}_1 + d_2\lambda_2{}^k\mathbf{b}_2 + \cdots + d_n\lambda_n{}^k\mathbf{b}_n. \tag{5}$$

Of course, this is the transformation analogue of the computation of $A^k\mathbf{x}$ in Section 4.3. We illustrate with an example.

EXAMPLE 3 Consider the vector space P_2 of polynomials of degree at most 2, and let B' be the ordered basis $(1, x, x^2)$ for P_2. Let $T: P_2 \rightarrow P_2$ be the linear transformation such that

$$T(1) = 3 + 2x + x^2, \qquad T(x) = 2, \qquad T(x^2) = 2x^2.$$

Find $T^4(x + 2)$.

Solution The matrix representation of T relative to B' is

$$R_{B'} = \begin{bmatrix} 3 & 2 & 0 \\ 2 & 0 & 0 \\ 1 & 0 & 2 \end{bmatrix}.$$

Using the methods of Chapter 4, we easily find the eigenvalues and eigenvectors of $R_{B'}$ and of T given in Table 6.1.

TABLE 6.1

Eigenvalues	Eigenvectors of $R_{B'}$	Eigenvectors of T
$\lambda_1 = -1$	$\mathbf{w}_1 = (-3, 6, 1)$	$p_1(x) = -3 + 6x + x^2$
$\lambda_2 = 2$	$\mathbf{w}_2 = (0, 0, 1)$	$p_2(x) = x^2$
$\lambda_3 = 4$	$\mathbf{w}_3 = (2, 1, 1)$	$p_3(x) = 2 + x + x^2$

Let B be the ordered basis $(-3 + 6x + x^2, x^2, 2 + x + x^2)$ consisting of these eigenvectors. We can find the coordinate vector $\mathbf{d}$ of $x + 2$ relative to the basis B by inspection. Since

$$x + 2 = 0(x^2 + 6x - 3) + (-1)x^2 + 1(x^2 + x + 2),$$

we see that

$$d_1 = 0, \quad d_2 = -1, \quad \text{and} \quad d_3 = 1.$$

Thus, Eq. (5) has the form

$$T^k(x + 2) = 2^k(-1)x^2 + 4^k(1)(x^2 + x + 2).$$

In particular,

$$T^4(x + 2) = -16x^2 + 256(x^2 + x + 2) = 240x^2 + 256x + 512. \qquad ■$$

SUMMARY

Let $T: V \to V$ be a linear transformation of a finite-dimensional vector space into itself.

1. If B and B' are ordered bases of V, then the matrix representations R_B and $R_{B'}$ of T relative to B and to B' are similar. That is, there is an invertible matrix C—namely, $C = C_{B',B}$—such that
$$R_{B'} = C^{-1}R_B C.$$

2. Conversely, two similar $n \times n$ matrices represent the same linear transformation of $\mathbb{R}^n$ into $\mathbb{R}^n$ relative to two suitably chosen ordered bases.

3. Similar matrices have the same eigenvalues with the same algebraic and geometric multiplicities.

4. If A and R are similar matrices with $R = C^{-1}AC$, and if $\mathbf{v}$ is an eigenvector of A, then $C^{-1}\mathbf{v}$ is an eigenvector of R corresponding to the same eigenvalue.

5. The transformation T is diagonalizable if V has a basis B consisting of eigenvectors of T. In this case, the matrix representation R_B is a diagonal matrix and computation of $T^k(\mathbf{v})$ by $R_B{}^k(\mathbf{v}_B)$ becomes relatively easy.

EXERCISES

In Exercises 1–14, find the matrix representations R_B and $R_{B'}$ and an invertible matrix C such that $R_{B'} = C^{-1}R_BC$ for the linear transformation T of the given vector space with the indicated ordered bases B and B'.

1. $T: \mathbb{R}^2 \to \mathbb{R}^2$ defined by $T(x, y) = (x - y, x + 2y)$; $B = ((1, 1), (2, 1))$, $B' = E$

2. $T: \mathbb{R}^2 \to \mathbb{R}^2$ defined by $T(x, y) = (2x + 3y, x + 2y)$; $B = ((1, -1), (1, 1))$, $B' = ((2, 3), (1, 2))$

3. $T: \mathbb{R}^3 \to \mathbb{R}^3$ defined by $T(x, y, z) = (x + y, x + z, y - z)$; $B = ((1, 1, 1), (1, 1, 0), (1, 0, 0))$, $B' = E$

4. $T: \mathbb{R}^3 \to \mathbb{R}^3$ defined by $T(x, y, z) = (5x, 2y, 3z)$; B and B' as in Exercise 3

5. $T: \mathbb{R}^3 \to \mathbb{R}^3$ defined by $T(x, y, z) = (z, 0, x)$; $B = ((3, 1, 2), (1, 2, 1), (2, -1, 0))$, $B' = ((1, 2, 1), (2, 1, -1), (5, 4, 1))$

6. $T: \mathbb{R}^2 \to \mathbb{R}^2$ defined as reflection of the plane through the line $5x = 3y$; $B = ((3, 5), (5, -3))$, $B' = E$

7. $T: \mathbb{R}^3 \to \mathbb{R}^3$ defined as reflection of $\mathbb{R}^3$ through the plane $x + y + z = 0$; $B = ((1, 0, -1), (1, -1, 0), (1, 1, 1))$, $B' = E$

8. $T: \mathbb{R}^3 \to \mathbb{R}^3$ defined as reflection of $\mathbb{R}^3$ through the plane $2x + 3y + z = 0$; $B = ((2, 3, 1), (0, 1, -3), (1, 0, -2))$, $B' = E$

9. $T: \mathbb{R}^3 \to \mathbb{R}^3$ defined as projection on the plane $x_2 = 0$; $B = E$, $B' = ((1, 0, 1), (1, 0, -1), (0, 1, 0))$

10. $T: \mathbb{R}^3 \to \mathbb{R}^3$ defined as projection on the plane $x + y + z = 0$; B and B' as in Exercise 7.

11. $T: P_2 \to P_2$ defined by $T(p(x)) = p(x + 1) + p(x)$; $B = (x^2, x, 1)$, $B' = (1, x, x^2)$

12. $T: P_2 \to P_2$ as in Exercise 11, but using $B = (x^2, x, 1)$ and $B' = (x^2 + 1, x + 1, 2)$

13. *(Calculus)* $T: P_3 \to P_3$ defined by $T(p(x)) = p'(x)$, the derivative of $p(x)$; $B = (x^3, x^2, x, 1)$, $B' = (1, x + 1, x^2 + 1, x^3 + 1)$

14. *(Calculus)* $T: W \to W$, where $W = \text{sp}(e^x, xe^x)$ and T is the derivative transformation; $B = (e^x, xe^x)$, $B' = (2xe^x, 3e^x)$

15. Let $T: P_2 \to P_2$ be the linear transformation and B' the ordered basis of P_2 given in Example 2. Find the matrix representation $R_{B'}$ of T by computing the matrix with column vectors $T(\mathbf{b}_1')_{B'}$, $T(\mathbf{b}_2')_{B'}$, $T(\mathbf{b}_3')_{B'}$.

16. Repeat Exercise 15 for the transformation in Example 3 and the basis $B' = (x + 1, -1, x^2 + x)$ of P_2.

In Exercises 17–22, find the eigenvalues λ_i and the corresponding eigenspaces of the linear transformation T. Determine whether the linear transformation is diagonalizable.

17. T defined on $\mathbb{R}^2$ by $T(x, y) = (2x - 3y, -3x + 2y)$

18. T defined on $\mathbb{R}^2$ by $T(x, y) = (x - y, -x + y)$

19. T defined on $\mathbb{R}^3$ by $T(x_1, x_2, x_3) = (x_1 + x_3, x_2, x_1 + x_3)$

20. T defined on $\mathbb{R}^3$ by $T(x_1, x_2, x_3) = (x_1, 4x_2 + 7x_3, 2x_2 - x_3)$

21. T defined on $\mathbb{R}^3$ by $T(x_1, x_2, x_3) = (5x_1, -5x_1 + 3x_2 - 5x_3, -3x_1 - 2x_2)$

22. T defined on $\mathbb{R}^3$ by $T(x_1, x_2, x_3) = (3x_1 - x_2 + x_3, -2x_1 + 2x_2 - x_3, 2x_1 + x_2 + 4x_3)$

23. Mark each of the following True or False

____ a) Two similar $n \times n$ matrices represent the same linear transformation of $\mathbb{R}^n$ into itself relative to the standard basis.

____ b) Two different $n \times n$ matrices represent different linear transformations of $\mathbb{R}^n$ into itself relative to the standard basis.

____ c) Two similar $n \times n$ matrices represent the same linear transformation of $\mathbb{R}^n$ into itself relative to two suitably chosen bases for $\mathbb{R}^n$.

____ d) Similar matrices have the same eigenvalues and eigenvectors.

____ e) Similar matrices have the same eigenvalues with the same algebraic and geometric multiplicities.

____ f) If A and C are $n \times n$ matrices and C is invertible and $\mathbf{v}$ is an eigenvector of A, then $C^{-1}\mathbf{v}$ is an eigenvector of $C^{-1}AC$.

_____ g) If A and C are $n \times n$ matrices and C is invertible and $\mathbf{v}$ is an eigenvector of A, then $C\mathbf{v}$ is an eigenvector of CAC^{-1}.

_____ h) Any two $n \times n$ diagonal matrices are similar.

_____ i) Any two $n \times n$ diagonalizable matrices having the same eigenvectors are similar.

_____ j) Any two $n \times n$ diagonalizable matrices having the same eigenvalues of the same algebraic multiplicities are similar.

24. Prove statement 2 of Theorem 6.5.

25. Prove statement 3 of Theorem 6.5.

26. Let A and R be similar matrices. Prove in two ways that A^2 and R^2 are similar matrices: using a matrix argument, and using a linear transformation argument.

27. Give a determinant proof that similar matrices have the same eigenvalues.

7 EIGENVALUES: FURTHER APPLICATIONS AND COMPUTATIONS

This chapter deals with further applications of eigenvalues and with the computation of eigenvalues. In Section 7.1, we discuss quadratic forms and their diagonalization. The principal axis theorem (Theorem 7.1) asserts that every quadratic form can be diagonalized. This is probably the most important result in the chapter, having applications to the vibration of elastic bodies, to quantum mechanics, and to electric circuits. Presentations of such applications are beyond the scope of this text, and we have chosen to present a more accessible application in Section 7.2: classification of conic sections and quadric surfaces. Although the facts about conic sections may be familiar to you, their easy derivation from the principal axis theorem should be seen and enjoyed.

Section 7.3 applies the parallel axis theorem to local extrema of functions and discusses maximization and minimization of quadratic forms on unit spheres. The latter topic is again important in vibration problems; it indicates that eigenvalues of maximum and of minimum magnitudes can be found for symmetric matrices by using techniques from advanced calculus for finding extrema of quadratic forms on unit spheres.

In Section 7.4, we sketch three methods for computing eigenvalues: the power method, Jacobi's method, and the QR method. We attempt to make as intuitively clear as we can how and why each method works, but proofs and discussions of efficiency are omitted.

7.1 DIAGONALIZATION OF QUADRATIC FORMS

Quadratic Forms

A quadratic form in one variable x is a polynomial $f(x) = ax^2$, where $a \neq 0$. A quadratic form in two variables x and y is a polynomial $f(x, y) = ax^2 + bxy + cy^2$, where at least one of a, b, or c is nonzero. The term *quadratic* means *degree 2*. The term *form* means *homogeneous*; that is, each summand contains a product of the *same number* of variables—namely, 2 for a quadratic form. Thus, $3x^2 - 4xy$ is a quadratic form in x and y, but $x^2 + y^2 - 4$ and $x - 3y^2$ are not quadratic forms.

Turning to the general case, a **quadratic form** $f(\mathbf{x}) = f(x_1, x_2, \ldots, x_n)$ in n variables is a polynomial that can be written, using summation notation, as

$$f(\mathbf{x}) = \sum_{\substack{i \leq j \\ i,j=1}}^{n} u_{ij} x_i x_j, \tag{1}$$

where not all u_{ij} are zero. To illustrate, the general quadratic form in x_1, x_2, and x_3 is

$$u_{11}x_1^2 + u_{12}x_1x_2 + u_{13}x_1x_3 + u_{22}x_2^2 + u_{23}x_2x_3 + u_{33}x_3^2. \tag{2}$$

A computation shows that

$$[x_1 \ x_2 \ x_3] \begin{bmatrix} u_{11} & u_{12} & u_{13} \\ 0 & u_{22} & u_{23} \\ 0 & 0 & u_{33} \end{bmatrix} \begin{bmatrix} x_1 \\ x_2 \\ x_3 \end{bmatrix}$$

$$= [x_1 \ x_2 \ x_3] \begin{bmatrix} u_{11}x_1 + u_{12}x_2 + u_{13}x_3 \\ u_{22}x_2 + u_{23}x_3 \\ u_{33}x_3 \end{bmatrix}$$

$$= x_1(u_{11}x_1 + u_{12}x_2 + u_{13}x_3) + x_2(u_{22}x_2 + u_{23}x_3) + x_3(u_{33}x_3)$$

$$= u_{11}x_1^2 + u_{12}x_1x_2 + u_{13}x_1x_3 + u_{22}x_2^2 + u_{23}x_2x_3 + u_{33}x_3^2,$$

which is again form (2). We can verify that the term involving u_{ij} for $i \leq j$ in the expansion of

$$[x_1 \ x_2 \ \ldots \ x_n] \begin{bmatrix} u_{11} & u_{12} & \cdots & u_{1n} \\ 0 & u_{22} & \cdots & u_{2n} \\ \vdots & \vdots & & \vdots \\ 0 & 0 & \cdots & u_{nn} \end{bmatrix} \begin{bmatrix} x_1 \\ x_2 \\ \vdots \\ x_n \end{bmatrix} \tag{3}$$

is precisely $u_{ij}x_ix_j$. Thus, matrix product (3) is a 1×1 matrix whose sole entry is equal to sum (1).

IN 1826 CAUCHY DISCUSSED QUADRATIC FORMS IN THREE VARIABLES—forms of the type $Ax^2 + By^2 + Cz^2 + 2Dxy + 2Exz + 2Fyz$. He showed that the characteristic equation formed from the determinant

$$\begin{vmatrix} A & D & E \\ D & B & F \\ E & F & C \end{vmatrix}$$

remains the same under any change of rectangular axes, what we would call an *orthogonal coordinate change*. Furthermore, he demonstrated a result already proved by Euler (see note on page 378): that one could find axes such that the new form has only the square terms. Three years later, Cauchy generalized the result to quadratic forms in n variables. In fact, he demonstrated that the roots $\lambda_1, \lambda_2, \ldots, \lambda_n$ of the characteristic equation are all real, and he showed how to find the linear substitution that converts the original form to the form $\lambda_1 x_1^2 + \lambda_2 x_2^2 + \cdots + \lambda_n x_n^2$.

In 1833 Jacobi proved the same result by a somewhat different method (see Section 7.4).

> Every quadratic form in n variables x_i can be written as $\mathbf{x}^T U \mathbf{x}$, where $\mathbf{x}$ is the column vector of variables and U is a nonzero upper-triangular matrix.

We will call the matrix U the **upper-triangular coefficient matrix** of the quadratic form.

EXAMPLE 1 Write $x^2 - 2xy + 6xz + z^2$ in matrix form (3).

Solution We obtain the matrix expression

$$[x \ y \ z] \begin{bmatrix} 1 & -2 & 6 \\ 0 & 0 & 0 \\ 0 & 0 & 1 \end{bmatrix} \begin{bmatrix} x \\ y \\ z \end{bmatrix}.$$

Just think of x as the first variable, y as the second, and z as the third. The summand $-2xy$, for example, gives the coefficient -2 in the row 1, column 2 position. ■

A matrix expression for a given quadratic form is by no means unique: $\mathbf{x}^T A \mathbf{x}$ gives a quadratic form for any nonzero $n \times n$ matrix A, and the form can be rewritten as $\mathbf{x}^T U \mathbf{x}$, where U is upper-triangular.

EXAMPLE 2 Expand

$$[x \ y \ z] \begin{bmatrix} -1 & 3 & 1 \\ 2 & 1 & 0 \\ -2 & 2 & 4 \end{bmatrix} \begin{bmatrix} x \\ y \\ z \end{bmatrix},$$

and find the upper-triangular coefficient matrix for the quadratic form.

Solution We find that

$$
[x \ y \ z] \begin{bmatrix} -1 & 3 & 1 \\ 2 & 1 & 0 \\ -2 & 2 & 4 \end{bmatrix} \begin{bmatrix} x \\ y \\ z \end{bmatrix} = [x \ y \ z] \begin{bmatrix} -x + 3y + z \\ 2x + y \\ -2x + 2y + 4z \end{bmatrix}
$$

$$= x(-x + 3y + z) + y(2x + y) + z(-2x + 2y + 4z)$$

$$= -x^2 + 3xy + xz + 2xy + y^2 - 2xz + 2yz + 4z^2$$

$$= -x^2 + 5xy + y^2 - xz + 2yz + 4z^2.$$

The upper-triangular coefficient matrix is

$$U = \begin{bmatrix} -1 & 5 & -1 \\ 0 & 1 & 2 \\ 0 & 0 & 4 \end{bmatrix}. \qquad ■$$

All the nice things that we will prove about quadratic forms come from the fact that any quadratic form $f(\mathbf{x})$ can be expressed as $f(\mathbf{x}) = \mathbf{x}^T A \mathbf{x}$, where A is a *symmetric* matrix.

EXAMPLE 3 Find the symmetric coefficient matrix of the form $x^2 - 2xy + 6xz + z^2$ discussed in Example 1.

Solution Rewriting the form as

$$x^2 - xy - yx + 3xz + 3zx + z^2,$$

we obtain the symmetric coefficient matrix

$$A = \begin{bmatrix} 1 & -1 & 3 \\ -1 & 0 & 0 \\ 3 & 0 & 1 \end{bmatrix}. \qquad ■$$

As illustrated in Example 3, we obtain the symmetric matrix A of the form

$$f(\mathbf{x}) = \sum_{\substack{i \leq j \\ i,j=1}}^{n} u_{ij} x_i x_j \qquad (4)$$

by writing each *cross term* $u_{ij} x_i x_j$ as $(u_{ij}/2) x_i x_j + (u_{ij}/2) x_j x_i$, and taking $a_{ij} = a_{ji} = u_{ij}/2$. The resulting matrix A is symmetric. We call it the **symmetric coefficient matrix** of the form.

Every quadratic form in n variables x_i can be written as $\mathbf{x}^T A \mathbf{x}$, where $\mathbf{x}$ is the column vector of variables and A is a symmetric matrix.

Diagonalization of Quadratic Forms

We saw in Section 5.3 that a symmetric matrix can be diagonalized by an orthogonal matrix. That is, if A is an $n \times n$ symmetric matrix, there exists an $n \times n$ orthogonal change-of-coordinates matrix C such that $C^{-1}AC = D$, where D is a diagonal matrix. Recall that the diagonal entries of D are $\lambda_1, \lambda_2, \ldots, \lambda_n$, where the λ_j are the (not necessarily distinct) eigenvalues of the matrix A, the jth column of C is a unit eigenvector corresponding to λ_j, and the column vectors of C form an orthonormal basis for $\mathbb{R}^n$.

Consider now a quadratic form $\mathbf{x}^T A \mathbf{x}$, where A is symmetric, and let C be an orthogonal diagonalizing matrix for A. Since $C^{-1} = C^T$, the substitution

$$\mathbf{x} = C\mathbf{t} \qquad \textbf{Diagonalizing substitution}$$

changes our quadratic form as follows:

$$\mathbf{x}^T A \mathbf{x} = (C\mathbf{t})^T A (C\mathbf{t}) = \mathbf{t}^T C^T A C \mathbf{t} = \mathbf{t}^T C^{-1} A C \mathbf{t}$$
$$= \mathbf{t}^T D \mathbf{t} = \lambda_1 t_1^2 + \lambda_2 t_2^2 + \cdots + \lambda_n t_n^2,$$

where the λ_j are the eigenvalues of A. We have thus diagonalized the quadratic form. The value of $\mathbf{x}^T A \mathbf{x}$ for any $\mathbf{x}$ in $\mathbb{R}^n$ is the same as the value of $\mathbf{t}^T D \mathbf{t}$ for $\mathbf{t} = C^{-1}\mathbf{x}$.

EXAMPLE 4 Find a substitution $\mathbf{x} = C\mathbf{t}$ that diagonalizes the form $3x_1^2 + 10x_1x_2 + 3x_2^2$, and find the corresponding diagonalized form.

Solution The symmetric coefficient matrix for the quadratic form is

$$A = \begin{bmatrix} 3 & 5 \\ 5 & 3 \end{bmatrix}.$$

We need to find the eigenvalues and eigenvectors for A. We have

$$|A - \lambda I| = \begin{vmatrix} 3 - \lambda & 5 \\ 5 & 3 - \lambda \end{vmatrix} = \lambda^2 - 6\lambda - 16 = (\lambda + 2)(\lambda - 8).$$

Thus, $\lambda_1 = -2$ and $\lambda_2 = 8$ are eigenvalues of A. Finding the eigenvectors that are to become the columns of the substitution matrix C, we compute

$$A - \lambda_1 I = A + 2I = \begin{bmatrix} 5 & 5 \\ 5 & 5 \end{bmatrix} \sim \begin{bmatrix} 1 & 1 \\ 0 & 0 \end{bmatrix}$$

and

$$A - \lambda_2 I = A - 8I = \begin{bmatrix} -5 & 5 \\ 5 & -5 \end{bmatrix} \sim \begin{bmatrix} 1 & -1 \\ 0 & 0 \end{bmatrix}.$$

Thus eigenvectors are $\mathbf{v}_1 = (-1, 1)$ and $\mathbf{v}_2 = (1, 1)$. Normalizing them to length 1 and placing them in the columns of the substitution matrix C, we obtain the diagonalizing substitution

$$\begin{bmatrix} x_1 \\ x_2 \end{bmatrix} = C \begin{bmatrix} t_1 \\ t_2 \end{bmatrix} = \begin{bmatrix} -1/\sqrt{2} & 1/\sqrt{2} \\ 1/\sqrt{2} & 1/\sqrt{2} \end{bmatrix} \begin{bmatrix} t_1 \\ t_2 \end{bmatrix}.$$

Our theory then tells us that making the variable substitution

$$x_1 = (1/\sqrt{2})(-t_1 + t_2),$$
$$x_2 = (1/\sqrt{2})(t_1 + t_2),$$ (5)

in the form $3x_1^2 + 10x_1x_2 + 3x_2^2$ will give the diagonal form

$$\lambda_1 t_1^2 + \lambda_2 t_2^2 = -2t_1^2 + 8t_2^2.$$

This can, of course, be checked by actually substituting the expressions for x_1 and x_2 in Eqs. (5) into the form $3x_1^2 + 10x_1x_2 + 3x_2^2$. ■

In the preceding example, we might like to clear denominators in our substitution given in Eqs. (5) by using $\mathbf{x} = (\sqrt{2}C)\mathbf{t}$, so that $x_1 = -t_1 + t_2$ and $x_2 = t_1 + t_2$. Using the substitution $\mathbf{x} = kC\mathbf{t}$ with a quadratic form $\mathbf{x}^T A\mathbf{x}$, we obtain

$$\mathbf{x}^T A\mathbf{x} = (kC\mathbf{t})^T A(kC\mathbf{t}) = k^2(\mathbf{t}^T C^T A C\mathbf{t}) = k^2(\mathbf{t}^T D\mathbf{t}).$$

Thus, with $k = \sqrt{2}$, the substitution $x_1 = -t_1 + t_2$, $x_2 = t_1 + t_2$ in Example 1 results in the diagonal form

$$2(-2t_1^2 + 8t_2^2) = -4t_1^2 + 16t_2^2.$$

In any particular situation, we would have to balance the desire for arithmetic simplicity against the desire to have the new orthogonal basis actually be orthonormal.

As an orthogonal matrix, C has determinant ± 1. (See Exercise 22 of Section 5.3.) We state without proof the significance of the sign of $\det(C)$. If $\det(C) = 1$, then the new ordered orthonormal basis B given by the column vectors of C has the **same orientation** in $\mathbb{R}^n$ as the standard ordered basis E; while if $\det(C) = -1$, then B has the **opposite orientation** to E. In order for $B = (\mathbf{b}_1, \mathbf{b}_2)$ in $\mathbb{R}^2$ to have the same orientation as $(\mathbf{e}_1, \mathbf{e}_2)$, there must be a *rotation* of the plane, given by a matrix transformation $\mathbf{x} = C\mathbf{t}$, which carries both $\mathbf{e}_1$ to $\mathbf{b}_1$ and $\mathbf{e}_2$ to $\mathbf{b}_2$. The same interpretation in terms of rotation is true in $\mathbb{R}^3$, with the additional condition that $\mathbf{e}_3$ be carried to $\mathbf{b}_3$. To illustrate for $\mathbb{R}^2$, Figure 7.1(a) shows an ordered orthonormal basis $B = (\mathbf{b}_1, \mathbf{b}_2)$, where $\mathbf{b}_1 = (-1/\sqrt{2}, 1/\sqrt{2})$ and $\mathbf{b}_2 = (-1/\sqrt{2}, -1/\sqrt{2})$, having the same orientation as E. Notice that

$$\det(C) = \begin{vmatrix} -1/\sqrt{2} & -1/\sqrt{2} \\ 1/\sqrt{2} & -1/\sqrt{2} \end{vmatrix} = \tfrac{1}{2} + \tfrac{1}{2} = 1.$$
$$\quad\;\; \mathbf{b}_1 \qquad \mathbf{b}_2$$

Counterclockwise rotation through an angle of 135° carries E into B, preserving the order of the vectors. However, we see that the basis $(\mathbf{b}_1, \mathbf{b}_2) = ((0, -1), (-1, 0))$ shown in Figure 7.1(b) does not have the same orientation as E, and this time

$$\det(C) = \begin{vmatrix} 0 & -1 \\ -1 & 0 \end{vmatrix} = -1.$$
$$\quad\; \mathbf{b}_1 \quad\; \mathbf{b}_2$$

For any orthogonal matrix C having determinant -1, multiplication of any *single* column of C by -1 gives an orthogonal matrix with determinant 1. For the diagonalization we are considering, where the columns are normalized eigenvectors, multiplication by -1 still gives a normalized eigenvector. While there will be some sign

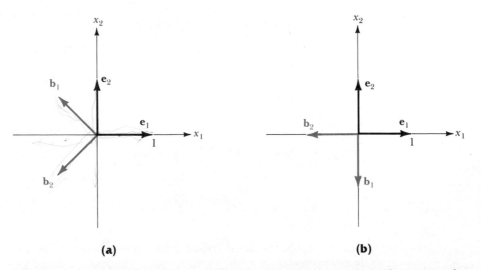

(a) **(b)**

FIGURE 7.1 **(a) Rotation of axes, hence $(\mathbf{b}_1, \mathbf{b}_2)$ and $(\mathbf{e}_1, \mathbf{e}_2)$ have the same orientation; (b) not a rotation of axes, hence, $(\mathbf{b}_1, \mathbf{b}_2)$ and $(\mathbf{e}_1, \mathbf{e}_2)$ have opposite orientation.**

changes in the diagonalizing substitution, the final diagonal form remains the same, because the eigenvalues are the same and their order has not been changed.

We summarize all our work in one main theorem, and then conclude with a final example. Section 7.2 will give an application of the theorem to geometry, and Section 7.3 will give an application to optimization.

THEOREM 7.1 Principal Axis Theorem

Every quadratic form $f(\mathbf{x})$ in n variables $x_1, x_2, \ldots, x_n$ can be diagonalized by a substitution $\mathbf{x} = C\mathbf{t}$, where C is an $n \times n$ orthogonal matrix. The diagonalized form appears as

$$\lambda_1 t_1^2 + \lambda_2 t_2^2 + \cdots + \lambda_n t_n^2,$$

where the λ_j are the eigenvalues of the symmetric coefficient matrix A of $f(\mathbf{x})$. The jth column vector of C is a normalized eigenvector $\mathbf{v}_j$ of A corresponding to λ_j. Moreover, C can be chosen so that $\det(C) = 1$.

We box a step-by-step outline for diagonalizing a quadratic form.

Diagonalizing a Quadratic Form $f(\mathbf{x})$

Step 1 Find the symmetric coefficient matrix A of the form $f(\mathbf{x})$.

Step 2 Find the (not necessarily distinct) eigenvalues $\lambda_1, \lambda_2, \ldots, \lambda_n$ of A.

Step 3 Find an orthonormal basis for $\mathbb{R}^n$ consisting of normalized eigenvectors of A.

Step 4 Form the matrix C, whose columns are the basis vectors found in step 3, in the order corresponding to the listing of eigenvalues in step 2. The transformation $\mathbf{x} = C\mathbf{t}$ is a **rotation** if $\det(C) = 1$. If a rotation is desired and $\det(C) = -1$, change the sign of all components of just one column vector in C.

Step 5 The substitution $\mathbf{x} = C\mathbf{t}$ transforms $f(\mathbf{x})$ to the diagonal form $\lambda_1 t_1^2 + \lambda_2 t_2^2 + \cdots + \lambda_n t_n^2$.

EXAMPLE 5 Find a variable substitution that diagonalizes the form $2xy + 2xz$, and give the resulting diagonal form.

Solution The symmetric coefficient matrix for the given quadratic form is

$$A = \begin{bmatrix} 0 & 1 & 1 \\ 1 & 0 & 0 \\ 1 & 0 & 0 \end{bmatrix}.$$

Expanding the determinant on the last column, we find that

$$|A - \lambda I| = \begin{vmatrix} -\lambda & 1 & 1 \\ 1 & -\lambda & 0 \\ 1 & 0 & -\lambda \end{vmatrix} = 1(\lambda) - \lambda(\lambda^2 - 1)$$

$$= -\lambda(-1 + \lambda^2 - 1) = -\lambda(\lambda^2 - 2).$$

The eigenvalues of A are then $\lambda_1 = 0$, $\lambda_2 = \sqrt{2}$, and $\lambda_3 = -\sqrt{2}$.

Computing eigenvectors, we have

$$A - \lambda_1 I = A = \begin{bmatrix} 0 & 1 & 1 \\ 1 & 0 & 0 \\ 1 & 0 & 0 \end{bmatrix} \sim \begin{bmatrix} 1 & 0 & 0 \\ 0 & 1 & 1 \\ 0 & 0 & 0 \end{bmatrix},$$

so $\mathbf{v}_1 = (0, -1, 1)$ is an eigenvector for $\lambda_1 = 0$. In addition,

$$A - \lambda_2 I = A - \sqrt{2}I = \begin{bmatrix} -\sqrt{2} & 1 & 1 \\ 1 & -\sqrt{2} & 0 \\ 1 & 0 & -\sqrt{2} \end{bmatrix} \sim \begin{bmatrix} 1 & -\sqrt{2} & 0 \\ -\sqrt{2} & 1 & 1 \\ 1 & 0 & -\sqrt{2} \end{bmatrix}$$

$$\sim \begin{bmatrix} 1 & -\sqrt{2} & 0 \\ 0 & -1 & 1 \\ 0 & \sqrt{2} & -\sqrt{2} \end{bmatrix} \sim \begin{bmatrix} 1 & -\sqrt{2} & 0 \\ 0 & 1 & -1 \\ 0 & 0 & 0 \end{bmatrix},$$

so $\mathbf{v}_2 = (\sqrt{2}, 1, 1)$ is an eigenvector for $\lambda_2 = \sqrt{2}$. In a similar fashion, we find that $\mathbf{v}_3 = (-\sqrt{2}, 1, 1)$ is an eigenvector for $\lambda_3 = -\sqrt{2}$. Thus,

$$C = \begin{bmatrix} 0 & \sqrt{2}/2 & -\sqrt{2}/2 \\ -1/\sqrt{2} & \frac{1}{2} & \frac{1}{2} \\ 1/\sqrt{2} & \frac{1}{2} & \frac{1}{2} \end{bmatrix}$$

is an orthogonal diagonalizing matrix for A. The substitution

$$x = \left(\tfrac{1}{\sqrt{2}}\right)(t_2 - t_3)$$

$$y = \tfrac{1}{2}(-\sqrt{2}t_1 + t_2 + t_3)$$

$$z = \tfrac{1}{2}(\sqrt{2}t_1 + t_2 + t_3)$$

will diagonalize $2xy + 2xz$ to become $\sqrt{2}t_2^2 - \sqrt{2}t_3^2$. ■

$$[t_1, t_2, t_3] \begin{bmatrix} 0 & 0 & 0 \\ 0 & \sqrt{2} & 0 \\ 0 & 0 & -\sqrt{2} \end{bmatrix} \begin{bmatrix} t_1 \\ t_2 \\ t_3 \end{bmatrix} = \sqrt{2}t_2^2 - \sqrt{2}t_3^2$$

SUMMARY

Let $\mathbf{x}$ *and* $\mathbf{t}$ *be* $n \times 1$ *column vectors with entries* x_i *and* t_i, *respectively.*

1. For every nonzero $n \times n$ matrix A, the product $f(\mathbf{x}) = \mathbf{x}^T A \mathbf{x}$ gives a quadratic form in the variables x_i.

2. Given any quadratic form $f(\mathbf{x})$ in the variables x_i, there exist an upper-triangular coefficient matrix U such that $f(\mathbf{x}) = \mathbf{x}^T U \mathbf{x}$ and symmetric coefficient matrix A such that $f(\mathbf{x}) = \mathbf{x}^T A \mathbf{x}$.

3. Any quadratic form $f(\mathbf{x})$ can be orthogonally diagonalized by using a variable substitution $\mathbf{x} = C\mathbf{t}$ in accordance with the boxed steps preceding Example 5.

EXERCISES

In Exercises 1–8, find the upper-triangular coefficient matrix U and the symmetric coefficient matrix A of the given quadratic form.

1. $3x^2 - 6xy + y^2$
2. $8x^2 + 9xy - 3y^2$
3. $x^2 - y^2 - 4xy + 3xz - 8yz$
4. $x_1^2 - 2x_2^2 + x_3^2 + 6x_4^2 - 2x_1x_4 + 6x_2x_4 - 8x_1x_3$
5. $[x \ \ y]\begin{bmatrix} -2 & 1 \\ 7 & 3 \end{bmatrix}\begin{bmatrix} x \\ y \end{bmatrix}$
6. $[x \ \ y]\begin{bmatrix} 7 & -10 \\ 15 & 2 \end{bmatrix}\begin{bmatrix} x \\ y \end{bmatrix}$
7. $[x \ \ y \ \ z]\begin{bmatrix} 8 & 3 & 1 \\ 2 & 1 & -4 \\ -5 & 2 & 10 \end{bmatrix}\begin{bmatrix} x \\ y \\ z \end{bmatrix}$
8. $[x_1 \ \ x_2 \ \ x_3 \ \ x_4]\begin{bmatrix} 2 & -1 & 3 & 0 \\ 4 & 2 & -1 & 3 \\ -6 & 3 & 0 & 7 \\ 10 & 2 & 1 & 5 \end{bmatrix}\begin{bmatrix} x_1 \\ x_2 \\ x_3 \\ x_4 \end{bmatrix}$

In Exercises 9–16, find an orthogonal substitution that diagonalizes the given quadratic form, and find the diagonalized form.

9. $2xy$
10. $3x^2 + 4xy$
11. $-6xy + 8y^2$
12. $x^2 + 2xy + y^2$
13. $3x^2 - 4xy + 3y^2$
14. $x_2^2 + 2x_1x_3$
15. $x_1^2 + x_2^2 + x_3^2 - 2x_1x_2 - 2x_1x_3 - 2x_2x_3$
   [SUGGESTION: Use Example 3 in Section 4.3.]
16. $x_1^2 + 2x_2^2 - 6x_1x_3 + x_3^2 - 4x_4^2$
17. Find a necessary and sufficient condition on a, b, and c such that the quadratic form $ax^2 + 2bxy + cy^2$ can be orthogonally diagonalized to kt^2.
18. Repeat Exercise 17, but require also that $k = 1$.

In Exercises 19–24, use MATCOMP or similar software to find a diagonal form into which the given form can be transformed by an orthogonal substitution. Do not give the substitution.

19. $3x^2 + 4xy - 5y^2$
20. $x^2 - 8xy + y^2$
21. $3x^2 + y^2 - 2z^2 - 4xy + 6yz$
22. $y^2 - 8z^2 + 3xy - 4xz + 7yz$
23. $x_1^2 - 3x_1x_4 + 5x_4^2 - 8x_2x_3$
24. $x_1^2 - 8x_1x_2 + 6x_2x_3 - 4x_3x_4$

7.2 APPLICATIONS TO GEOMETRY

Conic Sections in $\mathbb{R}^2$

Figure 7.2 shows three different types of plane curves obtained when a double right-circular cone is cut by a plane. These *conic sections* are the ellipse, the hyperbola, and the parabola. Figure 7.3(a) shows an ellipse in *standard position*, with center at the origin. The equation of this ellipse is

$$\frac{x^2}{a^2} + \frac{y^2}{b^2} = 1. \tag{1}$$

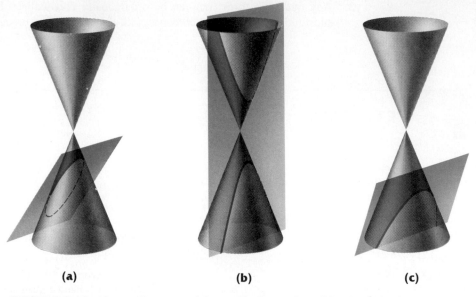

(a)   **(b)**   **(c)**

FIGURE 7.2   Sections of a cone: (a) an elliptic section; (b) a hyperbolic section; (c) a parabolic section.

The ellipse in Figure 7.3(b) with center (h, k) has equation

$$\frac{(x - h)^2}{a^2} + \frac{(y - k)^2}{b^2} = 1. \tag{2}$$

If we make the translation substitution $\bar{x} = x - h$, $\bar{y} = y - k$, then Eq. (2) becomes $\bar{x}^2/a^2 + \bar{y}^2/b^2 = 1$, which resembles Eq. (1).

   By completing the square, we can put any quadratic polynomial equation in x and y with no xy-term but with x^2 and y^2 having coefficients of the same sign into the form of Eq. (2), possibly with 0 or -1 on the right-hand side. The procedure should be familiar to you from work with circles. We give one illustration for an ellipse.

EXAMPLE 1   Complete the square in the equation

$$x^2 + 3y^2 - 4x + 6y = -1.$$

Solution   Completing the square, we obtain

$$(x^2 - 4x) + 3(y^2 + 2y) = -1$$
$$(x - 2)^2 + 3(y + 1)^2 = 4 + 3 - 1 = 6$$
$$\frac{(x - 2)^2}{6} + \frac{(y + 1)^2}{2} = 1,$$

which is of the form of Eq. (2). This is the equation of an ellipse with center $(h, k) = (2, -1)$. Setting $\bar{x} = x - 2$ and $\bar{y} = y + 1$, we obtain the equation

$$\frac{\bar{x}^2}{6} + \frac{\bar{y}^2}{2} = 1. \qquad \blacksquare$$

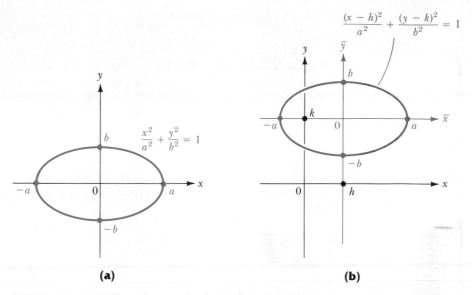

FIGURE 7.3   (a) Ellipse in standard position; (b) ellipse centered at (h, k).

If the constant on the right-hand side of the initial equation in Example 1 had been -7, we would have obtained $\bar{x}^2 + 3\bar{y}^2 = 0$, which describes the single point $(\bar{x}, \bar{y}) = (0, 0)$. This single point is regarded as a *degenerate* ellipse. If the constant had been -8, we would have obtained $\bar{x}^2 + 3\bar{y}^2 = -1$, which has no real solution; the ellipse is then called *empty*.

Thus, every polynomial equation

$$c_1 x^2 + c_2 y^2 + c_3 x + c_4 y = d, \qquad c_1 c_2 > 0, \tag{3}$$

describes an ellipse, possibly degenerate or empty.

ANALYTIC GEOMETRY is generally considered to have been founded by René Descartes (1596–1650) and Pierre Fermat (1601–1665) in the first half of the seventeenth century. But it was not until the appearance of a Latin version of Descartes' *Geometry* in 1661 by Frans van Schooten (1615–1660) that its influence began to be felt. This Latin version was published along with many commentaries; in particular, the *Elements of Curves* by Jan de Witt (1625–1672) gave a systematic treatment of conic sections. De Witt gave canonical forms of these equations similar to those in use today; for example, $y^2 = ax$, $by^2 + x^2 = f^2$, and $x^2 - by^2 = f^2$ represented the parabola, ellipse, and hyperbola, respectively. He then showed how, given an arbitrary second-degree equation in x and y, to find a transformation of axes that reduces the given equation to one of the canonical forms. This is, of course, equivalent to diagonalizing a particular symmetric matrix.

De Witt was a talented mathematician who, because of his family background, could devote but little time to mathematics. In 1653 he became in effect the prime minister of the Netherlands. Over the next decades, he guided the fortunes of the country through a most difficult period, including three wars with England. In 1672 the hostility of one of the Dutch factions culminated in his murder.

In a similar fashion, an equation

$$c_1 x^2 + c_2 y^2 + c_3 x + c_4 y = d, \qquad c_1 c_2 < 0, \tag{4}$$

describes a hyperbola. Notice that the coefficients of x^2 and y^2 have *opposite* signs. The standard form of the equation for a hyperbola centered at $(0, 0)$ is

$$\frac{x^2}{a^2} - \frac{y^2}{b^2} = 1 \qquad \text{or} \qquad \frac{-x^2}{a^2} + \frac{y^2}{b^2} = 1,$$

as shown in Figure 7.4. The dashed diagonal lines $y = \pm(b/a)x$ shown in the figure are the *asymptotes* of the hyperbola. By completing squares and translating axes, we can reduce Eq. (4) to one of the standard forms shown in Figure 7.4, in variables $\bar{x}$ and $\bar{y}$, unless the constant in the final equation reduces to zero. In that case we obtain

$$\frac{\bar{x}^2}{a^2} - \frac{\bar{y}^2}{b^2} = 0 \qquad \text{or} \qquad \bar{y} = \pm(b/a)\bar{x}.$$

These equations represent two lines that can be considered a degenerate hyperbola. Thus any equation of the form of Eq. (4) describes a (possibly degenerate) hyperbola.

Finally, the equations

$$c_1 x^2 + c_2 x + c_3 y = d \qquad \text{and} \qquad c_1 y^2 + c_2 x + c_3 y = d, \qquad c_1 \neq 0, \tag{5}$$

describe parabolas. If $c_3 \neq 0$ in the first equation in Eqs. (5) and $c_2 \neq 0$ in the second, these equations can be reduced to the form

$$\bar{x}^2 = a\bar{y} \qquad \text{and} \qquad \bar{y}^2 = a\bar{x} \tag{6}$$

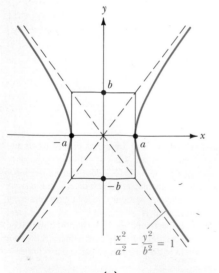

(a)

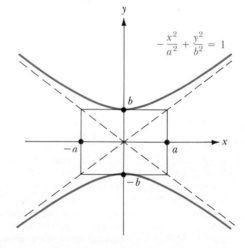

(b)

FIGURE 7.4   (a) The hyperbola $\dfrac{x^2}{a^2} - \dfrac{y^2}{b^2} = 1$; (b) the hyperbola $-\dfrac{x^2}{a^2} + \dfrac{y^2}{b^2} = 1$.

by completing the square and translating axes. Figure 7.5 shows two parabolas in standard position. If Eqs. (5) reduce to $c_1x^2 + c_2x = d$ and $c_1y^2 + c_3y = d$, each describes two parallel lines that can be considered degenerate parabolas.

In summary, every equation of the form

$$c_1x^2 + c_2y^2 + c_3x + c_4y - d \qquad (7)$$

with at least one of c_1 or c_2 nonzero describes a (possibly degenerate or empty) ellipse, hyperbola, or parabola.

Classification of Second-degree Curves

We can now apply our work in Section 7.1 on diagonalizing quadratic forms to classify the plane curves described by an equation of the type

$$ax^2 + bxy + cy^2 + dx + ey + f = 0 \qquad \text{for } a, b, c \text{ not all zero.} \qquad (8)$$

We make a substitution

$$\begin{bmatrix} x \\ y \end{bmatrix} = C\begin{bmatrix} t_1 \\ t_2 \end{bmatrix}, \qquad \text{where} \qquad \det(C) = 1, \qquad (9)$$

which orthogonally diagonalizes the quadratic-form part of Eq. (8), which is in color, and we obtain an equation of the form

$$\lambda_1 t_1^2 + \lambda_2 t_2^2 + gt_1 + ht_2 + k = 0. \qquad (10)$$

This equation has the form of Eq. (7) and describes an ellipse, hyperbola, or parabola. Remember that Eq. (9) corresponds to a rotation that carries the vector $\mathbf{e}_1$ to the first column vector $\mathbf{b}_1$ of C and carries $\mathbf{e}_2$ to the second column vector $\mathbf{b}_2$. We think of $\mathbf{b}_1$

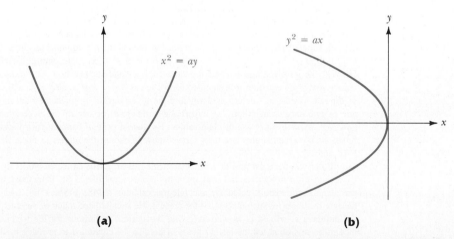

(a)   **(b)**

FIGURE 7.5   **(a) The parabola $x^2 = ay$, $a > 0$; (b) the parabola $y^2 = ax$, $a < 0$.**

as pointing out the t_1-axis and $\mathbf{b}_2$ as pointing out the t_2-axis. We summarize our work in a theorem and give just one illustration, leaving others to the exercises.

THEOREM 7.2 Classification of Second-degree Plane Curves

Every equation of the form of Eq. (8) can be reduced to an equation of the form of Eq. (10) by means of an orthogonal substitution corresponding to a rotation of the plane. The coefficients λ_1 and λ_2 in Eq. (10) are the eigenvalues of the symmetric coefficient matrix of the quadratic-form portion of Eq. (8). The curve describes a (possibly degenerate or empty)

$$
\begin{aligned}
\text{ellipse} \quad &\text{if } \lambda_1\lambda_2 > 0, \\
\text{hyperbola} \quad &\text{if } \lambda_1\lambda_2 < 0, \\
\text{parabola} \quad &\text{if } \lambda_1\lambda_2 = 0.
\end{aligned}
$$

EXAMPLE 2   Use rotation and translation of axes to sketch the plane curve $2xy + 2\sqrt{2}x = 1$.

Solution   The symmetric coefficient matrix of the quadratic form $2xy$ is

$$
A = \begin{bmatrix} 0 & 1 \\ 1 & 0 \end{bmatrix}.
$$

We easily find that the eigenvalues are $\lambda_1 = 1$ and $\lambda_2 = -1$, and that

$$
C = \begin{bmatrix} 1/\sqrt{2} & -1/\sqrt{2} \\ 1/\sqrt{2} & 1/\sqrt{2} \end{bmatrix}
$$

is an orthogonal diagonalizing matrix with determinant 1. The substitution

$$
\begin{aligned}
x &= (1/\sqrt{2})(t_1 - t_2) \\
y &= (1/\sqrt{2})(t_1 + t_2)
\end{aligned}
$$

then yields

$$
t_1{}^2 - t_2{}^2 + 2t_1 - 2t_2 = 1.
$$

THE EARLIEST CLASSIFICATION OF QUADRIC SURFACES was given by Leonhard Euler, in his precalculus text *Introduction to Infinitesimal Analysis* (1748). Euler's classification was similar to the conic-section classification of De Witt. Euler considered the second-degree equation in three variables $Ap^2 + Bq^2 + Cr^2 + Dpq + Epr + Fqr + Gp + Hq + Ir + K = 0$ as representing a surface in 3-space. As did De Witt, he gave canonical forms for these surfaces and showed how to rotate and translate the axes to reduce any given equation to a standard form such as $Ap^2 + Bq^2 + Cr^2 + K = 0$. An analysis of the signs of the new coefficients determined whether the given equation represented an ellipsoid, a hyperboloid of one or two sheets, an elliptic or hyperbolic paraboloid, or one of the degenerate cases. Euler did not, however, make explicit use of eigenvalues. He gave a general formula for rotation of axes in 3-space as functions of certain angles and then showed how to choose the angles to make the coefficients D, E, and F all zero.

Euler was the most prolific mathematician of all time. His collected works fill over 70 large volumes. Euler was born in Switzerland, but spent his professional life in St. Petersburg and Berlin. His texts on precalculus, differential calculus, and integral calculus (1748, 1755, 1768) had immense influence and became the bases for such texts up to the present. He standardized much of our current notation, introducing the numbers e, π, and i, as well as giving our current definitions for the trigonometric functions. Even though he was blind for the last 17 years of his life, he continued to produce mathematical papers almost up to the day he died.

Completing the square, we obtain

$$(t_1 + 1)^2 - (t_2 + 1)^2 = 1,$$

which describes the hyperbola shown in Figure 7.6. ■

Quadric Surfaces

An equation in three variables of the form

$$c_1 x^2 + c_2 y^2 + c_3 z^2 + c_4 x + c_5 y + c_6 z = d, \tag{11}$$

where at least one of c_1, c_2, or c_3 is nonzero, describes a *quadric surface* in space, which again might be degenerate or empty. Figures 7.7 through 7.15 show some of the quadric surfaces in standard position.

By completing the square in Eq. (11), which corresponds to translating axes, we see that Eq. (11) can be reduced to an equation involving $\bar{x}$, $\bar{y}$, and $\bar{z}$ in which a variable appearing to the second power does not appear to the first power. Notice that this is true for all the equations in Figures 7.7 through 7.15.

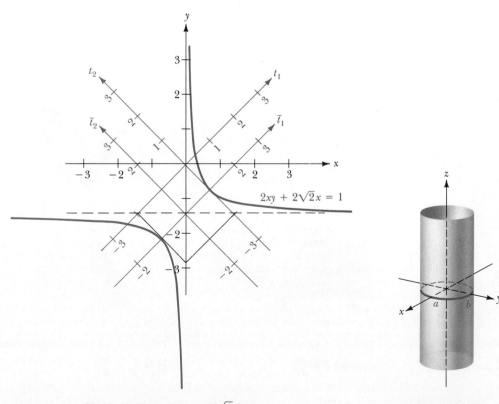

FIGURE 7.6 **The hyperbola $2xy + 2\sqrt{2}x = 1$.**

FIGURE 7.7 **The elliptic cylinder**
$$\frac{x^2}{a^2} + \frac{y^2}{b^2} = 1.$$

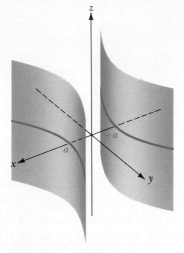

FIGURE 7.8 **The hyperbolic cylinder** $\dfrac{x^2}{a^2} - \dfrac{y^2}{b^2} = 1.$

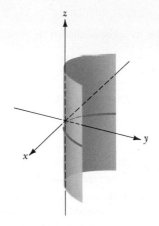

FIGURE 7.9 **The parabolic cylinder** $ay = x^2.$

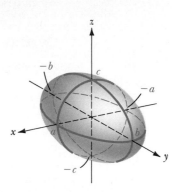

FIGURE 7.10 **The ellipsoid** $\dfrac{x^2}{a^2} + \dfrac{y^2}{b^2} + \dfrac{z^2}{c^2} = 1.$

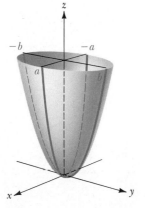

FIGURE 7.11 **The elliptic paraboloid** $z = \dfrac{x^2}{a^2} + \dfrac{y^2}{b^2}.$

FIGURE 7.12 **The hyperbolic paraboloid** $z = \dfrac{y^2}{b^2} - \dfrac{x^2}{a^2}.$

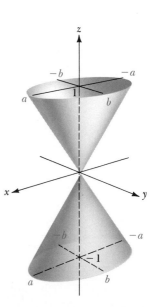

FIGURE 7.13 **The elliptic cone** $z^2 = \dfrac{x^2}{a^2} + \dfrac{y^2}{b^2}.$

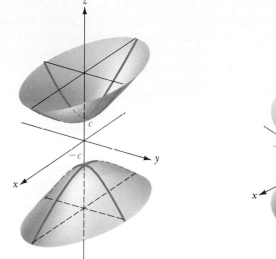

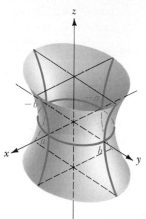

FIGURE 7.14 **The hyperboloid of two sheets** $\frac{z^2}{c^2} - 1 = \frac{x^2}{a^2} + \frac{y^2}{b^2}.$

FIGURE 7.15 **The hyperboloid of one sheet** $\frac{z^2}{c^2} + 1 = \frac{x^2}{a^2} + \frac{y^2}{b^2}.$

Again, degenerate and empty cases are possible. For example, the equation $x^2 + 2y^2 + z^2 = -4$ gives an empty ellipsoid. The elliptic cone, hyperboloid of two sheets, and hyperboloid of one sheet in Figures 7.13 through 7.15 differ only in whether a constant in their equations is zero, negative, or positive.

Now consider a general second-degree polynomial equation

$$ax^2 + by^2 + cz^2 + dxy + exz + fyz + px + qy + rz + s = 0, \qquad (12)$$

where the coefficient of at least one term of degree 2 is nonzero. Making a substitution

$$\begin{bmatrix} x \\ y \\ z \end{bmatrix} = C \begin{bmatrix} t_1 \\ t_2 \\ t_3 \end{bmatrix}, \qquad \text{where} \qquad \det(C) = 1, \qquad (13)$$

that orthogonally diagonalizes the quadratic-form portion of Eq. (12) (the portion in color), we obtain

$$\lambda_1 t_1^2 + \lambda_2 t_2^2 + \lambda_3 t_3^2 + p't_1 + q't_2 + r't_3 + s' = 0, \qquad (14)$$

which is of the form of Eq. (11). We state this as a theorem.

THEOREM 7.3 **Principal Axis Theorem for $\mathbb{R}^3$**

Every equation of the form of Eq. (12) can be reduced to an equation of the form of Eq. (14) by an orthogonal substitution (13) that corresponds to a rotation of axes.

Again, computation of the eigenvalues λ_1, λ_2, and λ_3 of the symmetric coefficient matrix of the quadratic-form portion of Eq. (12) may give us quite a bit of information as to the type of quadric surface. However, actually executing the substitution and completing squares, which can be tedious, may be necessary to distinguish among certain surfaces. We give a rough classification scheme in Table 7.1. Remember that empty or degenerate cases are possible, even where we do not explicitly give them.

TABLE 7.1

Eigenvalues of λ_1, λ_2, λ_3	Quadric Surface
All of the same sign	Ellipsoid
Two of one sign and one of the other sign	Elliptic cone, hyperboloid of two sheets or hyperboloid of one sheet
One zero, two of the same sign	Elliptic paraboloid or elliptic cylinder (degenerate case)
One zero, two of opposite signs	Hyperbolic paraboloid or hyperbolic cylinder (degenerate case)
Two zero, one nonzero	Parabolic cylinder or two parallel planes (degenerate case)

If you try to verify Table 7.1, you will wonder about an equation of the form $ax^2 + by + cz = d$ in the last case given. Exercise 11 indicates that, by means of a rotation of axes, this equation can be reduced to $at_1^2 + rt_2 = d$, which can be written as $at_1^2 + r(t_2 - d/r) = 0$, and consequently describes a parabolic cylinder. We conclude with four examples.

EXAMPLE 3   Classify the quadric surface $2xy + 2xz = 1$.

Solution   Example 5 of Section 7.1 shows that the orthogonal substitution

$$x = \left(\tfrac{1}{\sqrt{2}}\right)(t_2 - t_3)$$
$$y = \tfrac{1}{2}(-\sqrt{2}t_1 + t_2 + t_3)$$
$$z = \tfrac{1}{2}(\sqrt{2}t_1 + t_2 + t_3)$$

transforms $2xy + 2xz$ into $\sqrt{2}t_2^2 - \sqrt{2}t_3^2$. It can be checked that the matrix C corresponding to this substitution has determinant 1. Thus, the substitution corresponds to a rotation of axes and transforms the equation $2xy + 2xz = 1$ into $\sqrt{2}t_2^2 - \sqrt{2}t_3^2 = 1$, which we recognize as giving a hyperbolic cylinder.   ■

EXAMPLE 4   Classify the quadric surface $2xy + 2xz = y + 1$.

Solution   Using the same substitution as we did in Example 3, we obtain the equation

$$\sqrt{2}t_2^2 - \sqrt{2}t_3^2 = (1/2)(-\sqrt{2}t_1 + t_2 + t_3) + 1.$$

Translation of axes by completing squares yields an equation of the form

$$\sqrt{2}(t_2 - h)^2 - \sqrt{2}(t_3 - k)^2 = -(1/\sqrt{2})(t_1 - r),$$

which we recognize as a hyperbolic paraboloid. ▪

EXAMPLE 5 Classify the quadric surface $2xy + 2xz = x + 1$.

Solution Using again the substitution in Example 3, we obtain

$$\sqrt{2}t_2^2 - \sqrt{2}t_3^2 = (1/\sqrt{2})(t_2 - t_3) + 1$$

or

$$2t_2^2 - t_2 - 2t_3^2 + t_3 = \sqrt{2}.$$

Completing squares yields

$$2(t_2 - 1/4)^2 - 2(t_3 - 1/4)^2 = \sqrt{2},$$

which represents a hyperbolic cylinder. ▪

EXAMPLE 6 Classify the quadric surface

$$2x^2 - 3y^2 + z^2 - 2xy + 4yz - 6x + 8y - 8z = 17$$

as far as possible, by finding just the eigenvalues of the symmetric coefficient matrix of the quadratic-form portion of the equation.

Solution The symmetric matrix of the quadratic-form portion is

$$A = \begin{bmatrix} 2 & -1 & 0 \\ -1 & -3 & 2 \\ 0 & 2 & 1 \end{bmatrix}.$$

We find that

$$\det(A - \lambda I) = \begin{vmatrix} 2 - \lambda & -1 & 0 \\ -1 & -3 - \lambda & 2 \\ 0 & 2 & 1 - \lambda \end{vmatrix}$$

$$= (2 - \lambda)\begin{vmatrix} -3 - \lambda & 2 \\ 2 & 1 - \lambda \end{vmatrix} + \begin{vmatrix} -1 & 2 \\ 0 & 1 - \lambda \end{vmatrix}$$

$$= (2 - \lambda)(\lambda^2 + 2\lambda - 7) + \lambda - 1$$

$$= -\lambda^3 + 12\lambda - 15.$$

We can see that $\lambda = 0$ is not a solution of the characteristic equation $-\lambda^3 + 12\lambda - 15 = 0$. We could plot a rough sketch of the graph of $y = -\lambda^3 + 12\lambda - 15$, just to determine the signs of the eigenvalues. However, we prefer to use the available software program MATCOMP with the matrix A. We quickly find that the eigenvalues are approximately,

$$\lambda_1 = -3.9720, \qquad \lambda_2 = 1.5765, \qquad \lambda_3 = 2.3954.$$

According to Table 7.1, we have an elliptic cone, a hyperboloid of two sheets, or a hyperboloid of one sheet. ▪

SUMMARY

1. Given an equation $ax^2 + bxy + cy^2 + dx + ey + f = 0$, let A be the symmetric coefficient matrix of the quadratic-form portion of the equation (the portion in color), let λ_1 and λ_2 be the eigenvalues of A, and let C be an orthogonal matrix of determinant 1 with eigenvectors of A for columns. The substitution corresponding to matrix C followed by translation of axes reduces the given equation to a standard form for the equation of a conic section. In particular, the equation describes a (possibly degenerate or empty)

        ellipse   if $\lambda_1\lambda_2 > 0$,

        hyperbola if $\lambda_1\lambda_2 < 0$,

        parabola   if $\lambda_1\lambda_2 = 0$.

2. Proceeding in a way analogous to that described in summary item (1) but for the equation

   $$ax^2 + by^2 + cz^2 + dxy + exz + fyz + px + qy + rz + s = 0,$$

   one obtains a standard form for the equation of a (possibly degenerate or empty) quadric surface in space. Table 7.1 lists the information that can be obtained from the three eigenvalues λ_1, λ_2, and λ_3 alone.

EXERCISES

In Exercises 1–8, rotate axes, using a substitution

$$\begin{bmatrix} x \\ y \end{bmatrix} = C \begin{bmatrix} t_1 \\ t_2 \end{bmatrix},$$

complete squares if necessary, and sketch the graph (if it is not empty) of the given conic section.

1. $2xy = 1$

2. $2xy - 2\sqrt{2}y = 1$

3. $x^2 + 2xy + y^2 = 4$

4. $x^2 - 2xy + y^2 + 4\sqrt{2}x = 4$

5. $10x^2 + 6xy + 2y^2 = 4$

6. $5x^2 + 4xy + 2y^2 = -1$

7. $3x^2 + 4xy + 6y^2 = 8$

8. $x^2 + 8xy + 7y^2 + 18\sqrt{5}x = -9$

9. Show that the plane curve $ax^2 + bxy + cy^2 + dx + ey + f = 0$ is a (possibly degenerate or empty)

        ellipse   if $b^2 - 4ac < 0$,

        hyperbola if $b^2 - 4ac > 0$,

        parabola   if $b^2 - 4ac = 0$.

   [HINT: Diagonalize $ax^2 + bxy + cy^2$, using the quadratic formula, and check the signs of the eigenvalues.]

10. Use Exercise 9 to classify the conic section with the given equation.

   a) $2x^2 + 8xy + 8y^2 - 3x + 2y = 13$

   b) $y^2 + 4xy - 5x^2 - 3x = 12$

   c) $-x^2 + 5xy - 7y^2 - 4y + 11 = 0$

   d) $xy + 4x - 3y = 8$

   e) $2x^2 - 3xy + y^2 - 8x + 5y = 30$

   f) $x^2 + 6xy + 9y^2 - 2x + 14y = 10$

   g) $4x^2 - 2xy - 3y^2 + 8x - 5y = 17$

   h) $8x^2 + 6xy + 2y^2 - 5x = 25$

   i) $x^2 - 2xy + 4x - 5y = 6$

   j) $2x^2 - 3xy + 2y^2 - 8y = 15$

11. Show that the equation $ax^2 + by + cz = d$ can be transformed into an equation of the form

$at_1^2 + rt_2 = d$ by a rotation axes in space.
[HINT:

$$\text{If} \quad \begin{bmatrix} x \\ y \\ z \end{bmatrix} = C \begin{bmatrix} t_1 \\ t_2 \\ t_3 \end{bmatrix}, \quad \text{then}$$

$$\begin{bmatrix} t_1 \\ t_2 \\ t_3 \end{bmatrix} = C^{-1} \begin{bmatrix} x \\ y \\ z \end{bmatrix} = C^T \begin{bmatrix} x \\ y \\ z \end{bmatrix}.$$

Find an orthogonal matrix C^T such that
$\det(C^T) = 1$, $t_1 = x$, and $t_2 = (by + cz)/r$
for some r.]

*In Exercises 12–20, classify the quadric surface with
the given equation as one of the (possibly empty or
degenerate) types illustrated in Figures 7.7–7.15.*

12. $2x^2 + 2y^2 + 6yz + 10z^2 = 9$

13. $2xy + z^2 + 1 = 0$

14. $2xz + y^2 + 2y + 1 = 0$

15. $x^2 + 2yz + 4x + 1 = 0$

16. $3x^2 + 2y^2 + 6xz + 3z^2 = 1$

17. $x^2 - 8xy + 16y^2 - 3z^2 = 8$

18. $x^2 + 4y^2 - 4xz + 4z^2 = 8$

19. $-3x^2 + 2y^2 + 8yz + 16z^2 = 10$

20. $x^2 + y^2 + z^2 - 2xy + 2xz - 2yz = 9$

*In Exercises 21–27, use MATCOMP or similar
software to classify the quadric surface according to
Table 7.1.*

21. $x^2 + y^2 + z^2 - 2xy - 2xz - 2yz + 3x - 3z = 8$

22. $3x^2 + 2y^2 + 5z^2 + 4xy + 2yz - 3x + 10y = 4$

23. $2x^2 - 8y^2 + 3z^2 - 4xy + yz + 6xz - 3x - 8z = 3$

24. $3x^2 + 7y^2 + 4z^2 + 6xy - 8yz + 16x = 20$

25. $x^2 + 4y^2 + 16z^2 + 4xy + 8xz + 16yz - 8x + 3y = 8$

26. $x^2 + 6y^2 + 4z^2 - xy - 2xz - 3yz - 9x = 20$

27. $4x^2 - 3y^2 + z^2 + 8xz + 6yz + 2x - 3y = 8$

7.3 APPLICATIONS TO EXTREMA

Finding Extrema of Functions

We turn now to one of the applications of linear algebra to calculus. We simply state the facts from calculus that we need, trying to make them seem reasonable where possible.

There are many situations in which one desires to maximize or minimize a function of one or more variables. Applications involve maximizing profit, minimizing costs, maximizing speed, minimizing time, and so on. Such problems are of great practical importance.

Polynomial functions are especially easy to work with. Many important functions, such as trigonometric, exponential, and logarithmic functions, can't be expressed by a polynomial formula. However, each of those functions, and many others, can be expressed near a point in the domain of the function as a "polynomial" of infinite degree—an *infinite series*, as it is called. For example, it is shown in calculus that

$$\cos x = 1 - \left(\frac{1}{2!}\right)x^2 + \left(\frac{1}{4!}\right)x^4 - \left(\frac{1}{6!}\right)x^6 + \cdots \tag{1}$$

for any number x. [Recall that $2! = 2 \cdot 1$, $3! = 3 \cdot 2 \cdot 1$, and in general, $n! = n(n-1)(n-2) \cdots 3 \cdot 2 \cdot 1$.] We leave to calculus the discussion of the interpretation of the infinite sum

$$1 - \left(\frac{1}{2!}\right) + \left(\frac{1}{4!}\right) - \left(\frac{1}{6!}\right) + \cdots$$

as cos 1. From Eq. (1), it would seem that we should have

$$\cos(2x - y) =$$

$$1 - \left(\frac{1}{2!}\right)(2x - y)^2 + \left(\frac{1}{4!}\right)(2x - y)^4 - \left(\frac{1}{6!}\right)(2x - y)^6 + \cdots. \quad (2)$$

Notice that the term

$$-\left(\frac{1}{2!}\right)(2x - y)^2 = -\tfrac{1}{2}(4x^2 - 4xy + y^2)$$

is a quadratic form—that is, a form (homogeneous polynomial) of degree 2. Similarly,

$$\left(\frac{1}{4!}\right)(2x - y)^4 = \tfrac{1}{24}(16x^4 - 32x^3y + 24x^2y^2 - 8xy^3 + y^4)$$

is a form of degree 4.

Consider a function $g(x_1, x_2, \ldots, x_n)$ of n variables, which we denote as usual by $g(\mathbf{x})$. For many of the most common functions $g(\mathbf{x})$, it can be shown that, if $\mathbf{x}$ is near $\mathbf{0}$, so that all x_i are small in magnitude, then

$$g(\mathbf{x}) = c + f_1(\mathbf{x}) + f_2(\mathbf{x}) + f_3(\mathbf{x}) + \cdots + f_n(\mathbf{x}) + \cdots, \quad (3)$$

where each $f_i(\mathbf{x})$ is a form of degree i or is zero. Equation (2) illustrates this: forms of odd degree do not appear.

We wish to determine whether $g(\mathbf{x})$ in Eq. (3) has a *local extremum*—that is, a *local maximum* or *local minimum* at the origin $\mathbf{x} = \mathbf{0}$. A function $g(\mathbf{x})$ has a **local maximum at 0** if $g(\mathbf{x}) \leq g(\mathbf{0})$ for all $\mathbf{x}$ where $\|\mathbf{x}\|$ is sufficiently small. The notion of a **local minimum at 0** for $g(\mathbf{x})$ is analogously defined. For example, the function of one variable $g(x) = 1 + x^2$, whose graph is shown in Figure 7.16, has a local minimum of 1 at $x = 0$.

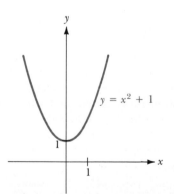

FIGURE 7.16 The graph of y = g(x) = x² + 1.

Now if $\mathbf{x}$ is really close to $\mathbf{0}$, all the forms $f_1(\mathbf{x}), f_2(\mathbf{x}), f_3(\mathbf{x}), \ldots$ in Eq. (3) have very small values, so the constant c, if $c \neq 0$, is the dominant term of Eq. (3) near zero. Notice that, for $g(x) = 1 + x^2$ in Figure 7.16, the function has values close to the constant 1 for x close to zero.

After a nonzero constant, the form in Eq. (3) that contributes the most to $g(\mathbf{x})$ for $\|\mathbf{x}\|$ close to $\mathbf{0}$ is the nonzero form $f_i(\mathbf{x})$ of lowest degree, because the *lower* the degree of a term of a polynomial, the *more* it contributes when the variables are all near zero. For example, x is greater than x^2 near zero; if $x = \frac{1}{100}$, then x^2 is only $\frac{1}{10,000}$. If $f_1(\mathbf{x}) = 0$ and $f_2(\mathbf{x}) \neq 0$, then $f_2(\mathbf{x})$ is the dominant form of Eq. (3) after a nonzero constant c, and so on.

We claim that, if $f_1(\mathbf{x}) \neq 0$ in Eq. (3), then $g(\mathbf{x})$ does not have a local maximum or minimum at $\mathbf{x} = \mathbf{0}$. Suppose that

$$f_1(\mathbf{x}) = d_1x_1 + d_2x_2 + \cdots + d_nx_n,$$

with some coefficient—say, d_1—nonzero. If k is small but of the same sign as d_1, then near the point $\mathbf{a} = (k, 0, 0, \ldots, 0)$ the function $g(\mathbf{x}) \approx c + d_1k > c$. On the other hand, if k is small but of opposite sign to d_1, then near this point $g(\mathbf{x}) < c$. Thus, $g(\mathbf{x})$ can have a local maximum or minimum of c at $\mathbf{0}$ only if $f_1(\mathbf{x}) = 0$ for all $\mathbf{x}$.

If $f_1(\mathbf{x}) = 0$ and $f_2(\mathbf{x}) \neq 0$ in Eq. (3), then, for $\mathbf{x}$ near $\mathbf{0}$, $f_2(\mathbf{x})$ is the dominant form in the equation; it can be expected to dominate the terms of higher degree for $\mathbf{x}$ close to $\mathbf{0}$. It seems reasonable that, if $f_2(\mathbf{x}) > 0$ for *all* $\mathbf{x} \neq \mathbf{0}$ but near $\mathbf{0}$, then for such $\mathbf{x}$

$$g(\mathbf{x}) = c + (\text{Little bit}),$$

and $g(\mathbf{x})$ has a local minimum of c at $\mathbf{0}$. On the other hand, if $f_2(\mathbf{x}) < 0$ for *all* such $\mathbf{x}$, then we expect that

$$g(\mathbf{x}) = c - (\text{Little bit})$$

for these values $\mathbf{x}$, and $g(\mathbf{x})$ has a local maximum of c at $\mathbf{0}$. This is proved in an advanced calculus course.

We know that we can orthogonally diagonalize the form $f_2(\mathbf{x})$ with a substitution $\mathbf{x} = C\mathbf{t}$ to become

$$\lambda_1t_1^2 + \lambda_2t_2^2 + \cdots + \lambda_nt_n^2, \tag{4}$$

where the λ_i are the eigenvalues of the symmetric coefficient matrix of $f_2(\mathbf{x})$. Form (4) is > 0 for *all* nonzero $\mathbf{t}$, and hence $f_2(\mathbf{x}) > 0$ for *all* nonzero $\mathbf{x}$, if and only if we have all $\lambda_i > 0$. Similarly, $f_2(\mathbf{x}) < 0$ for *all* nonzero $\mathbf{x}$ if and only if *all* $\lambda_i < 0$. It is also clear that, if some λ_i are positive and some are negative, then form (4) and hence $f_2(\mathbf{x})$ assume both positive and negative values arbitrarily close to zero.

DEFINITION 7.1 Definite Forms

A quadratic form $f(\mathbf{x})$ is **positive definite** if $f(\mathbf{x}) > 0$ for all nonzero $\mathbf{x}$ in $\mathbb{R}^n$, and it is **negative definite** if $f(\mathbf{x}) < 0$ for all such nonzero $\mathbf{x}$.

Our work in Section 7.1 and the preceding statement give us the following theorem and corollary.

THEOREM 7.4 Definite Forms

> A quadratic form is positive definite if and only if all the eigenvalues of its symmetric coefficient matrix are positive, and it is negative definite if and only if all those eigenvalues are negative.

COROLLARY   A Test for Local Extrema

Let $g(\mathbf{x})$ be a function of n variables given by Eq. (3). Suppose that the function $f_1(\mathbf{x})$ in Eq. (3) is zero. If $f_2(\mathbf{x})$ is positive definite, then $g(\mathbf{x})$ has a local minimum of c at $\mathbf{x} = \mathbf{0}$, whereas if $f_2(\mathbf{x})$ is negative definite, then $g(\mathbf{x})$ has a local maximum of c at $\mathbf{x} = \mathbf{0}$. If $f_2(\mathbf{x})$ assumes both positive and negative values, then $g(\mathbf{x})$ has no local extremum at $\mathbf{x} = \mathbf{0}$.

We have discussed local extrema only at the origin. To do similar work at another point $\mathbf{h} = (h_1, h_2, \ldots, h_n)$, we need only translate axes to this point, letting $\bar{x}_i = x_i - h_i$ so that Eq. (3) becomes

$$g(\bar{\mathbf{x}}) = c + f_1(\bar{\mathbf{x}}) + f_2(\bar{\mathbf{x}}) + \cdots + f_n(\bar{\mathbf{x}}) + \cdots. \tag{5}$$

Our discussion indicates a method for attempting to find local extrema of a function, which we box.

Finding an Extremum of $g(x)$

Step 1 Find a point $\mathbf{h}$ where the function $f_1(\bar{\mathbf{x}}) = f_1(\mathbf{x} - \mathbf{h})$ in Eq. (5) becomes the zero function. (This is the province of calculus.)

Step 2 Find the quadratic form $f_2(\bar{\mathbf{x}})$ at the point $\mathbf{h}$. (This also requires calculus.)

Step 3 Find the eigenvalues of the symmetric coefficient matrix of the quadratic form.

Step 4 If all eigenvalues are positive, then $g(\bar{\mathbf{x}})$ has a local minimum of c at $\mathbf{h}$. If all are negative, then $g(\bar{\mathbf{x}})$ has a local maximum of c at $\mathbf{h}$. If eigenvalues of both signs occur, then $g(\bar{\mathbf{x}})$ has no local extremum at $\mathbf{h}$.

Step 5 If any of the eigenvalues is zero, further study is necessary.

Exercises 17 through 22 illustrate some of the things that can occur if step 5 is the case. We shall not tackle steps 1 or 2, since they require calculus, but will simply start with equations of the form of Eqs. (3) or (5) that meet the requirements in steps 1 and 2.

EXAMPLE 1   Let

$$g(x, y) = 3 + (2x^2 - 4xy + 4y^2) + (x^3 + 4xy^2 + y^3).$$

Determine whether $g(x, y)$ has a local extremum at the origin.

Solution   The symmetric coefficient matrix for the quadratic-form portion $2x^2 - 4xy + 4y^2$ of $g(x, y)$ is

$$A = \begin{bmatrix} 2 & -2 \\ -2 & 4 \end{bmatrix}.$$

We find that

$$|A - \lambda I| = \begin{vmatrix} 2 - \lambda & -2 \\ -2 & 4 - \lambda \end{vmatrix} = \lambda^2 - 6\lambda + 4.$$

The solutions of the characteristic equation $\lambda^2 - 6\lambda + 4 = 0$ are found by the quadratic formula to be

$$\lambda = \frac{6 \pm \sqrt{36 - 16}}{2} = 3 \pm \sqrt{5}.$$

We see that $\lambda_1 = 3 + \sqrt{5}$ and $\lambda_2 = 3 - \sqrt{5}$ are both positive, so our form is positive definite. Thus, $g(x, y)$ has a minimum of 3 at $(0, 0)$.   ▪

EXAMPLE 2   Suppose that

$$g(x, y, z) = 7 + (2x^2 - 8y^2 + 3z^2 - 4xy + 2yz + 6xz) + (xz^2 - 5y^3)$$
$$+ \text{(higher-degree terms)}.$$

Determine whether $g(x, y, z)$ has a local extremum at the origin.

Solution   The symmetric coefficient matrix of the quadratic-form portion is

$$A = \begin{bmatrix} 2 & -2 & 3 \\ -2 & -8 & 1 \\ 3 & 1 & 3 \end{bmatrix}.$$

We could find $p(\lambda) = |A - \lambda I|$ and attempt to solve the characteristic equation, or we could try to sketch the graph of $p(\lambda)$ well enough to determine the signs of the eigenvalues. However, we prefer to use the available software MATCOMP, or similar software, and we find that the eigenvalues are approximately

$$\lambda_1 = -8.605, \qquad \lambda_2 = 0.042, \qquad \lambda_3 = 5.563.$$

Since both positive and negative eigenvalues appear, there is no local extremum at the origin.   ▪

Maximizing or Minimizing a Quadratic Form on the Unit Sphere

Let $f(\mathbf{x})$ be a quadratic form in the variables $x_1, x_2, \ldots, x_n$. We consider the problem of finding the maximum and minimum values of $f(\mathbf{x})$ for $\mathbf{x}$ on the *unit sphere*, where

$\|\mathbf{x}\| = 1$, that is, where $x_1^2 + x_2^2 + \cdots + x_n^2 = 1$. It is shown in advanced calculus that such extrema of $f(\mathbf{x})$ on the unit sphere always exist.

We need only orthogonally diagonalize the form $f(\mathbf{x})$, using an orthogonal transformation $\mathbf{x} = C\mathbf{t}$, obtaining as usual

$$\lambda_1 t_1^2 + \lambda_2 t_2^2 + \cdots + \lambda_n t_n^2. \tag{6}$$

Since our new basis for $\mathbb{R}^n$ is again orthonormal, the unit sphere has $\mathbf{t}$-equation

$$t_1^2 + t_2^2 + \cdots + t_n^2 = 1; \tag{7}$$

this is a very important point. Suppose that the λ_i are arranged so that

$$\lambda_1 \geq \lambda_2 \geq \cdots \geq \lambda_n.$$

On the unit sphere with Eq. (7), we see that formula (6) can be written as

$$\lambda_1(1 - t_2^2 - \cdots - t_n^2) + \lambda_2 t_2^2 + \cdots + \lambda_n t_n^2$$
$$= \lambda_1 - (\lambda_1 - \lambda_2)t_2^2 - \cdots - (\lambda_1 - \lambda_n)t_n^2. \tag{8}$$

Since $\lambda_1 - \lambda_i \geq 0$ for $i > 1$, we see that the maximum value assumed by formula (8) is λ_1 when $t_2 = t_3 = \cdots = t_n = 0$, and $t_1 = \pm 1$. Exercise 32 indicates similarly that the minimum value assumed by form (6) is λ_n, when $t_n = \pm 1$ and all other $t_i = 0$. We state this as a theorem.

THEOREM 7.5 Extrema of a Quadratic Form on the Unit Sphere

Let $f(\mathbf{x})$ be a quadratic form, and let $\lambda_1, \lambda_2, \ldots, \lambda_n$ be the eigenvalues of the symmetric coefficient matrix of $f(\mathbf{x})$. The maximum value assumed by $f(\mathbf{x})$ on the unit sphere $\|\mathbf{x}\| = 1$ is the maximum of the λ_i, and the minimum value assumed is the minimum of the λ_i. Each extremum is assumed at any eigenvector of length 1 corresponding to the eigenvalue that gives the extremum.

The preceding theorem is very important in vibration applications ranging from aerodynamics to particle physics. In such applications, one often needs to know the eigenvalue of maximum or of minimum magnitude for a symmetric matrix. These eigenvalues are frequently found by using advanced-calculus techniques to maximize or minimize the value of a quadratic form on a unit sphere, rather than using algebraic techniques like those presented in this text for finding eigenvalues. The principal axis theorem (Theorem 7.1) and the preceding theorem are the algebraic foundation for such an analytic approach. We illustrate the preceding theorem with an example that maximizes a quadratic form on a unit sphere by finding the eigenvalues, rather than illustrating the more important reverse procedure.

EXAMPLE 3   Find the maximum and minimum values assumed by $2xy + 2xz$ on the unit sphere $x^2 + y^2 + z^2 = 1$, and find all points where these extrema are assumed.

Solution   From Example 5 of Section 7.1, we see that the eigenvalues of the symmetric coefficient matrix A and an associated eigenvector are given by

$$\lambda_1 = 0, \qquad \mathbf{v}_1 = (0, -1, 1),$$
$$\lambda_2 = \sqrt{2}, \qquad \mathbf{v}_2 = (\sqrt{2}, 1, 1),$$
$$\lambda_3 = -\sqrt{2}, \qquad \mathbf{v}_3 = (-\sqrt{2}, 1, 1).$$

We see that the maximum value assumed by $2xy + 2xz$ on the unit sphere is $\sqrt{2}$, and it is assumed at the points $\pm\left(\sqrt{2}/2, \frac{1}{2}, \frac{1}{2}\right)$. The minimum value assumed is $-\sqrt{2}$, and it is assumed at the points $\pm\left(-\sqrt{2}/2, \frac{1}{2}, \frac{1}{2}\right)$. Notice that we normalized our eigenvectors to length 1 so that they lie on the unit sphere. ■

The extension of Theorem 7.5 to a sphere centered at the origin, with radius other than 1, is left as Exercise 33.

SUMMARY

1. Let $g(\mathbf{x}) = c + f_1(\mathbf{x}) + f_2(\mathbf{x}) + \cdots + f_n(\mathbf{x}) + \cdots$ near $\mathbf{x} = \mathbf{0}$, where $f_i(\mathbf{x})$ is a form of degree i or is zero.

   a) If $f_1(\mathbf{x}) = 0$ and $f_2(\mathbf{x})$ is positive definite, then $g(\mathbf{x})$ has a local minimum of c at $\mathbf{x} = \mathbf{0}$.

   b) If $f_1(\mathbf{x}) = 0$ and $f_2(\mathbf{x})$ is negative definite, then $g(\mathbf{x})$ has a local maximum of c at $\mathbf{x} = \mathbf{0}$.

   c) If $f_1(\mathbf{x}) \neq 0$ or if the symmetric coefficient matrix of $f_2(\mathbf{x})$ has both positive and negative eigenvalues, then $g(\mathbf{x})$ has no local extremum at $\mathbf{x} = \mathbf{0}$.

2. The natural analogue of summary item (1) holds at $\mathbf{x} = \mathbf{h}$; just translate the axes to the point $\mathbf{h}$, and replace x_i by $\bar{x}_i = x_i - h_i$.

3. A quadratic form in n variables has as maximum (minimum) value on the unit sphere $\|\mathbf{x}\| = 1$ in $\mathbb{R}^n$ the maximum (minimum) of the eigenvalues of the symmetric coefficient matrix of the form. The maximum (minimum) is assumed at each corresponding eigenvector of length 1.

EXERCISES

In Exercises 1–15, assume that $g(\mathbf{x})$, or $g(\bar{\mathbf{x}})$, is described by the given formula for values of $\mathbf{x}$, or $\bar{\mathbf{x}}$, near zero. Draw whatever conclusions are possible concerning local extrema of the function g.

1. $g(x, y) = -7 + (3x^2 - 6xy + 4y^2) + (x^3 - 4y^3)$

2. $g(x, y) = 8 - (2x^2 - 8xy + 3y^2) + (2x^2 y - y^3)$

3. $g(x, y) = 4 - 3x + (2x^2 - 2xy + y^2) + (2x^2y + y^3) + \cdots$

4. $g(x, y) = 5 - (8x^2 - 6xy + 2y^2) + (4x^3 - xy^2) + \cdots$

5. $g(\bar{x}, \bar{y}) = 3 - (4\bar{x}^2 - 8\bar{x}\bar{y} + 5\bar{y}^2) + (2\bar{x}^2\bar{y} - \bar{y}^3) + \cdots, \bar{x} = x + 5, \bar{y} = y$

6. $g(x, y) = -2 + (8x^2 + 4xy + y^2) + (x^3 + 5x^2y)$

7. $g(x, y) = 5 + (3x^2 + 10xy + 7y^2) + (7xy^2 - y^3)$

8. $g(x, y) = 4 - (x^2 - 6xy + 9y^2) + (x^3 - y^3) + \cdots$

9. $g(\bar{x}, \bar{y}) = 3 + (2\bar{x}^2 + 8\bar{x}\bar{y} + 8\bar{y}^2) + (4\bar{x}^3 - \bar{x}\bar{y}^2) + \cdots, \bar{x} = x - 3, \bar{y} = y - 1$

10. $g(\bar{x}, \bar{y}) = 4 - (\bar{x}^2 + 3\bar{x}\bar{y} - \bar{y}^2) + (4\bar{x}^2\bar{y} - 5\bar{x}\bar{y}^2), \bar{x} = x + 1, \bar{y} = y - 7$

11. $g(x, y, z) = 4 - (x^2 + 4xy + 5y^2 + 3z^2) + (x^3 - xyz)$

12. $g(x, y, z) = 3 + (2x^2 + 6xz - y^2 + 5z^2) + (x^2z - y^2z) + \cdots$

13. $g(x, y, z) = 5 + (4x^2 + 2xy + y^2 - z^2) + (xy^2 - 4xyz) + \cdots$

14. $g(\bar{x}, \bar{y}, \bar{z}) = 4 + (\bar{x}^2 + \bar{y}^2 + \bar{z}^2 - 2\bar{x}\bar{z} - 2\bar{x}\bar{y} - 2\bar{y}\bar{z}) + (3\bar{x}^3 - \bar{z}^3) + \cdots, \bar{x} = x + 1, \bar{y} = y - 2, \bar{z} = z + 5$

15. $g(\bar{x}, \bar{y}, \bar{z}) = 4 - 3\bar{z} + (2\bar{x}^2 - 2\bar{x}\bar{y} + 3\bar{x}\bar{z} + 5\bar{y}^2 + \bar{z}^2) + (2\bar{x}\bar{y}\bar{z} - \bar{z}^3) + \cdots, \bar{x} = x - 7, \bar{y} = y + 6, \bar{z} = z$

16. Define the notion of a local minimum c of a function $f(\mathbf{x})$ of n variables at a point $\mathbf{h}$.

In Exercises 17–22, let

$$g(x, y) = c + f_1(x, y) + f_2(x, y) + f_3(x, y) + \cdots$$

in accordance with the notation we have been using. Let λ_1 and λ_2 be the eigenvalues of the symmetric coefficient matrix of $f_2(x, y)$.

17. Give an example of a polynomial function $g(x, y)$ for which $f_1(x, y) = 0$, $\lambda_1 = 0$, $|\lambda_2| = 1$, and $g(x, y)$ has a local minimum of 10 at the origin.

18. Repeat Exercise 17, but make $g(x, y)$ have a local maximum of -5 at the origin.

19. Repeat Exercise 17, but make $g(x, y)$ have no local extremum at the origin.

20. Give an example of a polynomial function $g(x, y)$ such that $f_1(x, y) = f_2(x, y) = 0$, having a local maximum of 1 at the origin.

21. Repeat Exercise 20, but make $g(x, y)$ have a local minimum of 40 at the origin.

22. Repeat Exercise 20, but make $g(x, y)$ have no local extremum at the origin.

In Exercises 23–31, find the maximum and the minimum values of the quadratic form on the unit circle in $\mathbb{R}^2$, or unit sphere in $\mathbb{R}^n$, and find all points on the unit circle or sphere where the extrema are assumed.

23. xy, $n = 2$

24. $3x^2 + 4xy$, $n = 2$

25. $-6xy + 8y^2$, $n = 2$

26. $x^2 + 2xy + y^2$, $n = 3$

27. $3x^2 - 6xy + 3y^2$, $n = 2$

28. $y^2 + 2xz$, $n = 3$

29. $x^2 + y^2 + z^2 - 2xy + 2xz - 2yz$, $n = 3$

30. $x^2 + y^2 + z^2 - 2xy - 2xz - 2yz$, $n = 3$
    [SUGGESTION: Use Example 4 in Section 5.3.]

31. $x^2 + w^2 + 4yz - 2xw$, $n = 4$

32. Show that the minimum value assumed by a quadratic form in n variables on the unit sphere in $\mathbb{R}^n$ is the minimum eigenvalue of the symmetric coefficient matrix of the form, and show that this minimum value is assumed at any corresponding eigenvector of length 1.

33. Let $f(\mathbf{x})$ be a quadratic form in n variables. Describe the maximum and minimum values assumed by $f(\mathbf{x})$ on the sphere $\|\mathbf{x}\| = a^2$ for any $a > 0$. Describe the points of the sphere at which these extrema are assumed. [HINT: For any k in $\mathbb{R}$, how does $f(k\mathbf{x})$ compare with $f(\mathbf{x})$?]

⚠ *In Exercises 34–38, use MATCOMP or similar software, and follow the directions for Exercises 1–15.*

34. $g(x, y, z) = 7 + (3x^2 + 2y^2 + 5z^2 + 4xy + 2yz) + (xyz - x^2y)$

35. $g(x, y, z) = 5 - (x^2 + 6y^2 + 4z^2 - xy - 2xz - 3yz) + (xz^2 - 5z^3) + \cdots$

36. $g(x, y, z) = -1 - (2x^2 - 8y^2 + 3z^2 - 4xy + yz + 6xz) + (8x^3 - 4y^2z + 3z^3)$

37. $g(x, y, z) = 7 + (x^2 + 2y^2 + z^2 - 3xy - 4xz - 3yz) + (x^2y - 4yz^2)$

38. $g(x, y, z) = 5 + (x^2 + 4y^2 + 16z^2 + 4xy + 8xz + 16yz) + (x^3 + y^3 + 7xyz) + \cdots$

7.4 COMPUTING EIGENVALUES AND EIGENVECTORS

Computing the eigenvalues of a matrix is one of the toughest jobs in linear algebra. Many algorithms have been developed, but no one method can be considered the best for all cases. We have used the characteristic equation for computation of eigenvalues in most of our examples and exercises. The available program MATCOMP also uses it. Professionals frown on the use of this method in practical applications because small errors made in computing coefficients of the characteristic polynomial can lead to significant errors in eigenvalues. We describe three other methods in this section.

1. The *power method* is especially useful if one wants only the eigenvalue of largest (or of smallest) magnitude, as in many vibration problems.

2. *Jacobi's method* for symmetric matrices is presented without proof. This method is chiefly of historical interest, for the third and most recent of the methods we describe is more general and usually more efficient.

3. The *QR method* was developed by H. Rutishauser in 1958 and J. G. F. Francis in 1961 and is probably the most widely used method today. It finds all eigenvalues, both real and complex, of a real matrix. Details are beyond the scope of this text. We give only a rough idea of the method, with no proofs.

The available programs POWER, JACOBI, and QRFACTOR can be used to illustrate the three methods.

The Power Method

Let A be an $n \times n$ diagonalizable matrix with real eigenvalues. Suppose that one eigenvalue—say, λ_1—has greater magnitude than all the others. That is, $|\lambda_1| > |\lambda_i|$ for $i > 1$. We call λ_1 the **dominant eigenvalue** of A. In many vibration problems, we are interested only in computing this dominant eigenvalue and the eigenvalue of minimum absolute value.

Since A is diagonalizable, there exists a basis

$$\{\mathbf{b}_1, \mathbf{b}_2, \ldots, \mathbf{b}_n\}$$

for $\mathbb{R}^n$ composed of eigenvectors of A. We assume that $\mathbf{b}_i$ is the eigenvector corresponding to λ_i, and that the numbering is such that $|\lambda_1| > |\lambda_2| \geq |\lambda_3| \geq \cdots \geq |\lambda_n|$.

Let $\mathbf{w}_1$ be any nonzero vector in $\mathbb{R}^n$. Then

$$\mathbf{w}_1 = c_1\mathbf{b}_1 + c_2\mathbf{b}_2 + \cdots + c_n\mathbf{b}_n \tag{1}$$

for some constants c_i in $\mathbb{R}$. Applying A^s to both sides of Eq. (1) and remembering that $A^s\mathbf{b}_i = \lambda_i^s\mathbf{b}_i$, we see that

$$A^s\mathbf{w}_1 = \lambda_1^s c_1\mathbf{b}_1 + \lambda_2^s c_2\mathbf{b}_2 + \cdots + \lambda_n^s c_n\mathbf{b}_n. \tag{2}$$

Since λ_1 is dominant, we see that, for large s, the summand $\lambda_1^s c_1\mathbf{b}_1$ dominates the right-hand side of Eq. (2), as long as $c_1 \neq 0$. This is even more evident if we rewrite Eq. (2) in the form

$$A^s\mathbf{w}_1 = \lambda_1^s(c_1\mathbf{b}_1 + (\lambda_2/\lambda_1)^s c_2\mathbf{b}_2 + \cdots + (\lambda_n/\lambda_1)^s c_n\mathbf{b}_n). \tag{3}$$

If s is large, the quotients $(\lambda_i/\lambda_1)^s$ for $i > 1$ are close to zero, because $|\lambda_i/\lambda_1| < 1$. Thus, if $c_1 \neq 0$ and s is large enough, $A^s\mathbf{w}_1$ is very nearly parallel to $\lambda_1{}^s c_1\mathbf{b}_1$, which is an eigenvector of A corresponding to the eigenvalue λ_1. This suggests that we can approximate an eigenvector of A corresponding to the dominant eigenvalue λ_1 by multiplying an appropriate initial approximation vector $\mathbf{w}_1$ repeatedly by A.

A few comments are in order. In a practical application, we may have a rough idea of an eigenvector for λ_1 and be able to choose a reasonably good first approximation $\mathbf{w}_1$. In any case, $\mathbf{w}_1$ *should not* be in the subspace of $\mathbb{R}^n$ generated by the eigenvectors corresponding to the λ_j for $j > 1$. This corresponds to the requirement that $c_1 \neq 0$.

Repeated multiplication of $\mathbf{w}_1$ by A may produce very large (or very small) numbers. It is customary to scale after each multiplication, to keep the components of the vectors at a reasonable size. After the first multiplication, we find the maximum d_1 of the magnitudes of all the components of $A\mathbf{w}_1$ and apply A the next time to the vector $\mathbf{w}_2 = (1/d_1)A\mathbf{w}_1$. Similarly, we let $\mathbf{w}_3 = (1/d_2)A\mathbf{w}_2$, where d_2 is the maximum of the magnitudes of components of $A\mathbf{w}_2$, and so on. Thus we are always multiplying A times a vector $\mathbf{w}_j$ with components of maximum magnitude 1. This scaling also aids us in estimating the number-of-significant-figures accuracy we have attained in the components of our approximations to an eigenvector.

If $\mathbf{x}$ is an eigenvector corresponding to λ_1, then

$$\frac{A\mathbf{x} \cdot \mathbf{x}}{\mathbf{x} \cdot \mathbf{x}} = \frac{\lambda_1\mathbf{x} \cdot \mathbf{x}}{\mathbf{x} \cdot \mathbf{x}} = \lambda_1. \tag{4}$$

The quotient $(A\mathbf{x} \cdot \mathbf{x})/(\mathbf{x} \cdot \mathbf{x})$ is called a **Rayleigh quotient**. As we compute the $\mathbf{w}_j$, the Rayleigh quotients $(A\mathbf{w}_j \cdot \mathbf{w}_j)/(\mathbf{w}_j \cdot \mathbf{w}_j)$ should approach λ_1.

This *power method* for finding the dominant eigenvector should, mathematically, break down if we choose the initial approximation $\mathbf{w}_1$ in Eq. (1) in such a way that the coefficient c_1 of $\mathbf{b}_1$ is zero. However, due to roundoff error, it often happens that a nonzero component of $\mathbf{b}_1$ creeps into the $\mathbf{w}_j$ as they are computed, and the $\mathbf{w}_j$ then start swinging toward an eigenvector for λ_1 as desired. This is one case where roundoff error is helpful!

Equation (3) indicates that the ratio $|\lambda_2/\lambda_1|$, which is the maximum of the magnitudes $|\lambda_i/\lambda_1|$ for $i > 1$, should control the speed of convergence of the $\mathbf{w}_j$ to an eigenvector. If $|\lambda_2/\lambda_1|$ is close to 1, covergence may be quite slow.

We summarize the steps of the power method in the following box.

THE RAYLEIGH QUOTIENT is named for John William Strutt, the third Baron Rayleigh (1842–1919). Rayleigh was a hereditary peer who surprised his family by pursuing a scientific career instead of contenting himself with the life of a country gentleman. He set up a laboratory at the family seat in Terling Place, Essex, and spent most of his life there pursuing his research into many aspects of physics—in particular, sound and optics. He is especially famous for his resolution of the long-standing question in optics as to why the sky is blue, as well as for his codiscovery of the element argon, for which he won the Nobel prize in 1904. When he received the British Order of Merit in 1902 he said that "the only merit of which he personally was conscious was that of having pleased himself by his studies, and any results that may have been due to his researches were owing to the fact that it had been a pleasure to him to become a physicist."

Rayleigh used the Rayleigh quotient early in his career in an 1873 work in which he needed to evaluate approximately the normal modes of a complex vibrating system. He subsequently used it and related methods in his classic text *The Theory of Sound* (1877).

> ### The Power Method for Finding the Dominant Eigenvalue λ_1 of A
>
> **Step 1** Choose an appropriate vector $\mathbf{w}_1$ in $\mathbb{R}^n$ as first approximation to an eigenvector corresponding to λ_1.
>
> **Step 2** Compute $A\mathbf{w}_1$ and the Rayleigh quotient $(A\mathbf{w}_1 \cdot \mathbf{w}_1)/(\mathbf{w}_1 \cdot \mathbf{w}_1)$.
>
> **Step 3** Let $\mathbf{w}_2 = (1/d_1)A\mathbf{w}_1$, where d_1 is the maximum of the magnitudes of components of $A\mathbf{w}_1$.
>
> **Step 4** Repeat step 2, with all subscripts increased by 1. The Rayleigh quotients should approach λ_1, and the $\mathbf{w}_j$ should approach an eigenvector of A corresponding to λ_1.

EXAMPLE 1 Illustrate the power method for the matrix

$$A = \begin{bmatrix} 3 & -2 \\ -2 & 0 \end{bmatrix} \qquad \text{starting with} \qquad \mathbf{w}_1 = \begin{bmatrix} 1 \\ -1 \end{bmatrix}.$$

finding $\mathbf{w}_2$, $\mathbf{w}_3$, and the first two Rayleigh quotients.

Solution We have

$$A\mathbf{w}_1 = \begin{bmatrix} 3 & -2 \\ -2 & 0 \end{bmatrix}\begin{bmatrix} 1 \\ -1 \end{bmatrix} = \begin{bmatrix} 5 \\ -2 \end{bmatrix}.$$

We find as first Rayleigh quotient

$$(A\mathbf{w}_1 \cdot \mathbf{w}_1)/(\mathbf{w}_1 \cdot \mathbf{w}_1) = \tfrac{7}{2} = 3.5.$$

Since 5 is the maximum magnitude of a component of $A\mathbf{w}_1$, we have

$$\mathbf{w}_2 = \tfrac{1}{5}A\mathbf{w}_1 = \begin{bmatrix} 1 \\ -\tfrac{2}{5} \end{bmatrix}.$$

Then

$$A\mathbf{w}_2 = \begin{bmatrix} 3 & -2 \\ -2 & 0 \end{bmatrix}\begin{bmatrix} 1 \\ -\tfrac{2}{5} \end{bmatrix} = \begin{bmatrix} \tfrac{19}{5} \\ -2 \end{bmatrix}.$$

The next Rayleigh quotient is

$$(A\mathbf{w}_2 \cdot \mathbf{w}_2)/(\mathbf{w}_2 \cdot \mathbf{w}_2) = \tfrac{23/5}{29/25} = \tfrac{115}{29} \approx 3.966.$$

Finally,

$$\mathbf{w}_3 = \tfrac{5}{19}\begin{bmatrix} \tfrac{19}{5} \\ -2 \end{bmatrix} = \begin{bmatrix} 1 \\ -\tfrac{10}{19} \end{bmatrix}. \qquad ■$$

EXAMPLE 2 Use the available program POWER or similar software with the data in Example 1, and give the vectors $\mathbf{w}_j$ and the Rayleigh quotients until stabilization to all decimal places printed occurs.

Solution POWER does all its computations in double-precision arithmetic but prints the vectors $\mathbf{w}_j$ in single-precision to save space. It prints the Rayleigh quotients in double-precision. The data obtained using POWER are shown in Table 7.2. It is easy to solve the characteristic equation $\lambda^2 - 3\lambda - 4 = 0$ of A, and to see that the eigenvalues are really $\lambda_1 = 4$ and $\lambda_2 = -1$. POWER found the dominant eigenvalue 4, and it shows that $(1, -0.5)$ is an eigenvector for this eigenvalue. ■

If A has an eigenvalue of nonzero magnitude *smaller* than that of any other eigenvalue, the power method can be used with A^{-1} to find this smallest eigenvalue. The eigenvalues of A^{-1} are the reciprocals of the eigenvalues of A, and the eigenvectors are the same. To illustrate, if 5 is the dominant eigenvalue of A^{-1} and $\mathbf{v}$ is an associated eigenvector, then $\frac{1}{5}$ is the eigenvalue of A of smallest magnitude, and $\mathbf{v}$ is still an associated eigenvector.

TABLE 7.2 Power Method for $A = \begin{bmatrix} 3 & -2 \\ -2 & 0 \end{bmatrix}$

Vector Approximations	Rayleigh Quotients
$(1, -1)$	
$(1, -.4)$	3.5
$(1, -.5263158)$	3.96551724137931
$(1, -.4935065)$	3.997830802603037
$(1, -.5016286)$	3.999864369998644
$(1, -.4995932)$	3.999991522909338
$(1, -.5001018)$	3.999999470180991
$(1, -.4999746)$	3.999999966886309
$(1, -.5000064)$	3.999999997930394
$(1, -.4999984)$	3.99999999987065
$(1, -.5000004)$	3.999999999991916
$(1, -.4999999)$	3.999999999999495
$(1, -.5)$	3.999999999999968
$(1, -.5)$	3.999999999999998
$(1, -.5)$	4
$(1, -.5)$	4

Deflation for Symmetric Matrices

The method of deflation gives a way to compute eigenvalues of intermediate magnitude of a *symmetric* matrix by the power method. It is based on an interesting decomposition of a matrix product AB. Let A be an $m \times n$ matrix, and let B be an $n \times s$ matrix. We write AB symbolically as

$$AB = \begin{bmatrix} | & | & & | \\ \mathbf{c}_1 & \mathbf{c}_2 & \cdots & \mathbf{c}_n \\ | & | & & | \end{bmatrix} \begin{bmatrix} \rule{1cm}{0.4pt} & \mathbf{r}_1 & \rule{1cm}{0.4pt} \\ \rule{1cm}{0.4pt} & \mathbf{r}_2 & \rule{1cm}{0.4pt} \\ & \vdots & \\ \rule{1cm}{0.4pt} & \mathbf{r}_n & \rule{1cm}{0.4pt} \end{bmatrix},$$

where $\mathbf{c}_j$ is the jth column vector of A and $\mathbf{r}_i$ is the ith row vector of B. We claim that

$$AB = \mathbf{c}_1\mathbf{r}_1 + \mathbf{c}_2\mathbf{r}_2 + \cdots + \mathbf{c}_n\mathbf{r}_n, \tag{5}$$

where each $\mathbf{c}_i\mathbf{r}_i$ is the product of an $m \times 1$ matrix with a $1 \times s$ matrix. To see this, remember that the entry in the ith row and jth column of AB is

$$a_{i1}b_{1j} + a_{i2}b_{2j} + \cdots + a_{in}b_{nj}.$$

But $\mathbf{c}_1\mathbf{r}_1$ contributes precisely $a_{i1}b_{1j}$ to the ith row and jth column of the sum in Eq. (5), while $\mathbf{c}_2\mathbf{r}_2$ contributes $a_{i2}b_{2j}$, and so on. This establishes Eq. (5).

Let A be an $n \times n$ *symmetric* matrix with eigenvalues $\lambda_1, \lambda_2, \ldots, \lambda_n$, where

$$|\lambda_1| \geq |\lambda_2| \geq \cdots \geq |\lambda_n|.$$

We know from Section 5.3 that there exists an orthogonal matrix C such that $C^{-1}AC = D$, where D is a diagonal matrix, with $d_{ii} = \lambda_i$. Then

$$A = CDC^{-1} = CDC^T$$

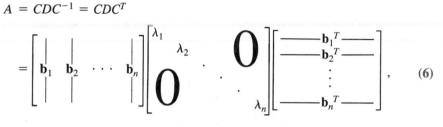

$$\tag{6}$$

where $\{\mathbf{b}_1, \mathbf{b}_2, \ldots, \mathbf{b}_n\}$ is an orthonormal basis for $\mathbb{R}^n$ consisting of eigenvectors of A and forming the column vectors of C. From Eq. (6), we see that

$$A = \begin{bmatrix} | & | & & | \\ \mathbf{b}_1 & \mathbf{b}_2 & \cdots & \mathbf{b}_n \\ | & | & & | \end{bmatrix} \begin{bmatrix} \text{---} \lambda_1\mathbf{b}_1{}^T \text{---} \\ \text{---} \lambda_2\mathbf{b}_2{}^T \text{---} \\ \vdots \\ \text{---} \lambda_n\mathbf{b}_n{}^T \text{---} \end{bmatrix}. \tag{7}$$

Using Eq. (5) and writing the $\mathbf{b}_i$ as column vectors, we see that A can be written in the following form:

Spectral Decomposition of A

$$A = \lambda_1\mathbf{b}_1\mathbf{b}_1{}^T + \lambda_2\mathbf{b}_2\mathbf{b}_2{}^T + \cdots + \lambda_n\mathbf{b}_n\mathbf{b}_n{}^T. \tag{8}$$

Equation (8) is known as the **spectral theorem** for symmetric matrices.

EXAMPLE 3 Illustrate the spectral theorem for the matrix

$$A = \begin{bmatrix} 3 & -2 \\ -2 & 0 \end{bmatrix}$$

of Example 1.

Solution Computation shows that the matrix A has eigenvalues $\lambda_1 = 4$ and $\lambda_2 = -1$, with corresponding eigenvectors

$$\mathbf{v}_1 - \begin{bmatrix} 1 \\ -\frac{1}{2} \end{bmatrix} \quad \text{and} \quad \mathbf{v}_2 = \begin{bmatrix} \frac{1}{2} \\ 1 \end{bmatrix}.$$

Eigenvectors $\mathbf{b}_1$ and $\mathbf{b}_2$ forming an orthonormal basis are

$$\mathbf{b}_1 = \begin{bmatrix} 2/\sqrt{5} \\ -1/\sqrt{5} \end{bmatrix} \quad \text{and} \quad \mathbf{b}_2 = \begin{bmatrix} 1/\sqrt{5} \\ 2/\sqrt{5} \end{bmatrix}.$$

We have

$$\lambda_1 \mathbf{b}_1 \mathbf{b}_1^T + \lambda_2 \mathbf{b}_2 \mathbf{b}_2^T = 4 \begin{bmatrix} 2/\sqrt{5} \\ -1/\sqrt{5} \end{bmatrix} [2/\sqrt{5} \quad -1/\sqrt{5}] - \begin{bmatrix} 1/\sqrt{5} \\ 2/\sqrt{5} \end{bmatrix} [1/\sqrt{5} \quad 2/\sqrt{5}]$$

$$= 4 \begin{bmatrix} \frac{4}{5} & -\frac{2}{5} \\ -\frac{2}{5} & \frac{1}{5} \end{bmatrix} - 1 \begin{bmatrix} \frac{1}{5} & \frac{2}{5} \\ \frac{2}{5} & \frac{4}{5} \end{bmatrix}$$

$$= \begin{bmatrix} \frac{16}{5} & -\frac{8}{5} \\ -\frac{8}{5} & \frac{4}{5} \end{bmatrix} - \begin{bmatrix} \frac{1}{5} & \frac{2}{5} \\ \frac{2}{5} & \frac{4}{5} \end{bmatrix} = \begin{bmatrix} 3 & -2 \\ -2 & 0 \end{bmatrix} = A. \qquad ■$$

Suppose now that we have found the eigenvalue λ_1 of maximum magnitude of A, and a corresponding eigenvector $\mathbf{v}_1$ by the power method. We compute the unit vector $\mathbf{b}_1 = \mathbf{v}_1/\|\mathbf{v}_1\|$. From Eq. (8), we see that

$$A - \lambda_1 \mathbf{b}_1 \mathbf{b}_1^T = 0\mathbf{b}_1 \mathbf{b}_1^T + \lambda_2 \mathbf{b}_2 \mathbf{b}_2^T + \cdots + \lambda_n \mathbf{b}_n \mathbf{b}_n^T \tag{9}$$

is a matrix with eigenvalues $\lambda_2, \lambda_3, \ldots, \lambda_n, 0$ in order of descending magnitude, and with corresponding eigenvectors $\mathbf{b}_2, \mathbf{b}_3, \ldots, \mathbf{b}_n, \mathbf{b}_1$. We can now use the power method on this matrix to compute λ_2 and $\mathbf{b}_2$. We then execute this *deflation* again, forming $A - \lambda_1 \mathbf{b}_1 \mathbf{b}_1^T - \lambda_2 \mathbf{b}_2 \mathbf{b}_2^T$ to find λ_3 and $\mathbf{b}_3$, and so on.

The available program POWER has an option to use this method of deflation. For the symmetric matrices of the small size that we use in our examples and exercises, POWER handles deflation well, provided that we compute each eigenvalue until stabilization to all places shown on the screen is achieved. In practice, scientists are wary of using deflation to find more than one or two further eigenvalues, since any error

---

THE TERM *SPECTRUM* was coined around 1905 by David Hilbert (1862–1943) for use in dealing with the eigenvalues of quadratic forms in infinitely many variables. The notion of such forms came out of his study of certain linear operators in spaces of functions. Hilbert was struck by the analogy one could make between these operators and quadratic forms in finitely many variables. Hilbert's approach was greatly expanded and generalized during the next decade by Erhard Schmidt, Frigyes Riesz (1880–1956), and Hermann Weyl. Interestingly enough, in the 1920s physicists called on spectra of certain linear operators to explain optical spectra.

David Hilbert was the most influential mathematician of the early twentieth century. He made major contributions in many fields, including algebraic forms, algebraic number theory, integral equations, the foundations of geometry, theoretical physics, and the foundations of mathematics. His speech at the 2nd International Congress of Mathematicians in Paris in 1900, outlining the important mathematical problems of the day, proved extremely significant in providing the direction for twentieth-century mathematics.

made in computation of an eigenvalue or eigenvector will propagate errors in the computation of subsequent ones.

EXAMPLE 4   Illustrate Eq. (9) for deflation with the matrix

$$A = \begin{bmatrix} 3 & -2 \\ -2 & 0 \end{bmatrix}$$

of Example 3.

Solution   From Example 3, we have

$$A = \lambda_1 \mathbf{b}_1 \mathbf{b}_1{}^T + \lambda_2 \mathbf{b}_2 \mathbf{b}_2{}^T = 4 \begin{bmatrix} \frac{4}{5} & -\frac{2}{5} \\ -\frac{2}{5} & \frac{4}{5} \end{bmatrix} - \begin{bmatrix} \frac{1}{5} & \frac{2}{5} \\ \frac{2}{5} & \frac{4}{5} \end{bmatrix}.$$

Thus,

$$A - \lambda_1 \mathbf{b}_1 \mathbf{b}_1{}^T = -\begin{bmatrix} \frac{1}{5} & \frac{2}{5} \\ \frac{2}{5} & \frac{4}{5} \end{bmatrix} = \begin{bmatrix} -\frac{1}{5} & -\frac{2}{5} \\ -\frac{2}{5} & -\frac{4}{5} \end{bmatrix}.$$

The characteristic polynomial for this matrix is

$$\begin{vmatrix} -\frac{1}{5} - \lambda & -\frac{2}{5} \\ -\frac{2}{5} & -\frac{4}{5} - \lambda \end{vmatrix} = (\lambda^2 + \lambda) = \lambda(\lambda + 1),$$

and the eigenvalues are indeed $\lambda_2 = -1$ and 0, as claimed after Eq. (9).   ■

EXAMPLE 5   Use the available program POWER or similar software to find the eigenvalues and eigenvectors, using deflation, for the symmetric matrix

$$\begin{bmatrix} 2 & -8 & 5 \\ -8 & 0 & 10 \\ 5 & 10 & -6 \end{bmatrix}.$$

Solution   Using POWER with deflation and finding eigenvalues as accurately as the printing on the screen permits, we obtain the eigenvectors and eigenvalues

$$\mathbf{v}_1 = (-.6042427, -.8477272, 1), \quad \lambda_1 = -17.49848531152027,$$

$$\mathbf{v}_2 = (-.760632, 1, .3881208), \quad \lambda_2 = \quad 9.96626448890372,$$

$$\mathbf{v}_3 = (1, .3958651, .9398283), \quad \lambda_3 = \quad 3.532220822616553.$$

Using MATCOMP as a check, we find the same eigenvalues and eigenvectors.   ■

Jacobi's Method for Symmetric Matrices

We present Jacobi's method for diagonalizing a symmetric matrix, omitting proofs. Let $A = [a_{ij}]$ be an $n \times n$ symmetric matrix, and suppose that a_{pq} is an entry of maximum magnitude among the entries of A that lie above the main diagonal. For example, in the matrix

$$A = \begin{bmatrix} 2 & -8 & 5 \\ -8 & 0 & 10 \\ 5 & 10 & -6 \end{bmatrix}, \tag{10}$$

the entry above the diagonal having maximum magnitude is 10. Then form the 2×2 matrix

$$\begin{bmatrix} a_{pp} & a_{pq} \\ a_{qp} & a_{qq} \end{bmatrix}. \tag{11}$$

From matrix (10), we would form the matrix consisting of the portion shown in color— namely,

$$\begin{bmatrix} 0 & 10 \\ 10 & -6 \end{bmatrix}.$$

Let $C = [c_{ij}]$ be a 2×2 orthogonal matrix that diagonalizes matrix (11). (Recall that C can always be chosen to correspond to a rotation of the plane, although this is not essential in order for Jacobi's method to work.) Now form an $n \times n$ matrix R, the same size as the matrix A, which looks like the identity matrix except that $r_{pp} = c_{11}$, $r_{pq} = c_{12}$, $r_{qp} = c_{21}$, and $r_{qq} = c_{22}$. For matrix (10), where 10 has maximum magnitude above the diagonal, we would have

$$R = \begin{bmatrix} 1 & 0 & 0 \\ 0 & c_{11} & c_{12} \\ 0 & c_{21} & c_{22} \end{bmatrix}.$$

This matrix R will be an orthogonal matrix, with $\det(R) = \det(C)$. Now form the new symmetric matrix $B_1 = R^T A R$, which has zero entries in the row p, column q position and in the row q, column p position. Other entries in B_1 can also be changed from those in A, but it can be shown that the maximum magnitude of off-diagonal entries has been reduced, assuming that no other above-diagonal entry in A had the magnitude of a_{pq}. Then repeat this process, starting with B_1 instead of A, to obtain another symmetric matrix B_2, and so on. It can be shown that the maximum magnitude of off-diagonal entries in the matrices B_i approaches zero as i increases. Thus the sequence of matrices

$$B_1, B_2, B_3, \ldots$$

will approach a diagonal matrix D whose eigenvalues $d_{11}, d_{22}, \ldots, d_{nn}$ are the same as those of A.

If one is going to use Jacobi's method much, one should find the 2×2 matrix C that diagonalizes a general 2×2 symmetric matrix

$$\begin{bmatrix} a & b \\ b & c \end{bmatrix};$$

that is, one should find formulas for computing C in terms of the entries a, b, and c. Exercise 23 develops such formulas.

Rather than give a tedious pencil-and-paper example of Jacobi's method, we choose to present data generated by the available program JACOBI for matrix (10)—

which is the same matrix as that of Example 5, where we used the power method. Observe how, in each step, the colored entries of maximum magnitude off the diagonal are reduced to "zero." While they may not remain zero in the next step, they never return to their original size.

EXAMPLE 6   Use the program JACOBI to diagonalize the matrix

$$\begin{bmatrix} 2 & -8 & 5 \\ -8 & 0 & 10 \\ 5 & 10 & -6 \end{bmatrix}.$$

Solution   The program JACOBI gives the following matrices:

$$\begin{bmatrix} 2 & -8 & 5 \\ -8 & 0 & 10 \\ 5 & 10 & -6 \end{bmatrix}$$

$$\begin{bmatrix} 2 & 8.786909 & 3.433691 \\ 8.786909 & -13.44031 & 0 \\ 3.433691 & 0 & 7.440307 \end{bmatrix}$$

$$\begin{bmatrix} -17.41676 & 0 & -1.415677 \\ 0 & 5.97645 & -3.128273 \\ -1.415677 & -3.128273 & 7.440307 \end{bmatrix}$$

$$\begin{bmatrix} -17.41676 & -.8796473 & -1.109216 \\ -.8796473 & 3.495621 & 0 \\ -1.109216 & 0 & 9.921136 \end{bmatrix}$$

$$\begin{bmatrix} -17.46169 & -.8789265 & 0 \\ -.8789265 & 3.495621 & 3.560333E\text{-}02 \\ 0 & 3.560333E\text{-}02 & 9.966067 \end{bmatrix}$$

$$\begin{bmatrix} -17.49849 & 0 & 1.489243E\text{-}03 \\ 0 & 3.532418 & 3.557217E\text{-}02 \\ 1.489243E\text{-}03 & 3.557217E\text{-}02 & 9.966067 \end{bmatrix}$$

$$\begin{bmatrix} -17.49849 & 8.233765E\text{-}06 & -1.48922E\text{-}03 \\ 8.233765E\text{-}06 & 3.532221 & 0 \\ -1.48922E\text{-}03 & 0 & 9.966265 \end{bmatrix}$$

$$\begin{bmatrix} -17.49849 & 8.233765E\text{-}06 & 0 \\ 8.233765E\text{-}06 & 3.532221 & -4.464509E\text{-}10 \\ 0 & -4.464508E\text{-}10 & 9.966265 \end{bmatrix}$$

The off-diagonal entries are now quite small, and we obtain the same eigenvalues from the diagonal that we did in Example 5 using the power method.   ▪

QR Algorithm

At the present time, an algorithm based on the *QR* factorization of an *invertible* matrix, discussed in Section 5.2, is often used by professionals to find eigenvalues of a matrix. A full treatment of the *QR* algorithm is beyond the scope of this text, but we give a brief description of the method. As with the Jacobi method, pencil-and-paper computations of eigenvalues using the QR algorithm are too cumbersome to include in this text. The program QRFACTOR can be used to illustrate the features of the QR algorithm that we now describe.

Let A be a nonsingular matrix. The QR algorithm generates a sequence of matrices $A_1, A_2, A_3, A_4, \ldots$, all having the same eigenvalues as A. To generate this sequence, let $A_1 = A$, and factor $A_1 = Q_1 R_1$, where Q_1 is the orthogonal matrix and R_1 is the upper-triangular matrix described in Section 5.2. Then let $A_2 = R_1 Q_1$, factor A_2 into $Q_2 R_2$, and set $A_3 = R_2 Q_2$. Continue in this fashion, factoring A_n into $Q_n R_n$ and setting $A_{n+1} = R_n Q_n$. Under fairly general conditions, the matrices A_i will approach an almost upper-triangular matrix of the form

$$\begin{bmatrix} X & X & X & X & X & \cdots & X & X \\ X & X & X & X & X & \cdots & X & X \\ 0 & 0 & X & X & X & \cdots & X & X \\ 0 & 0 & X & X & X & \cdots & X & X \\ 0 & 0 & 0 & 0 & X & \cdots & X & X \\ 0 & 0 & 0 & 0 & X & \cdots & X & X \\ & & & & & \vdots & & \\ 0 & 0 & 0 & 0 & 0 & \cdots & X & X \end{bmatrix}.$$

The colored entries just below the main diagonal may or may not be zero. If one of these entries is nonzero, the 2×2 submatrix having the entry in its lower left-hand corner, like the matrix shaded above, has a pair of complex conjugate numbers $a \pm bi$ as eigenvalues that are also eigenvalues of the large matrix, and of A. Entries on the diagonal that do not lie in such a 2×2 block are real eigenvalues of the matrix and of A.

The available program QRFACTOR can be used to illustrate this procedure. A few comments about the procedure and the program are in order.

From $A_1 = Q_1 R_1$, we have $R_1 = Q_1^{-1} A_1$. Then $A_2 = R_1 Q_1 = Q_1^{-1} A_1 Q_1$, so we see that A_2 is similar to $A_1 = A$ and therefore has the same eigenvalues as A. Continuing in this fashion, we see that each matrix A_i is similar to A. This explains why the eigenvalues don't change as the matrices of the sequence are generated.

Notice, too, that Q_1 and Q_1^{-1} are orthogonal matrices, so $Q_1^{-1}\mathbf{x}$ and $\mathbf{y}Q_1$ have the same magnitude as the vectors $\mathbf{x}$ and $\mathbf{y}$, respectively. It follows that, if E is the matrix of errors in the entries of A, then the error matrix $Q_1^{-1}EQ_1$ arising in the computation of $Q_1^{-1}AQ_1$ is of magnitude comparable to that of E. That is, the generation of the sequence of matrices A_i is *stable*. This is highly desirable in numerical computations.

Finally, it is often useful to perform a **shift**, adding a scalar multiple rI of the identity matrix to A_i before generating the next matrix A_{i+1}. Such a shift increases all eigenvalues by r (see Exercise 30 of Section 4.1), but we can keep track of the total change due to such shifts and adjust the eigenvalues of the final matrix found to obtain those of A. To illustrate one use of shifts, suppose that we wish to find eigenvalues of a singular matrix B. We can form an initial shift, perhaps taking $A = B + (.001)I$, to obtain an invertible matrix A to start the algorithm. For another illustration, we can find by using the program QRFACTOR that the QR algorithm applied to the matrix

$$\begin{bmatrix} 0 & 1 \\ 1 & 0 \end{bmatrix}$$

generates this same matrix repeatedly, even though the eigenvalues are 1 and -1 rather than complex numbers. This is an example of a matrix A for which the sequence of matrices A_i does not approach a form described earlier. However, a shift that adds $(.9)I$ produces the matrix

$$\begin{bmatrix} .9 & 1 \\ 1 & .9 \end{bmatrix},$$

which generates a sequence that quickly converges to

$$\begin{bmatrix} 1.9 & 0 \\ 0 & -.1 \end{bmatrix}.$$

Subtracting the scalar .9 from the eigenvalues 1.9 and $-.1$ of this last matrix, we obtain the eigenvalues 1 and -1 of the original matrix.

Shifts can also be used to speed convergence, which is quite fast when the ratios of magnitudes of eigenvalues are large. The program QRFACTOR displays the matrices A_i as they are generated, and it allows shifts. If we notice that we are going to obtain an eigenvalue whose decimal expansion starts with 17.52, then we can speed convergence greatly by adding the shift $(-17.52)I$. The resulting eigenvalue will be near zero, and the ratios of the magnitudes of other eigenvalues to its magnitude will probably be large. Using this technique with QRFACTOR, it is quite easy to find all eigenvalues, both real and complex, of most matrices of reasonable size that can be displayed conveniently.

Professional programs make many further improvements in the algorithm we have presented, in order to speed the creation of zeros in the lower part of the matrix.

SUMMARY

1. Let A be an $n \times n$ diagonalizable matrix with real eigenvalues λ_j and with a dominant eigenvalue λ_1 of algebraic multiplicity 1, so that $|\lambda_1| > |\lambda_j|$ for $j = 2$, $3, \ldots, n$. Start with any vector $\mathbf{w}_1$ in $\mathbb{R}^n$ that is not in the subspace generated by eigenvectors corresponding to the eigenvalues λ_j for $j > 1$. Form the vectors $\mathbf{w}_2 = (A\mathbf{w}_1)/d_1$, $\mathbf{w}_3 = (A\mathbf{w}_2)/d_2, \ldots$, where d_i is the maximum of the magnitudes of the components of $A\mathbf{w}_i$. The sequence of vectors

$$\mathbf{w}_1, \mathbf{w}_2, \mathbf{w}_3, \ldots$$

approaches an eigenvector of A corresponding to λ_1, and the associated Rayleigh quotients $(A\mathbf{w}_i \cdot \mathbf{w}_i)/(\mathbf{w}_i \cdot \mathbf{w}_i)$ approach λ_1. This is the foundation for the power method, which is summarized in the box preceding Example 1.

2. If A is diagonalizable and invertible, and if $|\lambda_n| < |\lambda_i|$ for $i < n$ with λ_n of algebraic multiplicity 1, then the power method may be used with A^{-1} to find λ_n.

3. Let A be an $n \times n$ symmetric matrix with eigenvalues λ_i such that $|\lambda_1| > |\lambda_2| \geq \cdots \geq |\lambda_n|$. If $\mathbf{b}_1$ is a unit eigenvector corresponding to λ_1, then $A - \lambda_1 \mathbf{b}_1 \mathbf{b}_1{}^T$ has eigenvalues $\lambda_2, \lambda_3, \ldots, \lambda_n, 0$ and if $|\lambda_2| > |\lambda_3|$, then λ_2 can be found by applying the power method to $A - \lambda_1 \mathbf{b}_1 \mathbf{b}_1{}^T$. This deflation can be continued with $A - \lambda_1 \mathbf{b}_1 \mathbf{b}_1{}^T - \lambda_2 \mathbf{b}_2 \mathbf{b}_2{}^T$ if $|\lambda_3| > |\lambda_4|$, and so on, to find more eigenvalues.

4. In the Jacobi method for diagonalizing a symmetric matrix A, one generates a sequence of symmetric matrices, starting with A. Each matrix of the sequence is obtained from the preceding one by multiplying it on the left by R^T and on the right by R, where R is an orthogonal "rotation" matrix designed to annihilate the two (symmetrically located) entries of maximum magnitude off the diagonal. The matrices in the sequence approach a diagonal matrix D having the same eigenvalues as A.

5. In the QR method, one begins with an invertible matrix A and generates a sequence of matrices A_i by setting $A_1 = A$ and $A_{n+1} = R_n Q_n$, where the QR factorization of A_n is $Q_n R_n$. The matrices A_i approach almost upper-triangular matrices having the real eigenvalues of A on the diagonal and pairs of complex conjugate eigenvalues of A as eigenvalues of 2×2 blocks appearing along the diagonal. Shifts may be used to speed convergence or to find eigenvalues of a singular matrix.

EXERCISES

In Exercises 1–4, use the power method to estimate the eigenvalue of maximum magnitude and a corresponding eigenvector for the given matrix. Start with first estimate

$$\mathbf{w}_1 = \begin{bmatrix} 1 \\ 1 \end{bmatrix},$$

and compute $\mathbf{w}_2$, $\mathbf{w}_3$, and $\mathbf{w}_4$. Also find the three Rayleigh quotients. Then find the exact eigenvalues, for comparison, using the characteristic equation.

1. $\begin{bmatrix} 3 & -3 \\ -5 & 1 \end{bmatrix}$ 2. $\begin{bmatrix} 3 & -3 \\ 4 & -5 \end{bmatrix}$

3. $\begin{bmatrix} -3 & 10 \\ -3 & 8 \end{bmatrix}$ 4. $\begin{bmatrix} -4 & 9 \\ -2 & 5 \end{bmatrix}$

In Exercises 5–8, find the spectral decomposition (8) of the given symmetric matrix.

5. $\begin{bmatrix} 2 & 3 \\ 3 & 2 \end{bmatrix}$ 6. $\begin{bmatrix} 3 & 5 \\ 5 & 3 \end{bmatrix}$

7. $\begin{bmatrix} 1 & 1 & 1 \\ 1 & 0 & 0 \\ 1 & 0 & 0 \end{bmatrix}$ 8. $\begin{bmatrix} 1 & -1 & -1 \\ -1 & 1 & -1 \\ -1 & -1 & 1 \end{bmatrix}$

[**Hint:** Use Example 4 in Section 5.3.]

In Exercises 9–12, find the matrix obtained by deflation after the (exact) eigenvalue of maximum magnitude and a corresponding eigenvector are found.

9. The matrix in Exercise 5
10. The matrix in Exercise 6
11. The matrix in Exercise 7
12. The matrix in Exercise 8

In Exercises 13–15, use the available program POWER or similar software to find the eigenvalue of maximum magnitude and a corresponding eigenvector of the given matrix.

13. $\begin{bmatrix} 1 & -44 & -88 \\ -5 & 55 & 113 \\ 1 & -24 & -48 \end{bmatrix}$

14. $\begin{bmatrix} 57 & -2 & 31 \\ -205 & 8 & -113 \\ -130 & 4 & -70 \end{bmatrix}$

15. $\begin{bmatrix} 3 & -22 & -46 \\ -3 & 23 & 47 \\ 1 & -10 & -20 \end{bmatrix}$

16. Use POWER with matrix inversion, or similar software, to find the eigenvalue of minimum magnitude and a corresponding eigenvector of the matrix in Exercise 13.

17. Repeat Exercise 16 for the matrix
$\begin{bmatrix} -1 & 3 & 1 \\ 3 & 2 & -11 \\ 1 & -11 & 7 \end{bmatrix}$.

In Exercises 18–21, use POWER and deflation to find all eigenvalues and corresponding eigenvectors for the given symmetric matrix. Always continue the method before deflating until as much stabilization as possible is achieved. Note the relationship between ratios of magnitudes of eigenvalues and the speed of convergence.

18. $\begin{bmatrix} 3 & 5 & -7 \\ 5 & 10 & 11 \\ -7 & 11 & 0 \end{bmatrix}$

19. $\begin{bmatrix} 0 & -1 & 4 \\ -1 & 2 & -1 \\ 4 & -1 & 0 \end{bmatrix}$

20. $\begin{bmatrix} 3 & 1 & -2 & 1 \\ 1 & 4 & 0 & -3 \\ -2 & 0 & 0 & 5 \\ 1 & -3 & 5 & 2 \end{bmatrix}$

21. $\begin{bmatrix} 5 & 7 & -2 & 6 \\ 7 & 4 & 11 & 3 \\ -2 & 11 & -8 & 0 \\ 6 & 3 & 0 & 6 \end{bmatrix}$

22. The eigenvalue option of the available program VECTGRPH is designed to strengthen your geometric understanding of the power method. There is no scaling, but you have the option of starting again if the numbers get so large that the vectors are off the screen. Read the directions in the program, and then run this eigenvalue option until you can reliably achieve a score of 85% or better.

23. Consider the matrix
$$A = \begin{bmatrix} a & b \\ b & c \end{bmatrix}.$$

The following steps will form an orthogonal diagonalizing matrix C for A.

Step 1 Let $g = (a - c)/2$.
Step 2 Let $h = \sqrt{g^2 + b^2}$.
Step 3 Let $r = \sqrt{b^2 + (g + h)^2}$.
Step 4 Let $s = \sqrt{b^2 + (g - h)^2}$.
Step 5 Let $C = \begin{bmatrix} -b/r & -b/s \\ (g + h)/r & (g - h)/s \end{bmatrix}$.

(If $b < 0$ and a rotation matrix is desired, change the sign of a column vector in C.) Prove this algorithm by showing the following.

a) The eigenvalues of A are
$$\lambda = \frac{a + c \pm \sqrt{(a - c)^2 + 4b^2}}{2}.$$
[HINT: Use the quadratic formula.]
b) The first row vector of $A - \lambda I$ is $(g \pm \sqrt{g^2 + b^2}, b)$.
c) The columns of the orthogonal matrix C in step 5 can be found using part (b).
d) The parenthetical statement following step 5 is valid.

24. Use the algorithm in Exercise 23 to find an orthogonal diagonalizing matrix with determinant 1 for each of the following.

a) $\begin{bmatrix} 0 & 1 \\ 1 & 0 \end{bmatrix}$ b) $\begin{bmatrix} 3 & -2 \\ -2 & 1 \end{bmatrix}$

In Exercises 25–28, use the available program JACOBI or similar software to find the eigenvalues of the matrix by Jacobi's method.

25. The matrix in Exercise 18
26. The matrix in Exercise 19
27. The matrix in Exercise 20
28. The matrix in Exercise 21

In Exercises 29–32, use the available program QRFACTOR to find all eigenvalues, both real and complex, of the given matrix. Be sure to make use of shifts to speed convergence whenever you can see approximately what an eigenvalue will be.

29. $\begin{bmatrix} -1 & 5 & 7 \\ 3 & -6 & 8 \\ 9 & 0 & -11 \end{bmatrix}$

30. $\begin{bmatrix} -3 & 6 & -7 & 2 \\ 13 & -18 & 4 & 6 \\ 21 & 32 & -16 & 9 \\ -4 & 8 & 7 & 11 \end{bmatrix}$

32. $\begin{bmatrix} -3 & 6 & 4 & -2 & 16 \\ 21 & -33 & 5 & 8 & -12 \\ 15 & -21 & 13 & 4 & 20 \\ -18 & 12 & 4 & 8 & 3 \\ -22 & 31 & 14 & 9 & 10 \end{bmatrix}$

31. $\begin{bmatrix} -12 & 3 & 15 & 2 & -21 \\ 47 & -34 & 87 & 24 & 7 \\ 35 & 72 & 33 & -57 & 82 \\ -145 & 67 & 32 & 10 & 46 \\ -9 & 22 & 21 & -45 & 8 \end{bmatrix}$

8

COMPLEX SCALARS

Much of the linear algebra we have discussed is equally valid in applications using complex numbers as scalars. In fact, the use of complex scalars actually extends and clarifies many of the results we have developed. For example, when we are dealing with complex scalars, every $n \times n$ matrix has n eigenvalues, counted with their algebraic multiplicities.

In Section 8.1, we review the algebra of complex numbers. Section 8.2 summarizes properties of the complex vector space $\mathbb{C}^n$ and of matrices with complex entries. Diagonalization of these matrices and Schur's unitary triangularization theorem are discussed in Section 8.3. Section 8.4 is devoted to Jordan canonical forms.

8.1 ALGEBRA OF COMPLEX NUMBERS

The Number i

Numbers exist only in our minds. There is no physical entity that *is* the number 1. If there were, 1 would be in a place of high honor in some great museum of science, and past it would file a steady stream of mathematicians, gazing at 1 in wonder and awe. As mathematics developed, new numbers were *invented* (some prefer to say *discovered*) to fulfill algebraic needs that arose. If we start just with the positive integers, which are the most *natural numbers*, inventing *zero* and *negative integers* enables us not merely to add any two integers, but also to subtract any two integers. Inventing *rational numbers* (fractions) enables us to divide an integer by any nonzero integer. It can be shown that no rational number squared is equal to 2, so *irrational numbers* have to be invented to solve the equation $x^2 = 2$, and our familiar decimal notation comes into use. This decimal notation provides us with other numbers, such as π and e, that are of practical use. The numbers (positive, negative, and zero) that we customarily write using decimal notation have unfortunately become known as the *real numbers*; they are no more real than any other numbers we may invent, because all numbers exist only in our minds.

The real numbers are still inadequate to provide solutions of even certain polynomial equations. The simple equation $x^2 + 1 = 0$ has no real number as a solution. In terms of our needs in this text, the matrix

$$\begin{bmatrix} 0 & -1 \\ 1 & 0 \end{bmatrix}$$

has no eigenvalues. Mathematicians have had to invent a solution i of the equation $x^2 + 1 = 0$ to fill this need. Of course, once a new number is invented, it must take its place in the algebra of numbers; that is, we would like to multiply it and add it to all the numbers we already have. Thus, we also allow numbers bi and finally $a + bi$ for all of our old real numbers a and b. As we will subsequently show, it then becomes possible to add, subtract, multiply, and divide any of these new *complex numbers*, except that division by zero is still not possible. It is unfortunate that i has become known as an *imaginary number*. The purpose of this introduction is to point out that i is no less real and no more imaginary than any other number.

With the invention of i and the consequent enlargement of our number system to the set $\mathbb{C}$ of all the complex numbers $a + bi$, a marvelous thing happens. Not only does the equation $x^2 + a = 0$ have a solution for all real numbers a, but *every* polynomial equation has a solution in the complex numbers. We state without proof the famous *Fundamental Theorem of Algebra*, which shows this.

Fundamental Theorem of Algebra

Every polynomial equation with coefficients in $\mathbb{C}$ has n solutions in $\mathbb{C}$, where n is the degree of the polynomial and the solutions are counted with their algebraic multiplicity.

Review of Complex Arithmetic

A **complex number** z is any expression $z = a + bi$, where a and b are real numbers and $i = \sqrt{-1}$. The scalar a is the **real part** of z, and b is the **imaginary part** of z. It is useful to represent the complex number $z = a + bi$ as the vector (a, b) in the x,y-plane $\mathbb{R}^2$, as shown in Figure 8.1. The x-axis is the **real axis**, and the y-axis is the **imaginary axis**. We let $\mathbb{C} = \{a + bi \mid a, b \in \mathbb{R}\}$ be the set of all complex numbers.

Complex numbers are added, subtracted, and multiplied by real scalars in the natural way:

$$\underset{z}{(a + bi)} \pm \underset{w}{(c + di)} = (a \pm c) + (b \pm d)i,$$

$$r(a + bi) = ra + (rb)i.$$

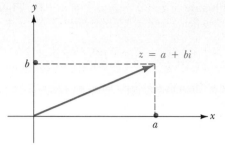

FIGURE 8.1 **The complex number z.**

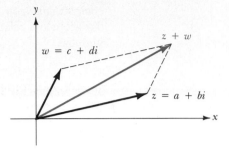

FIGURE 8.2 **Representation of z + w.**

These operations have the same geometric representations as in $\mathbb{R}^2$ and are illustrated in Figures 8.2, 8.3, and 8.4.

It is clear that the set $\mathbb{C}$ of complex numbers is a *real* vector space of dimension 2, isomorphic to $\mathbb{R}^2$.

The **modulus** (or **magnitude**) of the complex number $z = a + bi$ is $|z| = \sqrt{a^2 + b^2}$, which is the length of the vector in Figure 8.1. Notice that $|z| = 0$ if and only if $z = 0$, that is, if and only if $a = b = 0$.

Multiplication of two complex numbers is defined in the way it has to be to make the distributive laws $z_1(z_2 + z_3) = z_1 z_2 + z_1 z_3$ and $(z_1 + z_2)z_3 = z_1 z_3 + z_2 z_3$ valid. Namely, remembering that $i^2 = -1$, we have

$$(a + bi)(c + di) = (ac - bd) + (ad + bc)i.$$

Multiplication of complex numbers is commutative and associative. To divide one complex number by a nonzero complex number, we make use of the notion of a

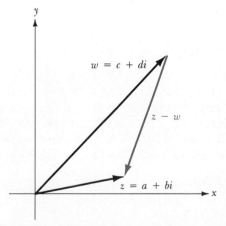

FIGURE 8.3 **Representation of z − w.**

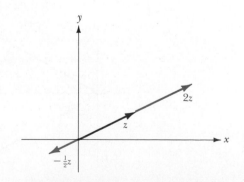

FIGURE 8.4 **Representation of rz.**

complex conjugate. The **conjugate** of $z = a + bi$ is $\bar{z} = a - bi$. Figure 8.5 illustrates the geometric relationship between $\bar{z}$ and z. Computing $z\bar{z} = (a + bi)(a - bi)$, we find that

$$z\bar{z} = a^2 + b^2 = |z|^2. \tag{1}$$

If $z \neq 0$, then Eq. (1) can be written as $z(\bar{z}/|z|^2) = 1$, so $1/z = \bar{z}/|z|^2$. More generally, if $z \neq 0$, then w/z is computed as

$$\frac{w}{z} = \frac{w\bar{z}}{z\bar{z}} = \frac{w\bar{z}}{|z|^2} = \frac{1}{|z|^2}(w\bar{z}). \tag{2}$$

We will see geometric representations of multiplication and division in a moment.

EXAMPLE 1   Find $(3 + 4i)^{-1} = 1/(3 + 4i)$.

Solution   Using Eq. (2), we have

$$\frac{1}{3 + 4i} = \left(\frac{1}{3 + 4i}\right)\left(\frac{3 - 4i}{3 - 4i}\right) = \frac{3 - 4i}{25} = \tfrac{1}{25}(3 - 4i) = \tfrac{3}{25} - \tfrac{4}{25}i. \qquad \blacksquare$$

EXAMPLE 2   Compute $(2 + 3i)/(1 + 2i)$.

Solution   Using the technique of Eq. (2), we obtain

$$\frac{2 + 3i}{1 + 2i} = \left(\frac{2 + 3i}{1 + 2i}\right)\left(\frac{1 - 2i}{1 - 2i}\right) = \tfrac{1}{5}(2 + 3i)(1 - 2i) = \tfrac{8}{5} - \tfrac{1}{5}i. \qquad \blacksquare$$

All of the familiar rules of arithmetic apply to the algebra of complex numbers. However, there is no notion of *order* for complex numbers; the notion $z_1 < z_2$ is defined only if z_1 and z_2 are real. The conjugation operation turns out to be very important; we summarize its properties in a theorem.

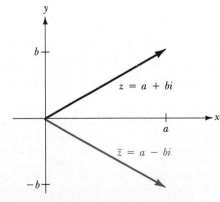

FIGURE 8.5   The conjugate of z.

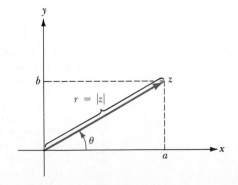

FIGURE 8.6   Polar form of z.

THEOREM 8.1   **Properties of Conjugation in ℂ**

Let $z = a + bi$ and $w = c + di$ be complex numbers. Then

1. $\overline{z + w} = \overline{z} + \overline{w}$,
2. $\overline{z - w} = \overline{z} - \overline{w}$,
3. $\overline{zw} = \overline{z}\,\overline{w}$,
4. $\overline{z/w} = \overline{z}/\overline{w}$,
5. $\overline{\overline{z}} = z$.

The proofs of the properties in Theorem 8.1 are very easy. We prove property 3 as an example and leave the rest as exercises. (See Exercises 12 and 13.)

EXAMPLE 3   Prove property 3 of Theorem 8.1.

Solution   Let $z = a + bi$ and $w = c + di$. Then

$$\overline{zw} = \overline{(a + bi)(c + di)} = (a - bi)(c - di)$$
$$= (ac - bd) - (ad + bc)i = \overline{z}\,\overline{w}. \qquad ■$$

Polar Form of Complex Numbers

Let us return to the vector representation of $z = a + bi$ in $\mathbb{R}^2$. Figure 8.6 indicates that, for $z \neq 0$, if θ is an angle from the positive x-axis to the vector representation of z, and if $r = |z|$, then $a = r \cos \theta$ and $b = r \sin \theta$. Thus,

$$z = r(\cos \theta + i \sin \theta). \quad \text{A polar form of } z \qquad (3)$$

The angle θ is an **argument** of z. Of particular importance is the restricted value

$$-\pi < \theta \leq \pi \quad \text{The principal argument of } z$$

denoted by $\mathrm{Arg}(z)$. We usually use this **principal argument** of z and refer to the corresponding form (3) as **the polar form** of z.

COMPLEX NUMBERS MADE THEIR INITIAL APPEARANCE on the mathematical scene in *The Great Art, or On the Rules of Algebra* (1545) by the 16th century Italian mathematician and physician, Gerolamo Cardano (1501–1576). It was in this work that Cardano presented an algebraic method of solution of cubic and quartic equations. But it was the quadratic problem of dividing 10 into two parts such that their product is 40 to which Cardano found the solution $5 + \sqrt{-15}$ and $5 - \sqrt{-15}$ by standard techniques. Cardano was not entirely happy with this answer, as he wrote, "So progresses arithmetic subtlety the end of which, as is said, is as refined as it is useless."

Twenty seven years later the engineer Raphael Bombelli (1526–1572) published an algebra text in which he dealt systematically with complex numbers. Bombelli wanted to clarify Cardano's cubic formula, which under certain circumstances could express a correct real solution to a cubic equation as the sum of two expressions each involving the square root of a negative number. Thus he developed our modern rules for operating with expressions of the form $a + b\sqrt{-1}$, including methods for determining cube roots of such numbers. Thus, for example, he showed that $\sqrt{2 + \sqrt{-121}} = 2 + \sqrt{-1}$ and therefore that the solution to the cubic equation $x^3 = 15x + 4$, which Cardano's formula gave as $x = \sqrt[3]{2 + \sqrt{-121}} + \sqrt[3]{2 - \sqrt{-121}}$, could be written simply as $x = (2 + \sqrt{-1}) + (2 - \sqrt{-1})$, or as $x = 4$, the obvious answer.

EXAMPLE 4 Find the principal argument and the polar form of the complex number $z = -\sqrt{3} - i$.

Solution Since $|z| = \sqrt{(-\sqrt{3})^2 + (-1)^2} = \sqrt{4} = 2$, we have $\cos \theta = -\sqrt{3}/2$ and $\sin \theta = -\frac{1}{2}$, as indicated in Figure 8.7. The principal argument of z is the angle θ between $-\pi$ and π satisfying these two conditions—that is, $\theta = -5\pi/6$. The required polar form is $z = 2(\cos(-5\pi/6) + i \sin(-5\pi/6))$. ■

If we multiply two complex numbers in their polar form, we quickly discover the geometric representation of multiplication. Let $z_1 = r_1(\cos \theta_1 + i \sin \theta_1)$ and $z_2 = r_2(\cos \theta_2 + i \sin \theta_2)$. Then

$$z_1 z_2 = r_1 r_2((\cos \theta_1 \cos \theta_2 - \sin \theta_1 \sin \theta_2) + i(\sin \theta_1 \cos \theta_2 + \cos \theta_1 \sin \theta_2))$$
$$= r_1 r_2(\cos (\theta_1 + \theta_2) + i \sin (\theta_1 + \theta_2)).$$

The last equation arises from the familiar trigonometric identities for the cosine and sine of a sum. This computation shows that, when two complex numbers are multiplied, the moduli are multiplied and the arguments are added. Notice, however, that the principal argument of a product may not be the sum of the principal arguments of the factors; the sum of the principal arguments may have to be adjusted to lie between $-\pi$ and π. (See Fig. 8.8.)

Geometric Representation of $z_1 z_2$

1. $|z_1 z_2| = |z_1||z_2|$.
2. $\mathrm{Arg}(z_1) + \mathrm{Arg}(z_2)$ is an argument of $z_1 z_2$.

(4)

Since z_1/z_2 is the complex number w such that $z_2 w = z_1$, we see that, in division, one divides the moduli and subtracts the arguments. (See Fig. 8.9.)

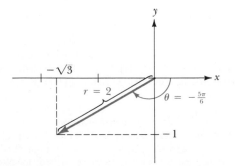

FIGURE 8.7 The polar form of $z = -\sqrt{3} - i$.

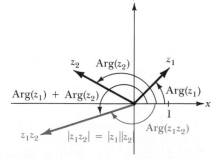

FIGURE 8.8 Representation of $z_1 z_2$.

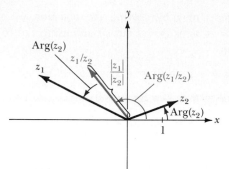

FIGURE 8.9 **Representation of z_1/z_2.**

Geometric Representation of z_1/z_2

1. $|z_1/z_2| = |z_1|/|z_2|$.
2. $\text{Arg}(z_1) - \text{Arg}(z_2)$ is an argument of z_1/z_2.

(5)

EXAMPLE 5 Illustrate relations (4) and (5) for the complex numbers $z_1 = \sqrt{3} + i$ and $z_2 = 1 + i$.

Solution For z_1, we have $r_1 = |z_1| = \sqrt{4} = 2$, $\cos \theta_1 = \sqrt{3}/2$, and $\sin \theta_1 = \frac{1}{2}$, so

$$z_1 = 2(\cos(\pi/6) + i \sin(\pi/6)).$$

For z_2, we have $r_2 = |z_2| = \sqrt{2}$ and $\cos \theta_2 = \sin \theta_2 = 1/\sqrt{2}$, so

$$z_2 = \sqrt{2}(\cos(\pi/4) + i \sin(\pi/4)).$$

Thus,

$$\begin{aligned}
z_1 z_2 &= 2\sqrt{2}(\cos(\pi/6 + \pi/4) + i \sin(\pi/6 + \pi/4)) \\
&= 2\sqrt{2}(\cos(5\pi/12) + i \sin(5\pi/12)),
\end{aligned}$$

and

$$\begin{aligned}
z_1/z_2 &= (2/\sqrt{2})(\cos(\pi/6 - \pi/4) + i \sin(\pi/6 - \pi/4)) \\
&= (2/\sqrt{2})(\cos(-\pi/12) + i \sin(-\pi/12)).
\end{aligned}$$

■

Relations (4) can be used repeatedly with $z_1 = z_2 = z = r(\cos \theta + i \sin \theta)$ to compute z^n for a positive integer n. We obtain the formula

$$z^n = r^n(\cos n\theta + i \sin n\theta).$$

(6)

We can find nth roots of z by solving the equation $w^n = z$. Writing $z = r(\cos \theta + i \sin \theta)$ and $w = s(\cos \phi + i \sin \phi)$ and using Eq. (6) with w instead of z, we obtain the equation

$$r(\cos \theta + i \sin \theta) = s^n(\cos n\phi + i \sin n\phi).$$

Therefore, $r = s^n$ and $n\phi = \theta \pm 2k\pi$ for $k = 0, 1, 2, \ldots$, so

$$w = s(\cos \phi + i \sin \phi),$$

where

$$s = r^{1/n} \quad \text{and} \quad \phi = \frac{\theta}{n} \pm \frac{2k\pi}{n}.$$

Exercises 27 indicates that these values for ϕ represent precisely n distinct complex numbers, as indicated in the following box.

---

nth roots of $z = r(\cos \theta + i \sin \theta)$

The nth roots of z are

$$r^{1/n}\left(\cos\left(\frac{\theta}{n} + \frac{2k\pi}{n}\right) + i \sin\left(\frac{\theta}{n} + \frac{2k\pi}{n}\right)\right) \tag{7}$$

for $k = 0, 1, 2, \ldots, n - 1$.

This illustrates the Fundamental Theorem of Algebra for the equation $w^n = z$ of degree n that has n solutions in $\mathbb{C}$.

EXAMPLE 6   Find the fourth roots of 16, and draw their vector representations in $\mathbb{R}^2$.

Solution   The polar form of $z = 16$ is $z = 16(\cos 0 + i \sin 0)$, where $\text{Arg}(z) = 0$. Applying formula (7), with $n = 4$, we find the following fourth roots of 16 (see Fig. 8.10):

k	$16^{1/4}\left(\cos\dfrac{2k\pi}{4} + i\sin\dfrac{2k\pi}{4}\right)$
0	$2(\cos 0 + i \sin 0) = 2$
1	$2\left(\cos\dfrac{\pi}{2} + i \sin\dfrac{\pi}{2}\right) = 2i$
2	$2(\cos \pi + i \sin \pi) = -2$
3	$2\left(\cos\dfrac{3\pi}{2} + i \sin\dfrac{3\pi}{2}\right) = -2i$

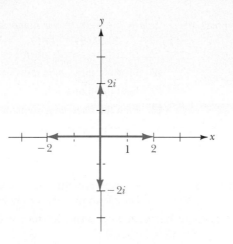

FIGURE 8.10 The fourth roots of 16.

SUMMARY

1. A complex number is a number of the form $z = a + bi$, where a and b are real numbers and $i = \sqrt{-1}$.

2. The modulus of the complex number $z = a + bi$ is
$$|z| = \sqrt{a^2 + b^2}.$$

3. The complex number $z = a + bi$ has the polar form $z = |z|(\cos \theta + i \sin \theta)$, and $\theta = \mathrm{Arg}(z)$ is the principal argument of z if $-\pi < \theta \leq \pi$.

4. Arithmetic computations in the set $\mathbb{C}$ of complex numbers, including division and extraction of roots, can be accomplished as illustrated in this section and can be represented geometrically as shown in the figures of this section.

EXERCISES

1. Find the sum $z + w$ and the product zw if
   a) $z = 1 + 2i$, $w = 3 - i$,
   b) $z = 3 + i$, $w = i$.

2. Find the sum $z + w$ and the product zw if
   a) $z = 2 + 3i$, $w = 5 - i$,
   b) $z = 1 + 2i$, $w = 2 - i$.

3. Find $|z|$ and $\bar{z}$, and verify that $z\bar{z} = |z|^2$, if
   a) $z = 3 + 2i$,   b) $z = 4 - i$.

4. Find $|z|$ and $\bar{z}$, and verify that $z\bar{z} = |z|^2$, if
   a) $z = 2 + i$,   b) $z = 3 - 4i$.

5. Show that z is a real number if and only if $z = \bar{z}$.

6. Express z^{-1} in the form $a + bi$, for a and b real numbers, if

   a) $z = -1 + i$,   b) $z = 3 + 4i$.

7. Express z/w in the form $a + bi$, for a and b real numbers, if

   a) $z = 1 + 2i$, $w = 1 + i$,
   b) $z = 3 + i$, $w = 3 + 4i$.

8. Find the modulus and principal argument for

   a) $\sqrt{3} - i$,   b) $-\sqrt{3} - i$.

9. Find the modulus and principal argument for

   a) $-2 + 2i$,   b) $-2 - 2i$.

10. Express $(\sqrt{3} + i)^6$ in the form $a + bi$ for a and b real numbers. [HINT: Write the given number in polar form.]

11. Express $(1 + i)^8$ in the form $a + bi$ for a and b real numbers. [HINT: Write the given number in polar form.]

12. Prove properties 1, 2, and 5 of Theorem 8.1.

13. Prove property 4 of Theorem 8.1.

14. Illustrate Eqs. (5) in the text for $z_1 = \sqrt{3} + i$ and $z_2 = -1 + \sqrt{3}i$.

15. Illustrate Eqs. (5) in the text for $z_1 = 2 + 2i$ and $z_2 = 1 + \sqrt{3}i$.

16. If $z^8 = 16$, find $|z|$.

17. Mark each of the following True or False.

   ___ a) The existence of complex numbers is more doubtful than the existence of real numbers.

   ___ b) Pencil-and-paper computations with complex numbers are more cumbersome than with real numbers.

   ___ c) The square of every complex number is a positive real number.

   ___ d) Every complex number has two distinct square roots in $\mathbb{C}$.

   ___ e) Every nonzero complex number has two distinct square roots in $\mathbb{C}$.

   ___ f) The Fundamental Theorem of Algebra asserts that the algebraic operations of addition, subtraction, multiplication, and division are possible with any two complex numbers, as long as we do not divide by zero.

   ___ g) The product of two complex numbers cannot be a real number unless both numbers are themselves real or unless both are of the form bi, where b is a real number.

   ___ h) If $(a + bi)^3 = 8$, then $a^2 + b^2 = 4$.

   ___ i) If $\text{Arg}(z) = 3\pi/4$ and $\text{Arg}(w) = -\pi/2$, then $\text{Arg}(z/w) = 5\pi/4$.

   ___ j) If $z + \bar{z} = 2z$, then z is a real number.

18. Find the three cube roots of 8.

19. Find the four fourth roots of -16.

20. Find the three cube roots of -27.

21. Find the four fourth roots of 1.

22. Find the six sixth roots of 1.

23. Find the eight eighth roots of 256.

24. A **primitive nth root of unity** is a complex number z such that $z^n = 1$ but $z^m \neq 1$ for $m < n$.

   a) Give a formula for one primitive nth root of unity.

   b) Find the primitive fourth roots of unity.

   c) How many primitive eighth roots of unity are there? [HINT: Argue geometrically in terms of polar forms.]

25. Let $z, w \in \mathbb{C}$. Show that $|z + w| \leq |z| + |w|$. [HINT: Remember that $\mathbb{C}$ is a real vector space of dimension 2, naturally isomorphic to $\mathbb{R}^2$.]

26. Show that the nth roots of $z \in \mathbb{C}$ can be represented geometrically as n equally spaced points on the circle $x^2 + y^2 = |z|^2$.

27. Show that the infinite list of values $\phi = \dfrac{\theta}{n} \pm \dfrac{2k\pi}{n}$ for $k = 0, 1, 2, \ldots$ yields just n distinct complex numbers $w = s(\cos \phi + i \sin \phi)$ of modulus s.

28. In calculus it is shown that

$$e^x = 1 + x + \frac{x^2}{2!} + \frac{x^3}{3!} + \frac{x^4}{4!} + \cdots$$

$$\sin x = x - \frac{x^3}{3!} + \frac{x^5}{5!} - \frac{x^7}{7!} + \frac{x^9}{9!} - \cdots$$

$$\cos x = 1 - \frac{x^2}{2!} + \frac{x^4}{4!} - \frac{x^6}{6!} + \frac{x^8}{8!} - \cdots.$$

   a) Proceeding formally, show that $e^{i\theta} = \cos \theta + i \sin \theta$. (This is **Euler's formula**.)

   b) Show that every complex number z may be written in the form $z = re^{i\theta}$, where $r = |z|$ and $\theta = \text{Arg}(z)$.

   c) Assuming that the usual laws of exponents hold for exponents that are complex numbers, use part (b) to derive again Eqs. (4) and (5) describing multiplication and division of complex numbers in polar form.

8.2 MATRICES AND VECTOR SPACES WITH COMPLEX SCALARS

Complex Matrices and Linear Systems

Both the real number system $\mathbb{R}$ and the complex number system $\mathbb{C}$ are algebraic structures known as *fields*. In a field, we can add any two elements and multiply any two elements to produce an element of the field. Addition and multiplication are commutative and associative operations, and multiplication is distributive over addition. The field contains an additive identity 0 and a multiplicative identity 1. Every element c in the field has an additive inverse $-c$ in the field—that is, an element that, when added to c, produces the additive identity. Similarly, every nonzero element d in the field has a multiplicative inverse $1/d$ in the field.

The part of our work in Chapters 1 and 2 that rests only on the field axioms of $\mathbb{R}$ is equally valid if we allow complex scalars. In particular, we can work with *complex matrices*—that is, with matrices having complex entries: adding matrices of the same size, multiplying matrices of appropriate sizes, and multiplying a matrix by a complex scalar. We can solve linear systems by using the same Gauss or Gauss–Jordan methods that we used in Chapter 1. All of our work in Chapter 1, Sections 1 through 5, makes perfectly good sense when applied to complex scalars. Pencil-and-paper computations are more tedious, however.

EXAMPLE 1 Solve the linear system

$$
\begin{aligned}
z_1 - z_2 + (1+i)z_3 &= i \\
iz_1 - 2iz_2 + iz_3 &= 2 - i \\
iz_2 - (1+i)z_3 &= 1 + 2i.
\end{aligned}
$$

Solution We use the Gauss–Jordan method as follows:

$$
\begin{bmatrix}
1 & -1 & 1+i & i \\
i & -2i & i & 2-i \\
0 & i & -1-i & 1+2i
\end{bmatrix}
\sim
\begin{bmatrix}
1 & -1 & 1+i & i \\
0 & -i & 1 & 3-i \\
0 & i & -1-i & 1+2i
\end{bmatrix}
$$

$$
\sim
\begin{bmatrix}
1 & 0 & 1+2i & 1+4i \\
0 & 1 & i & 1+3i \\
0 & 0 & -i & 4+i
\end{bmatrix}
\sim
\begin{bmatrix}
1 & 0 & 0 & 10+2i \\
0 & 1 & 0 & 5+4i \\
0 & 0 & 1 & -1+4i
\end{bmatrix}.
$$

Thus we obtain the solution $z_1 = 10 + 2i$, $z_2 = 5 + 4i$, $z_3 = -1 + 4i$. ▪

EXAMPLE 2 Find the inverse of the matrix

$$
A = \begin{bmatrix}
1 & 2i & 1+i \\
1 & 3i & i \\
0 & 1+i & -1
\end{bmatrix}.
$$

Solution We proceed precisely as in Chapter 1:

$$
\begin{bmatrix}
1 & 2i & 1+i & 1 & 0 & 0 \\
1 & 3i & i & 0 & 1 & 0 \\
0 & 1+i & -1 & 0 & 0 & 1
\end{bmatrix}
\sim
\begin{bmatrix}
1 & 2i & 1+i & 1 & 0 & 0 \\
0 & i & -1 & -1 & 1 & 0 \\
0 & 1+i & -1 & 0 & 0 & 1
\end{bmatrix}
$$

$$\sim \begin{bmatrix} 1 & 0 & 3+i & \bigm| & 3 & -2 & 0 \\ 0 & 1 & i & \bigm| & i & -i & 0 \\ 0 & 0 & -i & \bigm| & 1-i & -1+i & 1 \end{bmatrix}$$

$$\sim \begin{bmatrix} 1 & 0 & 0 & \bigm| & 1-4i & 4i & 1-3i \\ 0 & 1 & 0 & \bigm| & 1 & -1 & 1 \\ 0 & 0 & 1 & \bigm| & 1+i & -1-i & i \end{bmatrix}$$

Thus,

$$A^{-1} = \begin{bmatrix} 1-4i & 4i & 1-3i \\ 1 & -1 & 1 \\ 1+i & -1-i & i \end{bmatrix}.$$ ■

Complex Vector Spaces

The definition of a *complex vector space* is identical with that of a real vector space, except that the field of scalars used is $\mathbb{C}$ rather than $\mathbb{R}$. The set $\mathbb{C}^n$ of all *n*-tuples having entries in $\mathbb{C}$ is an example of a complex vector space. Another example is the set of all $m \times n$ matrices with complex entries. The vector space $\mathbb{C}^n$ has the same standard basis $\{\mathbf{e}_1, \mathbf{e}_2, \ldots, \mathbf{e}_n\}$ as $\mathbb{R}^n$, but of course now the field of scalars is $\mathbb{C}$. Thus, $\mathbb{C}^n$ is an *n*-dimensional *complex* vector space—that is, if we use *complex scalars*. In particular, $\mathbb{C} = \mathbb{C}^1$ is a one-dimensional complex vector space, even though we can regard it geometrically as a plane. That plane is a two-dimensional *real* vector space, but is a one-dimensional *complex* vector space. In general, $\mathbb{C}^n$ can be regarded geometrically in a natural way as a $2n$-dimensional real vector space, as we ask you to explain in Exercise 1.

All of our work in Chapter 2 regarding subspaces, generating sets, independence, and bases carries over to complex vector spaces, and the proofs are identical.

EXAMPLE 3 Determine whether the set

$$S = \{(1, 2i, 1+i), (1, 3i, i), (0, 1+i, -1)\}$$

is independent and is a basis for $\mathbb{C}^3$.

Solution The vectors given in S are the row vectors in matrix A of Example 2. Row reduction of the matrix in that example shows that the matrix has rank 3, so the vectors in S are independent and hence form a basis for the three-dimensional space $\mathbb{C}^3$. ■

EXAMPLE 4 Find the coordinate vector $\mathbf{v}_B$ in $\mathbb{C}^3$ of the vector $\mathbf{v} = (i, 2-i, 1+2i)$ relative to the ordered basis

$$B = ((1, i, 0), (-1, -2i, i), (1+i, i, -1-i)).$$

Solution To find the coordinate vector of $\mathbf{v}$ relative to B, we reduce the augmented matrix having the vectors in B as column vectors and having the vector $\mathbf{v}$ as the column to the right of the partition. This is precisely the augmented matrix that we reduced in Example 1, so we see that the coordinate vector $\mathbf{v}_B$, written as usual as a column vector, is

$$\mathbf{v}_B = \begin{bmatrix} 10 + 2i \\ 5 + 4i \\ -1 + 4i \end{bmatrix}.$$

▪

Euclidean Inner Product in $\mathbb{C}^n$

We now come to an essential difference in the structures of $\mathbb{C}$ and of $\mathbb{R}$. We have a natural idea of order for the elements of $\mathbb{R}$. We know what it means to say $x_1 < x_2$, and we have often used the fact that $x^2 \geq 0$ for all $x \in \mathbb{R}$. There is no idea of order in $\mathbb{C}$, extending the ordering of $\mathbb{R}$, for $i^2 = -1$ in $\mathbb{C}$. The nonzero numbers in $\mathbb{C}$ cannot be classified as either positive or negative on the basis of whether or not they are squares, because *all* numbers in $\mathbb{C}$ are squares. This is a very important difference between $\mathbb{R}$ and $\mathbb{C}$.

Let us see what problems this causes as we try to extend some more of the ideas in Chapter 2. We multiply matrices with complex entries by taking dot products of row vectors of the first with column vectors of the second, just as we do for matrices with real entries. Recall that, in $\mathbb{R}^n$, the length of a vector $\mathbf{v}$ is $\|\mathbf{v}\| = \sqrt{\mathbf{v} \cdot \mathbf{v}}$. However, this dot product cannot be used as an inner product to define length of vectors in $\mathbb{C}$, since, if $\mathbf{v} = (1, i)$, we would have $\mathbf{v} \cdot \mathbf{v} = 0$. The fix for this problem is simple. Recalling that $|a + bi|^2 = (a - bi)(a + bi)$, we make an adjustment and define the *Euclidean inner product* in $\mathbb{C}^n$.

DEFINITION 8.1 Euclidean Inner Product

Let $\mathbf{u} = (u_1, u_2, \ldots, u_n)$ and $\mathbf{v} = (v_1, v_2, \ldots, v_n)$ be vectors in $\mathbb{C}^n$. The **Euclidean inner product** of $\mathbf{u}$ and $\mathbf{v}$ is

$$\langle \mathbf{u}, \mathbf{v} \rangle = \bar{u}_1 v_1 + \bar{u}_2 v_2 + \cdots + \bar{u}_n v_n.$$

Notice that Definition 8.1 gives $\langle (1, i), (1, i) \rangle = (1)(1) + (-i)(i) = 1 + 1 = 2$. Since $|a + bi|^2 = (a - bi)(a + bi)$, we see at once that, for $\mathbf{v} \in \mathbb{C}^n$, we have

$$\langle \mathbf{v}, \mathbf{v} \rangle = |v_1|^2 + |v_2|^2 + \cdots + |v_n|^2, \tag{1}$$

just as in the $\mathbb{R}^n$ case.

We list properties of the Euclidean inner product in $\mathbb{C}^n$ as a theorem, leaving the proofs as exercises. (See Exercises 16 through 19.)

THEOREM 8.2 Properties of the Euclidean Inner Product

Let $\mathbf{u}$, $\mathbf{v}$, and $\mathbf{w}$ be vectors in $\mathbb{C}^n$, and let z be a complex scalar. Then

(i) $\langle \mathbf{u}, \mathbf{u} \rangle \geq 0$, and $\langle \mathbf{u}, \mathbf{u} \rangle = 0$ if and only if $\mathbf{u} = \mathbf{0}$,

(ii) $\langle \mathbf{u}, \mathbf{v} \rangle = \overline{\langle \mathbf{v}, \mathbf{u} \rangle}$,

(iii) $\langle (\mathbf{u} + \mathbf{v}), \mathbf{w} \rangle = \langle \mathbf{u}, \mathbf{w} \rangle + \langle \mathbf{v}, \mathbf{w} \rangle$,

(iv) $\langle \mathbf{w}, (\mathbf{u} + \mathbf{v}) \rangle = \langle \mathbf{w}, \mathbf{u} \rangle + \langle \mathbf{w}, \mathbf{v} \rangle$,

(v) $\langle z\mathbf{u}, \mathbf{v} \rangle = \bar{z}\langle \mathbf{u}, \mathbf{v} \rangle$, and $\langle \mathbf{u}, z\mathbf{v} \rangle = z\langle \mathbf{u}, \mathbf{v} \rangle$.

Property (i) of Theorem 8.2 and Eq. (1) preceding the theorem suggest that we define the **magnitude** or **norm** of a vector **v** in $\mathbb{C}^n$ as

$$\|\mathbf{v}\| = \sqrt{\langle \mathbf{v}, \mathbf{v} \rangle} = \sqrt{\bar{v}_1 v_1 + \bar{v}_2 v_2 + \cdots + \bar{v}_n v_n}. \quad \text{Magnitude of } \mathbf{v}$$

Vectors **u** and **v** in $\mathbb{C}^n$ are **perpendicular** (or **orthogonal**) if $\langle \mathbf{u}, \mathbf{v} \rangle = 0$; property (ii) of Theorem 8.2 tells us that $\langle \mathbf{u}, \mathbf{v} \rangle = 0$ if and only if $\langle \mathbf{v}, \mathbf{u} \rangle = 0$. The vectors are **parallel** if $\mathbf{u} = z\mathbf{v}$ for some scalar z in $\mathbb{C}$. Notice that $\|z\mathbf{v}\| = |z| \|\mathbf{v}\|$, just as in the real case. Vectors of magnitude 1 are **unit vectors**. Having made these definitions, we will feel free to consider without further definition such things as orthogonal subspaces and orthonormal bases.

EXAMPLE 5   Find a unit vector in $\mathbb{C}^3$ parallel to $\mathbf{v} = (1, i, 1 + i)$.

Solution   Since $\|\mathbf{v}\| = \sqrt{\langle \mathbf{v}, \mathbf{v} \rangle}$, we have

$$\|\mathbf{v}\| = \sqrt{1(1) + (-i)i + (1 - i)(1 + i)} = \sqrt{1 + 1 + 2} = 2.$$

Either of the vectors $\pm\frac{1}{2}(1, i, 1 + i)$ satisfies the requirement. ■

The Euclidean inner product given in Definition 8.1 reduces to the usual dot product of vectors if the vectors are in $\mathbb{R}^n$. However, we have to watch one feature very carefully:

> *The Euclidean inner product in $\mathbb{C}^n$ is not commutative.*

(See Property (ii) in Theorem 8.2.) For example,

$$\langle (1, i), (0, 1) \rangle = -i, \quad \text{but} \quad \langle (0, 1), (1, i) \rangle = i.$$

To illustrate the care that must be taken as a result of the noncommutativity of the inner product, we consider the Gram–Schmidt process applied in the vector space $\mathbb{C}^n$. Let **u** and **v** be vectors in $\mathbb{C}^n$. Then the vector

$$\mathbf{w} = \mathbf{u} - \left(\frac{\langle \mathbf{v}, \mathbf{u} \rangle}{\langle \mathbf{v}, \mathbf{v} \rangle} \right) \mathbf{v}$$

is orthogonal to **v**, since

$$\langle \mathbf{w}, \mathbf{v} \rangle = \langle \mathbf{u}, \mathbf{v} \rangle - \langle \overline{\mathbf{v}, \mathbf{u}} \rangle = \langle \mathbf{u}, \mathbf{v} \rangle - \langle \mathbf{u}, \mathbf{v} \rangle = 0.$$

We must not use $\mathbf{u} - \dfrac{\langle \mathbf{u}, \mathbf{v} \rangle}{\langle \mathbf{v}, \mathbf{v} \rangle} \mathbf{v}$, whose inner product with **v** is generally not zero.

EXAMPLE 6   Transform the basis

$$\{(1, i, i), (1, 0, -i), (1, 0, 1)\}$$

of $\mathbb{C}^3$ to an orthonormal one, using the Gram–Schmidt process.

Solution   First we transform the given basis to an orthogonal one. Since $\mathbf{v}_1 = (1, i, i)$ and $\mathbf{v}_2 = (1, 0, -i)$ are orthogonal, we work with $\mathbf{v}_3 = (1, 0, 1)$ and replace it by

$$\mathbf{v}_3 - \left(\frac{\langle \mathbf{v}_1, \mathbf{v}_3 \rangle}{\langle \mathbf{v}_1, \mathbf{v}_1 \rangle}\right)\mathbf{v}_1 - \left(\frac{\langle \mathbf{v}_2, \mathbf{v}_3 \rangle}{\langle \mathbf{v}_2, \mathbf{v}_2 \rangle}\right)\mathbf{v}_2$$

$$= \mathbf{v}_3 - \left(\frac{1-i}{3}\right)\mathbf{v}_1 - \left(\frac{1+i}{2}\right)\mathbf{v}_2$$

$$= (1, 0, 1) - \tfrac{1}{3}(1 - i, 1 + i, 1 + i) - \tfrac{1}{2}(1 + i, 0, 1 - i)$$

$$= \tfrac{1}{6}(1 - i, -2 - 2i, 1 + i).$$

In fact, we prefer to replace $\mathbf{v}_3$ by $(1 - i, -2 - 2i, 1 + i)$, which is just as good. An orthogonal basis is

$$\{(1, i, i), (1, 0, -i), (1 - i, -2 - 2i, 1 + i)\},$$

and an orthonormal basis is

$$\{\tfrac{1}{3}(1, i, i), \tfrac{1}{2}(1, 0, -i), \frac{1}{2\sqrt{3}}(1 - i, -2 - 2i, 1 + i)\}. \qquad ■$$

The Conjugate Transpose

In fixing up our old inner product to serve in $\mathbb{C}^n$ in Definition 8.1, we had to decide whether to take conjugates of the components u_i of the first vector in $\langle \mathbf{u}, \mathbf{v} \rangle$ or to take conjugates of the components v_i of the second vector. It is more convenient to take conjugates of components of the first vector for the following reason. The bulk of our work in $\mathbb{R}^n$ has been formulated in terms of column vectors, although we often use row notation to save space. For example, when working with matrix representations of linear transformations, we always write $A\mathbf{x}$, where $\mathbf{x}$ is a column vector. You may recall several instances where it has been convenient for us to use the fact that the inner product (dot product) of two vectors in $\mathbb{R}^n$ appears as sole entry in the matrix product of the first vector as a row vector and of the second vector as a column vector. That is, for column vectors $\mathbf{x}, \mathbf{y} \in \mathbb{R}^n$, we can write $\mathbf{x} \cdot \mathbf{y} = \mathbf{x}^T\mathbf{y}$, where $\mathbf{x}^T$ is a row vector and where no distinction is made between a 1×1 matrix and a scalar. Thinking in these terms, we can express the condition that the column vectors of a real square matrix A form an orthonormal basis as $A^TA = I$. In order to preserve these convenient algebraic formulas with as little change as possible, we choose to take conjugates of the components of the first vector in Definition 8.1, and we continue with a definition that allows us to recover these formulas.

DEFINITION 8.2   Conjugate Transpose, or Hermitian Adjoint

Let $A = [a_{ij}]$ be an $m \times n$ matrix with complex scalar entries.

  (i) The **conjugate** of A is the $m \times n$ matrix $\bar{A} = [\overline{a_{ij}}]$.

 (ii) The **conjugate transpose** (or **Hermitian adjoint**) of A is the matrix $A^* = [\overline{a_{ij}}]^T$.

For column vectors $\mathbf{v}$, $\mathbf{w} \in \mathbb{C}^n$, we have

$$\langle \mathbf{v}, \mathbf{w} \rangle = \mathbf{v}^*\mathbf{w}. \tag{2}$$

In addition, the condition for an $n \times n$ complex matrix A to have orthogonal unit column vectors can be written as

$$A^*A = I.$$

EXAMPLE 7 Find the conjugate transpose A^* of the matrix

$$A = \begin{bmatrix} 1 & i & 1+i \\ 2 & 0 & i \\ 2i & 1 & 1-i \end{bmatrix}.$$

Solution We form the transpose while taking the conjugate of each element, obtaining

$$A^* = \begin{bmatrix} 1 & 2 & -2i \\ -i & 0 & 1 \\ 1-i & -i & 1+i \end{bmatrix}. \qquad \blacksquare$$

Following are some properties of the conjugate transpose that can easily be verified. (See Exercise 32.)

THEOREM 8.3 **Properties of the Conjugate Transpose**

Let A and B be $m \times n$ matrices. Then

(i) $(A^*)^* = A$,

(ii) $(A + B)^* = A^* + B^*$,

(iii) $(zA)^* = \bar{z}A^*$ for any scalar $z \in \mathbb{C}$,

(iv) If A and B are square matrices, $(AB)^* = B^*A^*$.

EXAMPLE 8 Using the properties in Theorem 8.3, show that, for any $n \times n$ matrix A, we have $(A + A^*)^* = A + A^*$.

Solution Using properties (i) and (ii) of Theorem 8.3, we have

$$(A + A^*)^* = A^* + (A^*)^* = A^* + A = A + A^*,$$

which is what we wished to show. ▪

Unitary and Hermitian Matrices

Recall that a *real orthogonal matrix* is a square matrix having orthogonal unit column vectors. The complex analogue of such a matrix is known by another name.

DEFINITION 8.3 Unitary Matrix

A square matrix U with complex entries is **unitary** if its column vectors are orthogonal unit vectors—that is, if $U^*U = I$.

Our next example gives an important property of unitary matrices that is familiar to us from the real orthogonal case. Notice in this example how handy the notation $\mathbf{v}^*\mathbf{v}$ is for $\langle \mathbf{v}, \mathbf{v} \rangle$.

EXAMPLE 9   Let $T: \mathbb{C}^n \to \mathbb{C}^n$ be a linear transformation having a unitary matrix U as matrix representation with respect to the standard basis. Show that $\|T(\mathbf{z})\| = \|\mathbf{z}\|$ for all $\mathbf{z} \in \mathbb{C}^n$.

Solution   We know that $T(\mathbf{z}) = U\mathbf{z}$, since U is the standard matrix representation of T. Since $\|\mathbf{v}\|^2 = \langle \mathbf{v}, \mathbf{v} \rangle = \mathbf{v}^*\mathbf{v}$, we find by using property (iv) of Theorem 8.3 that

$$\|U\mathbf{z}\|^2 = (U\mathbf{z})^*U\mathbf{z} = \mathbf{z}^*(U^*U)\mathbf{z} = \mathbf{z}^*I\mathbf{z} = \mathbf{z}^*\mathbf{z} = \|\mathbf{z}\|^2.$$

Taking square roots, we find that $\|U\mathbf{z}\| = \|\mathbf{z}\|$, so $\|T(\mathbf{z})\| = \|\mathbf{z}\|$. ■

Real symmetric matrices play an important role in linear algebra. We saw in Chapter 7 that they are very handy in work with quadratic forms. We also stated without proof in Theorem 4.5 that every real symmetric matrix is diagonalizable. Symmetric matrices are defined in terms of the transpose operation. The useful analogue of symmetry for matrices with complex scalars involves the conjugate transpose operation.

DEFINITION 8.4 Hermitian Matrix

A square matrix H is **Hermitian** if $H^* = H$, that is, if H is equal to its conjugate transpose.

If a square matrix H actually has real entries, it is Hermitian if and only if it is symmetric. Notice that the condition $H = H^* = (\bar{H})^T$ implies that the entries on the diagonal of any Hermitian matrix are real numbers.

EXAMPLE 10   Example 8 shows that, for any square matrix A, the matrix $A + A^*$ is Hermitian. Illustrate that the entries on the diagonal of $A + A^*$ are real, using the matrix A in Example 7.

Solution   Example 7 shows that, for the matrix

$$A = \begin{bmatrix} 1 & i & 1+i \\ 2 & 0 & i \\ 2i & 1 & 1-i \end{bmatrix},$$

we have

$$A^* = \begin{bmatrix} 1 & 2 & -2i \\ -i & 0 & 1 \\ 1-i & -i & 1+i \end{bmatrix}.$$

Thus,

$$A + A^* = \begin{bmatrix} 2 & 2+i & 1-i \\ 2-i & 0 & 1+i \\ 1+i & 1-i & 2 \end{bmatrix},$$

which has real diagonal entries. ▪

Hermitian matrices provide the proper generalization of real symmetric matrices to enable us to prove in the next section that every Hermitian matrix is diagonalizable. The special case of this theorem for real matrices thus finally provides us with a proof of the fundamental theorem that every real symmetric matrix is diagonalizable. The fundamental theorem for real symmetric matrices is tough to prove if we stay within the real number system, but it is a corollary of a fairly easy theorem for complex matrices. This illustrates how we can obtain true insight into theorems in real analysis and linear algebra by studying analogous concepts that use complex numbers.

SUMMARY

1. $\mathbb{C}^n$ is an n-dimensional complex vector space.

2. If $\mathbf{u} = (u_1, u_2, \ldots, u_n)$ and $\mathbf{v} = (v_1, v_2, \ldots, v_n)$ are vectors in $\mathbb{C}^n$, then $\langle \mathbf{u}, \mathbf{v} \rangle = \overline{u_1}v_1 + \overline{u_2}v_2 + \cdots + \overline{u_n}v_n$ is the Euclidean inner product of $\mathbf{u}$ and $\mathbf{v}$ and satisfies the properties in Theorem 8.2. In general, $\langle \mathbf{u}, \mathbf{v} \rangle \neq \langle \mathbf{v}, \mathbf{u} \rangle$.

3. For $\mathbf{u}, \mathbf{v} \in \mathbb{C}^n$, the vector $\mathbf{u} - \left(\dfrac{\langle \mathbf{v}, \mathbf{u} \rangle}{\langle \mathbf{v}, \mathbf{v} \rangle} \right) \mathbf{v}$ is perpendicular to $\mathbf{v}$.

4. The conjugate transpose of an $m \times n$ matrix A is the $n \times m$ matrix $A^* = (\bar{A})^T$. The conjugate transpose operation satisfies the properties in Theorem 8.3.

5. A square matrix U is unitary if $U^*U = I$. A real unitary matrix is an orthogonal matrix.

6. A square matrix H is Hermitian if $H = H^*$. A real Hermitian matrix is a symmetric matrix.

EXERCISES

1. Explain how $\mathbb{C}^n$ can be viewed as a $2n$-dimensional real vector space and as an n-dimensional complex vector space. Give a basis in each case.

2. Is it appropriate to view $\mathbb{R}^n$ as a subspace of $\mathbb{C}^n$? Explain.

3. Find AB and BA if

$$A = \begin{bmatrix} 1 & i & i \\ 1+i & 1 & -i \\ i & 1+i & 1-i \end{bmatrix} \quad \text{and}$$

$$B = \begin{bmatrix} -1 & 1+i & i \\ 2+i & i & 1-i \\ i & 1 & i \end{bmatrix}.$$

4. Find A^2 and A^4 if $A = \begin{bmatrix} 1 & 1+i \\ -1+i & i \end{bmatrix}$.

5. Find A^{-1} if $A = \begin{bmatrix} 1 & i \\ 1+i & 2+i \end{bmatrix}$.

6. Find A^{-1} if $A = \begin{bmatrix} i & 1+i \\ 1 & i \end{bmatrix}$.

7. Find A^{-1} if $A = \begin{bmatrix} 1 & i & 1-i \\ 1 & 1 & 1+i \\ 0 & -1+i & 1 \end{bmatrix}$.

8. Find A^{-1} if $A = \begin{bmatrix} i & 1-i & 1+i \\ 0 & 1 & i \\ 1-i & -i & 1-i \end{bmatrix}$.

9. Solve the linear system $Az = \begin{bmatrix} i \\ 1+i \\ i \end{bmatrix}$ if A is the

   matrix in Exercise 7.

10. Solve the linear system $Az = \begin{bmatrix} -1+i \\ 2+i \\ 1 \end{bmatrix}$ if A is

    the matrix in Exercise 8.

11. Find the solution space of the homogeneous system

$$z_1 + iz_2 + (1-i)z_3 = 0$$
$$(1+2i)z_1 - z_2 + z_3 = 0$$
$$(1+i)z_1 + 2iz_2 + (3-2i)z_3 = 0.$$

12. Find the rank of the matrix

$$\begin{bmatrix} 1 & 1+i & i & 1-i \\ 1+i & 2+i & 1-i & 1 \\ 1+i & 1+4i & 1+3i & 2-i \end{bmatrix}.$$

13. Find the rank of the matrix

$$\begin{bmatrix} 2-i & i & 1-i & 1+3i \\ i & 1+i & -1+2i & i \\ 1-i & 1+2i & 1-3i & 2+3i \end{bmatrix}.$$

14. Find each of the indicated inner products.

    a) $\langle (2+i, 2, i), (1, 1+i, 2-i) \rangle$ in $\mathbb{C}^3$

b) $\langle (1-i, 1+i), (1+i, 1-i) \rangle$ in $\mathbb{C}^2$

c) $\langle (1+i, 1, 1-i, 1), (i, 1+i, i, 1-i) \rangle$ in $\mathbb{C}^4$

d) $\langle (2-i, 3+i, 1+i),$
   $\qquad (1+i, 2-i, 1+i) \rangle$ in $\mathbb{C}^3$

15. Compute $\langle \mathbf{u}, \mathbf{v} \rangle$ and $\langle \mathbf{v}, \mathbf{u} \rangle$ for each of the following.

    a) $\mathbf{u} = (1, i)$, $\mathbf{v} = (1+i, 1-i)$
    b) $\mathbf{u} = (1+i, 2-i, i)$, $\mathbf{v} = (1, 1-i, 1+i)$

16. Verify property (i) of Theorem 8.2.

17. Verify property (ii) of Theorem 8.2.

18. Verify properties (iii) and (iv) of Theorem 8.2.

19. Verify property (v) of Theorem 8.2.

20. Find the magnitude of each of the following vectors.

    a) $(1, i, i)$
    b) $(1+i, 1-i, 1+i)$
    c) $(1+i, 2+i, 3+i)$
    d) $(i, 1+i, 1-i, i)$
    e) $(1+i, 1-i, i, 1-i)$

21. Determine whether the given pairs of vectors are parallel, perpendicular, or neither.

    a) $(1, i)$, $(i, 1)$
    b) $(1+i, i, 1-i)$, $(1-i, 1, -1-i)$
    c) $(1+i, 2-i)$, $(3i, 3+i)$
    d) $(1, i, 1-i)$, $(1-i, 1+i, 2)$
    e) $(1+i, 1-i, 1)$, $(i, 1-i, -3-i)$

22. Find a unit vector parallel to $(1+i, 1-i, i)$.

23. Find a vector of length 2 parallel to $(i, 1-i, 1+i, 1-i)$.

24. Find a unit vector perpendicular to $(2-i, 1+i)$.

25. Find a vector perpendicular to both $(1, i, 1-i)$ and $(1+i, 1-i, 1)$.

In Exercises 26–29, transform the given basis of $\mathbb{C}^n$ into an orthogonal basis, using the Gram–Schmidt process.

26. $\{(1+i, 1-i), (1, 1)\}$ in $\mathbb{C}^2$

27. $\{(2+i, 1+i), (1+i, i)\}$ in $\mathbb{C}^2$

28. $\{(1-i, 1+i, 1+i), (1, 1, -1-i),$
    $(1, i, -i)\}$ in $\mathbb{C}^3$

29. $\{(1, i, i), (1+i, 1, i), (i, 1+i, 0)\}$ in $\mathbb{C}^3$

30. Find the conjugate transpose of each of the following matrices.

a) $\begin{bmatrix} 1 & 1+i & 2 \\ 1-i & 1+i & 2 \end{bmatrix}$

b) $\begin{bmatrix} i & 1+i \\ 2+i & 1-i \end{bmatrix}$

c) $\begin{bmatrix} 1+i & 1+i \\ 2-i & 1-i \\ 1 & 1-2i \end{bmatrix}$

d) $\begin{bmatrix} 1+2i & i & 1-i \\ i & 1-i & 1+i \\ 1+i & 2-i & 1+3i \end{bmatrix}$

31. Label each of the following matrices as Hermitian, unitary, both, or neither.

a) $(1/\sqrt{3}) \begin{bmatrix} 1 & 1+i \\ 1-i & -1 \end{bmatrix}$

b) $\frac{1}{2} \begin{bmatrix} 1 & i & 1-i \\ -i & 2 & 1 \\ 1+i & 1 & 1 \end{bmatrix}$

c) $\frac{1}{\sqrt{6}} \begin{bmatrix} \sqrt{2} & \sqrt{2}i & \sqrt{2}i \\ \sqrt{3}i & 0 & \sqrt{3} \\ -i & -2 & 1 \end{bmatrix}$

d) $\frac{1}{4} \begin{bmatrix} 1 & i & 1-i \\ i & 3 & 1 \\ 1-i & 1 & 2 \end{bmatrix}$

32. Prove the properties of the conjugate transpose operation given in Theorem 8.3. [HINT: From Section 1.1, we know that analogous properties of the *transpose* operation hold for real matrices and real scalars and can be derived using just field properties of $\mathbb{R}$, so they are also true for matrices with complex entries. Thus we can focus on the effect of the *conjugation*. From Theorem 8.1, we know that $\overline{z + w} = \bar{z} + \bar{w}$, $\overline{zw} = \bar{z}\bar{w}$, and $\bar{\bar{z}} = z$, for $z, w \in \mathbb{C}$. Use these properties of conjugation to complete the proof of Theorem 8.3.]

33. Mark each of the following True or False. Assume that all matrices and scalars are complex.

___ a) The definition of a determinant, properties of determinants (the transpose property, the row-interchange property, and so on), and techniques for computing them are developed using only field properties of $\mathbb{R}$ in Chapter 3, and thus they remain equally valid for square complex matrices.

___ b) Cramer's rule is valid for square linear systems with complex coefficients.

___ c) If A is any square matrix and $\det(A) \neq 0$, then $\det(iA) \neq \det(A)$.

___ d) If U is unitary, then $U^{-1} = U^T$.

___ e) If U is unitary, then $(\bar{U})^{-1} = U^T$.

___ f) The Euclidean inner product in $\mathbb{C}^n$ is not commutative.

___ g) For $\mathbf{u}, \mathbf{v} \in \mathbb{C}^n$, we have $\langle \mathbf{u}, \mathbf{v} \rangle = \langle \mathbf{v}, \mathbf{u} \rangle$ if and only if $\langle \mathbf{u}, \mathbf{v} \rangle$ is a real number.

___ h) For a square matrix A, we have $\det(\bar{A}) = \overline{\det(A)}$.

___ i) For a square matrix A, we have $\det(A^*) = \det(A)$.

___ j) If U is a unitary matrix, then $\det(U^*) = \pm 1$.

34. Show that, for vectors $\mathbf{v}_1, \mathbf{v}_2, \ldots, \mathbf{v}_n$ in $\mathbb{C}^n$, $\{\mathbf{v}_1, \mathbf{v}_2, \ldots, \mathbf{v}_n\}$ is a basis for $\mathbb{C}^n$ if and only if $\{\bar{\mathbf{v}}_1, \bar{\mathbf{v}}_2, \ldots, \bar{\mathbf{v}}_n\}$ is a basis for $\mathbb{C}^n$.

35. Show that an $n \times n$ matrix U is unitary if and only if the rows of U form an orthonormal basis for $\mathbb{C}^n$.

36. Show that, if A is a square matrix, then AA^* is a Hermitian matrix.

37. Show that the product of two commuting $n \times n$ Hermitian matrices is also a Hermitian matrix. What can you say about the sum of two Hermitian matrices?

38. Show that the product of two $n \times n$ unitary matrices is also a unitary matrix. What about the sum?

39. Let $T: \mathbb{C}^n \to \mathbb{C}^n$ be a linear transformation whose standard matrix representation is a unitary matrix U. Show that $\langle T(\mathbf{u}), T(\mathbf{v}) \rangle = \langle \mathbf{u}, \mathbf{v} \rangle$ for all $\mathbf{u}, \mathbf{v} \in \mathbb{C}^n$. [HINT: Remember that $\langle \mathbf{u}, \mathbf{v} \rangle = \mathbf{u}^*\mathbf{v}$.]

40. Show that for $\mathbf{u}, \mathbf{v} \in \mathbb{C}^n$, we have $(\mathbf{u}^*\mathbf{v})^* = \overline{\mathbf{u}^*\mathbf{v}} = \mathbf{v}^*\mathbf{u} = \mathbf{u}^T\bar{\mathbf{v}}$.

41. Describe the unitary diagonal matrices.

42. Show that, if U is unitary, then $\bar{U}$, U^T, and U^* are unitary matrices also.

43. A square matrix A is **normal** if $A^*A = AA^*$.

a) Show that every Hermitian matrix is normal.
b) Show that every unitary matrix is normal.
c) Show that, if $A^* = -A$, then A is normal.

44. Let A be an $n \times n$ matrix. Referring to Exercise 43, show that, if A is normal, then $\|A\mathbf{z}\| = \|A^*\mathbf{z}\|$ for all $\mathbf{z} \in \mathbb{C}^n$.

45. Prove the converse of the statement in Exercise 44.

| 8.3 | **EIGENVALUES AND DIAGONALIZATION** |

Recall the fundamental theorem of real symmetric matrices that we stated without proof as Theorem 5.8:

> Every real symmetric matrix is diagonalizable by a real orthogonal matrix.

Our main goal in this section is to extend this result to complex matrices, as follows:

> Every Hermitian matrix is diagonalizable by a unitary matrix.

We will prove this theorem, which has the theorem for real symmetric matrices as an easy corollary.

Eigenvalues for Complex Matrices

We begin by extending the notions of eigenvalues and eigenvectors to complex matrices. The definitions are identical to those for real matrices. If A is an $n \times n$ complex matrix and if $A\mathbf{v} = \lambda\mathbf{v}$, where $\lambda \in \mathbb{C}$ and $\mathbf{v} \in \mathbb{C}^n$, $\mathbf{v} \neq \mathbf{0}$, then λ is an **eigenvalue** of A and $\mathbf{v}$ is a corresponding **eigenvector**. The zero vector and the set of all eigenvectors of A corresponding to λ constitute the **eigenspace** E_λ. Computation of eigenvalues and eigenspaces of a complex matrix is the same as for real matrices, except that the arithmetic involves complex numbers and consequently is more laborious to do with pencil and paper. Every $n \times n$ complex matrix has n not necessarily distinct eigenvalues. This is a consequence of the Fundamental Theorem of Algebra, which we stated in Section 8.1. Recall that, for real matrices, there may exist no real eigenvalues.

EXAMPLE 1 Find the eigenvalues and eigenspaces of the matrix

$$A = \begin{bmatrix} 1 & 0 & i \\ 0 & 2 & 0 \\ -i & 0 & 1 \end{bmatrix}.$$

Solution The characteristic polynomial of A is

$$\det(A - \lambda I) = \begin{vmatrix} 1 - \lambda & 0 & i \\ 0 & 2 - \lambda & 0 \\ -i & 0 & 1 - \lambda \end{vmatrix} = (2 - \lambda)((1 - \lambda)^2 + i^2)$$

$$= (2 - \lambda)(1 - 2\lambda + \lambda^2 - 1) = -\lambda(2 - \lambda)^2.$$

The three roots of $-\lambda(2 - \lambda)^2 = 0$ are $\lambda_1 = 0$, $\lambda_2 = \lambda_3 = 2$.
For the eigenvalue $\lambda_1 = 0$, we have

$$A - \lambda_1 I = \begin{bmatrix} 1 & 0 & i \\ 0 & 2 & 0 \\ -i & 0 & 1 \end{bmatrix} \sim \begin{bmatrix} 1 & 0 & i \\ 0 & 1 & 0 \\ 0 & 0 & 0 \end{bmatrix},$$

which gives the eigenspace $E_0 = \mathrm{sp}\left(\begin{bmatrix} -i \\ 0 \\ 1 \end{bmatrix} \right)$. For the double root $\lambda_2 = \lambda_3 = 2$, we have

$$A - 2I = \begin{bmatrix} -1 & 0 & i \\ 0 & 0 & 0 \\ -i & 0 & -1 \end{bmatrix} \sim \begin{bmatrix} 1 & 0 & -i \\ 0 & 0 & 0 \\ 0 & 0 & 0 \end{bmatrix},$$

which gives the two-dimensional eigenspace $E_2 = \mathrm{sp}\left(\begin{bmatrix} i \\ 0 \\ 1 \end{bmatrix}, \begin{bmatrix} 0 \\ 1 \\ 0 \end{bmatrix} \right)$. ■

Definitions and theorems concerning eigenvalues and eigenvectors that depend only on the field axioms discussed at the beginning of Section 8.2 continue to make sense and to hold for complex matrices. In particular, an $n \times n$ complex matrix A is **diagonalizble** if and only if there exist an invertible matrix C and a diagonal matrix D such that $D = C^{-1}AC$. Just as for real matrices, two complex $n \times n$ matrices A and B are **similar** if there exists an invertible $n \times n$ matrix C such that $B = C^{-1}AC$. Similarity is an equivalence relation. Thus, A is similar to A; if A is similar to B, then B is similar to A; and if furthermore B is similar to D, then A is similar to D. All of these things are defined and proved using just field properties.

Consider again the equation $D = C^{-1}AC$, where D is a diagonal matrix. The equivalent equation, $CD = AC$, for an invertible matrix C shows that A is diagonalizable if and only if $\mathbb{C}^n$ has a basis of eigenvectors of A, and it shows that the matrix C must have such a basis of eigenvectors as its column vectors, while D has the corresponding eigenvalues on its diagonal. We obtain all of this from $CD = AC$ by considering the jth column vector of CD and comparing it to the jth column vector of AC, just as we did for the real case in Section 4.2. Such a basis for $\mathbb{C}^n$ of eigenvectors of A exists if and only if the algebraic multiplicity of each eigenvalue is equal to its geometric multiplicity (the dimension of the corresponding eigenspace).

EXAMPLE 2 Let $A = \begin{bmatrix} 1 & 0 & i \\ 0 & 2 & 0 \\ -i & 0 & 1 \end{bmatrix}$. Find a matrix C such that $C^{-1}AC$ is a diagonal matrix.

Solution From the preceding example, we see that A has an eigenvalue $\lambda_1 = 0$ of algebraic multiplicity 1 with eigenspace

$$E_0 = \mathrm{sp}\left(\begin{bmatrix} -i \\ 0 \\ 1 \end{bmatrix} \right)$$

and that it has the double eigenvalue $\lambda_2 = \lambda_3 = 2$ with eigenspace

$$E_2 = \text{sp}\left(\begin{bmatrix} i \\ 0 \\ 1 \end{bmatrix}, \begin{bmatrix} 0 \\ 1 \\ 0 \end{bmatrix}\right).$$

Thus, the algebraic multiplicity of each eigenvalue is equal to its geometric multiplicity, and the vectors shown form a basis for $\mathbb{C}^n$. Therefore, the matrix

$$C = \begin{bmatrix} -i & i & 0 \\ 0 & 0 & 1 \\ 1 & 1 & 0 \end{bmatrix}$$

is invertible and diagonalizes A; and we must also have $C^{-1}AC = D$, where

$$D = \begin{bmatrix} 0 & 0 & 0 \\ 0 & 2 & 0 \\ 0 & 0 & 2 \end{bmatrix}. \qquad ■$$

The proof of Theorem 4.3, which asserts that eigenvectors corresponding to distinct eigenvalues are independent, depends only on field properties and thus is valid in the complex case. Consequently, every matrix having only eigenvalues of algebraic multiplicity 1 is diagonalizable. We focus our attention on the geometric multiplicity of any eigenvalues of algebraic multiplicity greater than 1, when determining whether a matrix is diagonalizable.

EXAMPLE 3   Find all values of c for which the matrix

$$A = \begin{bmatrix} i & c & 1 \\ 0 & i & 2i \\ 0 & 0 & 1 \end{bmatrix}$$

is diagonalizable.

Solution   The eigenvalues of the upper-triangular matrix A are $\lambda_1 = \lambda_2 = i$ and $\lambda_3 = 1$. We focus our attention on the eigenvalue i of algebraic multiplicity 2. For A to be diagonalizable, its eigenspace E_i must have geometric multiplicity 2. The eigenspace is the nullspace of the matrix

$$A - \lambda_1 I = \begin{bmatrix} 0 & c & 1 \\ 0 & 0 & 2i \\ 0 & 0 & 1 - i \end{bmatrix},$$

and the nullspace has dimension 2 if and only if the matrix has rank 1, which is the case if and only if $c = 0$.   ■

Diagonalization of Hermitian Matrices

Diagonalization via a unitary matrix is of special importance, as we saw in the real case, where it becomes diagonalization by an orthogonal matrix. We call $n \times n$ matrices A and B **unitarily equivalent** if there is a unitary matrix U such that $B = U^{-1}AU$.

Since the inverse of a unitary matrix is unitary and since a product of unitary matrices is unitary, we can show that unitary equivalence is again an equivalence relation. Thus, A is unitarily equivalent to itself; and if A is unitarily equivalent to B, then B is unitarily equivalent to A; if furthermore B is unitarily equivalent to C, then A is unitarily equivalent to C.

Now we establish the main goal of this section: Hermitian matrices are unitarily equivalent to a diagonal matrix. That is, a Hermitian matrix can be diagonalized using a unitary matrix. This follows from a very important result known as *Schur's lemma* (or *Schur's unitary triangularization theorem*), which we state, deferring the proof until the end of this section.

THEOREM 8.4 Schur's Lemma

Let A be an $n \times n$ (complex) matrix. There is a unitary matrix U such that $U^{-1}AU$ is upper triangular.

Using Schur's lemma, we can prove that every Hermitian matrix is diagonalizable, and that the diagonalizing matrix can be chosen to be unitary. We express this by saying that every Hermitian matrix is **unitarily diagonalizable**.

THEOREM 8.5 Spectral Theorem for Hermitian Matrices

If A is a Hermitian matrix, there exists a unitary matrix U such that $U^{-1}AU$ is a diagonal matrix. Furthermore all eigenvalues of A are real.

PROOF By Schur's lemma, there exists a unitary matrix U such that $U^{-1}AU$ is an upper-triangular matrix. Since U is unitary, we have $U^*U = I$, so $U^{-1} = U^*$; and since A is Hermitian, we also know that $(A^*)^* = A$. Thus, we have

$$(U^{-1}AU)^* = (U^*AU)^* = U^*A^*(U^*)^* = U^*AU = U^{-1}AU,$$

which shows that the upper-triangular matrix $U^{-1}AU$ is also Hermitian. Since the conjugate transpose of an upper-triangular matrix is a lower-triangular matrix, we see that the entries above the diagonal in $U^{-1}AU$ must all be zero; therefore, $U^{-1}AU = D$, where D is a diagonal matrix. Thus, A is unitarily diagonalizable.

It remains to be shown that each eigenvalue of A is a real number. From the theory of diagonalization, we know that the entries on the diagonal of D are the eigenvalues of A. Now we showed in the preceding paragraph that the matrix $D = U^{-1}AU$ is Hermitian, so $D^* = D$. Forming the conjugate transpose for a diagonal matrix amounts simply to taking the conjugates of the entries on the diagonal. Since $D^* = D$, the entries on the diagonal of D remain unchanged under conjugation, so they must be real numbers. ▲

COROLLARY Fundamental Theorem of Real Symmetric Matrices

Every $n \times n$ real symmetric matrix has n real eigenvalues, counted with their algebraic multiplicity, and is diagonalizable by a real orthogonal matrix.

PROOF Since every real $n \times n$ symmetric matrix A is also Hermitian, Theorem 8.5 establishes that all of its eigenvalues in $\mathbb{C}$ actually lie in $\mathbb{R}$; therefore, the matrix has n real eigenvalues, counting them with their algebraic multiplicity. Furthermore, Theorem 8.5 asserts that A can be diagonalized by a unitary matrix U. We know that the column vectors of U are eigenvectors of A. Now the eigenvectors of A can be computed by row reductions of $A - \lambda_i I$, where the λ_i are eigenvalues of A. Since all the λ_i are real, the row reductions all take place in the field $\mathbb{R}$ of real numbers. The reduced echelon form of $A - \lambda_i I$ is thus a *real* matrix; it must have a nullspace of dimension (geometric multiplicity) equal to the algebraic multiplicity of λ_i, since A is diagonalizable. Thus, we can find bases for the eigenspaces E_{λ_i} consisting of vectors in $\mathbb{R}^n$. Using the Gram–Schmidt process, we can assume that the basis of each eigenspace is orthonormal. The matrix C having as column vectors the vectors in these orthonormal bases of eigenspaces E_{λ_i} is thus a real orthogonal matrix that diagonalizes A. ▲

EXAMPLE 4 Find a unitary matrix that diagonalizes the matrix A in Example 2.

Solution We found in Example 2 that the matrix

$$C = \begin{bmatrix} -i & i & 0 \\ 0 & 0 & 1 \\ 1 & 1 & 0 \end{bmatrix}$$

diagonalizes A. Notice that the inner product of any two distinct column vectors of this matrix is zero, so the column vectors are orthogonal. We need only normalize them to length 1 in order to obtain a unitary matrix that diagonalizes A. Thus, such a matrix is

$$U = (1/\sqrt{2}) \begin{bmatrix} -i & i & 0 \\ 0 & 0 & \sqrt{2} \\ 1 & 1 & 0 \end{bmatrix}.$$ ■

Recall that in Theorem 5.7 we showed that real eigenvectors of a symmetric matrix corresponding to distinct eigenvalues are orthogonal. Generalizing this, we can show that the eigenvectors of a Hermitian matrix corresponding to distinct eigenvalues are orthogonal. We ask you to show this in Exercise 21, using the fact that a Hermitian matrix can be unitarily diagonalized; but it is easy to demonstrate this orthogonality by using properties of matrices and the fact that the eigenvalues must be real.

THEOREM 8.6 Orthogonality of Eigenspaces of a Hermitian Matrix

The eigenvectors of a Hermitian matrix corresponding to distinct eigenvalues are orthogonal.

PROOF Let **v** and **w** be eigenvectors of a Hermitian matrix A corresponding to distinct eigenvalues λ_1 and λ_2, respectively. Using the facts that $A = A^*$ and that the eigenvalues are real, so that $\lambda_2^* = \lambda_2$, we have

$$\lambda_1(\mathbf{w}^*\mathbf{v}) = \mathbf{w}^*(\lambda_1\mathbf{v}) = \mathbf{w}^*(A\mathbf{v}) = \mathbf{w}^*(A^*\mathbf{v}) = (\mathbf{w}^*A^*)\mathbf{v}$$
$$= (A\mathbf{w})^*\mathbf{v} = (\lambda_2\mathbf{w})^*\mathbf{v} = \lambda_2(\mathbf{w}^*\mathbf{v}).$$

Therefore, $(\lambda_1 - \lambda_2)(\mathbf{w}^*\mathbf{v}) = 0$. Since $\lambda_1 \neq \lambda_2$, we must have $\mathbf{w}^*\mathbf{v} = 0$, so **w** and **v** are orthogonal. ▲

EXAMPLE 5   Find a unitary matrix C that diagonalizes the Hermitian matrix

$$A = \begin{bmatrix} -1 & i & 1+i \\ -i & 1 & 0 \\ 1-i & 0 & 1 \end{bmatrix}.$$

Solution   We find that

$$|A - \lambda I| = \begin{vmatrix} -1-\lambda & i & 1+i \\ -i & 1-\lambda & 0 \\ 1-i & 0 & 1-\lambda \end{vmatrix}$$

$$= (-1-\lambda)\begin{vmatrix} 1-\lambda & 0 \\ 0 & 1-\lambda \end{vmatrix} - i\begin{vmatrix} -i & 0 \\ 1-i & 1-\lambda \end{vmatrix} +$$

$$(1+i)\begin{vmatrix} -i & 1-\lambda \\ 1-i & 0 \end{vmatrix}$$

$$= (-1-\lambda)(1-\lambda)^2 - i(-i)(1-\lambda) + (1+i)(-1)(1-i)(1-\lambda)$$
$$= (1-\lambda)(\lambda^2 - 1 - 1 - 2) = (1-\lambda)(\lambda^2 - 4).$$

Thus, the eigenvalues are $\lambda_1 = 1$, $\lambda_2 = 2$, and $\lambda_3 = -2$. To find U, we need only compute one eigenvector of length 1 for each of these three distinct eigenvalues. The three eigenvectors we obtain must form an orthonormal set, according to Theorem 8.6.

For $\lambda_1 = 1$, we find that

$$A - I = \begin{bmatrix} -2 & i & 1+i \\ -i & 0 & 0 \\ 1-i & 0 & 0 \end{bmatrix},$$

so an eigenvector is

$$\mathbf{v}_1 = \begin{bmatrix} 0 \\ 1+i \\ -i \end{bmatrix}.$$

For $\lambda_2 = 2$, we find that

$$A - 2I = \begin{bmatrix} -3 & i & 1+i \\ -i & -1 & 0 \\ 1-i & 0 & -1 \end{bmatrix} \sim \begin{bmatrix} 1 & -i & 0 \\ 0 & -2i & 1+i \\ 0 & 1+i & -1 \end{bmatrix} \sim \begin{bmatrix} 1 & -i & 0 \\ 0 & 1 & \dfrac{i-1}{2} \\ 0 & 0 & 0 \end{bmatrix},$$

so a corresponding eigenvector is

$$\mathbf{v}_2 = \begin{bmatrix} 1 + i \\ 1 - i \\ 2 \end{bmatrix}.$$

Finally, for $\lambda_3 = -2$, we have

$$A + 2I = \begin{bmatrix} 1 & i & 1+i \\ -i & 3 & 0 \\ 1-i & 0 & 3 \end{bmatrix} \sim \begin{bmatrix} 1 & i & 1+i \\ 0 & 2 & -1+i \\ 0 & -1-i & 1 \end{bmatrix} \sim \begin{bmatrix} 1 & i & 1+i \\ 0 & 1 & \dfrac{-1+i}{2} \\ 0 & 0 & 0 \end{bmatrix},$$

and a corresponding eigenvector is

$$\mathbf{v}_3 = \begin{bmatrix} -3 - 3i \\ 1 - i \\ 2 \end{bmatrix}.$$

We normalize the vectors $\mathbf{v}_1$, $\mathbf{v}_2$, and $\mathbf{v}_3$ and form the column vectors in U from the resulting vectors of magnitude 1, obtaining

$$U = \begin{bmatrix} 0 & (1+i)/(2\sqrt{2}) & (-3-3i)/(2\sqrt{6}) \\ (1+i)/\sqrt{3} & (1-i)/(2\sqrt{2}) & (1-i)/(2\sqrt{6}) \\ -i/\sqrt{3} & 1/\sqrt{2} & 1/\sqrt{6} \end{bmatrix}. \qquad ■$$

A Criterion for Unitary Diagonalization

We have seen that every Hermitian matrix is unitarily diagonalizable, but of course a unitarily diagonalizable matrix need not be Hermitian. For example, the 1×1 matrix $[i]$ is already diagonal, so it is diagonalizable by the identity matrix. However, it is not Hermitian since $[i]^* = [-i]$. There is actually a way to determine whether a square matrix is unitarily diagonalizable, without having to find its eigenvalues and eigenvectors.

DEFINITION 8.5 Normal Matrix

A square matrix A is **normal** if it commutes with its conjugate transpose, so that $AA^* = A^*A$.

Exercises 25 and 26 ask you to prove the following theorem, which gives a criterion for A to be unitarily diagonalizable in terms of matrix multiplication!

THEOREM 8.7 Spectral Theorem for Normal Matrices

A square matrix A is unitarily diagonalizable if and only if it is a normal matrix.

EXAMPLE 6   Determine all values of a such that the matrix

$$A = \begin{bmatrix} i & a \\ 2 & i \end{bmatrix}$$

is unitarily diagonalizable.

Solution   In order for the matrix to be unitarily diagonalizable, we must have $AA^* = A^*A$ so that

$$\begin{bmatrix} i & a \\ 2 & i \end{bmatrix}\begin{bmatrix} -i & 2 \\ \bar{a} & -i \end{bmatrix} = \begin{bmatrix} -i & 2 \\ \bar{a} & -i \end{bmatrix}\begin{bmatrix} i & a \\ 2 & i \end{bmatrix}.$$

Equating entries, we obtain

row 1, column 1: $1 + |a|^2 = 1 + 4$,   so $|a| = 2$,

row 1, column 2: $2i - ai = -ai + 2i$,

row 2, column 1: $-2i + \bar{a}i = \bar{a}i - 2i$,

row 2, column 2: $4 + 1 = |a|^2 + 1$,   so $|a| = 2$.

Clearly these conditions are satisfied as long as $|a| = 2$, so a can be any number of the form $x + yi$, where $x^2 + y^2 = 4$.   ■

Proof of Shur's Lemma

We now prove by induction that, if A is an $n \times n$ matrix, there exists a unitary matrix U such that $U^{-1}AU = U^*AU$ is upper triangular. If $n = 1$, the lemma is trivial. We assume as induction hypothesis that the lemma is true for all matrices of size at most $(n - 1) \times (n - 1)$, and we proceed to show that it must hold for an $n \times n$ matrix A.

Let λ_1 be an eigenvalue of A, and let $\mathbf{v}_1$ be a corresponding eigenvector of norm 1. We can expand $\{\mathbf{v}_1\}$ to a basis for $\mathbb{C}^n$, and by the Gram–Schmidt process we can transform it into an orthonormal basis $\{\mathbf{v}_1, \mathbf{v}_2, \ldots, \mathbf{v}_n\}$. Let U_1 be the unitary matrix whose jth column vector is $\mathbf{v}_j$. Now the first column vector of AU_1 is $A\mathbf{v}_1 = \lambda_1\mathbf{v}_1$. Since the ith row vector of U_1^* is $\mathbf{v}_i^*$, and since the vectors $\mathbf{v}_i$ are mutually orthogonal, we see that the first column vector of the matrix $U_1^*AU_1$ is

$$U_1^*(\lambda_1\mathbf{v}_1) = \begin{bmatrix} \lambda_1 \\ 0 \\ \vdots \\ 0 \end{bmatrix}.$$

This shows that we can write $U_1^*AU_1$ symbolically as

$$U_1^*AU_1 = \begin{bmatrix} \lambda_1 & X & X & \cdots & X \\ 0 & & & & \\ \vdots & & & A_1 & \\ 0 & & & & \end{bmatrix}, \tag{1}$$

where we have denoted the $(n - 1) \times (n - 1)$ submatrix in the lower right-hand corner of $U_1^*AU_1$ by A_1. By our induction hypothesis, there exists an $(n - 1) \times (n - 1)$ unitary matrix C such that $C^*A_1C = B$, where B is upper triangular. Let

$$U_2 = \begin{bmatrix} 1 & 0 & 0 & \cdots & 0 \\ 0 & & & & \\ \vdots & & & C & \\ 0 & & & & \end{bmatrix}, \tag{2}$$

where we have used a symbolic notation similar to that in Eq. (1). Since C is unitary, it is clear that U_2 is a unitary matrix. Now let $U = U_1U_2$. Since $U^*U = U_2^*(U_1^*U_1)U_2 = U_2^*IU_2 = U_2^*U_2 = I$, we see that U is a unitary matrix. Now

$$U^*AU = U_2^*(U_1^*AU_1)U_2. \tag{3}$$

The matrix in parentheses in Eq. (3) is the matrix displayed in Eq. (1). From our definition of U_2 in Eq. (2), we see that the $(n - 1) \times (n - 1)$ block in the lower right-hand corner of U^*AU is $C^*A_1C = B$. Thus, we have

$$U^*AU = \begin{bmatrix} \lambda_1 & X & X & \cdots & X \\ 0 & & & & \\ \vdots & & & B & \\ 0 & & & & \end{bmatrix},$$

which is upper triangular because B is an upper-triangular matrix. This completes our induction argument. ▲

SUMMARY

1. An $n \times n$ matrix A is diagonalizable if and only if $\mathbb{C}^n$ has a basis consisting of eigenvectors of A. Equivalently, each eigenvalue has algebraic multiplicity equal to its geometric multiplicity.

2. Every Hermitian matrix is diagonalizable by a unitary matrix.

3. Every Hermitian matrix has real eigenvalues.

4. A square matrix A is unitarily diagonalizable if and only if it is normal, so that $AA^* = A^*A$.

5. Schur's lemma states that every square matrix is unitarily equivalent to an upper-triangular matrix.

EXERCISES

In Exercises 1–12, find a unitary matrix U and a diagonal matrix D such that $D = U^{-1}AU$ for the given matrix A.

1. $A = \begin{bmatrix} 1 & i \\ -i & 1 \end{bmatrix}$ 　　**2.** $A = \begin{bmatrix} 1 & 2i \\ -2i & 1 \end{bmatrix}$

3. $A = \begin{bmatrix} 1 & 1+i \\ 1-i & 2 \end{bmatrix}$ 　　**4.** $A = \begin{bmatrix} 9 & 3-i \\ 3+i & 0 \end{bmatrix}$

5. $A = \begin{bmatrix} 0 & i & 0 \\ -i & 0 & 0 \\ 0 & 0 & 1 \end{bmatrix}$ 　　**6.** $A = \begin{bmatrix} 1 & -i & 0 \\ i & 1 & 0 \\ 0 & 0 & 2 \end{bmatrix}$

7. $A = \begin{bmatrix} 1 & 2-2i & 0 \\ 2+2i & -1 & 0 \\ 0 & 0 & 3 \end{bmatrix}$

8. $A = \begin{bmatrix} 0 & 0 & 1+2i \\ 0 & 5 & 0 \\ 1-2i & 0 & 4 \end{bmatrix}$

9. $A = \begin{bmatrix} 1 & 0 & 2+2i \\ 0 & -3 & 0 \\ 2-2i & 0 & -1 \end{bmatrix}$

10. $A = \begin{bmatrix} 2 & 0 & 1-i \\ 0 & 3 & 0 \\ 1+i & 0 & 1 \end{bmatrix}$

11. $A = \begin{bmatrix} 3 & i & 1+i \\ -i & 1 & 0 \\ 1-i & 0 & 1 \end{bmatrix}$

12. $A = \begin{bmatrix} -3 & 5i & 1+i \\ -5i & 3 & 0 \\ 1-i & 0 & 3 \end{bmatrix}$

13. Find all $a \in \mathbb{C}$ such that the matrix $\begin{bmatrix} i & 4 \\ a & i \end{bmatrix}$ is unitarily diagonalizable.

14. Find all $a, b \in \mathbb{C}$ such that the matrix $\begin{bmatrix} i & a \\ b & i \end{bmatrix}$ is unitarily diagonalizable.

15. Find all $a \in \mathbb{C}$ such that the matrix $\begin{bmatrix} i & a \\ 1 & 3i \end{bmatrix}$ is unitarily diagonalizable.

16. Find all $a, b \in \mathbb{C}$ such that the matrix $\begin{bmatrix} a & -i \\ i & b \end{bmatrix}$ is unitarily diagonalizable.

17. Show that every 2×2 real matrix that is unitarily diagonalizable has one of the following forms:
$\begin{bmatrix} a & b \\ b & d \end{bmatrix}, \begin{bmatrix} a & b \\ -b & a \end{bmatrix}$, for $a, b, d \in \mathbb{R}$.

18. Determine whether the matrix $\begin{bmatrix} i & -1 & 1 \\ 1 & -i & -1 \\ -1 & 1 & i \end{bmatrix}$ is unitarily diagonalizable.

19. Mark each of the following True or False.

____ a) Every square matrix is unitarily equivalent to a diagonal matrix.

____ b) Every square matrix is unitarily equivalent to an upper-triangular matrix.

____ c) Every Hermitian matrix is unitarily equivalent to a diagonal matrix.

____ d) Every unitarily diagonalizable matrix is Hermitian.

____ e) Every real symmetric matrix is Hermitian.

____ f) Every diagonalizable matrix is normal.

____ g) Every unitarily diagonalizable matrix is normal.

____ h) Every real symmetric matrix is normal.

____ i) Every square matrix is diagonalizable, although perhaps not by a unitary matrix.

____ j) Every square matrix with eigenvalues of algebraic multiplicity 1 is diagonalizable by a unitary matrix.

20. Show that the eigenvalues of a Hermitian matrix are real, without using Theorem 8.5 or Schur's lemma. [HINT: Let $A\mathbf{v} = \lambda\mathbf{v}$, and use the fact that $\mathbf{v}^*A\mathbf{v} = \mathbf{v}^*A^*\mathbf{v}$.]

21. Argue directly from Theorem 8.5 that eigenvectors from different eigenspaces of a Hermitian matrix are orthogonal.

22. Suppose that A is an $n \times n$ matrix such that $A^* = -A$. Show that

a) A has eigenvalues of the form ri, where $r \in \mathbb{R}$,

b) A is diagonalizable by a unitary matrix. [HINT FOR BOTH PARTS: Work with iA.]

23. Show that an $n \times n$ matrix A is unitarily diagonalizable if and only if $\|A\mathbf{v}\| = \|A^*\mathbf{v}\|$ for all $\mathbf{v} \in \mathbb{C}^n$.

24. Show that a normal matrix is Hermitian if and only if all its eigenvalues are in $\mathbb{R}$.

25. a) Show that a diagonal matrix is normal.
    b) Show that, if A is normal and B is unitarily equivalent to A, then B is normal.
    c) Deduce from parts (a) and (b) that a unitarily diagonalizable matrix is normal.

26. a) Show that every normal matrix A is unitarily equivalent to a normal upper-triangular matrix B. (Use Schur's lemma and part (b) of Exercise 25.)

    b) Show that an $n \times n$ normal upper-triangular matrix B must be diagonal. [HINT: Let $C = B^*B = BB^*$. Equating the computations of c_{11} from B^*B and from BB^*, show that $b_{1j} = 0$ for $1 < j \le n$. Then equate the computations of c_{22} form B^*B and from BB^* to show that $b_{2j} = 0$ for $2 < j \le n$. Continue this process to show that B is lower triangular.]

    c) Deduce from parts (a) and (b) that a normal matrix is unitarily diagonalizable.

8.4   **JORDAN CANONICAL FORM**

Jordan Blocks

We have spent considerable time on diagonalization of matrices. The preceding section was concerned primarily with unitary diagonalization. As we have seen, diagonal matrices are easily handled. Unfortunately, not every $n \times n$ matrix A can be diagonalized, because we cannot always find a basis for $\mathbb{C}^n$ consisting of eigenvectors of A. We remind you of this with an example that is well worth studying.

EXAMPLE 1   Show that the matrix

$$J = \begin{bmatrix} 5 & 1 & 0 \\ 0 & 5 & 0 \\ 0 & 0 & 5 \end{bmatrix}$$

is not diagonalizable.

Solution   We see that 5 is the only eigenvalue of J, and that it has algebraic multiplicity 3. However,

$$J - 5I = \begin{bmatrix} 0 & 1 & 0 \\ 0 & 0 & 0 \\ 0 & 0 & 0 \end{bmatrix},$$

which shows that the eigenspace E_5 has dimension 2 and has basis $\{\mathbf{e}_1, \mathbf{e}_3\}$. Thus, the geometric multiplicity of the eigenvalue 5 is only 2, and J is not diagonalizable. We cannot find a basis for $\mathbb{C}^3$ consisting entirely of eigenvectors of J. ■

Let us examine the matrix J in Example 1 a bit more. Notice that, while $J - 5I$ has a nullspace of dimension 2, the matrix $(J - 5I)^2$ is the zero matrix and has all of $\mathbb{C}^3$ as nullspace. Moreover, multiplication on the left by $J - 5I$ carries $\mathbf{e}_2$ into $\mathbf{e}_1$ and carries both $\mathbf{e}_1$ and $\mathbf{e}_3$ into $\mathbf{0}$. We say that $J - 5I$ **annihilates** $\mathbf{e}_1$ and $\mathbf{e}_3$. The action of $J - 5I$ on these standard basis vectors is denoted schematically by the two strings

$$J - 5I: \quad \begin{array}{l} \mathbf{e}_2 \rightarrow \mathbf{e}_1 \rightarrow \mathbf{0}, \\ \mathbf{e}_3 \rightarrow \mathbf{0}. \end{array} \tag{1}$$

Diagram (1) also shows that $(J - 5i)^2$ maps each of these basis vectors into $\mathbf{0}$. Since $(J - 5I)\mathbf{e}_2 = \mathbf{e}_1$, we have $J\mathbf{e}_2 = 5\mathbf{e}_2 + \mathbf{e}_1$, while $J\mathbf{e}_1 = 5\mathbf{e}_1$ and $J\mathbf{e}_3 = 5\mathbf{e}_3$.

EXAMPLE 2 Let

$$J = \begin{bmatrix} \lambda & 1 & 0 & 0 & 0 \\ 0 & \lambda & 1 & 0 & 0 \\ 0 & 0 & \lambda & 1 & 0 \\ 0 & 0 & 0 & \lambda & 1 \\ 0 & 0 & 0 & 0 & \lambda \end{bmatrix}. \tag{2}$$

Discuss the action of $J - \lambda I$ on the standard basis vectors, drawing a schematic diagram similar to diagram (1). Describe also the action of J on the vectors in the standard basis.

Solution We find that

$$J - \lambda I = \begin{bmatrix} 0 & 1 & 0 & 0 & 0 \\ 0 & 0 & 1 & 0 & 0 \\ 0 & 0 & 0 & 1 & 0 \\ 0 & 0 & 0 & 0 & 1 \\ 0 & 0 & 0 & 0 & 0 \end{bmatrix}.$$

We see that multiplication of $\mathbf{e}_i$ on the left by $J - \lambda I$ yields $\mathbf{e}_{i-1}$ for $2 \leq i \leq 5$, while $(J - \lambda I)\mathbf{e}_1 = \mathbf{0}$. Schematically, we have just one *string*,

$$J - \lambda I: \quad \mathbf{e}_5 \rightarrow \mathbf{e}_4 \rightarrow \mathbf{e}_3 \rightarrow \mathbf{e}_2 \rightarrow \mathbf{e}_1 \rightarrow \mathbf{0}. \tag{3}$$

Left-multiplication by J yields

$$J\mathbf{e}_5 = \lambda\mathbf{e}_5 + \mathbf{e}_4, \ J\mathbf{e}_4 = \lambda\mathbf{e}_4 + \mathbf{e}_3, \ J\mathbf{e}_3 = \lambda\mathbf{e}_3 + \mathbf{e}_2,$$
$$J\mathbf{e}_2 = \lambda\mathbf{e}_2 + \mathbf{e}_1, \ J\mathbf{e}_1 = \mathbf{0}. \qquad ■$$

The matrix J in Example 2 is an example of a *Jordan block matrix*.

DEFINITION 8.6 Jordan Block

An $m \times m$ matrix is a **Jordan block** if it is structured as follows:

(i) All diagonal entries are equal.
(ii) Each entry immediately above a diagonal entry is 1.
(iii) All other entries are zero.

Thus, the matrix J of Example 2 is a Jordan block. However, the matrix of Example 1 is not a Jordan block, since the entry 5 at the bottom of the diagonal does not have a 1 just above it. A Jordan block has the properties described in the next theorem. These properties were illustrated in Example 2, and we leave a formal proof to you if you desire one. Notice that, for an $m \times m$ Jordan block

$$J = \begin{bmatrix} \lambda & 1 & 0 & \cdots & 0 & 0 \\ 0 & \lambda & 1 & \cdots & 0 & 0 \\ & & & \vdots & & \\ 0 & 0 & 0 & \cdots & \lambda & 1 \\ 0 & 0 & 0 & \cdots & 0 & \lambda \end{bmatrix},$$

we have just one string:

$$J - \lambda I: \qquad \mathbf{e}_m \to \mathbf{e}_{m-1} \to \cdots \to \mathbf{e}_2 \to \mathbf{e}_1 \to \mathbf{0}.$$

THEOREM 8.8 Properties of a Jordan Block

Let J be an $m \times m$ Jordan block with diagonal entries all equal to λ. Then the following properties hold:

(i) $(J - \lambda I)\mathbf{e}_i = \mathbf{e}_{i-1}$ for $1 < i \le m$, and $(J - \lambda I)\mathbf{e}_1 = \mathbf{0}$.

(ii) $(J - \lambda I)^m = O$, but $(J - \lambda I)^i \ne O$ for $i < m$.

(iii) $J\mathbf{e}_i = \lambda\mathbf{e}_i + \mathbf{e}_{i-1}$ for $1 < i \le m$, while $J\mathbf{e}_1 = \lambda\mathbf{e}_1$.

Jordan Canonical Forms

We have seen that not every $n \times n$ matrix is diagonalizable. It is our purpose in this section to show that every $n \times n$ matrix is similar to a matrix having all entries 0 except for those on the diagonal and entries 1 immediately above some diagonal entries; each 1 above a diagonal entry must have the same number on its left as below it on the diagonal. An example of such a matrix is

$$J = \begin{bmatrix} -i & 1 & 0 & 0 & 0 & 0 & 0 & 0 \\ 0 & -i & 1 & 0 & 0 & 0 & 0 & 0 \\ 0 & 0 & -i & 0 & 0 & 0 & 0 & 0 \\ 0 & 0 & 0 & -i & 1 & 0 & 0 & 0 \\ 0 & 0 & 0 & 0 & -i & 0 & 0 & 0 \\ 0 & 0 & 0 & 0 & 0 & 2 & 0 & 0 \\ 0 & 0 & 0 & 0 & 0 & 0 & 5 & 1 \\ 0 & 0 & 0 & 0 & 0 & 0 & 0 & 5 \end{bmatrix}. \tag{4}$$

As the shading indicates, this matrix J can be decomposed into four Jordan blocks, placed corner-to-corner along the diagonal.

DEFINITION 8.7 Jordan Canonical Form

An $n \times n$ matrix J is a **Jordan canonical form** if it consists of Jordan blocks, placed corner-to-corner along the diagonal, as in matrix (4), with only zero entries outside these Jordan blocks.

Every diagonal matrix is a Jordan canonical form, because each diagonal entry can be viewed as being the sole entry in a 1×1 Jordan block. Notice that matrix J in Eq. (4) contains the 1×1 Jordan block [2]. Notice, too, that the breaks between the Jordan blocks in matrix J occur at diagonal entries that have 0 rather than 1 immediately above them.

EXAMPLE 3   Is the matrix

$$\begin{bmatrix} 2 & 1 & 0 \\ 0 & 2 & 1 \\ 0 & 0 & 7 \end{bmatrix}$$

a Jordan canonical form? Why?

Solution   The matrix is not a Jordan canonical form. Since not all diagonal entries are equal, there should be at least two Jordan blocks present in order for the matrix to be a Jordan canonical form, and [7] should be a 1×1 Jordan block. However, the entry immediately above 7 is not 0. Consequently, this matrix is not a Jordan canonical form. ■

EXAMPLE 4   Describe the effect of the matrix J in Eq. (4) on each of the standard basis vectors in $\mathbb{C}^8$. Then give the eigenvalues and eigenspaces of J. Finally, find the dimension of the nullspace of $(J - \lambda I)^k$ for each eigenvalue λ of J and for each positive integer k.

Solution   We find that

$$J\mathbf{e}_3 = -i\mathbf{e}_3 + \mathbf{e}_2, \ J\mathbf{e}_2 = -i\mathbf{e}_2 + \mathbf{e}_1, \ J\mathbf{e}_1 = -i\mathbf{e}_1,$$
$$J\mathbf{e}_5 = -i\mathbf{e}_5 + \mathbf{e}_4, \ J\mathbf{e}_4 = -i\mathbf{e}_4,$$
$$J\mathbf{e}_6 = 2\mathbf{e}_6,$$
$$J\mathbf{e}_8 = 5\mathbf{e}_8 + \mathbf{e}_7, \ J\mathbf{e}_7 = 5\mathbf{e}_7.$$

The eigenvalues of J are $-i$, 2, and 5, which have algebraic multiplicities of 5, 1, and 2, respectively. The eigenspaces of J are $E_{-i} = \text{sp}(\mathbf{e}_1, \mathbf{e}_4)$, $E_2 = \text{sp}(\mathbf{e}_6)$, and $E_5 = \text{sp}(\mathbf{e}_7)$, as you can easily check.

The effect of $J - (-i)I$ on the first five standard basis vectors is given by the

---

THE JORDAN CANONICAL FORM appears in the *Treatise on Substitutions and Algebraic Equations*, the chief work of the French algebraist Camille Jordan (1838–1921). This text, which appeared in 1870, incorporated the author's group-theory work over the preceding decade and became the bible of the field for the remainder of the nineteenth century. The theorem containing the canonical form actually deals not with matrices over the real numbers, but with matrices with entries from the finite field of order p. And as the title of the book indicates, Jordan was not considering matrices as such, but the linear substitutions that they represented.

Camille Jordan, a brilliant student, entered the Ecole Polytechnique in Paris at the age of 17 and practiced engineering from the time of his graduation until 1885. He thus had ample time for mathematical research. From 1873 until 1912 he taught at both the Ecole Polytechnique and the Collège de France. Besides doing seminal work on group theory, he is known for important discoveries in modern analysis and topology.

two strings

$$J + iI: \quad \begin{array}{l} \mathbf{e}_3 \to \mathbf{e}_2 \to \mathbf{e}_1 \to \mathbf{0}, \\ \mathbf{e}_5 \to \mathbf{e}_4 \to \mathbf{0}. \end{array} \tag{5}$$

The 3×3 lower right-hand corner of $J + iI$ describes the action of $J + iI$ on $\mathbf{e}_6$, $\mathbf{e}_7$, and $\mathbf{e}_8$. Since this 3×3 matrix has a nonzero determinant, it causes $J + iI$ to carry these three vectors into three independent vectors, and the same is true of all powers of $J + iI$. Thus we can determine the dimension of the nullspace of $J + iI$ by schematic (5), and we easily see that

$J + iI$ has nullspace $\mathrm{sp}(\mathbf{e}_1, \mathbf{e}_4)$ of dimension 2,

$(J + iI)^2$ has nullspace $\mathrm{sp}(\mathbf{e}_1, \mathbf{e}_2, \mathbf{e}_4, \mathbf{e}_5)$ of dimension 4,

$(J + iI)^3$ has nullspace $\mathrm{sp}(\mathbf{e}_1, \mathbf{e}_2, \mathbf{e}_3, \mathbf{e}_4, \mathbf{e}_5)$ of dimension 5,

$(J + iI)^k$ has the same nullspace as that of $(J + iI)^3$ for $k > 3$.

By a similar argument, we find that

$(J - 2I)^k$ has nullspace $\mathrm{sp}(\mathbf{e}_6)$ of dimension 1 for $k \geq 1$,

$J - 5I$ has nullspace $\mathrm{sp}(\mathbf{e}_7)$ of dimension 1,

$(J - 5I)^k$ has nullspace $\mathrm{sp}(\mathbf{e}_7, \mathbf{e}_8)$ of dimension 2 for $k > 1$. ■

EXAMPLE 5   Suppose a 9×9 Jordan canonical form J has the following properties:

(i) $(J - 3iI)^k$ has rank 7 for $k = 1$, rank 5 for $k = 2$, and rank 4 for $k \geq 3$,

(ii) $(J + I)^j$ has rank 6 for $j = 1$ and rank 5 for $j \geq 2$.

Find the Jordan blocks that appear in J.

Solution   Since the rank of $J - 3iI$ is 7, the dimension of its nullspace is $9 - 7 = 2$, so $3i$ is an eigenvalue of geometric multiplicity 2. It must give rise to two Jordan blocks. In addition, $J - 3iI$ must annihilate two eigenvectors $\mathbf{e}_r$ and $\mathbf{e}_s$ in the standard basis. Since the rank of $(J - 3iI)^2$ is 5, its nullspace must have dimension 4, so in a schematic of the effect of $J - 3iI$ on the standard basis, we must have $(J - 3iI)\mathbf{e}_{r+1} = \mathbf{e}_r$ and $(J - 3iI)\mathbf{e}_{s+1} = \mathbf{e}_s$. Since $(J - 3iI)^k$ has rank 4 for $k \geq 3$, its nullity is 5, and we have just one more standard basis vector—either $\mathbf{e}_{r+2}$ or $\mathbf{e}_{s+2}$—that is annihilated by $(J - 3iI)^3$. Thus, the two Jordan blocks in J that have $3i$ on the diagonal are

$$J_1 = \begin{bmatrix} 3i & 1 & 0 \\ 0 & 3i & 1 \\ 0 & 0 & 3i \end{bmatrix} \quad \text{and} \quad J_2 = \begin{bmatrix} 3i & 1 \\ 0 & 3i \end{bmatrix}.$$

Since $J + 1I$ has rank 6, its nullspace has dimension $9 - 6 = 3$, so -1 is an eigenvalue of geometric multiplicity 3 and gives rise to three Jordan blocks. Since $(J + I)^j$ has rank 5 for $j \geq 2$, its nullspace has dimension 4, so $(J + I)^2$ annihilates a total of four standard basis vectors. Thus, just one of these Jordan blocks is 2×2, and the other two are 1×1. The Jordan blocks arising from the eigenvalue -1 are then

$$J_3 = \begin{bmatrix} -1 & 1 \\ 0 & -1 \end{bmatrix} \quad \text{and} \quad J_4 = J_5 = [-1].$$

The matrix J might have these blocks in any order down its diagonal. Symbolically, we might have

$$J = \begin{bmatrix} J_3 & & & & \\ & J_1 & & \mathbf{0} & \\ & & J_4 & & \\ \mathbf{0} & & & J_2 & \\ & & & & J_5 \end{bmatrix}, \quad J = \begin{bmatrix} J_4 & & & & \\ & J_5 & & \mathbf{0} & \\ & & J_2 & & \\ \mathbf{0} & & & J_1 & \\ & & & & J_3 \end{bmatrix},$$

or any other order. ■

Jordan Bases

If an $n \times n$ matrix A is similar to a Jordan canonical form J, we call J a **Jordan canonical form of** A. When this is the case, there exists an invertible matrix C such that $C^{-1}AC = J$. We know that similar matrices represent the same linear transformation, but with respect to different bases. Thus, if A is similar to J, there must exist a basis $\{\mathbf{b}_1, \mathbf{b}_2, \ldots, \mathbf{b}_n\}$ of $\mathbb{C}^n$ with the same schematic string properties relative to A that the standard ordered basis has relative to the matrix J. We proceed to define such a *Jordan basis*.

DEFINITION 8.8   Jordan Basis

Let A be an $n \times n$ matrix. An ordered basis $B = (\mathbf{b}_1, \mathbf{b}_2, \ldots, \mathbf{b}_n)$ of $\mathbb{C}^n$ is a **Jordan basis for** A if, for $1 \le j \le n$, we have either $A\mathbf{b}_j = \lambda \mathbf{b}_j$ or $A\mathbf{b}_j = \lambda \mathbf{b}_j + \mathbf{b}_{j-1}$, where λ is an eigenvalue of A that we say is **associated with** $\mathbf{b}_j$. If $A\mathbf{b}_j = \lambda \mathbf{b}_j + \mathbf{b}_{j-1}$, we require that the eigenvalue associated with $\mathbf{b}_{j-1}$ also be λ.

If an $n \times n$ matrix A has a Jordan basis B, then the matrix representation of the linear transformation $T(\mathbf{z}) = A\mathbf{z}$ relative to B must be a Jordan canonical form. We know then that $J = C^{-1}AC$, where C is the $n \times n$ matrix whose jth column vector is the jth vector $\mathbf{b}_j$ in B. In a moment we will prove that, for every square matrix, there is an associated Jordan basis, and consequently that every square matrix is similar to a Jordan canonical form. First, though, we make a few remarks concerning the computation of a Jordan canonical form of A.

Finding a Jordan Canonical Form of A

1. Find the eigenvalues of A.

2. For each eigenvalue λ, compute the rank of $(A - \lambda I)^k$ for consecutive values of k, starting with $k = 1$, until the same rank is obtained for two consecutive values of k.

3. From the data generated, find a Jordan canonical form for A, as in Example 5.

We now illustrate this technique.

EXAMPLE 6 Find a Jordan canonical form of the matrix

$$A = \begin{bmatrix} 2 & 5 & 0 & 0 & 1 \\ 0 & 2 & 0 & 0 & 0 \\ 0 & 0 & -1 & 0 & 0 \\ 0 & 0 & 0 & -1 & 0 \\ 0 & 0 & 0 & 0 & -1 \end{bmatrix}.$$

Solution Since A is an upper-triangular matrix, we see that the eigenvalues of A are $\lambda_1 = \lambda_2 = 2$ and $\lambda_3 = \lambda_4 = \lambda_5 = -1$. Now

$$A - \lambda_1 I = A - 2I = \begin{bmatrix} 0 & 5 & 0 & 0 & 1 \\ 0 & 0 & 0 & 0 & 0 \\ 0 & 0 & -3 & 0 & 0 \\ 0 & 0 & 0 & -3 & 0 \\ 0 & 0 & 0 & 0 & -3 \end{bmatrix}$$

has rank 4 and consequently has a nullspace of dimension 1. We find that

$$(A - 2I)^2 = \begin{bmatrix} 0 & 0 & 0 & 0 & -3 \\ 0 & 0 & 0 & 0 & 0 \\ 0 & 0 & 9 & 0 & 0 \\ 0 & 0 & 0 & 9 & 0 \\ 0 & 0 & 0 & 0 & 9 \end{bmatrix},$$

which has rank 3 and therefore has a nullspace of dimension 2. Furthermore,

$$(A - 2I)^3 = \begin{bmatrix} 0 & 0 & 0 & 0 & 9 \\ 0 & 0 & 0 & 0 & 0 \\ 0 & 0 & -27 & 0 & 0 \\ 0 & 0 & 0 & -27 & 0 \\ 0 & 0 & 0 & 0 & -27 \end{bmatrix}$$

has the same rank and nullity as $(A - 2I)^2$. Thus we have $A\mathbf{b}_1 = 2\mathbf{b}_1$ and $A\mathbf{b}_2 = 2\mathbf{b}_2 + \mathbf{b}_1$ for some Jordan basis $B = (\mathbf{b}_1, \mathbf{b}_2, \mathbf{b}_3, \mathbf{b}_4, \mathbf{b}_5)$ for A. There is just one Jordan block associated with $\lambda_1 = 2$, namely,

$$J_1 = \begin{bmatrix} 2 & 1 \\ 0 & 2 \end{bmatrix}.$$

For the eigenvalue $\lambda_3 = -1$, we find that

$$A - \lambda_3 I = A + I = \begin{bmatrix} 3 & 5 & 0 & 0 & 1 \\ 0 & 3 & 0 & 0 & 0 \\ 0 & 0 & 0 & 0 & 0 \\ 0 & 0 & 0 & 0 & 0 \\ 0 & 0 & 0 & 0 & 0 \end{bmatrix},$$

which has rank 2 and therefore has a nullspace of dimension 3. Since -1 is an eigenvalue of both algebraic multiplicity and geometric multiplicity 3, we realize that

$J_2 = J_3 = J_4 = [-1]$ are the remaining Jordan blocks. This is confirmed by the fact that

$$(A + I)^2 = \begin{bmatrix} 9 & 30 & 0 & 0 & 3 \\ 0 & 9 & 0 & 0 & 0 \\ 0 & 0 & 0 & 0 & 0 \\ 0 & 0 & 0 & 0 & 0 \\ 0 & 0 & 0 & 0 & 0 \end{bmatrix}$$

again has rank 2 and nullity 3. Thus, a Jordan canonical form for A is

$$J = \begin{bmatrix} 2 & 1 & 0 & 0 & 0 \\ 0 & 2 & 0 & 0 & 0 \\ 0 & 0 & -1 & 0 & 0 \\ 0 & 0 & 0 & -1 & 0 \\ 0 & 0 & 0 & 0 & -1 \end{bmatrix}. \qquad ■$$

EXAMPLE 7 Find a Jordan basis for matrix A in Example 6.

Solution For the part of a Jordan basis associated with the eigenvalue 2, we need to find a vector $\mathbf{b}_2$ in the nullspace of $(A - 2I)^2$ that is not in the nullspace of $A - 2I$; then we may take $\mathbf{b}_1 = (A - 2I)\mathbf{b}_2$. From the computation of $A - 2I$ and $(A - 2I)^2$ in Example 6, we see that we can take

$$\mathbf{b}_2 = \begin{bmatrix} 0 \\ 1 \\ 0 \\ 0 \\ 0 \end{bmatrix}, \quad \text{and then} \quad \mathbf{b}_1 = (A - 2I)\mathbf{b}_2 = \begin{bmatrix} 5 \\ 0 \\ 0 \\ 0 \\ 0 \end{bmatrix}.$$

For $\mathbf{b}_3$, $\mathbf{b}_4$, and $\mathbf{b}_5$, we need only take a basis for the nullspace of $A + I$. We see that we can take

$$\mathbf{b}_3 = \begin{bmatrix} 0 \\ 0 \\ 1 \\ 0 \\ 0 \end{bmatrix}, \quad \mathbf{b}_4 = \begin{bmatrix} 0 \\ 0 \\ 0 \\ 1 \\ 0 \end{bmatrix}, \quad \text{and} \quad \mathbf{b}_5 = \begin{bmatrix} -1 \\ 0 \\ 0 \\ 0 \\ 3 \end{bmatrix}. \qquad ■$$

In Example 7, it was easy to find vectors in a Jordan basis corresponding to the eigenvalue 2 whose geometric multiplicity is less than its algebraic multiplicity, because only one Jordan block corresponds to the eigenvalue 2. We now indicate how a Jordan basis can be constructed when more than one such block corresponds to a single eigenvalue λ. Let N_r be the nullspace of $(A - \lambda I)^r$ for $r \geq 1$, and suppose (for example) that $\dim(N_1) = 4$, $\dim(N_2) = 7$, and $\dim(N_r) = 8$ for $r \geq 3$. Then a Jordan basis for A contains four strings corresponding to λ, which we may represent as

$$\mathbf{b}_3 \rightarrow \mathbf{b}_2 \rightarrow \mathbf{b}_1 \rightarrow \mathbf{0},$$
$$\mathbf{b}_5 \rightarrow \mathbf{b}_4 \rightarrow \mathbf{0},$$
$$\mathbf{b}_7 \rightarrow \mathbf{b}_6 \rightarrow \mathbf{0},$$
$$\mathbf{b}_8 \rightarrow \mathbf{0}.$$

To find the first and longest of these strings, we *compute* a basis $\{\mathbf{v}_1, \mathbf{v}_2, \ldots, \mathbf{v}_8\}$ for the nullspace N_3 of $(A - \lambda I)^3$. The preceding strings show that multiplication of all of the vectors in N_3 on the left by $(A - \lambda I)^2$ must yield a space of dimension 1, so at least one of the vectors $\mathbf{v}_i$ has the property that $(A - \lambda I)^2\mathbf{v}_i \neq \mathbf{0}$. Let $\mathbf{b}_3$ be such a vector, and set $\mathbf{b}_2 = (A - \lambda I)\mathbf{b}_3$ and $\mathbf{b}_1 = (A - \lambda I)\mathbf{b}_2$. It is not difficult to show that $\mathbf{b}_1$, $\mathbf{b}_2$, and $\mathbf{b}_3$ must be independent. Thus we have found the first string.

Now $\mathbf{b}_1$ and $\mathbf{b}_2$ lie in N_2, and we can expand the independent set $\{\mathbf{b}_1, \mathbf{b}_2\}$ to a basis $\{\mathbf{b}_1, \mathbf{b}_2, \mathbf{w}_1, \ldots, \mathbf{w}_5\}$ of N_2. Again, the strings displayed earlier show that multiplication of the vectors in N_2 on the left by $A - \lambda I$ must yield a space of dimension 3, so there exist two vectors $\mathbf{w}_i$ and $\mathbf{w}_j$ such that the vectors $\mathbf{b}_1$, $(A - \lambda I)\mathbf{w}_i$, and $(A - \lambda I)\mathbf{w}_j$ are independent. Let $\mathbf{b}_5 = \mathbf{w}_i$ and $\mathbf{b}_4 = (A - \lambda I)\mathbf{b}_5$, while $\mathbf{b}_7 = \mathbf{w}_j$ and $\mathbf{b}_6 = (A - \lambda I)\mathbf{b}_7$. It can be shown that the vectors $\mathbf{b}_1, \mathbf{b}_2, \ldots, \mathbf{b}_7$ are independent. Finally, we expand the set $\{\mathbf{b}_1, \mathbf{b}_4, \mathbf{b}_6\}$ to a basis $\{\mathbf{b}_1, \mathbf{b}_4, \mathbf{b}_6, \mathbf{b}_8\}$ for N_1 to complete the portion of the Jordan basis corresponding to λ.

Although we know the techniques for finding bases for the nullspaces N_i and for expanding a given set of independent vectors to a basis, significant pencil-and-paper illustrations of this construction would be cumbersome, so we do not include them here. Any Jordan bases requested in the exercises can be found as in Example 7.

An application of the Jordan canonical form to differential equations is indicated in Exercise 32. We mention that computer-aided computation of a Jordan canonical form for a square matrix is not a stable process. Consider, for example, the matrix

$$A = \begin{bmatrix} 2 & c \\ 0 & 2 \end{bmatrix}.$$

If $c = 10^{-100}$, then the Jordan canonical form of A has 1 as its entry in the upper right-hand corner; but if $c = 0$, that entry is 0.

Existence of a Jordan Form for a Square Matrix

To demonstrate the existence of a Jordan canonical form similar to a given $n \times n$ matrix A, we need only show that we have a Jordan basis B for A. Let us formalize the concept of a *string* in a Jordan basis $B = (\mathbf{b}_1, \mathbf{b}_2, \ldots, \mathbf{b}_n)$. Let λ be an eigenvalue of A. If $A\mathbf{b}_i = \lambda\mathbf{b}_i$ and $A\mathbf{b}_k = \lambda\mathbf{b}_k + \mathbf{b}_{k-1}$ for $i < k < j$, while $A\mathbf{b}_j \neq \lambda\mathbf{b}_j + \mathbf{b}_{j-1}$, we refer to the sequence $\mathbf{b}_i, \mathbf{b}_{i+1}, \ldots, \mathbf{b}_{j-1}$ as a **string** of basis vectors **starting** at $\mathbf{b}_{j-1}$, **ending** at $\mathbf{b}_i$, and **associated with** λ. This string is represented by the diagram

$$A - \lambda I: \qquad \mathbf{b}_{j-1} \to \cdots \to \mathbf{b}_{i+1} \to \mathbf{b}_i \to \mathbf{0}.$$

THEOREM 8.9 Jordan Canonical Form of a Square Matrix

Let A be a square matrix. There exists an invertible matrix C such that the matrix $J = C^{-1}AC$ is a Jordan canonical form. This Jordan canonical form is unique, except for the order of the Jordan blocks of which it is composed.

PROOF We use a proof due to Filippov. First we note that it suffices to prove the theorem for matrices A having 0 as an eigenvalue. Observe that, if λ is an eigenvalue of A, then 0 is an eigenvalue of $A - \lambda I$. Now if we can find C such that $C^{-1}(A - \lambda I)C = J$ is a Jordan canonical form, then $C^{-1}AC = J + \lambda I$ is also a Jordan canonical form. Thus, we restrict ourselves to the case where A has an eigenvalue of 0.

In order to find a Jordan canonical form for A, it is useful to consider also the linear transformation $T: \mathbb{C}^n \to \mathbb{C}^n$, where $T(\mathbf{z}) = A\mathbf{z}$; a Jordan basis for A is considered to be a Jordan basis for T. We will prove the existence of a Jordan basis for any such linear transformation by induction on the dimension of the domain of the transformation.

If T is a transformation of a one-dimensional vector space $\mathrm{sp}(\mathbf{z})$, then $T(\mathbf{z}) = \lambda\mathbf{z}$ for some $\lambda \in \mathbb{C}$, and $\{\mathbf{z}\}$ is the required Jordan basis. (The matrix of T with respect to this ordered basis is the 1×1 matrix $[\lambda]$, which is already a Jordan canonical form.)

Now suppose that there exist Jordan bases for linear transformations on subspaces of $\mathbb{C}^n$ of dimension less than n, and let $T(\mathbf{z}) = A\mathbf{z}$ for $\mathbf{z} \in \mathbb{C}^n$ and an $n \times n$ matrix A. As noted, we can assume that zero is an eigenvalue of A. Then $\mathrm{rank}(A) < n$; let $r = \mathrm{rank}(A)$. Now T maps $\mathbb{C}^n$ onto the column space of A that is of dimension $r < n$. Let T' be the induced linear transformation of the column space of A into itself, defined by $T'(\mathbf{v}) = T(\mathbf{v})$ for $\mathbf{v}$ in the column space of A. By our induction hypothesis, there is a Jordan basis

$$B' = (\mathbf{u}_1, \mathbf{u}_2, \ldots, \mathbf{u}_r)$$

for this column space of A.

Let S be the intersection of the column space and the nullspace of A. We wish to separate the vectors in B' that are in S from those that are not. The nonzero vectors in S are precisely the eigenvectors in the column space of A with corresponding eigenvalue 0; that is, they are the eigenvectors of T' with eigenvalue 0. In other words, S is the nullspace of T'. Let J' be a matrix representation of T' relative to B'. Since J' is a Jordan canonical form, we see that the nullity of T' (and of J') is precisely the number of zero rows in J'. This is true because J' is an upper-triangular square matrix; it can be brought to echelon form by means of row exchanges that place the zero rows at the bottom while sliding the nonzero rows up. Thus, if $\dim(S) = s$, there are s zero rows in J'. Now in J' we have exactly one zero row for each Jordan block corresponding to the eigenvalue 0, namely, the row containing the bottom row of the block. Since the number of such blocks is equal to the number of strings in B' ending in S, we conclude that there are s such strings. Some of these strings may be of length 1 while others may be longer.

Figure 8.11 shows one possible situation when $s = 2$, where two vectors in S, namely, $\mathbf{u}_1$ and $\mathbf{u}_4$, are ending points of strings

$$\mathbf{u}_3 \to \mathbf{u}_2 \to \mathbf{u}_1 \to \mathbf{0} \qquad \text{and} \qquad \mathbf{u}_5 \to \mathbf{u}_4 \to \mathbf{0}$$

lying in the column space of A. These s strings of B' that end in S start at s vectors in the column space of A; these are the vectors $\mathbf{u}_3$ and $\mathbf{u}_5$ in the illustration in Figure

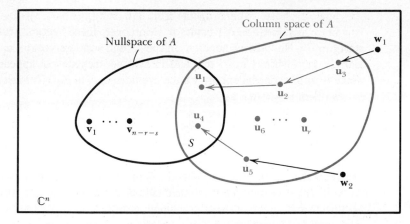

FIGURE 8.11 **Construction of a Jordan basis for A ($s = 2$).**

8.11. Since the vector at the beginning of the jth string is in the column space of A, it must have the form $A\mathbf{w}_j$ for some vector $\mathbf{w}_j$ in $\mathbb{C}^n$. Thus we obtain the vectors $\mathbf{w}_1$, $\mathbf{w}_2, \ldots, \mathbf{w}_s$ illustrated in Figure 8.11 for $s = 2$.

Finally, the nullspace of A has dimension $n - r$, and we can expand the set of s independent vectors in S to a basis for this nullspace. This gives rise to $n - r - s$ more vectors $\mathbf{v}_1, \mathbf{v}_2, \ldots, \mathbf{v}_{n-r-s}$. Of course, each $\mathbf{v}_i$ is an eigenvector with corresponding eigenvalue 0.

We claim that

$$(\mathbf{u}_1, \ldots, \mathbf{u}_r, \mathbf{w}_1, \ldots, \mathbf{w}_s, \mathbf{v}_1, \ldots, \mathbf{v}_{n-r-s})$$

can be reordered to become a Jordan basis B for A (and of course for T). We reorder it by moving the vectors $\mathbf{w}_i$, tucking each one in so that it starts the appropriate string in B' that was used to define it. For the situation in Figure 8.11, we obtain

$$(\mathbf{u}_1, \mathbf{u}_2, \mathbf{u}_3, \mathbf{w}_1, \mathbf{u}_4, \mathbf{u}_5, \mathbf{w}_2, \mathbf{u}_6, \ldots, \mathbf{u}_r, \mathbf{v}_1, \ldots, \mathbf{v}_{n-r-2})$$

as Jordan basis. From our construction, we see that B is a Jordan basis for A if it is a basis for C^n. Since there are $r + s + (n - r - s) = n$ vectors in all, we need only show that they are independent.

Suppose that

$$\sum_{i=1}^{r} a_i \mathbf{u}_i + \sum_{j=1}^{s} c_j \mathbf{w}_j + \sum_{k=1}^{n-r-s} d_k \mathbf{v}_k = \mathbf{0}. \tag{6}$$

Since the vectors $\mathbf{v}_k$ lie in the nullspace of A, if we apply A to both sides of this equation, we obtain

$$\sum_{i=1}^{r} a_i A\mathbf{u}_i + \sum_{j=1}^{s} c_j A\mathbf{w}_j = \mathbf{0}. \tag{7}$$

Since each $A\mathbf{u}_i$ is either of the form $\lambda\mathbf{u}_i$ or $\lambda\mathbf{u}_i + \mathbf{u}_{i-1}$, we see that the first sum is a linear combination of vectors $\mathbf{u}_i$. Moreover, these vectors $A\mathbf{u}_i$ *do not begin* any string in B'. Now the vectors $A\mathbf{w}_j$ in the second sum are vectors $\mathbf{u}_i$ that appear at the *start* of the s strings in B' that end in S. Thus they do not appear in the first sum. Since B' is an independent set, all the coefficients c_j in Eq. (7) must be zero. Equation (6) can then be written as

$$\sum_{i=1}^{r} a_i\mathbf{u}_i = \sum_{k=1}^{n-r-s} -d_k\mathbf{v}_k. \tag{8}$$

Now the vector on the left-hand side of this equation lies in the column space of A, while the vector on the right-hand side is in the nullspace of A. Consequently, this vector lies in S and is a linear combination of the s basis vectors $\mathbf{u}_i$ in S. Since the $\mathbf{v}_k$ were obtained by extending these s vectors to a basis for the nullspace of A, the vector $\mathbf{0}$ is the only linear combination of the $\mathbf{v}_k$ that lies in S. Thus, the vector on both sides of Eq. (8) is $\mathbf{0}$. Since the $\mathbf{v}_k$ are independent, we see that all d_k are zero. Since the $\mathbf{u}_i$ are independent, it follows that the a_i are all zero. Therefore, B is an independent set of n vectors and is thus a basis for $\mathbb{C}^n$. We have seen that, by our construction, it must be a Jordan basis. This completes the induction part of our proof, demonstrating the existence of a Jordan canonical form for every square matrix A.

Our work prior to this theorem makes clear that the Jordan blocks constituting a Jordan canonical form for A are completely determined by the ranks of the matrices $(A - \lambda I)^k$ for all eigenvalues λ of A and for all positive integers k. Thus, a Jordan canonical form J for A is unique except as to the order in which these blocks appear along the diagonal of J.   ▲

SUMMARY

1. A Jordan block is a square matrix with all diagonal entries equal, all entries immediately above diagonal entries equal to 1, and all other entries equal to 0.

2. Properties of a Jordan block are given in Theorem 8.8.

3. A square matrix is a Jordan canonical form if it consists of Jordan blocks placed corner to corner along its diagonal, with entries elsewhere equal to 0.

4. A Jordan basis (see Definition 8.8) for an $n \times n$ matrix A gives rise to a Jordan canonical form J that is similar to A.

5. A Jordan canonical form similar to an $n \times n$ matrix A can be computed if we know the eigenvalues λ_i of A and if we know the rank of $(A - \lambda_i I)^k$ for each λ_i and for all positive integers k.

6. Every square matrix has a Jordan canonical form; that is, it is similar to a Jordan canonical form.

EXERCISES

In Exercises 1–6, determine whether the given matrix is a Jordan canonical form.

1. $\begin{bmatrix} 0 & 0 & 0 \\ 0 & 0 & 0 \\ 0 & 0 & 0 \end{bmatrix}$
2. $\begin{bmatrix} 3 & 0 & 0 \\ 0 & 3 & 1 \\ 0 & 0 & 3 \end{bmatrix}$

3. $\begin{bmatrix} 3 & 1 & 0 & 0 \\ 0 & 3 & 1 & 0 \\ 0 & 0 & 2 & 1 \\ 0 & 0 & 0 & 2 \end{bmatrix}$
4. $\begin{bmatrix} 1 & 0 & 0 & 0 \\ 0 & 2 & 0 & 0 \\ 0 & 0 & 3 & 0 \\ 0 & 0 & 0 & 4 \end{bmatrix}$

5. $\begin{bmatrix} i & 1 & 0 & 0 \\ 0 & -i & 0 & 0 \\ 0 & 0 & 3 & 1 \\ 0 & 0 & 0 & 3 \end{bmatrix}$

6. $\begin{bmatrix} 2 & 1 & 0 & 0 \\ 0 & 2 & 0 & 0 \\ 0 & 0 & i & 0 \\ 0 & 0 & 0 & -1 \end{bmatrix}$

In Exercises 7–10:

a) *Find the eigenvalues of the given matrix J.*

b) *Give the rank and nullity of $(J - \lambda)^k$ for each eigenvalue λ of J and for every positive integer k.*

c) *Draw schemata of the strings of vectors in the standard basis arising from the Jordan blocks in J.*

d) *For each standard basis vector $\mathbf{e}_k$, express $J\mathbf{e}_k$ as a linear combination of vectors in the standard basis.*

7. $\begin{bmatrix} -2 & 1 & 0 & 0 \\ 0 & -2 & 1 & 0 \\ 0 & 0 & -2 & 1 \\ 0 & 0 & 0 & -2 \end{bmatrix}$

8. $\begin{bmatrix} i & 0 & 0 & 0 & 0 \\ 0 & i & 1 & 0 & 0 \\ 0 & 0 & i & 0 & 0 \\ 0 & 0 & 0 & -2 & 0 \\ 0 & 0 & 0 & 0 & -2 \end{bmatrix}$

9. $\begin{bmatrix} -1 & 0 & 0 & 0 & 0 \\ 0 & 2 & 1 & 0 & 0 \\ 0 & 0 & 2 & 0 & 0 \\ 0 & 0 & 0 & 2 & 1 \\ 0 & 0 & 0 & 0 & 2 \end{bmatrix}$

10. $\begin{bmatrix} i & 1 & 0 & 0 & 0 & 0 & 0 & 0 \\ 0 & i & 0 & 0 & 0 & 0 & 0 & 0 \\ 0 & 0 & i & 0 & 0 & 0 & 0 & 0 \\ 0 & 0 & 0 & i & 1 & 0 & 0 & 0 \\ 0 & 0 & 0 & 0 & i & 0 & 0 & 0 \\ 0 & 0 & 0 & 0 & 0 & 2 & 1 & 0 \\ 0 & 0 & 0 & 0 & 0 & 0 & 2 & 1 \\ 0 & 0 & 0 & 0 & 0 & 0 & 0 & 2 \end{bmatrix}$

In Exercises 11–14, find a Jordan canonical form for A from the given data.

11. A is 5×5, $A - 3I$ has nullity 2, $(A - 3I)^2$ has nullity 3, $(A - 3I)^3$ has nullity 4, $(A - 3I)^k$ has nullity 5 for $k \geq 4$.

12. A is 7×7, $A + I$ has nullity 3, $(A + I)^k$ has nullity 5 for $k \geq 2$; $A + iI$ has nullity 1, $(A + iI)^j$ has nullity 2 for $j \geq 2$.

13. A is 8×8, $A - I$ has nullity 2, $(A - I)^2$ has nullity 4, $(A - I)^k$ has nullity 5 for $k \geq 3$; $(A + 2I)^j$ has nullity 3 for $j \geq 1$.

14. A is 8×8; $A + iI$ has rank 4, $(A + iI)^2$ has rank 2, $(A + iI)^3$ has rank 1, $(A + iI)^k = O$ for $k \geq 4$.

In Exercises 15–22, find a Jordan canonical form and a Jordan basis for the given matrix.

15. $\begin{bmatrix} -10 & 4 \\ -25 & 10 \end{bmatrix}$
16. $\begin{bmatrix} 5 & -4 \\ 9 & -7 \end{bmatrix}$

17. $\begin{bmatrix} 4 & 0 & 0 \\ 2 & 1 & 3 \\ 5 & 0 & 4 \end{bmatrix}$
18. $\begin{bmatrix} -3 & 0 & 1 \\ 2 & -2 & 1 \\ -1 & 0 & -1 \end{bmatrix}$

19. $\begin{bmatrix} 2 & 5 & 0 & 0 & 0 \\ 0 & 2 & 0 & 0 & 0 \\ 0 & 0 & -1 & 0 & -1 \\ 0 & 0 & 0 & -1 & 0 \\ 0 & 0 & 0 & 0 & -1 \end{bmatrix}$

20. $\begin{bmatrix} i & 0 & 0 & 0 & 0 \\ 0 & i & 0 & 0 & 0 \\ 0 & 0 & 2 & 0 & 0 \\ 0 & 0 & 0 & 2 & 0 \\ 2 & 0 & -1 & 0 & 2 \end{bmatrix}$

21.
$$\begin{bmatrix} 2 & 0 & 0 & 0 & 1 \\ 0 & 2 & 0 & 0 & 0 \\ 0 & 0 & 2 & 0 & 1 \\ 0 & 0 & 0 & 2 & 0 \\ 0 & 0 & 0 & 0 & 2 \end{bmatrix}$$

22.
$$\begin{bmatrix} 1 & 2 & 0 & 0 & 1 \\ 0 & 1 & 0 & 0 & 0 \\ 0 & 0 & 1 & 0 & 1 \\ 0 & 0 & 0 & 1 & 2 \\ 0 & 0 & 0 & 0 & 1 \end{bmatrix}$$

23. Mark each of the following True or False.

____ a) Every Jordan block matrix has just one eigenvalue.

____ b) Every matrix having a unique eigenvalue is a Jordan block.

____ c) Every diagonal matrix is a Jordan canonical form.

____ d) Every square matrix is similar to a Jordan canonical form.

____ e) Every square matrix is similar to a unique Jordan canonical form.

____ f) Every 1×1 matrix is similar to a unique Jordan canonical form.

____ g) There is a Jordan basis for every square matrix A.

____ h) There is a unique Jordan basis for every square matrix A.

____ i) Every 3×3 diagonalizable matrix is similar to exactly six Jordan canonical forms.

____ j) Every 3×3 matrix is similar to exactly six Jordan canonical forms.

24. Let $A = \begin{bmatrix} 0 & 1 & 0 & 0 \\ 0 & 0 & 1 & 0 \\ 0 & 0 & 0 & 1 \\ 0 & 0 & 0 & 0 \end{bmatrix}$. Compute A^2, A^3, and A^4.

25. Let A be an $n \times n$ upper-triangular matrix with all diagonal entries 0. Compute A^m for all positive integers $m \geq n$. (See Exercise 24.) Prove that your answer is correct.

26. Let $A = \begin{bmatrix} 2 & 1 & 0 & 0 & 0 \\ 0 & 2 & 1 & 0 & 0 \\ 0 & 0 & 2 & 0 & 0 \\ 0 & 0 & 0 & 3 & 1 \\ 0 & 0 & 0 & 0 & 3 \end{bmatrix}$. Compute $(A - 2I)^3(A - 3I)^2$.

27. Let $A = \begin{bmatrix} 2 & 0 & 0 & 0 & 0 \\ 0 & 2 & 1 & 0 & 0 \\ 0 & 0 & 2 & 0 & 0 \\ 0 & 0 & 0 & 3 & 1 \\ 0 & 0 & 0 & 0 & 3 \end{bmatrix}$. Compute $(A - 2I)^2(A - 3I)^2$. Compare with Exercise 26.

28. Let $A = \begin{bmatrix} i & 0 & 0 & 0 & 0 \\ 0 & i & 1 & 0 & 0 \\ 0 & 0 & i & 0 & 0 \\ 0 & 0 & 0 & 2 & 0 \\ 0 & 0 & 0 & 0 & 2 \end{bmatrix}$. Find a polynomial in A (that is, a sum of terms $a_j A^j$ with a term $a_0 I$) that gives the zero matrix. (See Exercises 24–27.)

29. Repeat Exercise 28 for the matrix $A = \begin{bmatrix} -1 & 1 & 0 & 0 & 0 \\ 0 & -1 & 1 & 0 & 0 \\ 0 & 0 & -1 & 0 & 0 \\ 0 & 0 & 0 & i & 0 \\ 0 & 0 & 0 & 0 & i \end{bmatrix}$.

30. The Cayley–Hamilton theorem states that, if $p(\lambda) = a_n \lambda^n + \cdots + a_1 \lambda + a_0$ is the characteristic polynomial of a matrix A, then $p(A) = a_n A^n + \cdots + a_1 A + a_0 I = O$, the zero matrix. Prove it. [HINT: Consider $(A - \lambda_i I)^{n_i} \mathbf{b}$, where $\mathbf{b}$ is a vector in a Jordan basis corresponding to λ_i.] In view of Exercises 24–29, explain why you expect $p(J)$ to be O, where J is a Jordan canonical form for A. Deduce that $p(A) = O$.

31. Let $T: \mathbb{C}^n \to \mathbb{C}^n$ be a linear transformation. A subspace W of $\mathbb{C}^n$ is **invariant under** T if $T(\mathbf{w}) \in W$ for all $\mathbf{w} \in W$. Let A be the standard matrix representation of T.

a) Describe the one-dimensional invariant subspaces of T.

b) Show that every eigenspace E_λ of T is invariant under T.

c) Show that the vectors in any string in a Jordan basis for A generate an invariant subspace of T.

d) Is it true that, if S is a subspace of a subspace W that is invariant under T, then S is also invariant under T? If not, give a counterexample.

e) Is it true that every subspace of $\mathbb{R}^n$ invariant under T is contained in the nullspace of $(A - \lambda I)^n$, where λ is some eigenvalue of T? If not, give a counterexample.

32. In Section 4.3, we considered systems $\mathbf{x}' = A\mathbf{x}$ of differential equations, and we saw that, if $A = CJC^{-1}$, then the system takes the form $\mathbf{y}' = J\mathbf{y}$, where $\mathbf{x} = C\mathbf{y}$. (We used D in place of J in Section 4.3, since we were concerned only with diagonalization.) Let $\lambda_1, \lambda_2, \ldots, \lambda_n$ be the (not necessarily distinct) eigenvalues of an $n \times n$ matrix A, and let J be a Jordan canonical form for A.

a) Show that the system $\mathbf{y}' = J\mathbf{y}$ is of the form

$$\begin{aligned} y_1' &= \lambda_1 y_1 + c_1 y_2, \\ y_2' &= \lambda_2 y_2 + c_2 y_3, \\ &\vdots \\ y_{n-1}' &= \lambda_{n-1} y_{n-1} + c_{n-1} y_n, \\ y_n' &= \lambda_n y_n, \end{aligned}$$

where each c_i is either 0 or 1.

b) How can the system in part (a) be solved? [HINT: Start with the last equation.]

c) Given that, for

$$A = \begin{bmatrix} 2 & 2 & 3 \\ 0 & -1 & 0 \\ 2 & 2 & 1 \end{bmatrix},$$

$$C = \begin{bmatrix} \frac{1}{2} & 1 & \frac{3}{2} \\ 0 & -\frac{5}{4} & 0 \\ -\frac{1}{2} & 0 & 1 \end{bmatrix},$$

$$J = \begin{bmatrix} -1 & 1 & 0 \\ 0 & -1 & 0 \\ 0 & 0 & 4 \end{bmatrix},$$

we have $C^{-1}AC = J$, find the solution of the differential system $\mathbf{x}' = A\mathbf{x}$.

33. Let A be an $n \times n$ matrix with eigenvalue λ. Prove that the algebraic multiplicity of λ is at least as large as its geometric multiplicity.

9

SOLVING LARGE LINEAR SYSTEMS

The Gauss and Gauss–Jordan methods presented in Chapter 1 are fine for solving very small linear systems with pencil and paper. Some applied problems—in particular, those requiring numerical solution of differential equations—can lead to very large linear systems, involving thousands of equations in thousands of unknowns. Of course, such large linear systems must be solved by using a computer. That is the primary concern of this chapter. Although a computer can work tremendously faster than we can with pencil and paper, each individual arithmetic operation does take some time, and additional time is used whenever the value of a variable is stored or retrieved. In addition, indexing in subscripted arrays requires time. When solving a large linear system with a computer, we must use as efficient a computational algorithm as we can, so the number of operations required is as small as practically possible.

We begin this chapter with a discussion of the time required for a computer to execute operations and a comparison of the efficiency of the Gauss method including back substitution and the Gauss–Jordan method.

Section 9.2 presents the LU (lower- and upper-triangular) factorization of the coefficient matrix of a square linear system. This factorization will appear as we develop an efficient algorithm for solving, by repeated computer runs, many systems all having the same coefficient matrix.

Section 9.3 deals with problems of roundoff and discusses ill-conditioned matrices. We will see that there are actually very small linear systems, consisting of only two equations in two unknowns, for which good computer programs may give incorrect solutions.

9.1 CONSIDERATIONS OF TIME

Timing Data for One PC

The computation involved in solving a linear system is just a lot of arithmetic. Arithmetic takes time to execute even with a computer, although the computer is obviously much faster than an individual working with pencil and paper. In reducing a matrix by using elementary row operations, we spend most of our time adding a multiple of a row vector to another row vector. A typical step in multiplying row k of a matrix A by r and adding it to row i consists of

$$\text{replacing} \quad a_{ij} \quad \text{by} \quad a_{ij} + ra_{kj}. \tag{1}$$

In a computer language such as BASIC, operation (1) might become

$$A(I,J) = A(I,J) + R*A(K,J). \tag{2}$$

Computer instruction (2) involves operations of addition and multiplication, as well as indexing, retrieving values, and storing values. We use the terminology of C. B. Moler and call such an operation a **flop**. These flops require time to execute on any computer, although the times vary widely depending on the computer hardware and the language of the computer program. We experimented with our PC (personal computer), which is not noted for speed, using both interpretive BASIC and compiled BASIC. The computer can execute a compiled instruction faster than an interpretive one. We generated two random numbers, R and S, and then found the time required to add them 10,000 times, using the BASIC loop

$$\text{FOR I} = 1 \text{ TO N: C} = \text{R} + \text{S: NEXT}, \tag{3}$$

where we had earlier set N = 10,000. We obtained the data shown in the top row of Table 9.1. We also had the computer execute loop (3) with + replaced by − (subtraction), then by * (multiplication), and finally by / (division). We similarly timed the execution of 10,000 flops, using

$$\text{FOR I} = 1 \text{ TO N: A(K,J)} = \text{A(K,J)} + \text{R} * \text{A(M,J): NEXT}, \tag{4}$$

where we had set N = 10,000 and had also assigned values to all other variables except I. All our data are shown in Table 9.1.

TABLE 9.1 Time (in Seconds) for Executing 10,000 Operations

Routine	Interpretive BASIC		Compiled BASIC	
	Single-precision	Double-precision	Single-precision	Double-precision
Addition [using (3)]	37	44	8	9
Subtraction [using (3) with −]	37	48	8	9
Multiplication [using (3) with *]	39	53	9	11
Division [using (3) with /]	47	223	9	15
Flops [using (4)]	143	162	15	18

Table 9.1 gives us quite a bit of insight into the PC that generated the data. Here are some things we noticed immediately.

Point 1 Multiplication takes a bit, but surprisingly little, more time than addition.

Point 2 Division takes the most time of the four arithmetic operations. Indeed, our PC found double-precision division in interpretive BASIC very time-consuming. We should try to minimize divisions as much as possible. For example, when computing

$$\left(\tfrac{4}{3}\right)(3, 2, 5, 7, 8) + (-4, 11, 2, 1, 5)$$

to obtain a row vector with first entry zero, we should *not* compute the remaining entries as

$$\left(\tfrac{4}{3}\right)2 + 11, \qquad \left(\tfrac{4}{3}\right)5 + 2, \qquad \left(\tfrac{4}{3}\right)7 + 1, \qquad \left(\tfrac{4}{3}\right)8 + 5,$$

which requires four divisions. We should rather do a single division, finding $r = \tfrac{4}{3}$, and then compute

$$2r + 11, \qquad 5r + 2, \qquad 7r + 1, \qquad 8r + 5.$$

Point 3 Our program ran much faster in compiled BASIC than in interpretive BASIC. In the compiled version on our PC, the time for double-precision division was not so far out of line with the time for other operations.

Point 4 The indexing in the flops requires significant time in interpretive BASIC on our PC.

The software program TIMING available to schools using this text was used to generate the data in Table 9.1 on our PC. Exercise 22 asks students to obtain analogous data for their PC, using this program or similar software.

Counting Operations

We turn to counting the flops required to solve a square linear system $A\mathbf{x} = \mathbf{b}$ that has a unique solution. We assume that no row interchanges are necessary, which is frequently the case.

Suppose that we solve the system $A\mathbf{x} = \mathbf{b}$, using the Gauss method with back substitution. Form the augmented matrix

$$[A \mid \mathbf{b}] = \begin{bmatrix} a_{11} & a_{12} & \cdots & a_{1n} & b_1 \\ a_{21} & a_{22} & \cdots & a_{2n} & b_2 \\ & & \vdots & & \vdots \\ a_{n1} & a_{n2} & \cdots & a_{nn} & b_n \end{bmatrix}.$$

For the moment let us neglect the flops performed on $\mathbf{b}$ and count just the flops performed on A. In reducing the $n \times n$ matrix A to upper-triangular form U, we execute $n - 1$ flops in adding a multiple of the first row of $[A \mid \mathbf{b}]$ to the second row. (We do not need to compute the zero entry at the beginning of our new second row; we know it will be zero.) We similarly use $n - 1$ flops to obtain the new row 3, and so on. This gives a total of $(n - 1)^2$ flops executed using the pivot in row 1. The pivot in row 2 is then used for the $(n - 1) \times (n - 1)$ matrix obtained by crossing off the first row and column of the modified coefficient matrix. By the count just made for the $n \times n$ matrix, we realize that using this pivot in the second row will result in executing $(n - 2)^2$ flops. Continuing in this fashion, we see that reducing A to upper-triangular form U will require

$$(n - 1)^2 + (n - 2)^2 + (n - 3)^2 + \cdots + 1 \tag{5}$$

flops, together with some divisions. Let's count the divisions. We expect to use just one division each time a row is multiplied by a constant and added to another row (see point 2). There will be $n - 1$ divisions involving the pivot in row 1, $n - 2$ involving the pivot in row 2, and so on, for a total of

$$(n - 1) + (n - 2) + (n - 3) + \cdots + 1 \tag{6}$$

divisions.

There are some handy formulas for finding a sum of consecutive integers and a sum of squares of consecutive integers. It can be shown by induction (see Appendix A) that

$$1 + 2 + 3 + \cdots + n = \frac{n(n + 1)}{2} \tag{7}$$

and

$$1^2 + 2^2 + 3^2 + \cdots + n^2 = \frac{n(n + 1)(2n + 1)}{6}. \tag{8}$$

Replacing n by $n - 1$ in formula (8), we see that the number of flops given by sum (5) is equal to

$$\frac{(n - 1)n(2n - 1)}{6} = \frac{(n^2 - n)(2n - 1)}{6} = \tfrac{1}{6}(2n^3 - 3n^2 + n)$$

$$= \frac{n^3}{3} - \frac{n^2}{2} + \frac{n}{6}. \tag{9}$$

We assume that we are using a computer to solve a *large* linear system, involving hundreds or thousands of equations. With $n = 1000$, the value of Eq. (9) becomes

$$\frac{1,000,000,000}{3} - \frac{1,000,000}{2} + \frac{1,000}{6}. \tag{10}$$

The largest term in expression (10) is the first one, $1,000,000,000/3$, corresponding to the $n^3/3$ term in Eq. (9). The lower powers of n in the $n^2/2$ and $n/6$ terms contribute little in comparison to the $n^3/3$ for large n. It is customary to keep just the $n^3/3$ term as a measure of the **order of magnitude** of the expression in Eq. (9) for large values of n.

Turning to the count of the divisions in sum (6), we see from Eq. (7) with n replaced by $n - 1$ that we are using

$$\frac{(n - 1)n}{2} = \frac{n^2}{2} - \frac{n}{2} \tag{11}$$

divisions. For large n, this is of order of magnitude $n^2/2$, which is inconsequential in comparison to the order of magnitude $n^3/3$ for the flops. Exercise 1 shows that the number of flops performed on the column vector $\mathbf{b}$ in reducing $[A \mid \mathbf{b}]$ is again given by Eq. (11), so it can be neglected for large n in view of the order of magnitude $n^3/3$. The result is shown in the following box.

Flop Count for Reducing $[A \mid \mathbf{b}]$ to $[U \mid \mathbf{c}]$

If $A\mathbf{x} = \mathbf{b}$ is a square linear system in which A is an $n \times n$ matrix, the number of flops executed in reducing $[A \mid \mathbf{b}]$ to the form $[U \mid \mathbf{c}]$ is of order of magnitude $n^3/3$ for large n.

We turn now to finding the number of flops used in back substitution to solve the upper-triangular linear system $U\mathbf{x} = \mathbf{c}$. This system can be written out as

$$u_{11}x_1 + \cdots + \quad u_{1,n-1}x_{n-1} + \quad u_{1n}x_n = c_1$$

$$\vdots$$

$$u_{n-1,n-1}x_{n-1} + u_{n-1,n}x_n = c_{n-1}$$

$$u_{nn}x_n = c_n.$$

Solving for x_n requires one indexed division, which we consider to be a flop. Solving then for x_{n-1} requires an indexed multiplication and subtraction, followed by an indexed division, which we consider to be two flops. Solving for x_{n-2} requires two flops, each consisting of a multiplication combined with a subtraction, followed by an indexed division, and we consider this to contribute three more flops, and so on. We obtain the count

$$1 + 2 + 3 + \cdots + n = \frac{(n + 1)n}{2} = \frac{n^2}{2} + \frac{n}{2} \tag{12}$$

for the flops in this back substitution. Again, this is of lower order of magnitude than the order of magnitude $n^3/3$ for the flops required to reduce $[A \mid \mathbf{b}]$ to $[U \mid \mathbf{c}]$.

Flop Count for Back Substitution

If U is an $n \times n$ upper-triangular matrix, then back substitution to solve $U\mathbf{x} = \mathbf{c}$ requires the order of magnitude $n^2/2$ flops for large n.

Combining the results shown in the last two boxes, we arrive at the following flop count.

> **Flop Count for Solving $Ax = b$, Using the Gauss Method with Back Substitution**
>
> If A is an $n \times n$ matrix, the number of flops required to solve $Ax = \mathbf{b}$ using the Gauss method with back substitution is of order of magnitude $n^3/3$ for large n.

EXAMPLE 1   For the computer that produced the execution times shown in Table 9.1, find the approximate time required to solve a system of 100 equations in 100 unknowns, using single-precision arithmetic and (a) interpretive BASIC, (b) compiled BASIC.

Solution   From the flop count for the Gauss method, we see that solving such a system with $n = 100$ requires about $100^3/3 = 1,000,000/3$ flops. In interpretive BASIC, the time required for 10,000 flops in single precision was about 143 seconds. Thus the $1,000,000/3$ flops require about $(1,000,000/30,000)143 = 14,300/3$ seconds, or about 1 hour and 20 minutes. In compiled BASIC, where 10,000 flops took about 15 seconds, we find that the time is about $(1,000,000/30,000)15 = 1500/3 = 500$ seconds, or about 8 minutes. The PC used for Table 9.1 is regarded as slow. ■

It is interesting to compare the efficiency of the Gauss method with back substitution to that of the Gauss–Jordan method. Recall that, in the Gauss–Jordan method, $[A \mid \mathbf{b}]$ is reduced to a form $[I \mid \mathbf{c}]$, where I is the identity matrix. Exercise 2 shows that Gauss–Jordan flop count is of order of magnitude $n^3/2$ for large n. Thus the Gauss–Jordan method is not as efficient as the Gauss method with back substitution if n is large. One expects a Gauss–Jordan program to take about one and a half times as long to execute. The program TIMING can also be used to illustrate this. (See Exercises 23–26.)

Counting Flops for Matrix Operations

The exercises ask you to count the flops involved in matrix addition, multiplication, and exponentiation. Recall that a matrix product $C = AB$ is obtained by taking dot products of row vectors of an $m \times n$ matrix A with column vectors of an $n \times s$ matrix B. We indicate the usual way that a computer finds the dot product c_{ij} in C. First the computer sets $c_{ij} = 0$. Then it replaces c_{ij} by $c_{ij} + a_{i1}b_{1j}$, which gives c_{ij} the value $a_{i1}b_{1j}$. Then the computer replaces c_{ij} by $c_{ij} + a_{i2}b_{2j}$, giving c_{ij} the value $a_{i1}b_{1j} + a_{i2}b_{2j}$. This process continues in the obvious way until finally we have accumulated the desired value

$$c_{ij} = a_{i1}b_{1j} + a_{i2}b_{2j} + \cdots + a_{in}b_{nj}.$$

Each of these replacements of c_{ij} is accomplished by means of a flop, typically expressed by

$$C(I,J) = C(I,J) + A(I,K) * B(K,J) \tag{13}$$

in BASIC.

EXAMPLE 2 Find the number of flops required to compute the dot product of two vectors, each with n components.

Solution The preceding discussion shows that the dot product uses n flops of form (13), corresponding to the values $1, 2, 3, \ldots, n$ for K. ▪

SUMMARY

1. A flop is a rather vaguely defined execution by a computer, consisting typically of a bit of indexing, the retrieval and storage of a couple of values, and one or two arithmetic operations. Typical flops might appear in a computer program in instruction lines such as

$$C(I,J) = A(I,J) + B(I,J)$$

or

$$A(I,J) = A(I,J) + R*A(K,J).$$

2. A computer takes time to execute a flop, and it is desirable to use as few flops as possible when performing an extensive computation.

3. If the number of flops used to solve a problem is given by a polynomial expression in n, it is customary to keep only the term of highest degree in the polynomial as a measure of the *order of magnitude* for the number of flops when n is large.

4. The order of magnitude for the number of flops used in solving a system $A\mathbf{x} = \mathbf{b}$ for an $n \times n$ matrix A is

$$\frac{n^3}{3} \quad \text{for the Gauss method with back substitution}$$

and

$$\frac{n^3}{2} \quad \text{for the Gauss–Jordan method.}$$

5. The formulas

$$1 + 2 + 3 + \cdots + n = \frac{n(n+1)}{2}$$

and

$$1^2 + 2^2 + 3^2 + \cdots + n^2 = \frac{n(n+1)(2n+1)}{6}$$

are handy for determining the number of flops performed by a computer in matrix computations.

EXERCISES

In all of these exercises, assume that no row vectors are interchanged in a matrix.

1. Let A be an $n \times n$ matrix. Show that, in reducing $[A \mid \mathbf{b}]$ to $[U \mid \mathbf{c}]$ using the Gauss method, the number of flops performed on $\mathbf{b}$ is of order of magnitude $n^2/2$ for large n.

2. Let A be an $n \times n$ matrix. Show that, in solving $A\mathbf{x} = \mathbf{b}$ using the Gauss–Jordan method, the number of flops has order of magnitude $n^3/2$ for large n.

In Exercises 3–11, let A be an $n \times n$ matrix, and let B and C be $m \times n$ matrices. For each matrix, find the number of flops required for efficient computation of the matrix.

3. $B + C$	4. A^2	5. BA
6. A^3	7. A^4	8. A^5
9. A^6	10. A^{63}	11. A^{64}

12. Which of the following is more efficient with a computer?
    a) Solving a square system $A\mathbf{x} = \mathbf{b}$ by the Gauss–Jordan method, which makes each pivot 1 before creating the zeros in the column containing it.
    b) Using a similar technique to reduce the system to a diagonal system $D\mathbf{x} = \mathbf{c}$, where the entries in D are not necessarily all 1, and then dividing by these entries to obtain the solution.

13. Let A be an $n \times n$ matrix, where n is large. Find the order of magnitude for the number of flops if A^{-1} is computed using the Gauss–Jordan method on the augmented matrix $[A \mid I]$ without trying to reduce the number of flops used on I in response to the zeros that appear in it.

14. Repeat Exercise 13, using the Gauss method with back substitution rather than the Gauss–Jordan method.

15. Repeat Exercise 14, but this time cut down the number of flops performed on I during the Gauss reduction by taking into account the zeros in I.

*Exercises 16–20 concern band matrices. In a number of situations, square linear systems $A\mathbf{x} = \mathbf{b}$ occur in which the nonzero entries in the $n \times n$ matrix A are all concentrated near the main diagonal, running from the upper left-hand corner to the lower right-hand corner of A. Such a matrix is called a **band matrix**. For example, the matrix*

$$w = 2$$

$$\begin{bmatrix} 2 & 1 & 0 & 0 & 0 & 0 \\ 1 & 3 & 4 & 0 & 0 & 0 \\ 0 & 4 & 1 & 2 & 0 & 0 \\ 0 & 0 & 2 & 2 & 7 & 0 \\ 0 & 0 & 0 & 7 & 1 & 3 \\ 0 & 0 & 0 & 0 & 3 & 5 \end{bmatrix} \qquad (14)$$

$$w = 2$$

*is a symmetric 6×6 band matrix. We say that the **band width** of a band matrix $[a_{ij}]$ is w if w is the smallest integer such that $a_{ij} = 0$ for $|i - j| \geq w$. Thus matrix (14) has band width $w = 2$, as indicated. Such a matrix of band width 2 is also called **tridiagonal**. We usually assume that the band width is small compared with the size of n. As the band width approaches n, the matrix becomes **full**.*

In Exercises 16–20, assume that A is an $n \times n$ band matrix with band width w that is small in comparison to n.

16. What can be said concerning the band width of A^2? of A^3? of A^m?

17. Let A be tridiagonal, so $w = 2$. Find the order of magnitude of the number of flops required to reduce the partitioned matrix $[A \mid \mathbf{b}]$ to a form $[U \mid \mathbf{c}]$, where U is an upper-triangular matrix, taking into account the banded character of A.

18. Repeat Exercise 17 for band width w, expressing the result in terms of w.

19. Repeat Exercise 18, but include the flops used in back substitution to solve $A\mathbf{x} = \mathbf{b}$.

20. Explain why the Gauss method with back substitution is much more efficient than the Gauss–Jordan method for a banded matrix where w is very small compared with n.

21. Mark each of the following True or False.

___ a) A flop is a very precisely defined entity.

___ b) Computers can work so fast that it is not worthwhile to try to minimize the number of computations a computer makes to solve a problem, provided that the number of computations is only a few hundred.

___ c) Computers can work so fast that it is not worthwhile to try to minimize the number of computations required to solve a problem.

___ d) The Gauss method with back substitution and the Gauss–Jordan method for solving a large linear system both take about the same amount of computer time.

___ e) The Gauss–Jordan method for solving a large linear system takes about half again as long to execute as does the Gauss method on a computer.

___ f) Multiplying two $n \times n$ matrices requires more flops than does solving a linear system with an $n \times n$ coefficient matrix.

___ g) About n^2 flops are required to execute the back substitution in solving a linear system with an $n \times n$ coefficient matrix by using the Gauss method.

___ h) Executing the Gauss method with back substitution for a large linear system with an $n \times n$ coefficient matrix requires about n^2 flops.

___ i) Executing the Gauss method with back substitution for a large linear system with an $n \times n$ coefficient matrix requires about $n^3/3$ flops.

___ j) Executing the Gauss–Jordan method for a large linear system with an $n \times n$ coefficient matrix requires about $n^2/2$ flops.

⌷ *The software available to schools using our text includes a program TIMING that can be used to time*

algebraic operations and flops. The program can also be used to time the solution of square systems $A\mathbf{x} = \mathbf{b}$ by the Gauss method with back substitution and by the Gauss–Jordan method. For a user-specified integer $n \leq 60$, the program generates the $n \times n$ matrix A and column vector $\mathbf{b}$, where all entries are in the interval $[-20, 20]$. Use TIMING or similar software for Exercise 22–26.

22. Run the program TIMING, and obtain data for your PC analogous to those in Table 9.1.

23. Run the program TIMING in interpretive BASIC to time the Gauss method and the Gauss–Jordan method, starting with small values of n and increasing them until a few seconds' difference in times for the two methods is obtained. Does the time for the Gauss–Jordan method seem to be about $\frac{3}{2}$ the time for the Gauss method with back substitution? If not, why not? Experiment with both single-precision and double-precision timing.

24. Continuing Exercise 23, increase the size of n until the solutions take 2 or 3 minutes. Now does the Gauss–Jordan method seem to take about $\frac{3}{2}$ times as long? If this ratio is significantly different from that obtained in Exercise 23, explain why. (The two ratios may or may not appear approximately the same, depending on the speed of the computer used and on whether time in fractions of seconds is displayed.)

25. Repeat Exercise 23, using the compiled option in TIMING. Compare the ratios of times with those obtained in Exercise 23, and explain any difference.

26. Repeat Exercise 24, using the compiled option in TIMING. Compare the ratio of times with those in Exercise 24, and explain any difference.

9.2   THE *LU*-FACTORIZATION

Keeping a Record of Row Operations

We continue to work with a square linear system $A\mathbf{x} = \mathbf{b}$ having a unique solution that can be found by using Gauss elimination with back substitution, without having to interchange any rows. That is, the matrix A can be row-reduced to an upper-triangular matrix U, without making any row interchanges.

Situations occur in which it is necessary to solve many such systems, all having the same coefficient matrix A but different column vectors $\mathbf{b}$. In Section 1.3, we discussed solving such multiple systems by row-reducing a single augmented matrix

$$[A \mid \mathbf{b}_1, \mathbf{b}_2, \ldots, \mathbf{b}_s]$$

in which we line up all the different column vectors $\mathbf{b}_1, \mathbf{b}_2, \ldots, \mathbf{b}_s$ on the right of the partition. In practice, it may be impossible to solve all of these systems by using this single augmentation in a single computer run. Here are some possibilities in which a sequence of computer runs may be needed to solve all the systems:

1. Remember that we are concerned with *large* systems. If the number s of vectors $\mathbf{b}_j$ is large, there may not be room in the computer memory to accommodate all of these data at one time. Indeed, it might even be necessary to reduce the $n \times n$ matrix A in segments, if n is very large. If we can handle the s vectors $\mathbf{b}_j$ only in groups of m at a time, we must use at least s/m computer runs to solve all the systems.

2. Perhaps the vectors $\mathbf{b}_j$ are generated over a period of time, and we need to solve systems involving groups of the vectors $\mathbf{b}_j$ as they are generated. For example, we may want to solve $A\mathbf{x} = \mathbf{b}_j$ with r different vectors $\mathbf{b}_j$ each day.

3. Perhaps the vector $\mathbf{b}_{j+1}$ depends on the solution of $A\mathbf{x} = \mathbf{b}_j$. We would then have to solve $A\mathbf{x} = \mathbf{b}_1$, determine $\mathbf{b}_2$, solve $A\mathbf{x} = \mathbf{b}_2$, determine $\mathbf{b}_3$, and so on, until we finally solved $A\mathbf{x} = \mathbf{b}_s$.

From Section 9.1, we know that the magnitude of the number of flops required to reduce A to an upper-triangular matrix U is $n^3/3$ for large n. We want to avoid having to repeat all this work done in reducing A after the first computer run.

We assume that A can be reduced to U without interchanging rows—that is, that (nonzero) pivots always appear where we want them as we reduce A to U. This means that only elementary row-addition operations (those that add a multiple of a row vector to another row vector) are used. Recall that a row-addition operation can be accomplished by multiplying on the left by an $n \times n$ elementary matrix E, where E is obtained by applying the same row-addition operation to the identity matrix. There is a sequence $E_1, E_2, \ldots, E_h$ of such elementary matrices such that

$$E_h E_{h-1} \cdots E_2 E_1 A = U. \tag{1}$$

Once the matrix U has been found by the computer, the data in it can be stored on a disk or on tape, and simply read in for future computer runs. However, when we are solving $A\mathbf{x} = \mathbf{b}$ by reducing the augmented matrix $[A \mid \mathbf{b}]$, the sequence of elementary row operations described by $E_h E_{h-1} \cdots E_2 E_1$ in Eq. (1) must be applied to the *entire* augmented matrix, not merely to A. Thus we need to keep a *record* of this sequence of row operations to perform on column vectors $\mathbf{b}_j$ when they are used in subsequent computer runs.

Here is a way of recording the row-addition operations that is both efficient and algebraically interesting. As we make the reduction of A to U, we create a *lower-triangular matrix L*, which is a record of the row-addition operations performed. We start with the $n \times n$ identity matrix I, and as each row-addition operation on A is performed, we change one of the zero entries below the diagonal in I to give a record of that operation. For example, if during the reduction we add 4 times row 2 to row 6, we place -4 in the *second* column position in the *sixth* row of the matrix L that we are creating as a record. The general formulation is shown in the following box.

Creation of the Matrix L

Start with the $n \times n$ identity matrix I. If during the reduction of A to U, r times row i is added to row k, replace the zero in row k and column i of the identity matrix by $-r$. The final result obtained from the identity matrix is L.

EXAMPLE 1   Reduce the matrix

$$A = \begin{bmatrix} 1 & 3 & -1 \\ 2 & 8 & 4 \\ -1 & 3 & 4 \end{bmatrix}$$

to upper-triangular form U, and create the matrix L described in the preceding box.

Solution   We proceed in two columns, as follows:

Reduction of A to U	Creation of L from I
$A = \begin{bmatrix} 1 & 3 & -1 \\ 2 & 8 & 4 \\ -1 & 3 & 4 \end{bmatrix}$	$I = \begin{bmatrix} 1 & 0 & 0 \\ 0 & 1 & 0 \\ 0 & 0 & 1 \end{bmatrix}$

Add -2 times
row 1 to row 2.

$$\sim \begin{bmatrix} 1 & 3 & -1 \\ 0 & 2 & 6 \\ -1 & 3 & 4 \end{bmatrix} \qquad \begin{bmatrix} 1 & 0 & 0 \\ 2 & 1 & 0 \\ 0 & 0 & 1 \end{bmatrix}$$

Add 1 times
row 1 to row 3.

$$\sim \begin{bmatrix} 1 & 3 & -1 \\ 0 & 2 & 6 \\ 0 & 6 & 3 \end{bmatrix} \qquad \begin{bmatrix} 1 & 0 & 0 \\ 2 & 1 & 0 \\ -1 & 0 & 1 \end{bmatrix}$$

Add -3 times
row 2 to row 3.

$$\sim \begin{bmatrix} 1 & 3 & -1 \\ 0 & 2 & 6 \\ 0 & 0 & -15 \end{bmatrix} = U \qquad L = \begin{bmatrix} 1 & 0 & 0 \\ 2 & 1 & 0 \\ -1 & 3 & 1 \end{bmatrix}. \qquad ■$$

We now illustrate how the record kept in L in Example 1 can be used to solve a linear system $A\mathbf{x} = \mathbf{b}$ having the matrix A of Example 1 as coefficient matrix.

EXAMPLE 2   Use the record in L in Example 1 to solve the linear system $A\mathbf{x} = \mathbf{b}$ given by

$$
\begin{aligned}
x_1 + 3x_2 - x_3 &= -4 \\
2x_1 + 8x_2 + 4x_3 &= 2 \\
-x_1 + 3x_2 + 4x_3 &= 4.
\end{aligned}
$$

Solution   We use the record in L to find the column vector $\mathbf{c}$ that would occur if we were to reduce $[A \mid \mathbf{b}]$ to $[U \mid \mathbf{c}]$, using these same row operations:

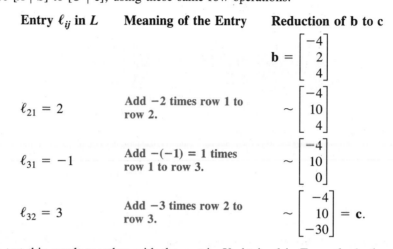

Entry ℓ_{ij} in L	Meaning of the Entry	Reduction of b to c
		$\mathbf{b} = \begin{bmatrix} -4 \\ 2 \\ 4 \end{bmatrix}$
$\ell_{21} = 2$	Add -2 times row 1 to row 2.	$\sim \begin{bmatrix} -4 \\ 10 \\ 4 \end{bmatrix}$
$\ell_{31} = -1$	Add $-(-1) = 1$ times row 1 to row 3.	$\sim \begin{bmatrix} -4 \\ 10 \\ 0 \end{bmatrix}$
$\ell_{32} = 3$	Add -3 times row 2 to row 3.	$\sim \begin{bmatrix} -4 \\ 10 \\ -30 \end{bmatrix} = \mathbf{c}.$

If we put this result together with the matrix U obtained in Example 1, the reduced partitioned matrix for the linear system becomes

$$
[U \mid \mathbf{c}] = \begin{bmatrix} 1 & 3 & -1 & \mid & -4 \\ 0 & 2 & 6 & \mid & 10 \\ 0 & 0 & -15 & \mid & -30 \end{bmatrix}.
$$

Back substitution then yields

$$
x_3 = \frac{-30}{-15} = 2,
$$

$$
x_2 = \frac{10 - 6(2)}{2} = \frac{-2}{2} = -1,
$$

$$
x_1 = -4 + 3 + 2 = 1.
$$ ■

We give another example.

EXAMPLE 3   Let

$$
A = \begin{bmatrix} 1 & -2 & 0 & 3 \\ -2 & 3 & 1 & -6 \\ -1 & 4 & -4 & 3 \\ 5 & -8 & 4 & 0 \end{bmatrix} \quad \text{and} \quad \mathbf{b} = \begin{bmatrix} 11 \\ -21 \\ -1 \\ 23 \end{bmatrix}.
$$

Generate the matrix L while reducing the matrix A to U. Then use U and the record in L to solve $A\mathbf{x} = \mathbf{b}$.

Solution   We work in two columns again. This time we fix up a whole column of A in each step:

Reduction of A   **Generation of L**

$$A = \begin{bmatrix} 1 & -2 & 0 & 3 \\ -2 & 3 & 1 & -6 \\ -1 & 4 & -4 & 3 \\ 5 & -8 & 4 & 0 \end{bmatrix} \qquad I = \begin{bmatrix} 1 & 0 & 0 & 0 \\ 0 & 1 & 0 & 0 \\ 0 & 0 & 1 & 0 \\ 0 & 0 & 0 & 1 \end{bmatrix}$$

$$\sim \begin{bmatrix} 1 & -2 & 0 & 3 \\ 0 & -1 & 1 & 0 \\ 0 & 2 & -4 & 6 \\ 0 & 2 & 4 & -15 \end{bmatrix} \qquad \begin{bmatrix} 1 & 0 & 0 & 0 \\ -2 & 1 & 0 & 0 \\ -1 & 0 & 1 & 0 \\ 5 & 0 & 0 & 1 \end{bmatrix}$$

$$\sim \begin{bmatrix} 1 & -2 & 0 & 3 \\ 0 & -1 & 1 & 0 \\ 0 & 0 & -2 & 6 \\ 0 & 0 & 6 & -15 \end{bmatrix} \qquad \begin{bmatrix} 1 & 0 & 0 & 0 \\ -2 & 1 & 0 & 0 \\ -1 & -2 & 1 & 0 \\ 5 & -2 & 0 & 1 \end{bmatrix}$$

$$\sim \begin{bmatrix} 1 & -2 & 0 & 3 \\ 0 & -1 & 1 & 0 \\ 0 & 0 & -2 & 6 \\ 0 & 0 & 0 & 3 \end{bmatrix} = U \qquad L = \begin{bmatrix} 1 & 0 & 0 & 0 \\ -2 & 1 & 0 & 0 \\ -1 & -2 & 1 & 0 \\ 5 & -2 & -3 & 1 \end{bmatrix}.$$

We now apply the record below the diagonal in L to the vector $\mathbf{b}$, working under the main diagonal down each column of the record in L in turn.

First column of L: $$\mathbf{b} = \begin{bmatrix} 11 \\ -21 \\ -1 \\ 23 \end{bmatrix} \sim \begin{bmatrix} 11 \\ 1 \\ -1 \\ 23 \end{bmatrix} \sim \begin{bmatrix} 11 \\ 1 \\ 10 \\ 23 \end{bmatrix} \sim \begin{bmatrix} 11 \\ 1 \\ 10 \\ -32 \end{bmatrix}.$$

Second column of L: $$\begin{bmatrix} 11 \\ 1 \\ 10 \\ -32 \end{bmatrix} \sim \begin{bmatrix} 11 \\ 1 \\ 12 \\ -32 \end{bmatrix} \sim \begin{bmatrix} 11 \\ 1 \\ 12 \\ -30 \end{bmatrix}.$$

Third column of L: $$\begin{bmatrix} 11 \\ 1 \\ 12 \\ -30 \end{bmatrix} \sim \begin{bmatrix} 11 \\ 1 \\ 12 \\ 6 \end{bmatrix}.$$

The partitioned matrix

$$\left[\begin{array}{cccc|c} 1 & -2 & 0 & 3 & 11 \\ 0 & -1 & 1 & 0 & 1 \\ 0 & 0 & -2 & 6 & 12 \\ 0 & 0 & 0 & 3 & 6 \end{array} \right]$$

yields, upon back substitution,

$$x_4 = \frac{6}{3} = 2,$$

$$x_3 = (12 - 12)/(-2) = 0,$$

$$x_2 = (1 - 0)/(-1) = -1,$$
$$x_1 = 11 - 2 - 6 = 3.$$ ▪

Two questions come to mind:

1. Why bother to put the entries 1 down the main diagonal in L, since they are not used?

2. If we add r times a row to another row while reducing A, why do we put $-r$ rather than r in the record in L, and then change back to r again when performing the operations on the column vector **b**?

We do these two things only because the matrix L formed in ths way has an interesting algebraic property, which we will discuss in a moment. In fact, when solving a large system using a computer, we certainly would not fuss with the entries 1 down the diagonal. Indeed, we can save memory space by not even generating a matrix L separate from the one being reduced. When creating a zero below the main diagonal, place the record $-r$ or r desired, as described in question 2 above, directly in the matrix being reduced at the position where a zero is being created! The computer already has space reserved for an entry there. Just remember that the final matrix contains the desired entries of U on and above the diagonal and the record for L or $-L$ below the diagonal. *We will always use $-r$ rather than r as record entry in this text.* Thus, for the 4×4 matrix in Example 3, we obtain

$$\begin{bmatrix} 1 & -2 & 0 & 3 \\ -2 & -1 & 1 & 0 \\ -1 & -2 & -2 & 6 \\ 5 & -2 & -3 & 3 \end{bmatrix}, \quad \text{Combined } L\backslash U \text{ display} \tag{2}$$

where the black entries on or above the main diagonal give the essential data for U, and the color entries are the essential data for L.

Let us examine the efficiency of solving a system $A\mathbf{x} = \mathbf{b}$ if U and L are already known. Each entry in the record in L requires one flop to execute when applying this record to reduce a column vector **b**. The number of entries is

$$1 + 2 + \cdots + n - 1 = \frac{(n-1)n}{2} = \frac{n^2}{2} - \frac{n}{2},$$

which is of order of magnitude $n^2/2$ for large n. We saw in Section 9.1 that back substitution requires about $n^2/2$ flops, too, giving a total of n^2 flops for large n to solve $A\mathbf{x} = \mathbf{b}$, once U and L are known. If instead we computed A^{-1} and found $\mathbf{x} = A^{-1}\mathbf{b}$, the product $A^{-1}\mathbf{b}$ would also require n^2 flops. But there are at least two advantages in using the *LU*-technique. First, finding U requires about $n^3/3$ flops for large n, whereas finding A^{-1} requires n^3 flops. (See Exercise 15 of Section 9.1.) If $n = 1000$, the difference in computer time is considerable. Second, more computer memory is used in reducing $[A \mid I]$ to $[I \mid A^{-1}]$ than is used in the efficient way we record L as we find U, illustrated in the combined $L\backslash U$ display (2).

We give a specific illustration in which keeping the record L is useful.

EXAMPLE 4   Let

$$A = \begin{bmatrix} 1 & 2 & -1 \\ -2 & -5 & 3 \\ -1 & -3 & 0 \end{bmatrix} \quad \text{and} \quad b = \begin{bmatrix} 9 \\ -17 \\ -44 \end{bmatrix}.$$

Solve the linear system $A^3 x = b$.

Solution   We view $A^3 x = b$ as $A(A^2 x) = b$, and substitute $y = A^2 x$ to obtain $Ay = b$. We can solve this equation for y. Then we write $A^2 x = y$ as $A(Ax) = y$ or $Az = y$, where $z = Ax$. We then solve $Az = y$ for z. Finally, we solve $Ax = z$ for the desired x. Since we are using the same coefficient matrix A each time, it is efficient to find the matrices L and U and then to proceed as in Example 3 to find y, z, and x in turn.

We find that the matrices U and L are given by

$$U = \begin{bmatrix} 1 & 2 & -1 \\ 0 & -1 & 1 \\ 0 & 0 & -2 \end{bmatrix} \quad \text{and} \quad L = \begin{bmatrix} 1 & 0 & 0 \\ -2 & 1 & 0 \\ -1 & 1 & 1 \end{bmatrix}.$$

Applying the record in L to b, we obtain

$$b = \begin{bmatrix} 9 \\ -17 \\ -44 \end{bmatrix} \sim \begin{bmatrix} 9 \\ 1 \\ -35 \end{bmatrix} \sim \begin{bmatrix} 9 \\ 1 \\ -36 \end{bmatrix}.$$

From the matrix U, we find that

$$y = \begin{bmatrix} -7 \\ 17 \\ 18 \end{bmatrix}.$$

To solve $Az = y$, we apply the record in L to y:

$$y = \begin{bmatrix} -7 \\ 17 \\ 18 \end{bmatrix} \sim \begin{bmatrix} -7 \\ 3 \\ 11 \end{bmatrix} \sim \begin{bmatrix} -7 \\ 3 \\ 8 \end{bmatrix}.$$

Using U, we obtain

$$z = \begin{bmatrix} 3 \\ -7 \\ -4 \end{bmatrix}.$$

Finally, to solve $Ax = z$, we apply the record in L to z:

$$z = \begin{bmatrix} 3 \\ -7 \\ -4 \end{bmatrix} \sim \begin{bmatrix} 3 \\ -1 \\ -1 \end{bmatrix} \sim \begin{bmatrix} 3 \\ -1 \\ 0 \end{bmatrix}.$$

Using U, we find that

$$x = \begin{bmatrix} 1 \\ 1 \\ 0 \end{bmatrix}. \qquad \blacksquare$$

The program LUFACTOR, available with this text, can be used to find the matrices L and U; it has an option for iteration to solve a system $A^m\mathbf{x} = \mathbf{b}$, as we did in Example 4.

The Factorization $A = LU$

This heading shows why the matrix L we described is algebraically interesting: we have $A = LU$.

EXAMPLE 5   Illustrate $A = LU$ for the matrices obtained in Example 3.

Solution   From Example 3, we have

$$
LU = \begin{bmatrix} 1 & 0 & 0 & 0 \\ -2 & 1 & 0 & 0 \\ -1 & -2 & 1 & 0 \\ 5 & -2 & -3 & 1 \end{bmatrix}\begin{bmatrix} 1 & -2 & 0 & 3 \\ 0 & -1 & 1 & 0 \\ 0 & 0 & -2 & 6 \\ 0 & 0 & 0 & 3 \end{bmatrix} = \begin{bmatrix} 1 & -2 & 0 & 3 \\ -2 & 3 & 1 & -6 \\ -1 & 4 & -4 & 3 \\ 5 & -8 & 4 & 0 \end{bmatrix} = A.
$$

■

It is not difficult to establish that we always have $A = LU$. Recall from Eq. (1) that

$$
E_h E_{h-1} \cdots E_2 E_1 A = U, \tag{3}
$$

for elementary matrices E_i corresponding to elementary row opertions consisting of multiplying a row by a constant and adding the result to a lower row. From Eq. (3), we obtain

$$
A = E_1^{-1} E_2^{-1} \cdots E_{h-1}^{-1} E_h^{-1} U. \tag{4}
$$

We proceed to show that the matrix L is equal to the product $E_1^{-1} E_2^{-1} \cdots E_{h-1}^{-1} E_h^{-1}$ appearing in Eq. (4). Now if E_i is obtained from the identity matrix by adding r (possibly $r = 0$) times a row to another row, then E_i^{-1} is obtained from the identity matrix by adding $-r$ times the same row to the other row. Since E_h corresponds to the last row operation performed, $E_h^{-1} I$ places the negative of the last multiplier in the nth row and $(n-1)$st column of this product, precisely where it should appear in L. Similarly,

$$
E_{h-1}^{-1}(E_h^{-1} I)
$$

has the appropriate entry desired in L in the nth row and $(n-2)$nd column;

$$
E_{h-2}^{-1}(E_{h-1}^{-1} E_h^{-1} I)
$$

has the appropriate entry desired in L in the $(n-1)$st row and $(n-2)$nd column; and so on. That is, the record entries of L may be created from the expression

$$
E_1^{-1} E_2^{-1} \cdots E_{h-1}^{-1} E_h^{-1} I,
$$

starting at the right with I and working to the left. This is the order shown by the color numbers below the main diagonal in the matrix

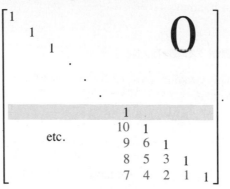

As we perform elementary row operations that correspond to adding scalar multiples of the shaded row to lower rows, none of the entries already created below the main diagonal will be changed, because the zeros to the right of the main diagonal lie over those entries. Thus,

$$E_1^{-1}E_2^{-1} \cdots E_{h-1}^{-1}E_h^{-1}I = L, \tag{5}$$

and $A = LU$ follows at once from Eq. (4).

We have shown how to find a solution of $Ax = \mathbf{b}$ from a factorization $A = LU$ by using the record in L to modify $\mathbf{b}$ to a vector $\mathbf{c}$ and then solving $Ux = \mathbf{c}$ by back substitution. Approximately n^2 flops are required. An alternate method to determine $\mathbf{x}$ from $LUx = \mathbf{b}$ is to view the equation as $L(Ux) = \mathbf{b}$. Letting $\mathbf{c} = Ux$, we first solve $Lc = \mathbf{b}$ for $\mathbf{c}$ by *forward substitution*, and then we solve $Ux = \mathbf{c}$ by back substitution. Since each of the forward and the back substitutions takes approximately $n^2/2$ flops, the total number required is again approximately n^2 flops. We illustrate this alternate method.

EXAMPLE 6 In Example 1, we found the factorization

$$
\begin{bmatrix} 1 & 3 & -1 \\ 2 & 8 & 4 \\ -1 & 3 & 4 \end{bmatrix} = \begin{bmatrix} 1 & 0 & 0 \\ 2 & 1 & 0 \\ -1 & 3 & 1 \end{bmatrix}\begin{bmatrix} 1 & 3 & -1 \\ 0 & 2 & 6 \\ 0 & 0 & -15 \end{bmatrix}.
$$

$$\underset{A}{} \qquad \underset{L}{} \qquad \underset{U}{}$$

Use the method of forward substitution and back substitution to solve the linear system $Ax = \begin{bmatrix} -4 \\ 2 \\ 4 \end{bmatrix}$, which we solved in Example 2.

Solution First, we solve $Lc = \mathbf{b}$ by forward substitution:

$$
\left[\begin{array}{ccc|c} 1 & 0 & 0 & -4 \\ 2 & 1 & 0 & 2 \\ -1 & 3 & 1 & 4 \end{array}\right]
$$

$$c_1 = -4,$$

$$2c_1 + c_2 = 2, c_2 = 2 + 8 = 10.$$

$$-c_1 + 3c_2 + c_3 = 4,$$

$$c_3 = c_1 - 3c_3 + 4 = -4 - 30 + 4 = -30.$$

Notice that this is the same **c** as was obtained in Example 2. The back substitution with $U\mathbf{x} = \mathbf{c}$ of Example 2 then yields the same solution **x**. ■

Factorization of an invertible square matrix A into LU, with L lower triangular and U upper triangular, is not unique. For example, if r is a nonzero scalar, then rL is lower triangular and $(1/r)U$ is upper triangular, and if $A = LU$, then we also have $A = (rL)((1/r)U)$. But let D be the $n \times n$ diagonal matrix having the same main diagonal as U; that is,

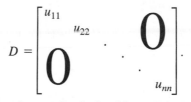

Let U^* be the upper-triangular matrix obtained from U by multiplying the ith row by $1/u_{ii}$ for $i = 1, 2, \ldots, n$. Then $U = DU^*$, and we have

$$A = LDU^*.$$

Now both L and U^* have all entries 1 on their main diagonals. This type of factorization is unique, as we now show.

THEOREM 9.1 Unique Factorization

Let A be an $n \times n$ matrix. A factorization of the form $A = LDU$, where

L is lower triangular with all main diagonal entries 1,
U is upper triangular with all main diagonal entries 1,
D is diagonal with all main diagonal entries nonzero,

is unique.

PROOF Suppose that $A = L_1D_1U_1$ and $A = L_2D_2U_2$ are two such factorizations. Observe that both L_1^{-1} and L_2^{-1} are also lower triangular, D_1^{-1} and D_2^{-1} are both diagonal, and U_1^{-1} and U_2^{-1} are both still upper triangular. Just think how the matrix reductions of $[L_1 \mid I]$ or $[D_1 \mid I]$ or $[U_1 \mid I]$ to find the inverses look. Furthermore, L_1^{-1}, L_2^{-1}, U_1^{-1}, and U_2^{-1} have all main diagonal entries equal to 1.

Now from $L_1D_1U_1 = L_2D_2U_2$, we obtain

$$L_2^{-1}L_1 = D_2U_2U_1^{-1}D_1^{-1}. \tag{6}$$

We see that $L_2^{-1}L_1$ is again lower triangular with entries 1 on its main diagonal, while $D_2U_2U_1^{-1}D_1^{-1}$ is upper triangular. Equation (6) then shows that both matrices must be I, so $L_1L_2^{-1} = I$ and $L_1 = L_2$. A similar argument starting over with $L_1D_1U_1 = L_2D_2U_2$ rewritten as

$$U_1U_2^{-1} = D_1 {}^{-1}L_1 {}^{-1}L_2D_2 \tag{7}$$

shows that $U_1 = U_2$. We then have $L_1D_1U_1 = L_1D_2U_1$, and multiplication on the left by L_1^{-1} and on the right by U_1^{-1} yields $D_1 = D_2$.   ▲

Systems Requiring Row Interchanges

Let A be an invertible square matrix whose row reduction to an upper-triangular matrix U requires at least one row interchange. Then not all elementary matrices corresponding to the necessary row operations add multiplies of row vectors to other row vectors. We can still write

$$E_hE_{h-1} \cdots E_2E_1A = U,$$

so

$$A = E_1^{-1}E_2^{-1} \cdots E_{h-1}^{-1}E_h^{-1}U,$$

but now that some of these E_i^{-1} interchange rows, their product may not be lower triangular. However, after we discover which row interchanges are necessary, we could start over, and make these necessary row interchanges in the matrix A before we start creating zeros below the diagonal. It is easy to see that the upper-triangular matrix U obtained would still be the same. Suppose, for example, that to obtain a nonzero element in pivot position in the ith row we interchange this ith row with a kth row further down in the matrix. The new ith row will be the same as though it had been put in the ith row position before the start of the reduction; in either case, it has been modified during the reduction only by the addition of multiples of rows *above* the ith row position. As multiples of it are now added to rows below the ith row position, the same rows (except possibly for order) below the ith row position are created, whether row interchange is performed during reduction or is completed before reduction starts.

Interchanging some rows before the start of the reduction amounts to multiplying A on the left by a sequence of elementary row-interchange matrices. Any product of elementary row-interchange matrices is called a **permutation matrix**. Thus we can form PA for a permutation matrix P, and PA will then admit a factorization $PA = LU$. We state this as a theorem.

THEOREM 9.2   *LU*-Factorization

Let A be an invertible square matrix. Then there exists a permutation matrix P, a lower-triangular matrix L, and an upper-triangular matrix U such that $PA = LU$.

EXAMPLE 7 Illustrate Theorem 9.2 for the matrix

$$A = \begin{bmatrix} 1 & 3 & 2 \\ -2 & -6 & 1 \\ 2 & 5 & 7 \end{bmatrix}.$$

Solution Starting to reduce A to upper-triangular form, we have

$$\begin{bmatrix} 1 & 3 & 2 \\ -2 & -6 & 1 \\ 2 & 5 & 7 \end{bmatrix} \sim \begin{bmatrix} 1 & 3 & 2 \\ 0 & 0 & 5 \\ 0 & -1 & 3 \end{bmatrix},$$

and we now find it necessary to interchange rows 2 and 3, which will then produce the desired U. Thus we may take

$$P = \begin{bmatrix} 1 & 0 & 0 \\ 0 & 0 & 1 \\ 0 & 1 & 0 \end{bmatrix},$$

and we have

$$PA = \begin{bmatrix} 1 & 3 & 2 \\ 2 & 5 & 7 \\ -2 & -6 & 1 \end{bmatrix} \sim \begin{bmatrix} 1 & 3 & 2 \\ 0 & -1 & 3 \\ 0 & 0 & 5 \end{bmatrix} = U.$$

The record matrix L for reduction of PA becomes

$$L = \begin{bmatrix} 1 & 0 & 0 \\ 2 & 1 & 0 \\ -2 & 0 & 1 \end{bmatrix},$$

and we confirm that

$$PA = \begin{bmatrix} 1 & 3 & 2 \\ 2 & 5 & 7 \\ -2 & -6 & 1 \end{bmatrix} = \begin{bmatrix} 1 & 0 & 0 \\ 2 & 1 & 0 \\ -2 & 0 & 1 \end{bmatrix}\begin{bmatrix} 1 & 3 & 2 \\ 0 & -1 & 3 \\ 0 & 0 & 5 \end{bmatrix} = LU. \qquad ▪$$

Suppose that, when solving a large square linear system $A\mathbf{x} = \mathbf{b}$ with a computer, keeping the record L, we find that row interchanges are advisable. (See partial pivoting in Section 9.3.) We could keep going and make the row interchanges to find an upper-triangular matrix; we could then start over, make those row interchanges first, just as in Example 7, to obtain a matrix PA, and then obtain U and L for the matrix PA instead. Of course, we would first have to compute $P\mathbf{b}$ when solving $A\mathbf{x} = \mathbf{b}$, since the system would then become $PA\mathbf{x} = P\mathbf{b}$, before we could proceed with the record L and back substitution. This is an undesirable procedure, because it requires reduction of both A and PA, taking a total of about $2n^3/3$ flops, assuming that A is $n \times n$ and n is large. Surely it would be better to devise a method of record-keeping that would also keep track of any row interchanges as they occurred, and would then apply this improved record to $\mathbf{b}$ and use back substitution to find the solution. We toss this suggestion out for enthusiastic programmers to consider on their own.

SUMMARY

1. If A is an $n \times n$ invertible matrix that can be row-reduced to an upper-triangular matrix U without row interchanges, there exists a lower-triangular $n \times n$ matrix L such that $A = LU$.

2. The matrix L in summary item (1) can be found as follows. Start with the $n \times n$ identity matrix I. If during the reduction of A to U, r times row i is added to row k, replace the zero in row k and column i of the identity matrix by $-r$. The final result obtained from the identity matrix is the matrix L.

3. Once A has been reduced to U and L has been found, a computer can find the solution of $A\mathbf{x} = \mathbf{b}$ for a new column vector $\mathbf{b}$, using about n^2 flops for large n.

4. If A is as described in summary item (1), then A has a unique factorization of the form $A = LDU$, where

   L is lower triangular with all diagonal entries 1,

   U is upper triangular with all diagonal entries 1, and

   D is a diagonal matrix with all diagonal entries nonzero.

5. For any invertible matrix A, there exists a permutation matrix P such that PA can be row-reduced to an upper-triangular matrix U and has the properties described for A in summary items (1), (2), (3), and (4).

EXERCISES

1. Discuss briefly the need to worry about the time required to create the record matrix L when solving

$$A\mathbf{x} = \mathbf{b}_1, \mathbf{b}_2, \ldots, \mathbf{b}_s$$

for a large $n \times n$ matrix A.

2. Is there any practical value in creating the record matrix L when one needs to solve only a *single* square linear system $A\mathbf{x} = \mathbf{b}$, as we did in Example 3?

In Exercises 3–7, find the solution of $A\mathbf{x} = \mathbf{b}$ from the given combined $L\backslash U$ display of the matrix A and the given vector $\mathbf{b}$.

3. $L\backslash U = \begin{bmatrix} 1 & 4 \\ 3 & -2 \end{bmatrix}$, $\mathbf{b} = \begin{bmatrix} 2 \\ -4 \end{bmatrix}$

4. $L\backslash U = \begin{bmatrix} -2 & 5 \\ -2 & 1 \end{bmatrix}$, $\mathbf{b} = \begin{bmatrix} 3 \\ -7 \end{bmatrix}$

5. $L\backslash U = \begin{bmatrix} 1 & -3 & 4 \\ -2 & -1 & 9 \\ 0 & 1 & -6 \end{bmatrix}$, $\mathbf{b} = \begin{bmatrix} 2 \\ -2 \\ 8 \end{bmatrix}$

6. $L\backslash U = \begin{bmatrix} 1 & -4 & 2 \\ 0 & 2 & -1 \\ 3 & 2 & -2 \end{bmatrix}$, $\mathbf{b} = \begin{bmatrix} -3 \\ 2 \\ 3 \end{bmatrix}$

7. $L\backslash U = \begin{bmatrix} 1 & 0 & 0 & 1 \\ 0 & -1 & 2 & 1 \\ -1 & -2 & 1 & 3 \\ 2 & 1 & -1 & 3 \end{bmatrix}$, $\mathbf{b} = \begin{bmatrix} 4 \\ 7 \\ -8 \\ 14 \end{bmatrix}$

In Exercises 8–13, let A be the given matrix. Find a permutation matrix P, if necessary, and matrices L and U such that $PA = LU$. Check the answer, using matrix multiplication. Then solve the system $A\mathbf{x} = \mathbf{b}$, using P, L, and U.

8. $A = \begin{bmatrix} 2 & -1 \\ 6 & -5 \end{bmatrix}$, $\mathbf{b} = \begin{bmatrix} 8 \\ 32 \end{bmatrix}$

9. $A = \begin{bmatrix} 2 & 1 & -3 \\ 6 & 3 & -8 \\ 2 & -1 & 5 \end{bmatrix}$, $\mathbf{b} = \begin{bmatrix} 6 \\ 17 \\ 0 \end{bmatrix}$

10. $A = \begin{bmatrix} 1 & 1 & -3 \\ 0 & 1 & 1 \\ 3 & -1 & 1 \end{bmatrix}$, $\mathbf{b} = \begin{bmatrix} -13 \\ 6 \\ -7 \end{bmatrix}$

11. $A = \begin{bmatrix} 1 & 2 & -1 \\ 3 & 7 & 2 \\ 4 & -2 & 1 \end{bmatrix}$, $\mathbf{b} = \begin{bmatrix} -3 \\ 1 \\ -2 \end{bmatrix}$

12. $A = \begin{bmatrix} 1 & -4 & 1 & -2 \\ 0 & 2 & -1 & 1 \\ 2 & -7 & -2 & 1 \\ 0 & 3 & 0 & -4 \end{bmatrix}$, $\mathbf{b} = \begin{bmatrix} -9 \\ 6 \\ 10 \\ -16 \end{bmatrix}$

13. $A = \begin{bmatrix} 1 & 2 & -3 & 0 \\ 2 & 5 & -6 & 0 \\ -1 & -2 & 1 & -1 \\ 4 & 10 & 9 & 1 \end{bmatrix}$, $\mathbf{b} = \begin{bmatrix} 8 \\ 17 \\ -8 \\ 33 \end{bmatrix}$

In Exercises 14–17, proceed as in Example 4 to solve the given system.

14. Solve $A^2\mathbf{x} = \mathbf{b}$ for $A = \begin{bmatrix} 1 & 1 \\ 3 & 0 \end{bmatrix}$, $\mathbf{b} = \begin{bmatrix} -11 \\ -6 \end{bmatrix}$.

15. Solve $A^4\mathbf{x} = \mathbf{b}$ for $A = \begin{bmatrix} -1 & 2 \\ 2 & -3 \end{bmatrix}$,

$\mathbf{b} = \begin{bmatrix} 144 \\ -233 \end{bmatrix}$.

16. Solve $A^2\mathbf{x} = \mathbf{b}$ for $A = \begin{bmatrix} 2 & -1 & 0 \\ 4 & -1 & 2 \\ -6 & 2 & 0 \end{bmatrix}$,

$\mathbf{b} = \begin{bmatrix} -2 \\ 14 \\ 12 \end{bmatrix}$.

17. Solve $A^3\mathbf{x} = \mathbf{b}$ if $A = \begin{bmatrix} -1 & 0 & 1 \\ 3 & 1 & 0 \\ -2 & 0 & 4 \end{bmatrix}$, $\mathbf{b} = \begin{bmatrix} 27 \\ 29 \\ 122 \end{bmatrix}$.

In Exercises 18–20, find the unique factorization LDU for the given matrix A.

18. The matrix in Exercise 8

19. The matrix in Exercise 10

20. The matrix in Exercise 11

21. Mark each of the following True or False.

____ a) Every matrix A has an *LU* factorization.

____ b) Every square matrix A has an *LU* factorization.

____ c) Every square matrix A has a factorization $P^{-1}LU$ for some permutation matrix P.

____ d) If an *LU* factorization of an $n \times n$ matrix A is known, using this factorization to solve a large linear system with coefficient matrix A requires about $n^2/2$ flops.

____ e) If an *LU* factorization of an $n \times n$ matrix A is known, using this factorization to solve a large linear system with coefficient matrix A requires about n^2 flops.

____ f) If $A = LU$, then this is the only factorization of this type for A.

____ g) If $A = LU$, then the matrix L can be regarded as a record of the steps in the row reduction of A to U.

____ h) All three types of elementary row operations used to reduce A to U can be recorded in the matrix L in an *LU* factorization of A.

____ i) Two of the three types of elementary row operations used to reduce A to U can be recorded in the matrix L in an *LU* factorization of A.

____ j) One can solve a linear system $LU\mathbf{x} = \mathbf{b}$ by means of a forward substitution followed by a back substitution.

In Exercises 22–24, use the available program LUFACTOR, or similar software, to find the combined L\U display (2) for the matrix A. Then solve $A\mathbf{x} = \mathbf{b}_i$ for each of the column vectors $\mathbf{b}_i$.

22. $A = \begin{bmatrix} 2 & 1 & -3 & 4 & 0 \\ 1 & 4 & 6 & -1 & 5 \\ -2 & 0 & 3 & 1 & 4 \\ 6 & 1 & 2 & 8 & -3 \\ 4 & 1 & 3 & 2 & 1 \end{bmatrix}$;

$\mathbf{b}_i = \begin{bmatrix} 13 \\ 7 \\ 4 \\ 27 \\ 17 \end{bmatrix}$, $\begin{bmatrix} 10 \\ 0 \\ 10 \\ 60 \\ 20 \end{bmatrix}$, and $\begin{bmatrix} -20 \\ 96 \\ 53 \\ 5 \\ 26 \end{bmatrix}$ for $i = 1$, 2, and 3, respectively.

23. $A = \begin{bmatrix} 1 & 3 & -5 & 2 & 1 \\ 4 & -6 & 10 & 8 & 3 \\ 3 & 6 & -1 & 4 & 7 \\ 2 & 1 & 11 & -3 & 13 \\ -6 & 3 & 1 & 4 & 21 \end{bmatrix}$;

$\mathbf{b}_i = \begin{bmatrix} 0 \\ 20 \\ -9 \\ -31 \\ -43 \end{bmatrix}$, $\begin{bmatrix} -48 \\ 218 \\ 0 \\ 100 \\ 143 \end{bmatrix}$, and $\begin{bmatrix} 87 \\ -151 \\ 102 \\ -46 \\ 223 \end{bmatrix}$

for $i = 1$, 2, and 3, respectively.

24. $A = \begin{bmatrix} -1 & 3 & 2 & 3 & 4 & 6 \\ 3 & 1 & 0 & -2 & 1 & 7 \\ 6 & 8 & -2 & 1 & 4 & 6 \\ 4 & -2 & 3 & 1 & 7 & 6 \\ 3 & -2 & 1 & 10 & 8 & 0 \\ 4 & -5 & 8 & -3 & 2 & 10 \end{bmatrix}$,

$\mathbf{b}_i = \begin{bmatrix} 19 \\ 1 \\ -8 \\ 82 \\ 76 \\ 103 \end{bmatrix}$, $\begin{bmatrix} 82 \\ 2 \\ 82 \\ 31 \\ 92 \\ -61 \end{bmatrix}$, and $\begin{bmatrix} 25 \\ -3 \\ -24 \\ -73 \\ -115 \\ 20 \end{bmatrix}$

for $i = 1, 2,$ and 3, respectively.

🖥 *In Exercises 25–29, use the program LUFACTOR to solve the indicated system.*

25. $A^5\mathbf{x} = \mathbf{b}$ for A and $\mathbf{b}$ in Exercise 14
26. $A^7\mathbf{x} = \mathbf{b}$ for A and $\mathbf{b}$ in Exercise 15
27. $A^6\mathbf{x} = \mathbf{b}$ for A and $\mathbf{b}$ in Exercise 16
28. $A^8\mathbf{x} = \mathbf{b}$ for A and $\mathbf{b}$ in Exercise 17
29. $A^3\mathbf{x} = \mathbf{b}_1$ for A and $\mathbf{b}_1$ in Exercise 24

🖥 *The available program TIMING has an option to time the formation of the combined L\U display from a matrix A. The user specifies the magnitude n for an n × n linear system Ax = b. The program then generates an n × n matrix A with entries in the interval [−20, 20] and creates the L\U display shown in matrix (2) in the text. The time required for the reduction is indicated. The user may then specify a number s for solution of Ax = $\mathbf{b}_1$, $\mathbf{b}_2$, . . . , $\mathbf{b}_s$. The computer generates a column vector and solves s systems, using the record in L and back substitution; the time to find these s solutions is also indicated. Recall that the reduction of A should require about $n^3/3$ flops for large n, and the solution for each column vector should require about n^2 flops. In Exercises 30–34, run the program TIMING or similar software, and see if the ratios of times obtained seems to conform to the ratios of the numbers of flops required, at least as n increases. Compare the time required to solve one system, including the computation of L\U, with the time required to solve a system of the same size using the Gauss method with back substitution. The times should be about the same.*

Choose single-precision or double-precision timing and interpretive BASIC or compiled BASIC.

30. $n = 4$; $s = 1, 3, 12$
31. $n = 10$; $s = 1, 5, 10$
32. $n = 16$; $s = 1, 8, 16$
33. $n = 24$; $s = 1, 12, 24$
34. $n = 30$; $s = 1, 15, 30$

9.3 PIVOTING, SCALING, AND ILL-CONDITIONED MATRICES

Some Problems Encountered with Computers

A computer can't do absolutely precise arithmetic with real numbers. For example, any computer can only compute using an approximation of the number $\sqrt{2}$, never with the number $\sqrt{2}$ itself. Computers work using base-2 arithmetic, representing numbers as strings of zeros and ones. But they encounter the same problems that we encounter if we pretend that we are a computer using base-10 arithmetic and are capable of handling only some fixed finite number of significant digits. We will illustrate computer problems in base-10 notation, since it is more familiar to all of us.

Suppose that our base-10 computer is asked to compute the quotient $\frac{1}{3}$, and suppose that it can represent a number in floating-point arithmetic using eight significant figures. It represents $\frac{1}{3}$ in **truncated** form as 0.33333333. It may represent $\frac{2}{3}$ as 0.66666667, **rounding off** to eight significant figures. For convenience, we will refer to all errors generated by either truncation or roundoff as **roundoff errors**.

Most people realize that in an extended arithmetic computation by a computer, the roundoff error can accumulate to such an extent that the final result of the computation becomes meaningless. We will see that there are at least two situations in which a computer cannot meaningfully execute even a single arithmetic operation. In order to avoid working with long strings of digits, we assume that our computer can compute only to three significant figures in its work. Thus, our three-figure, base-10 computer computes the quotient $\frac{2}{3}$ as 0.666. We box the first computer problem that concerns us, and give an illustration.

> Addition of two numbers of very different magnitude may result in the loss of some or even all the significant figures of the smaller number.

EXAMPLE 1   Evaluate our three-figure computer's computation of

$$45.1 + .0725.$$

Solution   Since it can handle only three significant figures, our computer represents the actual sum 45.1725 as 45.1. In other words, the second summand .0725 might as well be zero, so far as our computer is concerned. The datum .0725 is completely lost.   ■

The difficulty illustrated in Example 1 can cause serious problems in attempting to solve a linear system $A\mathbf{x} = \mathbf{b}$ with a computer, as we will illustrate in a moment. First we box another problem a computer may have.

> Subtraction of nearly equal numbers can result in a loss of significant figures.

EXAMPLE 2   Evaluate our three-figure computer's computation of

$$\frac{2}{3} - \frac{665}{1000}.$$

Solution   The actual difference is

$$.666666\cdots - .665 = .00166666\cdots.$$

However, our three-figure computer obtains

$$.666 - .665 = .001.$$

Two of the three significant figures with which the computer can work have been lost.   ■

The difficulty illustrated in Example 2 is encountered if one tries to use a computer to do differential calculus.

The cure for both difficulties is to have the computer work with more significant figures. But using more significant figures requires the computer to take more time to execute a program, and of course the same errors can occur "farther out." For example, the typical microcomputer of this decade, with software designed for routine computations, will compute $10^{10} + 10^{-10}$ as 10^{10}, losing the second summand 10^{-10} entirely.

Partial Pivoting

In row reduction of a matrix to echelon form, a technique called *partial pivoting* is often used. In **partial pivoting**, one interchanges the row in which the pivot is to occur with a row farther down, if necessary, so that the pivot becomes as large in absolute value as possible. To illustrate, suppose that row reduction of a matrix to echelon form leads to an intermediate matrix

$$\begin{bmatrix} 2 & 8 & -1 & 3 & 4 \\ 0 & -2 & 3 & -5 & 6 \\ 0 & 4 & 1 & 2 & 0 \\ 0 & -7 & 3 & 1 & 4 \end{bmatrix}.$$

Using partial pivoting, we would then interchange rows 2 and 4 to use the entry -7 of maximum magnitude among the possibilities $-2, 4$, and -7 for pivots in the second column. That is, we would form the matrix

$$\begin{bmatrix} 2 & 8 & -1 & 3 & 4 \\ 0 & -7 & 3 & 1 & 4 \\ 0 & 4 & 1 & 2 & 0 \\ 0 & -2 & 3 & -5 & 6 \end{bmatrix}$$

and continue with the reduction of this matrix.

We show by example the advantage of partial pivoting. Let us consider the linear system

$$\begin{aligned} .01x_1 + 100x_2 &= 100 \\ -100x_1 + 200x_2 &= 100. \end{aligned} \tag{1}$$

EXAMPLE 3 Find the actual solution of linear system (1). Then compare the result with that obtained by a three-figure computer using the Gauss method with back substitution, but without partial pivoting.

Solution First we find the actual solution:

$$\left[\begin{array}{cc|c} .01 & 100 & 100 \\ -100 & 200 & 100 \end{array} \right] \sim \left[\begin{array}{cc|c} .01 & 100 & 100 \\ 0 & 1,000,200 & 1,000,100 \end{array} \right],$$

and back substitution yields

$$x_2 = \frac{1,000,100}{1,000,200},$$

$$x_1 = \left(100 - \frac{100,010,000}{1,000,200} \right) 100 = \frac{1,000,000}{1,000,200}.$$

Thus, $x_1 \approx .9998$ and $x_2 \approx .9999$. On the other hand, our three-figure computer obtains

$$\begin{bmatrix} .01 & 100 \\ -100 & 200 \end{bmatrix} \begin{array}{|c} 100 \\ 100 \end{array} \sim \begin{bmatrix} .01 & 100 \\ 0 & 1,000,000 \end{bmatrix} \begin{array}{|c} 100 \\ 1,000,000 \end{array},$$

which leads to

$$x_2 = 1,$$
$$x_1 = (100 - 100)100 = 0.$$

The x_1-parts of the solution are very different. Our three-figure computer completely lost the second-row data entries 200 and 100 in the matrix when it added 10,000 times the first row to the second row. ■

EXAMPLE 4   Find the solution to system (1) that our three-figure computer would obtain using partial pivoting.

Solution   Using partial pivoting, the three-figure computer obtains

$$\begin{bmatrix} .01 & 100 \\ -100 & 200 \end{bmatrix} \begin{array}{|c} 100 \\ 100 \end{array} \sim \begin{bmatrix} -100 & 200 \\ .01 & 100 \end{bmatrix} \begin{array}{|c} 100 \\ 100 \end{array}$$

$$\sim \begin{bmatrix} -100 & 200 \\ 0 & 100 \end{bmatrix} \begin{array}{|c} 100 \\ 100 \end{array},$$

which yields

$$x_2 = 1,$$
$$x_1 = (100 - 200)/(-100) = 1.$$

This is close to the solution $x_1 \approx .9998$ and $x_2 \approx .9999$ obtained in Example 3, and it is much better than the erroneous $x_1 = 0$, $x_2 = 1$ obtained in Example 1 without pivoting. The data entries 200 and 100 in the second row of the initial matrix were not lost in this computation. ■

To understand completely the reason for the difference between the solutions in Example 3 and in Example 4, consider again the linear system

$$.01x_1 + 100x_2 = 100$$
$$-100x_1 + 200x_2 = 100,$$

which was solved in Example 3. Multiplication of the first row by 10,000 produced a coefficient of x_2 of such magnitude compared with the second equation coefficient 200 that the coefficient 200 was totally destroyed in the ensuing addition. (Remember that we are using our three-figure computer. In a less dramatic example, some significant figures of the smaller coefficient might still contribute to the sum.) If we use partial pivoting, the linear system becomes

$$-100x_1 + 200x_2 = 100$$
$$.01x_1 + 100x_2 = 100$$

and multiplication of the first equation by a number less than 1 does not threaten the significant digits of the numbers in the second equation. This explains the success of partial pivoting in this situation.

Suppose now that we multiply the first equation of system (1) by 10,000, which of course does not alter the solution of the system. We then obtain the linear system

$$100x_1 + 1,000,000x_2 = 1,000,000$$
$$-100x_1 + \quad\quad 200x_2 = \quad\quad 100.$$

$\quad$ (2)

If we solved system (2) using *partial pivoting*, we would not interchange the rows, since -100 is of no greater magnitude than 100. Addition of the first row to the second by our three-figure computer again totally destroys the coefficient 200 in the second row. Exercise 1 asks for a check that the erroneous solution $x_1 = 0$, $x_2 = 1$ is again obtained. We could avoid this problem by **full pivoting**. In full pivoting, columns are also interchanged, if necessary, to make pivots as large as possible. That is, if a pivot is to be found in the row i and column j position of an intermediate matrix G, then not only rows but also columns are interchanged as needed to put the entry g_{rs} of greatest magnitude, where $r \geq i$ and $s \geq j$, in the pivot position. Thus, full pivoting for system (2) will lead to the matrix

$$
\begin{array}{cc}
x_2 & x_1 \\
\end{array}
$$
$$
\left[\begin{array}{cc|c}
1,000,000 & 100 & 1,000,000 \\
200 & -100 & 100
\end{array} \right].
$$

$\quad$ (3)

Exercise 2 illustrates that row reduction by our three-figure computer of matrix (3) gives a reasonable solution of system (2). Notice, however, that we now have to do some bookkeeping and must remember that the entries in column 1 of matrix (3) are really the coefficients of x_2, not of x_1. For a matrix of any size, the search for elements of maximum magnitude and the bookkeeping required in full pivoting take a lot of computer time. Partial pivoting is frequently used, representing a compromise between time and accuracy. The program MATCOMP that we make available uses partial pivoting in its Gauss–Jordan reduction to reduced echelon form. Thus, if system (1) is modified so that the Gauss–Jordan method without pivoting fails to give a reasonable solution for a twenty-figure computer, MATCOMP could probably handle it. However, one can no doubt create a similar modification of the 2×2 system (2) for which MATCOMP would give an erroneous solution.

Scaling

We display again system (2):

$$100x_1 + 1,000,000x_2 = 1,000,000$$
$$-100x_1 + \quad\quad 200x_2 = \quad\quad 100.$$

We might recognize that the number 1,000,000 dangerously dominates the data entries in the second row, at least as far as our three-figure computer is concerned. We might multiply the first equation in system (2) by .0001 to cut those two numbers down to size, essentially coming back to system (1). Partial pivoting handles system (1) well. Multiplication of an equation by a nonzero constant for such a purpose is known as **scaling**. Of course, one could equivalently scale by multiplying the second equation in system (2) by 10,000, to bring its coefficients into line with the large numbers in the first equation.

We box one other computer problem that can sometimes be helped by scaling. In reducing a matrix to echelon form, we need to know whether the entry that appears in a pivot position as we start work on a column is truly nonzero, and indeed, whether there is any truly nonzero entry from that column to serve as pivot.

> Due to roundoff error, a computed number that should be zero is quite likely to be of small, nonzero magnitude.

Taking this into account, one usually programs row-echelon reduction so that entries of unexpectedly small size are changed to zero. MATCOMP finds the smallest nonzero magnitude m among 1 and all the coefficient data supplied for the linear system, and sets $E = rm$, where r is specified by the user. (Default r is .0001.) In reduction of the coefficient matrix, a computed entry of magnitude less than E is replaced by zero. The same procedure is followed in YUREDUCE. Whatever computed number we program a computer to choose for E in a program like MATCOMP, we will be able to devise some linear system for which E is either too large or too small to give the correct result.

An equivalent procedure to the one in MATCOMP that we just outlined is to *scale* the original data for the linear system in such a way that the smallest nonzero entry is of magnitude roughly 1, and then always to use the same value, perhaps 10^{-4}, for E.

When one of the authors first started experimenting with a computer, he was horrified to discover that a matrix inversion routine built into the mainframe BASIC program of a major computer company would refuse to invert a matrix if all the entries were small enough. An error message like

"Nearly singular matrix. Inversion impossible."

would appear. He considers a 2×2 matrix

$$\begin{bmatrix} a & b \\ c & d \end{bmatrix}$$

to be nearly singular if and only if lines in the plane having equations of the form

$$ax + by = r$$

$$cx + dy = s$$

are nearly parallel. Now the lines

$$10^{-8}x + \quad 0y = 1$$
$$0x + 10^{-9}y = 1$$

are actually perpendicular, the first one being vertical and the second horizontal. That is as far away from parallel as one can get! It annoyed the author greatly to have the matrix

$$\begin{bmatrix} 10^{-8} & 0 \\ 0 & 10^{-9} \end{bmatrix}$$

called "nearly singular." The inverse is obviously

$$\begin{bmatrix} 10^{8} & 0 \\ 0 & 10^{9} \end{bmatrix}.$$

A scaling routine was promptly written to be executed before calling the inversion routine. The matrix was multiplied by a constant that would bring the smallest nonzero magnitude to at least 1, and then the inversion subroutine was used, and the result rescaled to provide the inverse of the original matrix. For example, applied to the matrix just discussed, this procedure becomes

$$\begin{bmatrix} 10^{-8} & 0 \\ 0 & 10^{-9} \end{bmatrix} \quad \text{multiplied by } 10^9 \text{ becomes} \quad \begin{bmatrix} 10 & 0 \\ 0 & 1 \end{bmatrix},$$

$$\text{inverted becomes} \quad \begin{bmatrix} \frac{1}{10} & 0 \\ 0 & 1 \end{bmatrix},$$

$$\text{multiplied by } 10^9 \text{ becomes} \quad \begin{bmatrix} 10^8 & 0 \\ 0 & 10^9 \end{bmatrix}.$$

Having more programming experience now, this author is much more charitable and understanding. The user may also find unsatisfactory things about MATCOMP. We hope that this little anecdote has helped explain the notion of scaling.

Ill-conditioned Matrices

The line $x + y = 100$ in the plane has x-intercept 100 and y-intercept 100, as shown in Figure 9.1. The line $x + .9y = 100$ also has x-intercept 100, but it has y-intercept larger than 100. The two lines are almost parallel. The common x-intercept shows that the solution of the linear system

$$x + \quad y = 100$$
$$x + .9y = 100 \tag{4}$$

is $x = 100$, $y = 0$, as illustrated in Figure 9.2. Now the line $.9x + y = 100$ has y-intercept 100 but x-intercept larger than 100, so the linear system

$$x + y = 100$$
$$.9x + y = 100 \tag{5}$$

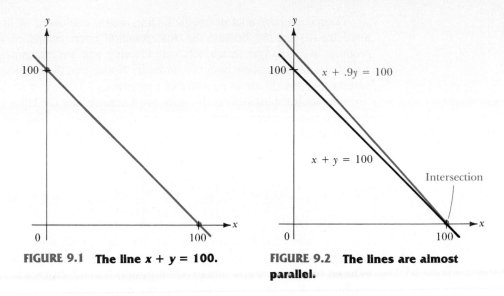

FIGURE 9.1 **The line $x + y = 100$.**

FIGURE 9.2 **The lines are almost parallel.**

has the very different solution $x = 0$, $y = 100$, as shown in Figure 9.3. Systems (4) and (5) are examples of **ill-conditioned** or **unstable** systems; small changes in the coefficients or in the constants on the right-hand sides can produce very great changes in the solutions. We say that a matrix A is **ill-conditioned** if a linear system $A\mathbf{x} = \mathbf{b}$ having A as coefficient matrix is ill-conditioned. For two equations in two unknowns, solving an ill-conditioned system corresponds to finding the intersection of two nearly parallel lines, as shown in Figure 9.4. Changing a coefficient of x or y slightly in one equation only changes the slope of that line slightly, but it may generate a big change in the location of the point of intersection of the two lines.

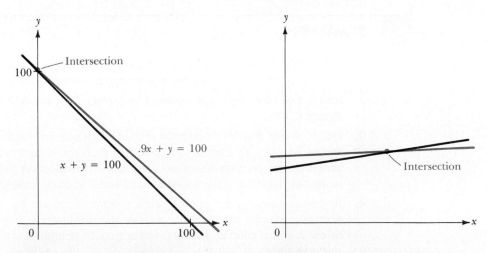

FIGURE 9.3 **A very different solution from the one in Figure 9.2.**

FIGURE 9.4 **The intersection of two nearly parallel lines.**

Computers have a lot of trouble finding accurate solutions of ill-conditioned systems like (4) and (5), because the small roundoff errors created by the computer can produce large changes in the solution. Pivoting and scaling usually don't help the situation; the systems are basically unstable. Notice that the coefficients of x and y in systems (4) and (5) are of comparable magnitude.

Among the most famous ill-conditioned matrices are the Hilbert matrices. These are very bad matrices named after a very good mathematician, David Hilbert! (See the historical note on page 398.) The entry in the ith row and jth column of a Hilbert matrix is $1/(i + j - 1)$. Thus, if we let H_n be the $n \times n$ Hilbert matrix, we have

$$H_2 = \begin{bmatrix} 1 & \frac{1}{2} \\ \frac{1}{2} & \frac{1}{3} \end{bmatrix}, \qquad H_3 = \begin{bmatrix} 1 & \frac{1}{2} & \frac{1}{3} \\ \frac{1}{2} & \frac{1}{3} & \frac{1}{4} \\ \frac{1}{3} & \frac{1}{4} & \frac{1}{5} \end{bmatrix}, \qquad \text{and so on.}$$

It can be shown that H_n is invertible for all n, so a square linear system $H_n\mathbf{x} = \mathbf{b}$ has a unique solution, but the solution may be very hard to find. When the matrix is reduced to echelon form, entries of surprisingly small magnitude appear. Scaling a row in which all entries are close to zero may help a bit.

Bad as the Hilbert matrices are, powers of them are even worse. The software we make available includes a program called HILBERT, which is modeled on YURE-DUCE. The computer generates a Hilbert matrix of the size we specify, up to 10×10. It will then raise the matrix to the power 2, 4, 8, or 16 if we so request. We may then proceed roughly as in YUREDUCE. Programs like YUREDUCE and HILBERT should help us understand this section, because we can watch and see just what is happening as we reduce the matrix. MATCOMP, which simply spits out answers, may produce an absurd result, but we have no way of knowing exactly where things went wrong.

SUMMARY

1. Addition of numbers of very different magnitudes by a computer can cause loss of some or all of the signficiant figures in the number of smaller magnitude.

2. Subtraction of nearly equal numbers by a computer can cause loss of significant figures.

3. Due to roundoff error, a computer may obtain a nonzero value for a number that should be zero. To attempt to handle this problem, a computer program might assign certain computed numbers the value zero whenever the numbers have a magnitude less than some predetermined small positive number.

4. In partial pivoting, the pivot in each column is created by changing rows, if necessary, so that the pivot has at least the maximum magnitude of any entry below it in that column. Partial pivoting may be helpful in avoiding the problem stated in summary item (1).

5. In full pivoting, columns are interchanged as well as rows, if necessary, to create pivots of maximum possible magnitude. Using full pivoting requires much more

computer time than does partial pivoting, and bookkeeping is necessary to keep track of the relationship between the columns and the variables. Partial pivoting is more commonly used.

6. *Scaling*, multiplication of a row by a nonzero constant, can be used to reduce the size of entries that threaten to dominate entries in lower rows, or to increase the size of entries in a row where some entries are very small.

7. A linear system $Ax = b$ is ill-conditioned or unstable if small changes in the numbers can produce a large change in the solution. The matrix A is then also called *ill-conditioned*.

8. Hilbert matrices, which are square matrices with entry $1/(i + j - 1)$ in the ith row and jth column, are examples of ill-conditioned matrices.

9. With the present technology, it appears hopeless to write a computer program that will successfully handle every linear system involving even very small coefficient matrices—say, of size at most 10×10.

EXERCISES

1. Find the solution by a three-figure computer of the system

$$100x_1 + 1,000,000x_2 = 1,000,000$$
$$-100x_1 + \quad\quad 200x_2 = \quad\quad 100$$

using just partial pivoting.

2. Repeat Exercise 1, but use full pivoting.

3. Find the solution without pivoting by a five-figure computer of the linear system in Exercise 1. Is the solution reasonably accurate? If so, modify the system a bit to obtain one for which a five-figure computer does not find a reasonable solution without pivoting.

4. Modify the linear system in Exercise 1 so that an eight-figure computer (roughly the usual single-precision computer) will not obtain a reasonable solution without partial pivoting.

5. Repeat Exercise 4 for an eighteen-figure computer (roughly the usual double-precision computer).

6. Find a linear system with two equations in x and y such that a change of .001 in the coefficient of x in the second equation produces a change of at least 1,000,000 in both of the values x and y in a solution.

7. Let A be an invertible square matrix. Show that the following scaling routine for finding A^{-1} using a computer is *mathematically* correct.

   a) Find the minimum absolute value m of all nonzero entries in A.
   b) Multiply A by an integer $n > 1/m$.
   c) Find the inverse of the resulting matrix.
   d) Multiply the matrix obtained in part (c) by n to get A^{-1}.

Use the programs MATCOMP, YUREDUCE, and HILBERT, or similar software, for exercises 8–16.

8. Use YUREDUCE in single-precision computational mode to solve the system

$$.00001x_1 + 10,000x_2 = 1$$
$$10,000x_1 + 20,000x_2 = 1,$$

without using any pivoting. Check the answer mentally to see if it is approximately correct. If it is, put more zeros in the large numbers and more zeros after the decimal point in the small one until a system is obtained in which the solution without pivoting is erroneous.

9. Use YUREDUCE to solve the system in Exercise 8, which did not give a reasonable solution without pivoting, but this time use partial pivoting.

10. Repeat Exercise 9, but this time use full pivoting. Compare the answer with that in Exercise 9.

11. Repeat Exercise 8 with the system that did not yield a reasonable solution, but use double-precision mode in YUREDUCE this time. Again, do not use any pivoting.

12. Modify the system in Exercise 8 to obtain one for which YUREDUCE in double-precision mode without pivoting yields an erroneous solution.

13. Take the system formed in Exercise 12, and solve it using YUREDUCE in double-precision and using (a) partial pivoting, and (b) full pivoting. Compare answers. Is the solution reasonable?

14. See how MATCOMP handles the linear system formed in Exercise 12.

15. Experiment with MATCOMP, and see if it gives a "nearly singular matrix" message when finding the inverse of a 2 × 2 diagonal matrix

$$rI = r\begin{bmatrix} 1 & 0 \\ 0 & 1 \end{bmatrix},$$

where r is a sufficiently small nonzero number. Use the default for roundoff control in MATCOMP.

16. Using MATCOMP, use the scaling routine suggested in Exercise 7 to find the inverse of the matrix

$$\begin{bmatrix} .000001 & -.000003 & .000011 & .000006 \\ -.000002 & .000013 & .000007 & .000010 \\ .000009 & -.000011 & 0 & -.000005 \\ .000014 & -.000008 & -.000002 & .000003 \end{bmatrix}$$

Then see if the same answer is obtained without using the scaling routine. Check the inverse in each case by multiplying by the original matrix.

17. Mark each of the following True or False.

___ a) Addition and subtraction never cause any problem when executed by a computer.

___ b) Given any present-day computer, one can find two positive numbers whose sum the computer will consider to be its representation of the larger of the two numbers.

___ c) A computer may have trouble representing as accurately as desired the sum of two numbers of extremely different magnitudes.

___ d) A computer may have trouble representing as accurately as desired the sum of two numbers of essentially the same magnitude.

___ e) A computer may have trouble representing as accurately as desired the sum of a and b, where a is approximately $2b$ or $-2b$.

___ f) Partial pivoting handles all problems due to roundoff error when solving a linear system by the Gauss method.

___ g) Full pivoting handles roundoff error problems better than partial pivoting, but it is generally not used because of the extra computer time required to implement it.

___ h) Given any present-day computer, one can find a system of two equations in two unknowns that the computer cannot solve accurately using the Gauss method with back substitution.

___ i) A linear system of two equations in two unknowns is unstable if the lines represented by the two equations are extremely close to being parallel.

___ j) The entry in the ith row and jth column of a Hilbert matrix is $1/(i + j)$.

⚠ *Use the program HILBERT or similar software in the remaining exercises. Since the Hilbert matrices are nonsingular, diagonal entries computed during the elimination should never be zero. Except in Exercise 25, enter 0 for r when r is requested during a run of HILBERT. Experiment with both single-precision and double-precision computing in the exercises that follow.*

18. Solve $H_2\mathbf{x} = \mathbf{b}, \mathbf{c}$, where

$$\mathbf{b} = \begin{bmatrix} 2 \\ 9 \end{bmatrix} \quad \text{and} \quad \mathbf{c} = \begin{bmatrix} 3 \\ 8 \end{bmatrix}.$$

Use just one run of HILBERT; that is, solve both systems at once. Notice that the components of $\mathbf{b}$ and $\mathbf{c}$ differ by just 1. Find the difference in the components of the two solution vectors.

19. Repeat Exercises 18, changing the coefficient matrix to $H_2{}^4$.

20. Repeat Exercise 18, using as coefficient matrix H_4 with

$$\mathbf{b} = \begin{bmatrix} 2 \\ 1 \\ 4 \\ 6 \end{bmatrix} \quad \text{and} \quad \mathbf{c} = \begin{bmatrix} 3 \\ 0 \\ 5 \\ 7 \end{bmatrix}.$$

21. Repeat Exercise 18, using as coefficient matrix $H_4{}^4$.

22. Find the inverse of H_2, and use the I menu option to test whether the computed inverse is correct.

23. Continue Exercise 22 by trying to find the inverses of H_2 raised to the powers 2, 4, 8, and 16. Was it always possible to reduce the power of the Hilbert matrix to diagonal form in both single-precision and double-precision computation? If not, what happened? Why did it happen? Was scaling (1 an entry) of any help? For those cases in which reduction to diagonal form was possible, did the computed inverse seem to be reasonably accurate when tested?

24. Repeat Exercises 22 and 23, using H_4 and the various powers of it. Are problems encountered for lower powers this time?

25. We have seen that, in reducing a matrix, we may wish to instruct the computer to assign the value zero to computed entries of sufficiently small magnitude, since entries that should be zero might otherwise be left nonzero due to roundoff error. Use HILBERT to try to solve the linear system

$$H_3{}^2\mathbf{x} = \mathbf{b}, \qquad \text{where} \qquad \mathbf{b} = \begin{bmatrix} -3 \\ 1 \\ 4 \end{bmatrix},$$

entering as r the number .0001 when requested. Will routine Gauss–Jordan elimination using HILBERT solve the system? Will HILBERT solve it using scaling (1 an entry)?

10 LINEAR PROGRAMMING

Linear programming was introduced in Section 1.7 and should be reviewed at this time. Linear programming deals with the problem of maximizing or minimizing a linear function

$$c_1x_1 + c_2x_2 + \cdots + c_nx_n,$$

subject to constraints that are linear equalities or linear inequalities in the variables x_i. In Section 1.7, we gave a geometric approach to the solution of such problems for the case $n = 2$. In this chapter we will see how solutions can be obtained algebraically. In Section 10.1, we describe Dantzig's simplex method for constraints of one type. The simplex method for general constraints is treated in Section 10.2, and duality is discussed in Section 10.3. Duality shows that every maximizing linear programming problem has an associated dual minimizing problem, and vice versa. Moreover, both problems can be solved at the same time.

We make no attempt to give a complete or rigorous treatment of the simplex method. Whole books have been written on the subject. We merely explain the technique while trying as best we can to show in an intuitive way why it works.

10.1 THE SIMPLEX METHOD FOR CONSTRAINTS $Ax \leq b$

The Standard Maximum Problem

In practice, linear programming problems that arise deal with maximizing or minimizing a linear function in which the variables can assume only nonnegative values. We start by presenting the simplex method for the *standard maximum problem*, which may be stated as follows.

486

Standard Maximum Problem

Maximize the linear function

$$P = c_1 x_1 + c_2 x_2 + \cdots + c_n x_n, \tag{1}$$

subject to the constraints

$$a_{11}x_1 + a_{12}x_2 + \cdots + a_{1n}x_n \leq b_1$$
$$a_{21}x_1 + a_{22}x_2 + \cdots + a_{2n}x_n \leq b_2$$
$$\vdots \tag{2}$$
$$a_{m1}x_1 + a_{m2}x_2 + \cdots + a_{mn}x_n \leq b_m,$$

where all b_i are nonnegative, and where

$$x_1 \geq 0, \, x_2 \geq 0, \, \ldots, \, x_n \geq 0. \tag{3}$$

The linear function in Eq. (1) is the **objective function**; in this maximizing situation, it is called the **profit function**. For a minimum problem, the objective function is referred to as the **cost function**. The solution set of constraints (2) and (3) is called the **feasible set**, and any point in the feasible set is a **feasible solution**.

In Section 1.7, we saw that the feasible set for a linear programming problem in two variables is a convex set obtained as an intersection of half-planes in $\mathbb{R}^2$. In the present situation, the feasible set is an intersection of half-spaces in $\mathbb{R}^n$, each given by an inequality (2) or (3). In Section 1.7, with $n = 2$, we saw that, if the maximum value of the objective function over the feasible set is attained, it is attained at a *corner point* of the feasible set in $\mathbb{R}^2$; the corner point appears as an intersction of two lines. Generalizing, a **corner point** is a *feasible solution* in $\mathbb{R}^n$ that can be obtained by turning n of the $n + m$ inequalities (2) and (3) into equations, replacing $\leq$ by $=$, and solving the resulting $n \times n$ linear system. If $n = 2$, a corner point becomes the intersection of two lines; if $n = 3$, it becomes the intersection of three planes; and so on. We state an important theorem without proof. An intuitive geometric argument in terms of convex sets and hyperplanes can be made, just as for the case $n = 2$ in Section 1.7.

THEOREM 10.1 **Achieving Optimal Values**

If a linear programming problem has an optimal solution, the optimal value of the objective function is achieved at some corner point of the feasible set.

Feasible solutions that are not corner points may also give the same optimal value. Any feasible solution giving the optimal value is called an **optimal solution**.

The standard maximum problem can be stated more concisely in matrix form. An inequality $A \leq B$ for two matrices of the same size means that $a_{ij} \leq b_{ij}$ for each index pair i, j. In a similar way, we can summarize our requirement $b_i \geq 0$ for all i by $\mathbf{b} \geq \mathbf{0}$.

Standard Maximum Problem (Matrix Form)

Maximize

$$\mathbf{c} \cdot \mathbf{x},$$

subject to

$$A\mathbf{x} \leq \mathbf{b},$$

where

$$\mathbf{b} \geq \mathbf{0} \quad \text{and} \quad \mathbf{x} \geq \mathbf{0}.$$

In this matrix form, $A = [a_{ij}]$ is an $m \times n$ matrix, $\mathbf{c}$ and $\mathbf{x}$ are n-component column vectors, and $\mathbf{b}$ is an m-component column vector. Notice that objective function (1) is written as a dot product.

Introducing Slack Variables

The first step in solving a linear programming problem by the simplex method is to convert the inequalities (2) into a linear system of equations. For the standard maximum problem, this is done by introducing the **slack variables** y_i defined by

$$a_{i1}x_1 + a_{i2}x_2 + \cdots + a_{in}x_n + y_i = b_i \tag{4}$$

for each $i = 1, 2, \ldots, m$. The inequalities (2) correspond to the requirements $y_i \geq 0$ for each i, or more briefly, to the requirement $\mathbf{y} \geq \mathbf{0}$. Notice that setting $y_i = 0$ in Eq. (4) gives the equation $a_{i1}x_1 + a_{i2}x_2 + \cdots + a_{in}x_n = b_i$, which is usually a bounding hyperplane of the feasible set. This is illustrated in Figure 10.1.

The linear system (4) of equations can be written in matrix form as

$$[A \quad I] \begin{bmatrix} \mathbf{x} \\ \mathbf{y} \end{bmatrix} = \mathbf{b}, \tag{5}$$

where the $m \times m$ identity matrix I has been appended to the matrix A to form the larger $m \times (n + m)$ coefficient matrix in Eq. (5). A solution vector

$$\begin{bmatrix} \mathbf{x} \\ \mathbf{y} \end{bmatrix} = \begin{bmatrix} x_1 \\ \vdots \\ x_n \\ y_1 \\ \vdots \\ y_m \end{bmatrix}$$

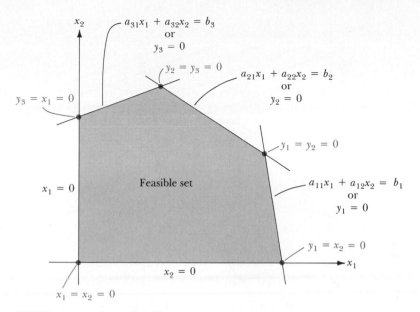

FIGURE 10.1 **Bounding lines and corner points.**

with nonnegative components will also be called a **feasible solution** for the linear programming problem.

The corner points of the feasible set are *represented by* those feasible solutions of Eq. (5) in which n of the $n + m$ components are zero, as illustrated in Figure 10.1 for the case $n = 2$, $m = 3$. The variables that are set equal to zero to produce a corner point are called **nonbasic variables** for that corner point. The remaining variables are **basic variables** there. Such a feasible solution of Eq. (5) representing a corner point is called a **basic feasible solution**.

In theory, the method of Section 1.7 could be used to solve linear programming problems: find all corner points, and substitute them in the objective function. In practical situations, there are too many corner points to make such a computation feasible. (See Exercise 1.)

We are interested in finding a corner point (basic feasible solution) at which the maximum of the objective function is attained. The simplex method achieves this by beginning with any corner point and moving from it to an adjacent corner point in a way that increases the objective function as quickly as possible. We continue until an optimal solution is found.

For the standard maximum problem, the origin $(0, 0, \ldots, 0)$ is always a corner point.

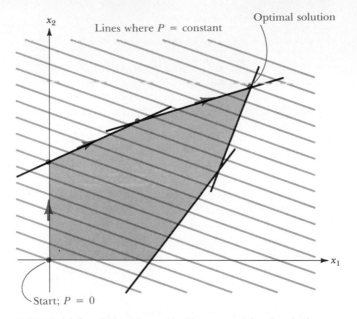

FIGURE 10.2   Going from (0, 0) to an optimal solution.

Figure 10.2 illustrates the simplex method for the case $n = 2$; start at the origin and follow the arrows along edges to the corner point that is an optimal solution. We will explain the simplex method, developed by George Dantzig, in the context of a simple problem.

PROBLEM   A boatyard produces three kinds of boats: cabin cruisers, racing sailboats, and cruising sailboats. Beginning with fiberglass hulls and decks only, the boatyard finishes the boats, taking 2, 1, and 3 weeks, respectively, to do each job. The boatyard is closed during the last week of December. The net unit profit on each kind of boat is $5,000, $4,000, and $6,000, respectively. Because of space considerations, the yard can finish no more than 25 boats in one year (51 weeks). Find the number of boats of each kind that will maximize the annual profit.

The data are summarized in Table 10.1. We wish to maximize the annual profit

$$P = 5x_1 + 4x_2 + 6x_3, \tag{6}$$

TABLE 10.1

	Cabin Cruiser	Racing Sailboat	Cruising Sailboat	Total for One Year
Number of boats per year	x_1	x_2	x_3	25
Unit time required (weeks)	2	1	3	51
Unit profit (thousands of dollars)	5	4	6	$5x_1 + 4x_2 + 6x_3$

subject to the constraints

$$x_1 + x_2 + x_3 \leq 25 \tag{7}$$
$$2x_1 + x_2 + 3x_3 \leq 51$$

and

$$x_1 \geq 0, \qquad x_2 \geq 0, \qquad x_3 \geq 0.$$

The two inequalities (7) are turned into the equations

$$x_1 + x_2 + x_3 + y_1 \qquad = 25 \tag{8}$$
$$2x_1 + x_2 + 3x_3 \qquad + y_2 = 51$$

by introducing the slack variables $y_1 \geq 0$ and $y_2 \geq 0$. The matrix equation (5) for system 8 is

$$\begin{bmatrix} 1 & 1 & 1 & 1 & 0 \\ 2 & 1 & 3 & 0 & 1 \end{bmatrix} \begin{bmatrix} x_1 \\ x_2 \\ x_3 \\ y_1 \\ y_2 \end{bmatrix} = \begin{bmatrix} 25 \\ 51 \end{bmatrix}. \tag{9}$$

The Initial Tableau

To begin the simplex method, we represent our problem in this *initial tableau*

	x_1	x_2	x_3	y_1	y_2	
y_1	1	1	1	1	0	25
y_2	2	1	3	0	1	51
P	-5	-4	-6	0	0	0,

(10)

which contains the augmented matrix of system (8). The last row in tableau (10) represents the objective function and Eq. (6), but in the form $P - 5x_1 - 4x_2 - 6x_3 - 0y_1 - 0y_2 = 0$. This bottom row is called the **objective row** and is always formed as illustrated.

At a corner point in the feasible set, $n = 3$ of the variables x_i, y_j are zero, and these are the nonbasic variables at that corner point. In the initial tableau, the corner point is the origin, where we have $x_1 = x_2 = x_3 = 0$. If we set $x_1 = x_2 = x_3 = 0$ in system (9) and solve for y_1 and y_2, we obtain the basic feasible solution

$$(x_1, x_2, x_3, y_1, y_2) = (0, 0, 0, 25, 51) \tag{11}$$

representing the corner point $(0, 0, 0)$. For this corner point, the nonbasic variables are x_1, x_2, x_3, and the basic variables are y_1, y_2. Notice that the column vectors under y_1 and y_2 above the objective row in initial tableau (10) make up the standard basis for $\mathbb{R}^m = \mathbb{R}^2$. In initial tableau (10) the first two rows are labeled at the left with the basic variables; their respective values, shown in Eq. (11), appear in the last column. The value of the objective function

$$P = 5x_1 + 4x_2 + 6x_3 + 0y_1 + 0y_2 \tag{12}$$

at the corner point represented by $(0, 0, 0, 25, 51)$ is 0 and appears at the bottom of the last column in the tableau. In summary, initial tableau (10) represents the situation at the corner point $(0, 0, 0)$—that is, at the origin.

In a similar manner, each subsequent simplex tableau exhibits data at a single corner point, as explained in the following box.

Interpreting a Simplex Tableau

1. The basic variables are those used both as row labels and as column labels. They can also be described as those labeling columns containing standard basis vectors—that is, columns with all zero entries except for the entry 1 in the row and column having a common label.

2. The values of each basic variable and of the objective function at the corner point appear in the final column of the tableau. Nonbasic variables have value zero.

Obtaining the Next Tableau

We wish to move to an adjacent corner point, where the value P of the objective function will be increased, and to continue until the maximum value is found. In order for us to move from $(0, 0, 0)$ to an *adjacent* corner point, *just one* of the basic variables in the variable list x_1, x_2, x_3, y_1, y_2 must become nonbasic, and vice versa. First we decide which nonbasic variable will become basic—that is, which variable will be increased from 0 to a positive quantity. Since we want to increase P as rapidly as possible, we choose a variable in Eq. (12)—in this case, x_3—corresponding to the largest coefficient. In initial tableau (10), this corresponds to a *negative entry of greatest magnitude* in the objective row. Thus, x_3 has been selected as the *entering variable* to become basic. The column labeled x_3 in the initial tableau is the **pivot column**.

Now we must decide which basic variable, y_1 or y_2, to choose as the *exiting variable*—that is, as the one to become nonbasic with value 0. This is done by computing the ratios of entries in the last column of tableau (10) by corresponding entries in the pivotal column. We choose the variable labeling the row corresponding to the *smallest nonnegative ratio*.* We can see the reason for this as follows. Since we are moving from corner point (11) by allowing x_3 to become positive, we must make one of the basic variables y_1, y_2 zero. Equations (8) with $x_1 = x_2 = 0$ can be written

---

*If a ratio is zero, one of the basic variables has value zero. The corresponding basic feasible solution

$$\begin{bmatrix} \mathbf{x} \\ \mathbf{y} \end{bmatrix}$$

is said to be *degenerate*. It is possible under these circumstances that the simplex method may enter an infinite cycle in which no progress is made. Such cyclic behavior is extremely rare in practice, and will not concern us further.

$$y_1 = 25 - x_3 = 25[1 - (1/25)x_3]$$
$$y_2 = 51 - 3x_3 = 51[1 - (1/17)x_3].$$

If y_1 is allowed to decrease to zero, then $x_3 = 25$, which makes y_2 negative and thus does not result in a feasible solution. Therefore, we choose y_2 to be zero, corresponding to $x_3 = 17$; this accommodates the requirements $y_1, y_2 \ge 0$. The largest *allowable* increase in x_3 corresponds to the smallest of the aforementioned ratios. The **pivotal row** for our initial tableau is then the row labeled y_2. The entry 3 in the pivotal row and pivotal column is the **pivot** for this step of the simplex method. (See the first tableau below.)

In choosing the pivotal row, we only consider nonnegative ratios, because a negative pivot would require the entering basic variable to become negative, causing a departure from the feasible set. In fact, if all the entries in the pivotal column should be zero or negative, then x_3 could be increased as much as we like without leaving the feasible set. The values of the objective function on the feasible set are then **unbounded**, and the linear programming problem has no optimal solution.

Using the pivot, we transform the initial tableau into a simplex tableau in which the nonbasic variables are x_1, x_2, y_2, and the basic variables are y_1, x_3. This is accomplished by using a Gauss–Jordan reduction to convert the pivot to 1 and to convert all the numerical entries above and below it to 0. Thus the column vector under x_3 becomes a standard basis vector. Such a reduction of an augmented matrix does not alter the solutions of the corresponding linear system (9). On the objective row, the reduction amounts to adding a multiple of a valid equation to the objective equation

$$P - 5x_1 - 4x_2 - 6x_3 + 0y_1 + 0y_2 = 0, \tag{13}$$

and the resulting equation is still a valid expression for our objective function. We obtain

	x_1	x_2	x_3	y_1	y_2				x_1	x_2	x_3	y_1	y_2	
y_1	1	1	1	1	0	25	$\sim$	y_1	$\frac{1}{3}$	$\frac{2}{3}$	0	1	$-\frac{1}{3}$	8
y_2	2	1	3	0	1	51		x_3	$\frac{2}{3}$	$\frac{1}{3}$	1	0	$\frac{1}{3}$	17
P	-5	-4	-6	0	0	0		P	-1	-2	0	0	2	102.

This new tableau exhibits the basic variables y_1 and x_3 as row labels in the left-most column, and their values 8 and 17 at the corner point where $x_1 = x_2 = y_2 = 0$ appear in the right-most column. Our choice of a negative entry in the objective row to determine the pivotal column and the test using ratios to find the pivotal row ensure that the entries in the last column of a tableau will always be nonnegative. Our new tableau shows the current corner point to be $(x_1, x_2, x_3) = (0, 0, 17)$, which is represented by the basic feasible solution

$$(x_1, x_2, x_3, y_1, y_2) = (0, 0, 17, 8, 0).$$

Our original objective equation (13) has been transformed into

$$P - x_1 - 2x_2 + 0x_3 + 0y_1 + 2y_2 = 102,$$

so $P = 102$ at the basic feasible solution where $x_1 = x_2 = y_2 = 0$. Of course, we could also evaluate the original objective function $P = 5x_1 + 4x_2 + 6x_3$ at the corner point $(0, 0, 17)$, obtaining again 102.

The process of finding a pivot corresponding to swapping a basic for a nonbasic variable is repeated until the objective row contains no negative entry. When this occurs, we have found an optimal solution of the linear programming problem. We summarize the steps involved in the following box.

Steps in the Simplex Method for the Standard Maximum Problem

1. Set up the initial tableau, exhibiting data at the origin.

2. *Test for an optimal solution:* If the entries in the objective row are nonnegative, an optimal solution has been found. Otherwise, continue.

3. *Find the pivotal column:* Choose the column in the tableau corresponding to a negative entry of largest magnitude in the bottom row.

4. *Test for unboundedness:* If all the entries in the pivotal column are zero or negative, the objective function is unbounded on the feasible set, and the problem has no optimal solution.

5. *Find the pivotal row:* Compute the ratios of entries in the last column by corresponding entries in the pivotal column that are positive. Choose the row corresponding to the smallest nonnegative ratio. (In case of a tie, either row may be chosen.)

6. The entry in the row and column described in steps 2 and 5 is the pivot. Use a Gauss–Jordan reduction on the tableau to convert this pivot to 1, with zeros above and below it. Relabel the rows appropriately. Go to step 2.

Following the rules in the box, we take our current simplex tableau and find the pivot. The pivot column is the column containing -2 in the objective row and is labeled x_2, the nonbasic variable to enter the basic category. The ratios of entries in the last column by entries in the pivot column are $8/\left(\frac{2}{3}\right) = 12$ and $17/\left(\frac{1}{3}\right) = 51$; thus the pivot row is the one labeled y_1, the variable to exit the basic category and become nonbasic. We use a Gauss–Jordan reduction with the indicated pivot, as follows:

	x_1	x_2	x_3	y_1	y_2	
y_1	$\frac{1}{3}$	$\frac{2}{3}$	0	1	$-\frac{1}{3}$	8
x_3	$\frac{2}{3}$	$\frac{1}{3}$	1	0	$\frac{1}{3}$	17
P	-1	-2	0	0	2	102

$\sim$

	x_1	x_2	x_3	y_1	y_2	
x_2	$\frac{1}{2}$	1	0	$\frac{3}{2}$	$-\frac{1}{2}$	12
x_3	$\frac{1}{2}$	0	1	$-\frac{1}{2}$	$\frac{1}{2}$	13
P	0	0	0	3	1	$126.$

Since all entries in the objective row of the new tableau are nonnegative, we have found the optimal solution. The maximum profit is $P = 126$, and it occurs at the corner point $(0, 12, 13)$ represented by

$$(x_1, x_2, x_3, y_1, y_2) = (0, 12, 13, 0, 0).$$

Since this result was obtained, the boatyard no longer produces cabin cruisers, because $x_1 = 0$. It now produces $x_2 = 12$ racing sailboats and $x_3 = 13$ cruising sailboats per year, earning a maximum annual profit of $126,000.

Minimizing Subject to $A\mathbf{x} \le \mathbf{b}$, $\mathbf{x} \ge \mathbf{0}$

Suppose that we wish to *minimize* a linear function $f(\mathbf{x})$ subject to constraints $A\mathbf{x} \le \mathbf{b}$, $\mathbf{x} \ge \mathbf{0}$, where $\mathbf{b} \ge \mathbf{0}$. A moment of thought shows that, if the minimum of $f(\mathbf{x})$ subject to these constraints exists, then

$$[\text{Minimum of } f(\mathbf{x})] = -[\text{Maximum of } -f(\mathbf{x})],$$

where the maximum of $-f(\mathbf{x})$ is subject to these same constraints. For example, in our boatyard problem we found that the maximum of the objective function $P = 5x_1 + 4x_2 + 6x_3$ subject to constraints (7) is 126, attained when $x_1 = 0$, $x_2 = 12$, and $x_3 = 13$. The minimum of $f(x) = -5x_1 - 4x_2 - 6x_3$ subject to these same constraints is -126 and is attained for the same values of x_1, x_2, and x_3. Thus, when we encounter a minimizing linear programming problem with constraints $A\mathbf{x} \le \mathbf{b}$, $\mathbf{x} \ge \mathbf{0}$, where $\mathbf{b} \ge \mathbf{0}$, we may simply maximize the negative of our objective function, and then change the sign of the maximum value to obtain the desired minimum value.

We give two examples. In Example 1, we maximize an objective function to emphasize the simplex method described earlier. Then in Example 2, we minimize the *same* objective function, using the technique just described.

EXAMPLE 1　Maximize $-2x_1 + x_2$, subject to the constraints

$$x_1 - 2x_2 \le 4, \qquad -3x_1 + x_2 \le 3, \qquad 4x_1 + 7x_2 \le 46.$$

Solution　We form the initial tableau and proceed to reduce it, following the boxed outline:

	x_1	x_2	y_1	y_2	y_3	
y_1	1	-2	1	0	0	4
y_2	-3	1	0	1	0	3
y_3	4	7	0	0	1	46
P	2	-1	0	0	0	0

$\sim$

	x_1	x_2	y_1	y_2	y_3	
y_1	-5	0	1	2	0	10
x_2	-3	1	0	1	0	3
y_3	25	0	0	-7	1	25
P	-1	0	0	1	0	3

$\sim$

	x_1	x_2	y_1	y_2	y_3	
y_1	0	0	1	$\frac{3}{5}$	$\frac{1}{5}$	15
x_2	0	1	0	$\frac{4}{25}$	$\frac{3}{25}$	6
x_1	1	0	0	$-\frac{7}{25}$	$\frac{1}{25}$	1
P	0	0	0	$\frac{18}{25}$	$\frac{1}{25}$	4.

Thus the maximum value of the objective function is 4 and is attained at $(x_1, x_2) = (1, 6)$. ∎

EXAMPLE 2 Minimize the same objective function $-2x_1 + x_2$ as in Example 1, subject to the same constraints.

Solution We maximize the function $2x_1 - x_2$, following again the steps in the boxed outline:

	x_1	x_2	y_1	y_2	y_3	
y_1	1	-2	1	0	0	4
y_2	-3	1	0	1	0	3
y_3	4	7	0	0	1	46
P	-2	1	0	0	0	0

$\sim$

	x_1	x_2	y_1	y_2	y_3	
x_1	1	-2	1	0	0	4
y_2	0	-5	3	1	0	15
y_3	0	15	-4	0	1	30
P	0	-3	2	0	0	8

$\sim$

	x_1	x_2	y_1	y_2	y_3	
x_1	1	0	$\frac{7}{15}$	0	$\frac{2}{15}$	8
y_2	0	0	$\frac{5}{3}$	1	$\frac{1}{3}$	25
x_2	0	1	$-\frac{4}{15}$	0	$\frac{1}{15}$	2
P	0	0	$\frac{6}{5}$	0	$\frac{1}{5}$	14.

Thus the maximum of $2x_1 - x_2$ is 14, attained at the point $(x_1, x_2) = (8, 2)$. Consequently, the minimum of our given objective function is -14, attained at this same point $(8, 2)$. ■

We close with an example illustrating the case in which the objective function can assume arbitrarily large values on the feasible set, so that no maximum value exists.

EXAMPLE 3 Maximize $-x_1 + 3x_2$, subject to the constraints

$$-2x_1 + x_2 \le 1, \qquad -x_1 + x_2 \le 4, \qquad x_1 \ge 0, \qquad x_2 \ge 0.$$

Solution We form the initial tableau and reduce it as usual:

	x_1	x_2	y_1	y_2	
y_1	-2	1	1	0	1
y_2	-1	1	0	1	4
P	1	-3	0	0	0

$\sim$

	x_1	x_2	y_1	y_2	
x_2	-2	1	1	0	1
y_2	1	0	-1	1	3
P	-5	0	3	0	3

$\sim$

	x_1	x_2	y_1	y_2	
x_2	0	1	-1	2	7
x_1	1	0	-1	1	3
P	0	0	-2	5	18.

The pivotal column labeled y_1 in this final tableau contains only negative entries, so the objective function is unbounded on the feasible set, and the problem has no optimal solution. (See step 4 of the boxed outline.) ■

SUMMARY

1. A linear programming problem consists of maximizing or minimizing a linear function, called the *objective function*, in nonnegative variables $x_1, x_2, \ldots, x_n$ that satisfy additional constraints in the form of linear equalities or inequalities.

2. The standard maximum problem is to maximize an objective function $c_1 x_1 + c_2 x_2 + \cdots + c_n x_n$ in which the variables x_i are required to be nonnegative and satisfy other linear inequality constraints of the form (Linear expression) $\leq b_i$, where $b_i \geq 0$. In matrix notation, the standard maximum problem is to maximize $\mathbf{c} \cdot \mathbf{x}$ subject to $A\mathbf{x} \leq \mathbf{b}$ and $\mathbf{x} \geq \mathbf{0}$, where $\mathbf{x}$ is the column vector of n variables, and A, $\mathbf{b}$, and $\mathbf{c}$ have appropriate size with $\mathbf{b} \geq \mathbf{0}$. A vector $\mathbf{x}$ that satisfies all of these conditions is a feasible solution for the problem, and a feasible solution that maximizes the linear function $\mathbf{c} \cdot \mathbf{x}$ is an optimal solution. The set of all feasible solutions is the feasible set. If an optimal solution exists, it occurs at a corner point of the feasible set.

3. For $i = 1, 2, \ldots, m$, the ith inequality represented by $A\mathbf{x} \leq \mathbf{b}$, namely,

$$a_{i1}x_1 + a_{i2}x_2 + \cdots + a_{in}x_n \leq b_i,$$

can be turned into an equation

$$a_{i1}x_1 + a_{i2}x_2 + \cdots + a_{in}x_n + y_i = b_i$$

by inserting the slack variable y_i. The standard matrix problem then takes the following form: maximize $\mathbf{c} \cdot \mathbf{x}$, subject to

$$[A \quad I]\begin{bmatrix} \mathbf{x} \\ \mathbf{y} \end{bmatrix} = \mathbf{b}, \qquad \mathbf{x} \geq \mathbf{0}, \qquad \mathbf{y} \geq \mathbf{0}.$$

4. A corner point of the feasible set is a feasible solution $\mathbf{x}$ obtained by setting n of the $n + m$ variables x_i, y_j equal to zero and solving the equations in summary item (3) for the other variables. The variables set equal to zero are called *nonbasic* at the corner point; the others are basic variables there.

5. A simplex tableau summarizes the status of the linear programming problem at a particular corner point of the feasible set. The simplex method consists of a Gauss–Jordan reduction, which transforms one simplex tableau into another in which the objective function has been increased. The steps for the simplex method are summarized in the box on page 494.

EXERCISES

1. Consider a standard maximum problem in seven variables with ten constraints in addition to the constraints $x_i > 0$. How many *corner points* must be found and checked to ensure that they are feasible to solve such a problem by the method in Section 1.7? (Use a counting argument involving the number of combinations of k things taken r at a time.)

2. Solve the boatyard problem stated in this section if the units of profit in Table 10.1 are changed to 8 units for a cabin cruiser, 3 units for a racing sailboat, and 6 units for a cruising sailboat.

3. Solve Exercise 15 of Section 1.7 (the coffee problem), using the simplex method.

4. At first, one might think that, if an objective function has some positive and some negative coefficients, a maximum would have to be achieved at a point where all variables with negative coefficients are zero, and a minimum at a point where all variables with positive coefficients are zero. Examples 1 and 2 show that this is not the case. Draw a sketch of the feasible set given by the constraints in Example 1, as in Section 1.7. Find the corner points, draw some lines where the objective function has constant values, and see graphically why one obtains the answers in Examples 1 and 2. Trace the path from the origin to the optimal solution in each example, and notice how the intermediate corner point appears in the second tableau in each case.

In Exercises 5–8, write the given linear programming problem as a standard maximum problem, using the matrix form boxed on page 488.

5. Maximize $x_1 + 2x_2 + 4x_3$, subject to the constraints

$$12 - x_3 \geq 3x_1 + 4x_2,$$
$$2x_1 + 2x_2 - x_3 \geq -4,$$
$$x_2 \leq x_1 + 1,$$
$$\text{all } x_i \geq 0.$$

6. Minimize $x_1 - 3x_2 - 8x_3$, subject to the constraints

$$2x_2 - 20 \leq 2x_1 - x_3,$$
$$-x_1 + 2x_2 - 2x_3 \leq 4,$$
$$16 - 4x_1 - 8x_2 + x_3 \geq 0,$$
$$\text{all } x_i \geq 0.$$

7. Minimize $2x_1 + x_2 + x_3 - x_4$, subject to the constraints

$$3x_2 + 5x_3 \leq 10 - x_1$$
$$-2x_2 - 5x_3 + x_4 \leq 5 - x_1,$$
$$\text{all } x_i \geq 0.$$

8. Maximize $5x_1 - x_2 - x_3 - 5x_4$, subject to the constraints

$$-x_1 + x_2 - x_3 + x_4 \geq -2,$$
$$4x_1 + x_2 - 5x_3 - 4x_4 \leq 3,$$
$$x_1 + 3x_2 \leq 5 - 3x_1 - x_2,$$
$$\text{all } x_i \geq 0.$$

In Exercises 9–16, solve the given linear programming problem, using the simplex method. All variables x_i given are assumed to be nonnegative.

9. Maximize $3x_1 + 2x_2$, subject to the constraints

$$x_1 + x_2 \leq 3,$$
$$2x_1 + x_2 \leq 5.$$

10. Minimize $-2x_1 + x_2$, subject to the constraints

$$2x_1 - x_2 \leq 8,$$
$$x_1 + x_2 \leq 10,$$
$$x_1 + 2x_2 \leq 14.$$

11. Maximize $4x_1 + 8x_2 - 4x_3$, subject to the constraints

$$x_1 + 2x_2 + x_3 \leq 2,$$
$$x_1 + x_2 + 2x_3 \leq 3.$$

12. Maximize $3x_1 + x_2 - 4x_3$, subject to the constraints

$$3x_1 + 2x_2 + x_3 \leq 5,$$
$$x_1 + x_2 - x_3 \leq 8,$$
$$2x_1 + x_2 + 3x_3 \leq 15.$$

13. Minimize $2x_1 - 4x_2$, subject to the constraints

$$x_1 + x_2 \leq 6,$$
$$-x_1 + 3x_2 \leq 18,$$
$$x_1 - x_2 \leq 8.$$

14. Minimize $-x_1 + 3x_2$, subject to the constraints

$$x_1 + 2x_2 \leq 10,$$
$$-x_1 + x_2 \leq 10.$$

15. Maximize $3x_1 + 2x_2 + x_3$, subject to the constraints

$$x_1 + x_2 - x_3 \leq 10,$$
$$2x_1 + x_2 + x_3 \leq 6.$$

16. Maximize $2x_1 + 2x_2 - x_3 + 3x_4$, subject to the constraints

$$4x_1 + 6x_2 + 2x_3 - 2x_4 \leq 7,$$
$$3x_1 + x_2 - 4x_3 + 5x_4 \leq 8.$$

17. Working graphically, construct a standard maximum problem in two variables with three constraints other than $x_i \geq 0$ having the following property:

Solving the problem using the simplex method and choosing the pivotal column in the initial tableau with negative objective row entry of maximum magnitude, as described in the text, requires three iterations, while another choice of initial pivotal column requires only two interations.

Check your problem by solving it both ways.

18. Continuing the idea of Exercise 17, draw a graph indicating that one can construct a standard maximum problem in two variables such that, with initial choice of pivotal column as described in the text, ten iterations will be required to solve the problem, but a different choice of pivotal column will solve the problem in a single iteration. (A great deal of effort has been expended in attempting to maximize the efficiency of the simplex method.)

19. Show graphically that the problem in Example 3 has no optimal solution.

20. Solve the problem given as text answer to Exercise 17, using the graphic program LINPROG, and observe graphically whether the text answer satisfies the requirements of the problem.

The available program SIMPLEX has three options:

1. An option allowing the user to specify the next step in a reduction of a tableau,

2. A quiz option on the simplex method as described in the text, and

3. An option to have the computer execute the simplex method and print the results from supplied data, with tableaux shown if desired.

In Exercises 21–24, use the first two options of SIMPLEX or similar software to find the indicated optimal value. Work at least the odd-numbered problems in the quiz format. It may be interesting to experiment with different choices for pivotal columns, and see how the numbers of iterations needed to obtain a solution compare.

21. The problem in Exercise 5

22. The problem in Exercise 6

23. The problem in Exercise 7

24. The problem in Exercise 8

10.2 THE SIMPLEX METHOD FOR GENERAL CONSTRAINTS

A constraint in a general linear programming problem has one of the forms

$$d_1x_1 + d_2x_2 + \cdots + d_nx_n \begin{cases} \leq b \\ = b \\ \geq b, \end{cases} \tag{1}$$

where we require that $b \geq 0$.

Recall that we always have the constraints $x_i \geq 0$. In the preceding section, we treated the case in which all other constraints have the ($\leq b$)-form in (1). We now discuss the situation in which any or all of the types in (1) may occur.

In the literature, one finds the simplex method divided geometrically into two parts:

Phase 1 Find a corner point of the feasible set.

Phase 2 Move in turn to adjacent corner points, until the optimal value (if it exists) of the objective function is obtained.

For the standard maximum problem in Section 10.1, phase 1 is easy, because the origin is *always* a corner point. This need not be true if we have constraints of the other types. For example, the origin $(0, 0)$ in $\mathbb{R}^2$ does not satisfy the constraints

$$x_1 + x_2 \geq 1, \qquad x_1 \geq 0, \qquad x_2 \geq 0.$$

We proceed to show how to execute phase 1 by passing to an augmented problem formed by introducing *surplus* and *artificial* variables, as well as the *slack* variables of the preceding section. We then show how to form an initial tableau that can be reduced just as in the preceding section. In the process, we make a smooth transition from phase 1 to phase 2.

Finding a Basic Feasible Solution

Recall these two features of the simplex method for the standard maximum problem:

1. *Converting to equations:* We convert the (≤ 1)-type of inequalities into equalities by inserting slack variables.

2. *Property of the initial tableau:* If m is the number of constraints other than the $x_i \geq 0$ constraints, the m columns above the objective row in the initial tableau labeled by the slack variables give the m column vectors in the standard basis for $\mathbb{R}^m$.

We show how to achieve these two features for a general linear programming problem wtih m constraints other than $x_i \geq 0$.

We introduce a slack variable y_i if the ith constraint is of the ($\leq b$)-type, just as in the preceding section. Such a slack variable converts the ith inequality into an equality, and it contributes one of the standard basis vectors of $\mathbb{R}^m$ in a column of the initial tableau.

Consider the first constraint listed of ($\geq b$)-type. To use a specific illustration, suppose that the kth constraint is $2x_1 - x_2 + 4x_3 \geq 7$. We introduce a **surplus variable** term $-y_k$ to convert this constraint into an inequality, requiring all four variables to be positive. That is, we form the equality

$$2x_1 - x_2 + 4x_3 - y_k = 7,$$

in which $y_k \geq 0$. We denote slack and surplus variables consecutively as y_1, y_2, y_3, and so on, numbered according to the order in which we list the constraints. We have

THE BASIC IDEAS OF LINEAR PROGRAMMING grew out of a World War II Air Force study of resource allocation problems, although there were many results over the years dealing with maximizing functions under certain constraints. George Dantzig, a member of this Air Force Project SCOOP (Scientific Computation of Optimum Programs), formulated the general linear programming problem presented here and devised the simplex method of solution in 1947. Since that time, it has been successfully applied to numerous problems, especially military and economic ones. The first successful solution of such a linear programming problem via the simplex method on an electronic computer took place in 1952 at the National Bureau of Standards. In recent years, faster methods of solving such problems have been developed.

Born in 1914, George Dantzig worked for the Rand Corporation after his Air Force stint. He is currently a professor of operations research at Stanford University.

converted the constraint into an equality with all variables positive, but in the initial tableau, the surplus variable column would be the *negative* of a standard basis vector of $\mathbb{R}^m$. Thus, we also introduce an **artificial variable** q_1, which we recognize should eventually take on the value zero in order to provide a feasible solution of the original problem. Our constraint then becomes

$$2x_1 - x_2 + 4x_3 - y_k + q_1 = 7,$$

and the column of the artificial variable q_1 in the initial tableau will provide the desired standard basis vector of $\mathbb{R}^m$.

We alter a constraint that is originally stated as an equality, in order to have it contribute a standard basis vector to the columns of the initial tableau, by inserting an artificial variable. For example, the constraint $x_1 - x_2 + 5x_3 = 4$ might become $x_1 - x_2 + 5x_3 + q_2 = 4$, and again we must eventually require that $q_2 = 0$ to provide a feasible solution of our original problem.

Each constraint now contains either a slack variable or an artificial variable. These variables are initially assigned the nonnegative values b_i on the right-hand side of their constraints, and all other variables (problem and surplus variables) are assigned the value zero. Thus, the slack and artificial variables are initially taken as basic variables and correspond to a basic feasible solution (corner point) of the *augmented problem*. We still may not have a basic feasible solution for the original problem. In other words, phase 1 continues until all artificial variables become zero.

Forming the Initial Tableau

The question now arises, after we have augmented the problem, "how do we form an initial tableau and reduce it in such a way that the artificial variables are driven to zero?" An ingenious technique is to modify the objective function so that it cannot be optimized without the artificial variables being zero. For example, suppose that we wish to maximize

$$P = 3x_1 - x_2 + 7x_3$$

and have introduced two artificial variables q_1 and q_2 for the reasons described earlier. We modify our objective function to become

$$P = 3x_1 - x_2 + 7x_3 - Mq_1 - Mq_2,$$

where we think of M as a positive number that is so huge in comparison to the magnitude of other data in our problem that this new objective function could not achieve a maximum without having $q_1 = q_2 = 0$. In pencil-and-paper computations, which can be terribly tedious, we generally carry M along as a particular but unspecified huge number—so huge that, for any nonzero k that we compute, the number kM will be of greater magnitude than any computed magnitude that doesn't involve M. Computer programs sometimes set M equal to something like 10,000 times the largest magnitude of the given data. Our program in SIMPLEX does not do this; it carries the designation M throughout its computations, just as we will do in our pencil-and-paper work.

The initial tableau for this general situation is formed just as for the standard maximum problem except for the objective row. We will illustrate using the plywood mill problem that we discussed in Section 1.7; we restate it here for easy reference.

TABLE 10.2

Plywood Type	Mill 1 per Day	Mill 2 per Day	6-month Demand
A	100 sheets	20 sheets	2000 sheets
B	40 sheets	80 sheets	3200 sheets
C	60 sheets	60 sheets	3600 sheets
Daily Costs	$3000	$2000	

PROBLEM   A lumber company owns two mills that produce sheets of hardwood veneer plywood. Each mill produces the same three types of plywood. Table 10.2 shows the daily production and the daily cost of operation of each mill. The rightmost column shows the plywood required by the lumber company for a period of 6 months. Find the number of days each mill should operate during the 6 months in order to supply the required sheets most economically.

If mill 1 operates for x_1 days and mill 2 operates for x_2 days, the cost is

$$C = 3000x_1 + 2000x_2.$$

To fulfill the company's requirements, these constraints must be satisfied:

$$100x_1 + 20x_2 \geq 2000$$
$$40x_1 + 80x_2 \geq 3200 \tag{2}$$
$$60x_1 + 60x_2 \geq 3600,$$

where $x_1 \geq 0$ and $x_2 \geq 0$. We wish to minimize C subject to these constraints.

We minimize C by maximizing $P = -C$, as explained in the preceding section. Converting our constraints to equalities with surplus and artificial variables as described earlier in this section, we obtain

$$100x_1 + 20x_2 - y_1 \qquad + q_1 \qquad = 2000$$
$$40x_1 + 80x_2 \qquad - y_2 \qquad + q_2 \qquad = 3200 \tag{3}$$
$$60x_1 + 60x_2 \qquad \qquad - y_3 \qquad + q_3 = 3600,$$

where all variables are to be nonnegative, and the q_i must eventually become zero. We take as objective function

$$P = -3000x_1 - 2000x_2 - Mq_1 - Mq_2 - Mq_3, \tag{4}$$

as explained previously. To form the objective row of the initial tableau, we first write Eq. (4) as

$$P + 3000x_1 + 2000x_2 + Mq_1 + Mq_2 + Mq_3 = 0, \tag{5}$$

just as we did for the standard maximum problem. However, formation of the objective row here is a bit more complicated than for the standard maximum problem. We give the initial tableau, and then explain how the objective row is obtained:

	x_1	x_2	y_1	y_2	y_3	q_1	q_2	q_3	
q_1	100	20	-1	0	0	1	0	0	2000
q_2	40	80	0	-1	0	0	1	0	3200
q_3	60	60	0	0	-1	0	0	1	3600
P	3000	2000	0	0	0	0	0	0	0
	$-200M$	$-160M$	M	M	M				$-8800M.$

Recall that the objective row should contain data that tell us which nonbasic variable should be increased to become basic. In Section 1.7, where we wished to maximize a function such as $P = 4x_1 + 3x_2$, a unit increase in x_1 provided 4 units increase in P, whereas a unit increase in x_2 provided only 3 units increase in P. Thus the rate of increase of P was measured by the coefficients of x_1 and x_2. But in our present situation, our objective function contains artificial variables. If we select x_1 as the entering basic variable, constraints (3) show that 1 unit increase in x_1 must cause 100 units decrease in q_1, 40 units decrease in q_2, and 60 units decrease in q_3. We find that this unit increase in x_1 and corresponding decreases in the q_i would produce a change of

$$3000 - 100M - 40M - 60M = 3000 - 200M$$

in the expression following P on the left-hand side of Eq. (5). This is the reason for the entry $3000 - 200M$ in the objective row in the column labeled x_1; it measures the change in P corresponding to a unit increase in x_1.

The entries in the other positions of the objective row are found in a similar way. In particular, Eqs. (3) show that an increase of 1 unit in a surplus variable y_i must be matched by an increase of 1 unit in the corresponding q_i, and that produces a change of M in the expression following P in the left-hand side of Eq. (5). Of course, the q_i are already basic variables in the initial tableau and are not candidates to enter the basic set. Our illustration using the plywood mill problem has no slack variables. Slack variables do not appear in modified constraints that contain artificial variables. However, when slack variables do appear in a problem, the objective row entry in a column labeled by a slack variable will be 0 in the initial tableau. Finally, the last entry in the objective row is the negative of the value of P in Eq. (4) at our initial basic solution (corner point). This last entry is thus vM, where v is the negative of the sum of the values of the *artificial* variables in the initial tableau. We summarize in the following box.

Objective Row of the Initial Tableau to Maximize P

1. A column labeled with a problem variable x_i has, in the objective row, the negative of the x_i-coefficient in the objective function minus M times the sum of the x_i-coefficients in constraints that involve artificial variables.

2. A column labeled with a surplus variable has M in the objective row.

3. A column labeled with a slack or an artificial variable has 0 in the objective row.

4. The final entry in the objective row is the negative of the sum of the values of the artificial variables multiplied by M.

This initial tableau may be reduced by using the same criteria for entering and exiting variables as were used in Section 1.7, because the same arguments go through. The reduction continues to generate the data in the objective row that are needed to decide what entering variable corresponds to the fastest increase in the objective function. We will not repeat the arguments here.

Returning to our illustration, we give again the initial tableau and proceed to give the successive reductions, with each pivot shown in color. The arithmetic is nothing people care to do in their heads.

	x_1	x_2	y_1	y_2	y_3	q_1	q_2	q_3	
q_1	100	20	−1	0	0	1	0	0	2000
q_2	40	80	0	−1	0	0	1	0	3200
q_3	60	60	0	0	−1	0	0	1	3600
P	3000	2000	0	0	0	0	0	0	0
	−200M	−160M	M	M	M				−8800M.

	x_1	x_2	y_1	y_2	y_3	q_1	q_2	q_3	
x_1	1	.2	−.01	0	0	.01	0	0	20
~ q_2	0	72	.4	−1	0	−.4	1	0	2400
q_3	0	48	.6	0	−1	−.6	0	1	2400
P	0	1400	30	0	0	−30	0	0	−60000
		−120M	−M	M	M	+2M			−4800M

	x_1	x_2	y_1	y_2	y_3	q_1	q_2	q_3	
x_1	1	0	−.0111	.0028	0	.0111	−.0028	0	13.33
~ x_2	0	1	.0056	−.0139	0	−.0056	.0139	0	33.33
q_3	0	0	.3333	.6667	−1	−.3333	−.6667	1	800
P	0	0	22.222	19.44	0	−22.22	−19.44	0	−106667
			−.333M	−.667M	M	+1.333M	+1.667M		−800M

	x_1	x_2	y_1	y_2	y_3	q_1	q_2	q_3	
x_1	1	0	−.0125	0	.0042	.0125	0	−.0042	10
~ x_2	0	1	.0125	0	−.0208	−.0125	0	.0208	50
y_2	0	0	.5	1	−1.5	−.5	−1	1.5	1200
P	0	0	12.5	0	29.17	−12.5	0	−29.17	−130000.
						+M	M	+M	

The objective row contains no more negative entries (remember that M is positive and huge), so we are done. The maximum of P is $-130,000$ when $x_1 = 10$ and $x_2 = 50$. Thus the minimum cost C is \$130,000 when $x_1 = 10$ and $x_2 = 50$. Notice that, in this problem, the artificial variables were not dirven to zero until the final tableau was computed.

The preceding tableaux indicate that paper-and-pencil execution of the simplex method can be a formidable task. However, a computer has no difficulty. The available program SIMPLEX has an option to allow the user to control the execution of the simplex method without having to do any actual computations.

We need only modify step 2 of the boxed outline on page 494 in Section 10.1 to obtain an outline for the simplex method for a general linear programming problem. Notice that each linear programming problem is described as a maximizing one before the initial tableau is formed.

Modified Step 2 in the Boxed Procedure on Page 494

Test for an optimal solution: If all entries in the objective row are nonnegative and if either no artificial variable is basic or all basic artificial variables have value zero, an optimal solution has been found. If the entries in the objective row are nonnegative and some artificial variable has positive value, the original linear programming problem has no optimal solution.

We now give an example involving three constraints: an equality, and one inequality of each type.

EXAMPLE 1   Maximize $P = x_1 + x_2$, subject to

$$-2x_1 + x_2 \leq 2$$
$$2x_1 + x_2 = 9$$
$$3x_1 + x_2 \geq 11,$$

where $x_1 \geq 0$ and $x_2 \geq 0$.

Solution   If we supply slack, surplus, and artificial variables, the constraints become

$$-2x_1 + x_2 + y_1 \qquad\qquad = 2$$
$$2x_1 + x_2 \qquad + q_1 \qquad = 9$$
$$3x_1 + x_2 \qquad -y_2 \quad + q_2 = 11.$$

Our profit function becomes

$$P = x_1 + x_2 - Mq_1 - Mq_2.$$

We now form the initial tableau and reduce it:

	x_1	x_2	y_1	y_2	q_1	q_2	
y_1	-2	1	1	0	0	0	2
q_1	2	1	0	0	1	0	9
q_2	3	1	0	-1	0	1	11
P	-1	-1	0	0	0	0	0
	$-5M$	$-2M$		M			$-20M$

	x_1	x_2	y_1	y_2	q_1	q_2	
y_1	0	$\frac{5}{3}$	1	$-\frac{2}{3}$	0	$\frac{2}{3}$	$\frac{28}{3}$
$\sim q_1$	0	$\frac{1}{3}$	0	$\frac{2}{3}$	1	$-\frac{2}{3}$	$\frac{5}{3}$
x_1	1	$\frac{1}{3}$	0	$-\frac{1}{3}$	0	$\frac{1}{3}$	$\frac{11}{3}$
P	0	$-\frac{2}{3}$	0	$-\frac{1}{3}$	0	$\frac{1}{3}$	$\frac{11}{3}$
		$-M/3$		$-2M/3$		$+5M/3$	$-5M/3$

	x_1	x_2	y_1	y_2	q_1	q_2	
y_1	0	2	1	0	1	0	11
$\sim y_2$	0	$\frac{1}{2}$	0	1	$\frac{3}{2}$	-1	$\frac{5}{2}$
x_1	1	$\frac{1}{2}$	0	0	$\frac{1}{2}$	0	$\frac{9}{2}$
P	0	$-\frac{1}{2}$	0	0	$\frac{1}{2}$	0	$\frac{9}{2}$
					$+M$	M	

	x_1	x_2	y_1	y_2	q_1	q_2	
y_1	0	0	1	-4	-5	4	1
$\sim x_2$	0	1	0	2	3	-2	5
x_1	1	0	0	-1	-1	1	2
P	0	0	0	1	2	-1	7
					$+M$	$+M$	

The maximum value of 7 is attained at the point $(x_1, x_2) = (2, 5)$. Notice that all artificial variables have been driven to zero in the third tableau. This tableau marks the end of phase 1, since we have attained a basic feasible solution to our *original* problem. ■

We conclude with an illustration of the case in which there is no solution, because the feasible set is empty, as shown by Exercise 8.

EXAMPLE 2   Maximize $P = 2x_1 + 3x_2$, subject to the constraint

$$x_1 + x_2 \leq -7,$$

where $x_1 \geq 0$ and $x_2 \geq 0$.

Solution We must write the constraint so that the right-hand side is positive. Multiplying it by -1, we obtain the constraint

$$-x_1 - x_2 \geq 7.$$

If we supply a surplus variable and an artificial variable, the constraint becomes

$$-x_1 - x_2 - y_1 + q_1 = 7.$$

We form the initial tableau and try to reduce it:

	x_1	x_2	y_1	q_1	
q_1	-1	-1	-1	1	7
P	-2	-3	0	0	0
	$+M$	$+M$	M		$-7M$

Since the entries in the objective row are all nonnegative (recall that M is huge and positive), no reduction is necessary. Since the artificial variable has value $7 > 0$, our boxed test for a solution given prior to Example 1 indicates that there is no solution to the problem.   ■

In the literature, one often finds all variables (problem variables, slack variables, surplus variables, and artificial variables) denoted by x_i. For example, x_1 through x_4 might be problem variables, x_5 and x_6 slack variables, x_7 a surplus variable, and x_8 and x_9 artificial variables. The simplex method treats all these variables in the same way, except that one must keep track of which are surplus and artificial variables in forming the objective row. In more advanced treatments, where the simplex method is carefully developed and proved, calling all variables x_i makes the notation cleaner. The development is usually in terms of matrices and generating vectors for their column spaces. Often tableaux are mentioned only in passing.

SUMMARY

1. Constraints other than $x_i \geq 0$ in a linear programming problem are modified as follows to obtain equalities and to create, in the columns of the initial tableau, a standard basis for the Euclidean space $\mathbb{R}^m$.

   a. A slack variable y_i is inserted in each $(\leq b)$-type constraint.

   b. An artificial variable q_i is inserted in each $(= b)$-type constraint.

   c. A surplus variable term $-y_i$ and an artifical variable q_j are inserted in each $(\geq b)$-type constraint.

   In each of these cases, we require that the constraint first be written so that $b_i \geq 0$. For a feasible solution of the original problem, we require that all variables be nonnegative and that the artificial variables be zero.

2. Formation of the objective row in the initial tableau is described in the box on page 503. Otherwise, the initial tableau is formed just as in the standard maximum problem.

3. The tableaux are reduced as described in the box on page 494 of Section 10.1 except for the test for an optimal solution. This test is described in the box that precedes Example 1.

EXERCISES

In Exercises 1–4, find the initial tableau for the given linear programming problem. All variables x_i given are assumed to be nonnegative.

1. Maximize $3x_1 - 2x_2 + 5x_3$, subject to the constraints

$$x_1 + 2x_2 + x_3 \leq 5,$$
$$4x_1 + 2x_2 - 3x_3 \geq 3,$$
$$5x_1 \qquad - x_3 \geq 4.$$

2. Maximize $x_1 + 2x_2 - 3x_3$, subject to the constraints

$$x_1 - x_2 \qquad = 1,$$
$$3x_1 \qquad + x_3 \geq 2,$$
$$4x_2 + x_3 \leq 10.$$

3. Minimize $-x_1 + 2x_2 - 5x_3$, subject to the constraints

$$3x_1 - x_2 \qquad = 5,$$
$$2x_2 + x_3 = 2,$$
$$x_1 + 4x_2 + 5x_3 \leq 20,$$
$$4x_1 - 2x_2 - x_3 \geq 3.$$

4. Minimize $20x_1 + 40x_2 + 10x_3$, subject to the constraints

$$x_1 + 3x_2 \qquad \geq 2,$$
$$5x_1 \qquad + 2x_3 \geq 5,$$
$$2x_2 + 4x_3 \geq 3,$$
$$x_3 \leq 12.$$

5. Use the simplex method to solve the vitamin problem in Exercise 17 of Section 1.7. Use a calculator.

6. Use the simplex method to solve the fertilizer problem in Exercise 19 of Section 1.7.

7. Use the simplex method to find the *minimum* value of the objective function in Example 1, subject to the constraints given there. Use a calculator.

8. Draw a sketch to show that there is no solution to the linear programming problem stated in Example 2.

9. (*A transportation problem*). A manufacturer has warehouses in towns W_1 and W_2, with 30 units of a product available at the warehouse in W_1 and 20 units of the product available at W_2. Buyer B_1 needs 10 units of this product, and buyer B_2 wants 40 units of it. The cost per unit for shipment from the warehouses to the buyers is as follows:

From W_1 to B_1: \$10 per unit,
From W_1 to B_2: \$ 5 per unit,
From W_2 to B_1: \$ 4 per unit,
From W_2 to B_2: \$ 8 per unit.

We wish to find how much of the product should be shipped from each warehouse to each buyer in order to minimize the cost of shipment.

a) Formulate this problem as a linear programming problem.
b) Set up the initial tableau for the problem.
c) Solve the problem by the simplex method.

In Exercises 10–18, use the user's choice option of the available program SIMPLEX to solve the indicated linear programming problem.

10.3 DUALITY

The Dual of a Standard Maximum Problem

Every linear programming problem has an associated dual problem. We begin our presentation with an economic motivation for the dual of a standard maximum problem.

Let us consider a modification of our coffee problem in Exercise 15 of Section 1.7. Suppose that a food packaging house has m different types of coffee beans in stock, with b_i lb of type i beans for $i = 1, 2, \ldots, m$. The house can create n different blends of coffee to sell to retailers. Suppose each pound of blend j requires a_{ij} lb of type i coffee, and can be sold to the retailers for c_j dollars. The house wishes to maximize its revenue R from the retailers for this coffee. Let x_j be the number of pounds of blend j created. The house wants to maximize the revenue

$$R = \mathbf{c} \cdot \mathbf{x} = c_1 x_1 + c_2 x_2 + \cdots + c_n x_n,$$

subject to the constraints

$$a_{11} x_1 + a_{12} x_2 + \cdots + a_{1n} x_n \leq b_1$$
$$a_{21} x_1 + a_{22} x_2 + \cdots + a_{2n} x_n \leq b_2$$
$$\vdots$$
$$a_{m1} x_1 + a_{m2} x_2 + \cdots + a_{mn} x_n \leq b_m,$$

that is, subject to $A\mathbf{x} \leq \mathbf{b}$, where $A = [a_{ij}]$ is an $m \times n$ matrix and $\mathbf{x} \geq \mathbf{0}$. We take this linear programming problem as our *primal* problem:

$$\text{maximize} \quad \mathbf{c} \cdot \mathbf{x} \quad \text{subject to} \quad A\mathbf{x} \leq \mathbf{b}, \quad \text{where} \quad \mathbf{x} \geq \mathbf{0}. \tag{1}$$

The *dual* problem arises if we consider the value $y_i \geq 0$ to the house of each pound of type i coffee beans. The total value of the stock of coffee beans becomes

$$V = \mathbf{b} \cdot \mathbf{y} = b_1 y_1 + b_2 y_2 + \cdots + b_m y_m. \tag{2}$$

Since each pound of blend j coffee that the house makes requires a_{ij} lb of type i beans for $i = 1, 2, \ldots, n$, the value of the beans in each pound of blend j is

$$a_{1j} y_1 + a_{2j} y_2 + \cdots + a_{mj} y_m. \tag{3}$$

Now we know that a pound of blend j can be sold to the retailers for c_j dollars, so we must have

$$a_{1j}y_1 + a_{2j}y_2 + \cdots + a_{mj}y_m \geq c_j \tag{4}$$

for $j = 1, 2, \ldots, m$. The inequalities (4) can be written as

$$A^T\mathbf{y} \geq \mathbf{c}. \tag{5}$$

The house would like to find (for tax purposes) the minimum value for this existing stock of coffee beans. That is, the house wishes to

$$\text{minimize} \quad \mathbf{b} \cdot \mathbf{y} \quad \text{subject to} \quad A^T\mathbf{y} \geq \mathbf{c}, \quad \text{where} \quad \mathbf{y} \geq \mathbf{0}. \tag{6}$$

This linear programming problem (6) is called the *dual* of the *primal* linear programming problem (1).

Let $R_{\max}$ be the optimal value for problem (1), and let $V_{\min}$ be the optimal value for problem (6). Since $R_{\max}$ will be the revenue obtained if the beans are retailed in blends in accordance with the solution of problem (1), the total value V of the beans is at least as large as $R_{\max}$. In particular, $V_{\min} \geq R_{\max}$. Exercise 10 requests an algebraic proof of this. The duality theorem, which we state later, asserts that actually $V_{\min} = R_{\max}$. This is not surprising, since the sale prices c_j that appear in the constraints in problem (6) are the only price information we have available to use in computing the values y_i. The values y_i that produce the optimal value $V_{\min}$ are called the **shadow prices** or **accounting prices**.

Here is an illustration showing how the dual of a standard maximum problem is formed.

EXAMPLE 1   Find the dual of the boatyard problem in Section 10.1, which is to maximize the profit function $P = 5x_1 + 4x_2 + 6x_3$, subject to the constraints

$$x_1 + x_2 + x_3 \leq 25$$

$$2x_1 + x_2 + 3x_3 \leq 51$$

$$x_1, x_2, x_3 \geq 0.$$

Solution   The given primal problem has matrix form (1), with

$$\mathbf{c} = \begin{bmatrix} 5 \\ 4 \\ 6 \end{bmatrix}, \qquad \mathbf{x} = \begin{bmatrix} x_1 \\ x_2 \\ x_3 \end{bmatrix}, \qquad \mathbf{b} = \begin{bmatrix} 25 \\ 51 \end{bmatrix}, \qquad A = \begin{bmatrix} 1 & 1 & 1 \\ 2 & 1 & 3 \end{bmatrix}.$$

Its dual has form (6)—that is,

$$\text{minimize} \quad \mathbf{b} \cdot \mathbf{y} = \begin{bmatrix} 25 \\ 51 \end{bmatrix} \cdot \begin{bmatrix} y_1 \\ y_2 \end{bmatrix} = 25y_1 + 51y_2,$$

$$\text{subject to} \quad A^T\mathbf{y} = \begin{bmatrix} 1 & 2 \\ 1 & 1 \\ 1 & 3 \end{bmatrix}\begin{bmatrix} y_1 \\ y_2 \end{bmatrix} \geq \begin{bmatrix} 5 \\ 4 \\ 6 \end{bmatrix} \quad \text{and} \quad \mathbf{y} \geq \mathbf{0}.$$

The constraints of the dual are then

$$y_1 + 2y_2 \geq 5$$

$$y_1 + y_2 \geq 4$$

$$y_1 + 3y_2 \geq 6$$

$$y_1, y_2 \geq 0.$$ ■

The initial tableau for the primal problem in Example 1 is

	x_1	x_2	x_3	y_1	y_2	C
y_1	1	1	1	1	0	25
y_2	2	1	3	0	1	51
P	-5	-4	-6	0	0	0

If we consider just the entries in color, ignoring the columns labeled with the slack variables, we obtain the **condensed tableau**

	x_1	x_2	x_3	C
y_1	1	1	1	25
y_2	2	1	3	51
P	-5	-4	-6	0

Reading from left to right across the rows of the condensed tableau, we obtain the data for the primal problem. The final row represents the objective function $P = 5x_1 + 4x_2 + 6x_3$. Recall that the negative entries are present because we want to view the entire last row as the equation $P - 5x_1 - 4x_2 - 6x_3 = 0$ so that the final entry gives the value of the objective function P. Mentally changing signs of elements in the last row and last column of either tableau and then reading from top to bottom down the colored columns, we can obtain the data for the dual problem. We place C in the upper right-hand corner of the tableau, analogous to our placement of P in the lower left-hand corner. (Recall that, for a maximizing problem, the objective function is generally referred to as the *profit*, while, for a minimizing problem, it is usually called the *cost*.) This condensed tableau exhibits the symmetry of a standard maximum primal problem with its dual. Notice that the basic variables of the primal problem are nonbasic variables of the dual, and vice versa. In implementing the simplex method, we choose to work with the original tableau.

The Dual of a General Linear Programming Problem

We now describe formation of the dual of any linear programming problem. To do this, we relax the condition that the constants on the right-hand side of our constraints be nonnegative, and we write all constraints other than $x_i \geq 0$ in the ($\leq$ type)-form. For example,

$$x_1 - 3x_2 \geq 5 \qquad \text{should be written} \qquad -x_1 + 3x_2 \leq -5.$$

Moreover, any equality constraint should be expressed as *two* inequality constraints. To illustrate, $2x_1 - 3x_2 = 5$ can be expressed as $2x_1 - 3x_2 \leq 5$ and $2x_1 - 3x_2 \geq 5$; or using only ($\leq$ type)-inequalities,

$$2x_1 - 3x_2 = 5 \qquad \text{becomes} \qquad 2x_1 - 3x_2 \leq 5 \qquad \text{and} \qquad -2x_1 + 3x_2 \leq -5.$$

Thus, any maximizing problem can be written in the form

$$\text{maximize} \quad \mathbf{c} \cdot \mathbf{x} \quad \text{subject to} \quad A\mathbf{x} \leq \mathbf{b}, \qquad \text{where} \qquad \mathbf{x} \geq \mathbf{0} \qquad (7)$$

and where entries in $\mathbf{b}$ may be positive, zero, or negative. In a similar manner, any minimizing problem can be written as

$$\text{minimize} \quad \mathbf{s} \cdot \mathbf{y} \quad \text{subject to} \quad B\mathbf{y} \geq \mathbf{r}, \qquad \text{where} \qquad \mathbf{y} \geq \mathbf{0}. \qquad (8)$$

Having written a linear programming problem in form (7) or (8), we may define its *dual* as follows.

DEFINITION 10.1 Dual Problems

The linear programming problems

$$\text{maximize} \quad \mathbf{c} \cdot \mathbf{x} \quad \text{subject to} \quad A\mathbf{x} \leq \mathbf{b}, \qquad \text{where} \qquad \mathbf{x} \geq \mathbf{0} \qquad (9)$$

and

$$\text{minimize} \quad \mathbf{b} \cdot \mathbf{y} \quad \text{subject to} \quad A^T\mathbf{y} \geq \mathbf{c}, \qquad \text{where} \qquad \mathbf{y} \geq \mathbf{0} \qquad (10)$$

are **dual problems**. If problem (9) is given first, it is the **primal** problem, and problem (10) is its **dual**. Similarly, if problem (10) is given first, it is the **primal** problem, and problem (9) is its **dual**.

EXAMPLE 2   Find the dual of the following minimizing linear programming problem:

$$\text{minimize the cost function} \qquad C = 3y_1 + 2y_2 + 5y_3,$$

$$\text{subject to} \qquad y_1 + 4y_2 + y_3 \geq 2$$

$$2y_2 + 10y_3 \geq 4$$

$$y_1 + y_3 \geq 5,$$

$$\text{where} \qquad y_1, y_2, y_3 \geq 0.$$

Solution   Our primal problem has the form of problem (10). Thus, its dual has the form of problem (9) and is:

$$\text{maximize the profit function} \qquad P = 2x_1 + 4x_2 + 5x_3,$$

$$\text{subject to} \qquad x_1 + x_3 \leq 3$$

$$4x_1 + 2x_2 \leq 2$$

$$x_1 + 10x_2 + x_3 \leq 5,$$

$$\text{where} \qquad x_1, x_2, x_3 \geq 0. \qquad \blacksquare$$

We now state the duality theorem. Our economic example at the beginning of the section anticipates the result.

THEOREM 10.2 Duality Theorem

If a linear programming problem has an optimal solution, its dual also has an optimal solution. Furthermore, the optimal values attained by the primal problem and by its dual are the same.

In a linear programming problem, one generally wants to know not only the optimal value attained but also a basic feasible solution (corner point) at which it is attained. The importance of duality lies partly in the fact that use of the simplex method to solve the primal problem solves at the same time the dual problem. Recall how a simplex tableau or condensed tableau for a primal problem exhibits the data for the dual problem also. We choose not to describe how to find the dual solution from the final primal tableau when some of the constraints are equalities. If equality constraints appear, we can always replace each of them by two inequality constraints and use the following boxed result.

Solution of the Dual Problem from the Final Primal Tableau

For a primal linear programming problem (9) not containing equality constraints, an optimal basic feasible solution of the dual problem (10) is provided by values in the bottom row of the final tableau for problem (9). The value of y_i in the dual solution appears in the objective row in the column labeled by the y_i at the top.

For an illustration, consider the boatyard problem stated in Example 1 of this section and solved in Section 10.1. The final tableau for this problem was found in Section 10.1 to be

	x_1	x_2	x_3	y_1	y_2	C
x_2	$\frac{1}{2}$	1	0	$\frac{3}{2}$	$-\frac{1}{2}$	12
x_3	$\frac{1}{2}$	0	1	$-\frac{1}{2}$	$\frac{1}{2}$	13
P	0	0	0	3	1	126.

From this tableau and the preceding boxed statement, we see that a solution of the dual problem is given by $y_1 = 3$, $y_2 = 1$. Looking back at Example 1, we see that the objective function to be minimized for the dual to the boatyard problem is $C = 25y_1 + 51y_2$, and we can check that $25(3) + 51(1) = 126$.

Remember that to solve problem (9) by the simplex method, we must write all constraints so that the constants on the right-hand sides are nonnegative.

EXAMPLE 3   Let the primal linear programming problem be

$$\text{maximize} \quad 3x_1 + 2x_2,$$

$$\text{subject to} \quad x_1 + 2x_2 \le 10$$

$$5x_1 + x_2 \ge 10$$

$$x_1 + 10x_2 \ge 20,$$

$$\text{where} \quad x_1 \ge 0, x_2 \ge 0.$$

Formulate the dual problem, and solve it by solving the primal problem and using the preceding boxed statement.

Solution   To form the dual problem, we rewrite the constraints of the pimal problem as

$$x_1 + 2x_2 \le 10$$

$$-5x_1 - x_2 \le -10$$

$$-x_1 - 10x_2 \le -20.$$

The dual problem is

$$\text{minimize} \quad 10y_1 - 10y_2 - 20y_3,$$

$$\text{subject to} \quad y_1 - 5y_2 - y_3 \ge 3$$

$$2y_1 - y_2 - 10y_3 \ge 2,$$

$$\text{where} \quad y_1 \ge 0, y_2 \ge 0, y_3 \ge 0.$$

We form the initial tableau as usual from the problem as originally stated, and we reduce it as usual:

	x_1	x_2	y_1	y_2	y_3	q_1	q_2	C
y_1	1	2	1	0	0	0	0	10
q_1	5	1	0	-1	0	1	0	10
q_2	1	10	0	0	-1	0	1	20
P	-3	-2	0	0	0	0	0	0
	$-6M$	$-11M$		M	M			$-30M$

	x_1	x_2	y_1	y_2	y_3	q_1	q_2	C
y_1	.8	0	1	0	.2	0	$-.2$	6
$\sim q_1$	4.9	0	0	-1	.1	1	$-.1$	8
x_2	.1	1	0	0	$-.1$	0	.1	2
P	-2.8	0	0	0	$-.2$	0	.2	4
	$-4.9M$			M	$-.1M$		$+1.1M$	$-8M$

	x_1	x_2	y_1	y_2	y_3	q_1	q_2	C
y_1	0	0	1	.1633	.1837	−.1633	−.1837	4.6939
~ x_1	1	0	0	−.2041	.0204	.2041	−.0204	1.6327
x_2	0	1	0	.0204	−.1020	−.0204	.1020	1.8367
P	0	0	0	−.5714	−.1429	.5714 +M	.1429 +M	8.5714

	x_1	x_2	y_1	y_2	y_3	q_1	q_2	C
y_2	0	0	6.125	1	1.125	−1	−1.125	28.75
~ x_1	1	0	1.25	0	.25	0	−.25	7.5
x_2	0	1	−.125	0	−.125	0	.125	1.25
P	0	0	3.5	0	.5	0 M	−.5 +M	25

From this final tableau, we see that the maximum value of the primal objective function $3x_1 + 2x_2$ is 25, attained at the basic feasible solution, $x_1 = 7.5$ and $x_2 = 1.25$. We also find that the minimum value of the dual objective function $10y_1 - 10y_2 - 20y_3$ is 25, attained at the basic feasible solution $y_1 = 3.5$, $y_2 = 0$, $y_3 = .5$. These are read off in the bottom row in the columns labeled y_1, y_2, and y_3, respectively. ■

We are accustomed to working with variables x_j as our problem variables in a simplex tableau. Given a primal linear programming problem stated with problem variables x_j, we may attempt to solve the problem directly by the simplex method, or we may choose to form the dual problem and solve the dual using the simplex method. In the latter case, we suggest that the dual be restated, using variables x_j as problem variables, and that the primal problem be restated, using variables y_i. Then the labels on our tableaux will be in their accustomed positions.

Duality is useful when a linear programming problem has a large number m of constraints compared with the number n of variables. Inserting a slack or surplus variable in each constraint, we find that the matrix of data in the initial tableau has $m + 1$ rows and at least $n + m + 1$ columns. On the other hand, the initial tableau of the dual problem has only $n + 1$ rows and again at least $n + m + 1$ columns. If m is much greater than n, it is advantageous for us to solve the dual problem instead. We close with an illustration of this.

EXAMPLE 4 Minimize the function $C = 2y_1 + y_2$, subject to the constraints

$$10y_1 + y_2 \geq 10$$

$$2y_1 + y_2 \geq 8$$

$$y_1 + y_2 \geq 6$$

$$y_1 + 2y_2 \geq 10$$

$$y_1 + 12y_2 \geq 12,$$

where $\mathbf{y} \geq \mathbf{0}$.

Solution In order to work with a smaller tableau, we choose to solve the dual problem. The dual problem is:

$$\text{maximize} \quad 10x_1 + 8x_2 + 6x_3 + 10x_4 + 12x_5,$$

$$\text{subject to} \quad 10x_1 + 2x_2 + x_3 + x_4 + x_5 \leq 2$$

$$x_1 + x_2 + x_3 + 2x_4 + 12x_5 \leq 1,$$

$$\text{where} \quad \mathbf{x} \geq \mathbf{0}.$$

We form the initial tableau and reduce it:

	x_1	x_2	x_3	x_4	x_5	y_1	y_2	C
y_1	10	2	1	1	1	1	0	2
y_2	1	1	1	2	12	0	1	1
P	-10	-8	-6	-10	-12	0	0	0

	x_1	x_2	x_3	x_4	x_5	y_1	y_2	C
~ y_1	9.9167	1.9167	.9167	.8333	0	1	$-.0833$	1.9167
x_5	.0833	.0833	.0833	.1667	1	0	.0833	.0833
P	-9	-7	-5	-8	0	0	1	1

	x_1	x_2	x_3	x_4	x_5	y_1	y_2	C
~ x_1	1	.1933	.0924	.0840	0	.1008	$-.0084$	.1933
x_5	0	.0672	.0756	.1597	1	$-.0084$	.0840	.0672
P	0	-5.2605	-4.1681	-7.2437	0	.9076	.9244	2.7395

	x_1	x_2	x_3	x_4	x_5	y_1	y_2	C
~ x_1	1	.1579	.0526	0	$-.5263$	.1053	$-.0526$	.1579
x_4	0	.4211	.4737	1	6.2632	$-.0526$	.5263	.4211
P	0	-2.2105	$-.7368$	0	45.3684	.5263	4.7368	5.7895

	x_1	x_2	x_3	x_4	x_5	y_1	y_2	C
~ x_1	1	0	$-.125$	$-.375$	-2.875	.125	$-.25$	0
x_2	0	1	1.125	2.375	14.875	$-.125$	1.25	1
P	0	0	1.75	5.25	78.25	.25	7.5	8.

Thus, the objective function $2y_1 + y_2$ in the original primal problem has minimum value 8 at the basic feasible solution $y_1 = .25$ and $y_2 = 7.5$.

In the next to the last tableau, we could have chosen the top row as pivot row, with .1579 as pivot entry. Notice that both of the ratios $.1579/.1579$ and $.4211/.4211$ determining the choice of pivot row are 1. The program SIMPLEX always chooses the candidate row closest to the top of the tableau as pivot row in such a tie case. Using SIMPLEX or pencil and paper, we see that the other choice leads to the solution $y_1 = 2$, $y_2 = 4$. The same value 8 is obtained for the objective function. ■

Solving the original primal problem in the preceding example, using the program SIMPLEX, we would work with a tableau containing thirteen columns and six rows of data; six iterations would be required. With the dual in the example, we work with eight columns and three rows of data, and four iterations are used. Thus it is advantageous for us to solve the dual in this case.

SUMMARY

1. To form the dual of a maximizing linear programming problem, write the problem as a maximizing one with all constraints in ($\leq$ type)-form, except for the non-negativity-of-variables constraints. An equality should be expressed as two such constraints. The constants on the right-hand sides of the constraints may be negative, positive, or zero. If the resulting problem is to maximize $\mathbf{c} \cdot \mathbf{x}$ subject to $A\mathbf{x} < \mathbf{b}$ and $\mathbf{x} > \mathbf{0}$, the dual problem is to minimize $\mathbf{b} \cdot \mathbf{y}$ subject to $A^T\mathbf{y} \geq \mathbf{c}$ and $\mathbf{y} \geq \mathbf{0}$. Similarly, the dual of the latter problem is the former. The problem that is stated first is regarded as the primal problem; the other is its dual.

2. The duality theorem asserts that, if a primal problem has an optimal solution, so does its dual. Moreover, the optimal value satisfying the primal problem equals the optimal value satisfying the dual.

3. The box following Theorem 10.2 explains how the basic feasible solution of the dual problem can be found from the final tableau for the primal problem, assuming that there are no equality constraints.

EXERCISES

In Exercises 1–8, formulate the dual of the given linear programming problem. Constraints asserting nonnegativity of variables are assumed but are omitted in the statement of the problem.

1. Maximize $3x_1 - 2x_2$,
   subject to $x_1 + 2x_2 \leq 10$,
   $3x_1 + x_2 \leq 15$.

2. Maximize $x_1 - 2x_2 + 4x_3$,
   subject to $2x_1 + x_2 + 4x_3 \leq 40$,
   $-x_1 + x_2 + x_3 \leq 10$.

3. Minimize $-4y_1 + 3y_2$,
   subject to $2y_1 + 3y_2 \geq 6$,
   $y_1 + 3y_2 \geq 4$.

4. Minimize $2y_1 + 2y_2 + y_3$,
   subject to $y_1 + 3y_2 + 4y_3 \geq 24$,
   $3y_1 + 2y_2 - 4y_3 \geq 4$,
   $2y_1 - 5y_2 + y_3 \geq 8$.

5. Maximize $3x_1 + 4x_2$,
   subject to $x_1 + 2x_2 \leq 8$,
   $x_1 - x_2 \geq 1$,
   $x_1 + x_2 \geq 2$.

6. Maximize $x_1 - x_2 + 3x_3$,
   subject to $2x_1 + x_2 + 4x_3 \leq 12$,
   $-2x_1 + x_2 + x_3 = 4$.

7. Minimize $y_1 - 2y_3$,
   subject to $y_1 + 3y_2 + y_3 = 12$,
   $y_2 + y_3 = 4$.

8. Minimize $y_1 + 3y_2 + y_4$,
   subject to $y_1 + 3y_2 + 9y_3 + 2y_4 \geq 18$,
                 $y_1 - 4y_2 + 2y_3 \qquad \leq 8$,
                 $\qquad\qquad y_3 + 3y_4 \leq 6$.

9. Consider a linear programming problem
   $$\text{maximize } \mathbf{c} \cdot \mathbf{x} \quad \text{subject to}$$
   $$A\mathbf{x} \leq \mathbf{b}, \qquad \mathbf{x} \geq \mathbf{0},$$
   where components of $\mathbf{b}$ may be positive, negative, or zero.

   a) State the vector form of the dual of this problem.
   b) Restate the answer to part (a) as a maximizing problem, as described in Section 10.1.
   c) Form the dual of the maximizing problem obtained in answer to part (b).
   d) Show that the minimizing problem found in part (c) is equivalent to the original given maximizing problem.

   [Some authors define the dual of any linear programming problem as follows:

   (i) Write the problem as a maximizing problem of the form given at the start of this exercise.
   (ii) Form the dual minimizing problem as described in this section.

   This exercise shows that, if one repeats by this method steps (i) and (ii) to form the dual of the dual just obtained, one essentially obtains the original problem. That is, the dual of the dual is the original problem.]

10. Let a primal problem to maximize $P = \mathbf{c} \cdot \mathbf{x}$ subject to $A\mathbf{x} \leq \mathbf{b}$ as in problem (9) have optimal solution $P_{\max}$. Let the dual problem (10) to minimize $C = \mathbf{b} \cdot \mathbf{y}$ subject to $A^T\mathbf{y} \geq \mathbf{c}$ have optimal solution $C_{\min}$. Proceeding algebraically, without using the duality theorem, show that $P_{\max} \leq C_{\min}$. [HINT: Let $\mathbf{x}$ and $\mathbf{y}$ be vectors with nonnegative components such that $A\mathbf{x} \leq \mathbf{b}$ and $A^T\mathbf{y} \geq \mathbf{c}$. Show from these inequalities that $\mathbf{c} \cdot \mathbf{x} \leq \mathbf{b} \cdot \mathbf{y}$.]

11. Solve the vitamin problem (Exercise 17 in Section 1.7) by using the simplex method to solve the dual problem.

12. Repeat Exercise 11 for the fertilizer problem (Exercise 19 in Section 1.7).

13. Repeat Exercise 11 for the coffee problem (Exercise 15 in Section 1.7).

14. Repeat Exercise 11 for the mill problem stated and solved in Section 1.7.

15. Suppose that a linear programming problem has the same number of constraints as problem variables, with some ($\leq$ type)-constraints and some ($\geq$ type)-constraints. (The constants on the right-hand sides of the constraints are nonnegative.) Under what circumstances might it be advantageous to solve the problem by using the simplex method on the dual problem, rather than on the given primal one?

In Exercises 16–19, solve the given linear programming problem by using the simplex method either on the given primal problem or on its dual, whichever appears to be easier. Constraints asserting nonnegativity of variables are assumed but are omitted in the statement of the problem.

16. Minimize $2y_1 + 3y_2$,
    subject to $2y_1 + y_2 \geq 6$,
                 $y_1 + y_2 \geq 5$,
                 $y_1 + 2y_2 \geq 8$,
                 $y_1 + 4y_2 \geq 12$.

17. Minimize $y_1 + 3y_2$,
    subject to $y_1 + 2y_2 \geq 4$,
                 $3y_1 + 4y_2 \geq 10$.

18. Maximize $2x_1 + 3x_2$,
    subject to $x_1 + 2x_2 \leq 8$,
                 $-x_1 + x_2 \leq 1$,
                 $x_1 \leq 4$.

19. Maximize $3x_1 + 5x_2$,
    subject to $-x_1 + x_2 \leq 2$,
                 $x_2 \leq 4$,
                 $x_1 + 3x_2 \leq 15$,
                 $x_1 + 2x_2 \leq 12$,
                 $x_1 + x_2 \leq 10$.

▦ *In Exercises 20–25, use the available program SIMPLEX to solve the given problem. Then form its dual, and use SIMPLEX again to solve the dual. Decide whether the primal or dual problem would be easier to solve if one were to use the simplex method with pencil and paper.*

20. The problem in Exercise 1
21. The problem in Exercise 2
22. The problem in Exercise 3
23. The problem in Exercise 4
24. The problem in Exercise 5
25. The problem in Exercise 8

A

MATHEMATICAL INDUCTION

Sometimes we want to prove that a statement about positive integers is true for all positive integers or perhaps for some finite or infinite sequence of consecutive integers. Such proofs are accomplished using *mathematical induction*. The validity of the method rests on the following axiom of the positive integers. The set of all positive integers is denoted by Z^+.

Induction Axiom

Let S be a subset of Z^+ satisfying

1. $1 \in S$,
2. If $k \in S$, then $(k + 1) \in S$.

Then $S = Z^+$.

This axiom leads immediately to the method of mathematical induction.

Mathematical Induction

Let $P(n)$ be a statement concerning the positive integer n. Suppose that

1. $P(1)$ is true,
2. If $P(k)$ is true, then $P(k + 1)$ is true.

Then $P(n)$ is true for all $n \in Z^+$.

Most of the time, we want to show that $P(n)$ holds for all $n \in Z^+$. If we wish only to show that it holds for $r, r + 1, r + 2, \ldots, s - 1, s$, then we show that $P(r)$ is true and that $P(k)$ implies $P(k + 1)$ for $r \le k \le s - 1$. Notice that r may be any integer—positive, negative, or zero.

EXAMPLE A.1   Prove the formula

$$1 + 2 + \cdots + n = \frac{n(n + 1)}{2} \tag{A.1}$$

for the sum of the arithmetic progression, using mathematical induction.

Solution   We let $P(n)$ be the statement that formula (A.1) is true. For $n = 1$, we obtain

$$\frac{n(n + 1)}{2} = \frac{1(2)}{2} = 1,$$

so $P(1)$ is true.

Suppose that $k \geq 1$ and $P(k)$ is true (our *induction hypothesis*), so

$$1 + 2 + \cdots + k = \frac{k(k + 1)}{2}.$$

To show that $P(k + 1)$ is true, we compute

$$1 + 2 + \cdots + (k + 1) = (1 + 2 + \cdots + k) + (k + 1)$$
$$= \frac{k(k + 1)}{2} + (k + 1) = \frac{k^2 + k + 2k + 2}{2}$$
$$= \frac{k^2 + 3k + 2}{2} = \frac{(k + 1)(k + 2)}{2}.$$

Thus, $P(k + 1)$ holds, and formula (A.1) is true for all $n \in Z^+$.   ▪

EXAMPLE A.2   Show that a set of n elements has exactly 2^n subsets for any nonnegative integer n.

Solution   This time we start the induction with $n = 0$. Let S be a finite set having n elements. We wish to show

$$P(n): S \text{ has } 2^n \text{ subsets.} \tag{A.2}$$

If $n = 0$, then S is the empty set and has only one subset—namely, the empty set itself. Since $2^0 = 1$, we see that $P(0)$ is true.

Suppose that $P(k)$ is true. Let S have $k + 1$ elements, and let one element of S be c. Then $S - \{c\}$ has k elements, and hence 2^k subsets. Now every subset of S either contains c or does not contain c. Those not containing c are subsets of $S - \{c\}$, so there are 2^k of them by the induction hypothesis. Each subset containing c consists of one of the 2^k subsets not containing c, with c adjoined. There are 2^k such subsets also. The total number of subsets of S is then

$$2^k + 2^k = 2^k(2) = 2^{k+1},$$

so $P(k + 1)$ is true. Thus, $P(n)$ is true for all nonnegative integers n.   ▪

EXAMPLE A.3   Let $x \in \mathbb{R}$ with $x > -1$ and $x \neq 0$. Show that $(1 + x)^n > 1 + nx$ for every positive integer $n \geq 2$.

Solution   We let $P(n)$ be the statement

$$(1 + x)^n > 1 + nx. \tag{A.3}$$

(Notice that $P(1)$ is false.) Then $P(2)$ is the statement $(1 + x)^2 > 1 + 2x$. Now $(1 + x)^2 = 1 + 2x + x^2$, and $x^2 > 0$, since $x \neq 0$. Thus, $(1 + x)^2 > 1 + 2x$, so

$P(2)$ is true.

Suppose that $P(k)$ is true, so

$$(1 + x)^k > 1 + kx. \tag{A.4}$$

Now $1 + x > 0$, since $x > -1$. Multiplying both sides of inequality (A.4) by $1 + x$, we obtain

$$(1 + x)^{k+1} > (1 + kx)(1 + x) = 1 + (k + 1)x + kx^2.$$

Since $kx^2 > 0$, we see that $P(k + 1)$ is true. Thus $P(n)$ is true for every positive integer $n \geq 2$. ▪

In a frequently used form of induction known as *complete induction*, the statement

If $P(k)$ is true, then $P(k + 1)$ is true

in the box on page A-1 is replaced by the statement

If $P(m)$ is true for $1 \leq m \leq k$, then $P(k + 1)$ is true.

Again, we are trying to show that $P(k + 1)$ is true, knowing that $P(k)$ is true. But if we have reached the stage of induction where $P(k)$ has been proved, we know that $P(m)$ is true for $1 \leq m \leq k$, so the strengthened hypothesis in the second statement is permissible.

EXAMPLE A.4   Recall that the set of all polynomials with real coefficients is denoted by P. Show that every polynomial in P of degree $n \in Z^+$ either is irreducible itself or is a product of irreducible polynomials in P. (An **irreducible polynomial** is one that cannot be factored into polynomials in P all of lower degree.)

Solution   We will use complete induction. Let $P(n)$ be the statement that is to be proved. Clearly $P(1)$ is true, since a polynomial of degree 1 is already irreducible.

Let k be a positive integer. Our induction hypothesis is then: every polynomial in P of degree less than $k + 1$ either is irreducible or can be factored into irreducible polynomials. Let $f(x)$ be a polynomial of degree $k + 1$. If $f(x)$ is irreducible, we have nothing more to do. Otherwise, we may factor $f(x)$ into polynomials $g(x)$ and $h(x)$ of lower degree than $k + 1$, obtaining $f(x) = g(x)h(x)$. The induction hypothesis indicates that each of $g(x)$ and $h(x)$ can be factored into irreducible polynomials, thus providing such a factorization of $f(x)$. This proves $P(k + 1)$. It follows that $P(n)$ is true for all $n \in Z^+$. ▪

EXERCISES

1. Show that

$$1^2 + 2^2 + 3^2 + \cdots + n^2 = \frac{n(n + 1)(2n + 1)}{6}$$

for $n \in Z^+$.

2. Show that

$$1^3 + 2^3 + 3^3 + \cdots + n^3 = \frac{n^2(n + 1)^2}{4}$$

for $n \in Z^+$.

3. Show that

$$1 + 3 + 5 + \cdots + (2n - 1) = n^2$$

for $n \in Z^+$.

4. Show that

$$\frac{1}{1 \cdot 2} + \frac{1}{2 \cdot 3} + \frac{1}{3 \cdot 4} + \cdots + \frac{1}{n(n + 1)}$$

$$= \frac{n}{n + 1}$$

for $n \in Z^+$.

5. Prove by induction that if $a, r \in \mathbb{R}$ and $r \neq 1$, then

$$a + ar + ar^2 + \cdots + ar^n$$
$$= a(1 - r^{n+})/(1 - r) \qquad \text{for } n \in Z^+.$$

6. Find the flaw in the following argument.

We prove that any two integers i and j in Z^+ are equal. Let

$$\max(i, j) = \begin{cases} i & \text{if} & i \geq j, \\ j & \text{if} & j > i. \end{cases}$$

Let $P(n)$ be the statement

$P(n)$: Whenever $\max(i, j) = n$, then $i = j$.

Notice that, if $P(n)$ is true for all positive integers n, then any two positive integers i and j are equal. We proceed to prove $P(n)$ for positive integers n by induction.

Clearly $P(1)$ is true, since, if $i, j \in Z^+$ and $\max(i, j) = 1$, then $i = j = 1$.

Assume that $P(k)$ is true. Let i and j be such that $\max(i, j) = k + 1$. Then $\max(i - 1, j - 1) = k$, so $i - 1 = j - 1$ by the induction hypothesis. Therefore, $i = j$ and $P(k + 1)$ is true. Consequently, $P(n)$ is true for all n.

7. Criticize the following argument.

Let us show that every positive integer has some interesting property. Let $P(n)$ be the statement that n has an interesting property. We use complete induction.

Of course $P(1)$ is true, since 1 is the only positive integer that equals its own square, which is surely an interesting property of 1.

Suppose that $P(m)$ is true for $1 \leq m \leq k$. If $P(k + 1)$ were not true, then $k + 1$ would be the smallest integer without an interesting property, which would, in itself, be an interesting property of $k + 1$. So $P(k + 1)$ must be true. Thus $P(n)$ is true for all $n \in Z^+$.

8. We have never been able to see any flaw in (a). Try your luck with it, and then answer (b).

a) A murderer is sentenced to be executed. He asks the judge not to let him know the day of the execution. The judge says, "I sentence you to be executed at 10 A.M. some day of this coming January, but I promise that you will not be aware that you are being executed that day until they come to get you at 8 A.M." The crimial goes to his cell and proceeds to prove, as follows, that he can't be executed in January.

Let $P(n)$ be the statement that I can't be executed on January $(31 - n)$. I want to prove $P(n)$ for $0 \leq n \leq 30$. Now I can't be executed on January 31, for since that is the last day of the month and I am to be executed that month, I would know that was the day before 8 A.M., contrary to the judge's sentence. Thus $P(0)$ is true. Suppose that $P(m)$ is true for $0 \leq m \leq k$, where $k \leq 29$. That is, suppose I can't be executed on January $(31 - k)$ through January 31. Then January $(31 - k - 1)$ must be the last possible day for execution, and I would be aware that was the day before 8 A.M., contrary to the judge's sentence. Thus I can't be executed on January $(31 - (k + 1))$, so $P(k + 1)$ is true. Therefore, I can't be executed in January.

(Of course, the criminal was executed on January 17.)

b) An instructor teaches a class 5 days a week, Monday through Friday. She tells her class that she will give one more quiz on one day during the final week of classes, but that the students will not know for sure the quiz will be that day until they come to the classroom. What is the last day of the week on which she can give the quiz in order to satisfy these conditions?

B

TWO DEFERRED PROOFS

PROOF OF THEOREM 3.2 ON EXPANSION BY MINORS

Our demonstration of the various properties of determinants in Section 3.2 depended on our ability to compute a determinant by expanding it by minors on *any row or column*, as stated in Theorem 3.2. In order to prove Theorem 3.2, we will need to look more closely at the form of the terms that appear in an expanded determinant.

Determinants of orders 2 and 3 can be written as

$$\begin{vmatrix} a_{11} & a_{12} \\ a_{21} & a_{22} \end{vmatrix} = (1)(a_{11}a_{22}) + (-1)(a_{12}a_{21})$$

and

$$\begin{vmatrix} a_{11} & a_{12} & a_{13} \\ a_{21} & a_{22} & a_{23} \\ a_{31} & a_{32} & a_{33} \end{vmatrix} = (1)(a_{11}a_{22}a_{33}) + (-1)(a_{11}a_{23}a_{32}) + (1)(a_{12}a_{23}a_{31})$$

$$+ (-1)(a_{12}a_{21}a_{33}) + (1)(a_{13}a_{21}a_{32}) + (-1)(a_{13}a_{22}a_{31}).$$

Notice that each determinant appears as a sum of products, each with an associated sign given by (1) or (−1), which is determined by the formula $(-1)^{1+j}$ as we expand the determinant across the first row. Furthermore, each product contains exactly one factor from each row and exactly one factor from each column of the matrix. That is, the row indices in each product run through all row numbers, and the column indices run through all column numbers. This is an illustration of a general theorem, which we now prove by induction.

THEOREM B.1 Structure of an Expanded Determinant

The determinant of an $n \times n$ matrix $A = [a_{ij}]$ can be expressed as a sum of signed products, where each product contains exactly one factor from each row and exactly one factor from each column. The expansion of det(A) on any row or column also has this form.

PROOF   We consider the expansion of $\det(A)$ on the first row and give a proof by induction. We have just shown that our result is true for determinants of orders 2 and 3. Let $n > 3$, and assume that our result holds for all square matrices of size smaller than $n \times n$. Let A be an $n \times n$ matrix. When we expand $\det(A)$ by minors across the first row, the only expression involving a_{1j} is $(-1)^{1+j}a_{1j}|A_{1j}|$. We apply our induction hypothesis to the determinant $|A_{1j}|$ of order $n - 1$: it is a sum of signed products, each of which has one factor from each row and column of A except for row 1 and column j. As we multiply this sum term by term by a_{1j}, we obtain a sum of products having a_{1j} as factor from row 1 and column j, and one factor from each other row and from each other column. Thus an expression of the stated form is indeed obtained as we expand $\det(A)$ by minors across the first row.

It is clear that essentially the same argument shows that expansion across any row or down any column yields the same type of sum of signed products.   ▲

Our illustration for 2×2 and 3×3 matrices indicates that we might always have the same number of products appearing in $\det(A)$ with a sign given by 1 as with a sign given by -1. This is indeed the case for determinants of order greater than 1, and the easy induction proof is left to the reader.

We now restate and prove Theorem 3.2.

Let $A = [a_{ij}]$ be an $n \times n$ matrix. Then

$$|A| = (-1)^{r+1}a_{r1}|A_{r1}| + (-1)^{r+2}a_{r2}|A_{r2}| + \cdots + (-1)^{r+n}a_{rn}|A_{rn}| \quad \textbf{(B.1)}$$

for any r from 1 to n, and

$$|A| = (-1)^{1+s}a_{1s}|A_{1s}| + (-1)^{2+s}a_{2s}|A_{2s}| + \cdots + (-1)^{n+s}a_{ns}|A_{ns}| \quad \textbf{(B.2)}$$

for any s from 1 to n.

PROOF OF THEOREM 3.2   We first prove Eq. (B.1) for any choice of r from 1 to n. Clearly, Eq. (B.1) holds for $n = 1$ and $n = 2$. Proceeding by induction, let $n > 2$ and assume that determinants of order less than n can be computed by using an expansion on *any row*. Let A be an $n \times n$ matrix. We show that expansion of det (A) by minors on row r is the same as expansion on row i for $i < r$. From Theorem B.1, we know that each of the expansions gives a sum of signed products, where each product contains a single factor from each row and from each column of A. We will compare the products containing as factors both a_{ij} and a_{rs} in each of the expansions. We consider two cases, as illustrated in Figures B.1 and B.2.

If $\det(A)$ is expanded on the ith row, the sum of signed products containing $a_{ij}a_{rs}$ is part of $(-1)^{i+j}a_{ij}|A_{ij}|$. In computing $|A_{ij}|$ we may, by our induction assumption, expand on the rth row. For $j < s$, terms of $|A_{ij}|$ involving a_{rs} are then $(-1)^{(r-1)+(s-1)}d$, where d is the determinant of the matrix obtained from A by crossing out rows i and r and columns j and s, as shown in Figure B.1. The exponent $(r - 1) + (s - 1)$ occurs because a_{rs} is in row $r - 1$ and column $s - 1$ of A_{ij}. Thus, the part of our expansion of $\det(A)$ across the ith row that contains $a_{ij}a_{rs}$ is equal to

FIGURE B.1 **The case $j < s$.**

FIGURE B.2 **The case $s < j$.**

$$(-1)^{i+j}(-1)^{(r-1)+(s-1)}a_{ij}a_{rs}d \qquad \text{for} \qquad j < s. \tag{B.3}$$

For $j > s$, we consult Figure B.2 and use similar reasoning to see that the part of our expansion of $\det(A)$ across the ith row, which contains $a_{ij}a_{rs}$, is equal to

$$(-1)^{i+j}(-1)^{(r-1)+s}a_{ij}a_{rs}d \qquad \text{for} \qquad j > s. \tag{B.4}$$

We now expand $\det(A)$ by minors on the rth row, obtaining $(-1)^{r+s}a_{rs}|A_{rs}|$ as the portion involving a_{rs}. Expanding $|A_{rs}|$ on the ith row, using our induction assumption, we obtain $(-1)^{i+j}a_{ij}d$ if $j < s$ and $(-1)^{i+(j-1)}a_{ij}d$ if $j > s$. Thus the part of the expansion of $\det(A)$ on the rth row, which contains $a_{ij}a_{rs}$, is equal to

$$(-1)^{r+s}(-1)^{i+j}a_{rs}a_{ij}d \qquad \text{for} \qquad j < s \tag{B.5}$$

or

$$(-1)^{r+s}(-1)^{i+(j-1)}a_{rs}a_{ij}d \qquad \text{for} \qquad j > s. \tag{B.6}$$

Expressions (B.3) and (B.5) are equal, since $(-1)^{r+s+i+j} = (-1)^{r+s+i+j-2}$, and expressions (B.4) and (B.6) are equal, since $(-1)^{r+s+i+j-1}$ is the algebraic sign of each. This concludes the proof that the expansions of $\det(A)$ by minors across rows i and r are equal.

A similar argument shows that expansions of $\det(A)$ down columns j and s are the same.

Finally, we must show that an expansion of $\det(A)$ on a row is equal to expansion on a column. It is sufficient for us to prove that the expansion of $\det(A)$ on the first row is the same as the expansion on the first column, in view of what we have proved above. Again, we use induction and dispose of the cases $n = 1$ and $n = 2$ as trivial to check. Let $n > 2$, and assume that our result holds for matrices of size smaller than $n \times n$. Let A be an $n \times n$ matrix. Expanding $\det(A)$ on the first row yields

$$a_{11}|A_{11}| + \sum_{j=2}^{n} (-1)^{1+j}a_{1j}|A_{1j}|.$$

For $j > 1$, we expand $|A_{1j}|$ on the first column, using our induction assumption, and obtain $|A_{1j}| = \sum_{i=2}^{n}(-1)^{(i-1)+1}a_{i1}d$, where d is the determinant of the matrix obtained from A by crossing out rows 1 and i and columns 1 and j. Thus the terms in the expansion of $\det(A)$ containing $a_{1j}a_{i1}$ are

$$(-1)^{1+j+i}a_{1j}a_{i1}d. \tag{B.7}$$

On the other hand, if we expand $\det(A)$ on the first column, we obtain

$$a_{11}|A_{11}| + \sum_{i=2}^{n} (-1)^{i+1}a_{i1}|A_{i1}|.$$

For $i > 1$, expanding on the first row, using our induction assumption, shows that $|A_{i1}| = \Sigma_{j=2}^{n}(-1)^{1+(j-1)}a_{1j}d$. This results in

$$(-1)^{i+1+j}a_{i1}a_{1j}d$$

as the part of the expansion of $\det(A)$ containing the sum $a_{i1}a_{1j}$, and this agrees with the expression in formula (B.7). This concludes our proof. ▲

PROOF OF THEOREM 3.5 ON THE VOLUME OF AN n-BOX IN $\mathbb{R}^m$

Theorem 3.5 of Section 3.3 asserts that the volume of an n-box in $\mathbb{R}^m$ determined by vectors $\mathbf{a}_1, \mathbf{a}_2, \ldots, \mathbf{a}_n$ can be computed as

$$\text{Volume} = \sqrt{\det(A^T A)}, \tag{B.8}$$

where A is the $m \times n$ matrix having $\mathbf{a}_1, \mathbf{a}_2, \ldots, \mathbf{a}_n$ as column vectors. Before proving this result, we must first give a proper definition of the *volume* of such a box. The definition proceeds inductively. That is, we define the volume of a 1-box directly, and then we define the volume of an n-box in terms of the volume of an $(n-1)$-box. Our definition of volume is a natural one, essentially taking the product of the measure of the base of the box and the altitude of the box. We will think of the base of the n-box determined by $\mathbf{a}_1, \mathbf{a}_2, \ldots, \mathbf{a}_n$ as an $(n-1)$-box determined by some $n-1$ of the vectors $\mathbf{a}_i$. In general, such a base can be selected in several different ways. We find it convenient to work with boxes determined by *ordered* sequences of vectors. We will choose one special base for the box and give a definition of its volume in terms of this order of the vectors. Once we obtain the expression $\det(A^T A)$ in Eq. (B.8) for the square of the volume, we can show that the volume does not change if the order of the vectors in the sequence is changed. We will show that $\det(A^T A)$ remains unchanged if the order of the columns of A is changed.

Observe that, if an n-box is determined by $\mathbf{a}_1, \mathbf{a}_2, \ldots, \mathbf{a}_n$, then $\mathbf{a}_1$ can be uniquely expressed in the form

$$\mathbf{a}_1 = \mathbf{b} + \mathbf{p}, \tag{B.9}$$

where $\mathbf{p}$ is the projection of $\mathbf{a}_1$ on $W = \text{sp}(\mathbf{a}_2, \ldots, \mathbf{a}_n)$ and $\mathbf{b} = \mathbf{a}_1 - \mathbf{p}$ is orthogonal to W. This follows from Theorem 5.1 and is illustrated in Figure B.3.

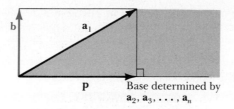

FIGURE B.3   The altitude vector b perpendicular to the base box.

DEFINITION B.1   Volume of an *n*-box

The **volume** of the 1-box determined by a nonzero vector $\mathbf{a}_1$ in $\mathbb{R}^m$ is $\|\mathbf{a}_1\|$. Let $\mathbf{a}_1, \mathbf{a}_2, \ldots, \mathbf{a}_n$ be an ordered sequence of n independent vectors, and suppose that the volume of an r-box determined by an ordered sequence of r independent vectors has been defined for $r < n$. The **volume of the *n*-box** in $\mathbb{R}^m$ determined by the ordered sequence of $\mathbf{a}_i$ is the product of the volume of the "base" determined by the ordered sequence $\mathbf{a}_2, \ldots, \mathbf{a}_n$ and the length of the vector $\mathbf{b}$ given in Eq. (B.9). That is,

$$\text{Volume} = (\text{Altitude } \|\mathbf{b}\|)(\text{Volume of the base}).$$

As a first step in finding a formula for the volume of an n-box, we establish a preliminary result on determinants.

THEOREM B.2   Property of det(*A^T A*)

Let $\mathbf{a}_1, \mathbf{a}_2, \ldots, \mathbf{a}_n$ be vectors in $\mathbb{R}^m$, and let A be the $m \times n$ matrix with jth column vector $\mathbf{a}_j$. Let B be the $m \times n$ matrix obtained from A by replacing the first column of A by the vector

$$\mathbf{b} = \mathbf{a}_1 - r_2\mathbf{a}_2 - \cdots - r_n\mathbf{a}_n$$

for scalars $r_2, \ldots, r_n$. Then

$$\det(A^TA) = \det(B^TB). \tag{B.10}$$

PROOF   The matrix B can be obtained from the matrix A by a sequence of $n - 1$ elementary *column-addition* operations. Each of the elementary column operations can be performed on A by multiplying A on the *right* by an elementary matrix formed by executing the same elementary column-addition operation on the $n \times n$ identity matrix I. Each elementary column-addition matrix therefore has the same determinant 1 as the identity matrix I. Their product is an $n \times n$ matrix E such that $B = AE$, and $\det(E) = 1$. Using properties of determinants and the transpose operation, we have

$$\det(B^TB) = \det((AE)^T(AE)) = \det(E^T(A^TA)E)$$
$$= 1 \cdot \det(A^TA) \cdot 1 = \det(A^TA). \quad \blacktriangle$$

We can now prove our volume formula in Theorem 3.5.

The volume of the n-box in $\mathbb{R}^m$ determined by the ordered sequence $\mathbf{a}_1, \mathbf{a}_2, \ldots, \mathbf{a}_n$ of n independent vectors is given by

$$\text{Volume} = \sqrt{\det(A^T A)},$$

where A is the $m \times n$ matrix with $\mathbf{a}_j$ as jth column vector.

PROOF OF THEOREM 3.5 Since our volume was defined inductively, we give an inductive proof. The theorem is valid if $n = 1$ or 2, by Eqs. (1) and (2), respectively, in Section 3.3. Let $n > 2$, and suppose that the theorem is proved for all k-boxes for $k \leq n - 1$. If we write $\mathbf{a}_1 = \mathbf{b} + \mathbf{p}$, as in Eq. (B.9), then, since $\mathbf{p}$ lies in $\text{sp}(\mathbf{a}_2, \ldots, \mathbf{a}_n)$, we have

$$\mathbf{p} = r_2 \mathbf{a}_2 + \cdots + r_n \mathbf{a}_n$$

for some scalars $r_2, \ldots, r_n$, so

$$\mathbf{b} = \mathbf{a}_1 - \mathbf{p} = \mathbf{a}_1 - r_2 \mathbf{a}_2 - \cdots - r_n \mathbf{a}_n.$$

Let B be the matrix obtained from A by replacing the first column vector $\mathbf{a}_1$ of A by the vector $\mathbf{b}$, as in the preceding theorem. Since $\mathbf{b}$ is orthogonal to each of the vectors $\mathbf{a}_2, \ldots, \mathbf{a}_n$, which determine the base of our box, we obtain

$$B^T B = \begin{bmatrix} \mathbf{b} \cdot \mathbf{b} & 0 & \cdots & 0 \\ 0 & \mathbf{a}_2 \cdot \mathbf{a}_2 & \cdots & \mathbf{a}_2 \cdot \mathbf{a}_n \\ \vdots & \vdots & & \vdots \\ 0 & \mathbf{a}_n \cdot \mathbf{a}_2 & \cdots & \mathbf{a}_n \cdot \mathbf{a}_n \end{bmatrix}. \tag{B.11}$$

From Eq. (B.11), we see that

$$\det(B^T B) = \|\mathbf{b}\|^2 \begin{vmatrix} \mathbf{a}_2 \cdot \mathbf{a}_2 & \cdots & \mathbf{a}_2 \cdot \mathbf{a}_n \\ \vdots & & \vdots \\ \mathbf{a}_n \cdot \mathbf{a}_2 & \cdots & \mathbf{a}_n \cdot \mathbf{a}_n \end{vmatrix}.$$

By our induction assumption, the square of the volume of the $(n - 1)$-box in $\mathbb{R}^m$ determined by the ordered sequence $\mathbf{a}_2, \ldots, \mathbf{a}_n$ is

$$\begin{vmatrix} \mathbf{a}_2 \cdot \mathbf{a}_2 & \cdots & \mathbf{a}_2 \cdot \mathbf{a}_n \\ \vdots & & \vdots \\ \mathbf{a}_n \cdot \mathbf{a}_2 & \cdots & \mathbf{a}_n \cdot \mathbf{a}_n \end{vmatrix}.$$

Applying Eq. (B.10) of Theorem B.2, we obtain

$$\det(A^T A) = \det(B^T B) = \|\mathbf{b}\|^2 (\text{Volume of the base})^2 = (\text{Volume})^2.$$

This proves our theorem. ▲

COROLLARY   Independence of Order

The volume of a box determined by independent vectors $\mathbf{a}_1$, $\mathbf{a}_2$, . . . , $\mathbf{a}_n$ and defined in Definition B.1 is independent of the order of the vectors; in particular, the volume is independent of a choice of base for the box.

PROOF   A rearrangement of the sequence $\mathbf{a}_1$, $\mathbf{a}_2$, . . . , $\mathbf{a}_n$ of vectors corresponds to the same rearrangement of the columns of matrix A. Such a rearrangement of the columns of A can be accomplished by multiplying A on the right by a product of $n \times n$ elementary *column-interchange* matrices, all having determinant 1. As in the proof of Theorem B.2, we see that, for the resulting matrix $B = AE$, we have $\det(B^T B) = \det(A^T A)$.   ▲

C AVAILABLE SOFTWARE

Following is a list of programs that make up the software available with this text, including a brief description of each.

LINPROG   A graphic program allowing the user to estimate solutions of two-variable linear programming problems. For use with Section 1.7.

MATCOMP   Performs matrix computations, solves linear systems, and finds real eigenvalues and eigenvectors. For use throughout the course.

YUREDUCE   Enables the user to select items from a menu for step-by-step row reduction of a matrix. For Sections 1.2–1.4 and Sections 8.2 and 9.3.

TIMING   Times operations in interpretive BASIC or compiled BASIC; also times various methods for solving linear systems. For Sections 9.1 and 9.2.

LUFACTOR   Gives the factorization $A = LU$, which can then be used to solve $A\mathbf{x} = \mathbf{b}$. For Section 9.2.

HILBERT   Enables the user to experiment with ill-conditioned Hilbert matrices. For Section 9.3.

VECTGRPH   Gives graded quizzes on vector geometry based on displayed graphics. Useful from Section 2.1 on.

EBYMTIME   Gives a lower bound for the time required to find the determinant of a matrix using only repeated expansion by minors. For Section 3.2.

ALLROOTS   Provides step-by-step execution of Newton's method to find both real and complex roots of a polynomial with real or complex coefficients; may be used in conjunction with MATCOMP to find complex as well as real eigenvalues of a small matrix. For Section 4.1.

YOUFIT   The user can experiment with a graphic to try to find the least-squares linear, quadratic, or exponential fit of data. The computer can then be asked to find the best fit. For Section 5.5.

QRFACTOR   Executes the Gram–Schmidt process, gives the QR factorization of a suitable matrix A, and can be used to find least-squares solutions. For Section 5.2. Also step-by-step computation by the QR algorithm of both real and complex eigenvalues of a matrix. The user specifies shifts and when to decouple. For Section 7.4

POWER   Executes, one step at a time, the power method (with deflation) for finding eigenvalues and eigenvectors. For Section 7.4.

JACOBI   Executes, one "rotation" at a time, the method of Jacobi for diagonalizing a symmetric matrix. For Section 7.4.

SIMPLEX   Allows the user to execute the simplex method by making choices from a menu of possible operations to be performed. A graded quiz option is included. Alternatively, the computer may be asked to simply execute the method and display the tableaux. For Chapter 10.

MATFILES   The user can create files of matrices that can be accessed from the programs MATCOMP, YUREDUCE, LUFACTOR, QRFACTOR, POWER, JACOBI, and SIMPLEX. Existing files can be viewed.

ANSWERS TO SOME ODD-NUMBERED EXERCISES

CHAPTER 1

Section 1.1

1. $\begin{bmatrix} -6 & 3 & 9 \\ 12 & 0 & -3 \end{bmatrix}$ **3.** $\begin{bmatrix} 2 & 2 & 1 \\ 9 & -1 & 2 \end{bmatrix}$ **5.** $\begin{bmatrix} 6 & -3 \\ -3 & 1 \\ -2 & 5 \end{bmatrix}$ **7.** Impossible **9.** $\begin{bmatrix} -130 & 140 \\ 110 & -60 \end{bmatrix}$

11. Impossible **13.** $\begin{bmatrix} 27 & -35 \\ -4 & 26 \end{bmatrix}$ **15.** $\begin{bmatrix} 20 & -2 & -10 \\ -2 & 1 & 3 \\ -10 & 3 & 10 \end{bmatrix}$ **17.** **(a)** $\begin{bmatrix} 4 & 0 & 0 \\ 0 & 1 & 0 \\ 0 & 0 & 1 \end{bmatrix}$;

(b) $\begin{bmatrix} 128 & 0 & 0 \\ 0 & -1 & 0 \\ 0 & 0 & 1 \end{bmatrix}$ **19.** $xy = [-14]$, $yx = \begin{bmatrix} -8 & 12 & -4 \\ 2 & -3 & 1 \\ -6 & 9 & -3 \end{bmatrix}$ **21.** T F F T F T T T F T

23. Let **u** have components $u_1, u_2, \ldots, u_n$, and let **v** have components $v_1, v_2, \ldots, v_n$. Then

$$\mathbf{u} \cdot \mathbf{v} = u_1 v_1 + u_2 v_2 + \cdots + u_n v_n = v_1 u_1 + v_2 u_2 + \cdots + v_n u_n = \mathbf{v} \cdot \mathbf{u}.$$

27. The (i, j)th entry in $(r + s)A$ is $(r + s)a_{ij} = ra_{ij} + sa_{ij}$, which is the (i, j)th entry in $rA + sA$. Since $(r + s)A$ and $rA + sA$ have the same size, they are equal.

33. Let $A = [a_{ij}]$ be an $m \times n$ matrix, $B = [b_{jk}]$ be an $n \times r$ matrix, and $C = [c_{kq}]$ be an $r \times s$ matrix. Then the ith row of AB is

$$\left[\sum_{j=1}^{n} a_{ij} b_{j1}, \sum_{j=1}^{n} a_{ij} b_{j2}, \ldots, \sum_{j=1}^{n} a_{ij} b_{jr} \right],$$

so the (i, q)th entry in $(AB)C$ is

$$\left(\sum_{j=1}^{n} a_{ij} b_{j1} \right) c_{1q} + \left(\sum_{j=1}^{n} a_{ij} b_{j2} \right) c_{2q} + \cdots + \left(\sum_{j=1}^{n} a_{ij} b_{jr} \right) c_{rq} = \sum_{k=1}^{r} \left[\sum_{j=1}^{n} (a_{ij} b_{jk} c_{kq}) \right].$$

Further, the qth column of BC has components

$$\sum_{k=1}^{r} b_{1k} c_{kq}, \sum_{k=1}^{r} b_{2k} c_{kq}, \ldots, \sum_{k=1}^{r} b_{nk} c_{kq},$$

so the (i, q)th entry in $A(BC)$ is

$$a_{i1}\left(\sum_{k=1}^{r} b_{1k}c_{kq}\right) + a_{i2}\left(\sum_{k=1}^{r} b_{2k}c_{kq}\right) + \cdots + a_{in}\left(\sum_{k=1}^{r} b_{nk}c_{kq}\right)$$

$$= \sum_{j=1}^{n} a_{ij}\left(\sum_{k=1}^{r} b_{jk}c_{kq}\right) = \sum_{j=1}^{n}\left(\sum_{k=1}^{r} (a_{ij}b_{jk}c_{kq})\right) = \sum_{k=1}^{r}\left(\sum_{j=1}^{n} (a_{ij}b_{jk}c_{kq})\right).$$

Since $(AB)C$ and $A(BC)$ are both $m \times s$ matrices with the same (i, q)th entry, they must be equal.

35. (a) $n \times m$; (b) $n \times n$; (c) $m \times m$ **39.** Since $(AA^T)^T = (A^T)^T A^T = AA^T$, we see that AA^T is symmetric.

41. (a) The jth entry in column vector Ae_j is $[a_{i1} \quad a_{i2} \quad \cdots \quad a_{in}] \cdot e_j = a_{ij}$. Therefore, Ae_j is the jth column vector of A. (b) (i) We have $Ae_j = 0$ for each $j = 1, 2, \ldots, n$; so, by part (a), the jth column of A is the zero vector for each j. That is, $A = O$. (ii) $Ax = Bx$ for each x

if and only if $(A - B)x = 0$ for each x,

if and only if $A - B = O$ by part (i),

if and only if $A = B$.

45. $\begin{bmatrix} 588 & -398 & -266 & 191 \\ 646 & 1038 & -55 & 290 \\ -581 & 2175 & 953 & -502 \\ -676 & 2762 & 2150 & -7 \\ 248 & 1424 & -602 & 105 \end{bmatrix}$ **47.** Undefined

49. $\begin{bmatrix} 36499 & 30055 & -6890 & 10010 & 50800 \\ 30055 & 32653 & -11723 & 3622 & 44680 \\ -6890 & -11723 & 17179 & 14629 & -15569 \\ 10010 & 3622 & 14629 & 22237 & 10107 \\ 50800 & 44680 & -15569 & 10107 & 75753 \end{bmatrix}$

51. $\begin{bmatrix} 26918 & -3837 & -13284 & 14372 & -6772 \\ 41446 & 353741 & 209672 & 310468 & 194032 \\ 934 & 34811 & 44048 & 53254 & 27054 \\ 20416 & 82936 & 44566 & 82892 & 52440 \\ -2966 & -64951 & -27916 & -52777 & -9719 \end{bmatrix}$

Section 1.2

1. (a) $\begin{bmatrix} 1 & 3 & 2 \\ 0 & 5 & 0 \\ 0 & 0 & 0 \end{bmatrix}$; (b) $\begin{bmatrix} 1 & 0 & 2 \\ 0 & 1 & 0 \\ 0 & 0 & 0 \end{bmatrix}$ **3.** (a) $\begin{bmatrix} -1 & 1 & 2 & 0 \\ 0 & 2 & -1 & 3 \\ 0 & 0 & 10 & 0 \\ 0 & 0 & 0 & 0 \end{bmatrix}$; (b) $\begin{bmatrix} 1 & 0 & 0 & 0 \\ 0 & 1 & 0 & 3 \\ 0 & 0 & 1 & 0 \\ 0 & 0 & 0 & 0 \end{bmatrix}$

5. (a) $\begin{bmatrix} 1 & -3 & 0 & 0 & -1 \\ 0 & 0 & 1 & 3 & -4 \\ 0 & 0 & 0 & 1 & 3 \\ 0 & 0 & 0 & 0 & 16 \end{bmatrix}$; (b) $\begin{bmatrix} 1 & -3 & 0 & 0 & 0 \\ 0 & 0 & 1 & 0 & 0 \\ 0 & 0 & 0 & 1 & 0 \\ 0 & 0 & 0 & 0 & 1 \end{bmatrix}$ **7.** $x = \begin{bmatrix} 5 - 6r \\ 2 - 4r \\ r \end{bmatrix}, \begin{bmatrix} -7 \\ -6 \\ 2 \end{bmatrix}$

9. $x = \begin{bmatrix} 1 - 2r \\ -2 - r - 3s \\ r \\ s \end{bmatrix}, \begin{bmatrix} -5 \\ 1 \\ 3 \\ -2 \end{bmatrix}$ **11.** $x = \begin{bmatrix} 0 \\ 2 \\ -5 \\ 2 \end{bmatrix}$ **13.** $x = 2, y = -4$ **15.** $x = -3, y = 2, z = 4$

17. $\mathbf{x} = \begin{bmatrix} 5 \\ 1 \end{bmatrix}$ **19.** Inconsistent **21.** $\mathbf{x} = \begin{bmatrix} -2 \\ 1 \end{bmatrix}$ **23.** $\mathbf{x} = \begin{bmatrix} -8 \\ -23-5s \\ -7+s \\ 2s \end{bmatrix}$ **25.** $\mathbf{x} = \begin{bmatrix} 2 \\ 0 \\ -1 \end{bmatrix}$

27. $\mathbf{x} = \begin{bmatrix} 3 \\ 1 \\ -1 \\ 2 \end{bmatrix}$ **29.** F F T T F T T T F T **31.** $x_1 = -1, x_2 = 3$ **33.** $x_1 = 2, x_2 = -3$

35. $x_1 = 1, x_2 = -1, x_3 = 1, x_4 = 2$ **37.** $x_1 = -3, x_2 = 5, x_3 = 2, x_4 = -3$ **39.** The system is consistent for all values of b_1 and b_2. **41.** The system is consistent if $b_3 + b_2 - b_1 = 0$. **43.** These products are equal for all values of r. **45.** $a = 1, b = -2, c = 1, d = 2$ **47.** $\mathbf{x} = \begin{bmatrix} -1 \\ 7 \end{bmatrix}$

49. $\mathbf{x} \approx \begin{bmatrix} -1.285714 \\ 3.142857 \\ 1.285714 \end{bmatrix}$ **51.** $\mathbf{x} = \begin{bmatrix} 1 \\ 1 \\ -1 \\ 1 \\ 2 \end{bmatrix}$ **53.** $\begin{bmatrix} 1 & 2 & 0 & 0 & 3 & -6 \\ 0 & 0 & 1 & 0 & 1 & -2 \\ 0 & 0 & 0 & 1 & -1 & 3 \end{bmatrix}$ **55.** $\mathbf{x} = \begin{bmatrix} 1 \\ 3 \\ 0 \end{bmatrix}$

Section 1.3

1. $\begin{bmatrix} x \\ y \end{bmatrix} = \begin{bmatrix} -1 \\ 1 \end{bmatrix}, \begin{bmatrix} x \\ y \end{bmatrix} = \begin{bmatrix} 2 \\ -5 \end{bmatrix}, \begin{bmatrix} x \\ y \end{bmatrix} = \begin{bmatrix} 3 \\ 0 \end{bmatrix}$ **3.** $\mathbf{x} = \begin{bmatrix} -3 \\ -5 \\ 2 \end{bmatrix}, \mathbf{x} = \begin{bmatrix} 0 \\ 0 \\ 1 \end{bmatrix}, \mathbf{x} = \begin{bmatrix} 2 \\ -1 \\ 3 \end{bmatrix}$

5. $\mathbf{x} = \begin{bmatrix} 7 + 2t - r - 5s \\ t \\ r \\ s \end{bmatrix} = \begin{bmatrix} 7 \\ 0 \\ 0 \\ 0 \end{bmatrix} + \begin{bmatrix} 2t - r - 5s \\ t \\ r \\ s \end{bmatrix}$ **7.** $\mathbf{x} = \begin{bmatrix} 16/5 \\ -2/5 \\ 0 \\ 0 \end{bmatrix} + \begin{bmatrix} s/30 \\ 13s/30 \\ -s/6 \\ s \end{bmatrix}$ **9.** 2

11. $E = \begin{bmatrix} 1 & 0 & 0 \\ 0 & 1 & 0 \\ -3 & 0 & 1 \end{bmatrix}$ **13.** $C = \begin{bmatrix} 1 & 0 & 0 \\ -3 & 1 & 0 \\ -4 & 0 & 1 \end{bmatrix}$ **15.** $\begin{bmatrix} 0 & 1 & 0 & 0 \\ 1 & 0 & 0 & 0 \\ 0 & 0 & 1 & 0 \\ 0 & 0 & 0 & 1 \end{bmatrix}$ **17.** $\begin{bmatrix} 5 & 0 & 0 & 0 \\ 0 & 0 & 1 & 0 \\ 0 & 1 & 0 & 0 \\ 0 & 2 & 0 & 1 \end{bmatrix}$

19. $\begin{bmatrix} 0 & 6 & 0 & 1 \\ -2 & 1 & 0 & 0 \\ 0 & -18 & 1 & -3 \\ 1 & 0 & 0 & 0 \end{bmatrix}$ **21.** F T T T F F T T T F **23.** In a system with more equations than unknowns, we naively expect that too many conditions are demanded of the unknowns to permit any solution; and in practice, this is usually the case (see Section 5.5). If the system does have a solution, some of the equations may be deleted to produce a square system with the same solution. **25.** $\begin{bmatrix} 0 & 1 & 0 \\ 0 & 0 & 1 \end{bmatrix} +$

$\begin{bmatrix} 0 & -1 & 0 \\ 0 & 0 & 1 \end{bmatrix} = \begin{bmatrix} 0 & 0 & 0 \\ 0 & 0 & 2 \end{bmatrix}$ is not in echelon form, even though the first two matrices are in echelon form.

35. $\mathbf{x} = \begin{bmatrix} 13 \\ -1 - 2r - 6t \\ 10 - 4r - 5s + 2t \\ r \\ s \\ 7 - 3t \\ t \end{bmatrix}$; the second system is inconsistent.

Section 1.4

1. (a) $A^{-1} = \begin{bmatrix} 1 & -1 \\ 0 & 1 \end{bmatrix}$; (b) The matrix A itself is an elementary matrix. **3.** Not invertible

5. (a) $A^{-1} = \begin{bmatrix} 1 & 0 & 1 \\ 0 & 1 & 1 \\ 0 & 0 & -1 \end{bmatrix}$; (b) $A = \begin{bmatrix} 1 & 0 & 0 \\ 0 & 1 & 0 \\ 0 & 0 & -1 \end{bmatrix}\begin{bmatrix} 1 & 0 & 0 \\ 0 & 1 & 1 \\ 0 & 0 & 1 \end{bmatrix}\begin{bmatrix} 1 & 0 & 1 \\ 0 & 1 & 0 \\ 0 & 0 & 1 \end{bmatrix}$ (Other answers are possible.)

7. (a) $A^{-1} = \begin{bmatrix} -7 & 5 & 3 \\ 3 & -2 & -2 \\ 3 & -2 & -1 \end{bmatrix}$;

(b) $A = \begin{bmatrix} 2 & 0 & 0 \\ 0 & 1 & 0 \\ 0 & 0 & 1 \end{bmatrix}\begin{bmatrix} 1 & 0 & 0 \\ 3 & 1 & 0 \\ 0 & 0 & 1 \end{bmatrix}\begin{bmatrix} 1 & 0 & 0 \\ 0 & \frac{1}{2} & 0 \\ 0 & 0 & 1 \end{bmatrix}\begin{bmatrix} 1 & \frac{1}{2} & 0 \\ 0 & 1 & 0 \\ 0 & 0 & 1 \end{bmatrix}\begin{bmatrix} 1 & 0 & 0 \\ 0 & 1 & 0 \\ 0 & -1 & 1 \end{bmatrix}\begin{bmatrix} 1 & 0 & 0 \\ 0 & 1 & 0 \\ 0 & 0 & -1 \end{bmatrix}\begin{bmatrix} 1 & 0 & 3 \\ 0 & 1 & 0 \\ 0 & 0 & 1 \end{bmatrix}\begin{bmatrix} 1 & 0 & 0 \\ 0 & 1 & -2 \\ 0 & 0 & 1 \end{bmatrix}$

9. $\begin{bmatrix} -1 & 0 & -1 & -1 \\ -3 & -1 & 0 & -1 \\ 5 & 0 & 4 & 3 \\ 3 & 0 & 3 & 2 \end{bmatrix}$ **11.** $\begin{bmatrix} 1 & 0 & 0 & 0 & 0 & 0 \\ 0 & -1 & 0 & 0 & 0 & 0 \\ 0 & 0 & \frac{1}{2} & 0 & 0 & 0 \\ 0 & 0 & 0 & \frac{1}{3} & 0 & 0 \\ 0 & 0 & 0 & 0 & \frac{1}{4} & 0 \\ 0 & 0 & 0 & 0 & 0 & \frac{1}{5} \end{bmatrix}$ **13.** (a) $A^{-1} - \begin{bmatrix} -7 & 3 \\ -5 & 2 \end{bmatrix}$;

(b) $\begin{bmatrix} x_1 \\ x_2 \end{bmatrix} = \begin{bmatrix} -37 \\ -26 \end{bmatrix}$ **15.** $x = -7r + 5s + 3t$, $y = 3r - 2s - 2t$, $z = 3r - 2s - t$

17. $\begin{bmatrix} 46 & 33 & 30 \\ 39 & 29 & 26 \\ 99 & 68 & 63 \end{bmatrix}$ **19.** $\begin{bmatrix} 3 & 2 & 1 \\ 0 & 3 & 2 \\ 1 & 3 & 4 \end{bmatrix}$ **21.** The matrix is invertible for any value of r except $r = 0$.

23. T T T F T T T F T F **25.** (a) No; (b) Yes

27. (a) Notice that $A(A^{-1}B) = (AA^{-1})B = IB = B$, so $X = A^{-1}B$ is a solution. To show uniqueness, suppose that $AX = B$. Then $A^{-1}(AX) = A^{-1}B$, $(A^{-1}A)X = A^{-1}B$, $IX = A^{-1}B$, and $X = A^{-1}B$; therefore, this is the only solution. (b) Let $E_1, E_2, \ldots, E_k$ be elementary matrices that reduce $[A \mid B]$ to $[I \mid X]$, and let $C = E_k E_{k-1} \cdots E_2 E_1$. Then $CA = I$ and $CB = X$. Thus, $C = A^{-1}$ and so $X = A^{-1}B$.

29. (a) $\begin{bmatrix} 0 & 0 \\ 1 & 0 \end{bmatrix}$ **39.** $\begin{bmatrix} 0.355 & -0.0645 & 0.161 \\ -0.129 & 0.387 & 0.0323 \\ -0.0968 & 0.290 & -0.226 \end{bmatrix}$ **41.** $\begin{bmatrix} 0.0275 & -0.0296 & -0.0269 & 0.0263 \\ 0.168 & -0.0947 & 0.0462 & -0.0757 \\ 0.395 & -0.138 & -0.00769 & -0.0771 \\ -0.0180 & 0.0947 & 0.00385 & 0.0257 \end{bmatrix}$

43. (See Exercise 9.) **45.** (See Exercise 41.)

47. $\begin{bmatrix} 0.291 & 0.199 & 0.0419 & -0.00828 & -0.272 \\ -0.0564 & 0.159 & 0.148 & -0.0737 & -0.0699 \\ 0.0276 & 0.145 & -0.00841 & -0.0302 & -0.0250 \\ -0.0467 & 0.122 & -0.029 & 0.133 & -0.00841 \\ 0.0116 & -0.128 & -0.0470 & -0.0417 & 0.178 \end{bmatrix}$

Section 1.5

1. $(24, -34)$ **3.** $(2, 7, -15)$ **5.** $(4, 2, 4)$ **7.** $\begin{bmatrix} 1 & 1 \\ 1 & -3 \end{bmatrix}$ **9.** $\begin{bmatrix} 1 & 1 & 1 \\ 1 & 1 & 0 \\ 1 & 0 & 0 \end{bmatrix}$ **11.** $\begin{bmatrix} 1 & -1 & 3 \\ 1 & 1 & 1 \\ 1 & 0 & 0 \end{bmatrix}$

13. Yes, because it can be written in the form $T(\mathbf{x}) = I\mathbf{x}$, where I is the 3×3 identity matrix.   **15.** No, because it does not satisfy Definition 1.10. For example, $T(2, 2, 2) = (1, 1, 1, 1) \neq (2, 2, 2, 2) = 2T(1, 1, 1)$.   **17.** The matrix associated with $T' \circ T$ is $\begin{bmatrix} 2 & 0 \\ 3 & 1 \end{bmatrix}$; $(T' \circ T)(x, y) = (2x, 3x + y)$.

19. Yes; $T^{-1}(x, y) = \left(\dfrac{3x + y}{4}, \dfrac{x - y}{4} \right)$.   **21.** Yes; $T^{-1}(x, y, z) = (z, y - z, x - y)$.

23. Yes; $T^{-1}(x, y, z) = \left(z, \dfrac{-x + 3y - 2z}{4}, \dfrac{x + y - 2z}{4} \right)$.   **25.** Yes: $(T' \circ T)^{-1}(x, y) = \left(\dfrac{x}{2}, \dfrac{-3x + 2y}{2} \right)$.

27. T F T T T F T T F T

37. In column vector notation, we have $T\left(\begin{bmatrix} x \\ y \end{bmatrix} \right) = \begin{bmatrix} -y \\ x \end{bmatrix} = \begin{bmatrix} 0 & -1 \\ 1 & 0 \end{bmatrix} \begin{bmatrix} x \\ y \end{bmatrix} = \begin{bmatrix} -1 & 0 \\ 0 & 1 \end{bmatrix} \begin{bmatrix} 0 & 1 \\ 1 & 0 \end{bmatrix} \begin{bmatrix} x \\ y \end{bmatrix}$, which represents a reflection through the line $y = x$ followed by a reflection through the y-axis.   **39.** In column vector notation, we have $T\left(\begin{bmatrix} x \\ y \end{bmatrix} \right) = \begin{bmatrix} -x \\ -y \end{bmatrix} = \begin{bmatrix} -1 & 0 \\ 0 & -1 \end{bmatrix} \begin{bmatrix} x \\ y \end{bmatrix} = \begin{bmatrix} -1 & 0 \\ 0 & 1 \end{bmatrix} \begin{bmatrix} 1 & 0 \\ 0 & -1 \end{bmatrix} \begin{bmatrix} x \\ y \end{bmatrix}$, which represents a reflection through the x-axis followed by a reflection through the y-axis.   **41.** In column vector notation, we have $T\left(\begin{bmatrix} x \\ y \end{bmatrix} \right) = A \begin{bmatrix} x \\ y \end{bmatrix} = \begin{bmatrix} 1 & 0 \\ 3 & 1 \end{bmatrix} \begin{bmatrix} 1 & 0 \\ 0 & 2 \end{bmatrix} \begin{bmatrix} 1 & 1 \\ 0 & 1 \end{bmatrix} \begin{bmatrix} x \\ y \end{bmatrix}$, which represents a horizontal shear followed by a vertical expansion followed by a vertical shear.

Section 1.6

1. Not a transition matrix   **3.** Not a transition matrix   **5.** A regular transition matrix   **7.** A transition matrix, but not regular   **9.** 0.28   **11.** 0.330   **13.** Not regular   **15.** Regular

17. Not regular   **19.** $\begin{bmatrix} \frac{3}{4} \\ \frac{1}{4} \end{bmatrix}$ **21.** $\begin{bmatrix} \frac{1}{4} \\ \frac{3}{8} \\ \frac{3}{8} \end{bmatrix}$ **23.** $\begin{bmatrix} \frac{12}{35} \\ \frac{8}{35} \\ \frac{15}{35} \end{bmatrix}$ **25.** T F F T T F T F F T   **27.** $A^{100} \approx \begin{bmatrix} \frac{12}{35} & \frac{12}{35} & \frac{12}{35} \\ \frac{8}{35} & \frac{8}{35} & \frac{8}{35} \\ \frac{15}{35} & \frac{15}{35} & \frac{15}{35} \end{bmatrix}$

29. 0.47   **31.** $\begin{bmatrix} .32 \\ .47 \\ .21 \end{bmatrix}$ **33.** The Markov chain is regular because T has no zero entry; $\mathbf{s} = \begin{bmatrix} \frac{11}{43} \\ \frac{17}{43} \\ \frac{15}{43} \end{bmatrix}$.   **35.** $\frac{1}{8}$

37. $\begin{bmatrix} \frac{1}{4} \\ \frac{1}{2} \\ \frac{1}{4} \end{bmatrix}$ **39.** The Markov chain is regular because T^2 has no zero entries; $\mathbf{s} = \begin{bmatrix} \frac{1}{4} \\ \frac{1}{2} \\ \frac{1}{4} \end{bmatrix}$.

41. $T = \begin{matrix} & \begin{matrix} \mathbf{D} & \mathbf{H} & \mathbf{R} \end{matrix} \\ \begin{bmatrix} 0 & 0 & 0 \\ 1 & \frac{1}{2} & 0 \\ 0 & \frac{1}{2} & 1 \end{bmatrix} & \begin{matrix} \mathbf{D} \\ \mathbf{H} \\ \mathbf{R} \end{matrix} \end{matrix}$. The recessive state is absorbing, since there is an entry 1 in the row 3, column 3 position.

45. $\begin{bmatrix} \frac{1}{5} \\ \frac{4}{5} \end{bmatrix}$ **47.** $\begin{bmatrix} \frac{5}{9} \\ \frac{4}{9} \end{bmatrix}$ **49.** $\begin{bmatrix} .4037267 \\ .3975155 \\ .1987578 \end{bmatrix}$ **51.** $\begin{bmatrix} .2115383 \\ .3461539 \\ .4423077 \end{bmatrix}$ **53.** $\begin{bmatrix} .2682278 \\ .2235232 \\ .1915913 \\ .1676424 \\ .1490154 \end{bmatrix}$

Section 1.7

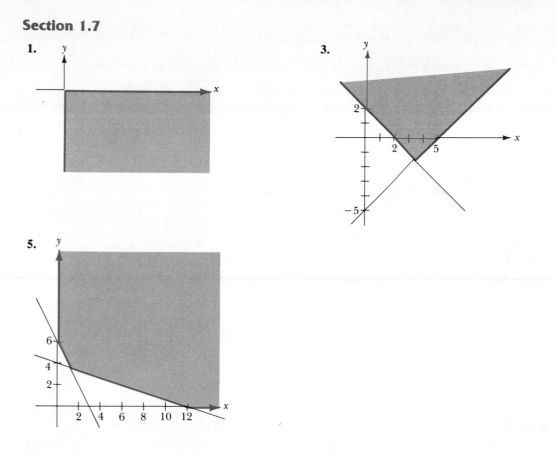

1.

3.

5.

7. **(a)** Minimum 7 at (2, 3); no maximum; **(b)** No minimum; no maximum; **(c)** No minimum; maximum -13 at (2, 3)   **9.** **(a)** No minimum; maximum 0 at (0, 0); **(b)** Minimum 0 at (0, 0); no maximum; **(c)** No minimum; no maximum

11. **(a)** Minimum 14 at (7, 0); no maximum; **(b)** Minimum 6 at (3, 1); no maximum; **(c)** Minimum 3 at (1, 2); no maximum   **13.** **(a)** Ratio $r \le \frac{1}{2}$; **(b)** Ratio $r \ge 5$ **15.** Maximum profit is \$400 per day, when 1000 lb of blend A and 2500 lb of blend B are produced daily.   **17.** The minimum cost is 46.25 cents, when $\frac{7}{2}$ oz of food 1 and $\frac{3}{4}$ oz of food 2 are used.   **19.** The minimum of \$7.25 is obtained by using 60 lb from bag 1 and 10 lb from bag 2.   **21.** T F F F F T F F T F   **23.** **(a)** Maximum $\approx$ 19; minimum $\approx$ 8; **(b)** Maximum $\approx$ 14.8; minimum $\approx$ 8; **(c)** Maximum $\approx$ 12; minimum $\approx$ 4; **(d)** Maximum $\approx$ 4; minimum $\approx$ -12.5 **25.** **(a)** No maximum; minimum $\approx$ 2; **(b)** Maximum $\approx$ 11.5; no minimum; **(c)** No maximum; minimum $\approx$ 2; **(d)** No maximum; minimum $\approx$ -5

CHAPTER 2

Section 2.1

1. $3\sqrt{2}$ **3.** $\sqrt{22}$ **5.** $(1, -3, -4)$ **7.** $(1, 4, 3)$ **9.** $(-9, 6, 19)$ **11.** $\left(-\frac{8}{5}, -\frac{4}{5}, \frac{12}{5}\right)$

13. $\left(\dfrac{2}{\sqrt{14}}, \dfrac{1}{\sqrt{14}}, \dfrac{-3}{\sqrt{14}}\right)$ **15.** 8   **17.** $\theta = \cos^{-1}\left(\dfrac{-3}{\sqrt{156}}\right) \approx 103.9°$ **19.** 11   **21.** $(1, -1, 1)$

23. $\theta = \cos^{-1}\left(\dfrac{34}{\sqrt{2449}}\right) \approx 46.6°$ **25.** Perpendicular   **27.** Parallel with opposite directions

29. Neither   **31.** 067.38; that is, 67.38° from north   **33.** (a) $50/(\cos \theta)$ lb; (b) 50 lb; (c) The rope will break.   **35.** T T F F T F T F F F

Section 2.2

1. (a) $(\mathbf{u} + \mathbf{v}) + \mathbf{w} = ((u_1 + v_1), \ldots, (u_n + v_n)) + (w_1, \ldots, w_n)$
$= ((u_1 + v_1) + w_1, \ldots, (u_n + v_n) + w_n)$
$= (u_1 + (v_1 + w_1), \ldots, u_n + (v_n + w_n))$
$= (u_1, \ldots, u_n) + (v_1 + w_1, \ldots, v_n + w_n)$
$= \mathbf{u} + (\mathbf{v} + \mathbf{w})$

3. $(4, 16, 8)$ **5.** $(-3, -5, -6)$ **7.** (a) Not closed under addition; (b) Not closed under scalar multiplication   **9.** (a) Not closed under addition; (b) Closed under scalar multiplication   **11.** (a) Not closed under addition. (b) Not closed under scalar multiplication   **13.** (a) Closed under addition; (b) Closed under scalar multiplication   **15.** Not a vector space   **17.** Not a vector space   **19.** Not a vector space   **21.** Not a vector space   **23.** Not a vector space   **25.** A vector space   **27.** Not a vector space   **29.** F F T T F T F T F T   **35.** (a) $(-1, 0)$ is the "zero vector," as shown in the solution of Exercise 16. (b) Part (v) of Theorem 2.2 in this vector space becomes $r(-1, 0) = (-1, 0)$, for all $r \in \mathbb{R}$. That is, $(0, 0)$ is not the zero vector $\mathbf{0}$ in this vector space.

Section 2.3

1. A subspace   **3.** Not a subspace   **5.** Not a subspace   **7.** Not a subspace   **9.** Not a subspace   **11.** A subspace   **13.** Not a subspace   **15.** Not a subspace   **17.** A subspace   **19.** Yes   **21.** Yes   **23.** Yes   **25.** (a) Since $1 = \sin^2 x + \cos^2 x$, we have $c = c(\sin^2 x) + c(\cos^2 x)$, which shows that $c \in \text{sp}(\sin^2 x, \cos^2 x)$. (b) Now $\cos 2x = \cos^2 x - \sin^2 x = (-1) \sin^2 x + (1) \cos^2 x$, which shows that $\cos 2x \in \text{sp}(\sin^2 x, \cos^2 x)$ (c) Now $\cos 4x = \cos^2 2x - \sin^2 2x = (1 - \sin^2 2x) - \sin^2 2x = \frac{1}{7}(7) + (-2) \sin^2 2x$, which shows that $\cos 4x$, and thus $8 \cos 4x$, is in $\text{sp}(7, \sin^2 2x)$.   **27.** $\mathbf{b} = 2\mathbf{v}_1 - \mathbf{v}_2 + \mathbf{v}_3$ **29.** Impossible

31. (a) We see that $\mathbf{v}_1, 2\mathbf{v}_1 + \mathbf{v}_2 \in \text{sp}(\mathbf{v}_1, \mathbf{v}_2)$; and therefore,

$$\text{sp}(\mathbf{v}_1, 2\mathbf{v}_1 + \mathbf{v}_2) \subseteq \text{sp}(\mathbf{v}_1, \mathbf{v}_2).$$

Furthermore, $\mathbf{v}_1 = 1\mathbf{v}_1 + 0(2\mathbf{v}_1 + \mathbf{v}_2)$ and $\mathbf{v}_2 = (-2)\mathbf{v}_1 + 1(2\mathbf{v}_1 + \mathbf{v}_2)$, showing that $\mathbf{v}_1, \mathbf{v}_2 \in \text{sp}(\mathbf{v}_1, 2\mathbf{v}_1 + \mathbf{v}_2)$; and therefore,

$$\text{sp}(\mathbf{v}_1, \mathbf{v}_2) \subseteq \text{sp}(\mathbf{v}_1, 2\mathbf{v}_1 + \mathbf{v}_2).$$

Thus, $\text{sp}(\mathbf{v}_1, \mathbf{v}_2) = \text{sp}(\mathbf{v}_1, 2\mathbf{v}_1 + \mathbf{v}_2)$.

35. T F T T T F F F T T   **37.** $W_2 = \text{sp}((1, 0, 1), (3, 0, -1))$ consists of all vectors in $\mathbb{R}^3$ whose middle component is zero, since $\begin{bmatrix} 1 & 3 & a \\ 0 & 0 & 0 \\ 1 & -1 & b \end{bmatrix}$ is the augmented matrix of a consistent system for all choices of a and b. The only vectors in $W_1 = \text{sp}((1, 2, 3), (2, 1, 1))$ with middle component zero are those of the form $r(1, 2, 3) + s(2, 1, 1)$, where $s = -2r$; and this gives vectors of the form $r(1, 2, 3) - 2r(2, 1, 1) = r(-3, 0, 1)$. Thus a generating set for $W_1 \cap W_2$ is $\{(-3, 0, 1)\}$.   **39.** $\{(0, 0)\}$

41. $\{(-1, 3, 0), (-1, 0, 3)\}$ **43.** $\{(-7, 1, 13, 0), (-6, -1, 0, 13)\}$ **45.** $\{(-60, 137, 33, 0, 1)\}$

47. Yes   **49.** $(-1)\begin{bmatrix} 0 \\ 1 \\ 1 \\ 1 \end{bmatrix} + 2\begin{bmatrix} 1 \\ 0 \\ 1 \\ 1 \end{bmatrix} + 0\begin{bmatrix} 1 \\ 1 \\ 0 \\ 1 \end{bmatrix} + 0\begin{bmatrix} 1 \\ 1 \\ 1 \\ 0 \end{bmatrix} = \begin{bmatrix} 2 \\ -1 \\ 1 \\ 1 \end{bmatrix}$

Section 2.4

1. Two nonzero vectors in $\mathbb{R}^2$ are dependent if and only if they are parallel.   **3.** Two nonzero vectors in $\mathbb{R}^3$ are dependent if and only if they are parallel.   **5.** Three vectors are dependent if and only if they all lie in one plane through the origin.   **7.** Dependent   **9.** Independent   **11.** Dependent   **13.** Independent
15. Independent   **17.** Dependent   **19.** Dependent   **21.** Independent   **23.** Dependent
25. Suppose that $r_1\mathbf{w}_1 + r_2\mathbf{w}_2 + r_3\mathbf{w}_3 = \mathbf{0}$, so that $r_1(2\mathbf{v}_1 + 3\mathbf{v}_2) + r_2(\mathbf{v}_2 - 2\mathbf{v}_3) + r_3(-\mathbf{v}_1 - 3\mathbf{v}_3) = \mathbf{0}$. Then $(2r_1 - r_3)\mathbf{v}_1 + (3r_1 + r_2)\mathbf{v}_2 + (-2r_2 - 3r_3)\mathbf{v}_3 = \mathbf{0}$. We solve the linear system

$$
\begin{aligned}
2r_1 \quad\quad - \; r_3 &= 0, \\
3r_1 + \; r_2 \quad\quad &= 0, \\
-2r_2 - 3r_3 &= 0.
\end{aligned}
$$

$$
\begin{bmatrix} 2 & 0 & -1 & \bigm| & 0 \\ 3 & 1 & 0 & \bigm| & 0 \\ 0 & -2 & -3 & \bigm| & 0 \end{bmatrix} \sim \begin{bmatrix} 1 & 0 & -\frac{1}{2} & \bigm| & 0 \\ 0 & 1 & \frac{3}{2} & \bigm| & 0 \\ 0 & -2 & -3 & \bigm| & 0 \end{bmatrix} \sim \begin{bmatrix} 1 & 0 & -\frac{1}{2} & \bigm| & 0 \\ 0 & 1 & \frac{3}{2} & \bigm| & 0 \\ 0 & 0 & 0 & \bigm| & 0 \end{bmatrix}.
$$

This system has the nontrivial solution $r_3 = 2$, $r_2 = -3$, and $r_1 = 1$. Thus, $(2\mathbf{v}_1 + 3\mathbf{v}_2) - 3(\mathbf{v}_2 - 2\mathbf{v}_3) + 2(-\mathbf{v}_1 - 3\mathbf{v}_3) = \mathbf{0}$ for all choices of vectors $\mathbf{v}_1, \mathbf{v}_2, \mathbf{v}_3 \in V$, so $\mathbf{w}_1, \mathbf{w}_2, \mathbf{w}_3$ are dependent. (Notice that, if the system had only the trivial solution, the dependence or independence would *not* be known.)

29. Let $A = \begin{bmatrix} 1 & 1 & 0 \\ 0 & 0 & 0 \\ 0 & 0 & 1 \end{bmatrix}$, $\mathbf{v} = \begin{bmatrix} 1 \\ 0 \\ 0 \end{bmatrix}$, $\mathbf{w} = \begin{bmatrix} 0 \\ 1 \\ 0 \end{bmatrix}$. Then $A\mathbf{v} = A\mathbf{w} = \begin{bmatrix} 1 \\ 0 \\ 0 \end{bmatrix}$. For the second part of the problem, let

$A = \begin{bmatrix} 1 & 0 & 0 \\ 0 & 1 & 0 \\ 0 & 0 & 0 \end{bmatrix}$, $\mathbf{v} = \begin{bmatrix} 1 \\ 0 \\ 0 \end{bmatrix}$, $\mathbf{w} = \begin{bmatrix} 0 \\ 1 \\ 0 \end{bmatrix}$. Then $A\mathbf{v} = \mathbf{v}$ and $A\mathbf{w} = \mathbf{w}$, so $A\mathbf{v}$ and $A\mathbf{w}$ are still independent.

33. T F T F T T T T F T

Section 2.5

1. A basis   **3.** Not a basis   **5.** A basis   **7.** Not a basis   **9.** Two vectors chosen *at random* from $\mathbb{R}^3$ are extremely likely to be nonzero and nonparallel, thus they are extremely likely to generate a plane through the origin in $\mathbb{R}^3$. A third vector chosen *at random* is extremely likely not to lie in that plane; thus the three vectors are extremely likely to be independent and to form a basis for $\mathbb{R}^3$.   **11.** $\{-3, 1\}$
13. $\{(-2, 3, 1), (3, -1, 2)\}$ **15.** $\{x^2 - 1, x^2 + 1, 2x - 3\}$ **17.** $\{1, \sin^2 x\}$ **19.** **(a)** Elementary row operations do not change the row space of a matrix. The row space of A is equal to the row space of an echelon form of A. Since the nonzero rows of an echelon form are independent and generate the row space, they form a basis for the rowspace of A.   **(b)** The basis obtained by using method (i) consists entirely of some of the rows of A itself, while the rows of an echelon form—obtained using method (ii)—may not be the same as rows of A.   **(c)** Reduce A to an echelon form H. A basis for the row space of A consists of the nonzero rows of H. A basis for the column space of A consists of the columns of A corresponding to columns of H containing pivots.   **21.** $\{(-1, -3, 11)\}$ **23.** A basis consists of the vectors $\begin{bmatrix} 1 \\ 2 \\ 3 \end{bmatrix}$ and $\begin{bmatrix} 3 \\ 0 \\ 2 \end{bmatrix}$.

25. The six matrices

$$\begin{bmatrix} 1 & 0 & 0 \\ 0 & 0 & 0 \end{bmatrix}, \begin{bmatrix} 0 & 1 & 0 \\ 0 & 0 & 0 \end{bmatrix}, \begin{bmatrix} 0 & 0 & 1 \\ 0 & 0 & 0 \end{bmatrix}, \begin{bmatrix} 0 & 0 & 0 \\ 1 & 0 & 0 \end{bmatrix}, \begin{bmatrix} 0 & 0 & 0 \\ 0 & 1 & 0 \end{bmatrix}, \begin{bmatrix} 0 & 0 & 0 \\ 0 & 0 & 1 \end{bmatrix}$$

form a basis.

27. F F T T F T F T T F   **35.** Let $W = \text{sp}(\mathbf{e}_1, \mathbf{e}_2)$ and $U = \text{sp}(\mathbf{e}_3, \mathbf{e}_4, \mathbf{e}_5)$ in $\mathbb{R}^5$. Then $W \cap U = \{\mathbf{0}\}$ and each $\mathbf{x} \in \mathbb{R}^5$ has the form $\mathbf{x} = \mathbf{w} + \mathbf{u}$, where $\mathbf{w} = x_1\mathbf{e}_1 + x_2\mathbf{e}_2$ and $\mathbf{u} = x_3\mathbf{e}_3 + x_4\mathbf{e}_4 + x_5\mathbf{e}_5$.   **37.** $\mathbf{v}_1, \mathbf{v}_2, \mathbf{v}_5$

39. $\mathbf{v}_1, \mathbf{v}_2, \mathbf{v}_3, \mathbf{v}_4, \mathbf{v}_6$

Section 2.6

1. 2   **3.** 1   **5.** $\{(1, 2, 1), (1, 0, 0), (0, 1, 0)\}$ **7.** $\{x^2 + 1, 1, x, x^3\}$ **9. (a)** 2;

(b) $\{(2, 0, -3, 1), (3, 4, 2, 2)\}$;   **(c)** The set consisting of $\begin{bmatrix} 2 \\ 3 \end{bmatrix}, \begin{bmatrix} 0 \\ 4 \end{bmatrix}$ or of $\begin{bmatrix} 1 \\ 0 \end{bmatrix}, \begin{bmatrix} 0 \\ 1 \end{bmatrix}$;   **(d)** The set

consisting of $\begin{bmatrix} 12 \\ -13 \\ 8 \\ 0 \end{bmatrix}, \begin{bmatrix} -4 \\ -1 \\ 0 \\ 8 \end{bmatrix}$ **11. (a)** 3;   **(b)** $\{(1, 0, -1, 0), (0, 1, 1, 0), (0, 0, 0, 1)\}$;   **(c)** The set con-

sisting of $\begin{bmatrix} 0 \\ 1 \\ 4 \\ 1 \end{bmatrix}, \begin{bmatrix} 6 \\ 2 \\ 1 \\ 3 \end{bmatrix}, \begin{bmatrix} 3 \\ 1 \\ 4 \\ 0 \end{bmatrix}$;   **(d)** The set consisting of the vector $\begin{bmatrix} 1 \\ -1 \\ 1 \\ 0 \end{bmatrix}$ **13. (a)** 3;

(b) $\{(6, 0, 0, 5), (0, 3, 0, 1), (0, 0, 3, 1)\}$;   **(c)** The set consisting of $\begin{bmatrix} 0 \\ 2 \\ 0 \end{bmatrix}, \begin{bmatrix} 1 \\ 1 \\ 2 \end{bmatrix}, \begin{bmatrix} 2 \\ 0 \\ 1 \end{bmatrix}$ or the set of column

vectors $\{\mathbf{e}_1, \mathbf{e}_2, \mathbf{e}_3\}$;   **(d)** The set consisting of $\begin{bmatrix} -5 \\ -2 \\ -2 \\ 6 \end{bmatrix}$ **15.** The rank is 3; therefore, the matrix is not

invertible.   **17.** The rank is 3; therefore, the matrix is invertible.   **19.** F F T F F T T F T T

29. No

Section 2.7

1. $(1, -1)$ **3.** $(2, 6, -4)$ **5.** $(2, 1, 3)$ **7.** $(4, 1, -2, 1)$ **9.** $\left(\frac{1}{2}, \frac{1}{2}, 1 - \frac{1}{2}, 0\right)$

11. T F F T T F F T T T

13. $C_{B',B} = \begin{bmatrix} -6 & 3 & 4 \\ 9 & -4 & -6 \\ 2 & -1 & -1 \end{bmatrix}$ $C_{B,B'} = \begin{bmatrix} 2 & 1 & 2 \\ 3 & 2 & 0 \\ 1 & 0 & 3 \end{bmatrix}$ $\begin{bmatrix} -6 & 3 & 4 \\ 9 & -4 & -6 \\ 2 & -1 & -1 \end{bmatrix}\begin{bmatrix} 2 & 1 & 2 \\ 3 & 2 & 0 \\ 1 & 0 & 3 \end{bmatrix} = \begin{bmatrix} 1 & 0 & 0 \\ 0 & 1 & 0 \\ 0 & 0 & 1 \end{bmatrix}$

15. $C_{B',B} = \begin{bmatrix} 0 & 0 & 0 & 1 \\ 0 & 0 & 1 & -1 \\ 0 & 1 & -1 & 0 \\ 1 & -1 & 0 & 0 \end{bmatrix}$ $C_{B,B'} = \begin{bmatrix} 1 & 1 & 1 & 1 \\ 1 & 1 & 1 & 0 \\ 1 & 1 & 0 & 0 \\ 1 & 0 & 0 & 0 \end{bmatrix}$

$\begin{bmatrix} 0 & 0 & 0 & 1 \\ 0 & 0 & 1 & -1 \\ 0 & 1 & -1 & 0 \\ 1 & -1 & 0 & 0 \end{bmatrix}\begin{bmatrix} 1 & 1 & 1 & 1 \\ 1 & 1 & 1 & 0 \\ 1 & 1 & 0 & 0 \\ 1 & 0 & 0 & 0 \end{bmatrix} = \begin{bmatrix} 1 & 0 & 0 & 0 \\ 0 & 1 & 0 & 0 \\ 0 & 0 & 1 & 0 \\ 0 & 0 & 0 & 1 \end{bmatrix}$

17.
$$\begin{bmatrix} \frac{1}{2} & -\frac{1}{2} & -\frac{1}{2} & -\frac{1}{2} \\ \frac{1}{2} & \frac{1}{2} & -\frac{1}{2} & -\frac{1}{2} \\ \frac{1}{2} & \frac{1}{2} & \frac{1}{2} & -\frac{1}{2} \\ \frac{1}{2} & \frac{1}{2} & \frac{1}{2} & \frac{1}{2} \end{bmatrix}$$

19. Let $\mathbf{v} \in \mathbb{R}^n$. Now $\mathbf{v} = \mathbf{v}_E = C_{B,E}\,\mathbf{v}_B$, so $\mathbf{v}_{B'} = C_{E,B'}\mathbf{v} = C_{E,B'}\,(C_{B,E}\,\mathbf{v}_B) = (C_{E,B'}\,C_{B,E})\mathbf{v}_B$. Thus the change-of-coordinates matrix from B to B' is $C_{B,B'} = C_{E,B'}\,C_{E,B}$.

21.
$$\begin{bmatrix} 0 & 0 & 1 & 0 \\ 0 & 1 & 0 & 2 \\ 1 & 0 & 0 & 0 \\ 0 & -1 & 0 & 0 \end{bmatrix}$$

23.
$$\begin{bmatrix} 0 & 0 & 1 & 0 \\ 1 & -1 & -1 & 1 \\ 1 & 0 & 0 & 0 \\ 0 & 0 & 3 & 1 \end{bmatrix}$$

25.
$$\begin{bmatrix} -\frac{1}{2} & \frac{1}{2} \\ \frac{1}{2} & \frac{1}{2} \end{bmatrix}$$

27.
$$\begin{bmatrix} -1 \\ -4 \\ -2 \end{bmatrix}$$

29. $C'C$

Section 2.8

1. $x_1 = 3 - 2t$
$x_2 = -3 + t$

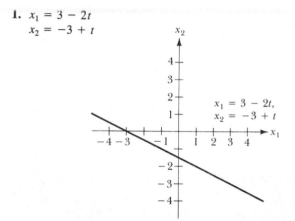

$x_1 = 3 - 2t,$
$x_2 = -3 + t$

3. $x_1 = \dfrac{d_1 c}{d_1{}^2 + d_2{}^2} + d_2 t,\; x_2 = \dfrac{d_2 c}{d_1{}^2 + d_2{}^2} - d_1 t$　**5. (a)** $x_1 = -2 + t, x_2 = 4 - t$;　**(b)** $x_1 = 3 - 3t, x_2 = -1 - 2t, x_3 = 6 - 7t$;　**(c)** $x_1 = 2 - 3t, x_2 = 5t, x_3 = 4 - 12t$　**7.** The lines contain exactly the same points $\{(5 - 3t, -1 + t) \mid t \in \mathbb{R}\}$.　**9. (a)** $\left(\frac{1}{2}, \frac{3}{2}\right)$; **(b)** $\left(\frac{3}{2}, -2, \frac{5}{2}\right)$; **(c)** $\left(-2, \frac{9}{2}, \frac{17}{2}\right)$
11. $\left(-\frac{3}{2}, -\frac{1}{2}, \frac{15}{4}\right)$　**13.** $\left(\frac{3}{2}, \frac{3}{2}, 1, \frac{7}{2}, -\frac{1}{2}\right)$　**15.** $(2, 3, 4)$　**17.** $x_1 + x_2 + x_3 = 1$　**19.** $2x_1 + x_2 - x_3 = 2$　**21.** $x_1 + 4x_2 + x_3 = 10$　**23.** $x_2 + x_3 = 3, -3x_1 - 7x_2 + 8x_3 + 3x_4 = 0$ (There are infinitely many other correct linear systems.)　**25.** $9x_1 - x_2 + 2x_3 - 6x_4 + 3x_5 = 6$　**27.** $x_1 + x_2 + x_3 + x_4 + x_5 + x_6 = 1$　**29.** The 0-flat $\mathbf{x} = \begin{bmatrix} -1 \\ 0 \\ 2 \end{bmatrix}$　**31.** The 1-flat $\mathbf{x} = \begin{bmatrix} -1 \\ -1 \\ 0 \end{bmatrix} + t\begin{bmatrix} 2 \\ 1 \\ 1 \end{bmatrix}$　**33.** The 1-flat $\mathbf{x} = \begin{bmatrix} -8 \\ -23 \\ -7 \\ 0 \end{bmatrix} + t\begin{bmatrix} 0 \\ -5 \\ 1 \\ 2 \end{bmatrix}$　**35.** The 0-flat $\mathbf{x} = \begin{bmatrix} -43 \\ -12 \\ 7 \\ 1 \end{bmatrix}$　**37.** T F F T F T F T F T

Section 2.9

1. An invertible transformation   3. A linear transformation; not invertible   5. A linear transformation; not invertible   7. Not a linear transformation   9. Not a linear transformation   11. Not a linear transformation   13. $A = \begin{bmatrix} 2 & -3 \\ 1 & 1 \end{bmatrix}$ 15. $A = \begin{bmatrix} 1 & 1 & 1 \\ 1 & -1 & -1 \\ -1 & -1 & 1 \end{bmatrix}$ 17. $A = \begin{bmatrix} 1 & 1 & 0 \\ 0 & 1 & 1 \\ 1 & 0 & 1 \\ 1 & 1 & 1 \end{bmatrix}$

19. Consider the parallelogram in $\mathbb{R}^2$ representing the sum of two vectors v_1 and v_2 emanating from the *origin*. Reflection in a line through the origin is a rigid motion of the plane corresponding to turning the plane over with the line acting as axis of rotation. The parallelogram is carried into another having vectors emanating from the origin, since the line of reflection passes through the origin. Calling the reflection T, we see that this image parallelogram represents the sum $T(v_1) + T(v_2)$, so $T(v_1 + v_2) = T(v_1) + T(v_2)$. That is, the vector diagonal of the original parallelogram is carried into the vector diagonal of the image parallelogram. Similarly, the rigid motion T carries $r(v)$ into $r(T(v))$, so T is a linear transformation.

21. $E_1 = \begin{bmatrix} 2 & 0 \\ 0 & 1 \end{bmatrix}$, $E_2 = \begin{bmatrix} 1 & 0 \\ -\sqrt{3}/2 & 1 \end{bmatrix}$, $E_3 = \begin{bmatrix} 1 & 0 \\ 0 & \frac{1}{2} \end{bmatrix}$, $E_4 = \begin{bmatrix} 1 & \sqrt{3} \\ 0 & 1 \end{bmatrix}$. $A = E_1^{-1}E_2^{-1}E_3^{-1}E_4^{-1} =$

$\begin{bmatrix} \frac{1}{2} & 0 \\ 0 & 1 \end{bmatrix}\begin{bmatrix} 1 & 0 \\ \sqrt{3}/2 & 1 \end{bmatrix}\begin{bmatrix} 1 & 0 \\ 0 & 2 \end{bmatrix}\begin{bmatrix} 1 & -\sqrt{3} \\ 0 & 1 \end{bmatrix}$. Rotation counterclockwise through $60°$ can be achieved by means of the sequence of transformations

the horizontal shear corresponding to E_4^{-1},

the vertical expansion corresponding to E_3^{-1},

the vertical shear corresponding to E_2^{-1},

the horizontal contraction corresponding to E_1^{-1}.

23. (a) $\begin{bmatrix} 1 & 0 & 0 \\ 0 & -1 & 0 \\ 0 & 0 & 1 \end{bmatrix}$; (b) $\begin{bmatrix} 1 & 0 & 0 \\ 0 & 1 & 0 \\ 0 & 0 & -1 \end{bmatrix}$ 25. T F T T F T F F F T

35. (a) Since

$$T_k(\mathbf{u} + \mathbf{v}) = (u_1, u_2, \ldots, u_k, 0, \ldots, 0) + (v_1, v_2, \ldots, v_k, 0, \ldots, 0)$$
$$= (u_1 + v_1, u_2 + v_2, \ldots, u_k + v_k, 0, \ldots, 0)$$
$$= T_k(\mathbf{u}) + T_k(\mathbf{v})$$

and

$$T_k(r\mathbf{u}) = (ru_1, ru_2, \ldots, ru_k, 0, \ldots, 0)$$
$$= r(u_1, u_2, \ldots, u_k, 0, \ldots, 0)$$
$$= rT_k(\mathbf{u}),$$

we know that T_k is a linear transformation. We see that $T_k[V] \subseteq W_k$, which means that it is a subspace of W_k, by Exercise 33.

(b) Since $T_k[V]$ is a subspace of $T_{k+1}[V]$, we have

$$d_1 \leq d_2 \leq d_3 \leq \cdots \leq d_n \leq n.$$

If the jth column of H has a pivot but the $(j + 1)$st column has no pivot, then $d_{j+1} = d_j$. However, if the jth and $(j + 1)$st columns both have a pivot, then $d_{j+1} = d_j + 1$. See parts (c) and (d) for illustrations.

(c) If $d_1 = d_2 = 1$ and $d_3 = d_4 = 2$, then H must have the form

$$H = \begin{bmatrix} p & X & X & X \\ 0 & 0 & p & X \\ 0 & 0 & 0 & 0 \\ 0 & 0 & 0 & 0 \\ \cdot & \cdot & \cdot & \cdot \\ \cdot & \cdot & \cdot & \cdot \\ \cdot & \cdot & \cdot & \cdot \\ 0 & 0 & 0 & 0 \end{bmatrix},$$

where a p denotes a pivot and an X denotes a possibly nonzero entry.

(d) If $d_1 = 1$, $d_2 = d_3 = d_4 = 2$, and $d_5 = d_6 = 3$, then H must have the form

$$H = \begin{bmatrix} p & X & X & X & X & X \\ 0 & p & X & X & X & X \\ 0 & 0 & 0 & 0 & p & X \\ 0 & 0 & 0 & 0 & 0 & 0 \\ 0 & 0 & 0 & 0 & 0 & 0 \\ \cdot & \cdot & \cdot & \cdot & \cdot & \cdot \\ \cdot & \cdot & \cdot & \cdot & \cdot & \cdot \\ \cdot & \cdot & \cdot & \cdot & \cdot & \cdot \\ 0 & 0 & 0 & 0 & 0 & 0 \end{bmatrix},$$

where a p denotes a pivot and an X denotes a possibly nonzero entry.

(e) The number of pivots in H is the number of distinct dimension numbers in the list $d_1, d_2, d_3, \ldots, d_n$. Moreover, a pivot occurs in the $(j + 1)$st column if and only if $d_{j+1} = d_j + 1$; the row and column positions of pivots are completely determined by the list $d_1, d_2, d_3, \ldots, d_n$. The pivot in the kth row occurs in the jth column if and only if d_j is the kth *distinct* positive integer in the list $d_1, d_2, d_3, \ldots, d_n$. Since the numbers d_i depend only on A (they are defined just in terms of the row space of A), so do the number and locations of the pivots; the number and locations are the same for all echelon forms of A.

(f) Let H be a *reduced* echelon form of A. Part (e) shows that the number of pivots and their locations depend only on A, and consequently the number of zero rows depends only on A and is the same for all choices of H. Suppose that the pivot in the kth row of H is in column j. Consider now a nonzero row vector in $\mathbb{R}^n$ that has entries zero in all components corresponding to columns of H containing pivots except for the jth component, where the entry is 1; entries in components not corresponding to pivot column locations in H may be arbitrary. We claim that there is a *unique* such vector in the row space of A—namely, the kth row vector of H. Such a vector must be a linear combination of the nonzero rows of H, which span the row space of A. The fact that there are zeros above as well as below pivots in the *reduced* row echelon form H shows that the only possible such linear combination of nonzero rows of H is 1 times the kth row plus 0 times the other rows. This gives a characterization of the kth nonzero row of H in terms of the row space of A and completes the demonstration that the reduced echelon form of A is unique.

Section 2.10

1. **(a)** $\mathbf{v} \cdot \mathbf{w} = (v_1, v_2, \ldots, v_n) \cdot (w_1, w_2, \ldots, w_n)$
$$= v_1 w_1 + v_2 w_2 + \cdots + v_n w_n = w_1 v_1 + w_2 v_2 + \cdots + w_n v_n$$
$$= (w_1, w_2, \ldots, w_n) \cdot (v_1, v_2, \ldots, v_n)$$
$$= \mathbf{w} \cdot \mathbf{v}.$$

(b) $\mathbf{u} \cdot (\mathbf{v} + \mathbf{w}) = (u_1, \ldots, u_n) \cdot ((v_1, \ldots, v_n) + (w_1, \ldots, w_n))$
$$= (u_1, \ldots, u_n) \cdot (v_1 + w_1, \ldots, v_n + w_n)$$
$$= u_1(v_1 + w_1) + \cdots + u_n(v_n + w_n)$$
$$= u_1 v_1 + u_1 w_1 + \cdots + u_n v_n + u_n w_n$$
$$= (u_1 v_1 + \cdots + u_n v_n) + (u_1 w_1 + \cdots + u_n w_n)$$
$$= \mathbf{u} \cdot \mathbf{v} + \mathbf{u} \cdot \mathbf{w}.$$

(c) $r(\mathbf{v} \cdot \mathbf{w}) = r((v_1, \ldots, v_n) \cdot (w_1, \ldots, w_n))$

$$= r(v_1 w_1 + \cdots + v_n w_n)$$

$= (rv_1)w_1 + \cdots + (rv_n)w_n$ ┃ $= v_1(rw_1) + \cdots + v_n(rw_n)$

$= (rv_1, \ldots, rv_n) \cdot (w_1, \ldots, w_n)$ ┃ $= (v_1, \ldots, v_n) \cdot (rw_1, \ldots, rw_n)$

$= (r\mathbf{v}) \cdot \mathbf{w}.$ ┃ $= \mathbf{v} \cdot (r\mathbf{w}).$

3. $15\sqrt{6}$ **5.** -540 **11.** No   **13.** No   **15.** No   **17.** Yes

19. P1: $\displaystyle\int_a^b f(x)g(x)\, dx = \int_a^b g(x)f(x)\, dx$, so $\langle f, g \rangle = \langle g, f \rangle$.

P2: $\displaystyle\langle f, g + h \rangle = \int_a^b f(x)((g(x) + h(x))\, dx$

$$= \int_a^b f(x)g(x)\, dx + \int_a^b f(x)h(x)\, dx$$

$$= \langle f, g \rangle + \langle f, h \rangle.$$

P3: $\displaystyle r\langle f, g \rangle = r\int_a^b f(x)g(x)\, dx$

$\displaystyle = \int_a^b (rf(x))g(x)\, dx$ ┃ $\displaystyle = \int_a^b f(x)(rg(x))\, dx$

$= \langle rf, g \rangle.$ ┃ $= \langle f, rg \rangle.$

P4: $\langle f, f \rangle = \int_a^b f(x)^2\, dx$. Now $f(x)$ is continuous. If $f(c) = r \neq 0$ for some c, where $a \leq c \leq b$, then there exists some small interval of length $h > 0$ containing c and contained in the interval $[a, b]$ from a to b such that $|f(x)| > |r|/2$ in this interval. Then $f(x)^2 > r^2/4$ in this interval, so $\langle f, f \rangle$ is greater than $r^2 h/4 > 0$. Of course, if $f = 0$, then $\langle f, f \rangle = 0$ as the integral of the zero function f^2 over the interval $[a, b]$. Thus, $\langle f, f \rangle \geq 0$ and $\langle f, f \rangle = 0$ if and only if $f = 0$.

21. $\langle \sin x, \cos x \rangle = \int_0^\pi \sin x \cos x\, dx = \dfrac{\sin^2 x}{2}\bigg|_0^\pi = 0$. Therefore, $\sin x$ and $\cos x$ are orthogonal functions.

23. -91 **25.** F T F F T T T F T F

CHAPTER 3

Section 3.1

1. -15 **3.** 15

5. $\mathbf{b} \cdot (\mathbf{b} \times \mathbf{c}) = \begin{vmatrix} b_1 & b_2 & b_3 \\ b_1 & b_2 & b_3 \\ c_1 & c_2 & c_3 \end{vmatrix} = b_1(b_2 c_3 - b_3 c_2) - b_2(b_1 c_3 - b_3 c_1) + b_3(b_1 c_2 - b_2 c_1) = 0,$

$\mathbf{c} \cdot (\mathbf{b} \times \mathbf{c}) = \begin{vmatrix} c_1 & c_2 & c_3 \\ b_1 & b_2 & b_3 \\ c_1 & c_2 & c_3 \end{vmatrix} = c_1(b_2 c_3 - b_3 c_2) - c_2(b_1 c_3 - b_3 c_1) + c_3(b_1 c_2 - b_2 c_1) = 0$

7. 120 **9.** -9 **11.** $\begin{vmatrix} a_1 & a_2 \\ b_1 & b_2 \end{vmatrix} = a_1 b_2 - a_2 b_1 = -(b_1 a_2 - b_2 a_1) = -\begin{vmatrix} b_1 & b_2 \\ a_1 & a_2 \end{vmatrix}$ **13.** $-6\mathbf{i} + 3\mathbf{j} + 5\mathbf{k}$

15. $0\mathbf{i} + 0\mathbf{j} + 0\mathbf{k}$ **17.** $22\mathbf{i} + 18\mathbf{j} + 2\mathbf{k}$ **19.** F T T F F T F T T F   **21.** 38 **23.** $\sqrt{62}$

25. $19/2$ **27.** $\sqrt{230}/2$ **29.** 16 **31.** $\sqrt{390}$

33. $\mathbf{a} \cdot (\mathbf{b} \times \mathbf{c}) = -6,$

$\mathbf{a} \times (\mathbf{b} \times \mathbf{c}) = 12\mathbf{i} + 4\mathbf{k}$

35. $\mathbf{a} \cdot (\mathbf{b} \times \mathbf{c}) = 19$,
   $\mathbf{a} \times (\mathbf{b} \times \mathbf{c}) = 3\mathbf{i} - 7\mathbf{j} + \mathbf{k}$

37. 20   39. 9   41. 1   43. $\frac{7}{3}$ 45. Not collinear   47. Collinear   49. Not coplanar

51. Not coplanar   53. 0   55. $\|\mathbf{a}\|^2\|\mathbf{b}\|^2$ 57. $\mathbf{i} \times (\mathbf{i} \times \mathbf{j}) = \mathbf{i} \times \mathbf{k} = -\mathbf{j}$, but $(\mathbf{i} \times \mathbf{i}) \times \mathbf{j} = \mathbf{0} \times \mathbf{j} = \mathbf{0}$.

59. $\mathbf{b} \times \mathbf{c} = \begin{vmatrix} \mathbf{i} & \mathbf{j} & \mathbf{k} \\ b_1 & b_2 & b_3 \\ c_1 & c_2 & c_3 \end{vmatrix} = (b_2c_3 - b_3c_2)\mathbf{i} - (b_1c_3 - b_3c_1)\mathbf{j} + (b_1c_2 - b_2c_1)\mathbf{k}$. Thus,

$$\mathbf{a} \cdot (\mathbf{b} \times \mathbf{c}) = a_1(b_2c_3 - b_3c_2) - a_2(b_1c_3 - b_3c_1) + a_3(b_1c_2 - b_2c_1)$$
$$= \begin{vmatrix} a_1 & a_2 & a_3 \\ b_1 & b_2 & b_3 \\ c_1 & c_2 & c_3 \end{vmatrix}.$$

Equation (4) of the text shows that this determinant is $\pm$(Volume of the box determined by $\mathbf{a}$, $\mathbf{b}$, and $\mathbf{c}$). Similarly,

$$(\mathbf{a} \times \mathbf{b}) \cdot \mathbf{c} = \mathbf{c} \cdot (\mathbf{a} \times \mathbf{b}) = \begin{vmatrix} c_1 & c_2 & c_3 \\ a_1 & a_2 & a_3 \\ b_1 & b_2 & b_3 \end{vmatrix}$$
$$= c_1(a_2b_3 - a_3b_2) - c_2(a_1b_3 - a_3b_1) + c_3(a_1b_2 - a_2b_1),$$

which is the same number.

Section 3.2

1. 13   3. −21   5. 19   7. 320   9. 0   11. −6   13. 2   15. 4   17. 54   19. 1/2

21. F T T F T T F T F   23. 0   25. 6   27. 5, −2   29. 1, 6, −2   31. 1/det(A)

33. Let A have column vectors $\mathbf{a}_1, \mathbf{a}_2, \ldots, \mathbf{a}_n$. Let

$$B = \begin{bmatrix} b_1 & & & \\ & b_2 & & \mathbf{0} \\ & & \ddots & \\ \mathbf{0} & & & b_n \end{bmatrix}. \quad \text{Then} \quad AB = \begin{bmatrix} | & | & & | \\ b_1\mathbf{a}_1 & b_2\mathbf{a}_2 & \cdots & b_n\mathbf{a}_n \\ | & | & & | \end{bmatrix}.$$

Applying the column scaling property n times, we have

$$\det(AB) = b_1 \begin{vmatrix} | & | & & | \\ \mathbf{a}_1 & b_2\mathbf{a}_2 & \cdots & b_n\mathbf{a}_n \\ | & | & & | \end{vmatrix} = b_1b_2 \begin{vmatrix} | & | & | & & | \\ \mathbf{a}_1 & \mathbf{a}_2 & b_3\mathbf{a}_3 & \cdots & b_n\mathbf{a}_n \\ | & | & | & & | \end{vmatrix}$$

$$= \cdots = (b_1b_2 \cdots b_n) \det(A) = \det(B) \det(A).$$

37. (a) The result is obvious for $n = 2$. Suppose that it is true for $n - 1$, where $n \geq 3$. Expand the determinant of an $n \times n$ matrix A by minors across the first row. There are n minors, each of which requires at least $(n - 1)!$ multiplications, by our induction assumption. Thus there are at least $n(n - 1)! = n!$ multiplications required. (b) The answer can't be given without our knowing the speed of the computer. For our computer, with $n = 50$, we obtained 4.50564×10^{54} years in interpretive BASIC and 8.036651×10^{53} years in compiled BASIC, using single precision.

Section 3.3

1. 27   3. −8   5. −195   7. −207   9. 496   13. Let $R_1, R_2, \ldots, R_k$ be square submatrices positioned corner-to-corner along the diagonal of a matrix A, where A has zero entries except possibly for

those in the submatrices R_i. Then $\det(A) = \det(R_1) \det(R_2) \cdots \det(R_k)$. **15.** $4\sqrt{5}$ **17.** 0 **19.** 38
21. 20 **23.** 12 **25.** $\frac{7}{6}$ **27.** The volume of an n-box, two of whose edges coincide, is zero.
29. Coplanar **33.** F T T F T T F T F T **35.** Both YUREDUCE and MATCOMP yield the value 19095.
37. With default roundoff, the smallest such value of m is 6, which gave 0 rather than 1 as the value of the
determinant. While this wrong result would lead you to believe that A^6 is singular, the error is actually quite
small compared with the size of the entries in A^6. We obtained 0 as $\det(A^m)$ for $6 \le m \le 17$, and we obtained
-1.470838×10^{10} as $\det(A^{18})$. MATCOMP gave us -1.598896×10^{13} as $\det(A^{20})$.

With roundoff control 0, we obtained $\det(A^6) = 1$ and $\det(A^7) = 0.999998$, which has an error of only
0.000002. The error became significantly larger with $m = 12$, where we obtained 11.45003 for the determinant. We obtained the same result with $m = 20$ as we did with the default roundoff.

With the default roundoff, computed entries of magnitude less than (0.0001)(Smallest nonzero magnitude
in A^m) are set equal to zero when A^m is reduced to echelon form. As soon as m is large enough (so that the
entries of A^m are large enough), this creates a false zero entry on the diagonal, which produces a false calculation of 0 for the determinant. With roundoff zero, this does not happen. But when m get sufficiently large,
roundoff error in the computation of A^m and in its reduction to echelon form creates even greater error in the
calculation of the determinant, no matter what roundoff control number is taken. Notice, however, that the
value given for the determinant is always small compared with the size of the entries of A^m.

Section 3.4

1. 16 **3.** 224 **5.** 11 **7.** 330 **9.** $\sqrt{3}$ **11.** $9\sqrt{3}$ **13.** $\sqrt{17}$ **15.** $9\pi\sqrt{17}$
17. T F T F T F T T F T

Section 3.5

1. $\begin{bmatrix} \frac{1}{2} & 0 \\ \frac{1}{2} & -1 \end{bmatrix}$ **3.** A is not invertible. **5.** $\frac{1}{15}\begin{bmatrix} 2 & 1 & -1 \\ -2 & 9 & 6 \\ 3 & -1 & 1 \end{bmatrix}$ **7.** $\begin{bmatrix} 6 & -5 \\ 3 & 4 \end{bmatrix}$ **9.** $\frac{1}{3}\begin{bmatrix} d & -b \\ -c & a \end{bmatrix}$.

11. $x_1 = 1, x_2 = 0$ **13.** $x_1 = 5, x_2 = -10$ **15.** $x_1 = \frac{1}{2}, x_2 = \frac{1}{2}, x_3 = -\frac{1}{2}$
17. $x_1 = \frac{1}{3}, x_2 = \frac{1}{3}, x_3 = 0$ **19.** 10 **21.** $(0, 3, 0, 0)$. **23.** F T T F T F F F T F
27. $\begin{bmatrix} 12 & -8 & -7 \\ -11 & 13 & 5 \\ 9 & -6 & -1 \end{bmatrix}$ **29.** $\begin{bmatrix} -4827 & 114 & -2211 & 2409 \\ -3218 & 76 & -1474 & 1606 \\ 8045 & -190 & 3685 & -4015 \\ 1609 & -38 & 737 & -803 \end{bmatrix}$

CHAPTER 4

Section 4.1

1. *Characteristic polynomial:* $\lambda^2 - 3\lambda + 2$
Eigenvalues: $\lambda_1 = 1, \lambda_2 = 2$
Eigenvectors: for $\lambda_1 = 1$: $\mathbf{v}_1 = \begin{bmatrix} -r \\ r \end{bmatrix}, r \ne 0,$
for $\lambda_2 = 2$: $\mathbf{v}_2 = \begin{bmatrix} 0 \\ s \end{bmatrix}, s \ne 0$

3. *Characteristic polynomial:* $\lambda^2 - 4\lambda + 3$
Eigenvalues: $\lambda_1 = 1, \lambda_2 = 3$
Eigenvectors: for $\lambda_1 = 1$: $\mathbf{v}_1 = \begin{bmatrix} -r \\ r \end{bmatrix}, r \ne 0,$
for $\lambda_2 = 3$: $\mathbf{v}_2 = \begin{bmatrix} -s \\ 2s \end{bmatrix}, s \ne 0$

5. *Characteristic polynomial*: $\lambda^2 + 1$
Eigenvalues: There are no real eigenvalues.

7. *Characteristic polynomial*: $-\lambda^3 + 2\lambda^2 + \lambda - 2$
Eigenvalues: $\lambda_1 = -1$, $\lambda_2 = 1$, $\lambda_3 = 2$

Eigenvectors: for $\lambda_1 = -1$: $\mathbf{v}_1 = \begin{bmatrix} 0 \\ r \\ 0 \end{bmatrix}$, $r \neq 0$,

for $\lambda_2 = 1$: $\mathbf{v}_2 = \begin{bmatrix} 0 \\ -s \\ s \end{bmatrix}$, $s \neq 0$,

for $\lambda_3 = 2$: $\mathbf{v}_3 = \begin{bmatrix} -t \\ -t \\ t \end{bmatrix}$, $t \neq 0$

9. *Characteristic polynomial*: $-\lambda^3 + 8\lambda^2 + \lambda - 8$
Eigenvalues: $\lambda_1 = -1$, $\lambda_2 = 1$, $\lambda_3 = 8$

Eigenvectors: for $\lambda_1 = -1$: $\mathbf{v}_1 = \begin{bmatrix} 0 \\ r \\ 0 \end{bmatrix}$, $r \neq 0$,

for $\lambda_2 = 1$: $\mathbf{v}_2 = \begin{bmatrix} 0 \\ -s \\ s \end{bmatrix}$, $s \neq 0$,

for $\lambda_3 = 8$: $\mathbf{v}_3 = \begin{bmatrix} -t \\ -t \\ t \end{bmatrix}$, $t \neq 0$

11. *Characteristic polynomial*: $-\lambda^3 - \lambda^2 + 8\lambda + 12$
Eigenvalues: $\lambda_1 = \lambda_2 = -2$, $\lambda_3 = 3$

Eigenvectors: for λ_1, $\lambda_2 = -2$: $\mathbf{v}_1 = \begin{bmatrix} -r \\ s \\ r \end{bmatrix}$,

r and s not both 0,

for $\lambda_3 = 3$: $\mathbf{v}_3 = \begin{bmatrix} 0 \\ -t \\ t \end{bmatrix}$, $t \neq 0$

13. *Characteristic polynomial*: $-\lambda^3 + 2\lambda^2 + 4\lambda - 8$
Eigenvalues: $\lambda_1 = -2$, $\lambda_2 = \lambda_3 = 2$

Eigenvectors: for $\lambda_1 = -2$: $\mathbf{v}_1 = \begin{bmatrix} -r \\ -3r \\ r \end{bmatrix}$, $r \neq 0$,

for λ_2, $\lambda_3 = 2$: $\mathbf{v}_2 = \begin{bmatrix} 0 \\ s \\ 0 \end{bmatrix}$, $s \neq 0$

15. *Characteristic polynomial*: $-\lambda^3 - 3\lambda^2 + 4$
Eigenvalues: $\lambda_1 = \lambda_2 = -2$, $\lambda_3 = 1$

Eigenvectors: for λ_1, $\lambda_2 = -2$: $\mathbf{v}_1 = \begin{bmatrix} 0 \\ r \\ 0 \end{bmatrix}$, $r \neq 0$,

for $\lambda_3 = 1$: $\mathbf{v}_3 = \begin{bmatrix} 3s \\ -s \\ 3s \end{bmatrix}$, $s \neq 0$

17. *Eigenvalues*: $\lambda_1 = -1$, $\lambda_2 = 5$

Eigenvectors: for $\lambda_1 = -1$: $\mathbf{v}_1 = \begin{bmatrix} r \\ r \end{bmatrix}$, $r \neq 0$,

for $\lambda_2 = 5$: $\mathbf{v}_2 = \begin{bmatrix} -s \\ s \end{bmatrix}$, $s \neq 0$

19. *Eigenvalues*: $\lambda_1 = 0$, $\lambda_2 = 1$, $\lambda_3 = 2$

Eigenvectors: for $\lambda_1 = 0$: $\mathbf{v}_1 = \begin{bmatrix} -r \\ 0 \\ r \end{bmatrix}$, $r \neq 0$,

for $\lambda_2 = 1$: $\mathbf{v}_2 = \begin{bmatrix} 0 \\ s \\ 0 \end{bmatrix}$, $s \neq 0$,

for $\lambda_3 = 2$: $\mathbf{v}_3 = \begin{bmatrix} t \\ 0 \\ t \end{bmatrix}$, $t \neq 0$

21. *Eigenvalues*: $\lambda_1 = -2$, $\lambda_2 = 1$, $\lambda_3 = 5$

Eigenvectors: for $\lambda_1 = -2$: $\mathbf{v}_1 = \begin{bmatrix} 0 \\ r \\ r \end{bmatrix}$, $r \neq 0$, for $\lambda_2 = 1$: $\mathbf{v}_2 = \begin{bmatrix} -6s \\ 5s \\ 8s \end{bmatrix}$, $s \neq 0$,

for $\lambda_3 = 5$: $\mathbf{v}_3 = \begin{bmatrix} 0 \\ -5t \\ 2t \end{bmatrix}$, $t \neq 0$

23. F F T T F T F T F T

29. *Eigenvalues:* $\lambda_1 = \dfrac{1 + \sqrt{5}}{2}$, $\lambda_2 = \dfrac{1 - \sqrt{5}}{2}$

Eigenvectors: for $\lambda_1 = \dfrac{1 + \sqrt{5}}{2}$: $\mathbf{v}_1 = r\begin{bmatrix} (1 + \sqrt{5})/2 \\ 1 \end{bmatrix}$, $r \neq 0$,

for $\lambda_2 = \dfrac{1 - \sqrt{5}}{2}$: $\mathbf{v}_2 = s\begin{bmatrix} (1 - \sqrt{5})/2 \\ 1 \end{bmatrix}$, $s \neq 0$

31. (a) Work with the matrix $A + 10I$, whose eigenvalues are approximately 30, 12, 7, and -9.5. (Other answers are possible.); (b) Work with the matrix $A - 10I$, whose eigenvalues are approximately -9.5, -8, -13, and -30. (Other answers are possible.)

35. When the determinant of an $n \times n$ matrix A is expanded according to the definition using expansion by minors, a sum of $n!$ (signed) terms is obtained, each containing a product of one element from each row and from each column. (This is readily proved by induction on n.) One of the $n!$ terms obtained by expanding $|A - \lambda I|$ to obtain $p(\lambda)$ is

$$(a_{11} - \lambda)(a_{22} - \lambda)(a_{33} - \lambda) \cdots (a_{nn} - \lambda).$$

We claim that this is the only one of the $n!$ terms that can contribute to the coefficient of λ^{n-1} in $p(\lambda)$. Any term contributing to the coefficient of λ^{n-1} must contain at least $n - 1$ of the factors in the preceding product; the other factor must be from the remaining row and column, and hence it must be the remaining factor from the diagonal of $A - \lambda I$. Computing the coefficient of λ^{n-1} in the preceding product, we find that, when $-\lambda$ is chosen from all but the factor $a_{ii} - \lambda$ in expanding the product, the resulting contribution to the coefficient of λ^{n-1} in $p(\lambda)$ is $(-1)^{n-1}a_{ii}$. Thus, the coefficient of λ^{n-1} in $p(\lambda)$ is

$$(-1)^{n-1}(a_{11} + a_{22} + a_{33} + \cdots + a_{nn}).$$

Now $p(\lambda) = (-1)^n(\lambda - \lambda_1)(\lambda - \lambda_2)(\lambda - \lambda_3) \cdots (\lambda - \lambda_n)$, and computation of this product shows in the same way that the coefficient of λ^{n-1} is also

$$(-1)^{n-1}(\lambda_1 + \lambda_2 + \lambda_3 + \cdots + \lambda_n).$$

Thus, $\text{tr}(A) = \lambda_1 + \lambda_2 + \lambda_3 + \cdots + \lambda_n$.

37. Since $\begin{vmatrix} 2 - \lambda & -1 \\ 1 & 3 - \lambda \end{vmatrix} = \lambda^2 - 5\lambda + 7$, we compute

$$A^2 - 5A + 7I = \begin{bmatrix} 2 & -1 \\ 1 & 3 \end{bmatrix}^2 - 5\begin{bmatrix} 2 & -1 \\ 1 & 3 \end{bmatrix} + 7\begin{bmatrix} 1 & 0 \\ 0 & 1 \end{bmatrix} = \begin{bmatrix} 3 & -5 \\ 5 & 8 \end{bmatrix} + \begin{bmatrix} -10 & 5 \\ -5 & -15 \end{bmatrix} + \begin{bmatrix} 7 & 0 \\ 0 & 7 \end{bmatrix} = \begin{bmatrix} 0 & 0 \\ 0 & 0 \end{bmatrix},$$

illustrating the Cayley–Hamilton theorem.

39. Look ahead at the proof of Theorem 4.3 in Section 4.2. The proof of this exercise is word-for-word the same, except that we must apply the transformation T instead of multiplying by A. **41.** (a) $F_8 = 21$; (b) $F_{30} = 832040$; (c) $F_{50} = 12586269025$; (d) $F_{77} = 5527939700884757$; (e) $F_{150} \approx 9.969216677189304 \times 10^{30}$

43. (a) 0, 1, 2, 1, -3, -7, -4, 10, 25; (b) $\begin{bmatrix} 2 & -3 & 1 \\ 1 & 0 & 0 \\ 0 & 1 & 0 \end{bmatrix}$; (c) $a_{30} = -191694$

45. $\lambda_1 = -6$, $\mathbf{v}_1 = \begin{bmatrix} -2r \\ 2r \\ r \end{bmatrix}$, $r \neq 0$; $\lambda_2 = 3$, $\mathbf{v}_2 = \begin{bmatrix} s \\ -s \\ s \end{bmatrix}$, $s \neq 0$; $\lambda_3 = 9$, $\mathbf{v}_3 = \begin{bmatrix} -2t \\ -t \\ t \end{bmatrix}$, $t \neq 0$.

47. $\lambda_1 = -2$, $\mathbf{v}_1 = \begin{bmatrix} r \\ r \\ r \\ 0 \end{bmatrix}$, $r \neq 0$; $\lambda_2 = 2$, $\mathbf{v}_2 = \begin{bmatrix} s \\ 0 \\ 0 \\ s \end{bmatrix}$, $s \neq 0$; $\lambda_3 = 4$, $\mathbf{v}_3 = \begin{bmatrix} 0 \\ t \\ 0 \\ 0 \end{bmatrix}$, $t \neq 0$; $\lambda_4 = 6$, $\mathbf{v}_4 = \begin{bmatrix} 0 \\ 0 \\ u \\ u \end{bmatrix}$, $u \neq 0$.

49. *Characteristic polynomial:* $-\lambda^3 + 19\lambda^2 + 18\lambda + 54$
 Eigenvalues: $\lambda_1 \approx -0.5165 + 1.5584i$, $\lambda_2 \approx -0.5165 - 1.5584i$, $\lambda_3 \approx 20.0331$

51. *Characteristic polynomial:* $\lambda^4 + 14\lambda^3 - 131\lambda^2 + 739\lambda - 21533$
 Eigenvalues: $\lambda_1 \approx -0.7828 + 9.4370i$, $\lambda_2 \approx -0.7828 - 9.4370i$, $\lambda_3 \approx -22.9142$, $\lambda_4 \approx 10.4799$

53. The characteristic polynomial is

$$\lambda^4 - 7\lambda^3 - 160\lambda^2 + 789\lambda + 4639.$$

Calling A the given matrix, we indeed find that

$$A^4 - 7A^3 - 160A^2 + 789A + 4639I = O.$$

Section 4.2

1. $\mathbf{v}_1$, $\mathbf{v}_3$, and $\mathbf{v}_5$ are eigenvectors of A_1, with corresponding eigenvalues -1, 2, and 2, respectively. $\mathbf{v}_2$, $\mathbf{v}_3$, and $\mathbf{v}_4$ are eigenvectors of A_2, with corresponding eigenvalues 5, 5, and 1, respectively.

3. *Eigenvalues:* $\lambda_1 = -1$, $\lambda_2 = 3$
 Eigenvectors: for $\lambda_1 = -1$: $\mathbf{v}_1 = \begin{bmatrix} -r \\ r \end{bmatrix}$, $r \neq 0$,
 for $\lambda_2 = 3$: $\mathbf{v}_2 = \begin{bmatrix} -2s \\ s \end{bmatrix}$, $s \neq 0$
 $C = \begin{bmatrix} -1 & -2 \\ 1 & 1 \end{bmatrix}$, $D = \begin{bmatrix} -1 & 0 \\ 0 & 3 \end{bmatrix}$

5. *Eigenvalues:* $\lambda_1 = -3$, $\lambda_2 = 1$, $\lambda_3 = 7$
 Eigenvectors: for $\lambda_1 = -3$: $\mathbf{v}_1 = \begin{bmatrix} r \\ 0 \\ 0 \end{bmatrix}$, $r \neq 0$,
 for $\lambda_2 = 1$: $\mathbf{v}_2 = \begin{bmatrix} s \\ s \\ s \end{bmatrix}$, $s \neq 0$,
 for $\lambda_3 = 7$: $\mathbf{v}_3 = \begin{bmatrix} t \\ t \\ 0 \end{bmatrix}$, $t \neq 0$
 $C = \begin{bmatrix} 1 & 1 & 1 \\ 0 & 1 & 1 \\ 0 & 1 & 0 \end{bmatrix}$, $D = \begin{bmatrix} -3 & 0 & 0 \\ 0 & 1 & 0 \\ 0 & 0 & 7 \end{bmatrix}$

7. *Eigenvalues:* $\lambda_1 = -1$, $\lambda_2 = 1$, $\lambda_3 = 2$
 Eigenvectors: for $\lambda_1 = -1$: $\mathbf{v}_1 = \begin{bmatrix} r \\ 0 \\ r \end{bmatrix}$, $r \neq 0$,

$$\text{for } \lambda_2 = 1: \mathbf{v}_2 = \begin{bmatrix} -s \\ 0 \\ 3s \end{bmatrix}, s \neq 0,$$

$$\text{for } \lambda_3 = 2: \mathbf{v}_3 = \begin{bmatrix} 0 \\ t \\ 0 \end{bmatrix}, t \neq 0$$

$$C = \begin{bmatrix} -1 & -1 & 0 \\ 0 & 0 & 1 \\ 1 & 3 & 0 \end{bmatrix}, D = \begin{bmatrix} -1 & 0 & 0 \\ 0 & 1 & 0 \\ 0 & 0 & 2 \end{bmatrix}$$

9. Yes, the matrix is symmetric.   **11.** Yes, the eigenvalues are distinct.   **13.** F T T F T F F T F T

15. Assume that A is a square matrix such that $D = C^{-1}AC$ is a diagonal matrix for some invertible matrix C. Since $CC^{-1} = I$, we have $(CC^{-1})^T = (C^{-1})^TC^T = I^T = I$, so $(C^T)^{-1} = (C^{-1})^T$. Then $D^T = (C^{-1}AC)^T = C^TA^T(C^T)^{-1}$, so A^T is similar to the diagonal matrix D^T.   **21.** Suppose that a linear combination $\mathbf{w}$ of vectors in the union $B_1 \cup B_2 \cup \cdots \cup B_m$ is zero. The linear combination $\mathbf{w}$ can be expressed in the form $\mathbf{w} = \mathbf{w}_1 + \mathbf{w}_2 + \cdots + \mathbf{w}_m$, where each $\mathbf{w}_i$ consists of the portion of the linear combination $\mathbf{w}$ that involves vectors in B_i. Since the eigenvalues λ_i are distinct for $i = 1, 2, \ldots, m$ and since $\mathbf{w}_i \in B_i$, Theorem 4.3 shows that the $\mathbf{w}_i$ are linearly independent. Thus, from $\mathbf{w} = \mathbf{w}_1 + \mathbf{w}_2 + \cdots + \mathbf{w}_m = \mathbf{0}$, we conclude that $\mathbf{w}_i = \mathbf{0}$ for $i = 1, 2, \ldots, m$. Since each $\mathbf{w}_i$ in turn is a linear combination of vectors in the basis B_i for E_{λ_i}, we conclude that the coefficients in the linear combination constituting each $\mathbf{w}_i$ must be zero. That is, all the coefficients of vectors in the original linear combination giving $\mathbf{w}$ must be zero. This shows that $B_1 \cup B_2 \cup \cdots \cup B_m$ is an independent set of vectors.   **25.** We indeed found that $C^{-1}AC = D$.   **27.** Not diagonalizable   **29.** Diagonalizable   **31.** Not diagonalizable   **33.** Diagonalizable   **35.** Not diagonalizable

Section 4.3

1. (a) $A = \begin{bmatrix} \frac{1}{2} & \frac{1}{2} \\ 1 & 0 \end{bmatrix}$; (b) Neutrally stable; (c) $\begin{bmatrix} a_{k+1} \\ a_k \end{bmatrix} = A^k \begin{bmatrix} 1 \\ 0 \end{bmatrix} = \frac{2}{3}(1)^k \begin{bmatrix} 1 \\ 1 \end{bmatrix} - \frac{1}{3}\left(-\frac{1}{2}\right)^k \begin{bmatrix} -1 \\ 2 \end{bmatrix}$. The sequence

starts $0, 1, \frac{1}{2}, \frac{3}{4}, \frac{5}{8}$ and $A^3 \begin{bmatrix} 1 \\ 0 \end{bmatrix} = \frac{2}{3}\begin{bmatrix} 1 \\ 1 \end{bmatrix} + \frac{1}{24}\begin{bmatrix} -1 \\ 2 \end{bmatrix} = \begin{bmatrix} \frac{5}{8} \\ \frac{3}{4} \end{bmatrix}$, which checks. (d) For large k, we have $\begin{bmatrix} a_{k+1} \\ a_i \end{bmatrix} \approx$

$\frac{2}{3}\begin{bmatrix} 1 \\ 1 \end{bmatrix} = \begin{bmatrix} \frac{2}{3} \\ \frac{2}{3} \end{bmatrix}$, so $a_k \approx \frac{2}{3}$.

3. (a) $A = \begin{bmatrix} \frac{1}{2} & \frac{1}{2} \\ 1 & 0 \end{bmatrix}$; (b) Neutrally stable; (c) $\begin{bmatrix} a_{k+1} \\ a_k \end{bmatrix} = A^k \begin{bmatrix} 0 \\ 1 \end{bmatrix} = \frac{1}{3}(1)^k \begin{bmatrix} 1 \\ 1 \end{bmatrix} + \frac{1}{3}\left(-\frac{1}{2}\right)^k \begin{bmatrix} -1 \\ 2 \end{bmatrix}$. The sequence

starts $1, 0, \frac{1}{2}, \frac{1}{4}, \frac{3}{8}$ and $A^3 \begin{bmatrix} 1 \\ 1 \end{bmatrix} = \frac{1}{3}\begin{bmatrix} 1 \\ 1 \end{bmatrix} - \frac{1}{24}\begin{bmatrix} -1 \\ 2 \end{bmatrix} = \begin{bmatrix} \frac{3}{8} \\ \frac{1}{4} \end{bmatrix}$, which checks. (d) For large k, we have

$\begin{bmatrix} a_{k+1} \\ a_k \end{bmatrix} \approx \frac{1}{3}\begin{bmatrix} 1 \\ 1 \end{bmatrix} = \begin{bmatrix} \frac{1}{3} \\ \frac{1}{3} \end{bmatrix}$, so $a_k \approx \frac{1}{3}$.   **5.** (a) $A = \begin{bmatrix} 1 & \frac{3}{4} \\ 1 & 0 \end{bmatrix}$;

(b) Unstable; (c) $\begin{bmatrix} a_{k+1} \\ a_k \end{bmatrix} = A^k \begin{bmatrix} 1 \\ 0 \end{bmatrix} = \frac{1}{4}\left(\frac{3}{2}\right)^k \begin{bmatrix} 3 \\ 2 \end{bmatrix} - \frac{1}{4}\left(-\frac{1}{2}\right)^k \begin{bmatrix} -1 \\ 2 \end{bmatrix}$. The sequence starts $0, 1, 1, \frac{7}{4}, \frac{5}{2}$ and

$A^3 \begin{bmatrix} 1 \\ 0 \end{bmatrix} = \frac{27}{32}\begin{bmatrix} 3 \\ 2 \end{bmatrix} + \frac{1}{32}\begin{bmatrix} -1 \\ 2 \end{bmatrix} = \begin{bmatrix} \frac{80}{32} \\ \frac{56}{32} \end{bmatrix} = \begin{bmatrix} \frac{5}{2} \\ \frac{7}{4} \end{bmatrix}$, which checks. (d) For large k we have $\begin{bmatrix} a_{k+1} \\ a_k \end{bmatrix} \approx$

$\frac{1}{4}(\frac{3}{2})^k \begin{bmatrix} 3 \\ 2 \end{bmatrix}$, so $a_k \approx \dfrac{3^k}{2^{k+1}}$; and a_k approaches ∞ as k approaches ∞.   **7.** $\begin{bmatrix} x_1 \\ x_2 \end{bmatrix} = \begin{bmatrix} -k_1 e^{-3t} + 4k_2 e^{4t} \\ k_1 e^{-3t} + 3k_2 e^{4t} \end{bmatrix}$

9. $\begin{bmatrix} x_1 \\ x_2 \end{bmatrix} = \begin{bmatrix} -2k_1 e^t + k_2 e^{4t} \\ k_1 e^t + k_2 e^{4t} \end{bmatrix}$ **11.** $\begin{bmatrix} x_1 \\ x_2 \\ x_3 \end{bmatrix} = \begin{bmatrix} k_1 e^{-3t} + k_2 e^t + k_3 e^{7t} \\ k_2 e^t + k_3 e^{7t} \\ k_2 e^t \end{bmatrix}$

13. $\begin{bmatrix} x_1 \\ x_2 \\ x_3 \end{bmatrix} = \begin{bmatrix} -k_1 e^{-t} - k_2 e^t \\ k_3 e^{2t} \\ k_1 e^{-t} + 3k_2 e^t \end{bmatrix}$

CHAPTER 5

Section 5.1

1. $\frac{2}{5}(3, 4)$ **3.** $\mathbf{p}_1 = (1, 0, 0)$, $\mathbf{p}_2 = (0, 2, 0)$, $\mathbf{p}_3 = (0, 0, 1)$ **5.** $\mathbf{p} = -\frac{1}{3}(2, -3, 1, 2)$
7. $\text{sp}((1, 0, 1), (-2, 1, 0))$ **9.** $\text{sp}((-12, 4, 5))$ **11.** $\text{sp} = ((2, -7, 1, 0), (-1, -2, 0, 1))$
13. $-5\mathbf{i} + 3\mathbf{j} + \mathbf{k}$ **15.** $\frac{1}{3}(5, 4, 1)$ **17.** $\frac{1}{7}(5, 3, 1)$ **19.** $\frac{1}{6}(2, -1, 5)$ **21.** $\frac{1}{3}(3, -2, -1, 1)$
23. $\frac{3}{2}x$ **25.** F T T T F T T F F T   **33.** $\sqrt{\frac{5}{2}}$ **35.** $5\sqrt{\frac{2}{7}}$ **37.** $\sqrt{\frac{29}{15}}$ **39.** $2\sqrt{\frac{7}{11}}$

Section 5.2

1. $(2, 3, 1) \cdot (-1, 1, -1) = -2 + 3 - 1 = 0$, so the generating set is orthogonal. $\mathbf{b}_W = \frac{1}{42}(136, 29, 103)$.
3. $(1, -1, -1, 1) \cdot (1, 1, 1, 1) = 1 - 1 - 1 + 1 = 0$,
$(1, -1, -1, 1) \cdot (-1, 0, 0, 1) = -1 + 0 + 0 + 1 = 0$, and
$(1, 1, 1, 1) \cdot (-1, 0, 0, 1) = -1 + 0 + 0 + 1 = 0$,

so the generating set is orthogonal. $\mathbf{b}_W = (2, 2, 2, 1)$.   **5.** $\left\{ \dfrac{1}{\sqrt{5}}(1, 0, -2), \dfrac{1}{\sqrt{70}}(6, -5, 3) \right\}$

7. $\left\{ (0, 1, 0), \dfrac{1}{\sqrt{2}}(1, 0, 1) \right\}$ **9.** $\left\{ \dfrac{1}{\sqrt{2}}(1, 0, 1), \dfrac{1}{\sqrt{3}}(-1, 1, 1), \dfrac{1}{\sqrt{6}}(1, 2, -1) \right\}$

11. $\left\{ \dfrac{1}{\sqrt{2}}(1, 0, 1, 0), (0, 1, 0, 0), \dfrac{1}{\sqrt{6}}(1, 0, -1, 2) \right\}$ **13.** $\left(\dfrac{9}{2}, -3, \dfrac{9}{2} \right)$

15. $\left(\dfrac{4}{3}, 0, \dfrac{-1}{3}, \dfrac{5}{3} \right)$

17. $\left\{ \dfrac{1}{\sqrt{2}}(1, 0, 1, 0), \dfrac{1}{\sqrt{6}}(-1, 2, 1, 0), \dfrac{1}{\sqrt{3}}(1, 1, -1, 0), (0, 0, 0, 1) \right\}$

19. $\{(3, -2, 0, 1), (-9, -8, 14, 11)\}$ **21.** $\left\{ \dfrac{1}{\sqrt{6}}(2, 1, 1), \dfrac{1}{\sqrt{2}}(0, -1, 1) \right\}$

23. $\{(2, 1, -1, 1), (1, 1, 3, 0), (-24, 9, 5, 44)\}$ **25.** F T T F F T F T T T

27. $Q = \begin{bmatrix} 1/\sqrt{2} & -1/\sqrt{3} & 1/\sqrt{6} \\ 0 & 1/\sqrt{3} & 2/\sqrt{6} \\ 1/\sqrt{2} & 1/\sqrt{3} & -1/\sqrt{6} \end{bmatrix}$, $R = \begin{bmatrix} \sqrt{2} & \sqrt{2} & \sqrt{2} \\ 0 & \sqrt{3} & -1/\sqrt{3} \\ 0 & 0 & 4/\sqrt{6} \end{bmatrix}$ **31.** $\left\{ \dfrac{1}{\sqrt{2}}, \dfrac{\sqrt{3}x}{\sqrt{2}}, \dfrac{3\sqrt{5}}{2\sqrt{2}}(x^2 - \frac{1}{3}) \right\}$

37. The answers checked when we used QRFACTOR.

Section 5.3

1. Let A be the given matrix. Then

$$A^T A = \frac{1}{\sqrt{2}} \begin{bmatrix} 1 & -1 \\ 1 & 1 \end{bmatrix} \frac{1}{\sqrt{2}} \begin{bmatrix} 1 & 1 \\ -1 & 1 \end{bmatrix} = \frac{1}{2} \begin{bmatrix} 2 & 0 \\ 0 & 2 \end{bmatrix} = \begin{bmatrix} 1 & 0 \\ 0 & 1 \end{bmatrix},$$

so A is orthogonal and $A^{-1} = A^T = \dfrac{1}{\sqrt{2}}\begin{bmatrix} 1 & -1 \\ 1 & 1 \end{bmatrix}$.

3. Let A be the given matrix. Then

$$A^T A = \tfrac{1}{7}\begin{bmatrix} 2 & 3 & -6 \\ -3 & 6 & 2 \\ 6 & 2 & 3 \end{bmatrix} \tfrac{1}{7}\begin{bmatrix} 2 & -3 & 6 \\ 3 & 6 & 2 \\ -6 & 2 & 3 \end{bmatrix} = \tfrac{1}{49}\begin{bmatrix} 49 & 0 & 0 \\ 0 & 49 & 0 \\ 0 & 0 & 49 \end{bmatrix} = \begin{bmatrix} 1 & 0 & 0 \\ 0 & 1 & 0 \\ 0 & 0 & 1 \end{bmatrix},$$

so A is orthogonal and $A^{-1} = A^T = \tfrac{1}{7}\begin{bmatrix} 2 & 3 & -6 \\ -3 & 6 & 2 \\ 6 & 2 & 3 \end{bmatrix}$.

5. $\begin{bmatrix} \frac{1}{2} & -\frac{1}{2} \\ \frac{1}{6} & \frac{1}{6} \end{bmatrix}$ 7. $\tfrac{1}{49}\begin{bmatrix} 1 & \frac{3}{2} & -3 \\ -3 & 6 & 2 \\ 6 & 2 & 3 \end{bmatrix}$ 9. $\pm\dfrac{1}{\sqrt{6}}\begin{bmatrix} -1 \\ 2 \\ -1 \end{bmatrix}$ 11. $\pm\sqrt{23}/6$ 13. $\dfrac{1}{\sqrt{2}}\begin{bmatrix} -1 & 1 \\ 1 & 1 \end{bmatrix}$

15. $\tfrac{1}{2}\begin{bmatrix} 1 & -2 & 1 \\ -\sqrt{2} & 0 & \sqrt{2} \\ 1 & 2 & 1 \end{bmatrix}$ 17. $\begin{bmatrix} 0 & \frac{1}{2} & -1/\sqrt{2} & \frac{1}{2} \\ -1/\sqrt{2} & -\frac{1}{2} & 0 & \frac{1}{2} \\ 1/\sqrt{2} & -\frac{1}{2} & 0 & \frac{1}{2} \\ 0 & \frac{1}{2} & 1/\sqrt{2} & \frac{1}{2} \end{bmatrix}$ 19. F T T T T T T T F T   23. $\begin{bmatrix} 2 & 1 \\ 1 & 1 \end{bmatrix}$

27. An orthogonal matrix A gives rise to an orthogonal linear transformation $T(\mathbf{x}) = A\mathbf{x}$ that preserves the magnitude of vectors. Thus, if $A\mathbf{v} = \lambda\mathbf{v}$, so that $T(\mathbf{v}) = \lambda\mathbf{v}$, we must have $\|\mathbf{v}\| = \|A\mathbf{v}\| = |\lambda|\|\mathbf{v}\|$. If $\mathbf{v}$ is an eigenvector, so that $\mathbf{v} \neq \mathbf{0}$, it follows that $|\lambda| = 1$; so $\lambda = \pm 1$.   **33.** No   **35.** Yes   **37.** Yes

Section 5.4

1. $P = \tfrac{1}{6}\begin{bmatrix} 4 & 2 & -2 \\ 2 & 1 & -1 \\ -2 & -1 & 1 \end{bmatrix}$, projection $= \tfrac{1}{2}\begin{bmatrix} 2 \\ 1 \\ -1 \end{bmatrix}$ 3. $P = \tfrac{1}{35}\begin{bmatrix} 34 & -3 & 5 \\ -3 & 26 & 15 \\ 5 & 15 & 10 \end{bmatrix}$, projection $= \tfrac{1}{35}\begin{bmatrix} 86 \\ 13 \\ 25 \end{bmatrix}$

5. $P = \tfrac{1}{6}\begin{bmatrix} 5 & -1 & 2 \\ -1 & 5 & 2 \\ 2 & 2 & 2 \end{bmatrix}$, projection $= \tfrac{1}{3}\begin{bmatrix} 2 \\ 8 \\ 5 \end{bmatrix}$ 7. $P = \tfrac{1}{21}\begin{bmatrix} 10 & -1 & 3 & 10 \\ -1 & 19 & 6 & -1 \\ 3 & 6 & 3 & 3 \\ 10 & -1 & 3 & 10 \end{bmatrix}$, projection $= \tfrac{1}{21}\begin{bmatrix} 41 \\ 40 \\ 27 \\ 41 \end{bmatrix}$

9. $\begin{bmatrix} 1 & 0 & 0 \\ 0 & 1 & 0 \\ 0 & 0 & 0 \end{bmatrix}$ 11. $\begin{bmatrix} 1 & 0 & 0 & 0 \\ 0 & 1 & 0 & 0 \\ 0 & 0 & 0 & 0 \\ 0 & 0 & 0 & 1 \end{bmatrix}$

13. If P is the projection matrix for a subspace W of $\mathbb{R}^n$ and if $\mathbf{b} \in \mathbb{R}^n$ is a column vector, then the projection of $\mathbf{b}$ on W is $P\mathbf{b}$. Since $P\mathbf{b}$ is in W, geometry indicates that the projection of $P\mathbf{b}$ on W is again $P\mathbf{b}$. Thus, $P(P\mathbf{b}) = P\mathbf{b}$, so $P^2\mathbf{b} = P\mathbf{b}$ and $(P^2 - P)\mathbf{b} = \mathbf{0}$ for all vectors $\mathbf{b} \in \mathbb{R}^n$. It follows from Exercise 41 of Section 1.1 that $P^2 - P = O$, so $P^2 = P$.

15. F T T F F T F F T T   17. I

19. (a) 0, 1;
    (b) 0 has geometric and algebraic multiplicity $n - k$,
        1 has geometric and algebraic multiplicity k;
    (c) Since the algebraic and geometric multiplicities of each eigenvalue are equal, P is a diagonalizable matrix.

21. The $n \times n$ identity matrix I for each positive integer n.   23. $\begin{bmatrix} \frac{9}{25} & \frac{12}{25} & 0 \\ \frac{12}{25} & \frac{16}{25} & 0 \\ 0 & 0 & 1 \end{bmatrix}$

25. $\frac{1}{49}\begin{bmatrix} 13 & -18 & 0 & -12 \\ -18 & 36 & 12 & 0 \\ 0 & 12 & 13 & -18 \\ -12 & 0 & -18 & 36 \end{bmatrix}$ **27.** $\frac{1}{3}\begin{bmatrix} 3 & 0 & 0 & 0 \\ 0 & 2 & -1 & 1 \\ 0 & -1 & 2 & 1 \\ 0 & 1 & 1 & 2 \end{bmatrix}$ **29.** $\begin{bmatrix} 10 \\ -4 \\ -2 \end{bmatrix}$ **31.** $\begin{bmatrix} 14 \\ 0 \\ -7 \\ 14 \end{bmatrix}$

33. Referring to Figure 5.11, we see that, for $\mathbf{p} = P\mathbf{b}$, the vector from the point $\mathbf{b}$ to the point $\mathbf{p}$ is $\mathbf{p} - \mathbf{b}$, which is also the required vector from the point $\mathbf{p}$ to the point $\mathbf{b}_r$. Thus, the vector $\mathbf{b}_r = \mathbf{b} + 2(\mathbf{p} - \mathbf{b}) = 2\mathbf{p} - \mathbf{b} = 2(P\mathbf{b}) - \mathbf{b} = (2P - I)\mathbf{b}$.

35. The projections are approximately $\begin{bmatrix} 1.151261 \\ -1.184874 \\ 3.89916 \end{bmatrix}$, $\begin{bmatrix} 3 \\ 3 \\ -1 \end{bmatrix}$, and $\begin{bmatrix} 1.932773 \\ -.806723 \\ 4.378151 \end{bmatrix}$, respectively.

37. The projections are approximately $\begin{bmatrix} 1.864516 \\ 1.496774 \\ -.135484 \\ 2.819355 \end{bmatrix}$, $\begin{bmatrix} 1.058064 \\ .787097 \\ -.941936 \\ 2.077419 \end{bmatrix}$, and $\begin{bmatrix} 4.116129 \\ 2.574194 \\ 1.116129 \\ 3.154839 \end{bmatrix}$, respectively.

Section 5.5

1. (a) $y = \frac{116.4}{59} + \frac{60.4}{59}x$; **(b)** ≈ 7.092 inches **3. (a)** $y = .528e^{.274x}$; **(b)** $\approx \$27,300$

5. $y = \frac{33}{35} + \frac{102}{35}x$ **7.** $y = -0.9 + 2.6x$

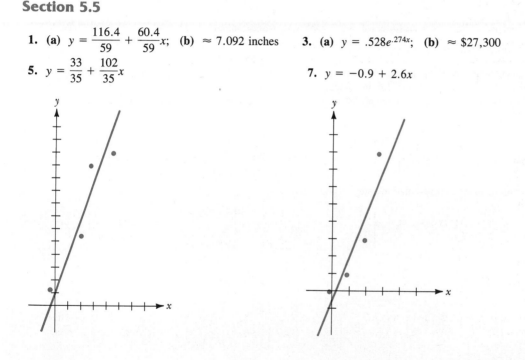

9. $y = 0.1 - 0.4x + x^2$ **11.** $y = 1.6 + 2x$ **13.** 4.5 min **15.** Let $t = x - c$, where $c = (\sum_{i=1}^{m} a_i)/m$. The data points $(a_1 - c, b_1), (a_2 - c, b_2), \ldots, (a_m - c, b_m)$ have the property that $\sum_{i=1}^{m} (a_i - c) = 0$. Exercise 14 then shows that these data points have least-squares linear fit given by $y = r_0 + r_1 t$, where r_0 and r_1 have the values given in Exercise 14. Making the substitution $t = x - c$, we see that the data points $(a_1, b_1), (a_2, b_2), \ldots, (a_m, b_m)$ have the least-squares linear fit given by $y = r_0 + r_1(x - c)$. **17.** $\bar{x} = \begin{bmatrix} -\frac{1}{5} \\ \frac{3}{5} \end{bmatrix}$ **19.** $\bar{x} = \begin{bmatrix} 0 \\ 2 \\ -\frac{1}{4} \end{bmatrix}$

21. FFTTFFTTFF **23.** $y = 0.4605263 + 0.3026316x$ **25.** $y = -1.5 + 2.6x + 0x^2$
27. $\bar{x}_1 = 0.4285714$, $\bar{x}_2 = -0.9$, $\bar{x}_3 = 0$
29. The computer gave the fit

$$y = -4.650418 + 3.703938x - 0.2053845x^2,$$

with least-squares sum 0.05979228.
31. We achieved a least-square sum of 0.09095781 with the exponential fit $y = 15e^{-0.51x}$. The computer achieved a least-squares sum of 43.36322 with the exponential fit $y = 35.6377e^{-0.8283194x}$. **33.** $y \approx 6.571 - 0.75x$
35. $y \approx 5.476 - 1.139x + 0.2738x^2 + 0.05556x^3$ **37.** $y = -1 + 2x + 3x^2$

CHAPTER 6

Section 6.1

1. $\mathbb{R}^2$ **3.** sp$((2, 3, 1), (0, 1, 1))$ **5.** $\mathbb{R}^2$ **7.** $\mathbb{R}$ **9.** $\{\mathbf{0}\}$ **11.** $\{\mathbf{0}\}$ **13.** $\{\mathbf{0}\}$
15. An isomorphism **17.** Not an isomorphism **19.** An isomorphism **21.** $\{\mathbf{0}\}$
23. $\lambda_1 = 1$, $\lambda_2 = 2$, $E_1 = $ sp$(-x + 3)$, $E_2 = $ sp(1) **25.** $\lambda_1 = \lambda_2 = 1$, $E_1 = $ sp$(\sin x + \cos x)$
27. Any $c \in \mathbb{R}$, $p(x) = 0$, **29.** FTTTFTTFFT
$c = -1$, $p(x) = r(-2x + 1)$, for any scalar r,
$c = 1$ and $p(x) = s(-x^2 + x) + t$, for any scalars s and t.
33. Since $(rT)(\mathbf{x}) = r(T(\mathbf{x})) = r(A\mathbf{x}) = (rA)\mathbf{x}$, for all $\mathbf{x} \in \mathbb{R}^n$, the standard matrix representation of rT is rA. For example, if

$$T(\mathbf{x}) = \begin{bmatrix} x_1 + 3x_2 \\ 2x_1 - x_2 \\ 3x_1 \end{bmatrix}, \quad \text{then} \quad (3T)(\mathbf{x}) = \begin{bmatrix} 3x_1 + 9x_2 \\ 6x_1 - 3x_2 \\ 9x_1 \end{bmatrix},$$

and the standard matrix representations are, respectively,

$$A = \begin{bmatrix} 1 & 3 \\ 2 & -1 \\ 3 & 0 \end{bmatrix} \quad \text{and} \quad 3A = \begin{bmatrix} 3 & 9 \\ 6 & -3 \\ 9 & 0 \end{bmatrix}.$$

Section 6.2

1. (a) $\begin{bmatrix} 2 & -3 \\ 1 & 1 \end{bmatrix}$; **(b)** $\begin{bmatrix} 3 & -1 \\ -1 & 2 \end{bmatrix}$ **3. (a)** $\begin{bmatrix} 1 & 1 & 1 \\ 1 & -1 & -1 \\ -1 & -1 & 1 \end{bmatrix}$; **(b)** $\begin{bmatrix} 6 & 3 & 5 \\ -2 & -1 & -1 \\ -4 & -3 & 1 \end{bmatrix}$

5. (a) $\begin{bmatrix} 1 & 1 & 0 \\ 0 & 1 & 1 \\ 1 & 0 & 1 \\ 1 & 1 & 1 \end{bmatrix}$; **(b)** $\begin{bmatrix} 3 & 2 & 1 \\ -1 & -1 & 0 \\ 0 & 0 & -1 \\ 0 & 1 & 1 \end{bmatrix}$

7. If A is the standard matrix representation of T, then $R_{B,B'}$ is obtained by row-reducing $[M_{B'} \mid AM_B]$ to $[I \mid R_{B,B'}]$. Thus the rank of $R_{B,B'}$ is equal to the rank of AM_B. The column space of AM_B is generated by the column vectors $A\mathbf{b}_1, A\mathbf{b}_2, \ldots, A\mathbf{b}_n$ of this matrix. The column space of A consists of all vectors $A\mathbf{x}$ for $\mathbf{x} \in \mathbb{R}^n$. Thus the column space of A contains the column space of AM_B. The reverse containment holds as well; since $B = \{\mathbf{b}_1, \mathbf{b}_2, \ldots, \mathbf{b}_n\}$ is a basis for $\mathbb{R}^n$, each $\mathbf{x} \in \mathbb{R}^n$ can be expressed in the form

$$\mathbf{x} = c_1\mathbf{b}_1 + c_2\mathbf{b}_2 + \cdots + c_n\mathbf{b}_n,$$

so $A\mathbf{x}$ is also a linear combination of the column vectors of AM_B. Thus, AM_B and A have the same column space. This shows that every matrix representation $R_{B,B'}$ of T has the same rank as the standard matrix representation A, and therefore each has the same nullity as A, by the rank equation. Since the column space of A is the same as the range of T and since the nullspace of A is the kernel of T, we see that T has the same rank and nullity as A—and consequently the same rank and nullity as any matrix representation $R_{B,B'}$.

9. Rank 2, nullity 0 **11.** Rank 3, nullity 0 **13.** Rank 2, nullity 0 **15.** $\begin{bmatrix} 0 & -1 \\ -1 & 0 \end{bmatrix}$

17. $\begin{bmatrix} \dfrac{1-m^2}{1+m^2} & \dfrac{2m}{1+m^2} \\ \dfrac{2m}{1+m^2} & \dfrac{m^2-1}{1+m^2} \end{bmatrix}$ **19.** $\begin{bmatrix} 1 & -2 & -1 \\ 0 & -1 & -1 \\ 0 & 0 & 1 \end{bmatrix}$ **21.** T T F T T F T F F T **23. (a)** $\begin{bmatrix} 2 & 1 \\ 0 & 2 \end{bmatrix}$;

(b) $32e^{2x} + 16xe^{2x}$ **25.** $u^2 - v^2 = 1$ **27.** $\begin{bmatrix} 3 & 1 & -2 \\ 1 & 2 & -1 \\ 4 & -1 & 2 \\ -1 & 2 & 0 \end{bmatrix}$ **29.** $M_{B'}^{-1}M_B R_{B,B} M_B^{-1} M_{B'}$

31. (a) $\begin{bmatrix} 2 & 0 & 0 \\ 0 & 2 & 0 \\ 8 & 0 & 3 \end{bmatrix}$; **(b)** $\begin{bmatrix} 3 & 0 & 8 \\ 0 & 2 & 0 \\ 0 & 0 & 2 \end{bmatrix}$; **(c)** $\begin{bmatrix} 8 & 0 & 3 \\ 0 & 2 & 0 \\ 2 & 0 & 0 \end{bmatrix}$; **(d)** $\begin{bmatrix} 0 & 0 & 2 \\ 0 & 2 & 0 \\ 3 & 0 & 8 \end{bmatrix}$ **33.** $-7\mathbf{b}_1' - 2\mathbf{b}_2' + 3\mathbf{b}_3' + 7\mathbf{b}_4'$

35. $\begin{bmatrix} 5 & 4 & -6 & 18 \\ -4 & -3 & -2 & 0 \\ 0 & 0 & 1 & -12 \\ 0 & 0 & 0 & 1 \end{bmatrix}$ **37.** $\begin{bmatrix} \frac{1}{2} & 0 & 0 \\ 0 & \frac{1}{4} & 0 \\ 0 & 0 & \frac{1}{8} \end{bmatrix}$ **39.** $6ae^{2x} + 5be^{4x} - 3ce^{8x}$

41. $(a+b)\sin 2x + (a-b)\cos 2x$

Section 6.3

1. $R_B = \begin{bmatrix} 6 & 7 \\ -3 & -3 \end{bmatrix}$, $R_{B'} = \begin{bmatrix} 1 & -1 \\ 1 & 2 \end{bmatrix}$, $C = \begin{bmatrix} -1 & 2 \\ 1 & -1 \end{bmatrix}$ **3.** $R_B = \begin{bmatrix} 0 & 1 & 0 \\ 2 & 0 & 1 \\ 0 & 1 & 0 \end{bmatrix}$, $R_{B'} = \begin{bmatrix} 1 & 1 & 0 \\ 1 & 0 & 1 \\ 0 & 1 & -1 \end{bmatrix}$,

$C = \begin{bmatrix} 0 & 0 & 1 \\ 0 & 1 & -1 \\ 1 & -1 & 0 \end{bmatrix}$

5. $R_B = \begin{bmatrix} \frac{13}{5} & \frac{4}{5} & 2 \\ -\frac{11}{5} & -\frac{3}{5} & -2 \\ -\frac{9}{5} & -\frac{2}{5} & -2 \end{bmatrix}$, $R_{B'} = \begin{bmatrix} -\frac{4}{3} & -\frac{1}{6} & -\frac{10}{3} \\ -\frac{4}{3} & -\frac{5}{3} & -\frac{16}{3} \\ 1 & \frac{1}{2} & 3 \end{bmatrix}$, $C = \begin{bmatrix} 0 & -\frac{9}{5} & -\frac{8}{5} \\ 1 & \frac{13}{5} & \frac{21}{5} \\ 0 & \frac{12}{5} & \frac{14}{5} \end{bmatrix}$

7. $R_B = \begin{bmatrix} 1 & 0 & 0 \\ 0 & 1 & 0 \\ 0 & 0 & -1 \end{bmatrix}$, $R_{B'} = \frac{1}{3}\begin{bmatrix} 1 & -2 & -2 \\ -2 & 1 & -2 \\ -2 & -2 & 1 \end{bmatrix}$, $C = \frac{1}{3}\begin{bmatrix} 1 & 1 & -2 \\ 1 & -2 & 1 \\ 1 & 1 & 1 \end{bmatrix}$ **9.** $R_B = \begin{bmatrix} 1 & 0 & 0 \\ 0 & 0 & 0 \\ 0 & 0 & 1 \end{bmatrix}$,

$R_{B'} = \begin{bmatrix} 1 & 0 & 0 \\ 0 & 1 & 0 \\ 0 & 0 & 0 \end{bmatrix}$, $C = \frac{1}{2}\begin{bmatrix} 1 & 0 & 1 \\ 1 & 0 & -1 \\ 0 & 2 & 0 \end{bmatrix}$ **11.** $R_B = \begin{bmatrix} 2 & 0 & 0 \\ 2 & 2 & 0 \\ 1 & 1 & 2 \end{bmatrix}$, $R_{B'} = \begin{bmatrix} 2 & 1 & 1 \\ 0 & 2 & 2 \\ 0 & 0 & 2 \end{bmatrix}$, $C = \begin{bmatrix} 0 & 0 & 1 \\ 0 & 1 & 0 \\ 1 & 0 & 0 \end{bmatrix}$

13. $R_B = \begin{bmatrix} 0 & 0 & 0 & 0 \\ 3 & 0 & 0 & 0 \\ 0 & 2 & 0 & 0 \\ 0 & 0 & 1 & 0 \end{bmatrix}$, $R_{B'} = \begin{bmatrix} 0 & 1 & -2 & -3 \\ 0 & 0 & 2 & 0 \\ 0 & 0 & 0 & 3 \\ 0 & 0 & 0 & 0 \end{bmatrix}$, $C = \begin{bmatrix} 0 & 0 & 0 & 1 \\ 0 & 0 & 1 & 0 \\ 0 & 1 & 0 & 0 \\ 1 & 1 & 1 & 1 \end{bmatrix}$ **15.** $\begin{bmatrix} 2 & 1 & -3 \\ -1 & 0 & 1 \\ 0 & 0 & 1 \end{bmatrix}$

17. $\lambda_1 = -1$, $\lambda_2 = 5$; $E_{-1} = \mathrm{sp}\left(\begin{bmatrix} 1 \\ 1 \end{bmatrix}\right)$, $E_5 = \mathrm{sp}\left(\begin{bmatrix} -1 \\ 1 \end{bmatrix}\right)$; diagonalizable **19.** $\lambda_1 = 0$, $\lambda_2 = 1$, $\lambda_3 = 2$;

$E_0 = \mathrm{sp}\left(\begin{bmatrix} -1 \\ 0 \\ 1 \end{bmatrix}\right)$, $E_1 = \mathrm{sp}\left(\begin{bmatrix} 0 \\ 1 \\ 0 \end{bmatrix}\right)$, $E_2 = \mathrm{sp}\left(\begin{bmatrix} 1 \\ 0 \\ 1 \end{bmatrix}\right)$; diagonalizable   **21.** $\lambda_1 = -2$, $\lambda_2 = \lambda_3 = 5$;

$E_{-2} = \mathrm{sp}\left(\begin{bmatrix} 0 \\ 1 \\ 1 \end{bmatrix}\right)$, $E_5 = \mathrm{sp}\left(\begin{bmatrix} 0 \\ -5 \\ 2 \end{bmatrix}\right)$; not diagonalizable   **23.** F T T F T T T F F T

CHAPTER 7

Section 7.1

1. $U = \begin{bmatrix} 3 & -6 \\ 0 & 1 \end{bmatrix}$, $A = \begin{bmatrix} 3 & -3 \\ -3 & 1 \end{bmatrix}$ **3.** $U = \begin{bmatrix} 1 & -4 & 3 \\ 0 & -1 & -8 \\ 0 & 0 & 0 \end{bmatrix}$, $A = \begin{bmatrix} 1 & -2 & \frac{3}{2} \\ -2 & -1 & -4 \\ \frac{3}{2} & -4 & 0 \end{bmatrix}$ **5.** $U = \begin{bmatrix} -2 & 8 \\ 0 & 3 \end{bmatrix}$,

$A = \begin{bmatrix} -2 & 4 \\ 4 & 3 \end{bmatrix}$ **7.** $U = \begin{bmatrix} 8 & 5 & -4 \\ 0 & 1 & -2 \\ 0 & 0 & 10 \end{bmatrix}$, $A = \begin{bmatrix} 8 & \frac{5}{2} & -2 \\ \frac{5}{2} & 1 & -1 \\ -2 & -1 & 10 \end{bmatrix}$

9. $\begin{bmatrix} x \\ y \end{bmatrix} = \begin{bmatrix} 1/\sqrt{2} & 1/\sqrt{2} \\ -1/\sqrt{2} & 1/\sqrt{2} \end{bmatrix}\begin{bmatrix} t_1 \\ t_2 \end{bmatrix}$, $-t_1^2 + t_2^2$ **11.** $\begin{bmatrix} x \\ y \end{bmatrix} = \begin{bmatrix} 3/\sqrt{10} & -1/\sqrt{10} \\ 1/\sqrt{10} & 3/\sqrt{10} \end{bmatrix}\begin{bmatrix} t_1 \\ t_2 \end{bmatrix}$, $-t_1^2 + 9t_2^2$

13. $\begin{bmatrix} x \\ y \end{bmatrix} = \begin{bmatrix} 1/\sqrt{2} & 1/\sqrt{2} \\ -1/\sqrt{2} & 1/\sqrt{2} \end{bmatrix}\begin{bmatrix} t_1 \\ t_2 \end{bmatrix}$, $5t_1^2 + t_2^2$ **15.** $\begin{bmatrix} x_1 \\ x_2 \\ x_3 \end{bmatrix} = \begin{bmatrix} 1/\sqrt{3} & -1/\sqrt{2} & -1/\sqrt{6} \\ 1/\sqrt{3} & 1/\sqrt{2} & -1/\sqrt{6} \\ 1/\sqrt{3} & 0 & 2/\sqrt{6} \end{bmatrix}\begin{bmatrix} t_1 \\ t_2 \\ t_3 \end{bmatrix}$, $-t_1^2 + 2t_2^2 + 2t_3^2$

17. $a + c = k$, $ac = b^2$ **19.** $3.472136t_1^2 - 5.472136t_2^2$
21. $4.69854t_1^2 + 1.323057t_2^2 - 4.021597t_3^2$ **23.** $5.5t_1^2 - 4t_2^2 + 4t_3^2 + .5t_4^2$

Section 7.2

1. $C = \begin{bmatrix} 1/\sqrt{2} & -1/\sqrt{2} \\ 1/\sqrt{2} & 1/\sqrt{2} \end{bmatrix}$
$t_1^2 - t_2^2 = 1$

3. $C = \begin{bmatrix} 1/\sqrt{2} & -1/\sqrt{2} \\ 1/\sqrt{2} & 1/\sqrt{2} \end{bmatrix}$
$2t_1^2 + 0t_2^2 = 4$

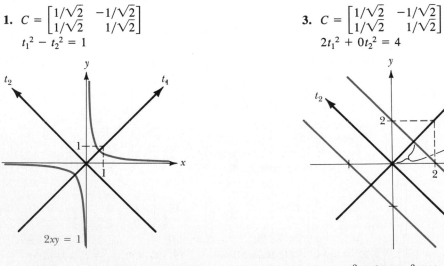

$2xy = 1$

$x^2 + 2xy + y^2 = 4$

5. $C = \begin{bmatrix} 3/\sqrt{10} & -1/\sqrt{10} \\ 1/\sqrt{10} & 3/\sqrt{10} \end{bmatrix}$

$11t_1^2 + t_2^2 = 4$

7. $C = \begin{bmatrix} 1/\sqrt{5} & -2/\sqrt{5} \\ 2/\sqrt{5} & 1/\sqrt{5} \end{bmatrix}$

$7t_1^2 + 2t_2^2 = 8$

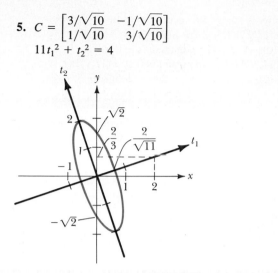

$10x^2 + 6xy + 2y^2 = 4$

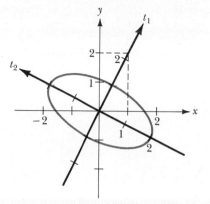

$3x^2 + 4xy + 6y^2 = 8$

9. The symmetric matrix of the quadratic-form portion is $\begin{bmatrix} a & b/2 \\ b/2 & c \end{bmatrix}$. Thus,

$$\det(A - \lambda I) = \begin{vmatrix} a - \lambda & b/2 \\ b/2 & c - \lambda \end{vmatrix} = \lambda^2 - (a + c)\lambda + ac - b^2/4.$$

The eigenvalues are given by $\lambda = (\frac{1}{2})(a + c \pm \sqrt{(a + c)^2 + b^2 - 4ac})$; they are real numbers with the same algebraic sign if $b^2 - 4ac < 0$, and with the opposite algebraic sign if $b^2 - 4ac > 0$. One of them is zero if $b^2 - 4ac = 0$. We obtain a (possibly degenerate) ellipse, hyperbola, or parabola accordingly.

11. Let $C^T = \begin{bmatrix} 1 & 0 & 0 \\ 0 & b/r & c/r \\ 0 & -c/r & b/r \end{bmatrix}$, where $r = \sqrt{b^2 + c^2}$. Then C is an orthogonal matrix such that $\det(C) = 1$ and

$$\begin{bmatrix} t_1 \\ t_2 \\ t_3 \end{bmatrix} = \mathbf{t} = C^T \begin{bmatrix} x \\ y \\ z \end{bmatrix} = \begin{bmatrix} x \\ (by + cz)/r \\ (-cy + bz)/r \end{bmatrix}.$$

Thus,

$$\begin{bmatrix} x \\ y \\ z \end{bmatrix} = C\mathbf{t} = \begin{bmatrix} t_1 \\ (bt_2 - ct_3)/r \\ (ct_2 + bt_3)/r \end{bmatrix}$$

represents a rotation of axes that transforms the equation $ax^2 + by + cz = d$ into the form $at_1^2 + rt_2^2 = d$.

13. Hyperboloid of two sheets   **15.** Hyperboloid of one sheet   **17.** Hyperbolic cylinder   **19.** Hyperboloid of one sheet   **21.** Elliptic cone or hyperboloid of one or two sheets   **23.** Elliptic cone or hyperboloid of one or two sheets   **25.** Parabolic cylinder or two parallel planes   **27.** Hyperbolic paraboloid or hyperbolic cylinder

Section 7.3

1. g has a local minimum of -7 at $(0, 0)$.   **3.** g has no local extremum at $(0, 0)$.   **5.** g has a local maximum of 3 at $(-5, 0)$.   **7.** g has no local extremum at $(0, 0)$.   **9.** The behavior of g at $(3, 1)$ is not determined.   **11.** g has a local maximum of 4 at $(0, 0, 0)$.   **13.** g has no local extremum at $(0, 0, 0)$.

15. g has no local extremum at $(7, -6, 0)$.   **17.** $g(x, y) = y^2 + 10$ **19.** $g(x, y) = y^2 + x^3$
21. $g(x, y) = x^4 + y^4 + 40$ **23.** The maximum is .5 at $\pm(1/\sqrt{2})(1, 1)$. The minimum is $-.5$ at $\pm(1/\sqrt{2})(1, -1)$.   **25.** The maximum is 9 at $\pm(1/\sqrt{10})(-1, 3)$. The minimum is -1 at $\pm(1/\sqrt{10})(3, 1)$.   **27.** The maximum is 6 at $\pm(1/\sqrt{2})(1, -1)$. The minimum is 0 at $\pm(1/\sqrt{2})(1, 1)$.
29. The maximum is 3 at $\pm(1/\sqrt{3})(1, -1, 1)$. The minimum is 0 at $\pm(1/\sqrt{2a^2 + 2b^2 - 2ab})(a - b, a, b)$.
31. The maximum is 2 at $(\pm 1/\sqrt{2a^2 + 2b^2})(a, b, b, -a)$. The minimum is -2 at $\pm(1/\sqrt{2})(0, -1, 1, 0)$.
33. The local maximum of f is $\lambda_1 a^2$, where λ_1 is the maximum eigenvalue of the symmetric coefficient matrix of the form f; it is assumed at any eigenvector corresponding to λ_1 and of length a. An analogous statement holds for the local minimum of f.   **35.** g has a local maximum of 5 at $(0, 0, 0)$.   **37.** g has no local extremum at $(0, 0, 0)$.

Section 7.4

1. $w_2 = \begin{bmatrix} 0 \\ -1 \end{bmatrix}$, $w_3 = \begin{bmatrix} 1 \\ -\frac{1}{3} \end{bmatrix}$, $w_4 = \begin{bmatrix} .75 \\ -1 \end{bmatrix}$

Rayleigh quotients: $-2, 1, 5.2$

Maximum eigenvalue 6, eigenvector $\begin{bmatrix} 1 \\ -1 \end{bmatrix}$

3. $w_2 = \begin{bmatrix} 1 \\ \frac{5}{7} \end{bmatrix}$, $w_3 = \begin{bmatrix} 1 \\ \frac{19}{29} \end{bmatrix}$, $w_4 = \begin{bmatrix} 1 \\ \frac{65}{103} \end{bmatrix}$

Rayleigh quotients: $6, 298/74 \approx 4, 4222/1202 \approx 3.5$

Maximum eigenvalue 3, eigenvector $\begin{bmatrix} 1 \\ \frac{3}{5} \end{bmatrix}$

5. $5 \begin{bmatrix} \frac{1}{2} & \frac{1}{2} \\ \frac{1}{2} & \frac{1}{2} \end{bmatrix} - \begin{bmatrix} \frac{1}{2} & -\frac{1}{2} \\ -\frac{1}{2} & \frac{1}{2} \end{bmatrix}$ **7.** $2 \begin{bmatrix} \frac{2}{3} & \frac{1}{3} & \frac{1}{3} \\ \frac{1}{3} & \frac{1}{6} & \frac{1}{6} \\ \frac{1}{3} & \frac{1}{6} & \frac{1}{6} \end{bmatrix} - \begin{bmatrix} \frac{1}{3} & -\frac{1}{3} & -\frac{1}{3} \\ -\frac{1}{3} & \frac{1}{3} & \frac{1}{3} \\ -\frac{1}{3} & \frac{1}{3} & \frac{1}{3} \end{bmatrix}$ **9.** $\begin{bmatrix} -\frac{1}{2} & \frac{1}{2} \\ \frac{1}{2} & -\frac{1}{2} \end{bmatrix}$ **11.** $\begin{bmatrix} -\frac{1}{3} & \frac{1}{3} & \frac{1}{3} \\ \frac{1}{3} & -\frac{1}{3} & -\frac{1}{3} \\ \frac{1}{3} & -\frac{1}{3} & -\frac{1}{3} \end{bmatrix}$

13. $\lambda = 12$, $\mathbf{v} = (-.7058823, 1, -.4117647)$ **15.** $\lambda = 6$, $\mathbf{v} = (-.9032258, 1, -.4193548)$
17. $\lambda = .1882784448992811$, $\mathbf{v} = (1, .2893009, .3203757)$
19. $\lambda_1 = 4.732050807568877$, $\mathbf{v}_1 = (1, -.7320508, 1)$
$\lambda_2 = 1.267949192431123$, $\mathbf{v}_2 = (.3660254, 1, .3660254)$
$\lambda_3 = -4$, $\mathbf{v}_3 = (-1, 0, 1)$
21. $\lambda_1 = 16.87586339619508$, $\mathbf{v}_1 = (.9245289, 1, .3678643, .7858846)$
$\lambda_2 = -15.93189429348535$, $\mathbf{v}_2 = (-.3162426, .6635827, -1, -.004253739)$
$\lambda_3 = 6.347821447472841$, $\mathbf{v}_3 = (-.5527083, .9894762, .8356429, -1)$
$\lambda_4 = -.291790550182573$, $\mathbf{v}_4 = (-1, .06924058, .3582734, .9206089)$

23. (a) The characteristic polynomial $A - \lambda I = \begin{vmatrix} a - \lambda & b \\ b & c - \lambda \end{vmatrix} = \lambda^2 + (-a - c)\lambda + (ac - b^2)$ has roots

$$\lambda = \tfrac{1}{2}(a + c \pm \sqrt{(a + c)^2 - 4(ac - b^2)}) = \tfrac{1}{2}(a + c \pm \sqrt{(a - c)^2 + 4b^2}).$$

(b) If we use part (a), the first row vector of $A - \lambda I$ is

$$(a - \lambda, b) = \left(\tfrac{1}{2}(a - c \mp \sqrt{(a - c)^2 + 4b^2}), b\right) = (g \mp \sqrt{g^2 + b^2}, b).$$

(c) From part (a), eigenvectors for the matrix are $(-b, g \pm \sqrt{g^2 + b^2}) = (-b, g \pm h)$. Normalizing, we obtain vectors $\dfrac{(-b, g \pm h)}{\sqrt{b^2 + (g \pm h)^2}}$. Using the upper choice of sign and setting $r = \sqrt{b^2 + (g + h)^2}$, we obtain $(-b/r, (g + h)/r)$ as the first column of C. Using the lower choice of sign and setting $s = \sqrt{b^2 + (g - h)^2}$, we obtain $(-b/s, (g - h)/s)$ as the second column of C.

(d) $\det(C) = \dfrac{-b(g-h)}{rs} + \dfrac{b(g+h)}{rs} = \dfrac{2bh}{rs}$; since $h, r, s \geq 0$, we see that the algebraic sign of $\det(C)$ is the same as that of b.

25. $\lambda_1 = -12.00517907692924$, $\lambda_2 = 7.906602974286551$, $\lambda_3 = 17.09857610264269$

27. $\lambda_1 = -5.210618568922174$, $\lambda_2 = 2.856693936892428$, $\lambda_3 = 3.528363748899602$, $\lambda_4 = 7.825560883130143$

29. 5.823349919059785, $-11.91167495952989 \pm 1.357830063519836i$

31. 57.22941613544168, -92.88108454947197, -54.25594801085533,
$47.45380821244281 \pm 44.48897425527453i$

CHAPTER 8

Section 8.1

1. (a) $z + w = 4 + i$, $zw = 5 + 5i$; **(b)** $z + w = 3 + 2i$, $zw = -1 + 3i$ **3. (a)** $|z| = \sqrt{13}$, $\bar{z} = (3 - 2i)$, $z\bar{z} = (3 + 2i)(3 - 2i) = 13 = |z|^2$; **(b)** $|z| = \sqrt{17}$, $\bar{z} = 4 + i$, $z\bar{z} = (4 - i)(4 + i) = 17 = |z|^2$ **7. (a)** $\frac{3}{2} + \frac{1}{2}i$, **(b)** $\frac{13}{25} + \left(-\frac{9}{25}\right)i$ **9. (a)** Modulus $2\sqrt{2}$, principal argument $3\pi/4$

11. 16 **17.** F T F F T F F T F T **19.** $\sqrt{2} + \sqrt{2}i$, $-\sqrt{2} + \sqrt{2}i$, $-\sqrt{2} - \sqrt{2}i$, $\sqrt{2} - \sqrt{2}i$

21. $1, i, -1, -i$ **23.** $2, \sqrt{2} + \sqrt{2}i, 2i, -\sqrt{2}, + \sqrt{2}i, -2, -\sqrt{2}, - \sqrt{2}i, -2i, \sqrt{2} - \sqrt{2}i$

Section 8.2

3. $AB = \begin{bmatrix} -3 + 2i & 2i & 2i \\ 2 & 2i & 1 \\ 2 + 3i & -1 + i & 2 + i \end{bmatrix}$, $BA = \begin{bmatrix} -2 + 2i & i & 2 - i \\ 2 + 3i & 1 + 3i & 0 \\ 2i & -1 + i & 0 \end{bmatrix}$ **5.** $\frac{1}{3}\begin{bmatrix} 2 + i & -i \\ -1 - i & 1 \end{bmatrix}$

7. $\frac{1}{10}\begin{bmatrix} 9 - 3i & 1 + 3i & -4 + 8i \\ -3 + i & 3 - i & -2 - 6i \\ -2 + 4i & 2 - 4i & 2 - 4i \end{bmatrix}$ **9.** $z = \frac{1}{10}\begin{bmatrix} -7 + 9i \\ 9 - 3i \\ 6 - 2i \end{bmatrix}$ **11.** $\text{sp}\left(\begin{bmatrix} 1 + i \\ 1 + 3i \\ 2 \end{bmatrix}\right)$ **13.** 3

15. (a) $\langle \mathbf{u}, \mathbf{v} \rangle = 0$, $\langle \mathbf{v}, \mathbf{u} \rangle = 0$; **(b)** $\langle \mathbf{u}, \mathbf{v} \rangle = 5 - 3i$, $\langle \mathbf{v}, \mathbf{u} \rangle = 5 + 3i$ **21 (a)** Perpendicular; **(b)** Parallel;
(c) Neither; **(d)** Parallel; **(e)** Perpendicular **23.** $\dfrac{2}{\sqrt{7}}(i, 1 - i, 1 + i, 1 - i)$

25. $(-3i, 1, 2 + 2i)$ **27.** $\{(2 + i, 1 + i), (1 - i, -2 + i)\}$

29. $\{(1, i, i), (1 + 3i, 3 - 2i, i), (1 + i, i, 1 - 2i)\}$

31. (a) Both; **(b)** Hermitian; **(c)** Unitary; **(d)** Neither

33. T T F F T T T T F F **41.** Diagonal matrices with entries of modulus 1 on the diagonal.

Section 8.3

1. $U = \dfrac{1}{\sqrt{2}}\begin{bmatrix} -i & i \\ 1 & 1 \end{bmatrix}$, $D = \begin{bmatrix} 0 & 0 \\ 0 & 2 \end{bmatrix}$ **3.** $U = \begin{bmatrix} (1 + i)/\sqrt{3} & (1 + i)/\sqrt{6} \\ -1/\sqrt{3} & 2/\sqrt{6} \end{bmatrix}$, $D = \begin{bmatrix} 0 & 0 \\ 0 & 3 \end{bmatrix}$

5. $U = \dfrac{1}{\sqrt{2}}\begin{bmatrix} -i & 0 & i \\ 1 & 0 & 1 \\ 0 & \sqrt{2} & 0 \end{bmatrix}$, $D = \begin{bmatrix} -1 & 0 & 0 \\ 0 & 1 & 0 \\ 0 & 0 & 1 \end{bmatrix}$ **7.** $U = \begin{bmatrix} (1 - i)/\sqrt{6} & 0 & (1 - i)/\sqrt{3} \\ -2\sqrt{6} & 0 & 1 \\ 0 & 1 & 0 \end{bmatrix}$, $D = \begin{bmatrix} -3 & 0 & 0 \\ 0 & 3 & 0 \\ 0 & 0 & 3 \end{bmatrix}$

9. $U = \begin{bmatrix} (1 + i)/\sqrt{6} & 0 & (1 + i)/\sqrt{3} \\ 0 & 1 & 0 \\ -2/\sqrt{6} & 0 & 1/\sqrt{3} \end{bmatrix}$, $D = \begin{bmatrix} -3 & 0 & 0 \\ 0 & -3 & 0 \\ 0 & 0 & 3 \end{bmatrix}$

11. $U = \begin{bmatrix} (-1-i)/\sqrt{8} & 0 & (3+3i)/\sqrt{24} \\ (1-i)/\sqrt{8} & (1+i)/\sqrt{3} & (1-i)/\sqrt{24} \\ 2/\sqrt{8} & -i/\sqrt{3} & 2/\sqrt{24} \end{bmatrix}$, $D = \begin{bmatrix} 0 & 0 & 0 \\ 0 & 1 & 0 \\ 0 & 0 & 4 \end{bmatrix}$ **13.** $\{a \in \mathbb{C} \mid |a| = 4\}$

15. $a = -1$ **19.** F T T F T F T T F F

Section 8.4

1. Yes   **3.** No   **5.** No

7. (a) $\lambda_1 = \lambda_2 = \lambda_3 = \lambda_4 = -2$.
  (b) $J + 2I$ has rank 3 and nullity 1,
    $(J + 2I)^2$ has rank 2 and nullity 2,
    $(J + 2I)^3$ has rank 1 and nullity 3,
    $(J + 2I)^k$ has rank 0 and nullity 4 for $k \geq 4$.
  (c) $J + 2I$: $\mathbf{e}_4 \rightarrow \mathbf{e}_3 \rightarrow \mathbf{e}_2 \rightarrow \mathbf{e}_1 \rightarrow \mathbf{0}$.
  (d) $J\mathbf{e}_1 = -2\mathbf{e}_1$, $J\mathbf{e}_2 = \mathbf{e}_1 - 2\mathbf{e}_2$, $J\mathbf{e}_3 = \mathbf{e}_2 - 2\mathbf{e}_3$, $J\mathbf{e}_4 = \mathbf{e}_3 - 2\mathbf{e}_4$.

9. (a) $\lambda_1 = -1$, $\lambda_2 = \lambda_3 = \lambda_4 = \lambda_5 = 2$.
  (b) $(J + I)^k$ has rank 4 and nullity 1 for $k \geq 1$,
    $(J - 2I)$ has rank 3 and nullity 2,
    $(J - 2I)^k$ has rank 1 and nullity 4 for $k \geq 2$.
  (c) $J + I$: $\mathbf{e}_1 \rightarrow \mathbf{0}$,
    $J - 2I$: $\mathbf{e}_3 \rightarrow \mathbf{e}_2 \rightarrow \mathbf{0}$, $\mathbf{e}_5 \rightarrow \mathbf{e}_4 \rightarrow \mathbf{0}$.
  (d) $J\mathbf{e}_1 = -\mathbf{e}_1$, $J\mathbf{e}_2 = 2\mathbf{e}_2$, $J\mathbf{e}_3 = \mathbf{e}_2 + 2\mathbf{e}_3$, $J\mathbf{e}_4 = 2\mathbf{e}_4$, $J\mathbf{e}_5 = \mathbf{e}_4 + 2\mathbf{e}_5$.

11. $\begin{bmatrix} 3 & 0 & 0 & 0 & 0 \\ 0 & 3 & 1 & 0 & 0 \\ 0 & 0 & 3 & 1 & 0 \\ 0 & 0 & 0 & 3 & 1 \\ 0 & 0 & 0 & 0 & 3 \end{bmatrix}$ **13.** $\begin{bmatrix} 1 & 1 & 0 & 0 & 0 & 0 & 0 & 0 \\ 0 & 1 & 0 & 0 & 0 & 0 & 0 & 0 \\ 0 & 0 & 1 & 1 & 0 & 0 & 0 & 0 \\ 0 & 0 & 0 & 1 & 1 & 0 & 0 & 0 \\ 0 & 0 & 0 & 0 & 1 & 0 & 0 & 0 \\ 0 & 0 & 0 & 0 & 0 & -2 & 0 & 0 \\ 0 & 0 & 0 & 0 & 0 & 0 & -2 & 0 \\ 0 & 0 & 0 & 0 & 0 & 0 & 0 & -2 \end{bmatrix}$ **15.** $\begin{bmatrix} 0 & 1 \\ 0 & 0 \end{bmatrix}$, $\left\{ \begin{bmatrix} 2 \\ 5 \end{bmatrix}, \begin{bmatrix} 1 \\ 3 \end{bmatrix} \right\}$ (Other

bases are possible.)

17. $\begin{bmatrix} 1 & 0 & 0 \\ 0 & 4 & 1 \\ 0 & 0 & 4 \end{bmatrix}$, $\left\{ \begin{bmatrix} 0 \\ 1 \\ 0 \end{bmatrix}, \begin{bmatrix} 0 \\ 5 \\ 5 \end{bmatrix}, \begin{bmatrix} 1 \\ 0 \\ 1 \end{bmatrix} \right\}$ (Other answers are possible.)

19. $\begin{bmatrix} 2 & 1 & 0 & 0 & 0 \\ 0 & 2 & 0 & 0 & 0 \\ 0 & 0 & -1 & 1 & 0 \\ 0 & 0 & 0 & -1 & 0 \\ 0 & 0 & 0 & 0 & -1 \end{bmatrix}$, $\left\{ \begin{bmatrix} 0 \\ 1 \\ 0 \\ 0 \\ 0 \end{bmatrix}, \begin{bmatrix} 5 \\ 0 \\ 0 \\ 0 \\ 0 \end{bmatrix}, \begin{bmatrix} 0 \\ 0 \\ -1 \\ 0 \\ 0 \end{bmatrix}, \begin{bmatrix} 0 \\ 0 \\ 0 \\ 0 \\ 1 \end{bmatrix}, \begin{bmatrix} 0 \\ 0 \\ 0 \\ 1 \\ 0 \end{bmatrix} \right\}$ (Other answers are possible.)

21. $\begin{bmatrix} 2 & 1 & 0 & 0 & 0 \\ 0 & 2 & 0 & 0 & 0 \\ 0 & 0 & 2 & 0 & 0 \\ 0 & 0 & 0 & 2 & 0 \\ 0 & 0 & 0 & 0 & 2 \end{bmatrix}$, $\{\mathbf{e}_1 + \mathbf{e}_3, \mathbf{e}_5, \mathbf{e}_2, \mathbf{e}_4, \mathbf{e}_1 - \mathbf{e}_3\}$ (Other answers are possible.)   **23.** T F T T F T T F F F

25. O **27.** O **29.** $A^4 + (3 - i)A^3 + (3 - 3i)A^2 + (1 - 3i)A - iI$

CHAPTER 9

Section 9.1

1. There are $n - 1$ flops performed on **b** while the first column of A is being fixed up, $n - 2$ flops while the second column of A is being fixed up, and so on. The total number is $(n - 1) + (n - 2) + \cdots + 2 + 1 = n(n - 1)/2$, which has order of magnitude $n^2/2$ for large n. **3.** mn, if we call each indexed addition a flop **5.** mn^2 **7.** $2n^3$ **9.** $3n^3$ **11.** $6n^3$ **13.** $3n^3/2$ **15.** n^3 **17.** $2n$ **19.** w^2n, counting each final division as a flop **21.** F T F F T T F F T **23.** (No text answer data are possible for this problem, since different computers run at different speeds. However, for n large enough to require 6 seconds or more to solve an $n \times n$ system, the computer did require roughly 50% more time when using the Gauss–Jordan method than when using the Gauss method with back substitution.) **25.** (See the answer to Exercise 23.)

Section 9.2

1. It is not significant; no arithmetic operations are involved, just storing of indexed values.

3. $\begin{bmatrix} -18 \\ 5 \end{bmatrix}$ **5.** $\begin{bmatrix} -27 \\ -11 \\ -1 \end{bmatrix}$ **7.** $\begin{bmatrix} 1 \\ -2 \\ 1 \\ 3 \end{bmatrix}$

9. $P = \begin{bmatrix} 0 & 0 & 1 \\ 0 & 1 & 0 \\ 1 & 0 & 0 \end{bmatrix}$, $L = \begin{bmatrix} 1 & 0 & 0 \\ 3 & 1 & 0 \\ 1 & \frac{1}{3} & 1 \end{bmatrix}$, $U = \begin{bmatrix} 2 & -1 & 5 \\ 0 & 6 & -23 \\ 0 & 0 & -\frac{1}{3} \end{bmatrix}$, $\mathbf{x} = \begin{bmatrix} 2 \\ -1 \\ -1 \end{bmatrix}$

11. $L = \begin{bmatrix} 1 & 0 & 0 \\ 3 & 1 & 0 \\ 4 & -10 & 1 \end{bmatrix}$, $U = \begin{bmatrix} 1 & 2 & -1 \\ 0 & 1 & 5 \\ 0 & 0 & 55 \end{bmatrix}$, $\mathbf{x} = \begin{bmatrix} -1 \\ 0 \\ 2 \end{bmatrix}$

13. $L = \begin{bmatrix} 1 & 0 & 0 & 0 \\ 2 & 1 & 0 & 0 \\ -1 & 0 & 1 & 0 \\ 4 & 2 & -\frac{3}{2} & 1 \end{bmatrix}$, $U = \begin{bmatrix} 1 & 2 & -3 & 0 \\ 0 & 1 & 0 & 0 \\ 0 & 0 & -2 & -1 \\ 0 & 0 & 0 & -\frac{1}{2} \end{bmatrix}$, $\mathbf{x} = \begin{bmatrix} 3 \\ 1 \\ -1 \\ 2 \end{bmatrix}$

15. $\begin{bmatrix} 0 \\ -1 \end{bmatrix}$ **17.** $\begin{bmatrix} -1 \\ 2 \\ 2 \end{bmatrix}$ **19.** $LDU = \begin{bmatrix} 1 & 0 & 0 \\ 0 & 1 & 0 \\ 3 & -4 & 1 \end{bmatrix}\begin{bmatrix} 1 & 0 & 0 \\ 0 & 1 & 0 \\ 0 & 0 & 14 \end{bmatrix}\begin{bmatrix} 1 & 1 & -3 \\ 0 & 1 & 1 \\ 0 & 0 & 1 \end{bmatrix}$

21. F F T F T F T F F T

23. Display: $\begin{bmatrix} 1 & 3 & -5 & 2 & 1 \\ 4 & -18 & 30 & 0 & -1 \\ 3 & .1666667 & 9 & -2 & 4.166667 \\ 2 & .2777778 & 1.407407 & -4.185185 & 5.413581 \\ -6 & -1.166667 & .6666667 & -4.141593 & 45.4764 \end{bmatrix}$.

For $\mathbf{b}_1$, $\mathbf{x} = \begin{bmatrix} 1 \\ -1 \\ 0 \\ 2 \\ -2 \end{bmatrix}$; for $\mathbf{b}_2$, $\mathbf{x} = \begin{bmatrix} -5 \\ -8 \\ 9 \\ 11 \\ 4 \end{bmatrix}$; for $\mathbf{b}_3$, $\mathbf{x} - \begin{bmatrix} -8 \\ 9 \\ -11 \\ 3 \\ 7 \end{bmatrix}$

25. $\begin{bmatrix} .4814815 \\ -1.592593 \end{bmatrix}$ **27.** $\begin{bmatrix} .8125 \\ 1.875 \\ -.3125 \end{bmatrix}$ **29.** $\begin{bmatrix} -.09131459 \\ -.08786787 \\ -.4790014 \\ -.01573263 \\ .08439226 \\ .2712737 \end{bmatrix}$

31. The ratios roughly conform on our computer. One solution that requires 100 flops took about $\frac{1}{3}$ of the time to form the L/U display, which requires about $1000/3$ flops, etc.

33. The ratios roughly conform on our computer, in the sense explained in the answer to Exercise 31.

Section 9.3

1. $x_1 = 0$, $x_2 = 1$ **3.** $x_1 = 1$, $x_2 = .9999$. Yes, it is reasonably accurate. $10x_1 + 1000000x_2 = 1000000$, $-10x_1 + 20x_2 = 10$ is a system that can't be solved without pivoting by a five-figure computer.   **5.** $x_1 + 10^{19}x_2 = 10^{19}$, $-x_1 + 2x_2 = 1$ **7.** We need to show that $(n(nA)^{-1})A = I$. It is easy to see that $(rA)B = A(rB) = r(AB)$ for any scalar r and matrices A and B such that AB is defined. Thus we have $(n(nA)^{-1})A = (nA)^{-1}(nA) = I$.   **9.** On our computer, $(x_1, x_2) = (-.0001, .0001)$, which is approximately correct.

11. On our computer, taking $r = 10^{-14}$ as the roundoff ratio, we obtain $(x_1, x_2) = (-.0001, .0001)$ in single-precision printing, which is approximately correct.   **13.** System: $10^{-11}x_1 + 10^{10}x_2 = 1$, $10^{10}x_1 + 2(10)^{10}x_2 = 1$;   **(a)** $(x_1, x_2) = (-10^{-10}, 10^{-10})$; approximately correct;   **(b)** Same answer as part (a)

15. Yes, on our computer, taking default value for roundoff ratio, and using the matrix $\begin{bmatrix} 10^{-13} & 0 \\ 0 & 10^{-13} \end{bmatrix}$

17. F T T T F F T T T F   **19.** We obtained $x_1 = -177552$, $x_2 = 331763.9$, using **b**; and we obtained $x_1 = -143177.3$, $x_2 = 267534.4$, using **c** and single precision. Using double precision, we obtained $x_1 = -176608$, $x_2 = 330000$, using **b**, and $x_1 = -142416$, $x_2 = 266112$, using **c**. Although components of **b** and **c** differ by only 1, components of the solutions obtained by using **b** and **c** differ by as much as 63,888.

21. We obtained the approximate values shown:

	Single Precision		Double Precision	
	c	d	c	d
x_1	-2.95×10^7	-4.80×10^7	-6.17×10^{13}	1.42×10^{14}
x_2	8.51×10^7	1.47×10^8	6.95×10^{14}	-1.60×10^{15}
x_3	1.21×10^8	1.64×10^8	-1.67×10^{15}	3.86×10^{15}
x_4	-2.15×10^8	-3.24×10^8	1.09×10^{15}	-2.51×10^{15}

Notice that totally different magnitudes are obtained, depending on whether we use single precision or double precision. The solutions obtained by using double precision are of the right order of magnitude, but experimentation with different schemes for matrix reduction shows that we can have no real confidence in their accuracy.

23. We produced the following data:

	Single Precision	Double Precision
$(H_2{}^4)^{-1}$:	$\begin{bmatrix} 11983.72 & -22391.05 \\ -22391.05 & 41838.45 \end{bmatrix}$	$\begin{bmatrix} 11920 & -22272 \\ -22272 & 41616 \end{bmatrix}$
$(H_2{}^4)^{-1}H_2{}^4$:	$\begin{bmatrix} 1 & 0 \\ 0 & 1 \end{bmatrix}$	$\begin{bmatrix} 1 & 9.1 \times 10^{-13} \\ 0 & 1 \end{bmatrix}$
$(H_2{}^8)^{-1}$:	$\begin{bmatrix} -2402679 & 4489447 \\ 4489447 & -8388608 \end{bmatrix}$	$\begin{bmatrix} 6.4 \times 10^8 & -1.2 \times 10^9 \\ -1.2 \times 10^9 & 2.2 \times 10^9 \end{bmatrix}$
$(H_2{}^8)^{-1}H_2{}^8$:	$\begin{bmatrix} 2 & 0 \\ 0 & 1 \end{bmatrix}$	$\begin{bmatrix} 1 & -1.2 \times 10^{-7} \\ 0 & 1 \end{bmatrix}$
$(H_2{}^{16})^{-1}$:	$\begin{bmatrix} -300334.9 & 561180.8 \\ 561180.9 & -1048576 \end{bmatrix}$	Unable to complete reduction with $r = .0001$ or $r = 0$
$(H_2{}^{16})^{-1}H_2{}^{16}$:	$\begin{bmatrix} 2 & -2 \\ 1 & 0 \end{bmatrix}$	None obtained

Scaling was necessary to complete reduction for the sixteenth power in single precision, and we were unable to complete reduction in double precision. (The only reason it is possible to do so in single precision is as a result of roundoff error.) The single-precision answers for the sixteenth power are of the wrong order of magnitude, since the inverse of the sixteenth power should be the square of the inverse for the eighth power.

25. We were unable to complete the reduction in either double or single precision without first scaling some small entries created up to 1. After scaling, we obtained these solution data:

Single precision: $(29137.48, -163717.3, 158334.5)$

Single precision: $(29133, -163692, 158310)$

CHAPTER 10

Section 10.1

1. 19,448   **3.** 1000 lb of blend A and 2500 lb of blend B

5. Maximize $x_1 + 2x_2 + 4x_3$, subject to

$$\begin{bmatrix} 3 & 4 & 1 \\ -2 & -2 & 1 \\ -1 & 1 & 0 \end{bmatrix}\begin{bmatrix} x_1 \\ x_2 \\ x_3 \end{bmatrix} \leq \begin{bmatrix} 12 \\ 4 \\ 1 \end{bmatrix}, \ \mathbf{x} \geq \mathbf{0}.$$

7. Maximize $-2x_1 - x_2 - x_3 + x_4$, subject to

$$\begin{bmatrix} 1 & 3 & 5 & 0 \\ 1 & -2 & -5 & 1 \end{bmatrix}\begin{bmatrix} x_1 \\ x_2 \\ x_3 \\ x_4 \end{bmatrix} \leq \begin{bmatrix} 10 \\ 5 \end{bmatrix}, \ \mathbf{x} \geq \mathbf{0}.$$

9. Value 8 at $(2, 1)$ **11.** Value 8 at $(0, 1, 0)$ **13.** Value -24 at $(6, 0)$ **15.** Value 12 at $(0, 6, 0)$

17. One such problem is to maximize $x_1 + 4x_2$, subject to $x_1 - x_2 \le 1$, $-x_1 + x_2 \le 3$, $x_2 \le 4$, $\mathbf{x} \ge \mathbf{0}$.

19. The feasible set is unbounded, as shown on the graph. The objective function $-x_1 + 3x_2$ assumes the arbitrarily large values $2x_1 + 12$ at points $(x_1, x_1 + 4)$, which are in the feasible set for all sufficiently large x_1. Thus the objective function can have no maximum value.

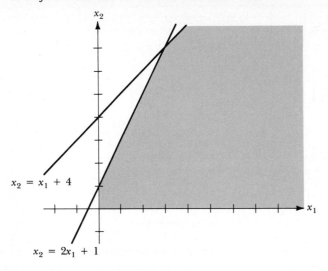

21. Value 30.4 at $(1.6, 0, 7.2)$ **23.** Value -13 at $(0, 0, 2, 15)$

Section 10.2

1.

	x_1	x_2	x_3	y_1	y_2	y_3	q_1	q_2	
y_1	1	2	1	1	0	0	0	0	5
q_1	4	2	-3	0	-1	0	1	0	3
q_2	5	0	-1	0	0	-1	0	1	4
P	-3	2	-5	0	0	0	0	0	0
	$-9M$	$-2M$	$+4M$		M	M			$-7M$

3.

	x_1	x_2	x_3	y_1	y_2	q_1	q_2	q_3	
q_1	3	-1	0	0	0	1	0	0	5
q_2	0	2	1	0	0	0	1	0	2
y_1	1	4	5	1	0	0	0	0	20
q_3	4	-2	-1	0	-1	0	0	1	3
P	-1	2	-5	0	0	0	0	0	0
	$-7M$	$+M$			M				$-10M$

5. 3.5 oz of food 1 and .75 oz of food 2   **7.** Value 4.5 at $(4.5, 0)$ **9. (a)** Let x_1 be shipped from W_1 to B_1, x_2 from W_1 to B_2, x_3 from W_2 to B_1, and x_4 from W_2 to B_2. Minimize $10x_1 + 5x_2 + 4x_3 + 8x_4$, subject to $x_1 + x_2 \le 30$, $x_3 + x_4 \le 20$, $x_1 + x_3 = 10$, $x_2 + x_4 = 40$.

(b)

	x_1	x_2	x_3	x_4	y_1	y_2	q_1	q_2	
y_1	1	1	0	0	1	0	0	0	30
y_2	0	0	1	1	0	1	0	0	20
q_1	1	0	1	0	0	0	1	0	10
q_2	0	1	0	1	0	0	0	1	40
P	10	5	4	8	0	0	0	0	0
	$-M$	$-M$	$-M$	$-M$					$-50M$

   (c) Value \$270 at (0, 30, 10, 10)

11. Value 4.5 at (4.5, 0)   **13.** 60 lb of bag 1, 10 lb of bag 2

15. Value 19.85714 at (2.571429, 0, 2.428572)   **17.** Value $-\frac{35}{3}$ at $\left(\frac{5}{3}, 0, 2\right)$

Section 10.3

1. Minimize $10y_1 + 15y_2$, subject to $y_1 + 3y_2 \geq 3$, $-2y_1 - y_2 \leq 2$.

3. Maximize $6x_1 + 4x_2$, subject to $-2x_1 - x_2 \geq 4$, $3x_1 + 3x_2 \leq 3$.

5. Minimize $8y_1 - y_2 - 2y_3$, subject to $y_1 - y_2 - y_3 \geq 3$, $2y_1 + y_2 - y_3 \geq 4$.

7. Maximize $12x_1 - 12x_2 + 4x_3 - 4x_4$, subject to $x_1 - x_2 \leq 1$, $3x_1 - 3x_2 + x_3 - x_4 \leq 0$, $-x_1 + x_2 - x_3 + x_4 \geq 2$.

9. (a) Minimize $\mathbf{b} \cdot \mathbf{y}$, subject to $A^T\mathbf{y} \geq \mathbf{c}$, $\mathbf{y} \geq \mathbf{0}$.

    (b) Maximize $-\mathbf{b} \cdot \mathbf{x}$, subject to $A^T\mathbf{x} \geq \mathbf{c}$, $\mathbf{x} \geq \mathbf{0}$.

    (c) Minimize $-\mathbf{c} \cdot \mathbf{y}$, subject to $-A\mathbf{y} \geq -\mathbf{b}$ or subject to $A\mathbf{y} \leq \mathbf{b}$, $\mathbf{y} \geq \mathbf{0}$.

    (d) Stating the answer to (c) as a maximum problem, we obtain: Maximize $\mathbf{c} \cdot \mathbf{x}$, subject to $A\mathbf{x} \leq \mathbf{b}$, $\mathbf{x} \geq \mathbf{0}$.

11. 3.5 oz of food 1 and .75 oz of food 2   **13.** 1000 lb of blend A, 2500 lb of blend B **15.** The matrix of data in the tableau will be smallest for the form that contains the fewest ($\geq$ type)-constraints, where constants on the right-hand sides are positive.

17. Value 4 at (4, 0)   **19.** Value 34 at (8, 2)   **21.** The simplex method is a bit easier with the primal problem, because of the smaller matrix of data in the tableau.

23. The simplex method is easier with the dual problem, because of the smaller matrix of data in the tableau.

25. The simplex method involves about the same amount of work with the primal as it does with the dual.

APPENDIX A

1. Let $P(n)$ be the equation to be proved. Clearly $P(1)$ is true, since $1 = \dfrac{1(1+1)(2+1)}{6}$. Assume that $P(k)$ is true. Then

$$1^2 + 2^2 + 3^2 + \cdots + k^2 + (k+1)^2 = \frac{k(k+1)(2k+1)}{6} + (k+1)^2$$

$$= \frac{(k+1)(2k^2 + k + 6k + 6)}{6}$$

$$= \frac{(k+1)(k+2)(2k+3)}{6}.$$

Thus, $P(k+1)$ is true, and $P(n)$ holds for all $n \in Z^+$.

3. Let $P(n)$ be the equation to be proved. We see that $P(1)$ is true. Assume that $P(k)$ is true. Then

$$1 + 3 + 5 + \cdots + (2k - 1) + (2k + 1) = k^2 + 2k + 1 = (k + 1)^2,$$

as required. Therefore, $P(n)$ holds for all $n \in Z^+$.

5. Let $P(n)$ be the equation to be proved. We see that $P(1)$ is true, since $a(1 - r^2)/(1 - r) = a(1 + r) = a + ar$. Assume that $P(k)$ is true. Then

$$\begin{aligned}
a + ar + ar^2 + \cdots + ar^k + ar^{k+1} &= a(1 - r^{k+1})/(1 - r) + ar^{k+1} \\
&= a(1 - r^{k+1} + r^{k+1}(1 - r))/(1 - r) \\
&= a(1 - r^{k+2})/(1 - r),
\end{aligned}$$

which establishes $P(k + 1)$. Therefore, $P(n)$ is true for all $n \in Z^+$.

7. The notion of an "interesting property" has not been made precise; it is not well defined. Moreover, we work in mathematics with two-valued logic: a statement is either true or false, *but not both*. The assertion that not having an interesting property would be an interesting property seems to contradict this two-valued logic. We would be saying that the integer both has and does not have an interesting property.

INDEX